ANNUAL REPORTS IN MEDICINAL CHEMISTRY
Volume 35

ANNUAL REPORTS IN MEDICINAL CHEMISTRY
Volume 35

*Sponsored by the Division of Medicinal Chemistry
of the American Chemical Society*

EDITOR-IN-CHIEF:
ANNETTE M. DOHERTY
INSTITUT DE RECHERCHE
JOUVEINAL / PARKE-DAVIS
FRESNES, FRANCE

SECTION EDITORS
*WILLIAM GREENLEE • WILLIAM K. HAGMANN
JACOB J. PLATTNER • DAVID W. ROBERTSON
GEORGE L. TRAINOR • JANET ALLEN*

EDITORIAL ASSISTANT
LISA BAUSCH / VIVIANE GUILLOT

ACADEMIC PRESS
San Diego   San Francisco   New York   Boston   London   Sydney   Tokyo

Academic Press
A Harcourt Science and Technology Company
525 B Street, Suite 1900, San Diego, California 92101-4495, USA
http://www.academicpress.com

Academic Press
Harcourt Place, 32 Jamestown Road, London NW1 7BY, UK
http://www.academicpress.com

International Standard Book Number: 0-12-040535-0

PRINTED IN THE UNITED STATES OF AMERICA
00 01 02 03 04 05 MB 9 8 7 6 5 4 3 2 1

CONTENTS

I. CENTRAL NERVOUS SYSTEM DISEASES

Section Editor: David W. Robertson, Pharmacia & Upjohn, Kalamazoo, Michigan

II. CARDIOVASCULAR AND PULMONARY DISEASES

Section Editor: William J. Greenlee, Schering Plough Research Institute, Kenilworth, New Jersey

III. CANCER AND INFECTIOUS DISEASES

Section Editor: Jacob J. Plattner, Chiron Corporation, Emeryville, California

IV. IMMUNOLOGY, ENDOCRINOLOGY AND METABOLIC DISEASES

Section Editor: William K. Hagmann, Merck Research Laboratories, Rahway, New Jersey

V. TOPICS IN BIOLOGY

Section Editor: Janet Allen, *Institut de Recherche Jouveinal/Parke-Davis, Fresnes, France*

CONTRIBUTORS

PREFACE

Annual Reports in Medicinal Chemistry continues to strive to provide timely and critical reviews of important topics in medicinal chemistry together with an emphasis on emerging topics in the biological sciences which are expected to provide the basis for entirely new future therapies.

Volume 35 retains the familiar format of previous volumes, this year with 31 chapters. Sections I - IV are disease-oriented and generally report on specific medicinal agents with updates from Volume 34 on anticoagulants, new therapeutics for Alzheimers disease and new antibacterials. As in past volumes, annual updates have been limited only to the most active areas of research in favor of specifically focussed and mechanistically oriented chapters, where the objective is to provide the reader with the most important new results in a particular field.

Sections V and VI continue to emphasize important topics in medicinal chemistry, biology, and drug design as well as the critical interfaces among these disciplines. Included in Section V, Topics in Biology, are chapters on pharmacogenetics, G protein - coupled receptors and immuno-modulatory proteins. Chapters in Section VI, Topics in Drug Design and Discovery, reflect the current focus on mechanism-directed drug discovery and newer technologies. These include chapters on privileged structures, ex vivo approaches to predicting oral bioavailability in humans, inhibition of cysteine proteases and principles for multivalent ligand design.

Volume 35 concludes with To Market, To Market - a chapter on NCE and NBE introductions worldwide in 1999 and, a chapter on genetically modified crops as a source of pharmaceuticals. In addition to the chapter reviews, a comprehensive set of indices has been included to enable the reader to easily locate topics in volumes 1-34 of this series.

Volume 35 *of Annual Reports in Medicinal Chemistry* was assembled with the superb editorial assistance of Ms. Viviane Guillot and Ms. Lisa Bausch and I would like to thank them for their hard work. I have continued to work with innovative and enthusiastic section editors and my sincere thanks goes to them again this year.

Annette M. Doherty
Fresnes, France
May 2000

SECTION I. CENTRAL NERVOUS SYSTEM DISEASES

Editor: David W. Robertson, Pharmacia & Upjohn Pharmaceuticals
Kalamazoo, MI 49001-0199

Chapter 1. Metabotropic Glutamate Receptor Modulators: Recent Advances and Therapeutic Potential

James A. Monn and Darryle D. Schoepp
Discovery Chemistry and Neuroscience Diseases Research Divisions
Eli Lilly and Company
Indianapolis, Indiana 46285

Introduction. Glutamate is considered the major excitatory neurotransmitter in the mammalian central nervous system, contributing input to the majority of excitatory synapses. Glutamatergic excitation per se is mediated by post-synaptic ion channel linked or "ionotropic" glutamate (iGlu) receptors, which include NMDA, AMPA and kainate receptor types. More recently, a family of G-protein coupled receptors (GPCRs), coined "metabotropic" glutamate (mGlu) receptors, have been identified (1). These possess structural homology to the calcium-sensing receptor of the parathyroid as well as GABA$_B$ receptors (i.e. type 3 GPCR superfamily). In general, mGlu receptors modulate glutamatergic excitations by pre-synaptic, post-synaptic, and glial mechanisms. Several recent reviews focused on mGlu receptor pharmacology, structure and function have been published (2-10).

Based in the proposal of Nakanishi (1), mGlu receptors have been categorized into three groups (Table 1) based on their degree of sequence homology (~70% homology within groups and ~40% homology between groups), signal transduction mechanisms when expressed in recombinant systems, and pharmacological properties (activation by group-selective agonists). A number of agonists/antagonists for mGlu receptors are selective for members within a group, but pharmacological agents that distinguish between receptor subtypes within groups are only now beginning to emerge. Thus, the "group" classification has been extensively used, allowing investigators to define pharmacological functions that are linked to receptors within each group. The agonist compounds in table 1 have been useful for this purpose.

Table 1. Classification of metabotropic glutamate receptors by "groups"			
	Subtypes	Transduction	Prototypic Agonists
Group I	mGlu1 mGlu5	Coupling to Gq and activation of phospholipase C	**1**, **6**
Group II	mGlu2 mGlu3	Coupling to Gi/Go and inhibition of adenylate cyclase	**7**, **8**, **11**, **12**, **16**
Group III	mGlu4 mGlu6 mGlu7 mGlu8	Coupling to Gi/Go and inhibition of adenylate cyclase	**18**, **19**, **24**

Pharmacological studies with group-selective agonists suggest that, in general, group I receptors reside post-synaptically and, upon activation, increase the excitability of neurons, while group II and group III agonists reside, at least in part, pre-synaptically and function to suppress glutamatergic neuronal excitations. The specific subtypes and cellular mechanisms by which group-selective agonists enhance or suppress glutamatergic excitations is an active area of research. Moreover, mGlu receptors can also modulate the actions (directly or indirectly) of other neurotransmitters, including purines, neuropeptides, GABA, serotonin and dopamine (11,12).

Structure and modeling of mGlu receptors - Metabotropic glutamate receptors have unique structural features that allow for multiple sites to which ligands can act to modulate receptor functions. They are characterized by a large extracellular amino terminal domain (ATD) to which glutamate binds and initiates receptor activation. This ATD region is homologous with a family of bacterial periplasmic proteins that function to transport amino acids across the bacterial cell wall. Members of this family include the leucine binding protein (LBP) and the leucine / isoleucine / valine binding protein (LIVBP) (13). Modeling of the glutamate recognition domain on mGlu receptors has been attempted using a number of approaches including comparative pharmacophore analysis of group-selective small molecule ligands (14-16), homology mapping and modeling to the known crystal structure of the LIVBP (13,17,18) and site-directed mutagenesis within the putative glutamate recognition domain (13,19). In addition to competitive agonists and antagonists, molecules have recently been identified that interact at domains distinct from the ATD and non-competitively block receptor activation. Importantly, these (vide infra) appear to be highly selective for specific mGlu receptor subtypes, and are thus useful tools to clearly understand the functional roles and therapeutic opportunities associated with mGlu receptor subtypes.

Development of new pharmacological tools for mGlu receptors - Efforts targeting mGlu receptors have continued to focus on identification of highly potent, group- or subtype-selective agonists and antagonists for use in target validation studies and subsequently as leads for development of potential clinical candidates. Lead generation strategies have largely centered on the synthesis of ω-acidic α-amino acids related to glutamate itself, incorporating either conformational constraints, bioisosteric replacements for the distal acid functionality and / or substitution (generally large arylalky groups) at either the α- or ω-positions of the glutamate chain. More recently, use of high throuput screening technologies has led to identification of non-amino acid antagonists that target sites distinct from the glutamate recognition domain.

1: n = 0
2: n = 1

3

4

5: R₁ = H, R₂ = Cl
6: R₁ = OH, R₂ = H

Group I mGlu receptor agonists - Pharmacological characterization of S-homoquisqualic acid (**2**) has revealed this molecule to act as a full efficacy agonist at recombinant mGlu5a receptors, though with approximately 600 times lower potency compared to quisqualic acid (**1**) itself. Interestingly, while **1** also activates mGlu1 receptors and has no mGlu2 activity, compound **2** was found to both antagonize glutamate-induced inositol phosphate production in mGlu1 expressing cells, and to act as a full agonist at mGlu2 receptors at concentrations comparable to its mGlu5 stimulating effect (20). Analogs **3** and **4**, which may be viewed as conformationally constrained derivatives of **2**, also displayed selective agonist activity at mGlu5 over mGlu1 receptors (21), with **3** showing approximately four-fold greater potency compared to **4**. Unlike **2** however, neither of these molecules produced agonist effects in mGlu2-expressing cells. The utility of **3** and **4** as pharmacological tools for studying selective mGlu5 activation is limited by their apparent AMPA-receptor agonist effects at concentrations within ten-fold of those required for mGlu5 activation (21). 2-Chloro-5-hydroxyphenylglycine (**5**), while significantly less potent than the prototypic group I mGlu agonist 3,5-dihydroxyglycine (**6**), displays greater than ten-fold agonist selectivity for mGlu5 versus mGlu1 receptors (22).

Group II agonists - LY354740 (**7**) remains one of the most potent and selective agonists developed to date for native group II (23,24) and recombinant mGlu2 and mGlu3 receptors (25-27). Owing to its ability to activate these targets following systemic administration in rodents, **7** has been used extensively in target validation studies and is currently undergoing clinical evaluation for anxiety. The preparation and use of [^{3}H]-**7** as a specific radioligand for group II mGlu receptors in rat cortex has been disclosed (28). Substitution of the C4-methylene group of **7** by either an oxygen or sulfur atom as in **8** and **9**, respectively (27), resulted in molecules possessing increased affinity at both mGlu2 (2-5 fold over **7**) and mGlu3 (15 fold over **7**). High agonist potency at these targets and low potency at other mGlu and iGlu receptor types was generally maintained, although **8** was found to possess sub-micromolar agonist activity at mGlu6. A number of patent applications describing substituted variants of **7** have published (29-35), albeit largely without biological data. Notably, the C3α-fluoro analog **10** was reported (34) as possessing mGlu2 receptor agonist activity with potency comparable to that described by others for **7**.

7: X = CH$_2$
8: X = O
9: X = S

10

11: X = Y = H
12: X = CO$_2$H, Y = H
13: X = Y = F

        The widely utilized group II-mGlu agonists L-CCG-I (**11**) and DCG-IV (**12**) have been profiled across all eight mGlu receptor subtypes (36). This has clearly established that **11** is not a selective agonist for group II mGlu receptors as it activates most of the other mGlu subtypes (with the exception of mGlu7) at concentrations only 5-10 times higher than that required for mGlu2 and mGlu3. This study also established the beneficial effect of the additional carboxylic acid group present in **12** on mGlu3 (but not mGlu2) receptor agonist potency relative to **11**, and that this molecule acts as an antagonist at higher concentrations at each of the other six mGlu receptor subtypes. A tritiated radioligand based on **12** has been prepared and evaluated for specific binding across rat brain regions (37). All eight stereoisomers of C3-difluoro-2-carboxycyclopropyl-glycine have been reported and **13** was found to possess both group I mGlu receptor-like and group II mGlu receptor-like agonist activity in a neonatal rat spinal cord assay (38). Subsequently, **13** has been found to act as an mGlu2 agonist with high potency (approximately three fold more potent than **11**), but also to display an interesting priming effect on α-aminopimelate-induced responses, suggesting that it (as well as **11**) is taken up into spinal cord storage sites and released upon α-aminopimelate challenge (39). This hypothesis has been further supported by the observation that an inhibitor of anion transport blocks this priming effect (40). Bicyclic amino acid **14** was prepared to test the hypothesis that the fully extended conformation of the glutamate backbone found in compound **7** is responsible for the high intrinsic potency and selectivity observed for this molecule (41). Interestingly, **14**, while displaying potent agonist activity at recombinant mGlu2 and mGlu3 receptors, also activated several other subtypes (mGlu1, 4, 5 and 6) at similar concentrations.

These data clearly establish that an extended glutamate conformation is capable of activating multiple mGlu receptor subtypes, and is not solely sufficient for high group II mGlu subtype potency and selectivity. The 1-amino analog **15** of the potent and selective mGlu2/3 agonist 2R,4R-APDC **16** has been identified as a modestly potent, highly subtype selective agonist for mGlu2 and mGlu3 receptors (42, 43). Of particular interest is this molecule's unique partial agonist activity at these targets that could engender it with properties distinct from full agonists. More recently, **17** has been reported as possessing selectivity toward mGlu3 receptors (44), although effects of this compound at mGlu3 are some 6,000 times weaker than it's inhibitory activity at it's primary target, glutamate carboxypeptidase II.

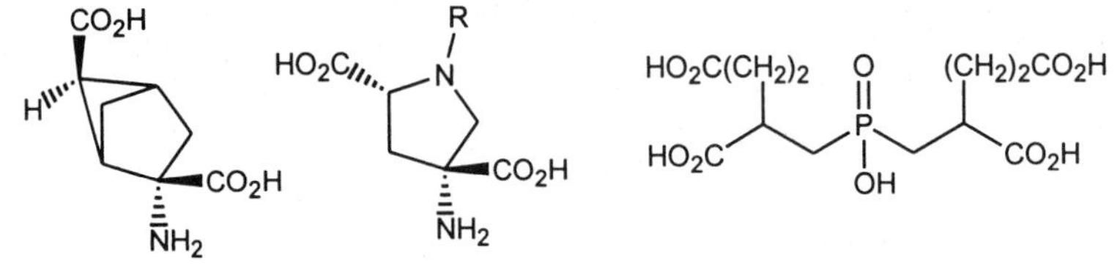

14   **15**: R = NH$_2$ **17**
              16: R = H

Group III agonists - Owing to their selectivity versus other mGlu and iGlu receptor subtypes and their commercial availability, L-AP4 and L-SOP (**18** and **19**) remain the most extensively utilized agonists for studying group III mGlu receptor function. More recent medicinal chemistry efforts, while generally lagging behind those applied to the group II mGlu receptors, have succeeded in generating additional group III-selective, and in some cases subtype-selective agonists. Extension of the chain linking the α-amino acid and hydroxyisoxazole in the prototypic iGlu$_{1-4}$ receptor agonist AMPA (**20**) resulted in **21**, a highly selective agonist for the mGlu6 receptor subtype, possessing no measurable activity at other mGlu or iGlu receptors (45). Incorporation of an additional carboxylic acid functionality at the 4-position of the prototypic mGlu receptor

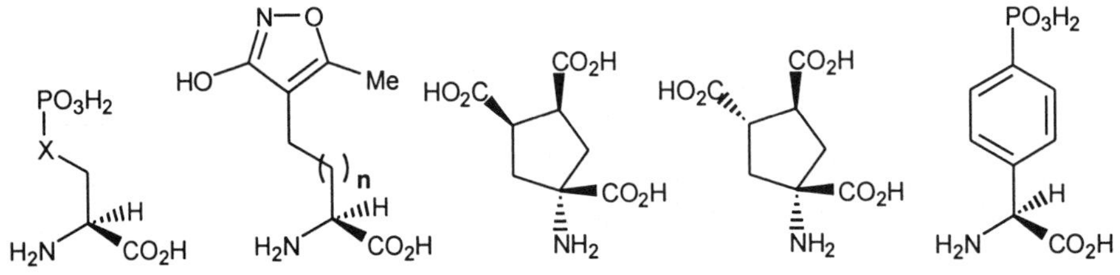

18: X = CH$_2$ **20**: n = 0   **22**   **23**   **24**
19: X = O   **21**: n = 1

agonist *trans*-1-aminocyclopentane-1,3-dicarboxylate afforded the 1-aminocyclo-pentane 1,3,4-tricarboxylates, or ACPTs (46). Of the isomers prepared, ACPT-I (**22**) and (+)-ACPT-III (**23**) show high agonist selectivity and low micromolar potency for mGlu4 receptors and show no appreciable agonist or antagonist activity at either mGlu1 or mGlu2. Recently, S-4-phosphonophenylglycine (**24**) has been described as a highly potent and selective group III mGlu agonist acting at all four of the group III mGlu receptor subtypes, showing greater than twenty-fold selectivity for mGlu8 (47).

Group I Antagonists - 4-Carboxyphenylglycine (4-CPG) is an antagonist of both mGlu1 and mGlu5 subtypes, with (depending on the agonist challenge) some selectivity for the former over the latter. Through introduction a methyl substituent *ortho* to the glycyl moiety (48), very high mGlu1 antagonist to mGlu5 antagonist selectivity was achieved for LY367385 (**25**). This selectivity was mimicked, but with greater potency in **26**. Attachment of relatively large alkyl chains to the α-position of 4-CPG afforded **27** which displays non-selective antagonist effects at both mGlu1 and mGlu5 receptors (49). In the same study 4-CPG and its corresponding α-methyl analog were found to be both significantly more potent and selective as mGlu1 antagonists, suggesting that mGlu1 is more sensitive to steric effects adjacent to the α-amino acid center than is mGlu5 (49). Attachment of a thioxanthylmethyl substituent to the same position afforded LY367366 (**28**), a molecule that, like **27** also lacks antagonist selectivity between mGlu1 and mGlu5, but which is some 40-50 times more potent than the simple alkyl-substituted variants (50).

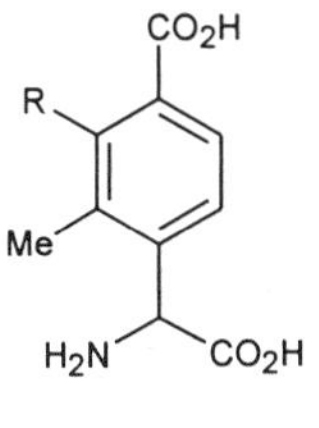

25: R = H
26: R = OH

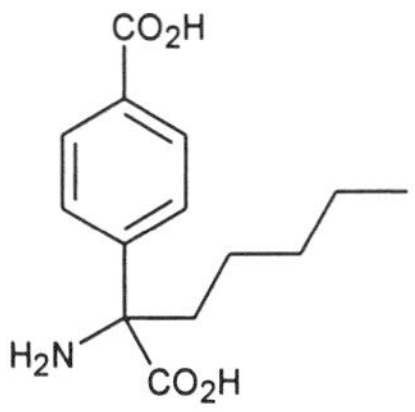

27

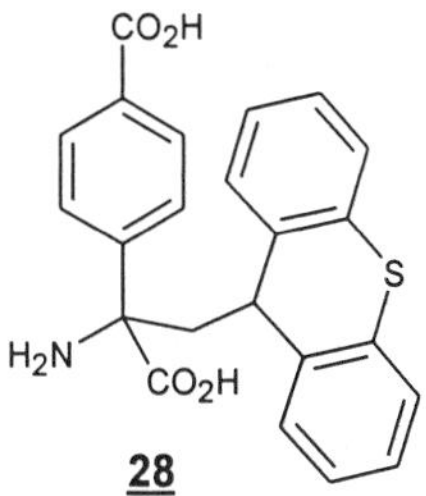

28

Constraining the glutamate pharmacophore into a conformationally-locked bicyclo[1.1.0]pentane ring system led to identification of **29** as a modestly potent antagonist in cells expressing mGlu1 receptors, while at slightly higher concentrations acting as a partial agonist at mGlu5 (51,52). Notably, **29** was found to possess no detectable activity at other mGlu (2 and 4) or iGlu targets. Extending the distance between the α-amino acid and distal acid functionalities while maintaining their co-linear relationship was achieved in compound **30**, a novel cubane derivative (53) which demonstrates relatively weak, but highly selective antagonist activity at mGlu1 receptors (no activity at mGlu5, very weak agonist effects at mGlu2 and mGlu4). The loss in potency of **30** compared to **29** has been ascribed to negative steric effects at the glutamate recognition domain. Attachment of a thioxanthylmethyl substituent at the α-position of 3-carboxycyclobutylglycine afforded **31**, a highly potent, though non-selective group I mGlu receptor antagonist (6,54).

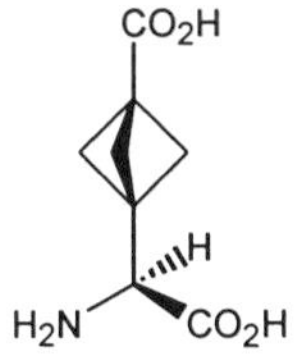

29

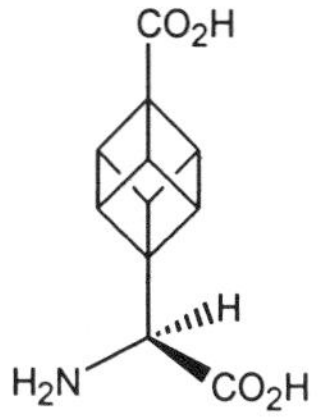

30

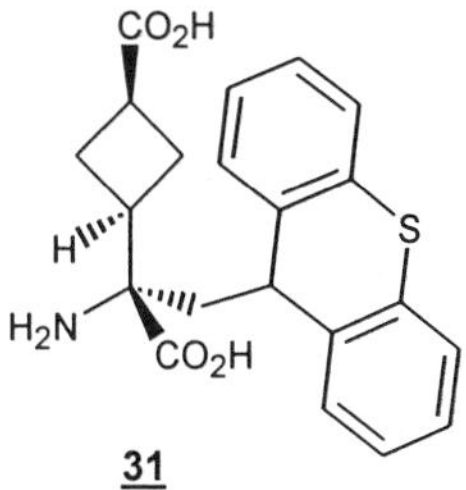

31

A number of non-competitive antagonists acting at mGlu1 or mGlu5 receptor subtypes have recently been disclosed. The first of these, **32-34**, were initially characterized as micromolar potency antagonists for recombinant mGlu1 receptors with a potency order of **33** > **34** > **32**, possessing no activity at mGlu2 or iGlu receptors up to 100 μM (55). Additionally, **32** was identified as an antagonist for mGlu5 inasmuch as it (at 100 μM) fully suppressed glutamate induced PI hydrolysis by this receptor subtype (55). Subsequent disclosures have revealed considerable selectivity by **32** for mGlu1 versus each of the other mGlu receptor subtypes and a non-competitive mechanism of action (56-58). Strong evidence has subsequently demonstrated the inhibition of mGlu1 by **32** involves its specific interaction with Thr[815] and Ala[818] residues located on the extracellular surface of the seventh transmembrane (TMVII) segment of mGlu1b (57). Consistent with an interaction of **32** in the 7TM domain, it also antagonized Ca^{2+} induced responses in cells expressing a chimeric receptor consisting of the ATD of the calcium-sensing receptor and the 7TM segment of mGlu1b, but not glutmate-induced responses in chimeras consisting of the mGlu1 ATD and 7TM segment of the Ca^{2+} receptor (59). High throughput screening (HTS) of a small molecule library against human mGlu5a and mGlu1b receptors and subsequent similarity mapping resulted in the identification of pyridine derivatives SIB1757 (**35**) and SIB1893 (**36**) as sub-micromolar potency, highly selective and non-competitive antagonists of recombinant mGlu5 receptors (60). Additional SAR resulted in the identification of **37**, which maintains the excellent mGlu5 receptor selectivity as found in **35** and **36**, but with an approximate ten-fold increase in antagonist potency. Importantly, **37** was found to suppress the excitatory effects of **6**, but not of **20** in the rat hippocampus following i.v. administration, establishing this molecule as a highly valuable tool for mGlu5 receptor target validation studies (61). Using [³H]-**37** as a specific radioligand probe in combination with site-directed mutagenesis techniques, it appears that this molecule may interact simultaneously with TMIII (Ser[658], Pro[655]) and TMVII (Ala[810]) amino acid residues (62). A number of other non-amino acid derivatives have been claimed as modulators of mGlu receptors in recently published patent applications (63-72), suggesting several pharmaceutical companies have successfully implemented HTS of mGlu targets as an approach toward lead generation.

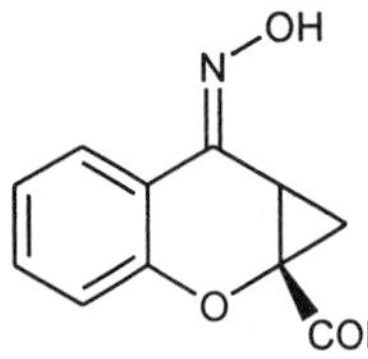

32: R = OEt
33: R = NHPh
34: R = NMePh

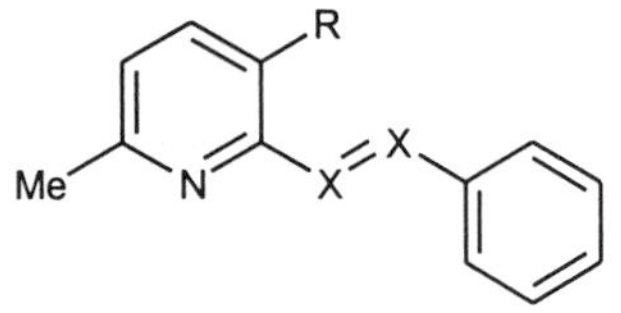

35: R = OH, X = N
36: R = H, X = CH

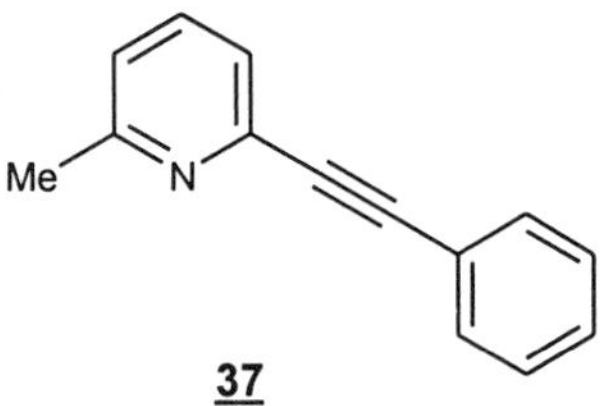

37

Group II mGlu Antagonists - An extensive SAR centered on α-substituted amino acids related to **11** has been disclosed (73,74). From these studies, the α-(9-xanthylmethyl)-substituted carboxycyclopropylglycine analog LY341495 (**38**) has been identified as a nanomolar potency competitive antagonist for both mGlu2 and mGlu3 receptors, although antagonist activity across all mGlu receptor subtypes is observed at higher concentrations (6,75). Significantly, while **38** displays low oral bioavailability in rats, it readily penetrates the blood brain barrier following intravenous or intraperitoneal administration, allowing this molecule to be utilized in group II mGlu-related target validation studies. A tritiated radioligand based on **38** has been prepared and characterized across mGlu subtypes (76,77). Somewhat higher group II mGlu selectivity, though with lower potency when compared to **38**, has been achieved with other substituted glutamate analogs, e.g. **39**-**41**. A lipophilic arylalkyl substituent attached to the glutamate pharmacophore appears to be essential for imparting both high antagonist potency and group II mGlu receptor selectivity (78-80). Interestingly, even **42**, an α-substituted variant of the group I selective agonist quisqualic acid, displays selective group II mGlu receptor antagonist activity (81).

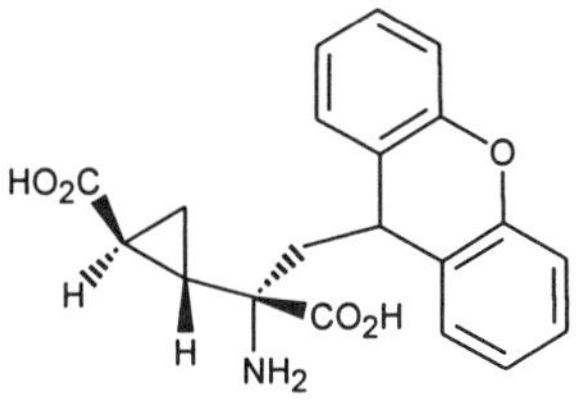

38 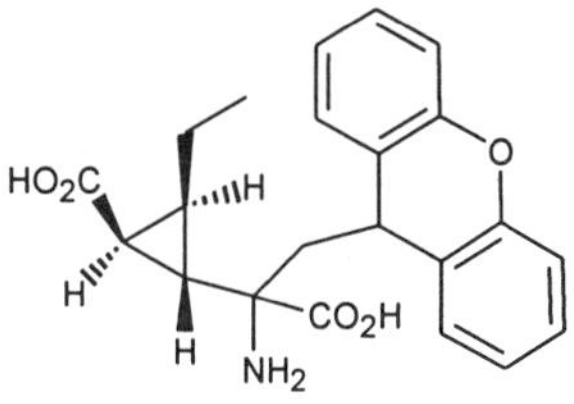   **39**

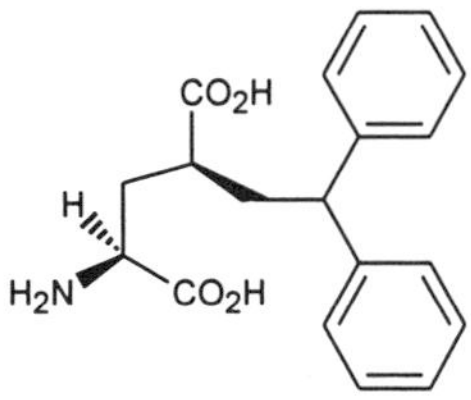

40 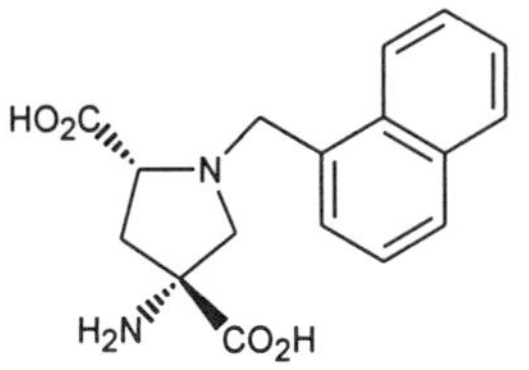   **41** 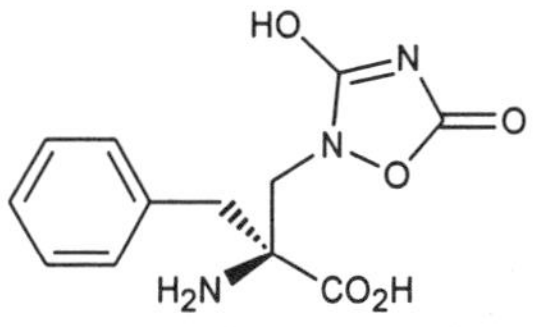   **42**

Group III mGlu Antagonists - No highly potent and selective antagonists for group III mGlu receptors have been reported. A number of molecules already cited (e.g. **12**, **38** and **39**) possess antagonist activity at cloned or native group III mGlu receptors, the potent of these being **38** (6). In addition, α-substituted ω-phosphono α-amino acids **43-46** display antagonism of group III agonist responses in a neonatal rat spinal cord preparation (82-85), though group II antagonist effects have been reported for **45** and **46** at somewhat higher concentrations (84,85), and **43**, **45** and **46** each suppress agonist responses in cells expressing recombinant mGlu2 (86,87).

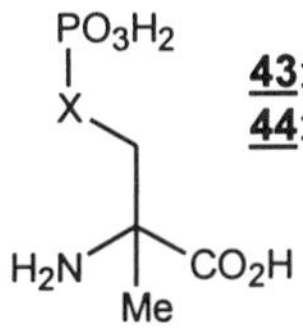

43: X = CH$_2$
44: X = O

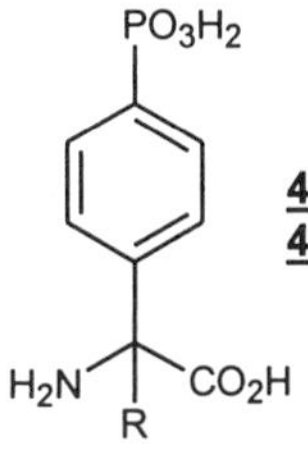

45: R = Me
46: R = Cyclopropyl

Therapeutic Opportunities - Dysfunctional glutamate neurotransmission (hypofunction or hyperfunction) has been associated with a variety of pathophysiological states and diseases including epilepsy, stroke / ischemia / trauma, pain, anxiety, psychosis, Parkinson's disease, drug withdrawal and diminished cognitive functioning. The modulatory effects of mGlu receptors on glutamatergic transmission suggest that mGlu receptor subtypes may be novel targets to treat these neurological and psychiatric disorders (4,88). In general, group II or group III agonists, as well as group I receptor antagonists, suppress glutamate excitation at the synaptic level. Accordingly, they have been reported to be anticonvulsant agents in animals and neuroprotective in tests (*in vitro* and/or *in vivo*) of excitotoxic injury and/or ischemic/traumatic injury (89,90). Compounds of these classes have also been found to suppress the induction and/or expression of pain states in animals (91). Group II agonists have been reported to have anxiolytic (92,93) anti-Parkinson's (94) and drug-withdrawal alleviating (94-96) properties in animal models. Further, compounds of this class block the behavioral effects of phencyclidine in rats (97,98), suggesting they might have antipsychotic activity in humans. In contrast, group I antagonists have been shown to block long term potentiation (LTP) and impair spatial learning in animals (99), suggesting group I receptor modulation might improve cognition in animals. Interestingly, one recent report showed that both group I and II mGlu agonists reverse cognitive disruptions induced by a nonselective mGlu antagonist (100). Many studies that support these hypotheses are hampered by the use of low potency, non-subtype selective, and non-systemically active compounds. Use of more potent, systemically active and subtype-selective agents will be critical to test hypotheses initially in animal models, then ultimately in humans. The therapeutic promise of this new area of drug research may then be realized.

References

1. S. Nakanishi, Science, 258, 597 (1992).
2. D. D. Schoepp, P. J. Conn, Trends Pharmacol. Sci., 14, 13 (1993).
3. J.-P. Pin, R. Duvoisin, Neuropharmacology, 34, 1 (1995).
4. P. J. Conn, J.-P. Pin, Annu. Rev. Pharmacol. Toxicol., 37, 205 (1997).
5. J. Bockaert, J.-P. Pin, EMBO J, 18, 1723 (1999).
6. D. D. Schoepp, D. E. Jane, J. A. Monn, Neuropharmacology, 38, 1431 (1999).
7. F. Bordi, A. Ugolini, Prog. Neurobiol., 59, 55 (1999).
8. R. Pellicciari, G. Constantino, Curr. Opin. Chem. Biol., 3, 433 (1999).
9. D. Ma, Bioorg. Chem., 27, 20 (1999).
10. J.-P. Pin, C. DeColle, A.-S. Bessis, F. Acher, Eur. J. Pharmacol., 375, 277 (1999).
11. R. Anwyl, Brain Res. Rev., 29, 83 (1999).
12. J. Cartmell, D. D. Schoepp, J. Neurochem. (2000) in press.
13. P. J. O'Hara, P. O. Sheppard, H. Thøgersen, D. Venezia, B. A. Haldeman, V. McGrane, K. M. Houamed, C. Thomsen, T. L. Gilbert, E. R. Mulvihill, Neuron, 11, 41 (1993).
14. N. Jullian, I. Brabet, J.-P. Pin, F. C. Acher, J. Med. Chem., 42, 1546 (1999).
15. G. Constantino, A. Macchiarulo, R. Pellicciari, J. Med. Chem., 42, 2816 (1999).
16. A.-S. Bessis, N. Jullian, E. Coudert, J.-P. Pin, F. Acher, Neuropharmacology, 38, 1543 (1999).
17. G. Constantino, R. Pellicciari, J. Med. Chem, 39, 3998 (1996).
18. G. Constantino, A. Macchiarulo, R. Pellicciari, J. Med. Chem., 42, 5390 (1999).
19. D. R. Hampson, X.-P. Huang, R. Pekhletski, V. Peltekova, G. Hornby, C. Thomsen, H. Thøgersen, J. Biol. Chem., 274, 33488 (1999).
20. H. Bräuner-Osborne, P. Krogsgaard-Larsen Br. J. Pharmacol., 123, 269 (1998).
21. L. Littman, C. Tokar, S. Venkatraman, R. J. Roon, J. F. Koerner, M. B. Robinson, R. L. Johnson, J. Med. Chem., 42, 1639 (1999).
22. A. J. Doherty, M. J. Palmer, J. M. Henley, G. L. Collingridge, D. E. Jane, Neuropharmacology, 36, 265 (1997).
23. J. A. Monn, M. J. Valli, S. M. Massey, R. A. Wright, C. R. Salhoff, B. G. Johnson, T. Howe, C. A. Alt, G. A. Rhodes, R. L. Robey, K. R. Griffey, J. P. Tizzano, M. J. Kallman, D. R. Helton, D. D. Schoepp, J. Med. Chem., 40, 528 (1997).
24. D. D. Schoepp, B. G. Johnson, R. A. Wright, C. R. Salhoff, J. A. Monn, Naunyn-Schmiedeberg's Arch. Pharm., 358, 175 (1998).
25. D. D. Schoepp, B. G. Johnson, R. A. Wright, C. R. Salhoff, N. G. Mayne, S. Wu, S. L. Cockerham, J. P. Burnett, R. Belagaje, D. Bleakman, J. A. Monn, Neuropharmacology, 36, 1 (1997).
26. S. Wu, R. A. Wright, P. K. Rockey, S. G. Burgett, P. R. Arnold, B. G. Johnson, D. D. Schoepp, R. Belagaje, Mol. Brain Res., 53, 88 (1998).
27. J. A. Monn, M. J. Valli, S. M. Massey, M. M. Hansen, T. J. Kress, J. P. Wepsiec, A. R. Harkness, J. L. Grutsch Jr., R. A. Wright, B. G. Johnson, S. L. Andis, A. E. Kingston, R. Tomlinson, R. Lewis, K. R. Griffey, J. P. Tizzano, D. D. Schoepp, J. Med. Chem., 42, 1027 (1999).
28. H. Schaffhauser, J. G. Richards, J. Cartmell, S. Chaboz, J. A. Kemp, A. Klingelschmidt, J. Messer, H. Stadler, T. Woltering, V. Mutel, Mol. Pharmacol., 53, 228 (1998).
29. C. Dominguez-Fernandez, D. R. Helton, S. M. Massey, J. A. Monn, U.S. Patent 5,916,920 (1999).
30. C. Dominguez-Fernandez, J. A. Monn, M. J. Valli, U.S. Patent 5,912,248 (1999).
31. S. M. Massey, J. A. Monn, M. J. Valli, U.S. Patent 5,958,960 (1999).
32. B. L. Chenard, Eur. Patent. Appl. 928792 A2 (1999).
33. G. Adam, P. N. Huguenin-Virchaux, V. Mutel, H. Stadler, T. J. Woltering, Ger. Offen. DE 19941675 (2000).
34. A. Nakazato, T. Kumagai, K. Sakagami, K. Tomisawa, PCT Int. Appl. WO 9938839 (1999).
35. A. Nakazato, T. Kumagai, K. Sakagami, K. Tomisawa, PCT Int. Appl. WO 0012464 (2000).

36. I. Brabet, M.-L. Parmentier, C. De Colle, J. Bockaert, F. Acher, J.-P. Pin, Neuropharmacology, 37, 1043 (1998).
37. V. Mutel, G. Adam, S. Chaboz, J. A. Kemp, A. Klingelschmidt, J. Messer, J. Wichmann, T. Woltering, J. G. Richards, J. Neurochem. 71, 2558 (1998).
38. A. Shibuya, A. Sato, T. Taguchi, Bioorg. Med. Chem. Lett., 8, 1979 (1998).
39. T. Saitoh, M. Ishida, H. Shinozaki, Br. J. Pharmacol., 123, 771 (1998).
40. M. Ishida, H. Shinozaki, Neuropharmacology, 38, 1531 (1999).
41. A. P. Kozikowski, D. Steensma, G. L. Araldi, W. Tückmantel, S. Wang, S. Pshenichkin, E. Surina, J. T. Wroblewski, J. Med. Chem., 41, 1641 (1998).
42. D. D. Schoepp, C. R. Salhoff, R. A. Wright, B. G. Johnson, J. P. Burnett, N. G. Mayne, R. Belagaje, S. Wu, J. A. Monn, Neuropharmacology, 35, 1661 (1996).
43. A. P. Kozikowski, G. L. Araldi, W. Tückmantel, S. Pshenichkin, E. Surina, J. T. Wroblewski, Bioorg. Med. Chem. Lett., 9, 1721 (1999).
44. F. Nan, T. Bzdega, S. Pshenichkin, J. T. Wroblewski, B. Wroblewska, J. H. Neale, A. P. Kozikowski, J. Med. Chem. 43, 772 (2000).
45. H. Ahmadian, B. Nielsen, H. Bräuner-Osborne, T. N. Johansen, T. B. Stensbøl, F. A. Sløk, N. Sekiyama, S. Nakanishi, P. Krogsgaard-Larsen, U. Madsen, J. Med. Chem., 40, 3700 (1997).
46. F. C. Acher, F. J. Tellier, R. Azerad, I. N. Brabet, L. Fagni, J.-P. Pin, J. Med. Chem., 40, 3119 (1997).
47. F. Gasparini, V. Bruno, G. Battaglia, S. Lukic, T. Leonhardt, W. Inderbitzin, D. Laurie, B. Sommer, M. A. Varney, S. D. Hess, E. C. Johnson, R. Kuhn, S. Urwyler, D. Sauer, C. Portet, M. Schmutz, F. Nicoletti, P. J. Flor, J. Pharmacol. Exp. Ther., 290, 1678 (1999).
48. B. P. Clark, S. R. Baker, J. Goldsworthy, J. R. Harris, A. E. Kingston, Bioorg. Med. Chem. Lett., 7, 2777 (1997).
49. A. J. Doherty, G. L. Collingridge, D. E. Jane, Br. J. Pharmacol., 126, 205 (1999).
50. V. Bruno, G. Battaglia, A. Kingston, M. J. O'Neill, M. V. Catania, R. Di Grezia, F. Nicoletti, Neuropharmacology, 38, 199 (1999).
51. R. Pellicciari, M. Raimondo, M. Marinozzi, B. Natalini, G. Constantino, C. Thomsen, J. Med. Chem., 39, 2874 (1996).
52. G. Mannaioni, S. Attucci, A. Missanelli, R. Pellicciari, R. Corradetti, F. Moroni, Neuropharmacology, 38, 917 (1999).
53. R. Pellicciari, G. Constantino, E. Giovagnoni, L. Mattoli, I. Brabet, J.-P. Pin, Bioorg. Med. Chem. Lett., 8, 1569 (1998).
54. S. R. Baker, B. P. Clark, J. R. Harris, K. I. Griffey, A. E. Kingston, J. P. Tizzano, Soc. Neurosci. Abs., 24, 576 (1998).
55. H. Annoura, A. Fukunaga, M. Uesugi, T. Tatsuoka, Y. Horikawa, Bioorg. Med. Chem. Lett., 6, 763 (1996).
56. G. Casabona, T. Knöpfel, R. Kuhn, F. Gasparini, P. Baumann, M. A. Sortino, A. Copani, F. Nicoletti, Eur. J. Neurosci., 9, 12 (1997).
57. S. Litschig, F. Gasparini, D. Rueegg, N. Stoehr, P. J. Flor, I. Vranesic, L. Prezeau, J.-P. Pin, C. Thompsen, R. Kuhn, Mol. Pharmacol., 55, 453 (1999).
58. E. Hermans, S. R. Nahorski, R. A. J. Challiss, Neuropharmacology, 37, 1645 (1998).
59. H. Bräuner-Osborne, A. A. Jensen, P. Krogsgaard-Larsen, NeuroReport, 10, 3923 (1999).
60. M. A. Varney, N. D. P. Cosford, C. Jachec, S. P. Rao, A. Sacaan, F.-F. Lin, L. Bleicher, E. M. Santori, P. J. Flor, H. Allgeier, F. Gasparini, R. Kuhn, S. D. Hess, G. Veliçelebi, E. C. Johnson, J. Pharmacol. Exp. Ther., 290, 170 (1999).
61. F. Gasparini, K. Lingenhöhl, N. Stoehr, P. J. Flor, M. Heinrich, I. Vranesic, M. Biollaz, H. Allgeier, R. Heckendorn, S. Urwyler, M. A. Varney, E. C. Johnson, S. D. Hess, S. P. Rao, A. I. Sacaan, E. M. Santori, G. Veliçelebi, R. Kuhn, Neuropharmacology, 38, 1493 (1999).
62. A. Pagano, D. Rüegg, S. Litschig, P. Floersheim, N. Stoehr, C. Stierlin, I. Vranesic, M. Heinrich, P. J. Flor, F. Gasparini, R. Kuhn, Neuropharmacology, 38, A33, abstract 107 (1999).
63. F. H. Fuller, B. C. Vanwagenen, M. F. Balandrin, E. G. Delmar, S. T. Moe, C.-P. Benjamin, PCT Int. Appl., WO9605818 (1996).
64. J. M. Lundbeck, A. Kanstrup, PCT Int. Appl., WO9705109 (1997).
65. P. H. Olesen, A. Kanstrup, PCT Int. Appl., WO9705137 (1997).
66. S. Hayashibe, H. Itahana, K. Yahiro, S.-i. Tsukamoto, M. Okada, PCT Int. Appl., WO9806724 (1998).
67. A. Stolle, H.-P. Antonicek, S. Lensky, A. Voerste, T. Müller, J. Baumgarten, K. Von Dem Bruch, G. Müller, U. Stropp, E. Horváth, J. M. V. De Vry, R. Schreiber, PCT Int. Appl., WO9936416 (1999).

68. A. Stolle, H.-P. Antonicek, S. Lensky, A. Voerste, T. Müller, J. Baumgarten, K. Von Dem Bruch, G. Müller, U. Stropp, E. Horváth, J. M. V. De Vry, R. Schreiber, PCT Int. Appl., WO9936417 (1999).

69. A. Stolle, H.-P. Antonicek, S. Lensky, A. Voerste, T. Müller, J. Baumgarten, K. Von Dem Bruch, G. Müller, U. Stropp, E. Horváth, J. M. V. De Vry, R. Schreiber, PCT Int. Appl., WO9936418 (1999).

70. A. Stolle, H.-P. Antonicek, S. Lensky, A. Voerste, T. Müller, J. Baumgarten, K. Von Dem Bruch, G. Müller, U. Stropp, E. Horváth, J. M. V. De Vry, R. Schreiber, PCT Int. Appl., WO9936419 (1999).

71. G. Adam, S. Kolczewski, V. Mutel, H. Stadler, J. Wichmann, T. J. Woltering, PCT Int. Appl., WO9908678 (1999).

72. G. Adam, Eur. Pat. Appl., EP0891978 (1999).

73. P. L. Ornstein, T. J. Bleisch, M. B. Arnold, R. A. Wright, B. G. Johnson, D. D. Schoepp, J. Med. Chem., 41, 346 (1998).

74. P. L. Ornstein, T. J. Bleisch, M. B. Arnold, J. H. Kennedy, R. A. Wright, B. G. Johnson, J. P. Tizzano, D. R. Helton, M. J. Kallman, D. D. Schoepp, M. Hérin, J. Med. Chem., 41, 358 (1998).

75. A. E. Kingston, P. L. Ornstein, R. A. Wright, B. G. Johnson, N. G. Mayne, J. P. Burnett, R. Belagaje, S. Wu, D. D. Schoepp, Neuropharmacology, 37, 1 (1998).

76. P. L. Ornstein, M. B. Arnold, T. J. Bleisch, R. A. Wright, W. J. Wheeler, D. D. Schoepp, Bioorg. Med. Chem. Lett., 8, 1919 (1998).

77. B. G. Johnson, R. A. Wright, M. B. Arnold, W. J. Wheeler, P. L. Ornstein, D. D. Schoepp, Neuropharmacology, 38, 1519 (1999).

78. I. Collado, J. Ezquerra, A. Mazón, C. Pedregal, B. Yruretagoyena, A. E. Kingston, R. Tomlinson, R. A. Wright, B. G. Johnson, D. D. Schoepp, Bioorg. Med. Chem. Lett., 8, 2849 (1998).

79. A. Escribano, J. Ezquerra, C. Pedregal, A. Rubio, B. Yruretagoyena, S. R. Baker, R. A. Wright, B. G. Johnson, D. D. Schoepp, Bioorg. Med. Chem. Lett., 8, 765 (1998).

80. M. J. Valli, D. D. Schoepp, R. A. Wright, B. G. Johnson, A. E. Kingston, R. Tomlinson, J. A. Monn, Bioorg. Med. Chem. Lett., 8, 1985 (1998).

81. A. P. Kozikowski, D. Steensma, M. Varasi, S. Pshenichkin, E. Surina, J. T. Wroblewski, Bioorg. Med. Chem. Lett., 8, 447 (1998).

82. D. E. Jane, P. L. St. J. Jones, P.C.-K. Pook, H.-W. Tse, J. C. Watkins, Br. J. Pharmacol., 112, 809 (1994).

83. N. K. Thomas, D. E. Jane, H.-W. Tse, J. C. Watkins, Neuropharmacology, 35, 637 (1996).

84. D. E. Jane, K. Pittaway, D. C. Sunter, N. K. Thomas, J. C. Watkins, Neuropharmacology, 34, 851 (1995).

85. D. E. Jane, N. K. Thomas, H.-W. Tse, J. C. Watkins, Neuropharmacology, 35, 1029 (1996).

86. J. Gomeza, S. Mary, I. Brabet, M.-L. Parmentier, S. Restituito, J. Bockaert, J.-P. Pin, Mol. Pharmacol., 50, 923 (1996).

87. J. Cartmell, G. Adam, S. Chaboz, R. Henningsen, J. A. Kemp, A. Klingelschmidt, V. Metzler, F. Monsma, H. Schaffhauser, J. Wichmann, V. Mutel, Br. J. Pharmacol., 123, 497 (1998).

88. T. Knöpfel, R. Kuhn, H. Allgeier, J. Med. Chem., 38, 1417 (1995).

89. F. Nicoletti, V. Bruno, A. Copani, G. Casabona, T. Knöpfel Trends. Neurosci., 19, 267 (1996).

90. F. Nicoletti, V. Bruno, M. V. Catania, G. Battaglia, A. Copani, G. Barbagallo, V. Cena, J. Sanchez-Prieto, P. F. Spano, M. Pizzi, Neuropharmacology, 38, 1477 (1999).

91. S. J. Boxall, S. W. N. Thompson, A. Dray, A. H. Dickenson, L. Urban, Neuroscience, 74, 13 (1996).

92. D. R. Helton, J. P. Tizzano, J. A. Monn, D. D. Schoepp, M. J. Kallman, J. Pharmacol. Exp. Ther., 284, 651 (1998).

93. A. Klodzinska, E. Chojnacka-Wójcik, A. Palucha, P. Branski, P. Popik, A. Pilc, Neuropharmacology, 38, 1831 (1999).

94. J. Konieczny, K. Ossowska, S. Wolfarth, A. Pilc, Naunyn-Schmiedebergs Arch. Pharmacol., 358, 500 (1998).

95. D. R. Helton, J. P. Tizzano, J. A. Monn, D. D. Schoepp, M. J. Kallman, Neuropharmacology, 36, 1511 (1997).

96. J. Vandergriff, K. Rasmussen, Neuropharmacology, 38, 217 (1999).

97. B. Moghaddam, B. W. Adams, Science, 281, 1349 (1998).

98. J. Cartmell, J. A. Monn, D. D. Schoepp, J. Pharmacol. Exp. Ther., 291, 161 (1999).

99. G. Riedel, K. G. Reimann, Acta Physiol. Scand., 157, 1 (1996).

100. C. Mathis, A. Ungerer, Exp. Brain Res., 129, 147 (1999).

Chapter 2. Recent Advances in Selective Serotonin Receptor Modulation

Albert J. Robichaud and Brian L. Largent
Medicinal Chemistry and CNS Diseases Research Departments
DuPont Pharmaceuticals Company, Wilmington, DE 19880

Introduction - First reported as a vasoconstrictor in the 1940's, serotonin (5-hydroxytryptamine, 5-HT) has been appreciated since that time as an important neurotransmitter in the brain. The receptors through which serotonin acts represent a diverse family of receptors. These receptors had been initially classified pharmacologically into multiple subtypes. The advent of molecular cloning provided additional discrimination within the serotonin receptor family ultimately revealing the existence of 7 subfamilies containing a total of 14 distinct receptors based on primary sequence, pharmacology and signal transduction pathways (1-6).

Serotonin receptor dysfunction has been implicated in a wide array of neuropsychiatric disorders including anxiety, depression and schizophrenia (4,5,7). To date, clinical use of therapeutics affecting serotonergic function is best exemplified by selective serotonin reuptake inhibitors (SSRI's) for depression and the triptans for migraine. Additionally, serotonin receptors have served as therapeutic targets for appetite suppression in obesity (fenfluramine).

A multitude of non-selective ligands for the various serotonin and other monamine receptors have been identified in the past decade, some of which have found clinical use. More recently selective serotonin receptor modulators have been the focus of many research groups. The goal of this report is to summarize the recent advances in identifying potent and selective serotonin receptor ligands. These tools will in large part dictate the direction and progress of research into serotonin based therapeutics.

5-HT1 RECEPTOR FAMILY

The 5-HT1 receptors are seven transmembrane receptors that couple to G-proteins (as are all but one of the serotonin receptor subfamilies). Since 5-HT1 receptors couple with either the G-protein Go or Gi they are generally regarded to be negatively coupled to adenylyl cyclase (8). Within the 5-HT1 receptor family, five receptor subtypes have been identified (5-HT1A, 5-HT1B, 5-HT1D, 5-HT1E, and 5-HT1F). 5-HT1 receptors have received much attention with respect to drug development. The 5-HT1A subtype has served as a target for anxiolytic treatment (Buspar™) and has been suggested to play a role in the antidepressant action of SSRI's. 5-HT1 receptors have played a more prominent role in the treatment of migraine with 5-HT1B, D, E, and F all being examined as targets for this indication (9).

5-HT1A receptor ligands - Among the serotonin receptors the 5-HT1A subtype is the most widely studied. The involvement of this receptor in the treatment of psychoses, depression (7,10,11) and anxiety (12) is well documented. Ligands with limited reported selectivity have been recently identified as 5-HT1A antagonists (13). Derivatives **1** and **2** display 5-HT1A Ki values of 2.4 and 7 nM, respectively, with >250-fold selectivity over dopamine D2 and adrenergic α1 receptors. In the area of selective 5-HT1A agonists the preparation of various novel compounds has been reported. The optimization of a series of piperidinyl benzamides afforded derivatives **3** and **4** which expressed subnanomolar affinity for 5-HT1A with >500-fold selectivity over dopamine D2 and adrenergic α1 receptors (14). A second report, focused on optimization of the pharmacokinetic profiles and *in vivo* efficacies, yielded ligands **5-7** (15). These selective 5-HT1A agonists demonstrated oral bioavailability and antidepressant activity. The results of the forced swimming test (FST) in rats showed derivatives **5-7** inhibited immobility in this assay more potently than the antidepressant imipramine (16-18).

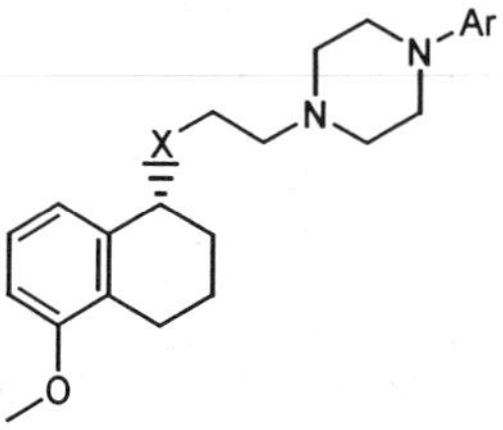

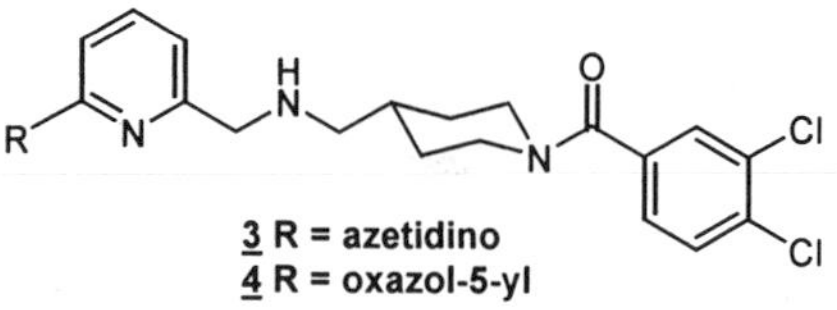

3 R = azetidino
4 R = oxazol-5-yl

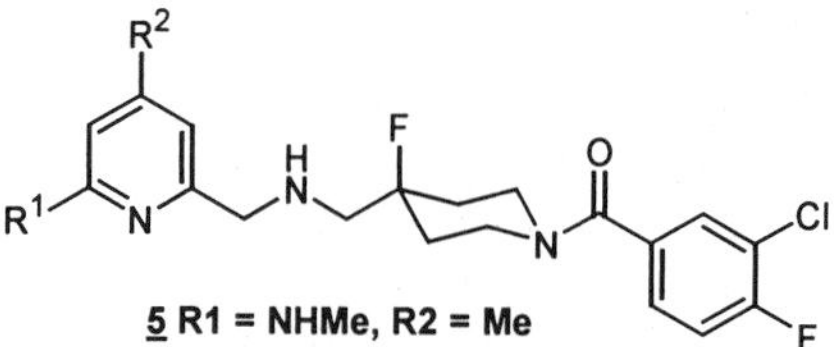

1 X = CH$_2$, Ar = Ph
2 X = NHCH$_2$, Ar = 2-MeOPh

5 R1 = NHMe, R2 = Me
6 R1 = NMe$_2$, R2 = H
7 R1 = furan-2-yl, R2 = H

5-HT1B receptor ligands - Progress in the identification of selective ligands for the 5-HT1B autoreceptor has recently been reported. Using computational chemistry models as a guide, the selective 5-HT1B inverse agonist **8** (SB-224289) was prepared (19). This potent ligand, with pKi value of 8.2, demonstrated 100-fold selectivity over numerous monoamine receptor sites and potent antagonism of 5-HT1B autoreceptors both *in vitro* and *in vivo*. Another report of a selective 5-HT1B antagonist is depicted by chromene **9** (20). This ligand shows moderate affinity (Ki 47 nM) for the rat 5-HT1B receptor with excellent selectivity over other monoamine sites. *In vitro* antagonism was demonstrated by potentiation of the K+-stimulated release of [3H]-5-HT from rat brain slices.

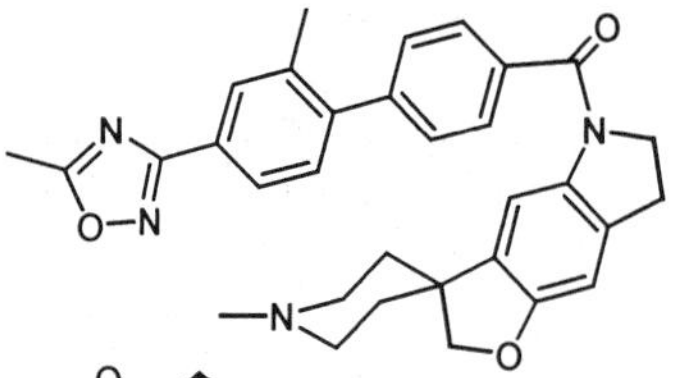

5-HT1D receptor ligands - The last decade has seen the development of multiple 5-HT1B/1D receptor agonists as antimigraine therapies. Termed "triptans", they have been extensively studied in attempts to elucidate their mechanism of action. These equipotent ligands at 5-HT1B and 5-HT1D receptors possess both a direct effect on vasoconstriction of cranial blood vessels (caused by 5-HT1B) (21), and inhibition of neurogenic inflammation (caused by 5-HT1B and 5-HT1D) (22). Efforts to prepare selective 5-HT1D agonists are aimed at effectively treating migraine headaches without the liability for vasoconstriction, a side effect that obviates their use with heart disease patients (23,24). In the last few years excellent progress in the preparation of selective 5-HT1D agonists has been realized. The first report of such a ligand is **10** (PNU-109291), which shows Ki = 0.9 nM, excellent (>100-fold) selectivity for all relevant receptor sites and >70% bioavailability (25). This selective ligand was shown to be >300 times more potent than sumatriptan, a non-selective 5-HT1D/1B agent, in an assay measuring inhibition of guinea pig dural plasma extravasation (26,27). Furthermore, comparison of **10** to sumatriptan in a carotid artery resistance assay in the cat was performed (25). While sumatriptan produced a dose-related increase in carotid resistance without affecting mean arterial pressure, the selective 5-HT1D agonist **10** was devoid of any changes in carotid resistance or mean arterial pressure, thus demonstrating it's lack of vasoconstrictive properties.

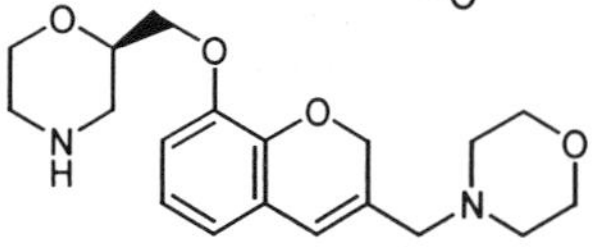

An extensive SAR optimization of a series of 3-substitiuted indoles has afforded several very potent, selective and orally bioavailable 5-HT1D agonists. Analog **11**, a reportedly potential development candidate in the treatment of migraine, has a 5-HT1D Ki value of 0.5 nM (28). This ligand is highly selective over other serotonin subtypes as well as >500-fold less potent in a broad screen of receptors and enzymes, and possesses a pharmacokinetic profile amenable to oral administration. Additional work aimed at optimization of this scaffold *via* modulation of the basicity of the piperazine nitrogen was undertaken (29). It was proposed that increasing the basicity would allow for better oral absorption *via* increased gut wall transport (30). A series of fluorinated derivatives was synthesized and evaluated, culminating in analogs such as **12**, where fluorination of the propyl linkage effects the desired result. While the potency and selectivity profile of these compounds are maintained, the oral bioavailability is increased significantly as compared to the non-fluorinated derivatives. Finally, a related class of pyrrolidinylethyl indoles was reported (31). Although optimization in this series afforded ligands such as **13**, with 5-HT1D Ki value of 0.45 nM and > 90-fold selectivity over 5-HT1B, efforts to effect any appreciable oral bioavailability in animals were not realized.

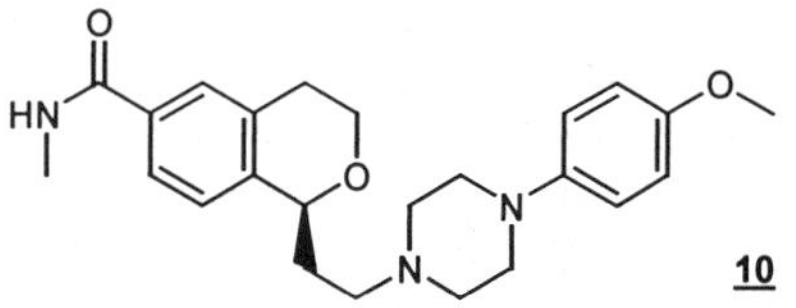

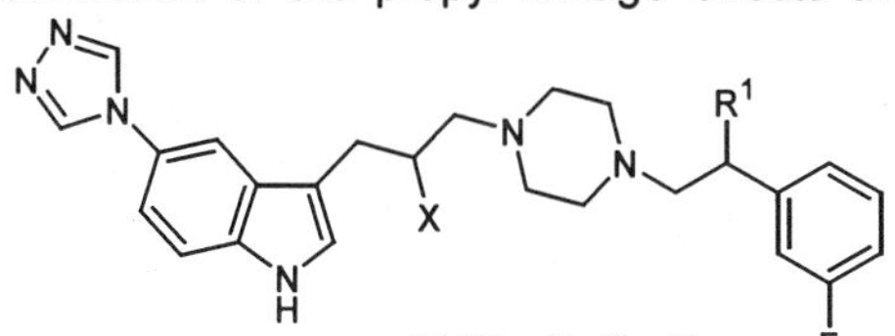

11 R1 = H X = H
12 R1 = Me X = F

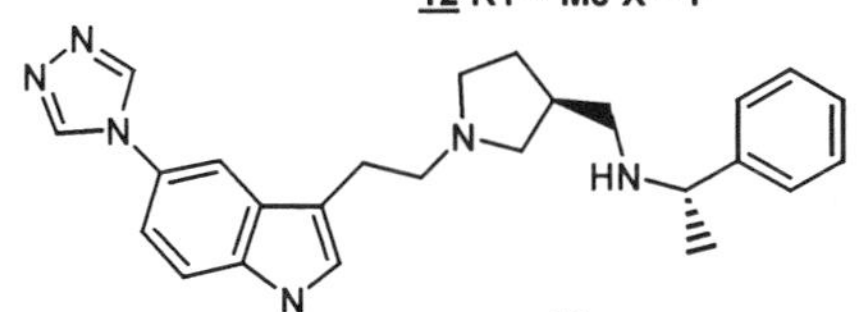

13

5-HT1E receptor ligands – The identification of this receptor subtype over a decade ago has yet to generate any significant effort in this area (32).

5-HT1F receptor ligands - Since last reviewed there have not been any reports of additional selective 5-HT1F ligands (33). However, the previously profiled 5-HT1F agonist **14** (LY334370) shows excellent potency (Ki 1.6 nM) with selectivity over 5-HT1A when measured functionally (34). This agent was reported to be a potent inhibitor of neurogenic dural inflammation and was undergoing clinical development as a possible therapy for migraine (35). Early results of phase II clinical trials of **14** indicated its efficacy in treating migraine with no cardiovascular side effects. However, a recent report indicated the suspension of **14** from phase III clinical trials due to non-mechanism based liver toxicity(36).

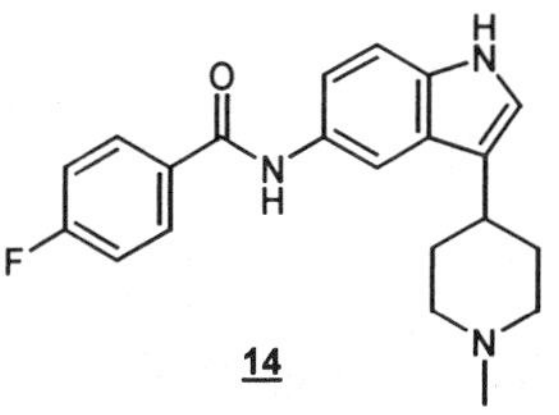

14

5-HT2 RECEPTOR FAMILY

The 5-HT2 receptor family represents a significant component, both in terms of function and clinical use, of the serotonin receptors. As a subfamily, these GPCRs couple though the Gq/11 family of G-proteins eliciting their second messenger effects predominantly through increases in activity of phospholiopase C (diacylglycerol pathway) and/or phospholipase A (arachidonic acid pathway). Three distinct subtypes of 5-HT2 receptors exist: 5-HT2A, 5-HT2B, and 5-HT2C. These subtypes share an overall amino acid homology of approximately 50% (5).

5-HT2A receptor ligands - The 5-HT2A receptor has been implicated as a therapeutic target for schizophrenia and depression (37-39). The most widely investigated selective 5-HT2A antagonist, **15** (M100907), is the standard for this class of selective agents, exhibiting a 5-HT2A Ki value of 0.4 nM with >100-fold selectivity over a wide range of receptors and enzyme sites (40). **15** has been the subject of intense scrutiny in the recent past culminating in a disappointing phase III clinical trial result (psychoses) and subsequent withdrawal from further development for chronic schizophrenia. Since the last review of this area there have not been any reports of selective 5-HT2A antagonists (33). There exist a plethora of combination (mixed dopamine, transporter, histamine, etc) and nonselective agents with affinity for the 5-HT2A receptor. However, identification of ligands with selectivities >50-fold versus other monoamine receptors and the various transporters has been rare. A recent example of a combination agent is portrayed by compound **16** (QF0610B). Possessing excellent selectivity over the 5-HT2C receptor, historically the most difficult to separate from, **16** is only modestly selective (10 - 20-fold selective) over the dopamine D2 receptor (41). Ligands of this type are currently being considered for use as atypical antipsychotic therapies (42-44).

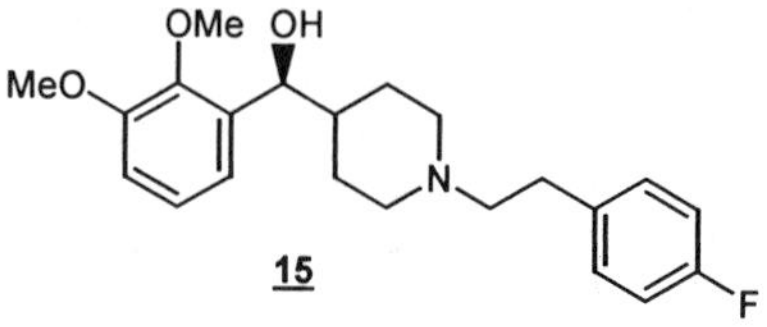

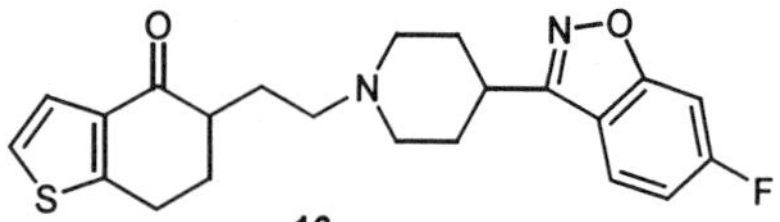

5-HT2B receptor ligands - Initially identified in rat stomach fundus (45), the 5-HT2B receptor has been implicated in the treatment of migraines and gastric motility (46-48). Additionally, 5-HT2B receptor activation has been reported too produce hyperphagia (49), anxiolysis (50) and cell proliferation (51) possibly causing the heart valvulopathies associated with chronic use of fenfluramine (52). Compound **17** (RS-127,445) has recently been described as a selective, high affinity, orally bioavailable 5-HT2B antagonist (53). This arylpyrimidine ligand (pKi 9.5) is 1,000-fold selective relative to numerous other receptors and ion channel binding sites.

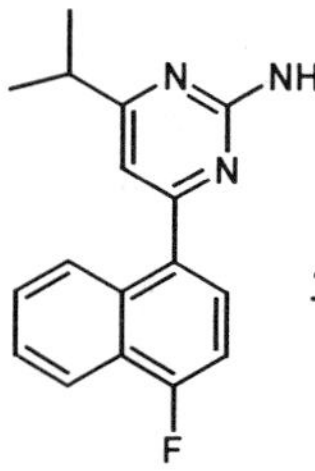

5-HT2C receptor ligands - The potential uses for 5-HT2C ligands include anxiety, depression, obesity and cognitive disorders (54-58). It has been shown that 5-HT2C receptor knockout mice exhibit obesity, epilepsy and cognitive dysfunction (59). The interest in selective receptor agonists and antagonists for the 5-HT2C subtype is currently very high. However, truly selective ligands have been elusive with a scant number of new developments since this area was last reviewed (19,60). With respect to 5-HT2C agonists, there have been no recent reports of selective ligands. However, in the antagonist area the search for selective compounds devoid of 5-HT2B activity has been realized. Optimization of a series of previously identified selective 5-HT2B/2C antagonists has afforded indolines **18** (SB-228357) and **19** (SB-243213) which show >100-fold selectivity over the closely related 5-HT2A and 5-HT2B receptor subtypes (61,62). In addition, these potent 5-HT2C antagonists (Ki 1.0 nM) were shown to be >100-fold selective over more than 50 receptor, ion channel and enzyme binding sites. The bispyridyl ether **19** is reported to be in phase I clinical development as an antidepressant/anxiolytic agent.

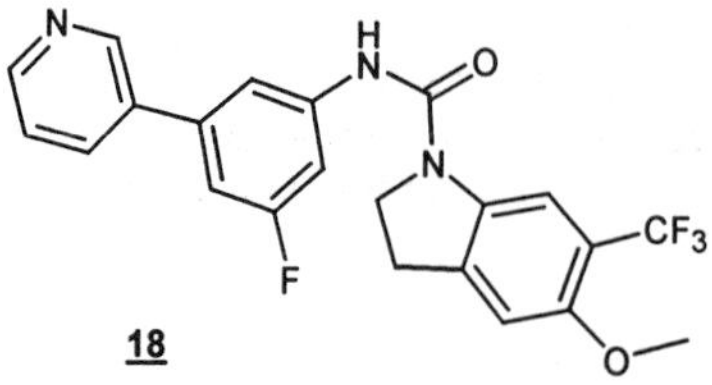

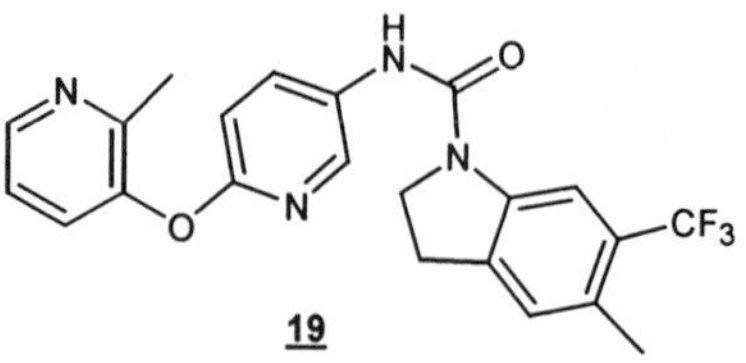

OTHER 5-HT RECEPTORS

5-HT3 receptor ligands - The 5-HT3 receptor is a ligand gated cation channel, distinct from all the other serotonin family members which are GPCRs. The recent focus on selective 5-HT3 antagonists has centered on their potential to affect various CNS related disorders and possibly irritable bowel syndrome (63-67). It has been shown that 5-HT3 receptors are localized to enteric neurons (68,69) and the wide clinical use of these agents to prevent chemotherapy induced emesis is well documented (70-72). Further possible therapeutic indications include the treatment of anxiety, psychoses, anorexia, as well as cognitive disorders and drug abuse. Several recent accounts of compounds with selectivities over a limited set of receptors have been reported. The new structural class represented by **20** has afforded analogs with excellent potency (5-HT3 Ki = 10-20 nM) and demonstrated selectivity (>100-fold) over 5-HT4 and dopamine D2 receptors (73). A second series of related compounds is exemplified by **21** and possesses a comparable profile (74). The related benzimidazole derivative **22** shows a structural similarity to a selective 5-HT4 antagonist described below, but with marked 5-HT3 selectivity (75). In this series, multiple derivatives showed nanomolar potencies with >100-fold selectivity over 5-HT4 and 5-HT1A.

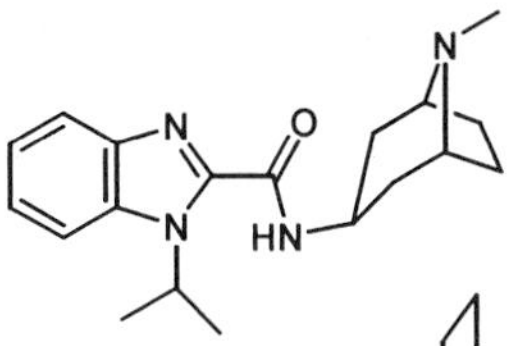

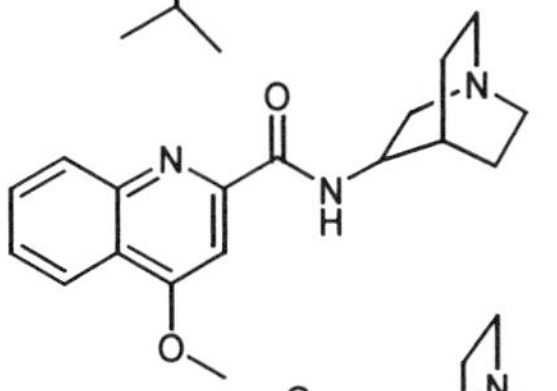

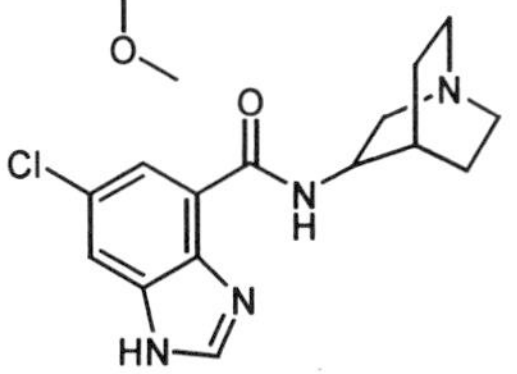

Two separate reports (76,77) of related fused heteroaromatic systems, represented by **23** and **24**, are examples of agonists based on quipazine **25**, the known but nonselective 5-HT3 ligand (78). Both series were shown to possess excellent potency with selectivity over various serotonin receptors. The pyrroloquinoxaline derivative **23** exhibited 5-HT3 Ki value of 0.80 nM in the receptor binding assay but failed to demonstrate *in vivo* anxiolytic activity. The functional efficacy of the 5-HT3 partial agonist **24** was shown to be dependent upon subtle changes to the tricyclic backbone (79).

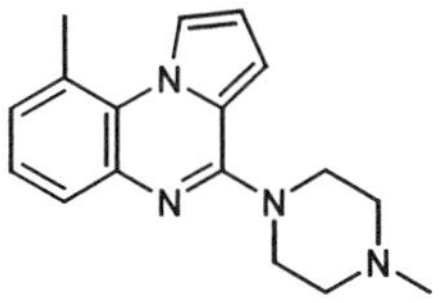

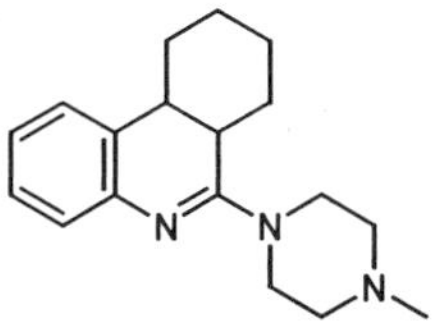

5-HT4 receptor ligands - The 5-HT4 receptor is a GCPR, which couples to the G-protein Gs and activates adenylyl cyclase (as do the 5-HT6 and 5-HT7 receptors). The localization of the 5-HT4 receptor to areas of the brain associated with dopamine function (such as the striatum and nucleus accumbens) (80) and the ability of 5-HT4 ligands to affect dopamine release (81) have suggested a role for 5-HT4 ligands in the treatment of Parkinson's disease and possibly schizophrenia. Peripherally, 5-HT4 receptors are localized in various organs, particularly along the alimentary tract giving rise to effects that suggest possible therapeutic roles in irritable bowel syndrome, gastroparesis, and urinary incontinence (82-85). However, the clinical interest in the development of selective 5-HT4 antagonists has been hampered by the paucity of selective ligands for this receptor subtype.

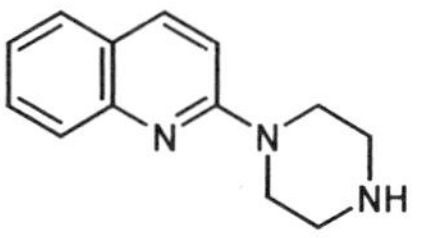

Benzimidazole derivatives of type **26-28**, with selectivity for 5-HT4 over 5-HT3, 5-HT2A and 5-HT1A receptors have recently been reported (86). These agents possess potent 5-HT4 antagonist activity in the guinea pig ileum (87), comparable to the known 5-HT4 receptor antagonist RS 39604 (88). Additional selectivity data for these subnanomolar 5-HT4 antagonists was not given. An interesting example of a class of compounds acting as either agonists or antagonists, dependant upon subtle changes in substitution, is represented by **29** and **30**. The potent antagonist properties of **29** compared to the high partial agonist **30** are suggested by the authors as a possible confirmation of two binding sites at the 5-HT4 receptor (89). These derivatives showed excellent selectivity (>100-fold) over 5-HT3 and dopamine D2 receptors, with no additional selectivity data offered. Progress in the design of selective 5-HT4 agonists has been reported (90) for benzamide derivatives **31** and **32**. These potent ligands were shown to be selective for the 5-HT4 receptor when compared to serotonin, dopamine and muscarinic receptors. The dihydrobenzofuran **31** was reported to exhibit greater potency as a 5-HT4 agonist than cisapride or mosapride, two clinically utilized gastroprokinetic agents which possess unwanted side effects due to their potent dopamine D2 antagonist activity (91,92).

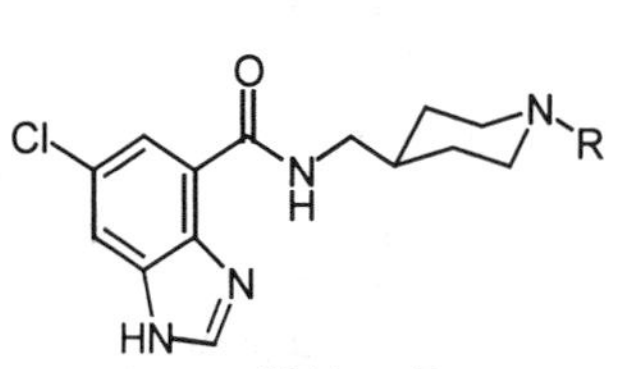

26 R = n-Bu
27 R = iBu
28 R = EtNHSO$_2$Me

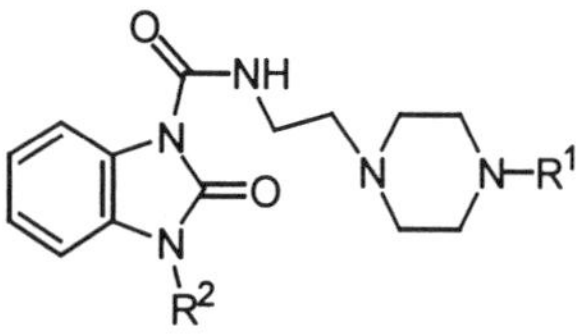

29 R1 = iPr   R2 = cyc-Pr
30 R1 =Me   R2 = iPr

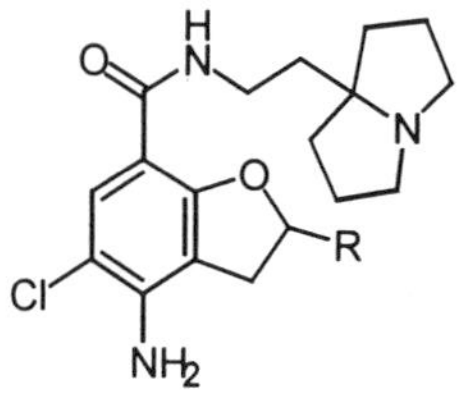

31 R = Me
32 R = Et

The preparation and SAR optimization of a series of carboxamide derivatives as 5-HT4 antagonists has been reported (93). The selectivity versus 5-HT3 and dopamine D2 is > 100-fold for both **33** and **34**. Interesting is the *in vivo* potency of **34** (EC$_{50}$ 1.2 nM), as measured by contractile effects in the guinea pig ascending colon, as compared to its *in vitro* receptor binding affinity (Ki 300 nM) at the 5-HT4 receptor (94).

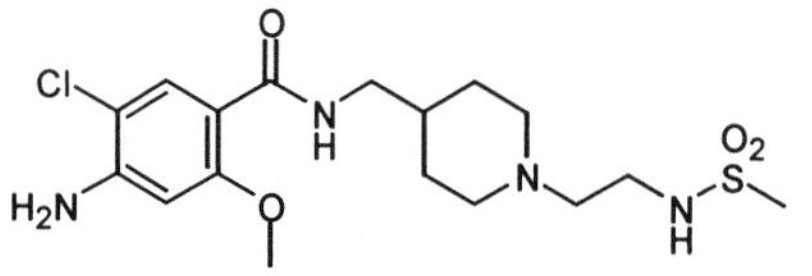

33 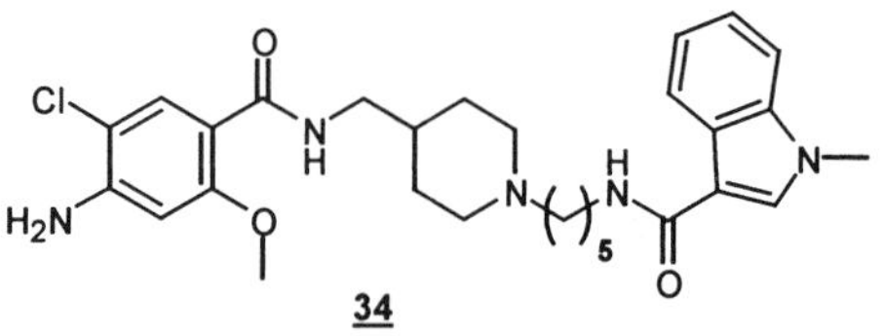   **34**

5-HT5 receptor ligands - To date there have been no reports of selective 5-HT5 receptor ligands. This subfamily of serotonin GPCRs has continued to be poorly characterized with respect to G-protein coupling mechanisms and function.

5-HT6 receptor ligands - The 5-HT6 receptor is another GCPR in the serotonin family which positively couples to adenylyl cyclase through the G-protein Gs (67,95,96). By mRNA, antibody mapping and radioligand binding, the receptor distribution in the CNS of humans and rats is most evident in the striatum with densities also noted in amygdala, nucleus accumbens, hippocampus, cortex, and olfactory tubercle (97-99). The localization of 5-HT6 receptors to limbic regions and the high affinity of clozapine, the most effectve antipsychotic on the market, have provided strong impetus to examine 5-HT6 receptors as a target for schizophrenia therapeutics (7,100).

There exists an array of nonselective agents such as antidepressants, antipsychotics and ergolines that bind to the 5-HT6 receptor with excellent affinity (the antipsychotic therapeutic clozapine being a notable example) (101,102). However, until recently, selective ligands for this serotonin subtype had not been identified. The first highly selective 5-HT6 antagonists (103) are exemplified by **35** (Ro 04-6790) and **36** (Ro 63-0563). These arylsulfonamides possess 5-HT6 pKi values of 7.26 and 7.91, respectively. The potential for these ligands as CNS agents may be limited due to the poor (<1%) brain penetrance, however intraperitoneal administration of **35** to rats produced sufficient brain levels to evoke an effect on stretching similar to that seen with antisense oligonucleotide treatment (103). The preparation and characterization of the orally bioavailable, brain penetrant 5-HT6 selective antagonist **37** (SB 271046) from a high throughput screening lead **38** has been reported (104). This optimized benzothiophene sulfonamide shows excellent potency (5-HT6 pKi 8.9) and selectivity, with moderate brain penetrance (10%) and a suitable pK profile for oral administration.

35 X = N
36 X = CH

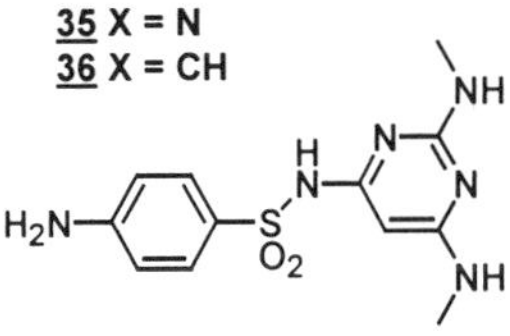

37

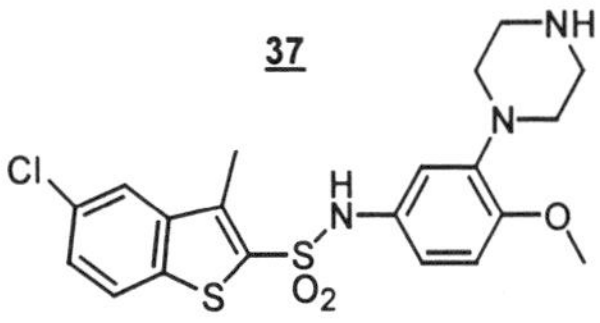

38

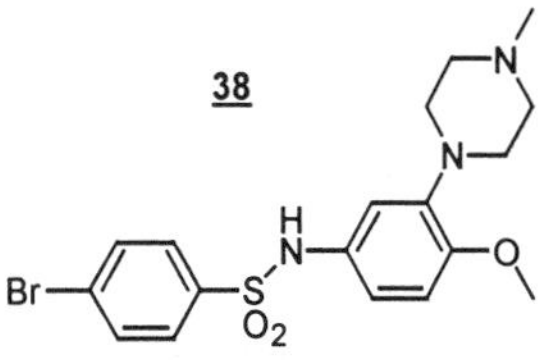

<u>5-HT7 receptor ligands</u> - The pharmacology and molecular biology of the 5-HT7 receptor, the newest member of the serotonin family, has recently been reported (105,106). Sequence alignments show a low overall homology (40%) with other 5-HT receptors but a high degree of inter-species homology (95%). This receptor subtype is positively coupled through G-protein Gs to adenylyl cyclase, and found mostly in brain where it is localized in the thalamus, hypothalamus and cortical regions (107-109). A possible therapeutic indication associated with the 5-HT7 receptor is migraine. Similarly to 5-HT2B receptors, 5-HT7 receptors have been localized to the cerebrovascular and may serve a role in vasodilation with subsequent sensory trigeminovascular afferents activation and pain associated with migraine.

Recent efforts in identifying 5-HT7 ligands has resulted in the first selective 5-HT7 antagonist **39** (5-HT7 pKi 7.5) reported to have >100-fold selectivity over a wide range of receptors (110). Additional optimization of this ligand, based partly on molecular modeling analysis, has afforded the ring constrained analog **40** which is reported (111) to possess improved selectivity and potency (5-HT7 pKi 8.9). Additionally, a series of tetrahydrobenzindole derivatives has been reported. SAR optimization has led to the identification of **41** (DR4004) as a potent (5-HT7 pKi 8.7) and selective (50-fold selective over other 5-HT and dopamine receptors) 5-HT7 antagonist (112).

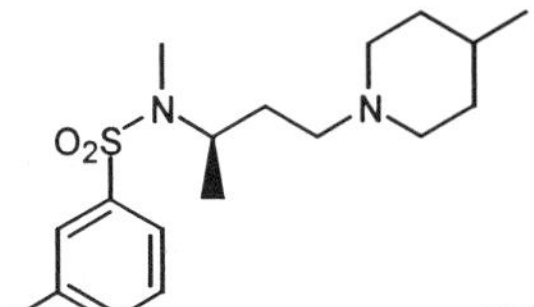

39

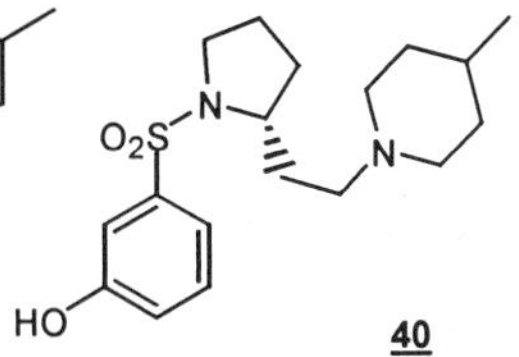

40

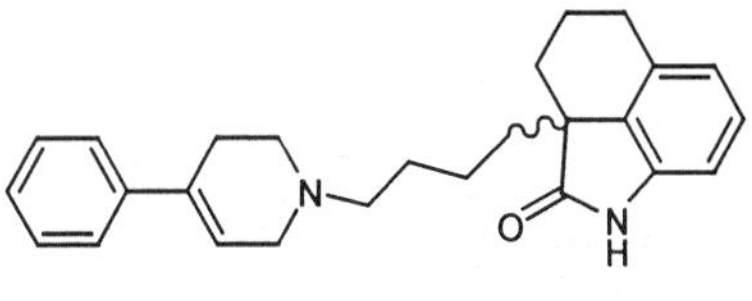

41

<u>Conclusion</u> -The advances in molecular biology coupled with the advent of high throughput synthesis and progress seen in asymmetric chemistry in the last decade have given rise to a multitude of novel selective serotonin subtype ligands. Utilization of these agents will enable a greater understanding of serotonin's role in the pathophysiology of various disease states. The increased focus on CNS therapeutics, and serotonin mediated pathways in particular, will fuel the progress of this research. The manifest goal of this effort is to develop selective serotonin therapeutics that hold improvements over current non-selective serotonergic agents with respect to efficacy, side-effect profile and safety margin.

References

1.   W.K. Kroeze and B.L. Roth, Biol. Psychiatry, 44, 1128 (1998).
2.   E. Hamel, Can.J.Neurol.Sci., 26, Suppl.3 (1999).
3.   M. Baez, J.D. Kursar, L.A. Helton, D.B. Wainscott and D.L. Nelson, Obes.Res., 3, 441S (1995).
4.   L. Uphouse, Neuroscience & Behavioral Rev., 21, 679 (1997).
5.   P.R. Hartig, "Molecular Biology and Transductional Characteristics of 5-HT Receptors", Springer,, 1997, Vol. 129(7).
6.   F.G. Boess and I.L. Martin, Neuropharmacology, 33, 275 (1993).
7.   H.Y. Meltzer, Neuropsychopharmacology, 21, 106S (1999).
8.   C.C. Gerhardt and H.v. Heerikhuizen, Euro.J.Pharmacology, 334, 1 (1997).
9.   D. Hoyer and G. Martin, Neuropharmacology, 36, 419 (1997).
10. C.L. Broekkamp, D. Leysen, B.W. Peeters and R.M. Pinder, J.Med.Chem., 38, 4615 (1995).
11. P. Blier and C.d. Montigny, Trends Pharmacol. Sci, 15, 220 (1994).
12. M. Hamon, Trends Pharmacol.Sci, 15, 36 (1994).
13. R. Perrone, F. Berardi, N. Colabufo, M. Leopoldo and V. Tortorella, J.Med.Chem., 42, 490 (1999).
14. B. Vacher, B. Bonnaud, P. Funes, N. Jubault, W. Koek, M. Assie and C. Cosi, J.Med. Chem., 41, 5070 (1998).
15. B. Vacher, B. Bonnaud, P. Funes, N. Jubault, W. Koek, M. Assie, C. Cossi and M. Kleven, J.Med.Chem., 42, 1648 (1999).
16. F. Borsini, Neurosci.Biobehav.Rev., 19, 377 (1995).
17. M.J. Detke, S. Weiland and I. Lucki, Psychopharmacology, 119, 47 (1995).
18. G.E. Simon, M. VonKorff, J.H. Heiligenstein, D.A. Revicki, L. Grothaus and W. Katon, JAMA, 275, 1897 (1996).
19. L.M. Gaster, F.E. Blaney, S. Davies, D.M. Duckworth, P. Ham, S. Jenkins, A. Jennings, G. Joiner, F. King, K. Mulholland, P. Wyman, J. Hagan, J. Hatcher, B. Jones, D. Middlemiss, G. Price, G. Riley, C. Roberts, C. Routledge, J. Selkirk and P. Slade, J.Med.Chem., 41, 1218 (1998).
20. S. Berg, L.-G. Larsson, L. Renyi, S.B. Ross, S.-O.Thorberg and G.Thorell-Svantesson, J.Med.Chem., 41, 1934 (1998).
21. G.W. Rebeck, K.I. Maynard, B.T. Hyman and M.A. Moskowitz, Proc.Natl.Acad.Sci.., 91, 3666 (1994).
22. M.A. Moskowitz and R. Macfarlane, Cerebrovasc.Brain Metab.Rev., 5, 159 (1993).
23. E. Hamel, L. Gregoire and B. Lau, Eur.J.Pharmacol., 75, 242 (1993).
24. E. Hamel, Serotonin, 1, 19 (1996).
25. M.D. Ennis, N.B. Ghazal, R.L. Hoffman, M.W. Smith, S.K. Schlachter, C.F. Lawson, W.B. Im, J.F. Pregenzer, K.A. Svensson, R.A. Lewis, E.D. Hall, D.M. Sutter and R.B. McCall, J.Med.Chem., 41, 2180 (1998).
26. P.P.A. Humphrey, W. Feniuk, M.J. Perren, H.E. Connor, A.W. Oxford, I.H. Coates and D. Butina, Br.J.Pharmacol., 94, 1123 (1988).
27. S. Markowitz, K. Saito and M.A. Moskowitz, J.Neurosci., 7, 4129 (1991).
28. M.S. Chambers, L.J. Street, S. Goodacre, S.C. Hobbs, P. Hunt, R.A. Jelley, W.G. Matassa, A.J. Reeve, F. Sternfeld, M.S. Beer, J.A. Stanton, D. Rathbone, A.P. Watt and A.M. MacLeod, J.Med.Chem., 42, 691 (1999).
29. M.B.v. Niel, I. Collins, M.S. Beer, H.B. Broughton, S.K.F. Cheng, S.C. Goodacre, A. Heald, K.L. Locker, A.M. MacLeod, D. Morrison, C.R. Moyes, D. O'Connor, A. Pike, M. Rowley, M.G.N. Russell, B. Sohal, J.A. Stanton, S. Thomas, H. Verrier, A.P. Watt and J.L. Castro, J.Med.Chem., 42, 2087 (1999).
30. J.L. Castro, I. Collins, M.G.N. Russell, A.P. Watt, B. Sohal, D. Rathbone, M.S. Beer and J.A. Stanton, J.Med.Chem., 41, 2667 (1998).
31. F. Sternfeld, R.A. Jelley, V.G. Matassa, A.J. Reeve, P.A. Hunt, M.S. Beer, A. Heald, J.A. Stanton, B. Sohal, A.P. Watt and L.J. Street, J.Med.Chem., 42, 677 (1999).
32. S. Leonhardt, K. Herrick-Davis and M. Titeler, Neurochem., 53, 465 (1989).
33. L.M. Gaster and F.D. King, Ann.Rev.Med.Chem., 35, 21 (1997).
34. L. Phebus, Soc. For Neuroscience, 22, 528 (1996).
35. C.D. Overshiner, Soc. for Neuroscience, 22, 528 (1996).
36. E.L.a.C.P. Release, , (3/19/99).
37. A. Carlsson, N. Waters and M.L. Carlsson, Biol. Psychiatry, 46, 1388 (1999).
38. G.J. Marek and G.K. Aghajanian, Biol. Psychiatry, 44, 1118 (1998).
39. E. Sibelle, Z. Sarnyai, D. Benjamin, J. Gal, H. Baker and M. Toth, Mol.Pharmacol., 52, 1056 (1997).
40. J.H. Kehne, J.Pharm.Exp.Ther., 277, 968 (1996).

41. E. Ravina, J. Negreira, J. Cid, C.F. Massguer, E. Rosa, M.E. Rivas, J.A. Fontenla, M.I. Loza, H. Tristan, M.I. Cadavid, F. Sanz, E. Lozoya, A. Carotti and A. Carrieri, J.Med.Chem., 42, 2774 (1999).

42. H.Y. Meltzer, S. Matsubara and J.C. Lee, J.Pharmacol.Exp.Ther., 251, 238 (1989).

43. H.Y. Meltzer, S. Matsubara and J.C. Lee, Psychopharmacol.Bull., 25, 390 (1989).

44. B.L. Roth, S. Tandra, L.H. Burgess, D.R. Sibley and H.Y. Meltzer, Psychopharmacology, 120, 365 (1995).

45. B.V. Clineschmidt, D.R. Reiss, D.J. Pettibone and J.L. Robinson, J.Pharmacol.Exp.Ther., 235, 696 (1985).

46. K. Schmuk, C. Ullmer, H.O. Kalkman, A. Probet and H. Lubbert, Eur.J.Neurosci., 8, 595 (1996).

47. H.O. Kalkman, Lif.Sci., 54, 641 (1994).

48. J.R. Fozard and H.O. Kalkman, Naunyn-Schmiedeberg's Arch., 350, 225 (1994).

49. G.A. Kennett, K. Ainsworth, B. Trail and T.P. Blackburn, Neuropharmacology, 36, 233 (1997).

50. G.A. Kennett, F. Bright, B. Trail, G.S. Baxter and T.P. Blackburn, Br.J.Pharmacol., 117, 1443 (1996).

51. J.-M. Launay, G. Birraux, D. Bondoux, J. Callebert, D.-S. Choi, S. Loric and L. Maroteaux, J.Biol.Chem., 271, 3141 (1996).

52. L.W. Fitzgerald, T.C. Burn, B.S. Brown, J.P. Patterson, P.A. Valentine, J.-H. Sun, J.R. Link, I. Abbaszade, J.M. Hollis, B.L. Largent, P.R. Hartig, G.F. Hollis, P.C. Meunier, A.J. Robichaud and D.W. Robertson, Mol.Pharmacol., 57, 75 (2000).

53. D.W. Bonhaus, L.A. Flippin, R.J. Greenhouse, S. Jaime, C. Rocha, M. Dawson, K.V. Natta, L.K. Chang, T. Pulido-Rios, A. Webber, E. Leung, R.M. Eglen and G.R. Martin, Br.J.Pharmacol., 127, 1075 (1999).

54. D. Leysen and J. Kelder, Trends in Drug Research II, 49(1998).

55. G.A. Kennett, Curr.Opin.Invest.Drugs, 2, 317 (1993).

56. G.A. Kennett, I.Drugs, 1, 456 (1998).

57. S.F. Leibowitz and J.T. Alexander, Biol.Psychiatry, 44, 851 (1998).

58. C.T. Dourish, Obes.Res., 3, 449S (1995).

59. L.H. Tecott, J. Psychopharmacol., 10, 223 (1996).

60. D.C.M. Leysen, I.Drugs, 2, 109 (1999).

61. S.M. Bromidge, S. Dabbs, D.T. Davies, D.M. Duckworth, I.T. Forbes, P. Ham, G.E.Jones, F.D. King, D.V. Saunders, S. Starr, K.M. Thewlis, P.A. Wyman, F.E. Blaney, C.B. Naylor, F. Bailey, T.P. Blackburn, V. Holland, G.A. Kennett, G.J. Riley and M.D. Wood, J.Med.Chem., 41, 1598 (1998).

62. S.M. Bromidge, S. Dabbs, D.T. Davies, S. Davies, D.M. Duckworth, I.T. Forbes, L.M. Gaster, P. Ham, G.E. Jones, F.D. King, K.R. Mulholland, D.V. Saunders, P.A. Wyman, F.E. Blaney, S.E. Clarke, T.P. Blackburn, V. Holland, G.A. Kennett, S. Lightowler, D.N. Middlemiss, B. Trail, G.J. Riley and M.D. Wood, J.Med.Chem., 43, 1123 (2000).

63. N. Kishibayashi, Y. Miwa, H. Hayashi, A. Ishii, S. Ichikawa, H. Nonaka, T. Yokoyama and F. Suzuki, J.Med.Chem., 36, 3286 (1993).

64. B. Costall and R.J. Naylor, Pharmacol.Toxicol., 70, 157 (1992).

65. A.J. Greenshaw and P.H. Silverstone, Drugs, 53, 20 (1997).

66. F.E. Bloom and M. Morales, Neurochem.Res., 23, 653 (1998).

67. R. Kohen, M.A. Metcalf, N. Khan, T. Druck, K. Huebner, J.E. Lachowicz, H.Y. Meltzer, D.R. Sibley, B.L. Roth and M.W. Hamblin, J. Neurochem., 66, 47 (1996).

68. M.D. Gershon, Adv.Exp.Med.Biol., 294, 221 (1991).

69. J.J. Galligan, Behav.Brain.Res., 73, 199 (1996).

70. K. Bunce, M. Tyers and P. Beranek, Trends Pharmacol.Sci., 12, 46 (1991).

71. C. Veyrat-Follet, R. Farinotti and J.L. Palmer, Drugs, 53, 206 (1997).

72. R.E. Gregory and D.S. Ettinger, Drugs, 55, 173 (1998).

73. A. Orjales, L. Alonso-Cires, P. Lopez-Tudanca, I. Tapia, R. Mosquera and L. Labeaga, Eur.J.Med.Chem., 34, 415 (1999).

74. A. Orjales, L. Alonso-Cires, P. Lopez-Tudanca, I. Tapia, L. Labeaga and R. Mosquera, Drug Design and Discovery, 10, 271 (2000).

75. M. Lopez-Rodriguez, B. Benhamu, M.J. Morcillo, I.D. Tejada, L. Orensanz, M.J. Alfaro and M.I. Martin, J.Med.Chem., 42, 5020 (1999).

76. G. Campiani, E. Morelli, S. Gemma, V. Nacci, S. Butini, M. Hamon, E. Novellino, G. Greco, A. Cagnotto, M. Goegan, L. Cervo, F.D. Valle, C. Fracasso, S. Caccia and T. Mennini, J.Med.Chem., 42, 4362 (1999).

77. A. Cappelli, M. Anzini, S. Vomero, L. Canullo, L. Mennuni, F. Makovec, E. Doucet, M. Hamon, C. Menziani, P.D. Benedetti, G. Bruno, M.R. Romeo, G. Giorgi and A. Donatti, J.Med.Chem., 42, 1556 (1999).

78. S.J. Ireland and M.B. Tyers, Br.J.Pharmacol., 90, 229 (1987).

79. A. Cappelli, M. Anzini, S. Vomero, L. Mennuni, F. Makovec, E. Doucet, M. Hamon, G. Bruno, M.R. Romeo, M.C. Menziani, P.D. Benedetti and T. Langer, J.Med.Chem., 41, 728 (1998).

80. V. Compan, A. Daszuta, P. Salin, M. Sebben, J. Bockaert and A. Dumuis, Eur.J.Neurosci., 8(12), 2591 (1996).

81. L.J. Steward, J. Ge, R.L. Stowe, D.C. Brown, R.K. Bruton, P.R. Stokes and N.M. Barnes, Br.J.Pharmacol., 117, 55 (1996).

82. S.S. Hedge and R.M. Eglen, FASEB J., 10, 1398 (1996).

83. E. Leung, M.T. Pulido-Rios, D.W. Bonhaus, L.A. Pekins, K.D. Keitung, S.A. Hsu, R.D. Clark, E.H. Wong and R.M. Eglen, Naunyn Schiedebergs Arch.Pharmacol., , 145 (1996).

84. J.G. Jin, A.E. Foxx-Orenstein and J.R. Grider, J.Pharmacol.Exp.Ther., 288, 93 (1999).

85. L.A. Houghton, N.A. Jackson, P.J. Whorwell and S.M. Cooper, Aliment.Pharmacol.Ther., 13, 1437 (1999).

86. M.L. Lopez-Rodriguez, B. Benhamu, A. Viso, M.J. Morcillo, M. Murcia, L. Orensanz, M.J. Alfaro and M.I. Martin, Biorg.Med.Chemistry, 7, 2271 (1999).

87. R.M. Eglen, S.R. Swank, L.K.M. Walsh and R.L. Whiting, Br.J.Pharmacol., 101, 513 (1990).

88. S.S. Hegde, D.W. Bonhaus, L.G. Johnson, E. Leung, R.D. Clark and R.M. Eglen, Br.J.Pharmacol., 115, 1087 (1995).

89. I. Tapia, L. Alonso-Cires, P.L. Lopez-Tudanca, R. Mosquera, L. Labeaga, A. Innerarity and A. Orjales, J.Med.Chem., 42, 2870 (1999).

90. T. Kakigama, T. Usui, K. Tsukamoto and T. Kataoka, Chem.Pharm.Bull., 46, 42 (1998).

91. R.M. Pinder, R.N. Brogden, P.R. Sawyer, T.M. Speigh and G.S. Avery, Drugs, 12, 81 (1976).

92. R.W. McCallum, C. Prakash, D.M. Campoli-Richards and K.L. Goa, Drugs, 36, 652 (1988).

93. K. Itoh, K. Kanzaki, T. Ikebe, T. Kuroita, H. Tomozane, S. Sonda, N. Sato, K. Haga and T. Kawakita, Eur.J.Med.Chem., 34, 977 (1999).

94. K. Itoh, H. Tomozane, H. Hakira, S. Sonda, K. Asano, M. Fujimura, N. Sato, K. Haga and T. Kawakita, Eur.J.Med.Chem., 34, 1101 (1999).

95. A.J. Sleight, F.G. Boess, M. Bos and A. Bourson, Ann. NY Acad.Sci., 861, 91 (1998).

96. M. Ruat, E. Traiffort, J.M. Arrang, J. Tardivel-Lacombe, J. Diaz, R. Leurs and J. Schwartz, Biochem.Biophys.Res. Commun., 193, 268 (1993).

97. C. G'erard, S.e. Mestikawy, C. Lebrand, J. Adrien, M. Ruat, E. Traiffort, M. Hamon and M.P. Martres, Synapse, 23, 164 (1996).

98. C. G'erard, M.P. Martres, K. Lef'evre, M.C. Miquel, D. Verg'e, L. Lanfumey, E. Doucet, M. Hamon and S.e. Mestikawy, Brain Res., 746, 207 (1997).

99. F.G. Boess, C. Riemer, M. Bos, J. Bently, A. Bourson and A.J. Sleight, Mol. Pharmacol., 54, 577 (1998).

100. C.E. Glatt, A.M. Snowman, D.R. Sibley and S.H. Snyder, Mol.Med., 1, 398 (1995).

101. F.J. Monsma, Y. Shen, R.P. Ward, M.W. Hamblin and D.R. Sibley, Mol.Pharmacol., 43, 320 (1993).

102. B.L. Roth, S.C. Craigo, M.S. Choudhary, M.S. Uluer, M.J. Monsma, F.J. Shen, H.Y. Meltzer and D.R. Sibley, J.Pharm.Exp.Ther., 268, 1403 (1994).

103. A.J. Sleight, F.G. Boess, M. Bos, B. Levet-Trafit, C. Riemer and A. Bourson, Br.J.Pharmacol,, 124, 556 (1998).

104. S.M. Bromidge, A.M. Brown, E. Clarke, K. Dodgson, T. Gager, H.L. Grassam, P.M. Jeffrey, G.F. Joiner, F.D. King, D.N. Middlemiss, S.F. Moss, H. Newman, G. Riley, C. Routledge and P. Wyman, J.Med.Chem., 42, 202 (1999).

105. R.M. Eglen, J.R. Jasper, D.J. Chang and G.R. Martin, TiPS, 18, 104 (1997).

106. Y. Shen, F.J.M. Jr., M.A. Metcalf, P.A. Jose, M.W. Hamblin and D.R. Sibley, J.Biol.Chem., 268, 18200 (1993).

107. Y. Shen, F.J. Monsma, M.A. Metcalf, P.A. Jose, M.W. Hamblin and D.R. Sibley, J.Biol.Chem., 268, 18200 (1993).

108. M. Ruat, E. Traiffort, R. Leurs, J. Tardivel-Lacombe, J. Diaz, J. Arrang and J. Schwartz, Proc.Natl.Acad.Sci.U.S.A., 90, 8547 (1993).

109. J.A. Bard, J. Zgombick, N. Adham, P. Vaysse, T.A. Branchek and R.L. Weinshank, J.Biol.Chem., 268, 23422 (1993).

110. I.T. Forbes, S. Dabbs, D.M. Duckworth, A.J. Jennings, F.D. King, P.J. Lovell, A.M. Brown, L. Collin, J.J. Hagan, D.N. Middlemiss, G.J. Riley, D.R. Thomas and N. Upton, J.Med.Chem., 41, 655 (1998).

111. P.J. Lovell, S.M. Bromidge, S. Dabbs, D.M. Duckworth, I.T. Forbes, A.J. Jennings, F.D. King, D.N. Middlemiss, S.K. Rahman, D.V. Saunders, L.L. Collin, J. Hagan, G. Riley and D.R. Thomas, J.Med.Chem., 43, 342 (2000).

112. C. Kikuchi, H. Nagaso, T. Hiranuma and M. Koyama, J.Med.Chem., 42, 533 (1999).

Chapter 3. Recent Advances in Development of Novel Analgesics

Elizabeth A. Kowaluk, Kevin J. Lynch and Michael F. Jarvis
Abbott Laboratories, Abbott Park, IL 60064

Introduction — Nonsteroidal anti-inflammatory drugs (NSAIDs, e.g. aspirin, ibuprofen and naproxen), opioids (e.g. morphine) and a variety of adjuvant agents (e.g. tricyclic antidepressants, lidocaine and certain anticonvulsant agents) have been mainstays of pain therapy for decades. All these agents suffer from drawbacks in clinical use. The NSAIDs are associated with gastrointestinal side effects, increased bleeding time and do not effectively ameliorate severe pain. The opioids can produce tolerance and dependence, constipation, respiratory depression and sedation. Adjuvant agents, which non-selectively block sodium channels, are associated with CNS and cardiovascular side effects. Currently available analgesics also have limited utility in treatment of neuropathic pain (pain arising from nerve injury). Thus, there is a significant unmet medical need for safer and more effective analgesic agents, and this need is likely to be met by identification of novel ligands for newly identified molecular targets.

A number of advances have been made in understanding of the neurobiology of pain in recent years (1, 2). It is now appreciated that distinct mechanisms contribute to physiological pain, to pain arising from tissue damage (inflammatory or nociceptive pain) and to pain arising from injury to the nervous system (neuropathic pain) (3, 4). Tissue injury results in release of mediators which sensitize peripheral nerve terminals (peripheral sensitization), leading to phenotypic alterations of sensory neurons and increased excitability of spinal cord dorsal horn neurons (central sensitization) (5, 6). Nerve injury may be associated with abnormal discharges of the injured neurons, leading also to central sensitization and phenotypic changes in spinal cord neurons (7, 8). In addition, the response of the nervous system to pain is influenced by descending modulatory systems (9). Major advances have been made in elucidation of the molecular mechanisms of pain. A variety of receptors, transmitters, second messenger systems, transcription factors and other signaling molecules that are involved in pain pathways have been identified (1,2). Some receptors have been identified which appear to be expressed primarily on nociceptors (e.g. SNS/PN3 and P2X$_3$) , offering the possibility of selective analgesia with minimal side effects.

The introduction of selective COX-2 inhibitors into clinical practice during the past 18 months represents the first step towards novel, safer and more effective pain therapies. COX-2 inhibitors are described in this report, along with efforts to identify safer opioids with a focus largely on novel, potential near-term analgesics. In addition, this report highlights some newly emerging molecular targets that may represent future advances in the search for novel pain therapies.

COX-2 Inhibitors — Cyclooxygenase (COX) pathway products, particularly the prostaglandin PGE$_2$, mediate inflammation and contribute to pain (hyperalgesia) by sensitizing peripheral nociceptors. The enzyme, COX, exists as a constitutive isoform (COX-1), which mediates the formation of prostanoids involved in homeostatic functions (e.g. in gastric mucosa, platelets), and an isoform which is upregulated by inflammatory stimuli (COX-2) (10, 11). Traditionally available NSAIDs (e.g. ibuprofen and naproxen) are nonselective COX inhibitors (12). Hence, their COX-2 mediated anti-inflammatory and analgesic effects are tempered by COX-1 mediated gastrointestinal and hematologic adverse events. Much attention has consequently been focused on developing selective COX-2 inhibitors as safer NSAIDs. The crystal structures of COX-1 and COX-2 have been solved (13), revealing subtle differences in the conformation and size of the active site, and supporting the potential for selective COX-2 inhibition.

In late 1998 and 1999, two selective COX-2 inhibitors, celecoxib (**1**) and rofecoxib (**2**) have been approved for clinical use in the US and certain other countries (14 – 17). These compounds are approved for use in osteoarthritis and rheumatoid arthritis, and in some cases, acute pain (18 – 21). Both compounds show an incidence of gastrointestinal injury no greater than placebo, and no inhibition of platelet function (15, 17, 22, 23). Celecoxib is 375-fold selective for COX-2 (IC_{50} = 40 nM) *versus* COX-1 *in vitro* (14). Rofecoxib exhibited 1000-fold selectivity for COX-2 (IC_{50} = 18 – 46 nM in a variety of cell-based assays) compared to COX-1 (16). The COX-2 selective inhibitor, parecoxib is in clinical development as a parenteral formulation and has shown analgesic efficacy in patients after dental extraction (24). Additional compounds in development include valdecoxib (**3**) and MK-663, which are believed to be more potent and selective than the first-generation compounds, as well as JTE-522 (**4**) and GR-253035 (25, 26). A combined COX-2/5-lipoxygenase inhibitor, CI-1004 is also being investigated in clinical trials (25).

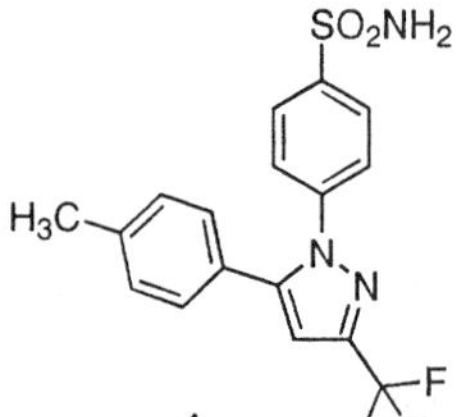

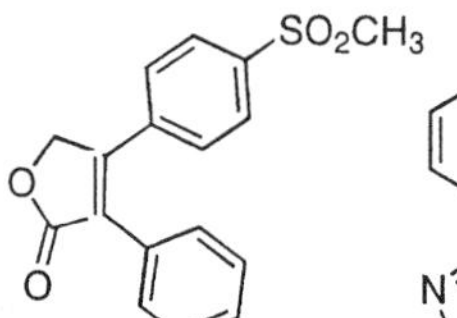

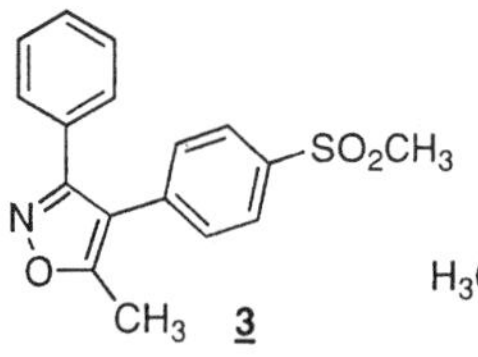

The development of EP$_1$ receptor antagonists represents an alternative approach to interfering with the effects of prostanoids in order to elicit analgesia. Activation of the EP$_1$ receptor by PGE$_2$, initiates a cascade of events that results in sensitization of peripheral afferents and neurons in the spinal cord dorsal horn. The selective EP$_1$ receptor antagonist , ZM325802 (**5**), dose-dependently relieved thermal hyperalgesia, but not cold allodynia, in the rat chronic constriction injury model of neuropathic pain after oral dosing (27).

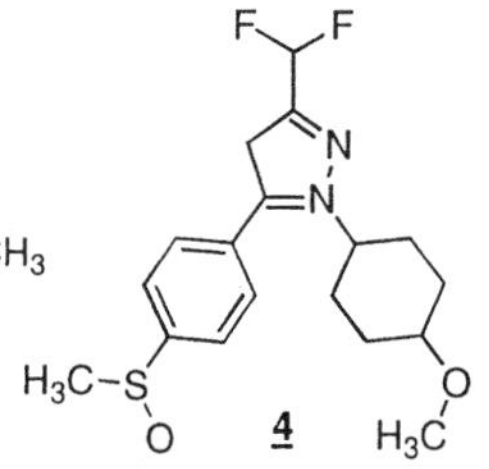

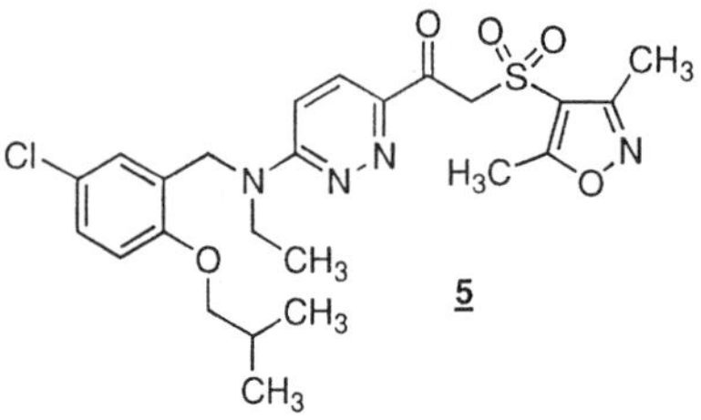

Opioids - The broad spectrum analgesic efficacy of the opioids, like morphine, coupled with the fact that these agents do not show analgesic ceiling effects makes opioid compounds the mainstay in the control of moderate to severe pain (28). The analgesic actions of opioid drugs are mediated at multiple sites of action including peripheral sites, the spinal cord, and supraspinal sites such as the brainstem and midbrain (2). This multitude of opiate interactions also contributes to the side-effects associated with opioid analgesic therapy including dependence, tolerance, immunosuppression, respiratory depression and constipation (2, 29). Opioid dose titration can be achieved to manage some nociceptive condidtions (29), however this strategy does not provide full efficacy in all chronic pain syndromes such as cancer and neuropathic pain (30). The cloning and characterization of the major opioid receptor subtypes (μ, OP1; δ, OP2; and κ, OP3) has stimulated significant basic and clinical research to discover new opioids with improved target selectivity, safety and efficacy. Moreover, each of these opioid subtypes have been further subdivided into putative subtypes, and an "orphan" member of this family, ORL-1 (OP4), has also been described (28).

Other trends that may show therapeutic promise include targeting peripheral sites of opioid action and the use of drug combinations such as morphine and dextromethorphan (e.g. MorphiDex) to respectively limit centrally mediated opioid side-effects or to reduce side-effects associated with effective doses of individual analgesics administered alone. Opioid agents in clinical development include MorphiDex (late stage clinical development); BCH-3963 (LEF-576) a peripherally-selective μ-receptor analgesic; ADL 2-1294, a topically active formulation of the μ-receptor agonist, loperamide, targeted for use in burns and abrasions. An opthalmic formulation of ADL 2-1294 is also under clinical evaluation for inflammatory corneal pain and post-surgical pain.

EMD 61753 (asimadioline, **6**) is a potent ($IC_{50} = 6$ nM) and fully effective κ-agonist that has limited systemic availability to the CNS (31). In the persistent phase of the formalin test, **6** reduces noccieptive behaviors (31, 32). In a recent clinical trial, **6** actually increased post-operative pain by a putative non-κ-opioid mechanism (33). ADL 10-0101 is another peripherally-selective κ-agonist currently in clinical development. TRK-820 is a potent and highly selective κ-agonist that has been reported to be approximately 100-fold more potent than morphine in the mouse abdominal constriction assay (34). TRK-820 does not affect the development of morphine tolerance indicating that this compound may be pharmacologically distinct from previous κ-agonists (35).

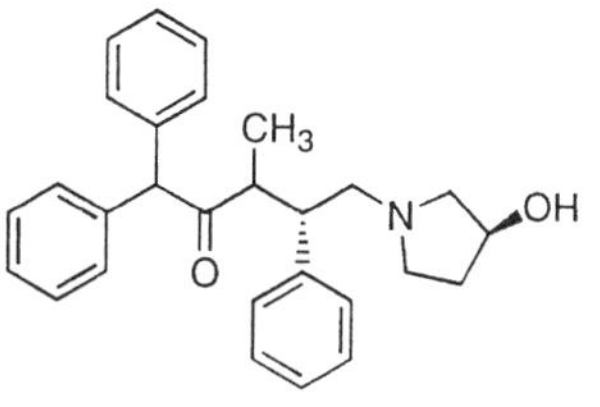

6 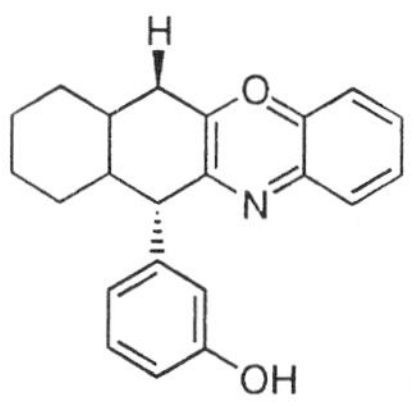   7 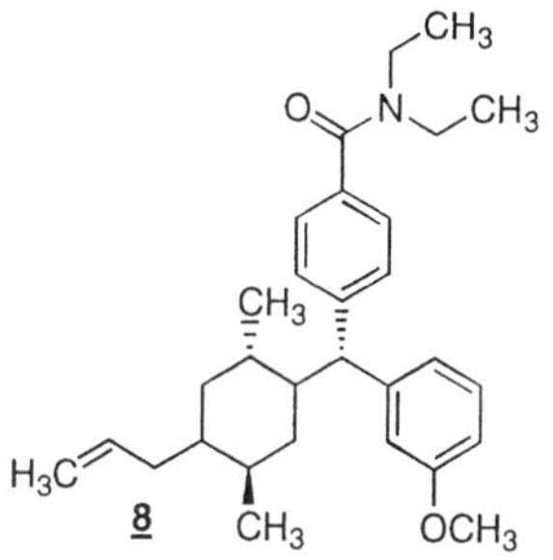   8

More recent preclinical results indicate that δ-opioid agonists may be safe and effective analgesics (36). Compound **7** (TAN-67; δ $K_i = 1.1$ nM, μ $K_i = 2320$ nM, κ $K_i = 1800$ nM) is representative of these efforts (15). Interestingly, (-)TAN-67 reduces acute nociception in the mouse tail flick test while the (+) enantiomer is pronociceptive and can block the antinociceptive actions of orphanin FQ (37). In addition, SNC-80 (**8**) is a highly selective and potent δ-opioid that effectively blocks acute nociception when administered into the CNS (spinal and supraspinal sites) (38). SNC-80 blocks acute nociception following systemic administration in rodents and monkeys at 3-10 fold lower doses than are required to produce gastrointestinal and CNS side-effects (32).

Excitatory Amino Acid Receptor Antagonists - The excitatory amino acid (EAA), glutamate, functions as a primary excitatory neurotransmitter in the CNS, and activation of EAA specific ionotropic and metabotropic receptor superfamiles in the spinal cord underlies the process of central sensitization involved in chronic pain (8). Activation of the heteromultimeric NMDA receptors (NR1/NR2B/NR2D subunits) expressed in spinal cord (39) contributes to the expression of tactile and thermal hyperalgesia. A number of competitive and noncompetitive NMDA receptor antagonists including (±)-3-(2-carboxypiperazin-4-yl)-propyl-1-phosphonic acid ((±)-CPP), MK-801, ketamine, and dextromethorphan have been shown to block hyperalgesia in animal models (40) and to attenuate the process of central sensitization (41). As such, NMDA antagonists appear to be more effective as compared to AMPA and kainate receptor antagonists (42). A problematic issue associated with NMDA receptor antagonists is their psychotomimetic effects that include both dysphoria and cognitive impairment. Further development of ligands selective for specific NMDA subunits like NR2B may provide enhanced therapeutic utility (39).

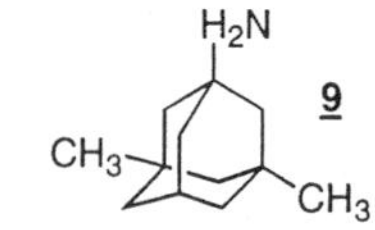

Memantine (**9**) is a low affinity (NR1/NR2B IC_{50} = 820 nM) noncompetitive NMDA antagonist that has shown analgesic efficacy in man. Memantine attenuated ongoing neuropathic pain symptoms in both diabetic and post-herpetic neuralgia patients. MRZ-2/579, a low affinity noncompetitive NMDA antagonist, has been reported to attenuate carrageenan-induced thermal hyperalgesia at doses that do not affect sensory-motor function (43). GV 196771A (**10**) is a compound which modulates NMDA receptor function by blocking the glycine binding site of the NMDA receptor complex (44). Like memantine and dextromethorphan, **10** produces antihyperalgeisa in animal models at doses that do not elicit CNS side effects. Additionally, **10** is only weakly active in cerebral stroke models suggesting differences in the physiological substrates of nociception and neuroprotection. NMDA receptor antagonists can provide opioid sparing effects and may prevent the tolerance related to prolonged opioid use (40). As noted above, late-stage clinical trials of a dextromethorphan/morphine combination (MorphiDex) are ongoing. Another noncompetitive NMDA receptor antagonist, CNS-5161 (**11**) has shown antinociceptive effects in healthy volunteers as part of a Phase I study (45). Antagonists of the kainic acid subtype of the glutamate receptor include LY 293558 (**12**) which is active when delivered at a dose of 1.2 mg/kg in a capsaicin model of allodynia/ hyperalgesia (46).

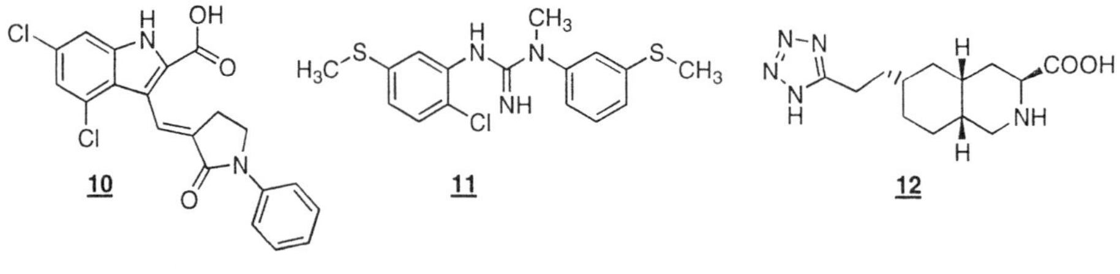

Metabotropic glutamate receptors (mGluRs) may also contribute to nociception (see chapter 1). Intrathecal delivery of antisense oligonucleotides directed against the mGlu1 receptor attenuated hyperalgesia and allodynia in a rat model of neuropathic pain (47). Additionally, the mGlu5 antagonist, 2-methyl-6-(phenylethynyl)-pyridine (MPEP) attenuates carrageenan-induced hyperalgesia following systemic or local administration (48).

Calcium Channel Modulators - Control of N-type calcium channels has been shown to provide an avenue for development of novel analgesics, as exemplified by ziconotide (SNX-111). Ziconotide is a 25 amino-acid polycationic peptide originally isolated from the venom of a cone snail. Its delivery is limited to the epidural and intrathecal routes, as systemic administration has led to a risk for orthostatic hypotension (49). An NDA has been filed with the US FDA for the use of ziconotide in intractable cancer pain and chronic neuropathic pain.

While gabapentin is approved for use as an anticonvulsant, it is also extensively used for the treatment of neuropathic pain. This compound may act as a calcium channel modulator, as recent reports indicate that gabapentin interacts with domains on the $\alpha2\delta$ subunit of voltage-sensitive calcium channels, and it inhibits neuronal Ca^+

currents (50). Nonetheless, it remains unclear whether this mechanism fully accounts for the actions of gabapentin. A gabapentin analog, pregabalin (S-(+)-3-isobutylGABA, (CI-1008)) is in phase III clinical trials for neuropathic pain.

<u>Adenosine kinase inhibitors</u> - Selective adenosine kinase (AK) inhibitors are designed to potentiate the concentration and actions of extracellular endogenously released adenosine (ADO). ADO is involved in modulating endogenous antinociceptive and anti-inflammatory pathways, and exerts its actions by activating specific cell-surface receptors (P1 receptors) (51). AK rapidly phosphorylates ADO, maintaining intracellular ADO concentrations at low levels. Since ADO uptake is driven by its concentration gradient, AK inhibition has the net effect of decreasing cellular uptake of ADO (52), thus potentiating the local concentration and the effects of ADO in the extracellular compartment. AK inhibitors have been shown to have antinociceptive effects in a animal models of inflammatory and neuropathic pain (51). Efficacy of AK inhibitors has also been demonstrated in animal models of inflammation (53). Potent AK inhibitors have recently been synthesized that demonstrate high specificity for this enzyme as compared to other ADO metabolic enzymes, transporters, and receptors (54). A wide variety of potent substituted 5-aryl-pyrrolo[2,3-*d*]-pyrimidine and 3-aryl-pyrazolo[3,4-*d*]-pyrimidine nucleoside analog AK inhibitors have been disclosed (e.g. 55), including GP515 (<u>13</u>) (IC_{50} = 5 nM) (56) and GP3269 (<u>14</u>) (IC_{50} = 11 nM)(57). Changing the ribose ring to a carbocyclic ring and truncation of the 5'-methylene atom led to an alternative series of novel, potent AK inhibitors, exemplified by A-134974 (<u>15</u>) (IC_{50} = 0.06 nM) (58). Recently, a series of novel non-nucleoside AK inhibitors has been disclosed (59).

           13   14   15

<u>Cannabinoids</u> - The use of marijuana (cannabis) to relieve pain has been in practice for centuries (60). However, clinical evaluation of the major active cannabinoid, tetrahydrocannabinol (THC), has produced equivocal results in cancer pain patients. Further, the analgesic actions of THC could not be clearly separated from the other well-described psychotropic actions of THC (60). Investigation of the pharmacological actions of the cannabinoids has been greatly aided by the recent discovery of specific cannabinoid receptor subtypes (CB$_1$ and CB$_2$), elucidation of their signal transduction pathways, and the identification of putative endogenous ligands, such as anandamide (60 – 62). High densities of CB$_1$ receptors are found in the CNS, while CB$_2$ receptors are localized primarily to immune cells and peripheral nerve terminals (62). These advances in cannabinoid pharmacology suggest the possibility of identifying receptor subtype selective ligands (63, 64).

Cannabimimetics have been shown to produce antinociception in animal pain models *via* spinal (65) and supraspinal actions on CB$_1$ receptors, and by peripheral actions at CB$_2$ receptors on sensory afferents and, indirectly, on immune cells (52). Recent compounds in preclinical development include agonists with improved oral bioavailability and/or enhanced receptor subtypes selectivity. CT-3 (<u>16</u>) is an orally active and nonselecitve analog of THC that dose-dependently reduces acute nociception in the rat (66). O-1057 (<u>17</u>) is a potent, and moderately CB$_1$ receptor selective analog of CT-3 that has improved water solubility and acute antinociceptive actions (67). HU-308 (**18**) is a novel, highly CB_2 receptor selective agonist (Ki: CB_1 > 10 mM, CB_2 = 23 nM) that has antinociceptive effects in the persistent phase of the mouse formalin test, but was inactive in the acute phase of the formalin test (68). While no effects of HU-308 were observed on motor function, antinociceptive doses of the compound also reduced gastrointestinal motility and blood pressure (68).

Sodium Channel Modulators- As a long time target for the action of local anesthetics, voltage-gated sodium channels have again come into the spotlight as molecular targets for novel analgesics. Successful use of intravenous lidocaine to treat some patients with neuropathic pain has pointed out the potential effectiveness of sodium channel blockers in this application (69, 70). Unfortunately, the CNS and cardiovascular liabilities encountered due to the lack of molecular specificity of these compounds has limited their usefulness in this role. Nonetheless, there are some new compounds on the horizon that, while not demonstrating subtype selectivity, do appear promising as analgesics. Co102862 (**19**) has shown antiallodynic efficacy in animal models of neuropathic pain and is also effective in the rat formalin test of acute pain (71). This compound exhibits no reduction in seizure threshold and no local anesthetic activity. Additionally, the novel sodium channel blocker, RS-132943 ((S)-3-(4-bromo-2,6-dimethylphenoxymethyl)-1-methylpiperidine hydrochloride) has been shown to be

orally bioavailable and to provide dose-dependent relief of thermal hyperalgesia (ED_{50} = 16 mg/kg, p.o.) and cold allodynia (ED_{50} = 55 mg/kg, p.o.) in preclinical neuropathic pain models (72).

The cloning and characterization of several sensory nerve-specific sodium channel subtypes has raised interest in the possibility of developing subtype-specific inhibitors which might overcome the cardiovascular and proconvulsant liabilities of nonselective agents. Two of these channels, SNS/PN3 (73, 74) and SNS2/NaN (75, 76), contribute to the tetrodotoxin (TTX) resistant sodium currents primarily restricted to sensory neurons. Several studies have implicated PN3/SNS in pain signaling. Inflammatory mediators such as PGE_2, and serotonin increase the magnitude of TTX-resistant sodium currents (77). PN3/SNS immunoreactivity is increased in the carrageenan inflammatory pain model, and has also been shown to increase proximal to the site of nerve injury in rats and humans (78, 79). Antisense oligonucleotides to PN3/SNS prevented thermal hyperalgesia or mechanical allodynia from developing in animal models of neuropathic pain (80), and were also effective at reducing prostaglandin-induced hyperalgesia (81). PN3/SNS knockout mice demonstrated a diminished response to noxious mechanical stimuli and delayed inflammatory hyperalgesia (82). While the PN3/SNS subtype has been a major focus of research, other sodium channels may also be appealing targets for pharmaceutical intervention of pain. The significant upregulation of a previously silent type III TTX-sensitive sodium channel after peripheral axonal injury has been recently described (83).

Vanilloid receptor modulators - While the potential of the vanilloids (e.g capsaicin) for use as topical analgesics has been appreciated for some time, their utilization has been hampered by the well-known initial burning sensation these compounds elicit. Nonetheless, the administration of high concentrations of topical capsaicin in combination with local anesthetic has proven to provide extended relief from neuropathic pain (84). In addition, the ultrapotent natural capsaicin analog, resiniferatoxin, has been shown to be effective in providing prolonged regional analgesia (greater than 7 days) upon epidural administration in animals (85). These compounds appear to function by chronically desensitizing sensory neurons in the treated area.

Cloning of the vanilloid receptor VR1 in 1997 aided in understanding the mechanism by which vanilloids exert their effects (86). This receptor is expressed by small-diameter nociceptive neurons in sensory ganglia. In addition to capsaicin, this nonselective ion channel is sensitive to elevations in temperature and to acidic pH, suggesting a role in pain signaling (87). The endogenous cannabinoid anandamide has also been found to be an agonist for the vanilloid receptor, raising the possibility that endogenous lipids may serve to modulate the action of this receptor (88, 89).

The role of VR1 in nociception has been more clearly elucidated with the creation of VR1 knockout mice (90). These animals were found to have normal behavioral responses to noxious mechanical stimuli, but greatly altered thermal nociceptive responses. Interestingly, these mice exhibited impaired behavioral responses to tissue injury induced thermal hyperalgesia, but disruption of VR1 had no apparent effect on thermal sensitization following nerve injury. The nociceptive phenotype of these mice suggest that antagonists to VR1 may be effective analgesics in inflammatory and traumatic pain and will no doubt spur an increased interest in this area of analgesic development.

P2X$_3$ Receptor Antagonists - It is now appreciated that ATP serves as an extracellular signaling molecule, and that it acts as an excitatory neurotransmitter by activating specific ligand-gated ion channel receptors, termed P2X receptors. One member of this family, P2X$_3$, has been cloned and found to be localized primarily to sensory neurons, suggesting a role in pain transmission (91, 92). The P2X$_3$ mRNA occurs only in the trigeminal, dorsal root and nodose ganglia (93), and the receptor is selectively expressed in sensory C-fiber neurons that project to the periphery and spinal cord, and which are predominantly nociceptors. Moreover, immunohistochemical studies have localized the P2X$_3$ receptor protein to the peripheral and central terminals of these neurons (94, 95). Thus, P2X$_3$ receptors on primary afferent nerve endings at the periphery may serve as a target for extracellular ATP released from damaged cells, initiating a nociceptive signal (94, 96). In addition, P2X$_3$ receptors located presynaptically at the central terminals of primary afferent neurons may have a facilitatory role to enhance neurotransmission, leading to a further increase in pain sensation (94, 96).

A role of ATP in pain transmission is consistent with the observed induction of pain by ATP upon application to human skin (97, 98), and with reports that intradermal and intrathecal application of ATP and ATP analogs (e.g. α,β–methylene-ATP (α,β-meATP)) into the rat hindpaw (99) evokes acute nociceptive behavioral responses. In addition, recent reports indicate that intradermally and intrathecally administered P2X receptor antagonists are analgesic in animal models (100). P2X$_3$ antagonists may, therefore, be useful analgesics with high selectivity, given the highly restricted localization of the P2X$_3$ receptor.

The pharmacological tools available for examining P2 receptor function remain suboptimal. All known P2 receptor agonists are bioisosteres of the purine nucleotide pharmacophore (101, 102). Their use in both *in vitro* and *in vivo* pharmacological studies is complicated by a lack of selectivity, by chemical and biological instability and by susceptibility to degradation by ecto-nucleotidases. Traditionally used antagonist ligands like suramin, PPADS and reactive blue also have limited selectivity and efficacy, complicating their use *in vitro* and *in vivo* to characterize P2 receptor function. More recently, 2',3'-*O*-(2,4,6-trinitrophenyl)-ATP (TNP-ATP) has been identified as a nanomolar antagonist of $P2X_1$, $P2X_3$, and heteromeric $P2X_{2/3}$ receptors (103); however, the identification of highly selective $P2X_3$ antagonists has not yet been reported.

Conclusion — The adverse physiological, psychological and economic effects of inadequate pain management have become increasingly recognized in recent years. This has been accompanied by a growing awareness on the part of patients and caregivers that pain need not be tolerated, and an increased emphasis amongst physicians on the proactive treatment of pain. The unmet need for new analgesics remains substantial. Recent advances in the neurobiology of pain, together with the development of new preclinical and clinical pain paradigms, have revealed new opportunities for the development of analgesics, and raised the exciting possibility of entirely novel classes of analgesics in the future.

References

1. R. Dubner and M. Gold, Proc. Natl. Acad. Sci., 96, 7627 (1999).
2. J.M. Besson, Lancet, 353, 1610 (1999).
3. R. Dubner, A.I. Basbaum in "Textbook of Pain" 3rd edition, P.D. Wall, R.D. Melzack (eds.), Churchill Livingstone, Edinburgh, 1994, p. 225.
4. C.J. Woolf in "Textbook of Pain" 3rd edition, P.D. Wall, R.D. Melzack (eds.), Churchill Livingstone, Edinburgh, 1994, p. 101.
5. C. J. Woolf and M. Costigan, Proc. Natl. Acad. Sci. 96, 7723 (1999).
6. A. Neumann, T.P. Doubell, T. Leslie, C.J. Woolf, Nature, 384, 360 (1996).
7. M. Devor in "Textbook of Pain" 3rd edition, P.D. Wall, R.D. Melzack (eds.), Churchill Livingstone, Edinburgh, 1994, p. 79.
8. C.J. Woolf and R.J. Mannion, Lancet, 353, 1959 (1999).
9. M.O. Urban and G.F.Gebhart, Proc. Natl. Acad. Sci 96, 7687 (1999).
10. J.R. Vane and R.M. Botting, Inflamm. Res. 44, 1 (1995).
11. J.A. Mitchell and T.D.Warner, Br. J. Pharmacol. 128, 1121 (1999).
12. E.A. Meade, W.L.Smith, D.L.DeWitt, J. Biol. Chem. 268, 6610 (1993).
13. R.G. Kurumbail, A.M.Stevens, J.K.Gierse, J.J.McDonald, R.R.Stegeman, J.Y. Pak, D. Gildehaus, J.M. Miyashiro, T.D.Penning, K. Seibert, P.C.Isakson and W.C. Stallings, Nature, 384, 644 (1996).
14. T.D. Penning, J.T. Talley, S.R. Bertenshaw, J.S. Carter, P.W. Collins, S. Docter, M.J. Graneto, L.F. Lee, J.W. Malecha, J.M. Miyashiro, R.S. Rogers, D.J. Rogier, S.S. Yu, G.D. Anderson, E.G. Burton, J.N. Cogburn, S.A. Gregory, A.M. Koboldt, W.E. Perkins, K. Seibert, A.W. Veenhuizen, Y.Y. Zhang and P.C. Isakson, J. Med. Chem., 40, 1347 (1997).
15. J. Wallace and B. Chin, Curr, Opin. CPNS Invest. Drugs, 1, 132 (1999).
16. C.C Chan, S. Boyce, C. Brideau, S. Charleson, W. Cromlish, D. Ethier, J. Evans, A. W. Ford-Hutchinson, M. J. Forrest, J. Y. Gauthier, R. Gordon, M. Gresser, J. Guay, S. Kargman, B. Kennedy, Y. LeBlanc, S. Leger, J. Mancini, G. P. O'Neill, M. Ouellet, D. Patrick, M. D. Percival, H. Perrier, P. Prasit, I. Rodger, P. Tagari, M. Therien, P. Vickers, D. Visco, Z. Wang, J. Webb, E. Wong, L-J. Xu, R.N.Young, R. Zamboni and D. Riendeau. J. Pharmacol. Exp. Ther. 290, 551 (1999).
17. F. Kamali, Curr. Opin. CPNS Invest. Drugs, 1, 126 (1999).
18. D.R. Mehlisch, R.C. Hubbard, P. Isakson, A. Karim, M. Weaver and S. Mills, Clin Pharmacol. Ther. 60, PIII-2 (1997).
19. L.S. Simon, F.L. Lanza, P.E. Lipsky, R.C. Hubbard, S. Talwalker, B.D. Schwartz, P.S. Isakson, G.S. Geis, Arth. Rheum., 41, 1591 (1998).
20. E.W. Ehrich, A. Dallob, I. DeLepeleire, D. Riendeau, W. Yuan, A. Porras, J. Wittreich, J.R. Seibold, P. DeSchepper, D. Mehlisch and B.J. Gertz, Clin. Pharmacol. Ther. 65, 336 (1999).
21. E. Ehrich, T. Schnitzer, A. Weaver, A. McKay Brabham, M. Schiff, A. Ko, C. Brett, J. Bolognese and B. Seidenberg, Rheumatol. Eur. 27 (Suppl 2), 198 (1998).
22. F.L. Lanza, M.F. rack, D.A. Callison, R.C. Hubbard, S.S. Yu, S. Talwalker and G.S. Geis, Gastroenterology, 112 (Suppl A), 194 (1997).
23. R. Hunt, B. Bowen, C. James, S. Sridhar, T. Simon, E. Mortensen, A. Cagliola, H. Quan and J. Bolognese. Am. J. Gastroenterol. 93, 1671 (1998).
24. M. Kuss, D. Mehlisch, A. Bauman, D. Baum, R. Hubbard, Abstract , 9th World Congress on Pain, Vienna, p. 450 (1999).

25. T.L.McCarty and A. Marfat, Curr. Opin. Anti-Inflamm. Immunomod. Invest. Drugs. $\underline{1}$,1464 (1999).
26. D.P.Rotella, Curr. Opin. Drug Discovery Development, $\underline{1}$, 165 (1998).
27. M.M.S. Lo, Y.K. Xie, A.J. barker, G. Breault, E.J. Griffen, J.S.Shaw, Abstract , 9th World Congress on Pain, Vienna, p. 272 (1999).
28. G.W. Pasternak, S.R. Letchworth, Curr. Opin. CPNS Invest. Drug, $\underline{1}$, 54-64, (1999).
29. H. McQuay, Lancet, $\underline{353}$, 2229-2232, (1999).
30. R.K. Portenoy, P. Lesage, Lancet, $\underline{353}$, 1695-1700, (1999).
31. A.Barber, R. Gottschlich, Curr. Opin. Invest. Drugs, $\underline{6}$, 1351-1368, (1997).
32. J.S. Bryans, Invest. Drugs, $\underline{2}$, 1170-1182, (1999).
33. H. Machelska, M. Pfluger, W. Weber, M. Piranvisseh-Volk, J. Daubert, R. Dehaven, C. Stein. J. Pharmacol. Exp. Ther. 290: 354-361, (1999).
34. H. Nagase, J. Hayakawa, K. Kawamura, K. Kawai, Y. Takezawa, H. Matsuura, C. Tajima, T. Endo, Chem. Pharm. Bull., $\underline{46}$, 366-369, (1998).
35. M. Tsuji, M. Yamazaki, H. Takeda, T. Matsumiya, H. Nagase, L.F. Tseng, M. Narita, T. Suzuki, Eur. J. Pharmacol., $\underline{394}$, 91-95, (2000).
36. G. Dondio, S. Ronzoni, P. Petrillo, Exp. Opin. Ther. Patents, $\underline{7}$, 1075 (1997).
37. J. Kamei, M. Ohsawa, T. Kashiwazaki, H. Nagase, Eur. J. Pharmacol., $\underline{370}$, 109-116, (1999).
38. E.J. Bilisky, S.N. Calderon, T. Wang, R.N. Bernstein, P. Davis, V.J. Hruby, R.W. McNutt, R.B. Rothman, K.C. Rice, F. Porreca, J. Pharmacol. Exp. Ther., $\underline{273}$, 359-366, (1995).
39. P.L. Chazot, L.M. Hawkins, Invest. Drugs, $\underline{2}$, 1313-1326, (1999).
40. B.H. Herman, F.Vocci, P.Bridge, Neuropsychopharmacology, $\underline{13}$, 269 (1995).
41. J.D. Kristensen, B.Svensson, T.Gordh, Pain, $\underline{51}$, 249 (1992).
42. R.M. Eglen, J.C. Hunter, A. Dray, Trends Pharmacol. Sci. $\underline{20}$, 337-342, (1999).
43. C. Parsons, New Developments in Glutamate Pharmacology, Orlando, FL, March 4-5, (1999).
44. A. Quartaroli, American Pain Society Annual Meeting (1997).
45. J.T.M. Linders Curr. Opin. CPNS Invest. Drugs, $\underline{1}$, 167-170, (1999).
46. B. Sang, Abst. Soc. Neurosci. $\underline{23}$, 401.14 (1997).
47. M.E. Fundytus, Metabotropic Glutamate Receptors-Third International Meeting, Taormina, Sicily, Italy, September 19-24, (1999).
48. K. Walker, Metabotropic Glutamate Receptors-Third International Meeting, Taormina, Sicily, Italy, September 19-24, (1999).
49. S.S. Bowersox, T. Gadbois, T. Singh, M. Pettus, Y.X. Wang, R.R. Luther, J. Pharmacol. Exp. Ther., $\underline{279}$, 1243 (1996).
50. N.S. Gee, J.P. Brown, V.U.K. Dissanayake, J. Offord, R. Thurlow, G.N. Woodruff, J. Biol Chem., $\underline{271}$, 5768 (1996).
51. E.A. Kowaluk and M.F.Jarvis, Exp. Opin. Invest. Drugs, $\underline{9}$, 551 (2000).
52. L.P. Davies, D.D. Jamieson, J.A. Baird-Lambert and R. Kazlauskas, Biochem. Pharmacol., $\underline{33}$, 347 (1984).
53. G.S. Firestein,. Drug Dev. Res., $\underline{39}$, 371 (1996).
54. E.A. Kowaluk, S.S. Bhagwat and M.F. Jarvis, Curr. Pharm Design, $\underline{4}$,403 (1998).
55. B.G. Ugarkar, M.D. Erion, J.E. Gomez Galeno, A.J. Castellino, C.E. Browne, US patent 5721356 (1998).
56. J.B. Wiesner, B.G. Ugarkar, A.J. Castellino, J. Barankiewicz, D.P. Dumas, H.E. Gruber et al., J. Pharmacol. Exp. Ther. $\underline{289}$, 1669 (1999).
57. M.D. Erion, J.B. Wiesner, J. Dare, J.Kopcho and B.G. Ugarkar, Nucleosides And Nucleotides, $\underline{16}$, 1013 (1997).
58. M. Cowart and S.S. Bhagwat, US patent 5,665,721 (1997).
59. S.S. Bhagwat, C.-H. Lee, M.D. Cowart, J. McKie, A.L. Grillot, WO patent application 09846605 (1998).
60. B.R. Martin, A.H. Lichtman, Neurobio. Disease, $\underline{5}$, 447-461, (1998).
61. E. Pop, Curr. Opin. CPNS Invest. Drugs, $\underline{1}$, 587-596, (1999).
62. R.G. Pertwee, Curr. Med. Chem. $\underline{6}$, 635-664, (1999).
63. D.B. Jack, Drug News Persp. $\underline{10}$, 440 (1997).
64. R.G. Pertwee, Pharmacol. Ther., $\underline{2}$, 129 (1997).
65. W.J. Martin, C.M. Loo, A.I. Basbaum, Pain, $\underline{82}$, 199-205, (1999).
66. E.Z. Dajani, K.R. Larsen, J. Taylor, N.E. Dajani, T.G. Shahwan, S.D. Neeleman, M.S. Taylor, M.T. Dayton, G.N. Mir, J. Pharmacol. Exp. Ther., $\underline{291}$, 31-38, (1999).
67. R.G. Pertwee, T.M. Gibson, L.A. Stevenson, R.A. Ross, W.K. Banner, B. Saha, R.K. Razdan, B.R. Martin, Brit. J. Pharmacol.,129, 1577-1584, (2000).
68. L. Hanus, A. Breuer, S. Tchilibon, S. Shiloah, D. Goldenberg, M. Horowitz, R.G. Pertwee, R.A. Ross, R. Mechoulam, E. Fride, Proc. Natl. Acad. Sci. U.S.A., $\underline{96}$, 14228-14233, (1999).

69. P. Petersen, J. Kastrup, I. Zeeberg, G. Boysen, Neurol. Res. 8, 189 (1986)
70. F.W. Bach, T.S. Jensen, J. Kastrup, B. Stigsby, A. Dejgard, Pain, 40, 29 (1990)
71. R.B. Carter, Presented at "Novel Targets in the Treatment of Pain" Sept. 16-17, 1999, Washington, D.C.
72. K.R. Gogas, L.A. Flippin, X.-F. Lin, D.G. Loughhead, R. O'Ryan, R.J. Weikert, P.E. Haroldsen, L. Lin, K.R. Bley, S.M. Amagasu, D.D.Blissard, L.O. Jacobson, R.S. Lewis, D. Walingora, R.M. Englen, J.C. Hunter. 9th World Congress on Pain, Aug. 22-27, 1999 Vienna, Austria
73. A.N. Akopian, L. Sivilotti, J.N. Wood, Nature, 379, 257 (1996)
74. L. Sangameswaran, S.G. Delgado, L.M. Fish, B.D. Koch, L.B. Jakeman, G.R. Stewart, P. Sze, J.C. Hunter, R.M. Eglen, R.C. Herman, J. Biol. Chem., 271, 5953 (1996)
75. S.D. Dib-Hajj, L. Tyrell, J.A. Black, S.G. Waxman, Proc. Natl. Acad. Sci. USA, 95, 8963 (1998)
76. S. Tate, S. Benn, C. Hick, D. Trezise, V. John, R.J. Mannion, M. Costigan, C. Plumpton, D. Grose, Z.Gladwell, G. Kendall, K. Dale, C. Bountra, C.J. Woolf, Nat. Neurosci. 1, 653 (1998)
77. M.S. Gold, D.B. Reichling, M.J. Shuster, J.D. Levine, Proc. Natl. Acad. Sci. USA 93, 1108 (1996)
78. M. Tanaka, T.R. Cummins, K. Ishikawa, S.D. Dib-Hajj, J.A. Black, S.G. Waxman, Neuroreport, 9, 967 (1998)
79. K. Coward, C. Plumpton, P. Facer, R. Birch, T. Carlstedt, S. Tate, C. Bountra, P. Anand, Pain, 85, 41 (2000)
80. F. Porreca, J. Lai, D. Bian, S. Wegert, M.H. Ossipov, R.M. Eglen, L. Kassotakis, S. Novakovic, D.K. Rabert, L. Sangameswaran, Proc. Natl. Acad. Sci. USA, 96, 7640 (1999)
81. S.G. Khasar, M.S. Gold, J.D. Levine, Neurosci. Lett. 256, 17 (1998)
82. A.N. Akopian, V. Souslova, S. England, K. Okuse, N. Ogata, J. Ure, A. Smith, B.J. Kerr, S.B. McMahon, S. Boyce, R. Hill, L.C. Stanfa, A.H. Dickenson, J.N. Wood, Nat. Neurosci., 2,541 (1999)
83. J.A. Black, T.R. Cummins, C. Plumpton, Y.H. Chen, W. Hormuzdiar, J.J. Clare, S.G. Waxman, J. Neurophysiol., 82, 2776 (1999)
84. W.R. Robbins, P.S. Staats, J. Levine, H.L. Fields, R.W. Allen, J. N. Campbell, M. Pappagallo, Anesth. Analg., 86, 579 (1998)
85. T. Szabo, Z. Olah, M.J. Iadarola, P.M. Blumberg, Brain Res., 840, 92 (1999)
86. M.J. Caterina, M.A. Schumacher, M. Tominga, T.A. Rosen, J.D. Levine, D. Julius, Nature, 389, 816 (1997)
87. M. Tominaga, M.J. Caterina, A.B. Malmberg, T.A. Rosen, H. Gilbert, K. Skinner, B.E. Raumann, A.I. Basbaum, D. Julius, Neuron, 21, 531 (1998)
88. P.M. Zygmunt, J. Petersson, D.A. Andersson, H. Chuang, M. Sørgård, V. DiMarzo, D. Julius, E.D. Högestätt, Nature, 400, 452 (1999)
89. D. Smart, M.J. Gunthorpe, J.C. Jerman, S. Nasir, J. Gray, A.I. Muir, J.K. Chambers, A.D. Randall, J.B. Davis, Br. J. Pharmacol., 129, 227 (2000)
90. M.J. Caterina, A. Leffler, A.B. Malmberg, W.J. Martin, J. Trafton, K.R. Petersen-Zeitz, M. Koltzenburg, A.I. Basbaum, D. Julius, Science, 288, 306 (2000)
91. C. Chen, A.N. Akopian, L. Sivilotti, D. Colquhoun, G. Burnstock and J.N. Wood, Nature 377, 428 (1995).
92. C. Lewis, S. Niedhart, C. Holy, R.A. North, G. Buell and A. Surprenant, Nature, 377, 432 (1995).
93. G. Collo, R.A. North, E. Kawashima, E. Merol-Pich, S. Neidhart, A. Surprenant and G. Buell, J. Neuroscience, 16, 2495 (1996).
94. G. Burnstock, Lancet, 347, 1604 (1996).
95. S.P. Cook, L. Vulchanova, K.M. Hargreaves, R. Elde and E.W. McCleskey, Nature, 387, 505 (1997).
96. A. Brake, M. Schumacher and D. Julius. Chemistry and Biology, 3, 229 (1996).
97. T. Bleehen and C.A. Keele, Pain, 3, 367 (1977).
98. S. Hamilton and S.B. McMahon. Abstract , 9th World Pain Congress, Vienna, Austria p. 236 (1999).
99. P.A. Bland-Ward and P.P.A. Humphrey, Br. J. Pharmacol, 122, 365 (1997).
100. B. Driessen, W. Reiman, N. Selve, E. Friderichs and R. Bultman, Brain Res., 666, 182 (1994).
101. K.A. Jacobson, B. Fischer, M. Maillard, J.L. Boyer, C.H.V. Hoyle, T.K. Harden and G. Burnstock in: "Adenosine and Adenine Nucleotides: From Molecular Biology to Integrative Physiology". L. Belardinelli and A. Pelleg (eds.), Kluwer, Boston, 1995 p. 149.
102. N.J. Cusack, Drug Dev. Res., 28, 244 (1993).
103. C. Virginio, G. Robertson, A. Surprenant and R.A. North, Mol. Pharm., 53, 969 (1998).

Chapter 4. Secretase Inhibitors as Therapeutics for Alzheimer's Disease

Richard E. Olson and Lorin A. Thompson
DuPont Pharmaceuticals Company, Wilmington, DE 19880

Introduction – Alzheimer's disease (AD), the most common form of dementia, was the 12th leading cause of death in the U.S. in 1998, claiming 22,824 individuals (1). The disease affects 7-10% of those over 65 years of age; the prevalence among those 80 years of age or older may be >40% (2-5). Beginning with mild cognitive impairment, the clinical course proceeds through significant, then profound dementia, loss of motor function, and death (6,7). In addition to the devastating impact on individuals and families, the financial burden of this protracted illness has been estimated at over $100 billion in the U.S. each year (8). In view of the aging population, the societal burden of AD is likely to increase significantly.

Alzheimer's disease is defined by the co-occurrence of two histologic lesions in brain – extracellular deposits (senile plaques) whose principal component is a fibrillar form of a predominantly 40-42 amino acid peptide known as β–amyloid (Aβ), and intracellular tangles of a hyperphosphorylated form of the microtubule associated protein tau. Tau dysfunction appears to have a role in the neurodegeneration of Alzheimer's disease (9,10). However, considerable evidence points to a causative role for the accumulation of Aβ peptide, potentially related to the neurotoxicity shown by Aβ in *in vitro* and *in vivo* models (11). A variety of approaches towards mitigating the impact of Aβ on the brain have been pursued, including inhibition of Aβ production, the use of antiinflammatory agents (12,13), inhibition of Aβ aggregation (14), and immunization against Aβ (15,16). Therapeutic strategies aimed at reducing amyloid production and aggregation have been reviewed recently in this series (17,18). This chapter will summarize progress in characterizing the proteolytic enzymes responsible for the production of Aβ, and recent work towards identifying secretase inhibitors for treatment of this devastating illness.

THE AMYLOID HYPOTHESIS

The study of the molecular genetics of early onset (age <60), or familial, AD (FAD) has been critical to development of the amyloid hypothesis, which links genetics, cell biology and clinical pathology in a compelling argument for a central role for Aβ in the development of AD (19). Except for the age of onset, the clinical manifestations of familial and sporadic AD are indistinguishable (20-22).

APP Processing and FAD Mutations - Characterization of Aβ in plaques stimulated the cloning of the β-amyloid precursor protein (APP), a 695-770 residue, type-I transmembrane protein encoded by a gene on chromosome 21 (23). Although the biological functions of APP have not been fully elucidated, it may play a role in memory and cell adhesion (24-27). APP is subject to a complex proteolytic cascade initiated by either α- or β-secretase (Figure 1). α-Secretase cleavage is mediated by metalloproteases including TACE (28-32), and accounts for the majority of APP processing in most cell types (33). This cleavage occurs in the center of the Aβ sequence after Aβ Lys16, precluding the formation of Aβ. Cleavage by α- or β–secretases releases the soluble ectodomains APPsα and APPsβ respectively, leaving membrane-associated C-terminal fragments (CTFs) termed C83 and C99 (Figure 1). Subsequent cleavage by γ-secretase(s) within the presumed transmembrane domain (TMD) releases the 24-26 residue p3 and the 40-42 residue Aβ peptide from C83 and C99, respectively. Although the Aβ$_{40}$ cleavage predominates, Aβ$_{42}$ is less soluble than Aβ$_{40}$ and is the major form of the peptide found in plaques. Aβ$_{42}$ may provide a nidus for extracellular amyloid deposition (34,35). Intracellular accumulation of Aβ$_{42}$ has recently been described in cell culture and in brain, and may have a role in plaque formation (36,37). While the processing of APP by β- and γ-secretases represents a minor proteolytic pathway, genetic analysis of a fraction of early onset families has revealed mutations in APP near the β-and γ-secretase cleavage sites (Figure 1), implicating inappropriate APP proteolysis in the disease process (38). In the brain and body fluids of FAD patients bearing the K670N/M671L mutation (Swedish/sw) higher levels of both Aβ$_{40}$ and Aβ$_{42}$ are observed implicating increased β-secretase activity. Mutations near the γ-secretase cleavage site (APP716, 717) increase Aβ$_{42}$, indicating an effect on γ-secretase activity (19,39). The AD pathology observed in Down's syndrome patients has been attributed to increased APP levels associated with the extra copy of chromosome 21 (40).

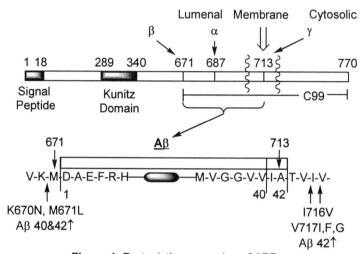

Figure 1. Proteolytic processing of APP$_{770}$

The majority of FAD cases are associated with mutations in genes on chromosomes 14 and 1 encoding presenilin 1 (PS1) and presenilin 2 (PS2), respectively. The presenilins are homologous transmembrane proteins found mainly in the endoplasmic reticulum and Golgi apparatus (22,41-43). FAD PS mutations increase the production of Aβ$_{42}$ in transfected cells and *in vivo* (22,41,44). APP and PS mutations have additive effects on Aβ$_{42}$ production in cells and transgenic mice (45,46). Thus, a common feature of the APP and PS FAD mutations, as well the trisomy 21 of Down's syndrome, is to increase the levels of Aβ$_{42}$ (44). Studies of normal and AD brain have demonstrated a correlation of AD severity with levels of Aβ in brain (47-51).

β-SECRETASE

Cloning and Isolation – Serine proteases, metalloproteases, the aspartyl protease cathepsin D, and the proteosome have all been considered as candidate β–secretases (52). The isolation of a novel membrane-anchored aspartyl protease termed "BACE" (beta-site cleaving enzyme) with a number of the features expected for the β-secretase enzyme was reported in late 1999 (53). The candidate gene was discovered using an expression cloning strategy to search a library for genes that increased the secretion of Aβ from HEK293 cells overexpressing APPsw. Soon after this report additional strategies were described which resulted in the identification of the same protein, also referred to as "asp2" and "memapsin 2". These included the location of ESTs encoding novel aspartyl proteases (52,54,55), and direct purification of the enzyme from human brain using an inhibitor-functionalized affinity column (56). BACE is a 501 amino acid aspartyl protease containing an unusual 80 amino acid C-terminal extension with a predicted TMD 21 amino acids from the C-terminus (Figure 2). The protein has an N-terminal 21 amino acid signal peptide, and a prodomain at residues 22-45. The mature, active enzyme begins at Glu46 of the predicted amino acid sequence. Two copies of the conserved D(T/S)G sequence indicative

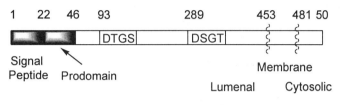

Figure 2. BACE

of aspartyl proteases are present. The highest levels of BACE mRNA are found in pancreas and brain; however, high BACE enzymatic activity was present only in the brain (56). The enzyme has a pH optimum of 4.5-5. It is not inhibited by the aspartyl protease inhibitor pepstatin.

A number of features of this protein make it likely to be the secretase relevant to *in vivo* cleavage of APP in AD brain. The enzyme is capable of cleaving native APP and APP-related peptides at the correct KM/DA site to generate the N-terminus of Aβ. The enzyme also cleaves APPsw approximately 10-fold faster than APPwt. The protein is expressed intracellularly in the expected sub-cellular organelles, including the Golgi appatatus, the trans-Golgi network, and in endosomes (57,58). The protein is predicted to be a type I transmembrane protein with 4 putative N-linked glycosylation sites and an active site on the lumenal side of the cell membrane. Antisense oligonucleotides targeting the BACE mRNA reduce secretion of Aβ with an efficiency of 70-80% (53,54). Transfection of BACE into several different cell lines stably expressing APPwt or APPsw results in an increase in APPsβ and a decrease in APPsα. The effect on Aβ production, however, seems to be sensitive to the cell line employed and the exact APP construct, and may also vary with the expression level of the protein (53,54,56).

β-Secretase Inhibitors – Prior to the recent characterization of the enzyme, a few inhibitors of Aβ production were reported to be β-secretase inhibitors on the basis of their ability to reduce the amount of C99 present in cells or to reduce the amount of APPsβ secreted by cells (17). For example, SIB1281 (**1**, IC_{50} = 0.8 μM) decreases the formation of APPsβ (59,60). Compound **2**, which incorporates a reduced amide isostere based on the Swedish mutation peptide sequence (~40% inhibition at 50 μM) and the tripeptide aldehyde Cbz-Val-Leu-Leu-H (IC_{50} = 0.7 μM) both inhibit the formation of $Aβ_{1-40}$ and $Aβ_{1-42}$ with approximately equal efficacy, leading to the suggestion that these inhibitors act at the β-secretase level (61). Reports of potent inhibitors with demonstrated activity against the enzyme are limited to peptide-based structures incorporating at least 7-8 amino acid residues. Effective inhibitors have been designed using the hydroxyethylene isostere transition state mimetic (62). The octapeptide OM99-2 (**3**) is based on the Swedish mutant

APP sequence (EVNL/DAEF) and incorporates an additional D to A replacement at $P_{1'}$ (55). The length of the peptide is important, as OM99-2 (K_i = 10 nM) is 7-fold more active than the corresponding N-terminally truncated OM99-1 (**4**). The 14-mer inhibitor **5** (IC_{50} ~ 30 nM) was based on the Swedish mutant APP sequence from P_{10} to $P_{4'}$ and incorporated (*S*)-statine in place of Asp at P_1, and Val at $P_{1'}$. These peptidomimetic scaffolds may hold promise as leads for the development of smaller, nonpeptide inhibitors. The effects of long-term suppression of β-secretase, or the 50% homologous BACE-2, a potential cross-reactivity of β-secretase inhibitors, are unknown (63).

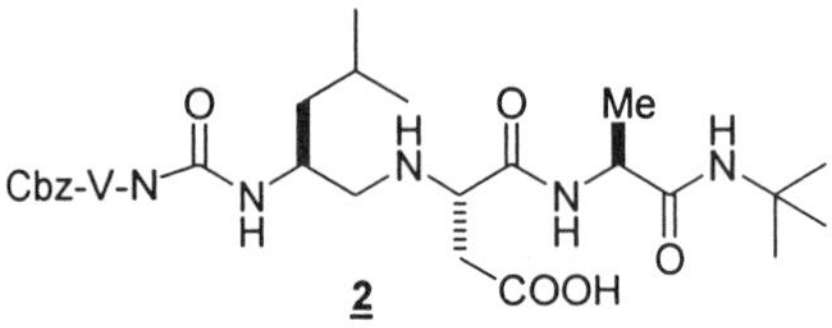

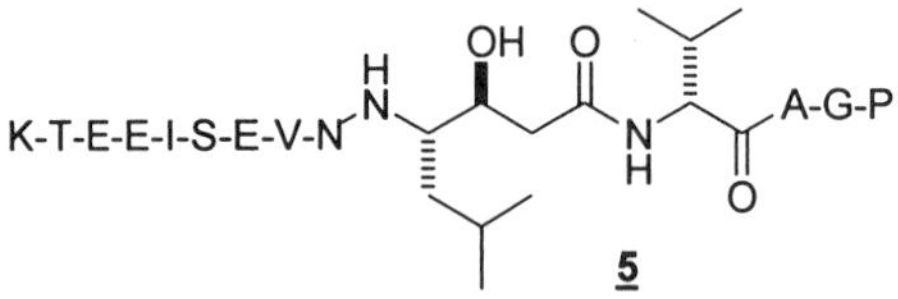

Letters Denote Amino Acids

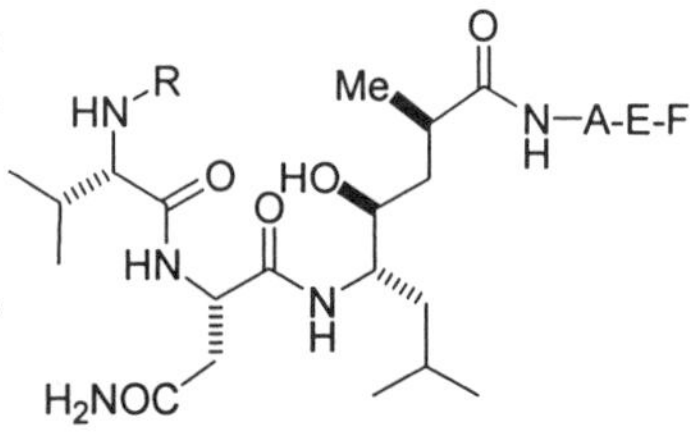

3 R = Glu
4 R = H

γ-SECRETASE

Characterization of γ-Secretase – γ-Secretase appears to be a member of a small group of proteases capable of intramembrane cleavage (64). A cleavage mechanism involving prior removal of the cleavage site from the membrane environment has also been proposed (72). Mutation studies demonstrate that γ-secretase has loose sequence specificity, tolerating many hydrophobic residues near the scissile bonds (65-67). In a study using a C99 construct as substrate, systematic substitution of TMD residues carboxy to $A\beta_{42}$ with phenylalanine changed $A\beta_{40}$ vs $A\beta_{42}$ cleavage specificity in a manner consistent with a helical substrate model. Phe-substitution as distal as $A\beta_{51}$ ($P_{9'}$ of the $A\beta_{42}$ scissile bond) affected cleavage specificity, indicating that γ-secretase may interact with the entire TMD carboxyl to the cleavage sites (68,69). It is unclear whether $A\beta_{40}$ and $A\beta_{42}$ are generated by one or multiple γ-secretase enzymes. Evidence for multiple enzymes includes different cellular localizations for $A\beta_{40}$ and $A\beta_{42}$ generation (36), different potencies for the inhibition of $A\beta_{40}$ and $A\beta_{42}$ formation by the same inhibitor (70-73) and the existence of γ-secretase activities which are both sensitive and insensitive to high levels of pepstatin (67). Alternatively, differential inhibitor access to cellular processing compartments, and selective inhibition of a carboxypeptidase acting on $A\beta_{43}$ have been suggested as causes for different inhibitory potencies for $A\beta_{40}$ and $A\beta_{42}$ production (74,75). A recent study showed an identical rank order of potencies for inhibition of $A\beta_{40}$ and $A\beta_{42}$

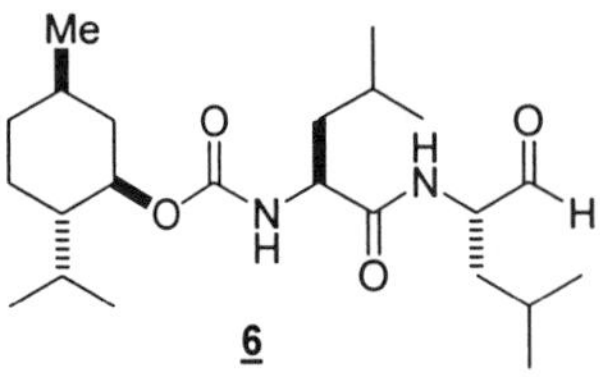

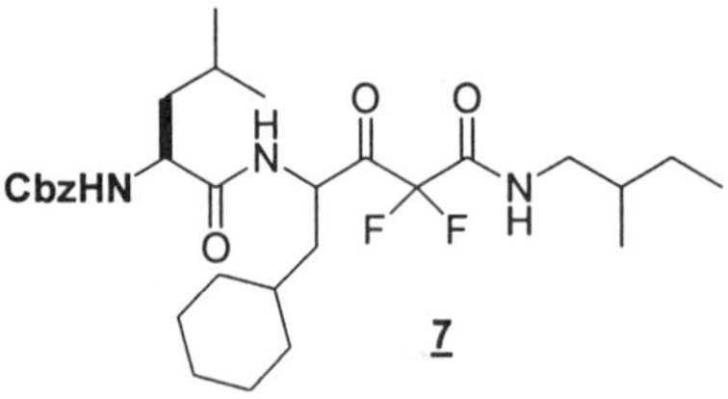

for a series of structually diverse inhibitors, including <u>6</u> and <u>7</u>, suggesting a single γ-secretase (74). Sub-inhibitory concentrations of γ-secretase inhibitors are reported to cause increased Aβ secretion (71,73,74,76).

<u>Role of Presenilins</u> - The presenilin proteins have been found to affect APP processing at the level of γ-secretase (41,42). Cultured neurons from PS1-/- mice transfected with human APP were found to generate greatly reduced (<20%) amounts of Aβ, and to accumulate the C-terminal APP γ-secretase substrate C83 (77,78). This finding has been confirmed in brain tissue from PS-/- mice (79). Most of the over 40 known FAD PS defects are missense mutations, occurring in topologically diverse regions of the proposed 6-8 TMD (Figure 3) PS structure (44,80). A model suggesting the alignment of certain TMD FAD residues on the same faces of PS TMD helices has been suggested (22,81). Presenilin holoproteins are subject to a proteolytic cleavage in the cytoplasmic loop bridging putative TMD's 6 and 7, which generates N- (30 kDa) and C-terminal (20 kDa) fragments. These recombine in long-lived heterodimers and represent the predominant PS form present under

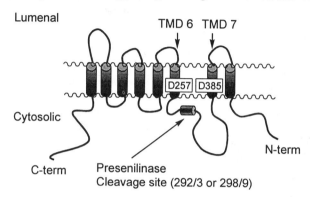

Figure 3. Proposed 8 TMD structure of presenilin 1. normal conditions. Endoproteolysis may not be required for Aβ generation (82). The splicing error-derived FAD PS1Δ9 results in a proteolysis-defective, yet functional, PS (83). Presenilins interact with other proteins including glycogen synthase kinase-3β, β-catenin, and E-cadherin, and are components of high molecular weight intracellular complexes (84-87).

The PS genes are known regulators of Notch signaling and their structure and function appear to be conserved across *C. elegans* and humans (88-90). Mice lacking PS1 die of defects possibly related to disturbed Notch signaling during embryogenesis (77,91,92). PS1 deficient mouse neurons demonstrate defects in Notch processing (93), and a general role for PS proteins in Notch activation is now appreciated (94). Presenilin-mediated activation of Notch appears to involve an intramembrane proteolysis which releases an intracellular domain (NICD) involved in Notch signaling (95-98). A series of γ-secretase inhibitors (including <u>9</u>, below) show similar potencies with respect to Aβ inhibition and NICD release (93). However, a recent study using PS mutants demonstrated a functional separation between PS-mediated APP processing and Notch proteolysis (99).

The loss of γ-secretase function caused by PS deficiency implicates PS proteins as γ-secretases or as γ-secretase regulators. Presenilins do not share homology with known proteases. However, the mutation of either of two conserved transmembrane 6 and 7 Asp residues in PS1 to Ala, Asn or Glu results in reduced Aβ production, buildup of C83 and C99, and decreased PS endoproteolysis in cells and cell-free microsomes (100). Coexpression of PS1 and PS2 mutants bearing the relevant Asp mutations in CHO cells causes Aβ production to drop to undetectable levels, consistent with a role for PS2 in the residual amyloidogenic processing observed in PS1-mutant systems (77,81,101). The proteolytic release of the Notch intracellular domain is deficient in the presence of Asp-mutant PS1 (102). These results have formed the basis of a hypothesis that proposes presenilins as unique aspartyl proteases capable of intramembrane cleavage of APP and Notch (81). A model aligning an helical APP substrate within the TMD helices of PS1 has been suggested (81). The cellular localization of PS1 mediated APP and Notch processing remains controversial, although co-localization of PS1 with APP and Notch has been reported (41,94,102-105).

γ-Secretase Inhibitors - Screening strategies employing cell-based assays have identified γ- rather than β-secretase inhibitors (51), and a number of γ-secretase inhibitors with hydrophobic semi-peptidic structures have been described. Cinnamamide peptidyl aldehyde **8** (IC$_{50}$ = 9.6 μM) was identified using a combinatorial approach (71). The difluoroketone Aβ$_{42/43}$ substrate analog **9** showed CTF accumulation and inhibition of Aβ$_{40}$ secretion at 25 μM (72,73). The tolerance for various hydrophobic groups both in the difluoroketone series and in peptide aldehydes may reflect the loose sequence specificity reported for γ-secretase. The hydroxyethylene-based peptidomimetic **10** demonstrates potent inhibition of both Aβ$_{40}$ and Aβ$_{42}$ production (IC$_{50}$ ~0.3 nM) through a γ-secretase mechanism in a cell-free preparation. Immunoprecipitation of PS1 from this system removes γ-secretase activity, consistent with earlier studies showing a critical role for PS1 in γ-secretase processing (106). A recent patent application discloses bis-benzazepines such as **11** as inhibitors of γ-secretase activity through a high affinity (53 pM) interaction with a cellular binding site (107).

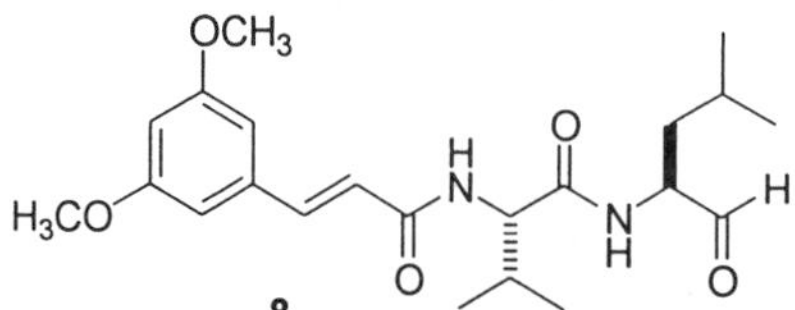

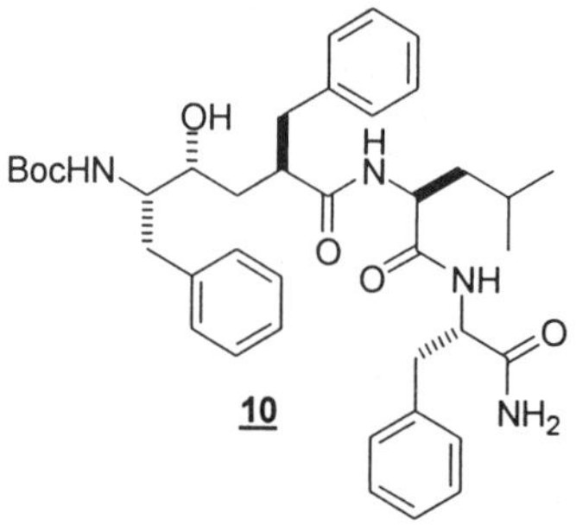

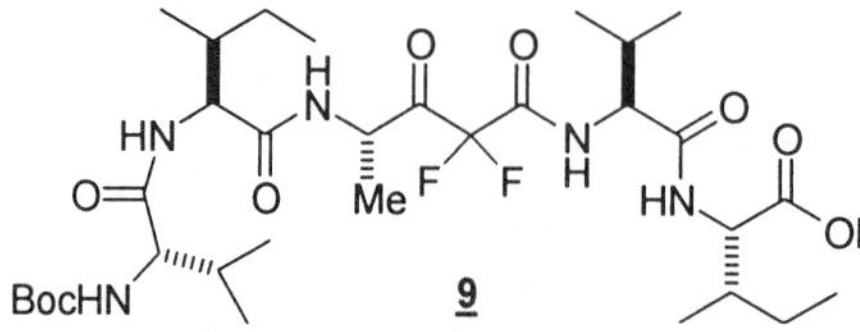

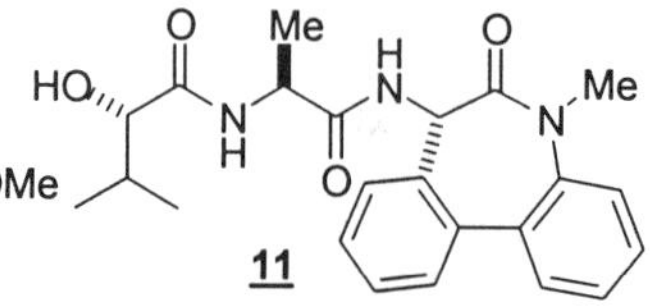

The potential side-effects of γ-secretase inhibition remain to be elucidated. Should γ-secretase inhibitors modulate Notch processing, the role of Notch signaling in adult organisms will need clarification. An effect on hematopoeisis has been suggested (108,109). Potential modulation of the function of other PS interacting proteins will need consideration (41).

Inhibitors of Aβ Production with Unreported Mechanisms - Compounds reported as inhibiting Aβ production in cell-based assays without detailed characterization of the mechanism of action have been reviewed recently (110,111). A series of amino acid-substituted aminobenzazepines (**12-15**) have been reported to inhibit Aβ formation in cells (112-115). Malate hydroxamate **16** and succinamide **17** were also disclosed as inhibitors of Aβ production (115,116).

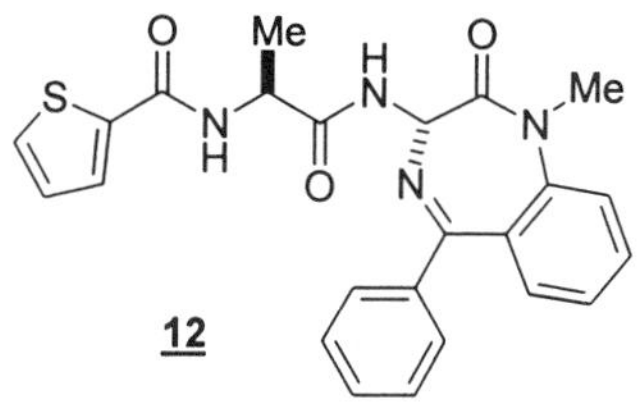

Conclusion - Recent years have seen tremendous progress in our understanding of the cell biology and biochemistry of APP processing. In addition, potent β- and γ-secretase inhibitors have been identified. Reports that γ-secretase inhibitors are approaching clinical evaluation may indicate that the benefits of lowering Aβ production will soon be tested in man (94,117).

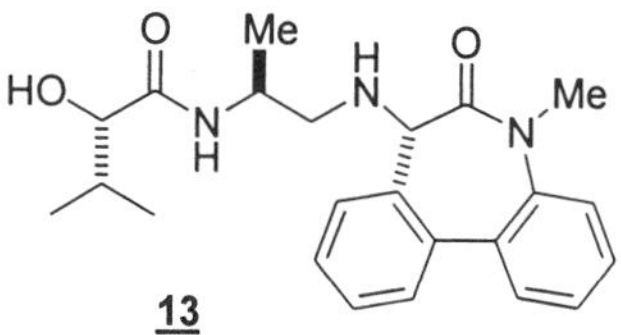

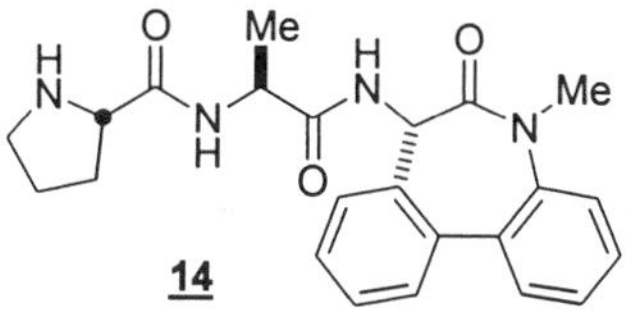

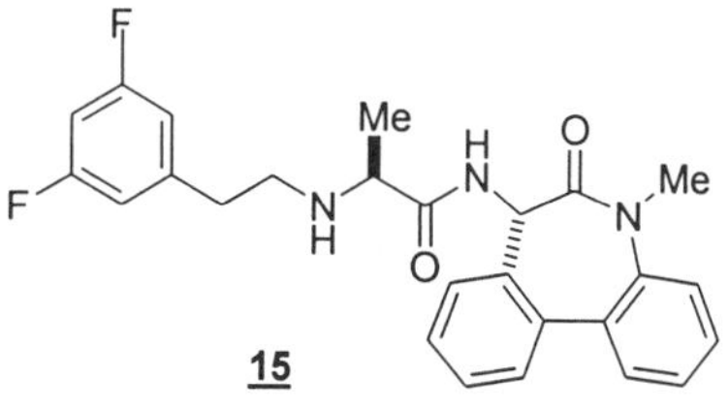

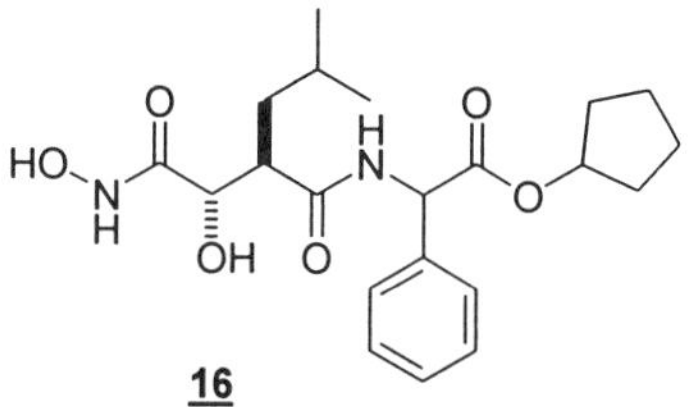

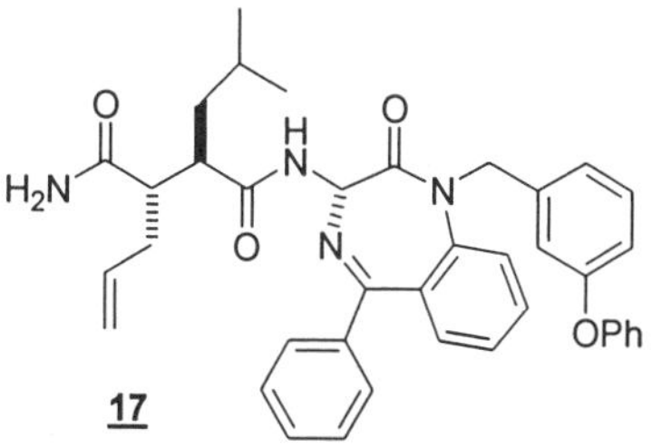

References

1. J.A. Martin, B.L. Smith, T.J. Mathews and S.J. Ventura, Natl. Vital Stat. Rep., 47, 1 (1999).
2. S.S. Sisodia, J. Clin. Invest., 104, 1169 (1999).
3. A.F. Jorm and D. Jolley, Neurology, 51, 728 (1998).
4. R. Katzman, New Engl. J. Med., 314, 964 (1986).
5. K. Heininger, Human Psychopharmacol, 14, 363 (1999).
6. P. Celsis, Ann Med, 32, 6 (2000).
7. W. Pendlebury and P. Solomon, Clin. Symposia, 48, 1 (1996).
8. C.E. Milligan, Nat. Med., 6, 385 (2000).
9. M. Tolnay and A. Probst, Neuropathol. Appl. Neurobiol., 25, 171 (1999).
10. R. Brandt and J. Eidenmuller, Mol. Biol. Alzheimer's Dis., , 17 (1998).
11. B.A. Yankner, Neuron, 16, 921 (1996).
12. G. Halliday, S.R. Robinson, C. Shepherd and J. Kril, Clin. Exp. Pharmacol. Physiol., 27, 1 (2000).
13. B.L. Flynn and K.A. Theesen, Ann. Pharmacother., 33, 840 (1999).
14. M.A. Findeis, Curr. Opin. Cent. Peripher. Nerv. Syst. Invest. Drugs, 1, 333 (1999).
15. K. Duff, Trends Neurosci., 22, 485 (1999).
16. D. Schenk, R. Barbour, W. Dunn, G. Gordon, H. Grajeda, T. Guido, K. Hu, J. Huang, K. Johnson-Wood, K. Khan, D. Kholodenko, M. Lee, Z. Liao, I. Lieberburg, R. Motter, L. Mutter, F. Soriano, G. Shopp, N. Vasquez, C. Vandevert, S. Walker, M. Wogulis, T. Yednock, D. Games and P. Seubert, Nature, 400, 173 (1999).
17. V. John, L.H. Latimer, J.S. Tung and M.S. Dappen, Annu. Rep. Med. Chem., 32, 11 (1997).
18. C.E. Augelli-Szafran, L.C. Walker and H. LeVine, III, Annu. Rep. Med. Chem., 34, 21 (1999).
19. D.J. Selkoe, Nature, 399, A23 (1999).
20. G.D. Schellenberg, Proc. Natl. Acad. Sci. U. S. A., 92, 8552 (1995).

21. D. Dawbarn and S.J. Allen In "The Molecular and Cellular Neurobiology Series"; Collingridge, G. L., Davies, R. W., Hunt, S. P., Eds.; BIOS Scientific Publishers Oxford, 1995, p 9.

22. J. Hardy, Trends Neurosci., 20, 154 (1997).

23. D.J. Selkoe, Annu. Rev.Cell Biol., , 10373 (1994).

24. E. Storey and R. Cappai, Neuropathol. Appl. Neurobiol., 25, 81 (1999).

25. H. Meziane, J.C. Dodart, C. Mathis, S. Little, J. Clemens, S.M. Paul and A. Ungerer, Proc. Natl. Acad. Sci. U. S. A., 95, 12683 (1998).

26. S.S. Sisodia and M. Gallagher, Proc. Natl. Acad. Sci. U. S. A., 95, 12074 (1998).

27. G.R. Dawson, G.R. Seabrook, H. Zheng, D.W. Smith, S. Graham, G. O'Dowd, B.J. Bowery, S. Boyce, M.E. Trumbauer, H.Y. Chen, L.H.T. Van Der Ploeg and D.J.S. Sirinathsinghji, Neuroscience (Oxford), 90, 1 (1999).

28. J.D. Buxbaum, K.-N. Liu, Y. Luo, J.L. Slack, K.L. Stocking, J.J. Peschon, R.S. Johnson, B.J. Castner, D.P. Cerretti and R.A. Black, J. Biol. Chem., 273, 27765 (1998).

29. S. Lammich, E. Kojro, R. Postina, S. Gilbert, R. Pfeiffer, M. Jasionowski, C. Haass and F. Fahrenholz, Proc. Natl. Acad. Sci. U. S. A., 96, 3922 (1999).

30. S. Parvathy, I. Hussain, E.H. Karran, A.J. Turner and N.M. Hooper, Biochemistry, 37, 1680 (1998).

31. D.M. Skovronsky, D.B. Moore, M.E. Milla, R.W. Doms and V.M.Y. Lee, J. Biol. Chem., 275, 2568 (2000).

32. S. Parvathy, I. Hussain, E.H. Karran, A.J. Turner and N.M. Hooper, Biochemistry, 38, 9728 (1999).

33. S.F. Lichtenthaler, C.L. Masters and K. Beyreuther, Handb. Exp. Pharmacol., 140, 359 (2000).

34. J.T. Jarrett, E.P. Berger and P.T. Lansbury Jr., Biochemistry, 32, 4693 (1993).

35. T. Iwatsubo, A. Okada, N. Suzuki, H. Mizusawa, N. Nukina and Y. Ihara, Neuron, 13, 45 (1994).

36. C.A. Wilson, R.W. Doms and V.M.Y. Lee, J. Neuropathol. Exp. Neurol., 58, 787 (1999).

37. G.K. Gouras, J. Tsai, J. Naslund, B. Vincent, M. Edgar, F. Checler, J.P. Greenfield, V. Haroutunian, J.D. Buxbaum, H. Xu, P. Greengard and N.R. Relkin, Am. J. Pathol., 156, 15 (2000).

38. P.H. St George-Hyslop, Biol. Psychiatry, 47, 183 (2000).

39. D.J. Selkoe, Science, 275, 630 (1997).

40. J.K. Teller, C. Russo, L.M. DeBusk, G. Agelini, D. Zaccheo, F. Dagna-Bricarelli, P. Scartezzini, S. Bertolini, D.M.A. Mann, M. Tabaton and P. Gambetti, Nat. Med., 2, 93 (1996).

41. G. Thinakaran, J. Clin. Invest., 104, 1321 (1999).

42. H. Steiner, A. Capell and C. Haass, Biochem. Soc. Trans., 27, 234 (1999).

43. F. Checler, Mol. Neurobiol., 19, 255 (1999).

44. J. Hardy, Proc. Natl. Acad. Sci. U. S. A., 94, 2095 (1997).

45. M. Citron, C.B. Eckman, T.S. Diehl, C. Corcoran, B.L. Ostaszewski, W. Xia, G. Levesque, P.S.G. Hyslop, S.G. Younkin and D.J. Selkoe, Neurobiol. Dis., 5, 107 (1998).

46. D.R. Borchelt, T. Ratovitski, J. Van Lare, M.K. Lee, V. Gonzales, N.A. Jenkins, N.G. Copeland, D.L. Price and S.S. Sisodia, Neuron, 19, 939 (1997).

47. J. Wang, D.W. Dickson, J.Q. Trojanowski and V.M.Y. Lee, Exp. Neurol., 158, 328 (1999).

48. C.A. McLean, R.A. Cherny, F.W. Fraser, S.J. Fuller, M.J. Smith, K. Beyreuther, A.I. Bush and C.L. Masters, Ann. Neurol., 46, 860 (1999).

49. L.-F. Lue, Y.-M. Kuo, A.E. Roher, L. Brachova, Y. Shen, L. Sue, T. Beach, J.H. Kurth, R.E. Rydel and J. Rogers, Am. J. Pathol., 155, 853 (1999).

50. J. Naslund, V. Haroutunian, R. Mohs, K.L. Davis, P. Davies, P. Greengard and J.D. Buxbaum, JAMA, 283, 1571 (2000).

51. D.J. Selkoe, JAMA, 283, 1615 (2000).

52. I. Hussain, D. Powell, D.R. Howlett, D.G. Tew, T.D. Meek, C. Chapman, I.S. Gloger, K.E. Murphy, C.D. Southan, D.M. Ryan, T.S. Smith, D.L. Simmons, F.S. Walsh, C. Dingwall and G. Christie, Mol. Cell. Neurosci., 14, 419 (1999).

53. R. Vassar, B.D. Bennett, S. Babu-Khan, S. Kahn, E.A. Mendiaz, P. Denis, D.B. Teplow, S. Ross, P. Amarante, R. Loeloff, Y. Luo, S. Fisher, J. Fuller, S. Edenson, J. Lile, M.A. Jarosinski, A.L. Biere, E. Curran, T. Burgess, J.-C. Louis, F. Collins, J. Treanor, G. Rogers and M. Citron, Science, 286, 735 (1999).

54. R. Yan, M.J. Bienkowski, M.E. Shuck, H. Miao, M.C. Tory, A.M. Pauley, J.R. Brashler, N.C. Stratman, W.R. Mathews, A.E. Buhl, D.B. Carter, A.G. Tomasselli, L.A. Parodi, R.L. Heinrikson and M.E. Gurney, Nature, 402, 533 (1999).

55.   X. Lin, G. Koelsch, S. Wu, D. Downs, A. Dashti and J. Tang, Proc. Natl. Acad. Sci. U. S. A., 97, 1456 (2000).
56.   S. Sinha, J.P. Anderson, R. Barbour, G.S. Basi, R. Caccavello, D. Davis, M. Doan, H.F. Dovey, N. Frigon, J. Hong, K. Jacobson-Croak, N. Jewett, P. Keim, J. Knops, I. Lieberburg, M. Power, H. Tan, G. Tatsuno, J. Tung, D. Schenk, P. Seubert, S.M. Suomensaari, S. Wang, D. Walker, J. Zhao, L. McConlogue and V. John, Nature, 402, 537 (1999).
57.   D.G. Cook, M.S. Forman, J.C. Sung, S. Leight, D.L. Kolson, T. Iwatsubo, V.M.Y. Lee and R.W. Doms, Nat. Med., 3, 1021 (1997).
58.   J. Knops, S. Suomensaari, M. Lee, L. McConlogue, P. Seubert and S. Sinha, J. Biol. Chem., 270, 2419 (1995).
59.   B. Munoz, I.A. McDonald and E. Albrecht, PCT Int. Appl. WO 9620725 A2 (1996).
60.   I.A. McDonald, E. Albrecht, B. Munoz, B.A. Rowe, R.S. Siegel and S.L. Wagner, PCT Int. Appl. WO 9620949 A1 (1996).
61.   G. Abbenante, D.M. Kovacs, D.L. Leung, D.J. Craik, R.E. Tanzi and D.P. Fairlie, Biochem. Biophys. Res. Commun., 268, 133 (2000).
62.   A.K. Ghosh, D. Shin, D. Downs, G. Koelsch, X. Lin, J. Ermolieff and J. Tang, J. Am. Chem. Soc., 122, 3522 (2000).
63.   H. Potter and D. Dressler, Nat. Biotechnol., 18, 125 (2000).
64.   M.S. Brown, J. Ye, R.B. Rawson and J.L. Goldstein, Cell, 100, 391 (2000).
65.   S.F. Lichtenthaler, N. Ida, G. Multhaup, C.L. Masters and K. Beyreuther, Biochemistry, 36, 15396 (1997).
66.   E. Tischer and B. Cordell, J. Biol. Chem., 271, 21914 (1996).
67.   M.P. Murphy, L.J. Hickman, C.B. Eckman, S.N. Uljon, R. Wang and T.E. Golde, J. Biol. Chem., 274, 11914 (1999).
68.   S.F. Lichtenthaler, R. Wang, H. Grimm, S.N. Uljon, C.L. Masters and K. Beyreuther, Proc. Natl. Acad. Sci. U. S. A., 96, 3053 (1999).
69.   K.S. Kosik, Proc. Natl. Acad. Sci. U. S. A., 96, 2574 (1999).
70.   H.W. Klafki, D. Abramowski, R. Swoboda, P.A. Paganetti and M. Staufenbiel, J. Biol. Chem., 271, 28655 (1996).
71.   J.N. Higaki, S. Chakravarty, C.M. Bryant, L.R. Cowart, P. Harden, J.M. Scardina, B. Mavunkel, G.R. Luedtke and B. Cordell, J. Med. Chem., 42, 3889 (1999).
72.   M.S. Wolfe, M. Citron, T.S. Diehl, W. Xia, I.O. Donkor and D.J. Selkoe, J. Med. Chem., 41, 6 (1998).
73.   M.S. Wolfe, W. Xia, C.L. Moore, D.D. Leatherwood, B. Ostaszewski, T. Rahmati, I.O. Donkor and D.J. Selkoe, Biochemistry, 38, 4720 (1999).
74.   J.T. Durkin, S. Murthy, E.J. Husten, S.P. Trusko, M.J. Savage, D.P. Rotella, B.D. Greenberg and R. Siman, J. Biol. Chem., 274, 20499 (1999).
75.   H. Hamazaki, FEBS Lett., 424, 136 (1998).
76.   L. Zhang, L. Song and E.M. Parker, J. Biol. Chem., 274, 8966 (1999).
77.   B. De Strooper, P. Saftig, K. Craessaerts, H. Vanderstichele, G. Guhde, W. Annaert, K. Von Figura and F. Van Leuven, Nature, 391, 387 (1998).
78.   C. Haass and D.J. Selkoe, Nature, 391, 339 (1998).
79.   W. Xia, J. Zhang, B.L. Ostaszewski, W.T. Kimberly, P. Seubert, E.H. Koo, J. Shen and D.J. Selkoe, Biochemistry, 37, 16465 (1998).
80.   T. Nakai, A. Yamasaki, M. Sakaguchi, K. Kosaka, K. Mihara, Y. Amaya and S. Miura, J. Biol. Chem., 274, 23647 (1999).
81.   M.S. Wolfe, J.D.L. Angeles, D.D. Miller, W. Xia and D.J. Selkoe, Biochemistry, 38, 11223 (1999).
82.   H. Steiner, H. Romig, B. Pesold, U. Philipp, M. Baader, M. Citron, H. Loetscher, H. Jacobsen and C. Haass, Biochemistry, 38, 14600 (1999).
83.   H. Steiner, H. Romig, M.G. Grim, U. Philipp, B. Pesold, M. Citron, R. Baumeister and C. Haass, J. Biol. Chem., 274, 7615 (1999).
84.   G. Yu, F. Chen, G. Levesque, M. Nishimura, D.-M. Zhang, L. Levesque, E. Rogaeva, D. Xu, Y. Liang, M. Duthie, P.H. St. George-Hyslop and P.E. Fraser, J. Biol. Chem., 273, 16470 (1998).
85.   A. Georgakopoulos, P. Marambaud, S. Efthimiopoulos, J. Shioi, W. Cui, H.-C. Li, M. Schutte, R. Gordon, G.R. Holstein, G. Martinelli, P. Mehta, V.L. Friedrich, Jr. and N.K. Robakis, Mol. Cell, 4, 893 (1999).
86.   A. Takashima, M. Murayama, O. Murayama, T. Kohno, T. Honda, K. Yasutake, N. Nihonmatsu, M. Mercken, H. Yamaguchi, S. Sugihara and B. Wolozin, Proc. Natl. Acad. Sci. U. S. A., 95, 9637 (1998).
87.   C. Czech, G. Tremp and L. Pradier, Prog. Neurobiol., 60, 363 (2000).

88. R. Baumeister, Eur. Arch. Psychiatry Clin. Neurosci., <u>249</u>, 280 (1999).
89. S.S. Sisodia, S.H. Kim and G. Thinakaran, Am. J. Hum. Genet., <u>65</u>, 7 (1999).
90. J.S. Nye, Curr. Biol., <u>9</u>, R118 (1999).
91. P.C. Wong, H. Zheng, H. Chen, M.W. Becher, D.J.S. Sirinathsinghji, M.E. Trumbauer, H.Y. Chen, D.L. Price, L.H.T. Van der Ploeg and S.S. Sisodia, Nature, <u>387</u>, 288 (1997).
92. J. Shen, R.T. Bronson, D.F. Chen, W. Xia, D.J. Selkoe and S. Tonegawa, Cell, <u>89</u>, 629 (1997).
93. B. De Strooper, W. Annaert, P. Cupers, P. Saftig, K. Craessaerts, J.S. Mumm, E.H. Schroeter, V. Schrijvers, M.S. Wolfe, W.J. Ray, A. Goate and R. Kopan, Nature, <u>398</u>, 518 (1999).
94. D.J. Selkoe, Curr. Opin. Neurobiol., <u>10</u>, 50 (2000).
95. E.H. Schroeter, J.A. Kisslinger and R. Kopan, Nature, <u>393</u>, 382 (1998).
96. Y. Ye, N. Lukinova and M.E. Fortini, Nature, <u>398</u>, 525 (1999).
97. G. Struhl and I. Greenwald, Nature, <u>398</u>, 522 (1999).
98. W. Song, P. Nadeau, M. Yuan, X. Yang, J. Shen and B.A. Yankner, Proc. Natl. Acad. Sci. U. S. A., <u>96</u>, 6959 (1999).
99. A. Capell, H. Steiner, H. Romig, S. Keck, M. Baader, M.G. Grim, R. Baumeister and C. Haass, Nat. Cell Biol., <u>2</u>, 205 (2000).
100. M.S. Wolfe, W. Xia, B.L. Ostaszewski, T.S. Diehl, W.T. Kimberly and D.J. Selkoe, Nature, <u>398</u>, 513 (1999).
101. W.T. Kimberly, W. Xia, T. Rahmati, M.S. Wolfe and D.J. Selkoe, J. Biol. Chem., <u>275</u>, 3173 (2000).
102. W.J. Ray, M. Yao, J. Mumm, E.H. Schroeter, P. Saftig, M. Wolfe, D.J. Selkoe, R. Kopan and A.M. Goate, J. Biol. Chem., <u>274</u>, 36801 (1999).
103. W.G. Annaert, L. Levesque, K. Craessaerts, I. Dierinck, G. Snellings, D. Westaway, P.S. George-Hyslop, B. Cordell, P. Fraser and B. De Strooper, J. Cell Biol., <u>147</u>, 277 (1999).
104. W.J. Ray, M. Yao, P. Nowotny, J. Mumm, W. Zhang, J.Y. Wu, R. Kopan and A.M. Goate, Proc. Natl. Acad. Sci. U. S. A., <u>96</u>, 3263 (1999).
105. L. Pradier, N. Carpentier, L. Delalonde, N. Clavel, M.-D. Bock, L. Buee, L. Mercken, B. Tocque and C. Czech, Neurobiol. Dis., <u>6</u>, 43 (1999).
106. Y.-M. LI, M.-T. Lai, M. Xu, Q. Huang, J. DiMuzio-Mower, M.K. Sardana, X.-P. Shi, K.-C. Yin, J.A. Shafer and S.J. Gardell, Proc. Natl. Acad. Sci. U. S. A, Early Edition, (2000).
107. J.E. Audia, P.A. Hyslop, J.S. Nissen, R.C. Thompson, J.S. Tung and L.I. Tanner, PCT Int. Appl. WO 0019210 A2 (2000).
108. J. Hardy and A. Israel, Nature, <u>398</u>, 466 (1999).
109. B. Varnum-Finney, L.E. Purton, M. Yu, C. Brashem-Stein, D. Flowers, S. Staats, K.A. Moore, I. Le Roux, R. Mann, G. Gray, S. Artavanis-Tsakonas and I.D. Bernstein, Blood, <u>91</u>, 4084 (1998).
110. E. Thorsett and I. Lieberburg, Curr. Opin. Cent. Peripher. Nerv. Syst. Invest. Drugs, <u>1</u>, 327 (1999).
111. C.L. Moore and M.S. Wolfe, Expert Opin. Ther. Pat., <u>9</u>, 135 (1999).
112. J.E. Audia, B.A. Dressman and Q. Shi, PCT Int. Appl. WO 9966934 A1 (1999).
113. J.E. Audia, W.J. Porter, R.C. Thompson, S.C. Wilkie, D.R. Stack and Q. Shi, PCT Int. Appl. WO 9967219 A1 (1999).
114. R.C. Thompson, S. Wilkie, D.R. Stack, E.E. Vanmeter, Q. Shi, T.C. Britton, J.E. Audia, J.K. Reel, T.E. Mabry, B.A. Dressman, C.L. Cwi, S.S. Henry, S.L. McDaniel, R.D. Stucky and W.J. Porter, PCT Int. Appl. WO 9967221 A1 (1999).
115. J.E. Audia, R.C. Thompson, S.C. Wilkie, T.C. Britton, W.J. Porter, G.W. Huffman and L.H. Latimer, Wo 9967220 A1 (1999).
116. R.E. Olson, T.P. Maduskuie and L.A. Thompson, PCT Int. Appl. WO 0007995 A1 (2000).
117. S.L. Wagner and B. Munoz, J. Clin. Invest., <u>104</u>, 1329 (1999).

Chapter 5. Targeting Nicotinic Acetylcholine Receptors: Advances in Molecular Design and Therapies

Jeffrey D. Schmitt and Merouane Bencherif
Targacept Inc.
P.O. Box 1487, Winston-Salem, North Carolina, 27101

Introduction – Cholinergic pharmacology research, initiated over one hundred years ago by the discovery of a 'receptive substance' in muscle tissue, has exploded in recent years. Indeed, mounting evidence suggests that modulation of nicotinic acetylcholine receptors (nAChRs) may benefit numerous central nervous system (CNS) and peripheral nervous system (PNS) disorders (1,2). The nAChR approach is particularly attractive where therapeutic intervention is currently limited. Such indications include senile dementia of the Alzheimer's type, Parkinson's disease, Huntington's chorea, tardive dyskinesia, hyperkinesia, mania, depression, attention deficit disorder, anxiety, dyslexia, schizophrenia, Tourette's syndrome and smoking cessation. Nicotinic systems also appear to be involved in the pathophysiology of pain and in non-CNS disorders such as ulcerative colitis. Advances in nAChR therapeutics have also been augmented by new synthetic methodologies that tackle the challenging heterocyclic chemistry required to develop nAChR selective drug candidates.

nAChRs are members of the ligand-gated ion-channel neurotransmitter family of receptors. These receptors consist of five individual subunit proteins assembled around a central ion-conducting pore (3). Advances in molecular biology and selective probe design have resulted in the discovery of nine human subunit genes expressed in the CNS, designated $\alpha 2$-$\alpha 7$ and $\beta 2$-$\beta 4$. Five additional nAChR subunits expressed in the PNS are designated $\alpha 1$, $\beta 1$, γ, δ, and ε (4). Naming of subunits has been largely chronological, with the exception that α subunits contain vicinal disulphides in their extracellular domain. nAChR subunits are known to oligomerize into both homopentameric and heteropentameric subtypes, each of which possesses distinct ligand specificity, pharmacological profile and distribution in the nervous system (5). The predominant homopentameric subtype, $\alpha 7$, is characterized by its high affinity for α-bungarotoxin (6). In the CNS and ganglia, α and β subunits assemble to form subtypes with the following stoichiometry: $\{\alpha\}_2\{\beta\}_3$. Of major interest is the $\alpha 4\beta 2$ subtype which exhibits high-affinity binding to nicotine, cytisine and epibatidine (7,8). nAChR induced dopamine release in the CNS is evidenced (vide infra) to be elicited by heterogenous $\alpha 3\beta 2\alpha 4\beta 2\beta x$ ('x' implies an unknown subtype) receptors (9,10). Ganglionic nAChRs are known consist of the following subtypes: $\alpha 3\beta 4$, $\alpha 3\alpha 5\beta 4$ and $\alpha 3\alpha 7\beta x$ (11,12). Finally, receptors at the neuromuscular junction take the form of $\alpha 1\beta 1\delta\gamma$ or $\alpha 1\beta 1\delta\varepsilon$ (13).

STRUCTURE-ACTIVITY RELATIONSHIPS OF NOVEL nAChR LIGANDS

Despite the fact that nAChR ligands have been explored for the better part of a century (14), especially in the realm of nicotinic insecticides, only in recent years has the focus turned to their use as therapeutics. Subsequently, the scientific community is only starting to unravel the determinants of CNS/PNS and subtype selectivity (15).

Substitution of nicotine's (1, R= H) pyridyl moiety can dramatically effect both nAChR affinity (16) and subtype specificity (17). Modification of the 6-pyridyl position (R) results in a moderate increase in affinity (^{3}H-nicotine; rat cerebrum) in the case of halogen substitution, moderate diminution in affinity with small moieties of moderate oxidation state or a dramatic decrease with carboxylates and amides (18). A new procedure for the synthesis of chiral pyridyl pyrrolidines (**2**) using an inverse electron-demand Diels-Alder reaction has been described. This methodology may lead to other

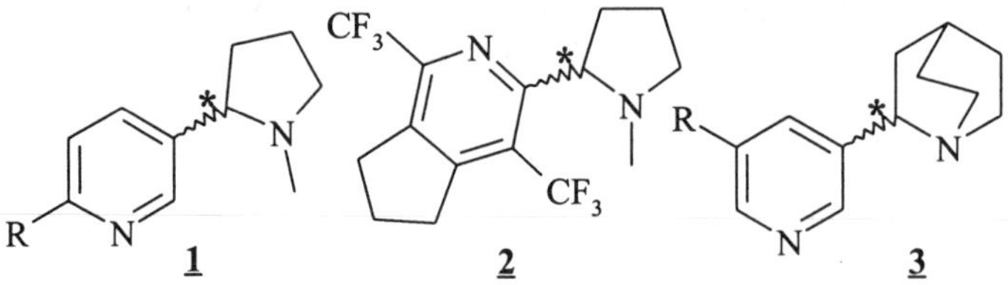

synthetic in-roads to the difficult pyridine nucleus (19). Optimization of the racemic pyridyl quinuclidine RJR-2429 (**3**, R= H, *R/S*) has resulted in the determination that the *R* enantiomer of **3** has a marginally higher affinity at $\alpha 4\beta 2$ sites than the *S* form. Additionally, the *S* isomer of **3** is 20-fold less potent at ganglionic subtypes relative to the *R* isomer, an observation which can be explained on the basis of differing Hahn receptor models for $\alpha 3$ versus $\alpha 4$ containing receptors. Additionally, modification of the 5-pyridyl position with alkoxy ethers results in a dramatic reduction in interaction at muscle and ganglionic receptors (20,21). A series of pyridyl ethers has been recently reported in the patent literature (22). These compounds, exemplified by **4** and **5**, exhibit high affinity binding (K_i < 100 nM) at ^{3}H-cytisine rat forebrain sites and moderate affinity (K_i < 700 nM) at ^{3}H-epibatidine sites.

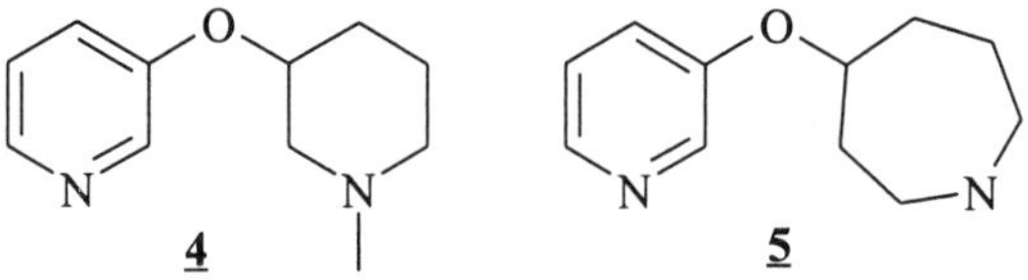

New pyrido-azepines have been evaluated for affinity at $\alpha 4\beta 2$ sites (23). Compounds of genus **6** have are very sensitive to ring size (m,n) and the position of

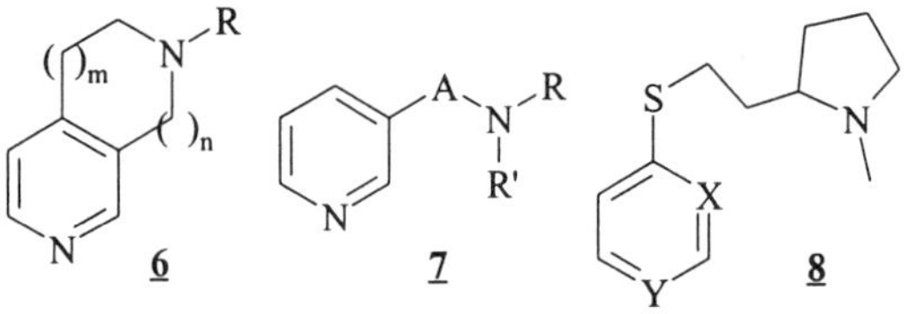

the cationic nitrogen relative to pyridine ring. The methylation state of the cationic nitrogen (R= H *vs.* R= CH$_3$) is also shown to dramatically impact receptor affinity. In the same publication, these investigators also explored three families of acyclic pyridyl amines of the genus **7** (where A is -(CH$_2$)$_{1-3}$-, -CH=CHCH$_2$-, -CαCCH$_2$-, -O(CH$_2$)-; R and R' are various alkyl and cycloalkyl substituents). A family of second-generation SIB-1553A (**8**, X= CH, Y= C-OH) analogs has been disclosed in the patent literature (24). These 2- and 4-pyridyl substituted thioethers (**8-10**, X= N or Y= N) are said to bind tightly to ^{3}H-nicotine sites (rat cerebrum) and induce calcium flux through $\alpha 4\beta 2$ nAChRs with potencies comparable to nicotine. This SAR information is further substantiation that SIB-1553A-like compounds interact with nAChRs in a manner different from other nicotinic alkaloids (16).

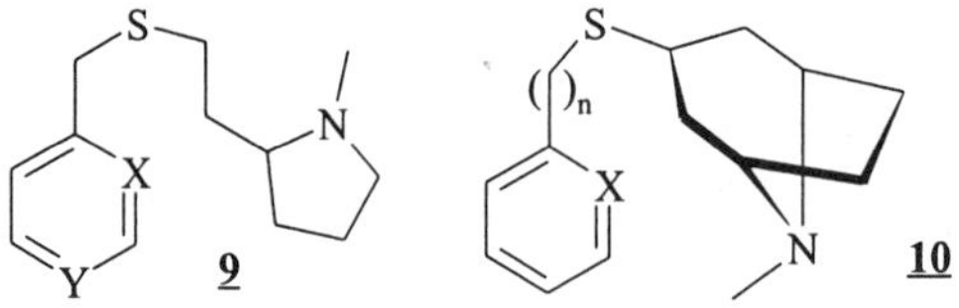

A number of 3-pyridyl ether nicotinic ligands were recently disclosed in the patent literature (25). Compounds of genus **11** are shown to bind to ^{3}H-cytisine (K_i= 50-200 nM) and ^{3}H-epibatidine (K_i= 300-700 nM) sites in rat forebrain while exhibiting much lower affinity at ^{125}I-α-bungarotoxin (α7) sites (>15 uM). Numerous investigators have

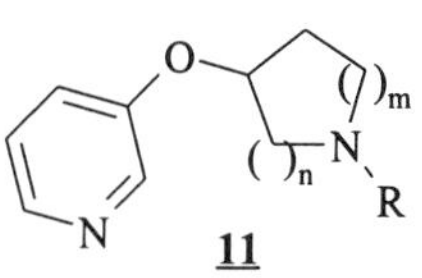

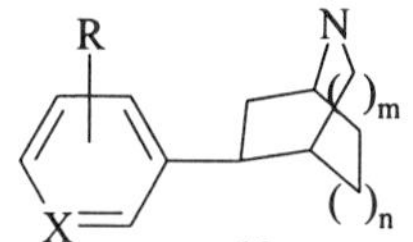

recently developed synthetic strategies for the synthesis of epibatidine (**12**, m= 0, n= 1, R= 6-Cl) analogs. The *exo*- and *endo*- forms of the 2-aza-bicyclo[2.2.1]heptane (m= 1, n= 0, R= 6-Cl) system have been described as well (26). A synthesis using N-protected 7-azabicyclo[2.2.1]hept-2-ene typifies another method of generating epibatidine variants, including those where X= CF and X= COH (27). Analogs possessing a 3-substituted 8-azabicylo[3.2.1]octane moiety (**13**) have been synthesized from tropinone (28). Ionic hydrogenation methodology affords products like **15**, from common pyridyl tropinone intermediates. Unfortunately no biological data are given for these compounds. Rapoport and coworkers have significantly advanced synthesis of chiral nicotinic alkaloids, employing glutamic acid as a key intermediate (29). Systems accessible *via* this methodology are exemplified by **14**. Further development of epiboxidine (**16**, R= CH$_3$) analogs has led to the synthesis of the phenylisoxazole (**16**, R= phenyl). This analog shows moderate binding to ^{3}H-cytisine sites in rat cortex (IC$_{50}$ = 147 nM) and a dramatic decrease in toxicity relative to epibatidine (mouse LD$_{50}$> 130 mg/kg (**16**, R= phenyl) *vs.* 0.196 mg/kg (**12**, m= 0, n= 1, R= 6-Cl).

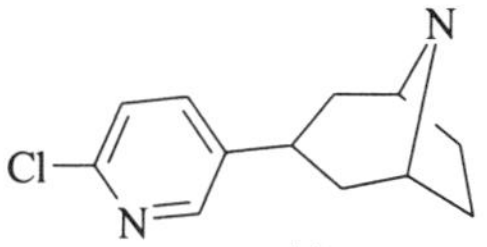

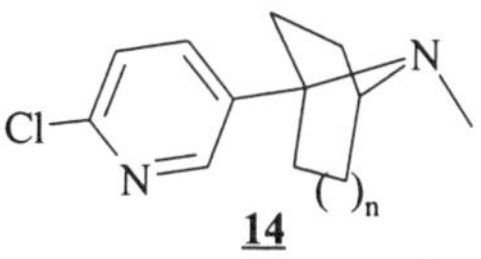

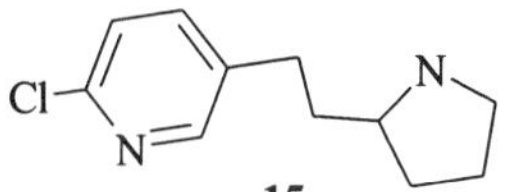

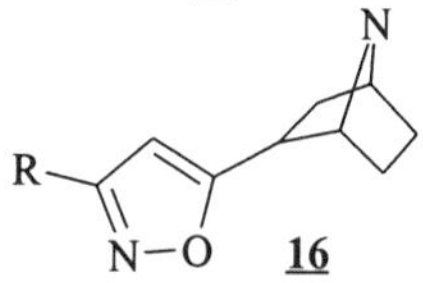

In the patent literature more than fifty diazacycloalkanes of genus **17** are described and evaluated for their interaction at ^{3}H-cytisine, ^{3}H-nicotine and ^{3}H-α-bungarotoxin sites (30). Results indicate that proper choice of substituents on the pyridine moiety can significantly effect subtype specificity. There has been considerable recent advance in characterizing the interaction of nicotinic insecticides with mammalian nAChRs. The *des*-nitro form of imidacloprid (**18**), as well as many related compounds, have been shown to interact with numerous nAChR subtypes (22). Additionally, the facile synthesis and preliminary pharmacological evaluation of numerous 6-chloro-3-pyridinylmethyl analogs have been reported by the same investigators (32). This work expands the current lexicon of cationic moieties recognized by nAChRs. Another exciting finding relates to nicotinic/muscarinic acetylcholine receptor specificity, as it is not

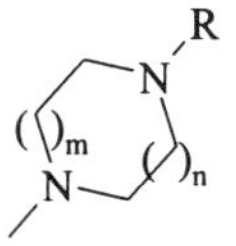

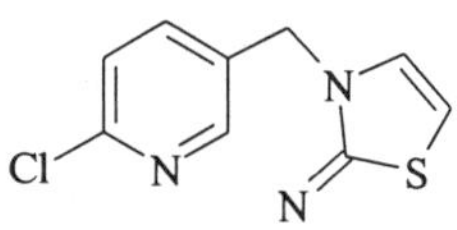

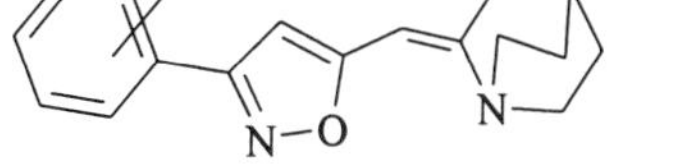

19: R= 3,4-di(OH)

20: R=3,4-di(OCH$_2$(4-OCH$_3$)Phenyl))

surprising that numerous nicotinic ligands interact with muscarinic acetylcholine receptors (mAChRs) and *vice versa*. The nAChR/mAChR selectivity ratio for ligand displacement, as exemplified by **19** and **20**, depends on the size of substitution on the hydrogen-bond acceptor geometrically opposite the cationic moiety, smaller substitutions favor nAChRs whilst larger substitutions favor mAChRs (33).

A number of studies have indicated a role for α7 nAChRs in neuroprotection,

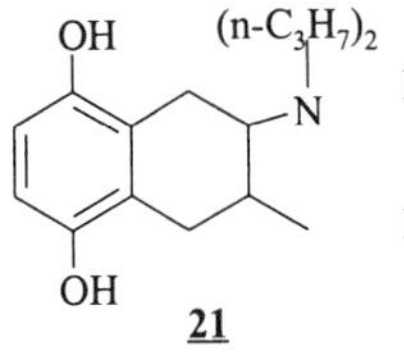

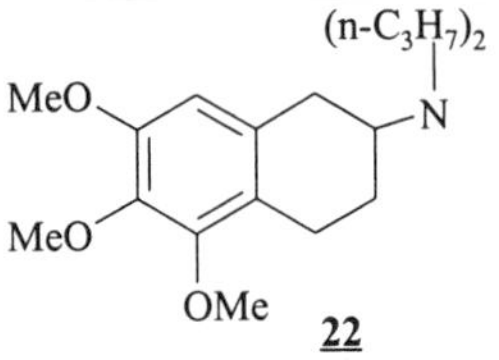

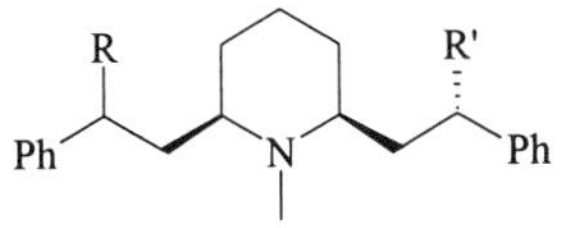

including modulation of -amyloid cytotoxicity (34). Recent evidence suggests intriguing tight interactions between Aβ1-42, a protein key to the pathophysiology of AD, and α7 receptors. In the recent patent literature a class of compounds, exemplified by **21**, has been shown to interact with α7 nAChRs and subsequently to inhibit the specific association of β-amyloid (35). Related aminotetralin derivatives like **22** have been shown to inhibit nicotine-induced relaxation of amphibian muscle (36). Notable advancement has been made in understanding of the determinants of lobeline (**23**) high-affinity nAChR binding (^{3}H-nicotine; rat cerebrum) and analgesia (37). Two salient findings are reported: 1) that chlorine substitution for the hydroxyl (**24**) leads to a compound with equal affinity (**23** K_i = 4 nM; **24** K_i = 5 nM); and 2) that the presence of two oxygens is not necessary for receptor interaction (**25** K_i = 340 nM).

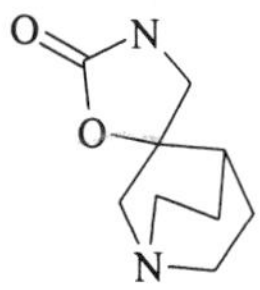

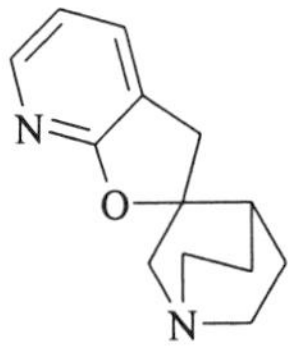

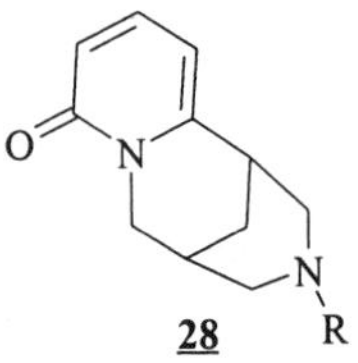

In an effort to extend SAR information on the potent 7 nAChR ligand AR-R17779 (**26**), a number of spiro(furanopyridyl)quinuclidines of genus **27** have been reported in the patent literature (38). The compounds described therein are said to have affinities at 7 and 4 2 nAChRs < 1μM. Finally, a number of N-substituted derivatives of cytisine (**28**, R= H) have been synthesized and evaluated for nAChR receptor affinity, in addition to broader spectrum *in vitro*/*vivo* activity (39). Relative to ^{3}H-cytisine binding (rat cortex), bulk in the R position of **28** has a deleterious effect on receptor affinity, with the interesting exception of two examples of cytisine dimers tethered by either a two (K_i = 96 nM) or three (K_i = 30 nM) carbon chain.

CLINICAL APPLICATIONS OF nAChR LIGANDS

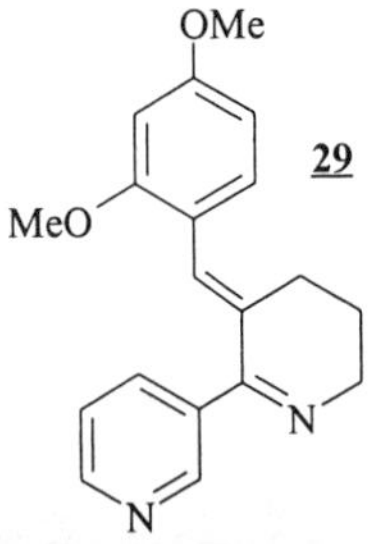

Neurodegenerative Disease – Alzheimer's Disease (AD) is a debilitating neurodegenerative disease characterized by progressive intellectual and personality decline, as well as a loss of memory, perception, reasoning, orientation and judgment. AD is becoming a major public health burden with the increasingly aging population. A salient characteristic of AD is a decline in cholinergic system function, and specifically, a severe depletion of cholinergic neurons. A 'nicotinic' hypothesis for the etiology and treatment of Alzheimer's is further supported by the early and significant depletion of high affinity nicotinic receptors in Alzheimer's patients' brains. Nicotine has long been known to improve learning and memory in animal models and to improve attention and learning in humans (40). Additionally, nicotine's neuroprotective effects *in vitro* and *in vivo* suggest that nicotinic ligands may exhibit long-term therapeutic benefits. These observations, together with the prominence of acetylcholinesterase inhibitor therapies-- which themselves target cholinergic neurotransmission-- have motivated the development of several nicotinic candidates, now in various stages of clinical trials (GTS-21(**29**); SIB-1553A (**8**, X= CH, Y= C-OH)). These efforts have been all the more stimulated by the lack of success in muscarinic therapies. Compound **29** has been evaluated for liver toxicity and induction of 9 different cytochrome P450 isoforms in rats. Oral administration of **29** for 7-days at 3, 60 and 300 mg/kg/day failed to change liver weight, and significant induction was only seen with CYP1A2 at the highest dosing (41). Compound **29** also appears readily absorbed *via* oral administration, but despite these encouraging results, **29** suffers from high first-pass metabolism (42). Progress in clinical development of **8** has not yet been reported.

Parkinson's Disease (PD) is a debilitating neurodegenerative disease, presently of unknown etiology, characterized by tremors and muscular rigidity. Degeneration of dopaminergic neurons believed to modulate the process of dopamine secretion, is evidenced to be a salient feature of this disease (as well as Lewy body dementia) and, like AD, there is concurrent loss of associated nAChR's. Intriguing are consistent findings of a negative correlation between smoking and the incidence of Parkinson's disease. In addition, nicotine exhibits neuroprotective effects in PD animal models and has been shown to modulate a number of neurotransmitters involved in PD pathophysiology. These neurotransmitters include acetylcholine (cognitive impaitment), norepinephrine (dysautonomia), 5-HT (depression) and glutamate (dyskinesia). There

is presently one nicotinic cholinergic candidate in clinical development: altinicline, or SIB-1508Y (**30**). This compound exhibits selectivity for the human recombinant $\alpha 4 \beta 2$ nicotinic receptor subtype. In rodent models, **30** increases locomotor activity and circling behavior in unilaterally 6-OHDA lesioned animals. **30** Also potentiates L-dopa in reserpine treated animals and has cognitive effects in rodents. Compound **30** also improves a visual memory task in monkeys following chronic low-dose 1-methyl-4-phenyl-1,2,3,6-tetrahydropyridine (MPTP) (43).   However, in a similar model **30** failed to improve motoric deficit but showed effects when combined with L-dopa. Two Phase II studies were conducted that assessed **30** in treating cognitive and motor deficits in PD patients. The initial study was a placebo-controlled, multi-dose, four-week study designed to test the safety of oral administration of **30** as a monotherapy in early stage PD patients. Recently SIBIA has reported that **30** did not show a statistically significant improvement in either motor or cognitive symptoms in early-stage PD patients as compared to placebo. However, it was mentioned that an unusually high placebo response rate was observed and that **30** *did* lead to improvement with respect to baseline values. The second study was a double-blind, randomized, placebo controlled, crossover study measuring safety, tolerability and efficacy of **30** in mid-to-late-stage PD patients receiving half their usual dose of L-DOPA. The combination of **30** (one of three doses in random order) and L-dopa (half the normally effective dose of L-DOPA) for four days was reportedly well-tolerated in later stage PD patients. Acute motor and cognitive performance were measured over an eight-hour period. Although, a trend towards significance was observed, with significant differences between the various doses of **30**, the effects were not significant compared to placebo. It should be noted that GI side effects were observed in the phase I study, and doses had to be decreased (44). In summary, side effects requiring decreased dosing and unusually high placebo effects are reported to be the main problems revealed in this study.

Neuroimaging – There has been a concerted effort to develop non-invasive techniques to probe neuro-receptors *in vivo*. Positron emission tomography (PET) and single photon emission computed tomography (SPECT) of high affinity ligands are routinely used to map and monitor alterations in densities of a variety of receptor targets having relevance to human diseases. The wide distribution and function of nAChRs in the body reinforces the view that nicotinic cholinergic signaling is involved in the regulation of the key neurochemicals as well as neuronal processes involved in sensory processing and cognition. Major cholinergic systems and subsystems have been described in rodent and primate brains (4). There is an early and significant depletion of high affinity nicotinic receptors in the brains of Alzheimer's patients. For example, individuals with β-amyloid plaques show a greater depletion of high affinity nAChRs in the entorhinal cortex than those without such plaques (45). Additionally, selective loss of nAChR subtypes has been reported in post-mortem studies using brain tissues from Alzheimer, Parkinson, and schizophrenic patients (46-52).

Radiolabeled nicotine was the first probe developed for measurement of nAChRs *in vivo* using PET imaging in primates and human. Despite its success, ^{11}C-nicotine has numerous limitations. While effects on regional cerebral blood flow can be corrected using dual tracer approaches (i.e., using ^{15}O- followed by ^{11}C-nicotine), marginal affinity and high non-specific binding of ^{11}C-nicotine are major shortcomings.

33: R=^{18}F, R'=H
34: R=H, R'=^{18}F

31   32

For these reasons, efforts have been directed toward development of other radio-tracers that image specific nAChR subtypes within the brain (53). A high specific activity is required for facile visualization of low density receptor subtypes, and the administered dose must be well below the toxic range and still provide high contrast images. Attempts were made to use the Abbott compound ^{11}C-ABT-418 (**31**) as a probe for neuronal nAChRs in primates, but the results have been disappointing (54, 55). The novel ^{11}C–labeled nicotinic agonist (*R,S*)-1-[^{11}C]methyl-2-(3-pyridyl)azetidine (**32**; [^{11}C]MPA) was reported to be superior to either ^{11}C-nicotine or **31**, based on affinity and selectivity (54). The ^{18}F–radiolabeled analogue of the compound A85380, 2-[^{18}F]fluoro-3-(2(*S*)-azetidinylmethoxy)pyridine (**33**; 2-[^{18}F]-A85380) was evaluated *in vivo* by PET in Rhesus monkey brain (56). The ligand exhibits picomolar affinity at α4β2 nAChRs, crosses the blood-brain barrier, and distributes in a pattern consistent with that previously reported for this receptor subtype. This compound shows an adequate specific/non-specific binding ratio and the study authors suggest that it shows promise for use with PET in human subjects. Another radiolabled analogue of A-85380 6-[^{18}F]Fluoro-3-(2(*S*)-azetidinylmethoxy)pyridine (**34**; 6-[^{18}F]fluoro-A-85380 or 6-[18F]FA), is being evaluated as a new tracer for PET (57). ^{123}I-Analogs of A-85380 are also being evaluated as a probes using SPECT (58).

Neurological And Neuropsychiatric Disorders – Tourette's Syndrome (TS), first described by Gilles de la Tourette in 1872, is the most debilitating of tic disorders. It is an autosomal dominant neuropsychiatric disorder characterized by a range of symptoms, including multiple motor and phonic tics. Motor tics generally include eye blinking, head jerking, shoulder shrugging and facial grimacing; while phonic or vocal tics include throat clearing, sniffling, yelping, tongue clicking and corpolalia.   The pathophysiology of TS presently is unknown, but it is believed that neurotransmission dysfunction is involved (59). It has been proposed that stimulation of nicotinic cholinergic systems may be beneficial in suppressing symptoms associated with TS. Studies with the nicotine patch have yielded promising results and nicotine appears to be effective in reducing Tourette's syndrome in populations whose symptoms are not well controlled with dopamine receptor blockers (60). Recent successful attempts using nicotinic antagonists in the management of Tourette's syndrome suggest that more targeted nicotinic therapies could be beneficial in this disorder (61).

Attention Deficit Hyperactivity Disorder (ADHD) is a disorder that affects mainly children who suffer from difficulty in concentrating, listening, learning and completing tasks, and who are restless, fidgety, impulsive and easily distracted.   Early clinical studies with the nicotine patch show promising results in adult ADHD patients. A recent study using nicotine skin patch treatment in eight Alzheimer patients showed significant attention improvement during the 4-week treatment (62).

There is a decreased number of hippocampal $\alpha7$ nAChRs in the postmortem brain tissue of schizophrenics (52); and the human $\alpha7$ receptor message, located on chromosome-15, has been linked to sensory inhibition/familial schizophrenia (63, 64). A recent study reports that the duplication of the $\alpha7$ subunit encoding gene which is expressed in the brain of normal subjects is missing in some schizophrenic subjects (65). Nicotine has been shown to improve sensory gating deficits in schizophrenic patients (66), in inbred DBA/2 mouse strains, in isolation-reared rats and in rats with amphetamine or cocaine-induced deficits (67, 68). Blockade of the $\alpha7$ nAChR induces gating deficits similar to those in schizophrenia (69) and selective antagonism blocks the positive effects of nicotine. Two $\alpha7$-selective ligands, DMXB or GTS-21 (**29**) and AR-R17779 (**26**), have been reported to reverse sensory gating deficits and to restore pre-pulse inhibition in DBA/2 mice, respectively (67, 70, 71). On a related front, there is an epidemiological association between nicotine use and major depression. Cessation of nicotine use is also reportedly harder in people with a history of major depression. Nicotine has been shown to be effective in rodent models for depression (forced swim test), and the nicotine patch significantly improved symptoms and REM sleep in depressed patients (72). SIB-1508Y (**65**) in phase II clinical trials for Parkinson's, has been reported to ameliorate mood and affect in those patients.

Autosomal Dominant Nocturnal Frontal Lobe Epilepsy (ADNFLE) is a form of epilepsy characterized by nocturnal brief partial seizures. The gene responsible for this form of epilepsy has been discovered on chromosome-20 q13.2-q13.3 (73, 74), a locus that contains the gene encoding the $\alpha4$ nAChR subunit.   Specific mis-sense mutations have been identified on the $\alpha4$ gene and electrophysiological studies with recombinant human $\alpha4\beta2$ nAChR indicate that these mutations alter its calcium permeability and desensitization (75).

Although the anxiolytic effects of nicotine in humans remains controversial, in animals models both anxiolytic effects at low doses and anxiogenic effects at higher nicotine doses have been reported (76). Additional conditions that could involve nicotinic pharmacology include obesity and obsessive compulsive disorder, a condition that has been associated with abnormal serotoninergic tone, and shows co-morbidity with Tourette's syndrome (77).

Pain – Current approaches for systemic analgesics include: non-steroidal anti-inflammatories (NSAIDS), opiates, and the emerging COX-2 inhibitors. Significant opportunities exist for oral and injectable non-opioid analgesics due to the shortcomings of existing therapies. Although nicotine has been known to produce analgesia in a number of animal and human models, the effects are modest and limited by toxic side-effects. Other nicotinic ligands targeting relevant CNS receptors can produce analgesia by an opioid-independent mechanism with potency much greater than that of morphine (78). Further development of such analgesics will require significant reduction of the cardio-respiratory side effects associated with the prototypical potent nicotinic analgesic epibatidine. An $\alpha 4\beta 2$-selective ligand RJR-2403 (**35**), initially developed for Alzheimer's disease (79, 80), has successfully completed Phase I clinical trials (81) and exhibits broad analgesic effects (82-84). These reports summarize the effects of **35** in both mice and rats on acute thermal pain (hot-plate and tail-flick), mechanical (paw pressure test), chemical (PPQ test), persistent pain (formalin test), chronic pain (inflammatory pain: arthritic models), and neuropathic pain (chronic constrictive injury model, Chung model, and capsaicin model). Interestingly, some gender-differences were noted in acute administration with a greater sensitivity in females rats. Another candidate in development, ABT-594 (**36**), was reported to exhibit potent antinociceptive effects (85). The *in vitro* nAChR profile indicates a rank order potency $\alpha 4\beta 2 > \alpha 3\beta 4 > \alpha 3\beta 2 > \alpha 7$ (with a relative EC$_{50}$ ratio of 1:3:10:100), and **36** is effective in a number of animal pain models. Both **35** and **36** provide analgesic effects through non-opiate mechanism as assessed by the lack of effect of the opiate antagonist naloxone, and some evidence suggests that they exert their effects through activation of inhibitory noradrenergic descending pathways (83). However, determination of the precise site(s) of action is complicated by the ubiquity of cholinergic neurotransmission in pain pathways, which include ganglia, spinal and supra-spinal sites. The development of safe long-lasting nicotinic analgesics remains the subject of intense research effort.

Non-CNS Diseases – Ulcerative Colitis (UC) is currently incurable except by surgical removal of the colon. Treatment is limited to anti-inflammatories, immunosuppressants, and antibiotics, all of which have numerous side effect liabilities. An effective non-surgical approach for treating the underlying cause of UC and preventing or delaying recurrence is needed. Transdermal nicotine reverses symptoms and pathology in 40-70% of UC patients, but side effects limit its use and have motivated the exploration of alternative administration routes such as by enema (86). The promise of highly targeted nicotinic therapies having lower side-effects has been highlighted by a recent long term cross over study. This work indicates that patients with mild to moderately active UC suffered fewer relapses when treated with mesalazine plus transdermal nicotine than with the combination of mesalazine plus oral corticosteroids (87). Proposed mechanisms explaining nicotinic effects include increased mucus secretion, inhibition of interleukins and/or eicosanoids, and/or normalization of vagal dysfunction. However the precise mechanism(s) remain elusive.

Other pathological non-CNS conditions where nicotine efficacy has been reported but remains to be confirmed include pouchitis (88) and influenza virus-induced pneumonitis (89). These effects are presumably elicited through immunosuppressive and anti-inflammatory mechanisms. Auto-immune diseases with high serum antibody titers against the muscle–type nicotinic receptor (Myasthenia Gravis) have been extensively studied and reviewed (90, 91). A high titer of antibodies against the nicotinic acetylcholine receptor (nAChR) neuronal subunits $\alpha 3$ and $\beta 4$ was reported in the serum of patients afflicted with pemphigus vulgaris and pemphigus foliaceus. Functional nicotinic receptors and expression of specific subunits including $\alpha 3$, $\beta 2$ and $\beta 4$ have been reported in keratinocytes (92, 93).

Conclusion – Several recent reviews have offered perspectives on how nAChR-targeted compounds might be developed to modify clinically-relevant end-points, and intense effort has been directed toward new nicotinic agonists that are potent and selective for specific nAChR subtypes. However, the relationship between nAChR subtypes and clinical end-points is complicated by the broad pharmacological role of nAChRs in modulating the release of various neurotransmitters (acetylcholine, dopamine, serotonin, norepinephrine, GABA, glutamate, takykinins, nitric oxide), peptides (e.g., CGRP, NGF, BDNF, bFGF, transthyretin), cytokines, eicosanoids, and gene expression (e.g., TH, PNMT). The dissection of these pharmacological end-points and the identification of their role in disease processes is a major challenge. Yet, the recent emergence of novel ligands that discriminate between the major neuronal nAChR subtypes ($\alpha 7$, $\alpha 4\beta 2$, $\alpha 4\beta 4$, $\alpha 3\beta 2$, $\alpha 3\beta 2\alpha 4\beta 2\beta x$ and $\alpha 3b4$) suggest that the successful clinical development of potent nAChR-based therapeutics is imminent and may expand the therapeutic arsenal available for treatment of a variety of disorders.

References

1. M.W. Holladay, M.J. Dart and J.K. Lynch, J. Med. Chem., 40, 4169 (1997).
2. M.W. Decker and S.P. Arneric in "Neuronal Nicotinic Receptors: Pharmacology and Therapeutic Opportunities," S.P. Arneric and J.D. Brioni, Eds. Wiley-Liss, New York, N.Y., 1998, p. 395.
3. H.A. Lester, Annu. Rev. Biophys. Biomol. Struct., 21, 267 (1992).
4. C. Gotti, D. Fornasari and F. Clementi, Prog.Neurobiology, 53, 199 (1997).
5. J.D. Brioni, M.W. Decker, J.P. Sullivan and S.P. Arneric, Adv. Pharmacol., 37, 153 (1997).
6. P. Seguela, J. Wadiche, K. Dineley-Miller, J.A. Dani and J.W. Patrick, J. Neurosci., 13, 596 (1993).
7. P. Whiting and J. Lindstrom, J. Neurosci., 6, 3061 (1986).
8. P. Whiting, F. Esch, S. Shimasaki and J. Lindstrom, FEBS Lett., 219, 459 (1987).
9. R.J. Lukas, J. Pharmacol. Exp. Ther., 265, 294 (1993).
10. J.R. Forsayeth and E. Kobrin, J. Neurosci., 17, 1531 (1997).
11. S.W. Halvorsen and D.K. Berg, J. Neurosci., 10, 1711 (1990).
12. S. Couturier, D. Bertrand, J.M. Matter, M.C. Hernandez, S. Bertrand, N. Millar, S. Valera, T. Barkas and M. Ballivet, Neuron, 5, 847 (1990).
13. J.A. Reynolds and A. Karlin, Biochemistry, 17, 2035 (1978).
14. P.A. Crooks in "Analytical Determination of Nicotine Related Compounds and Their Metabolism," J.W. Gorrod, and P. Jacob III, Eds. Elsevier Science B.V., Amsterdam Neth., 1999, p. 69.
15. K.L. Swanson, M. Alkondon, E.F.R. Pereira and E.X. Albuquerque in "The Toxic Action of Marine and Terrestrial Alkaloids," M.S. Blum, Ed., Alaken, Fort Collins Colo., 1995, p. 191.
16. J.D. Schmitt, Curr. Med. Chem., 7, 749 (2000).
17. N.D.P. Cosford, L. Bleicher, A. Herbaut, J.S. McCallum, J.-M. Vernier, H. Dawson, J.P. Whitten, P. Adams, L. Chavez-Noriega, L.D. Correa, J.H. Crona, L.S. Mahaffy, F. Menzaghi, T.S. Rao, R. Reid, A.I. Saccan, E. Santori, K.A. Stauderman, K. Whelan, G.K. Lloyd, and I.A. McDonald, J. Med. Chem., 39, 3235 (1996).
18. M. Dukat, M. Dowd, M.I. Damaj, B. Martin, M.A. El-Zahabi and R.A. Glennon, Eur. J. Med. Chem., 34, 31 (1999).
19. D. Che, J. Siegl, and G. Seitz, Tetrahedron: Asymm. 10, 573 (1999).
20. W.S. Caldwell, Proceedings from IBC 2nd International symposium on Nicotinic Acetylcholine Receptors. Advances in Molecular Pharmacology and Drug Development, Annapolis, MD, May 13-14, 1999.
21. M. Bencherif, J.D. Schmitt, B.S. Bhatti, P.A. Crooks, W.S. Caldwell, M.E. Lovette, K. Fowler, L. Reeves, and P.M. Lippiello, J. Pharmacol. Exp. Ther., 284, 886 (1998).
22. D. Peters, G.M. Olsen, S.F. Nielsen and E.O. Nielsen, PCT Patent Appl. WO 99/24422.
23. Y. Cheng, M. Dukat, M. Dowd, W. Fiedler, B. Martin, M.I. Damaj, and R.A. Glennon, Eur. J. Med. Chem., 34, 1 (1999).
24. J. Vernier, N.D. Cosford, and I.A. McDonald, PCT Patent Appl. WO 99/32117 (1999).
25. D. Peters, G.M. Olsen, S.F. Nielsen, and E.O. Neilsen, PCT Patent Appl. WO 99/24422 (1999).
26. J.R. Malpass, and C.D. Cox, Tetrahedron Lett., 40, 1419(1999).
27. D. Yohannes, and M.W. Bundesmann, Patent EP 0955301A2 (1999).
28. H.F. Olivo, D. A. Colby, and M.S. Hemenway, J. Org. Chem., (1999).

29. Y. Xu, J. Choi, M.I. Calaza, S. Turner, and H. Rapoport, J. Org. Chem., 64, 4069 (1999).
30. D. Peters, G.M. Olsen, S.F. Nielsen and E.O. Neilsen, PCT Patent Appl. WO 99/21834 (1999).
31. M. Tomizawa, and J.E. Casida, Brit. J. Pharm., 127, 115 (1999).
32. B. Latli, K. D'Amour, and J.E. Casida, J. Med. Chem. 42, 2227 (1999).
33. K.I. Choi, H.H. Cha, Y.S. Cho A.N. Pae and C. Jin, Bioorg. Med. Chem. Lett, 9, 2795 (1999).
34. H. Wang, D.H. S. Lee, M.R. D'Andrea, P.A. Peterson, R.P. Shank and A.B. Reitz, J. Biol. Chem., 275, 5626 (2000).
35. A.B. Reitz, D.A. Demeter, D.H. Lee, H. Wang, R.H. Chen, T.M. Ross, M.K. Scott, and C. Plata-Salaman, PCT Patent Appl. WO 99/62505 (1999).
36. M.O. Babaogul, T.R. Aydos, H.S. Orer, and M. Ilhan, Drug Res., 19(II), 566 (1999).
37. D. Flammia, M. Dukat, M.I. Damaj, B. Martin, and R.A. Glennon, J. Med. Chem., 42, 3726 (1999).
38. E. Phillips, R. Mack, J. Macor and S. Semus, PCT Patent Appl. WO 99/03859.
39. C.C. Boido, and F. Sparatore, Il Farmaco, 54, 438 (1999).
40. E. Phillips, R. Mack, J. Macor and S. Semus, Patent WO 99/03859 (1999).
41. R. Azuma, M. Komuro, S. Black and J.M. Mathews, Toxicol. Lett., 110, 137 (1999).
42. R. Azuma, M. Komuro, B.H. Korsch, J.C. Andre, O. Onnagawa, S.R. Black, and J.M. Mathews, Xenobiotica, 29, 747 (1999).
43. J.S. Schneider, J.P. Tinker, M. Van Velson, F. Menzaghi, and G.K. Lloyd, J. Pharmacol. Exp. Ther., 290, 731 (1999).
44. D.E. McClure, IBC's 2nd International Symposium on Nicotinic Acetylcholine receptors, Annapolis, MD, May 13-14, 1999.
45. E.K. Perry, J.A. Court, S. Lloyd, M. Johnson, M.H. Griffiths, D. Spurden, M.A. Piggott, J. Turner and R.H. Perry, Ann. N.Y. Acad. Sci., 777: 388 (1996).
46. A. Nordberg and W. Bengt, Neurosci. Lett., 72, 115 (1986).
47. K.J. Kellar, P.J. Whitehouse, A.M. Martion-Barrows, K. Marcus and D.L. Price, Brain Res. 436, 62 (1987).
48. D.M. Araujo, P.A. Lapchak, Y. Robitaille, S. Gauthier and R. Quirion, J. Neurochem., 50, 1914 (1988).
49. P.J. Whitehouse, A.M. Martino, K.A. Marcus, R.M. Zweig, H.S. Singer, D.L. Price, and K.J. Kellar, Arch. Neurol. 45: 722-4, (1988).
50. P.J. Whitehouse, Neurol., 36, 720 (1998).
51. E.D. London, M.J. Ball and S.B. Waller, Neurochem. Res., 14, 745 (1989).
52. R. Freedman, M. Hall, L.E. Adler, and S. Leonard, Biol. Psychiatry. 38, 22-33 (1995).
53. Z.-Z. Guan, X. Zhang, R. Rivka and A. Nordberg, J. Neurochem. 74, 237 (2000)
54. H. Valette, M. Bottlaender, F. Dolle, L. Dolci, A. Syrota and C. Crouzel, Nucl. Med. Commun. 18, 164 (1997).
55. W. Sihver, K.J. Fasth, M. Ogren, H. Lundqvist, M. Bergstrom, Y. Watanabe, B. Langstrom, and A. Nordberg, Nucl. Med. Biol. 26, 633, (1999)
56. S.I. Chefer, A.G. Horti, A.O. Koren, D. Gundisch, J.M. Links, V. Kurian, R.F. Dannals, A.G. Mukhin, and E.D. London, Neuroreport, 10, 2715 (1999).
57. U. Scheffel, A.G. Horti, A.O. Koren, H.T. Ravert, J.P. Banta, P.A. Finley, E.D. London, and R.F. Dannals, Nucl. Med. Biol., 27, 51 (2000).
58. J.L. Musachio, V.L. Villemagne, U.A. Scheffel, R.F. Dannals, A.S. Dogan, F. Yokoi, and D.F. Wong. Nucl. Med. Biol., 26, 201, (1999).
59. M. Ernst, A.J. Zametkin, P.H. Jons, J.A. Matochik, D. Pascualvaca and R.M. Cohen, J. Am. Acad. Child. Adolesc. Psychiatry, 38, 86 (1999).
60. A.A. Silver and P.R. Sanberg, Lancet, 342, 182 (1993).
61. P.R. Sanberg, R.D. Shytle, and A.A. Silver, Lancet, 352, 705 (1998).
62. H.K. White and E.D. Levin, Psychopharm. (Berlin) 143, 158 (1999).
63. N. Baker, L.E. Adler, R.D. Franks, M.C. Waldo, S. Berry, H. Nagamoto, A. Muckle, and R. Freedman, Biol. Psych., 22, 603 (1987).
64. R. Freedman, L.E. Adler, N. Baker, M.C. Waldo, and G. Mizner, Somat. Cell. Mol. Gen., 13, 479 (1987).
65. L.E. Adler, L.J. Hoffer, J. Griffith, M.C. Waldo and R. Freedman, Biol. Psych., 32, 607 (1992).
66. S. Leonard, C. Breese, C. Adams, K. Benhammou, J. Gault, K. Stevens, M. Lee, L. A. Adler, R. Olincy, R. Ross and R. Freedman, Eur. J. Pharmacol., 393, 237 (2000).
67. K. Stevens, Proceedings from IBC 2nd International symposium on Nicotinic Acetylcholine Receptors. Advances in Molecular Pharmacology and Drug Development, Annapolis, MD, May 13-14, 1999.
68. K.E. Stevens, W.R. Kem , and R. Freedman, Biol. Psychiatry, 46, 1443 (1999).
69. V. Luntz-Leybman, P.C. Bickford, R. Friedman,. Brain Res. 587, 130, (1992).
70. K.E. Stevens, W.R. Kem, V.M. Mahnir and R. Freedman , Psychopharm.(Berlin), 136, 320 (1998).
71. J.J. Gordon, Proceedings from IBC 2nd International Symposium on Nicotinic Acetylcholine Receptors. Advances in Molecular Pharmacology and Drug Development, Annapolis, MD, May 13-14, 1999.
72. R.J. Salin-Pascual, M. Rosas, A. Jimenez-Genchi, B.L. Rivera-Meza and V.J Delgado-Parra, Clin. Psych., 57, 387 (1996).
73. O.K. Steilein, J.C Mulley, P. Propping, R.H. Wallace, H.A. Phillips and G.R. Sutherland, Nat. Genetics, 11, 201 (1995).

74. O.K. Steilein, A. Magnusson, J. Stoodt, S. Bertrand, S. Weiland and S. Berkovic, Hum. Mol. Genetic., <u>6</u>, 943 (1997).

75. S. Bertrand, S. Weiland, S.F. Berkovic. and O.K. Steilein., Br. J. Pharmacol., <u>125</u>, 751 (1998).

76. S.E. File, S. Cheeta and P.J. Kenny, Eur. J. Pharmacol., <u>393</u>, 231 (2000).

77. E.C. Miguel, M.C. Rosario-Campos, H.S. Prado, R. Valle, S.L. Rauch, B.J. Coffey, L. Baer, C.R. Savage, R.L. O'Sullivan, M.A. Jenike and J.F. Leckman, J. Clin. Psych., <u>61</u>, 150 (2000).

78. T.F. Spande, H.M. Garraffo, M.W. Edwards, H.J.C. Yeh, L. Pannel and J.W. Daly, J. Am. Chem. Soc., <u>114</u>, 3475 (1992).

79. M. Bencherif, M.E. Lovette, K.W. Fowler, S. Arrington, L. Reeves, W.S. Caldwell and P.M. Lippiello, J. Pharmacol. Exp. Ther., <u>279</u>, 1413 (1996).

80. P.M. Lippiello, M. Bencherif, J.A. Gray, S. Peters, G. Grigoryan, S. Hodges and A.C. Collins, J. Pharmacol. Exp. Ther. <u>279</u>, 1422 (1996).

81. M. Bencherif, G. Byrd, W.S. Caldwell., J.R. Hayes and P.M. Lippiello, CNS Drug Rev., <u>3</u>, 325 (1997).

82. M.I. Damaj, W. Glassco, M.D. Aceto, and B.R. Martin, J. Pharmacol. Exp. Ther. <u>291</u>, 390 (1999).

83. A. Chiari, J.R. Tobin, H.L. Pan, D.D. Hood, and J.C. Eisenach, Anesthesiology <u>91</u>, 1447 (1999).

84. P.M. Lavand'homme, and J.C. Eisenach. Anesthesiology, <u>91</u>, 1455 (1999).

85. A.W. Bannon, M.W. Decker, P. Curzon, M.J. Buckley, D.J. Kim, R.J. Radek, J.K. Lynch, J.T. Wasicak, N.H. Lin, W.H. Arnold, M.W. Holladay, M. William and S.P. Arneric, J. Pharmacol. Exp. Ther., <u>285</u>, 787 (1998).

86. J.T. Green, J. Rhodes, G.A. Thomas, B.K. Evans, C. Feyerabend, M.A. Russell and W.J. Sandborn, Ital. J. Gastroenterol. Hepatol., <u>30</u>, 260 (1998).

87. M. Guslandi, Int. J. Colorectal Dis., <u>14</u>, 261 (1999).

88. M.N. Merrett, N. Mortensen, M. Kettlewell and D.O. Jewell, Gut, <u>38</u>, 362 (1996).

89. M.L. Sopori and W.J. Kozak Neuroimmunol. <u>83</u>, 148 (1998).

90. J.M. Lindstrom, Muscle Nerve, <u>23</u>, 453 (2000).

91. B.M. Conti-Tronconi, K.E. McLane, M.A. Raftery , S.A. Grando and M.P. Protti, Crit. Rev. Biochem. Mol. Biol., 29, <u>69</u> (1994).

92. S.A. Grando, R.M. Horton, E.F. Pereira, B.M. Diethelm-Okita, P.M. George, E.X. Albuquerque and B.M. Conti-Fine, J. Invest. Dermatol., <u>105</u>, 774 (1995).

93. S.A. Grando, R.M. Horton, T.M. Mauro, D.A. Kist, T.X. Lee, M.V. and M.V. Dahl, J. Invest. Dermatol., <u>107</u>, 412 (1996).

SECTION II. CARDIOVASCULAR AND PULMONARY DISEASES

Editor: William J. Greenlee, Schering Plough Research Institute, Kenilworth, New Jersey

Chapter 6. Recent Developments in Antitussive Therapy

John A. Hey and Neng-Yang Shih
Schering Plough Research Institute
2015 Galloping Hill Road
Kenilworth, NJ 07033

Introduction - Cough is a forceful defensive reflex maneuver that leads to expulsion of irritants, fluids, mucus or foreign material from the respiratory tract. Specifically, the reflex triggers a complex, multiphasic motor pattern characterized by sequential coordination of large increases in motor output to an array of inspiratory and expiratory skeletal muscles. This highly coordinated musculoskeletal activation process consists of three sequential phases, namely deep inspiration, compression (i.e. contraction against a closed glottis) and vigorous expulsion. The expulsion is ultimately attained through the combined forceful contraction of thoracic, abdominal and diaphragm muscles through the generation of rapid airflow (1, 2). While generally beneficial, cough is a prominent pathophysiological feature associated with many airway and lung diseases such as asthma, upper respiratory viral and bacterial infections, post-nasal drip syndrome, gastroesophageal reflux disease, pulmonary neoplasm, chronic bronchitis and chronic obstructive pulmonary disease (1, 3). Treatment may be complicated by the determination whether cough is mainly productive (sputum generating) or non-productive (dry cough). Chronic persistent non-productive cough (i.e. cough greater than three weeks duration) occurs in a post-viral airway disease setting in about 30% of the patients with acute respiratory infection. Furthermore, uncontrolled chronic cough is known to lead to significant morbidity and is associated with rib fractures, ruptured abdominal muscles, pneumothorax, marked decrement in the amount and quality of sleep, and loss of consciousness (1).

Drugs used for the treatment of cough are among the most widely used prescription and OTC drugs in the world (4). In the prescription market in 1999, sales of antitussives in the USA, Canada and Europe totaled approximately $750 million (5). The pharmacotherapy of cough is broadly divided into two major categories based on their purported site of action. These include drugs that act by a central mechanism to block the frequency (and/or amplitude) of the efferent cough motor output (e.g. opiates) or act in the periphery to inhibit the generation of the tussigenic sensory impulses (e.g. local anesthetics) (6, 7). Current antitussive therapy is dominated by the use of older drugs such as dextromethorphan and codeine. These drugs however carry significant side effect liabilities including among others, sedation, abuse liability and respiratory depression. Thus, because persistent cough is underserved by current therapeutic options available, there is an increasing need for safe and efficacious antitussive alternative(s) to existing medications. This report highlights some of the advances in the development of novel antitussives, and briefly reviews their chemistry, mechanism and site of action.

NEURAL REGULATION OF THE COUGH REFLEX

<u>Afferent Mechanisms of the Cough Reflex</u> - Coughing is elicited by stimulation of specialized sensory nerves in the airways. A variety of different stimuli including airway inflammation, chemical irritants (e.g. capsaicin) and mechanical stimuli can trigger coughing in humans and animals. The polymodal sensory nerves that are prominently involved in neural regulation of cough are myelinated Aδ rapidly adapting receptors (RAR) and unmyelinated bronchial and pulmonary C-fibers (8). Their nerve endings appear to be prominently localized within the epithelial layer of the trachea and lower airways (2). The afferents transmitting the tussigenic impulses are carried by the vagus nerve to the nucleus tractus solitarius (NTS) in the medulla oblongata.

The unmyelinated C-fiber afferents contain neuropeptides of the tachykinin family, such as substance P and neurokinin A (NKA) as well as calcitonin gene related peptide (CGRP), which are synthesized in the nodose and jugular nerve cell bodies and undergo retrograde transport to the nerve terminals in the airways. In humans, the major subgroup of afferent nerves involved in the generation of cough is located primarily throughout the trachea and intra- and extrapulmonary bronchi. In addition to relaying impulse information to the CNS cough center, these nerves exert a local pro-inflammatory response due to the release of tachykinins from the peripheral nerve terminals. This neurogenic inflammatory response is characterized by smooth muscle contraction, edema and stimulation of neighboring RARs leading to amplification of the cough response (2, 8).

<u>Efferent Mechanisms of the Cough Reflex</u> - Tussigenic sensory impulses, in the form of propagating action potentials conducting up the axon, reach the NTS in the lower brainstem. When a critical threshold is exceeded, a cough response is produced. The «brainstem cough center», or central organization of the neural substrates involved in the coordination of the cough motor response, are poorly defined and refer primarily to the functional neural network in the medulla oblongata. The large forces generated during the expiratory motor phase of coughing collapse bronchi, which in turn increases the shear forces that promote clearance of airway material.

SITE OF ACTION OF ANTITUSSIVE DRUGS

<u>Central Site of Action</u> - Centrally active antitussive agents act preferentially by depressing the cough center at the level of the lower brainstem without affecting peripheral sensory or motor endplate effector responses, Figure 1 (6, 7). The μ opiate receptor agonists codeine and dextromethorphan, which are considered among the most effective antitussive agents available, are prototype centrally active antitussive agents. Whereas these agents are generally effective against cough of various etiologies, they also have prominent dose-dependent sedative and respiratory side effects that often limit their clinical utility.

<u>Peripheral Site of Action</u> – Antitussive agents may also act in the periphery to inhibit the sensory impulses that lead to the activation of the cough reflex, Figure 1. Recently there has been a strong interest in developing broadly efficacious antitussive drugs that act exclusively by a peripheral action in order to minimize the potential for undesirable side effects and liabilities characteristic of centrally active antitussive drugs such as codeine. One challenging aspect to the development of peripherally acting antitussives is the requirement that any novel agent display a high degree of selectivity for pulmonary afferents over other sensory afferents so as to avoid potential side effects related to nonspecific sensory function impairment. To

date, however, most of the peripherally active antitussives display limited efficacy and/or poor oral bioavailability. In general, peripherally active agents act by decreasing the sensitivity of airway cough receptors to tussigenic stimuli leading to an inhibition of pulmonary vagal discharge. When applied topically, local anesthetic antitussives, such as benzonatate, are generally considered to block voltage-dependent Na^+ channels, which in turn inhibits the generation and transmission of impulses to the CNS, leading to an attenuation of the cough response. Unfortunately, the local anesthetics are of minimal clinical value because their use is limited to topical delivery in lozenge form, and they have marginal efficacy at tolerated doses, short duration of action, and dose dependent potential for CNS-related side effects.

Figure 1

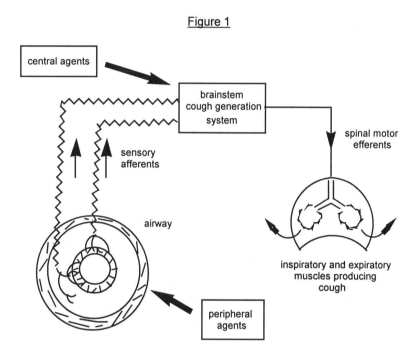

CENTRALLY ACTING AGENTS

Several classes of drugs inhibit the cough reflex primarily by a central mechanism in humans and experimental animals. Included among these are opioids, $GABA_B$ agonists, and neurokinin antagonists (6, 9, 10, 11).

Opioids - Studies detailing the selective δ opioid receptor agonist SB 227122 (**1**) have recently been described (12). Compound **1** binds to the human δ receptor with high affinity (K_i = 6.9 nM) while its activity at the μ and κ opioid receptors is significantly weaker (K_i = 2030 nM and >5000 nM respectively). In vivo, **1** dose-dependently inhibited citric acid induced cough in the guinea pig with an ED_{50} = 7.3 mg/kg when administered parenterally. This compares favorably with the activity of codeine (a μ receptor agonist) and BRL 52974 (a κ receptor agonist) in the same model (ED_{50} = 5.2 and 5.3 mg/kg respectively). Activation of the μ and κ receptors is associated with many of the side effects seen with the opioid antitussives such as respiratory depression, constipation and dependence (μ receptor), and diuresis and sedation (κ receptors). Therefore, it may be expected that a selective δ receptor agonist might be devoid of such side effects.

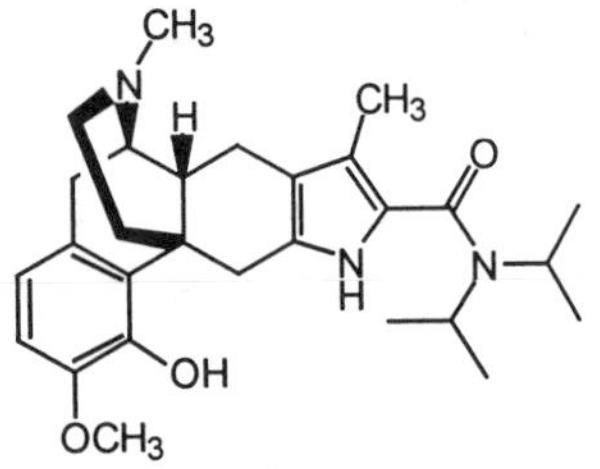

1

Several patent applications have claimed a variety of opioid-like compounds for the treatment of cough e.g. octahydroisoquinolines of general structure **2** (13,14,15,16). In a capsaicin-induced cough model in the mouse, maximum antitussive activity of **2** was obtained with a small hydrophobic group (R^1 = CH$_3$) (13).

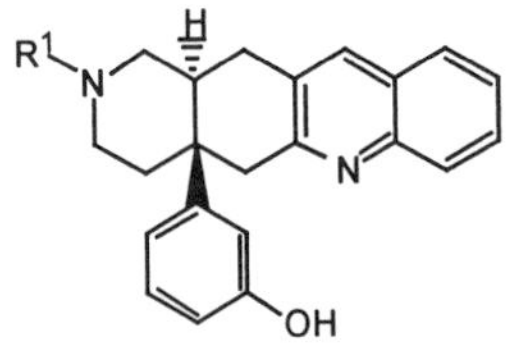

R^1	Dose (mg/kg, s.c.)	% Inhibition
CH$_3$	1	72
H	10	34
CH$_2$cyclopropyl	1	30
CH$_2$CH$_2$Ph	1	44

2

As mentioned previously, selective δ agonists may offer a therapeutic advantage over nonselective opioid agonists due to their potentially improved side-effect profile. A series of selective δ agonists exemplified by **3** were recently described (14). Larger hydrophobic groups present as R^1 appear to impart greater binding affinity for this receptor than the smaller methyl group implying the presence of a hydrophobic pocket. Among compounds that were assayed for in vivo antitussive activity, using the capsaicin-induced cough model in the rat, compounds **4** and **5** were the most potent (14).

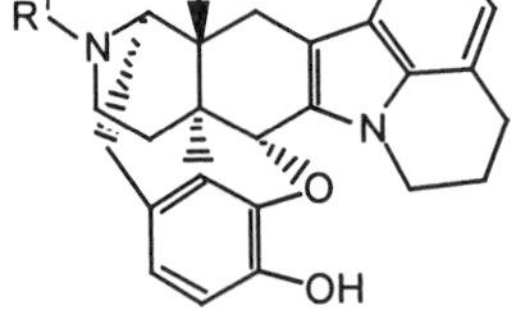

R^1	K_e (nM)
CH$_2$CHCH$_2$CH$_2$	0.05
CH$_3$	0.21
CH$_2$CH$_2$Ph	0.06
CH$_2$CH$_2$-2-thienyl	0.03

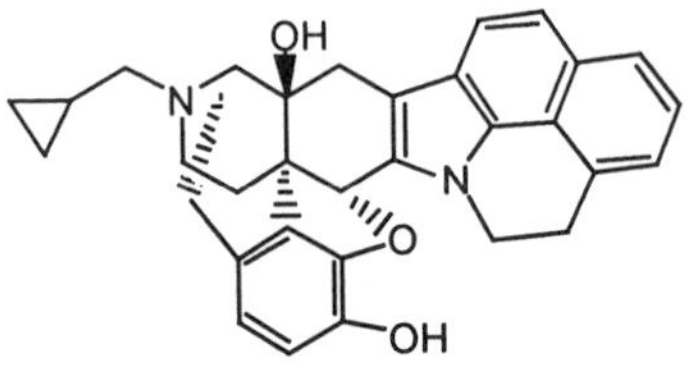

3

4

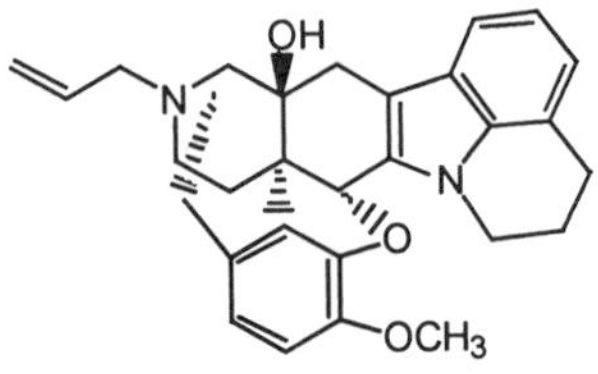

5

Analogs of compound **4** in which the indole group was replaced with an amide moiety have also been described (15, 16). It appears from the binding data that these compounds are, in general, more selective for the κ receptor than for the μ or δ receptors. The antitussive activity of two representative examples, compounds **6** and **7**, was 2.6 μg/kg and 3.7 μg/kg, respectively, determined in the tracheal stimulated antitussive model.

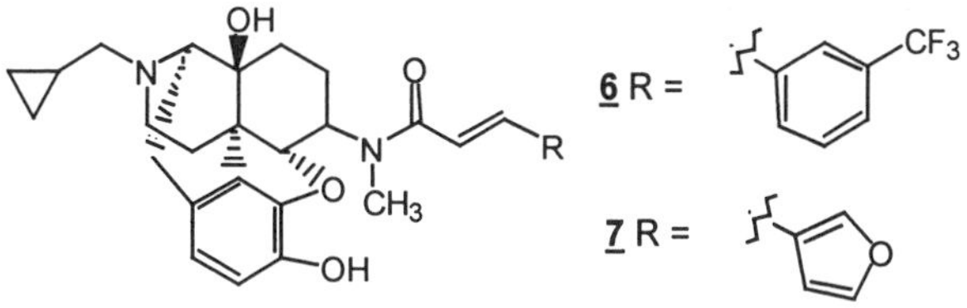

While the presence of functional μ opioid receptors have been reported on peripheral afferent terminals in the airways (17), it does not appear that they contribute significantly to the clinical antitussive action of μ opioid agonists since aerosol administration of codeine, morphine or BW 443C, a peptide opioid agonist, failed to reduce capsaicin-induced cough in human subjects (18).

GABA$_B$ Agonists - GABA (γ-aminobutyric acid) is an inhibitory neurotransmitter present in both the peripheral and central nervous systems. It binds to two different receptors termed GABA$_A$ and GABA$_B$. Experimental evidence indicates that activation of central GABA$_B$ receptors can influence the cough reflex. For example, the GABA$_B$ agonists baclofen (**8**) potently inhibits capsaicin-induced cough in the guinea pig by a central mechanism, ED$_{50}$ = 0.04 mg/kg, s.c. (9, 19). This level of activity compared favorably with codeine and dextromethorphan in this model. The ability of baclofen (**8**) to increase the cough threshold against capsaicin was further demonstrated in a double-blind, placebo-controlled study in healthy human volunteers. Pretreatment with baclofen (10 mg, 3 times/day) for fourteen days suppressed capsaicin-induced cough (20).

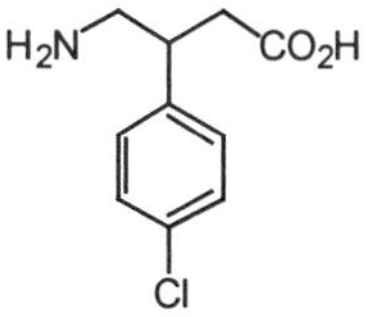

The activity of GABA is modulated by its rapid uptake by high-affinity presynaptic transporters. Therefore, an alternative approach to activation of the GABA receptor would be to administer a compound that could inhibit the reuptake of GABA. The use of compounds that inhibit the uptake of GABA have been claimed for the treatment of cough (21). Among a number of compounds specified in this application, compound **9** was reported to inhibit citric acid induced cough in the guinea pig by 42% (10 mg/kg, p.o.).

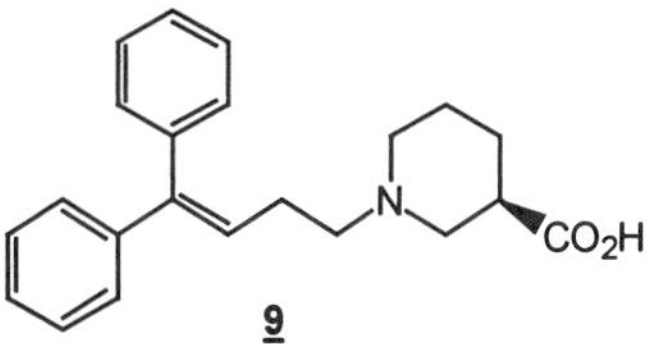

Neurokinin Antagonists - Activation of the NK_1 and NK_2 receptors by endogenous tachykinins has been associated with a variety of pathological conditions affecting the lung and bronchi including plasma extravasation, bronchoconstriction and cough (22). Several studies in a number of different animal models have also demonstrated that antagonists of the NK_2 receptor can modulate the cough reflex by a central action (11). For example, SR 48968 (**10**), a non-peptide NK_2 receptor antagonist, was shown to inhibit cough in a dose dependent manner in the citric acid induced cough model in the guinea pig (ED_{50} = 0.1 mg/kg, i.p.) (23). By comparison, under similar conditions codeine was significantly less active (ED_{50} = 8 mg/kg, p.o.). The activity of SR 48968 was apparently not mediated via an opioid mechanism since naloxone, a nonspecific opioid antagonist that blocks the antitussive effect of codeine, had no effect.

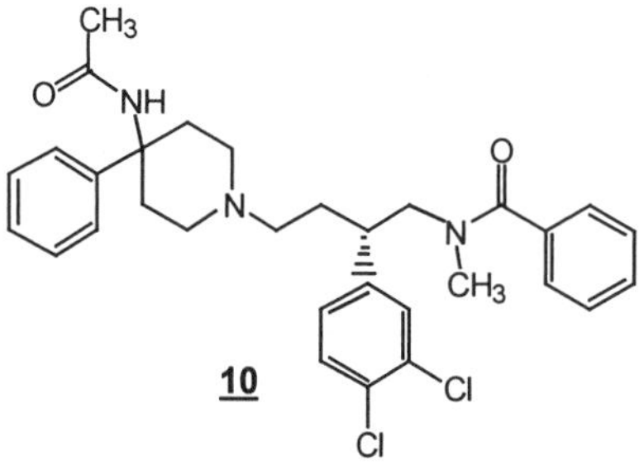

10

The antitussive activity of selective NK_1 antagonists is still an area of active investigation (11, 24). The NK_1 antagonist CP 99,994 (**11**) has been shown to inhibit cough in both the guinea pig and cat by a central mechanism (24). However, the NK_1 antagonist SR 140333 was not active in the citric acid-induced cough model in the guinea pig (25).

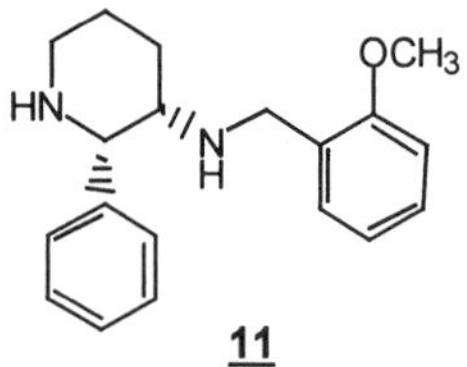

11

In addition to their central activity, neurokinin antagonists can also act at peripheral sites. NK_1 and NK_2 antagonists have been shown to inhibit the pro-inflammatory, bronchoconstrictor and RAR stimulation (tussigenic Aδ fibers) activity of sensory tachykinins by a blocking action at the level of the effector cell.

Miscellaneous - CH-170 (**12**) is a xanthine derivative that was in preclinical development as a potential antitussive (26). In pharmacological studies, **12** demonstrated antitussive activity weaker than that of codeine in several in vivo models e.g. the citric acid induced cough model [ED_{50} (CH-170) = 22.2 mpk, ED_{50} (codeine) = 12.2 mpk]. Development of **12** was suspended when it was discovered that the demethylated metabolite, CH-13584 (**13**), was a more potent antitussive. In several in vivo studies, **13** showed similar activity to both codeine and dextromethorphan as an antitussive [ED_{50} in the 4-8 mg/kg range, p.o.] and displayed mucolytic activity (27).

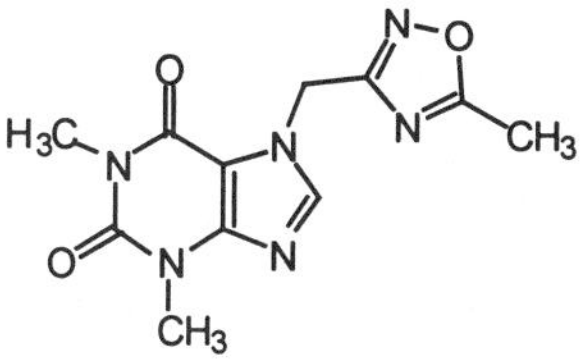

12 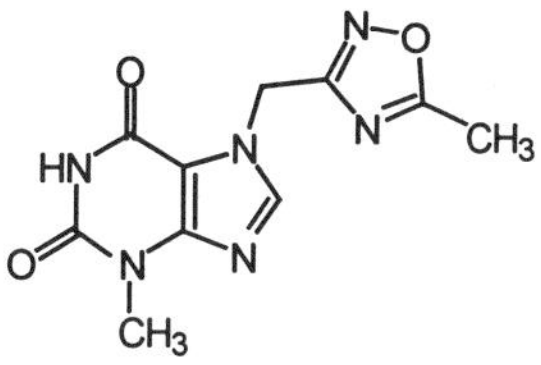   13

A number of studies were undertaken to determine the mechanism of action of **13** but these led to no firm conclusions. It is clear from these studies that **13** does not behave like theophylline although it structurally quite similar (28), nor does it exert its antitussive activity via an opioid mechanism since it does not show appreciable binding to any of the opioid receptors. However, the antitussive activity of CH-13584 is blocked by the non-specific opioid antagonist naloxone. This implies that **13** may act by facilitating the release of endogenous opioids which then exert an antitussive effect (29).

Selective nociceptin ORL-1 ligands have recently been claimed for the treatment of cough (30). A representative compound, **14,** had a K_I = 18 nM in a nociceptin binding assay.

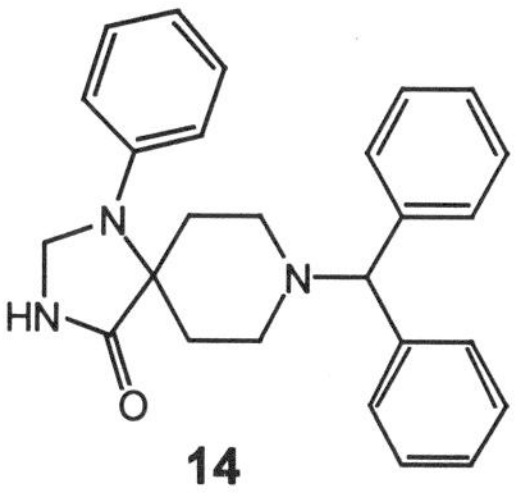

14

PERIPHERALLY ACTING AGENTS

Other classes of drugs appear to inhibit the cough primarily by a peripheral mechanism. Representatives among these antitussive agents are α_2-adrenoceptor agonists, potassium channel openers and $GABA_B$ agonists.

α_2-Adrenoceptor Agonists - α_2-Adrenoceptors are present on nerves that innervate the airways and smooth muscle. Like GABA, activation of these receptors has been shown to cause inhibition of sensory nerve transmission which in turn can modulate the cough reflex. Inhaled clonidine (**15**), a non-specific α_2-adrenoceptor agonist, caused a concentration dependent inhibition of citric acid induced cough in the guinea pig at a dose of 10-1000 μM (31). However, in contrast to its effect in the guinea pig, inhaled clonidine had no effect on capsaicin-induced cough in healthy human volunteers indicating a difference in the innervation of the airways of the two species (31).

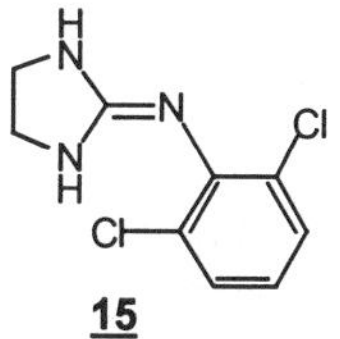

15

A series of bicyclic heteroaromatic analogs of clonidine has been reported to be α_2-adrenoceptor agonists useful for the treatment of a number of diseases including cough (32). Compound **16**, a typical example of the general structure, was reported to be more selective for α2-adrenoceptors over α1 in a binding assay and to display agonist activity similar to clonidine. In vivo, **16** was orally active and did not produce centrally mediated effects at doses that were peripherally effective (33).

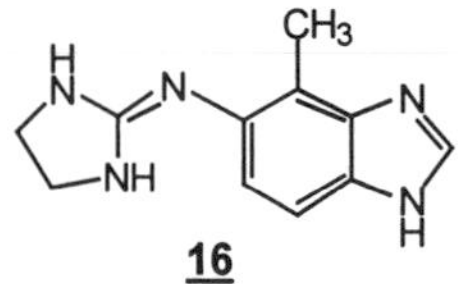

16

Potassium Channel Openers - Sensory nerves play an important role in airway disease by mediating central reflexes such as cough and local axon reflexes which result in the release of neuropeptides such as Substance P and NKA. NS1619 (**17**), a Ca^{+2} activated potassium channel opener (BK_{Ca}), inhibits the activity of sensory fiber mediated reflexes in guinea pig airways (34). In conscious guinea pigs, cough due to citric acid inhalation was reduced by 60% in animals that were previously exposed to **17** (300 µM). Selective BK_{Ca} channel openers may represent a new class of antitussive agents that act by reducing peripheral cough reflexes.

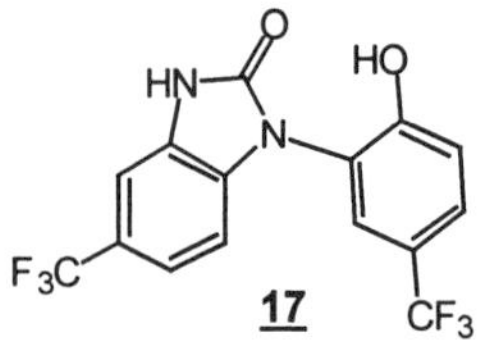

17

GABA_B Agonists - A peripheral site of action to inhibit the efferent function of sensory afferents may also contribute to the antitussive activity of GABA_B agonists. Specifically, 3-aminopropylphosphinic acid (3-APPI, **18**), a GABA_B agonist that does not cross the blood-brain-barrier, inhibits tachykinergic-mediated airway pathophysiological responses such as bronchospasm, vagally-induced airway microvascular leakage and cough (inhibition of capsaicin-induced cough in guinea pigs, ED_{50} = 0.36 mg/kg, s.c.) in experimental animals by attenuating the release of pro-inflammatory neurogenic neuropeptides from pulmonary afferent nerve terminals (9, 19).

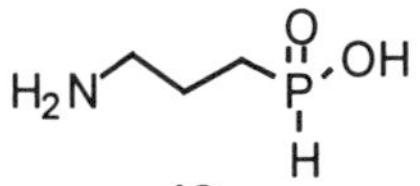

18

Miscellaneous - Moguisteine (**19**) is a non-narcotic antitussive agent currently in Phase II clinical trials (35). Although its mechanism of action is not known, it is thought to act by decreasing the sensitivity of pulmonary RARs in the lung. It was equipotent to codeine in several animal models of cough. In eight volunteers, acetylene induced coughing was suppressed by 61% after treatment with moguisteine (50 mg tablet, p.o.).

DF-1012 (**20**) has demonstrated potent, long-acting and orally effective antitussive activity in the guinea pig in several different cough models (36). It is currently in Phase II clinical trials.

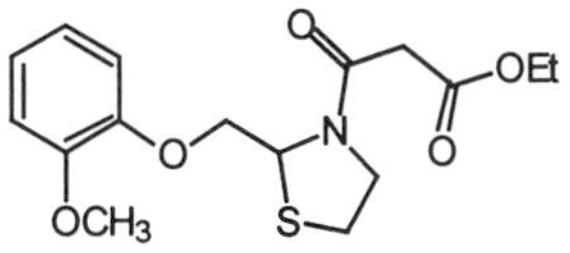

19 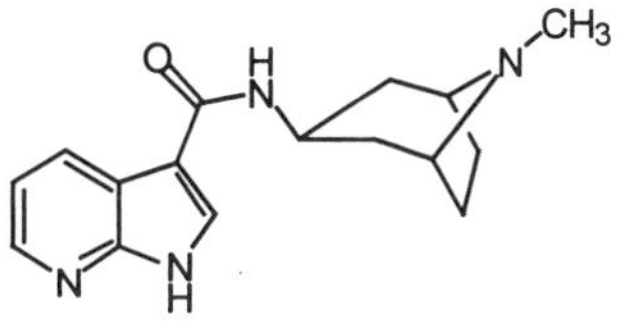   **20**

Conclusion - Cough is a serious symptom of a wide range of pathophysiological conditions. Current therapies rely heavily on the use of centrally acting agents such as codeine and dextromethorphan and to a lesser extent on the peripherally acting local anesthetics. While codeine is an effective antitussive agent, it is plagued by a number of unwanted side effects such as sedation, respiratory depression, tolerance, and abuse potential. Dextromethorphan is generally less effective than codeine and still suffers from some of the same liabilities. The ideal antitussive should possess the following pharmacodynamic and pharmacokinetic properties: efficacy equivalent or better than codeine, oral activity, a pharmacokinetic profile that supports once daily dosing, no GI side effects, no sedative or respiratory depression liability, and no tolerance or abuse liability. It remains to be determined whether any of the new approaches to antitussive therapy will demonstrate a significant therapeutic advance over existing therapies.

REFERENCES

1. R.S. Irwin, M.J. Rosen and S.S. Braman, Arch. Int. Med. 137, 1186 (1977).
2. J.G. Widdicombe, Monaldi Arch. Chest Dis. 54, 275 (1999).
3. J.G. Widdicombe, Eur. Respir. J., 8, 1193 (1995)
4. N.B. Choudry and R.W. Fuller, Eur. Respir. J., 5, 296 (1992)
5. MAT 1999 (Nov.), sales for US, Germany, France, UK, Italy, Spain and Canada.
6. J.-A. Karlsson and R.W. Fuller, Pul. Pharmacol. Ther., 12, 215 (1999).
7. D.C. Bolser, Pul. Pharmacol., 9, 357 (1996).
8. J.G. Widdicombe, Res. Physiol., 114, 5 (1998).
9. D.C. Bolser, F.C. DeGennaro, S. O'Reilly, R.W. Chapman, W. Kreutner, R.W. Egan and J.A. Hey, Br. J. Pharmacol., 113, 1344 (1994).
10. R. Yasumitsu, Y. Hirayama, T. Imai, K. Miyayasu and J. Hiroi, Eur. J. Pharmacol., 300, 215 (1996).
11. D.C. Bolser, F.C. DeGennaro, S. O'Reilly, R.L. McLeod and J.A. Hey, Br. J. Pharmacol., 121, 165 (1997).
12. C.J. Kotzer, D.W.P. Hay, G. Dondio, G. Giardina, P. Petrillo and D.C. Underwood, J. Pharacol. Exp. Ther., 292, 803 (2000).
13. H. Nagase, K. Kawai, A. Mizusuna and J. Kamei World Patent Application WO9902157 (1999).
14. H. Nagase, K. Kawai, T. Endo and S. Ueno World Patent Application WO9711948 (1997).
15. H. Nagase, K. Kawai, T. Endo, S. Ueno and Y. Negishi, World Patent Application WO9501178 (1995).
16. H. Nagase, J. Hayakawa, K. Kawamura, K. Kawa and T. Endo, World Patent Application WO9503308 (1995).
17. J.J. Addock, C. Schneider and T.W. Smith, Br. J. Pharmacol., 93, (1988).
18. N.B. Choudry, S.J. Gray, J. Posner and R. W. Fuller, Br. J. Clin. Pharmacol., 32, 633 (1991).
19. D.C. Bolser, S.M. Aziz, F.C. DeGennaro, W. Kreutner, R.W. Egan, M.I. Siegel and R.W. Chapman, Br. J. Pharmacol., 110, 491 (1993).
20. P.V. Dicpinigaitis and J.B. Dobkin, Chest, 111, 996 (1997).
21. W.E. Bondinell, D.C. Underwood and C.J. Kotzer, World Patent Application WO9743902 (1997).
22. C. Swain and N.M.J. Rupniak in «Annual Reports in Medicinal Chemistry,» Vol. 34, A. M. Doherty, Ed., Academic Press New York, N. Y., 1999, p.51.
23. C. Advenier, V. Girard, E. Naline, P. Vilain, X. Emonds-Alt and E.Ur. J. Pharmacol., 250, 169 (1993).

24. K. Sekizawa, Y.X. Jia, T. Ebihara, Y. Hirose, Y. Hirayama and H. Sasaki, Pul. Pharmacol., 9, 323 (1996).
25. V. Girard, E. Naline, P. Vilain, X. Emonds-Alt and C. Advenier, Eur. Respir. J., 8, 1110 (1995).
26. D. Korbonits, G. Heja and P. Kormoczy, Eur. J. Pharmacol., 183, 730 (1990).
27. E.G. Mikus, J. Revesz, E. Minker, D. Korbonits, V. Saano, M. Pascal and P. Aranyi, Arzneim.-Forsch/Drug Res., 47, 395 (1997).
28. E.G. Mikus, Z. Kapui, D. Korbonits, K. Boer, E. Borobkay, J. Gyurky, J. Revesz, F. Lacheretz, M. Pascal and P. Aranyi, Arzneim.-Forsch/Drug Res., 47, 1358 (1997).
29. Z. Kapui, E.G. Mikus, J. Bence, K. Gerber, K. Boer, D. Korbonits, A. Borsodi and P. Aranyi, Arzneim.-Forsch/Drug Res., 47, 1147 (1998).
30. D. Tulshian, G.D. Ho, L.S. Silverman, J.J. Matasi, R.L. McLeod, J.A. Hey, R.W. Chapman and A. Bercovici, F. M. Cuss, World Patent Application WO00006545 (2000).
31. F. O'Connell, V.E. Thomas, R.W. Fuller, N.B. Pride and J-A. Karlsson, J. Appl. Physiol. 76, 1082 (1994).
32. T.L. Cupps, S.E. Bogdan, G.E. Mieling, N. Nikolaides, R.T. Henry, R.J. Sheldon, World Patent Application WO9846595 (1998).
33. K. Rasmussen, J. Lillibridge and M. Soehner, FASEB J, 10, A426 (1996).
34. A.J. Fox, P.J. Barnes, P. Venkatesan and M.G. Belvisi, J. Clin. Invest., 99, 513 (1997).
35. R. Ishii, M. Furuta, M. Hashimoto, T. Naruse, L. Gallico and R. Ceserani, Eur. J. Pharmacol., 362, 207 (1998).
36. Pharmaprojects (May, 1999).

Chapter 7. Advances in the Understanding and Treatment of Congestive Heart Failure (HF)

John E. Macor and Mark C. Kowala
Discovery Chemistry and Metabolic and Cardiovascular Drug Discovery
Pharmaceutical Research Institute
Bristol-Myers Squibb, Princeton, NJ 08543

Introduction - Heart failure (HF) is a chronic, expensive and fatal disease. Heart failure occurs when cardiac output fails to meet end organ needs for blood flow. However, from this simple description, it has become apparent that heart failure at present is a disease whose etiology is poorly understood, and likely, heart failure occurs and progresses via a multitude of factors. Not surprisingly, many different classes of drugs have been used in the treatment of HF with varied results. The disease as viewed now is best treated with several drugs of multiple actions, but the exact combination of agents is an ever shifting landscape.

Chronic heart failure (HF) affects approximately 5 million people in the US (approximately 2% of the population), and the disease has reached epidemic proportions with over 400,000 new cases being diagnosed each year (1, 2). Congestive heart failure is one of the single largest causes of death in industrialized countries. Hospitalization costs from heart failure alone account for as much as 1.5% of total health care budgets, with out-patient management costs requiring as much as an additional 1% of total health care costs (3, 4). Surprisingly, therapeutic intervention of heart failure has mostly used drugs whose efficacy had been established elsewhere in the realm of cardiovascular disease (i.e., hypertension). Few drugs, if any, have been designed and pursued specifically for the treatment of heart failure alone.

There have been a number of recent reviews discussing the treatment of heart failure (5, 6). These serve as good in-depth discussions of the treatment and management of the disease. A description of the detailed pathogenesis of heart failure is beyond the scope of this chapter. Instead, the major physiological and biochemical mechanisms of heart failure that current drug therapies modify are summarized. Included in this discussion will be descriptions of the drugs used to treat HF, recent clinical trials with those drugs, and new molecular entities into the different classes of drugs. The final part of the review will focus on novel, untested molecules and approaches to the treatment of congestive heart failure.

Heart failure can be described as *a clinical syndrome that arises when the heart is unable to pump sufficient blood to meet the metabolic needs of the body at normal filling pressures, provided that the venous return is normal* (7). Heart failure frequently begins with a loss of myocardial cells as a result of myocardial infarction, or when there is an increased cardiac overload from either uncontrolled hypertension or heart valve disease. The reduced cardiac output during heart failure triggers the activation of compensatory neurohormonal responses such as stimulation of α_1- and β-adrenergic receptors by norepinephrine, activation of the renin-angiotensin-aldosterone system, increased production of arginine vasopressin, endothelins and atrial natriuretic peptides. During heart failure, cardiac myocytes hypertrophy, synthesize collagen and undergo apoptosis. The left ventricle wall thickens and becomes fibrotic as part of an active remodeling process. Subsequent left ventricular (LV) dilatation leads to poor contraction, reduced stroke volume and decreased cardiac output. These pathophysiological changes occur as the heart adapts to increased cardiac overload and to the neurohormones acting directly on the cardiac myocytes. Resulting hypoperfusion of the kidney increases renin synthesis and promotes angiotensin II mediated vasoconstriction in the peripheral arteries. Angiotensin II increases renal Na^+ and water retention. It also stimulates aldosterone production, which, in turn, inhibits natriuresis (Na^+ excretion) and diuresis (water excretion), leading to hypertension. The severe syndrome of congestive heart failure involves shortness of breath (dyspenia), pulmonary edema, pulmonary hypertension and edema in the periphery. The prognosis of heart failure patients is poor due to progressive pump failure, increased risk of myocardial ischemia and the unpredictable occurrence of sudden death as a result of ventricular arryhthmias. In fact, the life expectancy of a person diagnosed with HF is similar to that of a person diagnosed with cancer (8).

RATIONALE FOR DRUGS TREATING HEART FAILURE

As mentioned earlier, many different classes of drugs have been tested clinically and preclinically in the treatment of HF. This section will review the different types of drugs that have been evaluated in the treatment of HF, starting with established agents (i.e., ACE inhibitors) and followed by more speculative agents. Older drugs [i.e., digoxin (9) and amiodorone (10)] that have been used extensively, but with few documented mortality benefits, will not be reviewed here.

Renin-Angiotensin - The activity of the renin-angiotensin system is elevated in patients with heart failure. Reduced cardiac output leads to renal hypoperfusion and to increased synthesis and release of renin into the circulation. Renin cleaves angiotensinogen to angiotensin I, and angiotensin converting enzyme (ACE) cleaves angiotensin I to angiotensin II (11 - 13). Increased production of angiotensin II has multiple physiological and biochemical actions. Angiotensin II binding to AT_1 receptors on smooth muscle cells causes contraction, vasoconstriction and increases peripheral vascular resistance, which deleteriously augments cardiac afterload (14). As a vasoconstrictor, angiotensin II causes hypertension that induces left ventricular remodeling and hypertrophy. Angiotensin II binding to cardiomyocyte AT_1 receptors directly stimulates hypertrophy leading to left ventricular enlargement, reduced myocyte contractility and reduced cardiac output (15, 16). Angiotensin II increases Na^+ transport in the renal proximal tubule, and stimulates aldosterone secretion from the adrenal cortex (17). Increased levels of aldosterone promote Na^+ and water retention producing increased cardiac preload and afterload, thereby diminishing cardiac performance. Thus, interruption of the renin-angiotensin system has been one of the most effective treatments for HF. This has been accomplished through the use of ACE inhibitors, angiotensin (AT_1) receptor antagonists and vasopeptidase inhibitors (ACE/NEP inhibitors).

ACE Inhibitors - A number of clinical trials have established the benefits of ACE inhibitors which improve cardiac function, and more importantly, reduce mortality compared to placebo. ACE inhibitors were originally designed for the treatment of hypertension, but the CONSENSUS trial in 1987 and SOLVD trial in 1991 with enalapril, and the SAVE trial with captopril showed that the drugs had significant benefits in the treatment of HF (18 - 20). Ten ACE inhibitors are presently listed in the 2000 Physician's Desk Reference (54th Edition, Medical Economics Company, Montvale, NJ), and the use of an ACE inhibitor as first-line treatment for HF is in the general practice. However, a recent review of the prescribing practices of physicians indicates that the full benefits of ACE inhibitors in HF may not yet have been achieved. It appears that ACE inhibitors are being under prescribed in HF, and when prescribed, the doses used to treat HF are often lower than those doses which showed benefits in the HF clinical trials (21, 22).

Angiotensin Receptor Blockers (AT$_1$ antagonists) - Like ACE inhibitors before them, AT$_1$ receptor antagonists have found their way from use as anti-hypertensive agents to use in the treatment of HF. AT$_1$ receptor antagonists were believed to offer an advantage over ACE inhibitors in the treatment of HF due to their lack of side effects, i.e., lack of the cough that is commonly encountered during the use of ACE inhibitors. The ELITE trial using losartan (**1**) in over 770 elderly patients suffering from systolic left ventricular dysfunction suggested that losartan had clinical benefits superior to the ACE inhibitor captopril in terms of both reduced mortality and reduced hospitalizations for those patients on **1** (23). However, when the study was repeated with a larger study population (ELITE II), these results were not replicated. In fact, there was a trend toward better benefit from the ACE inhibitor used than from **1** (24). As a result of this study, the use of AT$_1$ receptor antagonists in HF rather than ACE inhibitors has been questioned. However, studies are ongoing which are examining the potential additive benefit of using both ACE inhibitors and AT$_1$ blockers as combination therapy for HF (25). At present, there are five AT$_1$ receptor antagonists commercially available for hypertension [losartan (**1**), irbesartan (**2**), valsartan (**3**), eprosartan (**4**) and candesartan (**5**)], and should HF become an indication for this class of drug, one likely would view the class as highly competitive.

Atrial Natriuretic Peptides - During heart failure, atrial concentrations of atrial natriuretic peptides (ANPs) are increased (26). Atrial natriuretic peptides are synthesized and stored in specialized atrial cells, which are released into the circulation following atrial stretching (27). ANPs cause natriuresis and diuresis and promote vascular smooth muscle relaxation (28). They also reduce the production of renin, aldosterone and arginine vasopressin (29). All of these effects of the ANPs have the potential to decease blood pressure, blood volume and cardiac afterload. Atrial natriuretic peptides are inactivated by neutral endopeptidases. Thus, inhibition of these enzymes (NEP inhibitors) leads to an increase in the concentration of the ANPs, and this has been viewed as a new potential treatment of HF.

ACE/NEP Inhibitors (Vasopeptidase Inhibitors) - A very exciting second generation family of ACE inhibitors are the dual metalloprotease inhibitors (ACE/NEP or vasopeptidase inhibitors) such as omapatrilat (**6**) and sampatrilat (**7**) (30). These compounds are dual ACE and neutral endopeptidase (NEP) inhibitors. NEP hydrolyzes the vasorelaxant atrial natriuretic peptides (ANPs), and inhibition of NEP produces a more sustained effect from ANPs resulting in vasodilation. Recent reviews on vasopeptidase inhibitors have summarized the rationale and advances in the field (31 - 33). Most of the clinical study thus far has been done with omapatrilat (**6**, ACE IC$_{50}$ = 5 nM; NEP IC$_{50}$ = 8 nM) (34 - 36). The IMPRESS trial included over 570 patients with HF, and omapatrilat produced significant advantages over the potent ACE inhibitor lisinopril (37). These results have prompted another clinical trial (OVERTURE) which will examine the long-term effects of **6** in the treatment of HF in over 4000 patients. Because of its unique mechanism of action and excellent clinical results in hypertension and HF, **6** was granted priority review by the FDA, and it is anticipated that **6** will be launched sometime in 2000/2001. A second vasopeptidase inhibitor related to omapatrilat (BMS-189921, **8**) has been described recently as has mixanpril (**9**) (38, 39). However, the clinical development of these compounds has not yet been disclosed.

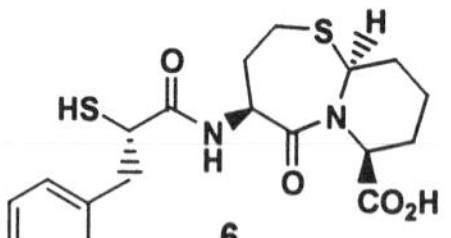

Recently, an analog of omapatrilat was published which utilized a novel zinc-binding moiety: an N-formyl hydroxylamine (**10**) (40). This compound was potent against both ACE and NEP with IC_{50}'s < 10 nM against both enzymes. Not only does **10** represent a new ACE/NEP combination, it also provides evidence for the use of a N-formyl hydroxylamine as a zinc-binding pharmacophore.

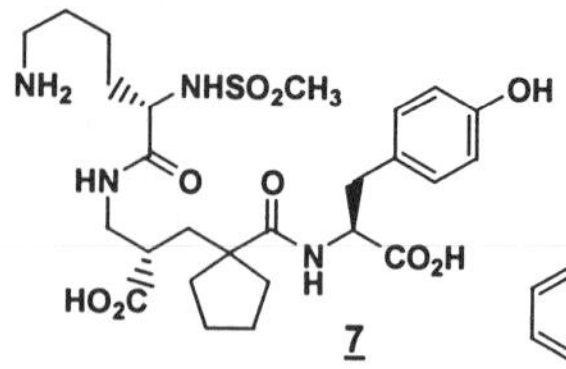

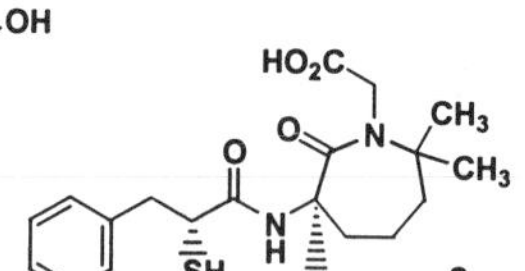

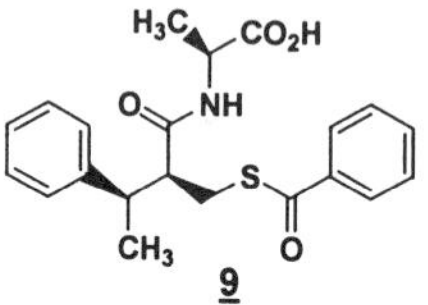

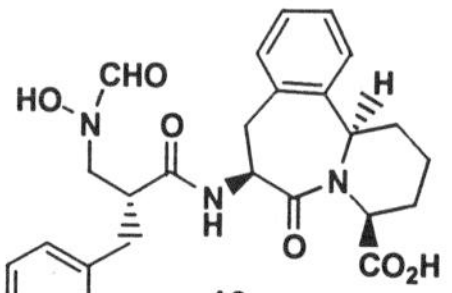

Autonomic Nervous System - During heart failure, one of the earliest physiological adaptations is an increase in adrenergic activity. Augmented sympathetic activation of α_1- and β-adrenergic receptors by norepinephrine increases myocardial contractility, heart rate, arterial vasoconstriction and cardiac afterload. Venous constriction increases cardiac preload (41). Increasing plasma concentrations of norepinephrine contribute to cardiac myocyte hypertrophy, either by directly stimulating α_1-adrenergic receptors on cardiac myocytes or by activating the renin angiotensin aldosterone system (42). Activation of β-adrenergic receptors increases heart rate and stroke volume (43). This places great energy demand on the failing heart and may promote cardiac ischemia. In patients with heart failure, enhanced α_1-adrenergic stimulation increases Na^+ transport in the renal proximal tubule and increases water retention (44). For these reasons, α_1- and β-adrenergic antagonists have been studied as first-line treatment for HF.

α- and β-Blockers - In clinical trials with carvedilol (**11**), a mixed α- and β-adrenergic receptor antagonist, a clear benefit for treating HF was seen with reduced mortality in the drug-treated group (67% reduction compared with placebo) (45). This benefit was so striking that clinical trials were suspended within the placebo group for ethical reasons. Complicating the interpretation of these trials is the observation that **11** also functions as an anti-oxidant. Nonetheless, these results have heightened interest in this class of established adrenergic drugs (46). A review of some of the clinical trials with **11** in HF has been recently published (47). Nebivolol (**12**) is a selective β_1-adrenergic receptor antagonist which has shown efficacy in symptomatic treatment of HF in trials with a small number of patients (48). The CIBIS-II study with over 2600 patients with NYHA class III or IV symptoms of HF and LV ejection fractions < 35% using the β-adrenergic antagonist bisoprolol (**13**) showed a significant reduction in all-cause mortality compared to placebo over a 1.3 year period (49). Furthermore, in the MERIT-HF trial with almost 4000 patients with NYHA Class II-IV symptoms of HF, and LV ejection fractions of < 40%, metoprolol (**14**) showed a 34% reduction in all-cause mortality compared to 11% in the placebo group (50). Recent reviews outlining the evolution of the use of β-adrenergic antagonists in clinical trials for HF have been published (51 - 53).

<u>Aldosterone</u> - During heart failure, angiotensin II stimulates the release of aldosterone from the adrenal cortex. When aldosterone binds to its mineralocorticoid receptor, it enhances renal Na^+ and water retention and exacerbates the loss of potassium and magnesium. Aldosterone is associated with baroreceptor dysfunction and impaired arterial compliance. Aldosterone promotes collagen synthesis and cardiac fibrosis, which may lead to sudden cardiac death as myocardial fibrosis predisposes patients to variations in ventricular arryhthmias (54, 55).

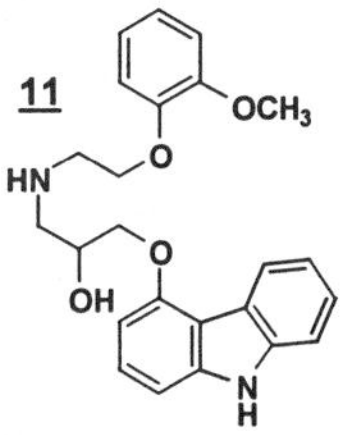

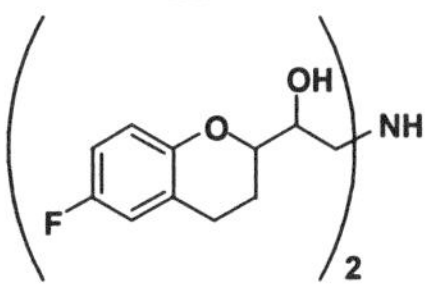

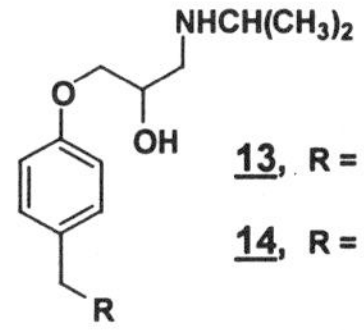

Mineralocorticoid receptor antagonists - Diuretics have been the historical first-line treatment for HF for many years. Because of the heart's inability to adequately perfuse the body, diuretics have been used in HF to relieve the edema that classically accompanies HF. A review of the use of diuretics in HF has recently been published (56). The RALES HF trial with spironolactone (**15**) used in conjunction with an ACE-inhibitor was actually concluded prematurely because of the approximately 30% mortality-reduction benefit seen in the spironolactone (**15**)/ACE- inhibitor group versus ACE-inhibitor alone (57). Presently, a large Phase III trial whose goal is to replicate these initial findings with **15** is ongoing (58). The exact reason why **15** showed such a positive benefit in the treatment of HF is still under study and debate. It is a competitive antagonist of the mineralocorticoids, particularly aldosterone, and it is thought that blockade of mineralocorticoid receptors may be the source of its activity. These receptors are responsible for enhancing reabsorption of Na^+ and secretion of K^+. Thus, the action of **15** is that of a diuretic which preserves K^+ concentrations (potassium sparing diuretic). Because of the new clinical indication for **15**, effort has been rekindled in the area of potassium sparing diuretics and/or mineralocorticoid receptor antagonists. Eplerenone (**16**) is a close structural analog to **15** which is presently under clinical evaluation (Phase III) for HF (59).

<u>Endothelins</u> - Preproendothelin-1, -2, and -3 are produced by endothelial cells, and a furin-like protease cleaves these peptides into big endothelin-1, -2, and -3, respectively (60). Endothelin converting enzyme (ECE) cleaves big endothelin-1, -2, and -3 into the biologically active endothelin-1, -2, and -3, respectively (61). *In vitro*, endothelin activation of endothelin type A (ET_A) receptors on cardiomyocytes increases protein synthesis and stimulates mitogen activated protein kinase, causing cellular hypertrophy and thickening of the left ventricle (62 - 64). Endothelin stimulation of ET_A and ET_B receptors on smooth muscle cells stimulates

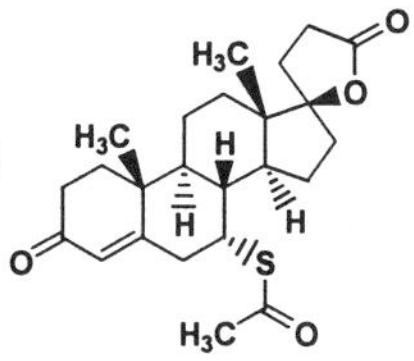

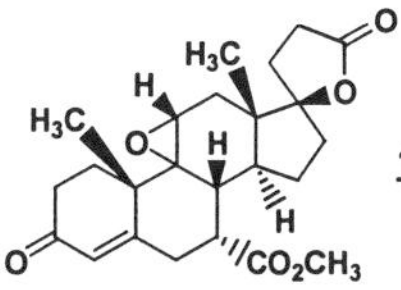

vasoconstriction and increases peripheral vascular resistance, thus augmenting cardiac overload (65). ET is also implicated in causing secondary pulmonary hypertension in patients with left ventricular HF (66). Disruption of endothelin signaling has been viewed as a novel approach to treating HF.

Endothelin (ET_A / ET_B) Receptor Antagonists – Endothelin (ET) has been implicated in the progression of HF, especially in the hypertrophic aspect of the disease. Thus, a number of endothelin receptor antagonists have been tested in clinical trials for the treatment of HF. A recent review on endothelin antagonists has been published (67). The most clinically advanced ET receptor antagonist reported is bosentan (**17**), which has shown positive results in HF trials when given in addition to a standard therapy of ACE inhibitors. Bosentan (1 g, bid) significantly reduced hemodynamic parameters, with a significant number of patients reporting an improved general status using **17** (68). However, mortality benefits were not determined as the trials were interrupted by elevation of liver transaminase levels on patients treated with **17**. Another phase III trial with **17** has begun (ENABLE, June 1999). A number of other endothelin receptor antagonists have since been studied in the treatment of HF, including BMS-193884 (**18**), TBC-11251 (**19**), LU-135252 (**20**), tarasentan (**21**), and tezosentan (**22**). Clear hemodynamic benefits have been observed with many of these compounds in HF trials, but mortality benefits have not yet been demonstrated due to the insufficient length of these trials and the small number of patients in these trials. However, because the role of endothelin as a potent mitogen, ET_A receptor antagonists represent a potential new class of drugs for the treatment of HF.

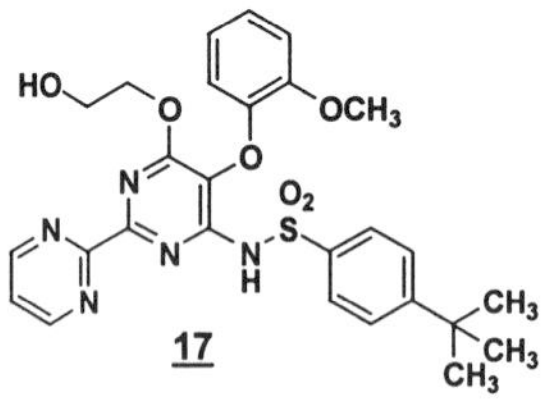

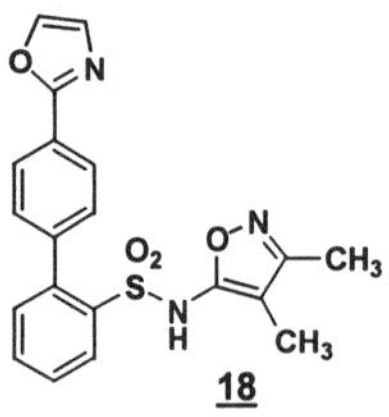

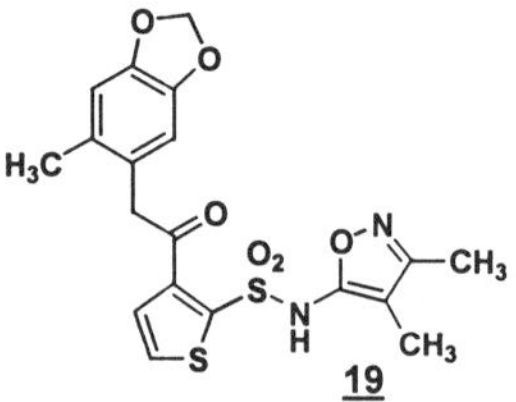

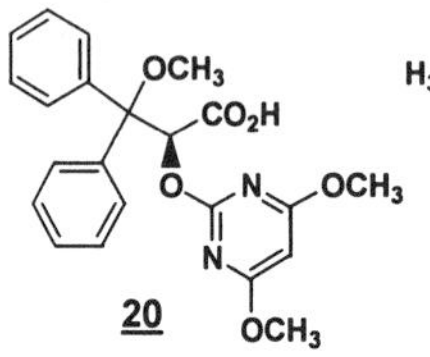

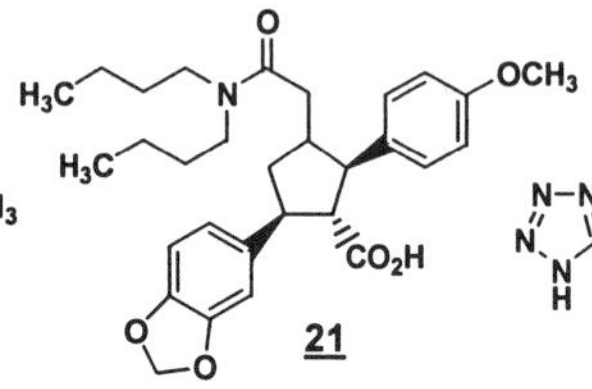

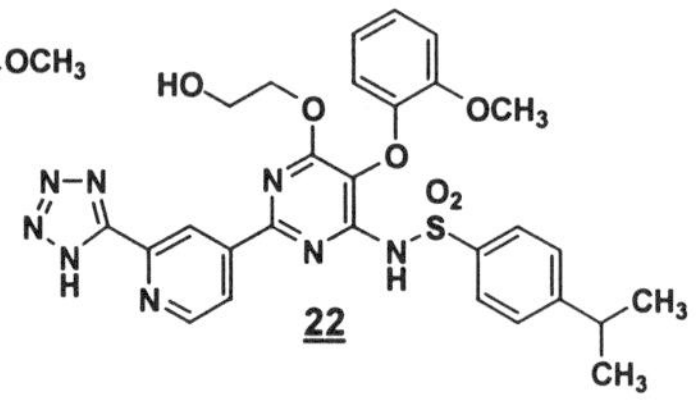

Endothelin Converting Enzyme (ECE-1) Inhibitors – Agents incorporating ECE-1 inhibition are relatively new, and little clinical data on their use in treating HF is available. A recent publication disclosed a small molecule ECE-1 inhibitor (**23**, CGS-35066) claimed to be selective for ECE-1 versus NEP (rhECE-1 IC_{50} = 22 nM, NEP IC_{50} = 2300 nM) (69). Combined ECE-1/NEP inhibitors have also been disclosed, and CGS-31447 (**24**) is a potent inhibitor at both enzymes (rhECE-1 IC_{50} = 17 nM, NEP IC_{50} = 5 nM) (70). A recent review of ECE-1 inhibitors has been published (71).

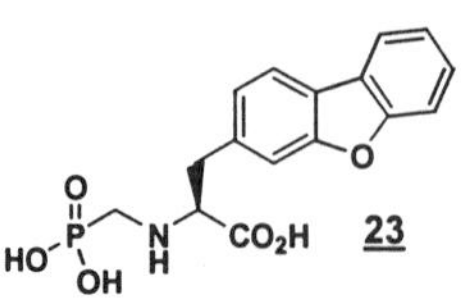

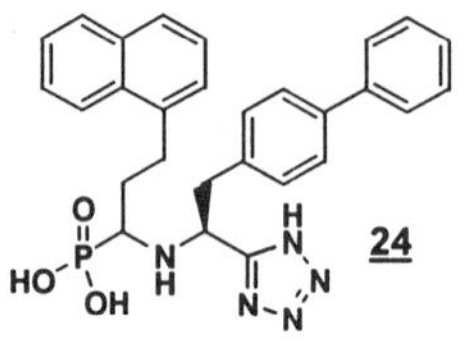

Calcium Channel Antagonists - The benefit of the use of calcium channel blockers in the treatment of HF is equivocal with little evidence at present suggesting a positive role for the agents. Recent reviews on the use of calcium channel blockers in HF have appeared (72, 73).

NEWER APPROACHES

Arginine vasopressin - In patients with severe heart failure, arginine vasopressin (AVP) secretion from the pituitary increases, where it reduces renal natriuresis and diuresis and acts as a strong vasoconstrictor (74, 75). All these actions increase blood volume, peripheral vascular resistance and cardiac afterload. Thus, a number of groups have focused on agents which block the actions of arginine vasopressin. YM-087 (**25**) and OPC-31260 (**26**) are vasopressin receptor antagonists presently in clinical trials for hypertension and HF (76 - 79).

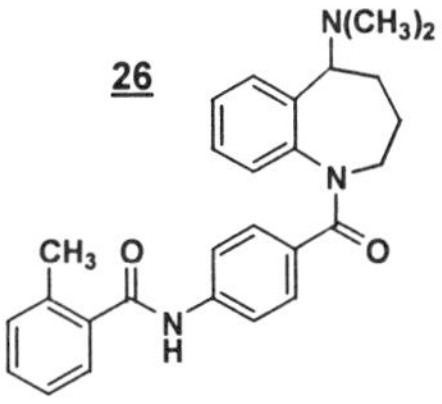

TNFα - Plasma levels of TNFα are elevated in patients with advanced heart failure, and this cytokine enhances left ventricular remodeling, dilatation, dysfunction and pulmonary edema. TNFα may not initiate heart failure, but it may promote the progression of cardiac dysfunction (80).

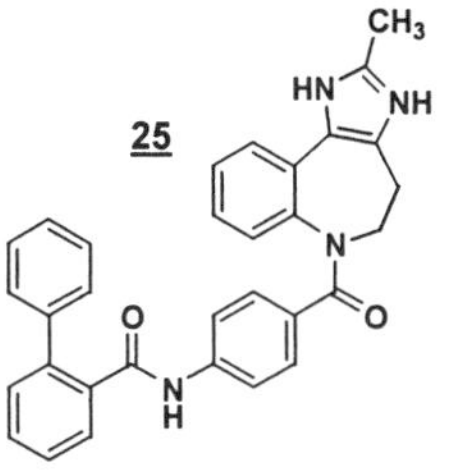

As discussed above, there exists an inflammatory component to HF (81). Modulation of the action or production of TNF-α has been proposed as a potential treatment for HF. Etanercept is a TNF receptor fusion protein that was designed for treating conditions mediated by TNF-α. Etanercept was tested in a small number of heart failure patients in Phase I studies and was found to be well-tolerated. It produced a dose-dependent improvement of cardiac function and quality of life (82). A 900-patient Phase II trial studying the effect of etanercept in HF has been started. Positive results in this and succeeding trials with the fusion protein would certainly generate interest in a small molecule mimic (i.e., TNF-α antagonist or modulator).

Dual Angiotensin Endothelin Receptor Antagonists – As discussed above, both angiotensin (AT$_1$) receptor antagonists and endothelin (ET$_A$) receptor antagonists have shown promise in the treatment of heart failure. A small series of compounds typified by L-746,072 (**27**) were published in 1995 as single molecule antagonists of AT$_1$, AT$_2$, ET$_A$ and ET$_B$ receptors (83). No further reports on their development have appeared. A recent patent application described a series of biphenyl sulfonamides (**28**), which are claimed to block both AT$_1$ and ET$_A$ receptors (84). No biological data was given.

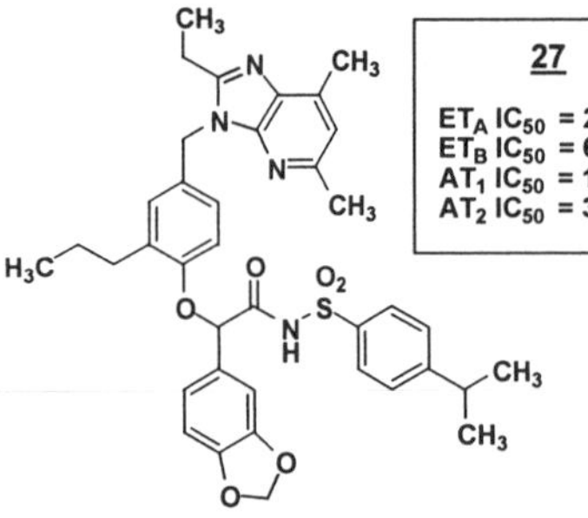

27

$ET_A IC_{50} = 24$ nM
$ET_B IC_{50} = 60$ nM
$AT_1 IC_{50} = 13$ nM
$AT_2 IC_{50} = 32$ nM

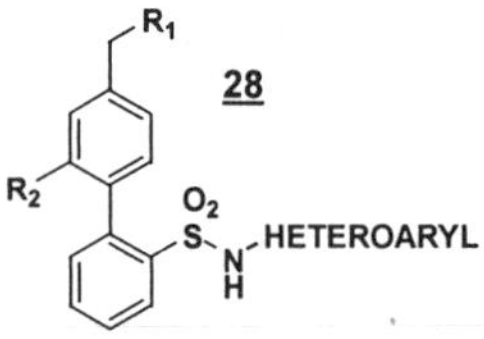

28

Phosphodiesterase Type III inhibitors – The PDEIII inhibitors pimobendan (**29**) and MCI-154 (**30**) have both been studied clinically for the treatment of HF. Whereas **29** showed positive improvement in hemodynamic parameters in patients with acute HF, further development of the drug has been delayed because another clinical trial (PICO) suggested that the drug may be associated with a higher risk of death than placebo (85, 86). MCI-154 (**30**) increased cardiac output and decreased pulmonary capillary wedge pressure in recent Phase III trials, but a mortality benefit has not yet been disclosed (87).

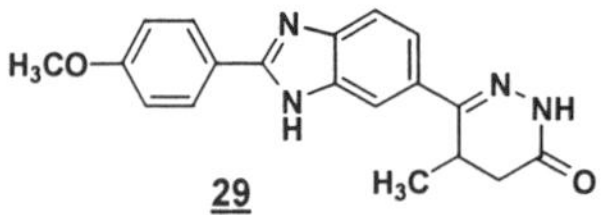

29

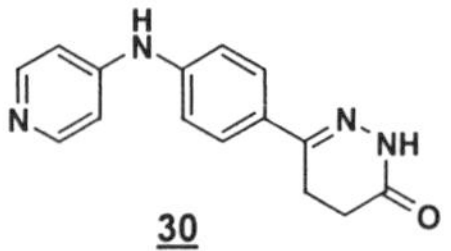

30

FUTURE PROSPECTS

Despite the fact that HF is a terminal disease similar in prognosis to cancer, few (if any) drugs have been specifically designed and targeted for direct treatment of HF. Treatment of HF continues to be an off-shoot of other cardiovascular modalities without regard to the economic potential of a specific, disease-arresting or modifying treatment in HF. A clear understanding of the etiology of the disease will likely lead to more targeted, efficacious and life-extending drug treatments for HF. Clinical trials of new treatments for HF need to show not only a significant improvement in cardiac functioning, but they must also demonstrate a clear mortality benefit on top of the present pharmacological platform of ACE inhibitors and potassium sparing diuretics. Drugs which can clear this hurdle will provide physicians with a better armament of choices for treating HF patients with more than just symptomatic relief.

References

1. R. Garg, M. Packer, B. Pitt, and S. Yusuf, J. Am. Coll. Cardiol., 22 (Suppl. A), 3A (1993).
2. M. W. Rich, J. Am. Geriatric. Soc., 45, 968 (1997).
3. J. McMurray, W. Hart and G. Rhodes, Br. J. Med. Econ., 6, 99 (1993).
4. E. Levy, Eur. Heart J., 19 (Suppl. 9), P2 (1998).
5. R. N. Doughty, Cardiovascular, Pulmonary & Renal Investigational Drugs, 2, 17 (2000).
6. M. W. Rich, Drugs of Today, 35, 171 (1999).
7. R. C. Schlant, E. H. Sonnenblick, and A. M. Katz, "The Heart", Vol. 1, W. R. Alexander, R. C. Schlant and V. Fuster (eds.), p. 687 (1998).
8. K. B. Kannel, Cardiol. Clin., 7, 1 (1989).
9. J. Soler-Soler and G. Permanyer-Miralda, Eur. Heart J., 19 (Suppl. P), P26 (1998).
10. S. Singh, Int. J. Clin. Pract., 52, 432 (1998).
11. J. C. Forfar, J. Cardiol., 67, 3C (1991).
12. J. H. Laragh and J. E. Scaley. "Handbook of Physiology", J. Orloff and R. W. Berliner (eds.), Section 8, "Renal Physiology", 831 (1973).

13. V. Dzau, W. S. Collucci, N. K. Hollenberg and G. H. Williams, Circulation, 63, 645 (1981).
14. C. Curtiss, J. N. Cohn, T. Vrobel and J. A. Franciosa, Circulation, 58, 763 (1971).
15. J. Sadoshima, R. Malhotra, and S. Izumo, J. Cardiac Failure, 2, S1 (1996).
16. P. B. Timmermans, P. C. Wong, A. T. Chiu, W. F. Herblin, D. J. Carini, R. J. Lee, R. R. Wexler, J.-A. Saye and R. D. Smith, Pharmacol. Rev., 45, 205 (1993).
17. B. D. Meyers, W. M. Deen and B. M. Brenner, Circ. Res., 37, 101 (1983).
18. The CONSENSUS Trial Study Group, N. Eng. J. Med., 316, 1429 (1987).
19. The SOLVD Investigators, N. Eng. J. Med., 325, 293 (1991).
20. The SAVE Investigators, N. Eng. J. Med., 327, 669 (1992).
21. F. Andersson, C. Cline, T. Ryden-Bergsten and L. Erhardt, PharmacoEconomics, 15, 535 (1999).
22. A. R. Houghton and A. J. Cowley, Int. J. Cardiol., 59, 7 (1997).
23. Scrip World Pharmaceutical News, 20, 2216 (1997).
24. B. Pitt, P. Poole-Wilson, R. Segal, F. A. Martinez, K. Dickstein, A. J. Camm, M. A. Konstam, G. Riegger, G. H. Klinger, D. Sharma and B. Thiyagarajan, J. Cardiac Failure, 5, 146 (1999).
25. M. Burnier and M. Maillard, IDrugs, 3, 304 (2000).
26. H. Nakaoka, K Imakata, M. Amano, J. Fuji, M. Ishibashi and T. Yamaji, N. Eng. J. Med., 313, 892 (1985).
27. B. Weber, M. Brurnier, J. Nussberger and H. R. Brunner, Horm. Res., 34, 161 (1990).
28. E. R. Bates, Y. Shenker and R. J. Grekin, Circulation, 73, 1155 (1986).
29. K. Atarashi, P. J. Mulrow, R. Franco-Saenz, R. M. Snajclar and J. P. Rapp, Science, 224, 992 (1984).
30. C. J. Brearly, M. J. Allen, V. C. Grimwood, M. MacMahon, European Meeting On Hypertension, Abstract 107, Milan (1995).
31. M. Weber, Am. J. Hypertens., 12 (Number 11, Pt. 2), 139S (1999).
32. J. Robl and D. Ryono, Exp. Opin. Ther. Pat., 9, 1665 (1999).
33. B.-M. Löffler, Curr. Opin. Cardiovasc., Pulm. Renal Invest. Drugs, 1, 352 (1999).
34. C. A. Fink, Exp. Opin. Ther. Pat. 6, 1147 (1996).
35. J. A. Robl, C.-Q. Sun, J. Stevenson, D. E. Ryono, L. M. Simpkins, M. P. Cimarusti, T. Dejneka, W. A. Slusarchyk, S. Chao, L. Stratton, R. N. Misra, M. S. Bednarz, M. M. Asaad, H. S. Cheung, B. E. Abboa-Offei, P. L. Smith, P. D. Mathers, M. Fox, T. R. Schaeffer, A. A. Seymour and N. C. Trippodo, J. Med. Chem., 40, 1570 (1997)
36. A. Graul, P. Leeson and J. Castaner, Drugs Future, 24, 269 (1999).
37. J. L. Rouleau, M. A. Pfeffer, D. J. Stewart, E. K. Kerut, C. B. Porter, J. O. Parker, L. K. Smith, G. Proulx, C. Qian and A. J. Block, Abstract 4132 from the 72nd Am. Heart Assoc. Annual Meeting, Atlanta (1999).
38. J. A. Robl, R. Sulsky, E. Sieber-McMaster, D. E. Ryono, M. P. Cimarusti, L. M. Simpkins, D. S. Karanewsky, S. Chao, M. M. Asaad, A. A. Seymour, M. Fox, P. L. Smith and N. C. Trippodo, J. Med. Chem., 42, 305 (1999).
39. M. C. Fournie-Zaluski, W. Gonzalez, S. Turcaud, I. Pham , B. P. Roques, J. B. Michel, Proc. Natl. Acad. Sci., 91, 4072 (1994).
40. J. A. Robl, L. M. Simpkins and M. M. Asaad, Bioorg. Med. Chem. Lett., 10, 257 (2000).
41. R. W. Schrier and W. T. Abraham, N. Eng. J. Med., 341, 577 (1999).
42. R. W. Schrier, N. Eng. J. Med., 319, 1065 (1988).
43. G. S. Francis, S. R. Goldsmith, T. B. Levine, M. T. Olivari and J. N. Cohn, Ann. Int. Med., 101, 370 (1984).
44. E. Bell-Reuss, D. L. Trevino and C. W. Gottschalk, J. Clin. Invest., 57, 1104 (1976).
45. Scrip, 23, 1999 (1995).
46. R. R. Ruffolo, G. Z. Feuerstein and E. H. Ohlstein, Am. J. Hypertens., 11 (Suppl. 1, Pt. 2), 9S (1998).
47. J. McGowan, R. Murphy, and J. G. F. Cleland, Heart Failure Review, 4, 89 (1999).
48. T. Wisenbaugh, I. Katz, J. Davis, R. Essop, J. Skoularigis, S. Middlemost, C. Rothlisberger, D. Skudicky and P. Sareli, J. Am. Coll. Cardiol., 21 1094 (1993).
49. P. Lechat and the CIBIS-II Investigators and Committees, Lancet, 353, 9 (1999).
50. B. Fagerberg and the MERIT-HF Study Group, Lancet, 353, 2001 (1999).
51. J. R. Teerlink and B. M. Massie, Am. J. Cardiol., 84, 94R (1999).
52. R. Moskowitz and M. Kukin, Expert Opin. Invest. Drugs, 8, 1795 (1999).
53. H. Krum, Drugs, 58, 203 (1999).
54. C. G. Brilla, L. S. Matsubara and K. T. Weber, J. Mol. Cell. Cardiol., 25, 563 (1993).
55. C. S. Barr, A. Naas, M. Freeman, C. C. Lang and A. D. Struthers, Lancet, 343, 327 (1994).
56. F. Folliath, Eur. Heart J., 19 (Suppl. P), P5 (1998).

57. B. Pitt, F. Zannad, W. J. Remme, R. Cody, A. Castaigne, A. Perez, J. Palensky and J. Wittes, N. Eng. J. Med., 341, 709 (1999).
58. GD Searle and Co., Corporate Annual Report (1993).
59. GD Searle & Co, Press Release, January 26 (2000).
60. S. Laporte, J-B. Denault, P. D'Orleans-Juste and R. Leduc, J. Cardiovasc. Pharmacol., 22, S7 (1993).
61. T. J. Opgenorth, J. R. Wu-Wong and K. Shiosaki, FASEB J., 6, 2653 (1992).
62. K. Ponicke, I. Heinroth-Hoffman, K. Becker and O. E. Brode, Br. J. Pharmacol., 121, 118 (1997).
63. M. A. Bogoyevitch, P. E. Glennon, M. B. Andersson, A. Clerk, A. Lazou, C. J. Marshall, P. J. Parker and P. H. Sugden, J. Biol. Chem., 269, 1110 (1994).
64. H. Ito, Y. Hirata, M. Hiroe, S. Tsujino, T. Adachi, T. Takamoto, M. Nitta, K. Taniguchi and F. Marumo, Cir. Res., 69, 209 (1991).
65. M. Yanagisawa, H. Kurihara, S. Kimura, Y. Tomobe, M. Kobayashi, Y. Mitsui, Y. Yazaki, K. Goto and T. Masaki, Nature, 332, 411 (1988).
66. T. Sasaki, T. Noguchi, K. Komamura, T. Nishikimi, H. Yoshikawa and K. Miyatake, J. Card. Fail., 5, 38 (1999).
67. J. R. Wu-Wong, Curr. Opin. Cardiovasc. Pulm. Renal Invest. Drugs, 1, 346 (1999).
68. V. Breu, "Endothelin Inhibitors"; Advances in Therapeutic Application and Development IBC's Third International Symposium, Philadelphia, USA, Iddb Meeting Report, June 26-27 (1997).
69. S. De Lomaert, L. Blanchard, L. B. Stamford, J. Tan, E. M. Wallace, Y. Satoh, J. Fitt, D. Hoyer, D. Simonsbergen, J. Moliterni, N. Marcopoulos, P. Savage, M. Chou, A. J. Trapani and A. Y. Jeng, J. Med. Chem., 43, 488 (2000).
70. E. M. Wallace, J. A. Moliterni, M. A. Moskal, A. D. Neubert, N. Marcopulos, L. B. Stamford, A. J. Trapani. P. Savage, M. Chou and A. Y. Jeng, J. Med. Chem., 41, 1513 (1998).
71. B.-M. Löffler, Curr. Opin. Cardiovasc. Pulm. Renal Invest. Drugs, 1, 352 (1999).
72. N. Mahon and W. J. McKenna, Prog. Cardiovasc. Dis., 41, 191 (1998).
73. U. Elkayam, Cardiology, 89 (Suppl. 1), 38 (1998).
74. S. R. Goldsmith, G. S. Francis, A. W. Cowley Jr. and T. B. Levine, J. Am. Coll. Cardiol., 1, 1385, (1983).
75. A. W. Cowley Jr., "Cardiovascular Physiology IV, International Review of Physiology", A. C Guyton and J. E. Halls (eds.), p. 189 (1982).
76. T. Yatsu, Y. Tomura, A. Tahara, K. Wada, T. Kusayama, J. Tsukada, T. Tokioka, W. Uchida, O Inagaki, Y. Iizumi, A. Tanaka and K. Honda, Eur. J. Pharmacol., 376, 239 (1999).
77. Y. Yamamura, S. Nakamura, S. Itoh, T. Hirano, T. Onogawa, T. Yamashita, Y. Yamada, K. Tsujimae, M. Aoyama, K. Kotosai, H. Ogawa, H. Yamashita, K. Kondo, M. Tominaga, G. Tsujimoto and T. Mori, J. Pharmacol. Exp. Ther., 287, 860 (1998).
78. F-D-C Report 'The Pink Sheet,' 60, 19 (1998).
79. Pharmaceutical Marketing, 6, 50 (1995).
80. G. Torre-Amione, B. Bozkurt, A. Deswal and D. L. Mann, Curr. Opin. Cardiol., 14, 206 (1999).
81. A. Blum and H. Miller, Am. Heart J., 135 (Suppl. 2, Pt.1), 181 (1998).
82. A. Deswal, B. Bozkurt, Y. Seta, S. Parilti-Eiswirth, F. A. Hayes, C. Blosch and D. L. Mann, Circulation, 99, 3224 (1999).
83. T. F. Walsh, K. J. Fitch, D. L. Williams, K. L. Murphy, N. A. Nolan, D. J. Pettibone, R. S. L. Chang, S. S. O'Malley, B. V. Clineschmidt, D. F. Veber, W. J. Greenlee, Bioorg. Med. Chem. Lett., 5, 1155 (1995).
84. N. Murugesan, J. Tellew, J. Macor and Z. Gu, Patent Application WO 00/01389 (2000).
85. S. H. Kubo, Cardiology, 88 [Suppl. 2], 21 (1997).
86. Scrip World Pharmaceutical News, 18, 2056 (1995).
87. H. Takaoka, M. Takeuchi and K. Hata, Am Heart Assoc Ann. Meeting, Abstract 1168 (1994).

Chapter 8. Recent Advances in Endothelin Antagonism

Gang Liu
Pharmaceutical Products Research Division
Abbott Laboratories, Abbott Park, IL 60064

Introduction – The endothelins (ET-1, ET-2 and ET-3), a family of 21-amino acid peptides discovered in 1988 by Yanagisawa, *et al.* (1,2), are potent, long-acting constrictors of vascular smooth muscle and are also potent mitogens (3-5). Aided by known endothelin antagonists, investigators have identified the possible involvement of abnormally high plasma or tissue levels of ET-1, the predominant isoform of ET, in a diverse array of clinical syndromes, including systemic and pulmonary hypertension, cardiac and renal failure, cerebral vasospasm, restenosis, and prostate cancer (6-12). A number of reviews on endothelin systems have appeared recently (13-15), and endothelin inhibitors, including endothelin converting enzyme (ECE) inhibitors and endothelin receptor antagonists, were last reviewed in this forum several years ago (16). This chapter will focus on advances in discovering novel endothelin inhibitors since 1998.

ENDOTHELIN BIOSYNTHESIS AND ECE INHIBITORS

Mature human ET-1 is post-translationally derived from big ET-1, a 38-amino acid inactive precursor peptide, through specific hydrolysis at Trp^{21}-Val^{22} bond in a rate-limiting fashion (17) by endothelin-converting enzyme-1 (ECE-1), a novel zinc metalloprotease which shares a 37% homology with neutral endopeptidase (NEP).-24.11 (18,19). At least three ECE-1 isoforms, ECE-1a, ECE-1b, and ECE-1c, differing in their N-terminal cytosolic domains as a result of alternative splicing of a single gene have been reported (20-22). ECE-1 is ubiquitously distributed in the human vascular endothelium, where both big ET-1 and mature ET-1 have been observed (23,24). The significance of ECE-1 in ET-1 biosynthesis has been supported by the recent observation that disruption of the ECE-1 gene in mice results in term embryos exhibiting craniofacial and cardiac abnormalities virtually identical to those lacking the ET-1 or ET_A receptor genes (25).

The ECE-1 inhibitors discovered so far can be classified as metal chelators, peptide inhibitors, natural products, and low molecular weight organic compounds featuring a phosphorus-containing functionality, a sulfhydryl, a hydroxamic acid, or a carboxylic acid as the zinc-binding group (26). Recent developments in the field have included the isolation of new natural product inhibitors and improvement over existing small molecule ECE inhibitors (27). TMC-66, **1**, a new member of the benzo[a]naphthacenequinone class of antibiotics, was isolated from the fermentation broth of *Streptomyces* sp. A5008 (28). It has a selective inhibitory activity for rat ECE with an IC_{50} value of 2.9 μM, and no inhibitory activity against NEP up to 100 μM. WS 75624 A (**2**), is one of the two structural isomers which were isolated from the fermentation broth of *Saccharothrix* sp. No. 75624 and reported as nonpeptide inhibitors of ECE-1 (29). A series of novel biaryl pyridine carboxylic acids has been prepared and evaluated as inhibitors of human ECE-1, exemplified with imidazole analog **3** (IC_{50} = 2.4 μM), and benzyl substituted thiazole **4** (IC_{50} = 0.86 μM). The relatively flat SAR along with the efficient zinc binding properties of these analogs raised questions about whether the inhibitors act at the catalytic site (30).

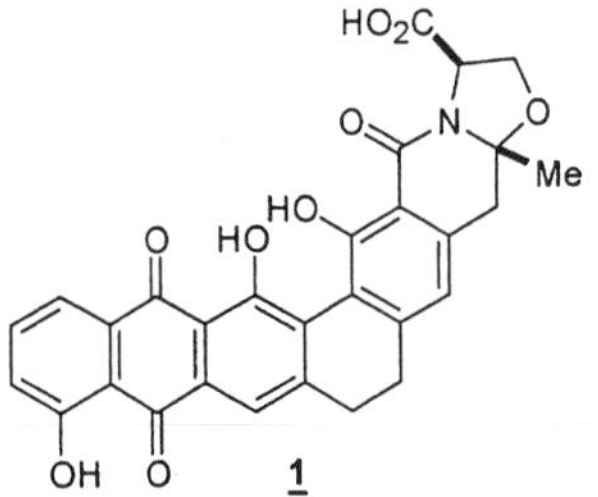

2: R = ... R' = OMe

3: R = ... R' = H

4: R = ... R' = H

1

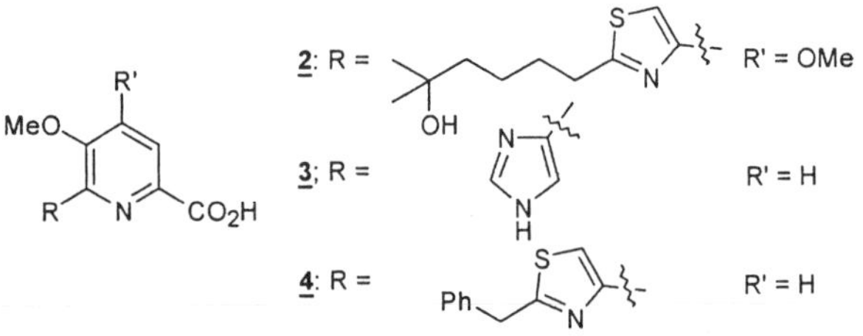

A nonpeptide ECE inhibitor, CGS 26303, **5**, was identified through screens of NEP inhibitors (31). It has moderate ECE-1 inhibitory activity (IC_{50} = 410 nM and 1 nM for ECE and NEP, respectively). Substitution of the α-aminophosphonate moiety of **5** with a naphthylethyl group produced a dramatic increase of ECE-1 inhibitory activity (IC_{50} = 17 nM) while maintaining potent NEP inhibition (IC_{50} = 5 nM). This led to the discovery of CGS 31447, **6**, a new class of potent, dual ECE/NEP inhibitors (32). Using arylacetylenes as novel P_1' biphenyl surrogates decreased NEP inhibition while maintaining the potent ECE-1 inhibition of **5**, leading to ECE-1 selective inhibitors (33). Compound **7**, a phosphonate amide, exhibited IC_{50} values of 33 nM and 6.5 μM for ECE-1 and NEP, respectively. Similarly, a tripeptide **8** inhibited ECE-1 and NEP with IC_{50}s of 8 and 5800 nM, respectively. Probably due to its peptidic nature, compound **8** (10 mg/kg i.v.) exhibited significant inhibition of the big ET-1 pressor response in rats for only 15 min. Further optimization of **5** by replacing the P_1' biphenyl substituent with a conformationally restricted 3-dibenzofuranyl group led to more potent and selective ECE-1 inhibitors, such as CGS 34043 (**9**) (IC_{50} (nM), ECE-1 6.0, and NEP 114) and CGS 35066 (**10**) (IC_{50} (nM), ECE-1 22, and NEP 2300). These compounds (10 mg/kg i.v.) also produced sustained (>90 min) blockade of the hypertensive effects induced by big ET-1 in rats similar to SB209670 (34).

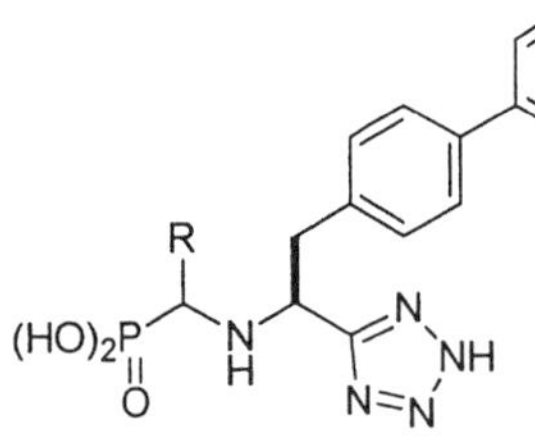

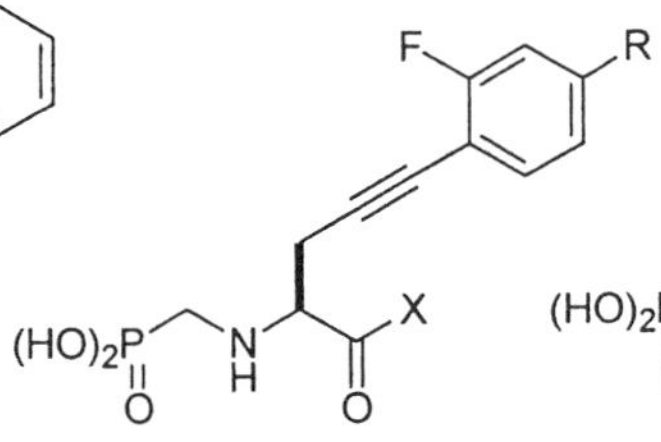

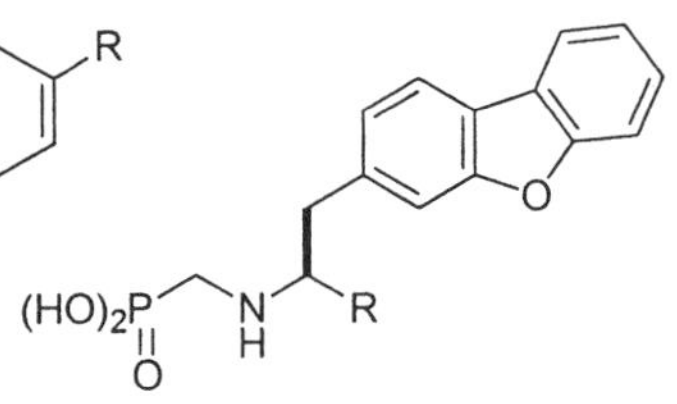

5: R = H
6: R = CH$_2$CH$_2$(1-Naph)

7: R = H, X = NH(CH$_2$)$_2$Ph(p-Ph)
8: R = F, X = Leu-Ala-OH

9: R = tetrazole
10: R = COOH

SLV 306, **11**, is another dual ECE/NEP inhibitor reported to show significant efficacy in a rat model of chronic heart failure when administered at 30 mg/kg/day for 5 months (35). Triple inhibitors of ECE, NEP and ACE could offer a superior therapeutic profile in the treatment of various cardiovascular and renal disorders by inhibiting the effects of ET-1 while concomitantly potentiating those of ANP. Benzofused macrocyclic lactams, such as CGS 26582, **12**, with IC_{50} values of 620, 4, and 175 nM, respectively, were reported as triple ECE-1/NEP/ACE inhibitors (36). Compound **12** (30 mg/kg i.v.) suppressed the increase in mean arterial blood pressure induced by big ET-1 in conscious rats by 44%. SCH54470, **13**, another triple inhibitor (IC_{50} = 70, 90 and 2.5 nM for ECE-1, NEP, and ACE, respectively), was reported to be effective in a remnant kidney model of chronic renal failure (37).

Although ECE-1 constitutes a prime therapeutic target for the regulation of ET-1 production *in vivo*, it remains possible that other enzymes could partially contribute to the proteolytic cleavage of big ET-1 under physiological and pathological conditions. The emergence of potent and selective ECE-1 inhibitors with good *in vivo* efficacy represents novel pharmacological tools to address this question. More animal studies are also needed to assess the therapeutic potential of ECE-1 inhibitors.

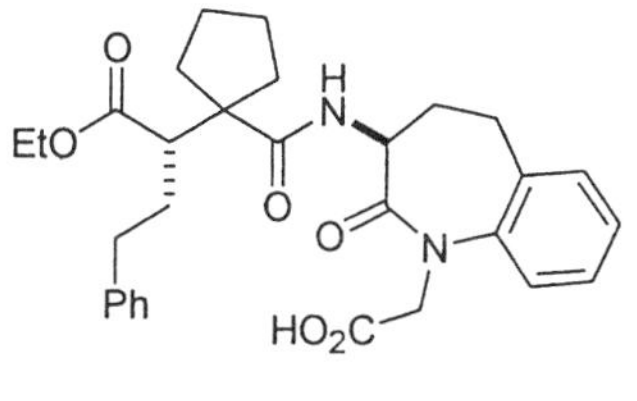

11 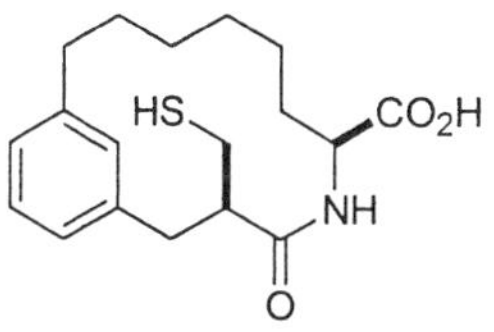   **12** 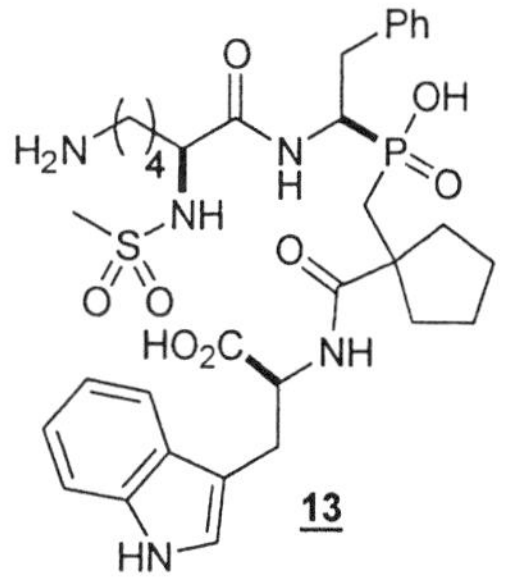   **13**

ENDOTHELIN RECEPTORS AND ET RECEPTOR ANTAGONISTS

The biological effects of the ETs are elicited through endothelin receptors, a family of G-protein-coupled receptors (GPCR) with seven transmembrance (TM) domains. Two subtypes of human endothelin receptors, ET_A and ET_B, encoded by genes located on chromosomes 4 and 13, respectively, have been cloned and are approximately 55% homologous (38, 39). Expressed mainly in vascular smooth muscle cells, ET_A receptor mediates the vasoconstrictive and mitogenic effects of ET-1 both *in vitro* and *in vivo* (3-5). Activation of the endothelial ET_B receptor may attenuate the vasoconstrictive effects of local ET-1 by mediating production of nitric oxide (40) and prostacyclin, and by clearing ET-1 from the circulation (41), whereas activation of smooth muscle ET_B receptors produces vasoconstriction. The positive clinical data from both non-selective antagonist bosentan (42) and ET_A selective antagonist sitaxsentan (43) in the treatment of pulmonary and cardiovascular diseases illustrated the ongoing debate concerning the merits of these two approaches. A wide variety of ET receptor antagonists, peptidic and nonpeptidic, have been reported since 1994. More recently, the existing ET antagonists have been further explored to improve their potency, selectivity and pharmacological properties. Meanwhile, novel molecular templates acting as ET receptor antagonists have been identified, some of which have been further optimized to potential development candidates.

Implementation of a pyridylcarbamoyl group and an isopropylpyridylsulfonamide substituent as key components in the scaffold of bosentan, a heteroarylsulfonamido pyrimidine-based ET antagonist, resulted in the identification of Ro 48-5695 (**14**) with IC_{50} values of 0.3 nM and 0.7 nM for ET_A and ET_B receptors, respectively (44,45). Compound **14** has high functional antagonistic potency *in vitro* by shifting the concentration-response curve of ET-1 induced contraction of the isolated rat aorta to the right (pA$_2$ = 9.3). A water soluble analog of Bosentan was developed for parenteral use by incorporating a tetrazolylpyridyl group to the positions 2 of the central pyrimidine core to give Ro 61-1790, **15**, which is selective for ET_A receptor (K_i = 0.13 nM) over ET_B receptor (K_i = 175 nM). Compound **15** also exhibited high potency in functional assays *in vitro* and *in vivo* (46). A similar agent, tesosentan (Ro 61-0612), **16**, with $t_{1/2}$ of 10 min (i.v.), can offer a unique approach for treatment of acute pathological conditions (47). TA-0201, **17**, from another series of pyrimidine sulfonamides, was reported to be highly potent against the ET_A receptor (K_i = 0.015 nM), selective for ET_A over ET_B receptor (K_i = 41 nM), and potent in functional assays *in vitro* (pA$_2$ = 9.0). In anesthetized rats, compound **17** (0.01-1 mg/kg) inhibited the pressor response to exogeneous big ET-1 (1 nmol/kg, i.v.) after both i.v. and p.o. administration, in a dose-dependent manner for up to 8 hr (48). A similar pyrazine sulfonamide, ZD1611, **18**, was also reported to have good potency against ET_A

receptor (pIC_{50} = 8.6) vs. ET_B receptor (pIC_{50} = 5.6). In pithed rats, compound **18** (0.2 mg/kg, i.v.) produced a 50% reversal of an established big ET-1-induced pressor response in the presence of continuous big ET-1 infusion (49).

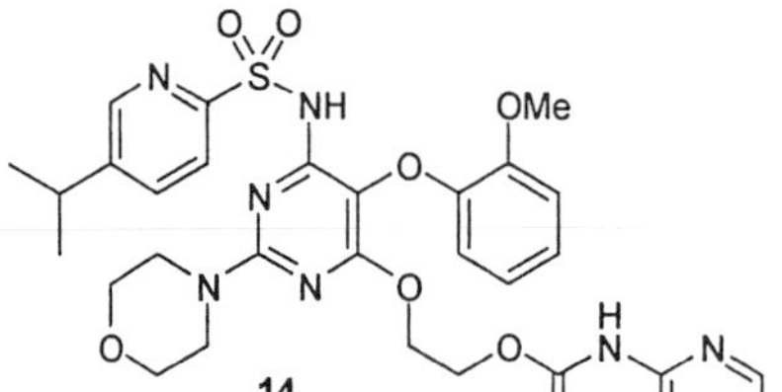

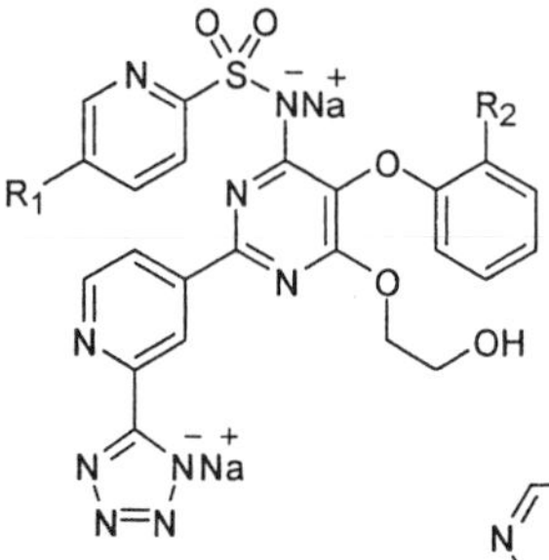

15 R_1 = Me, R_2 = OMe
16 R_1 = i-Pr, R_2 = ONa

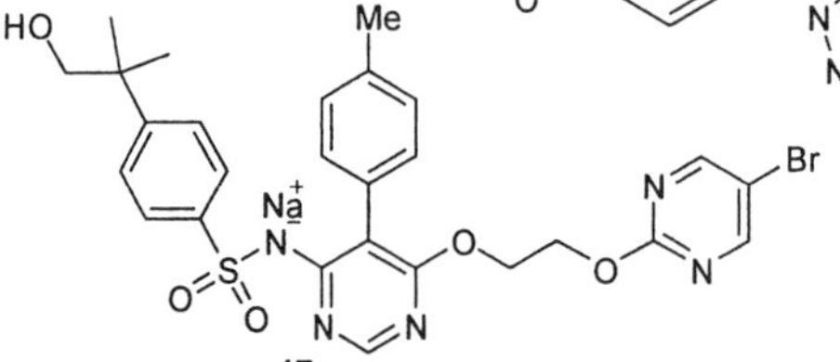

Isoxazoleamino sulfonamide-based ET_A receptor selective antagonist BMS-187308, **19**, was developed based on BMS-182874 by modifying the *N,N*-dimethylaminonaphthalene portion of the molecule with a biaryl moiety (50). The combination of the optimal 4'-isobutyl substituent with the 2'-amino function maintained the ET_A selective character of the compound while improving the affinities for the receptors by 20-fold (K_i = 4.7 nM for ET_A and 1.7 µM for ET_B). Compound **19** has 48% bioavailability and good oral activity in inhibiting the pressor effect caused by an ET-1 infusion in rats and monkeys. Further efforts on reducing the metabolism of **19** produced BMS-193884, **20**, with K_i of 1.4 nM for ET_A and 19 µM for ET_B, respectively. Compound **20** improved the hemodynamics in a porcine model of heart

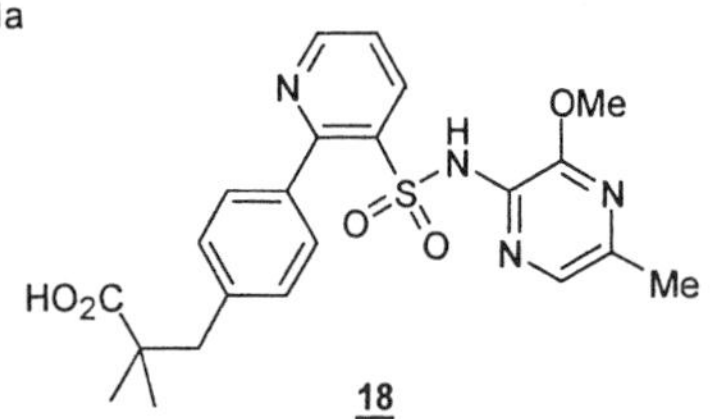

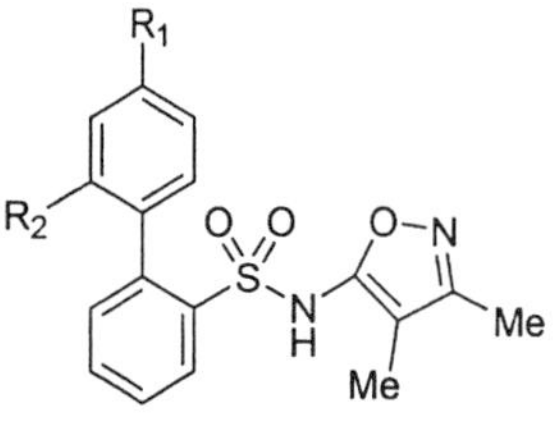

19 R_1 = i-PrCH$_2$, R_2 = NH$_2$
20 R_1 = 2-oxazolyl, R_2 = H

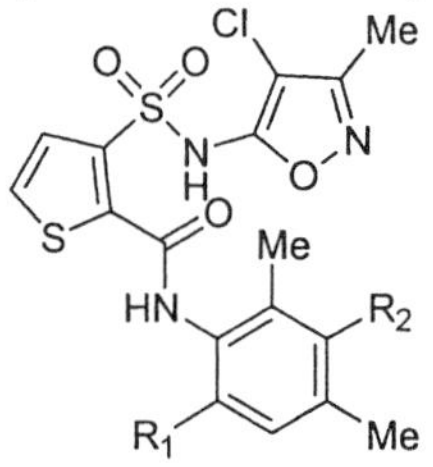

21 R_1 = Me, R_2 = OH
22 R_1 = acetyl, R_2 = H

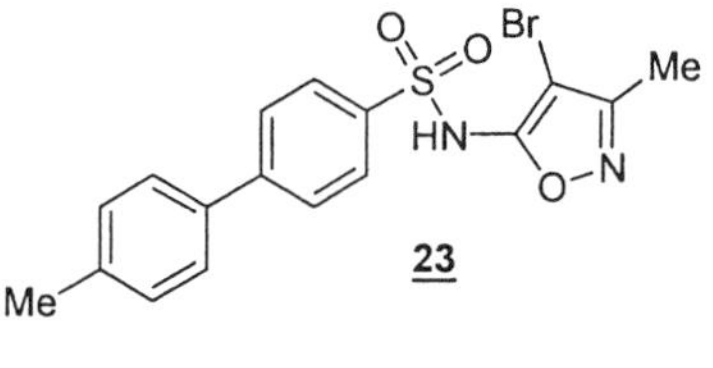

23

failure (51). The SAR of sitaxsentan (TBC 11251), an anilino thiophenesulfonamide-based ET_A-selective antagonist, was further explored (52). Changing the ketone linkage of TBC 11251 to a carboxyamide analog, in conjunction with fine-tuning the anilino ring substituents rendered TBC 2576, **21**, which has ~10-fold increase of ET_A binding affinity (IC_{50} = 0.11 nM) than sitaxsentan, over 10,000-fold ET_A/ET_B selectivity ratio, good aqueous solubility, and long lasting *in vivo* activity in a rat model of acute hypoxia-induced pulmonary hypertension at 5 mg/kg (53). Introduction of an acetyl group at the 6-position of the anilino ring further improved the *in vitro* activity of this class of compounds (TBC3214, **22**, IC_{50} (ET_A) = 0.04 nM, 442,000-fold selective, $t_{1/2}$ = 4.0 h, F = 25% in rats), and *in vivo* activity was demonstrated in the same model as well (54). The reversal of the receptor selectivity was achieved with TBC-10950, **23**, a *para*-tolyl substituted benzenesulfonamide, which exhibited high affinity for the ET_B receptor (IC_{50} = 17 nM), 290-fold selectivity over the ET_A receptor, and *in vitro* functional activity in PI hydrolysis with an IC_{50} of 2.0 µM (55).

The pyrrolidine-3-carboxylic acid-based ET antagonist series, represented by atrasentan (ABT-627), has been further modified by taking advantage of the critical role of the *N*-linked side chain plays in determining receptor binding and selectivity (56). Replacing the *N,N*-dialkylacetamide side chain of atrasentan with a 2,6-diethylacetanilide or a bis-*o*-tolylmethylamine acetamide group resulted in A-192621 (**24**) and A-308165 (**25**), respectively, which exhibited complete reversal of receptor selectivity, preferring ET_B (IC_{50} (nM) 6.4 and 1.9) over ET_A (IC_{50} (μM) 8.2 and 52) (57,58). The orally available **24** has helped elucidating the role of ET_B receptors in the maintenance of vascular tone. Replacing the *p*-anisyl group of atrasentan with a 2,2-dimethylpentyl group was found to further increase the selectivity for ET_A receptor. This change combined with fine-tuning of the physical properties provided A-216546, **26**, a potent (K_i = 0.46 nM), highly ET_A selective (>28,000-fold) antagonist as a potential clinical backup for atrasentan (59).

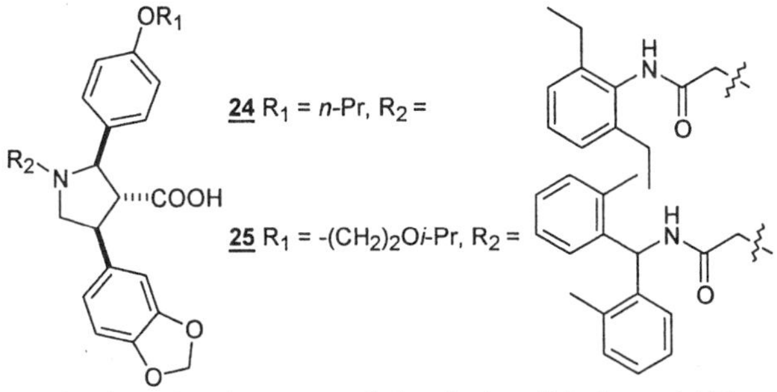

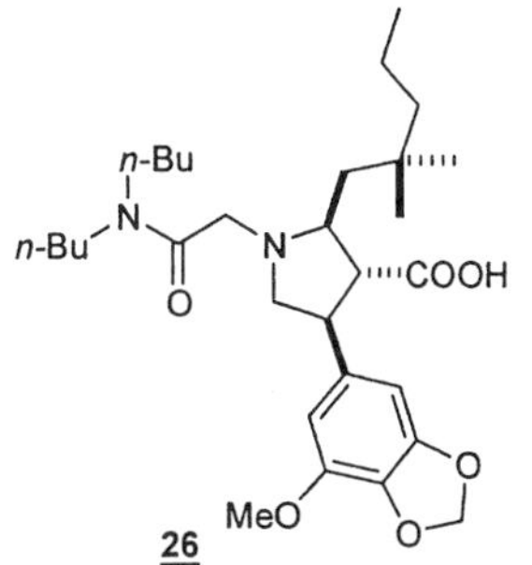

24 R_1 = *n*-Pr, R_2 =

25 R_1 = -(CH$_2$)$_2$O*i*-Pr, R_2 =

26

Further development of the butenolide-based ET_A selective antagonist CI-1020 has led to the identification of **27**, a sodium salt of a sulfonic acid, which offered much improved aqueous solubility (60,61). Compound **27** has an ET_A IC_{50} of 0.38 nM, ET_A selectivity of 4,200-fold, ET_A functional activity of pK_B = 7.8, and an ED_{50} of 0.3 μg/kg/h (i.v.) in preventing acute hypoxia-induced pulmonary hypertension in rats. EMD 122801, **28**, with IC_{50}s of 0.30 nM for ET_A and 340 nM for ET_B receptor, respectively, was derived from the reference compound CI-1020 by replacing the metabolically undesirable methylenedioxyphenyl with a benzothiadiazole group using a Kohonen neural network approach (62,63). Further exploration of the SAR led to the highly potent ET_A selective antagonist EMD 122946, **29**, (IC_{50} = 32 pM for ET_A and 160 nM for ET_B, respectively) by introducing a fluorine atom to the 3-position of the *p*-anisyl group (64). Compound **29** also displayed potent functional activity (pA_2 = 9.5) and inhibited the ET-1 induced pressor response in pithed rats with an ED_{50} of 0.3 mg/kg. In conscious spontaneously hypertensive rats and in DOCA-salt hypertensive rats, the compound lowered MAP with an ED_{50} of 0.06 mg/kg.

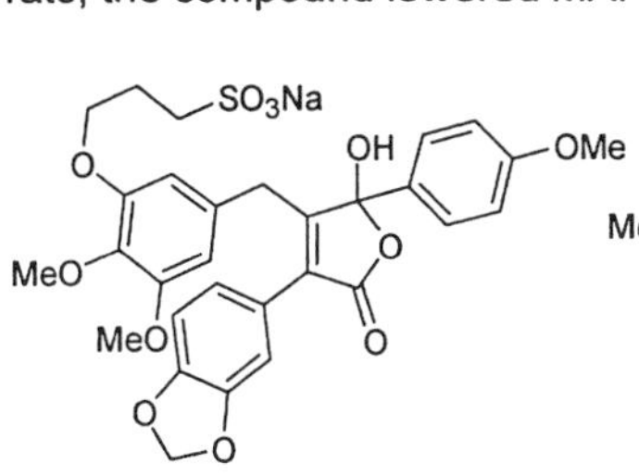

27

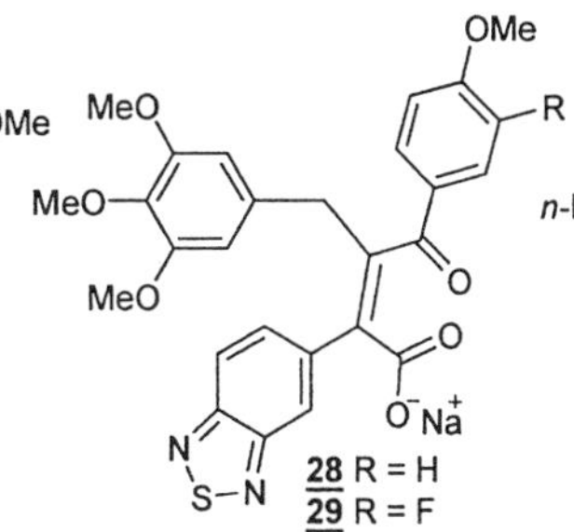

28 R = H
29 R = F

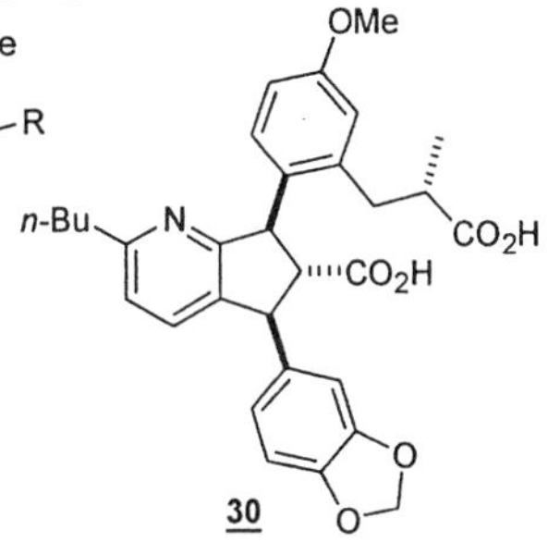

30

Indane-carboxylic acid-based ET receptor antagonists, typified with enrasentan (SB-217242) (65), was expanded to cyclopenteno[1,2-b]pyridine-6-carboxylic acid, such as J-104132 (L753037), **30**. Compound **30** is a potent, orally active, mixed ET receptor antagonist with K_i of 0.034 nM and 0.104 nM for ET_A and ET_B receptor, respectively. Functionally, compound **30** inhibits ET-1-induced contractions in rabbit iliac artery (pA_2 = 9.7) and BQ-3020-induced contractions in pulmonary artery (pA_2 = 10.14). In conscious, normotensive rats, pressor responses to 0.5 nmol/kg i.v. ET-1 are inhibited by **30** after i.v. (0.1 mg/kg) or p.o. (1 mg/kg) administration (66). The SAR of potent, ET_A selective antagonist LU 127043 was further explored (67). Its active enantiomer, LU 135252, **31**, (K_i = 1.4 nM and 184 nM for ET_A and ET_B receptor, respectively) demonstrated the importance of the substituent in the β-position of the propionic acid. Modification of the β-alkoxy substituent led to the discovery of a potent, mixed ET_A/ET_B receptor antagonist LU 302872, **32**, (K_i = 2.15 nM and 4.75 nM for the ET_A and ET_B receptor, respectively) (68). The compound is orally active (F = 50-70 % in dogs), and antagonizes the big ET-1-induced blood pressure increase in rats and bronchospasm in guinea pigs at a dose of 10 mg/kg.

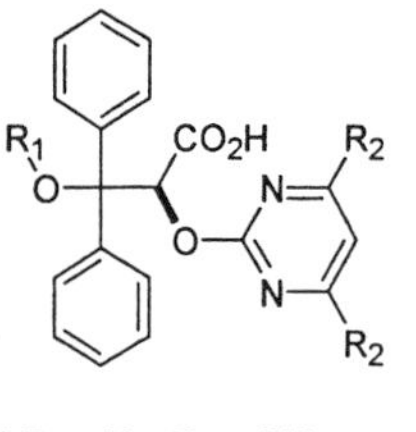

31 R_1 = Me, R_2 = OMe
32 R_1 = 3,4-DiOMePhCH$_2$CH$_2$, R_2 = Me   **33** 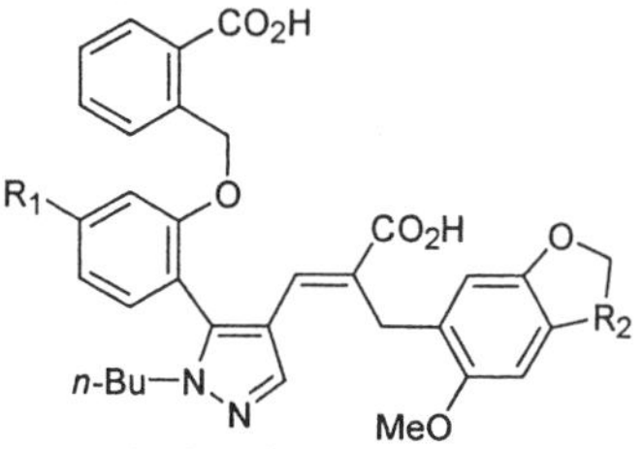   **34** R_1 = OMe, R_2 = O
          35 R_1 = Cl, R_2 = CH$_2$

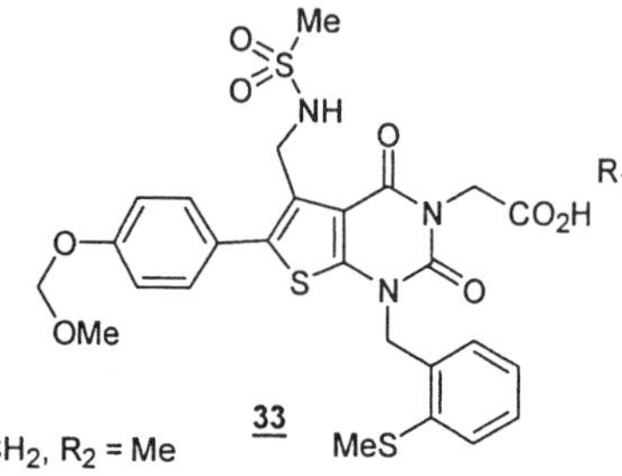

The power of rational drug design to optimize a peptide lead compound was demonstrated by the discovery of a series of thieno[2,3-d]pyrimidinone-3-acetic acids as nonpeptide ET antagonists. Based on the structural information of a cyclic hexapeptide ET antagonist TAK-044 (69), crucial functional moieties for ET receptor binding, i.e. a carboxyl group and aromatic rings, were introduced onto a bicyclic heterocycle "scaffold" which mimics the fixed main chain of TAK-044, and provided this series of nonpeptide ET receptor ligands. Simultaneous optimization of each substituent in the molecule led to the discovery of compound **33**, which binds to human ET_A and ET_B receptor subtypes with affinities (IC_{50}) of 7.6 and 100 nM, respectively (70).

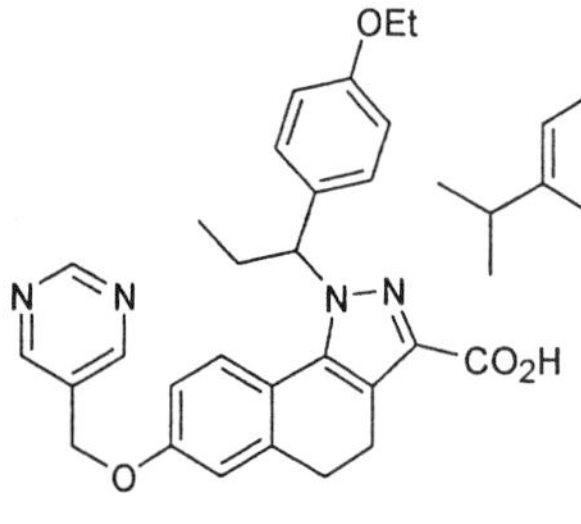

36 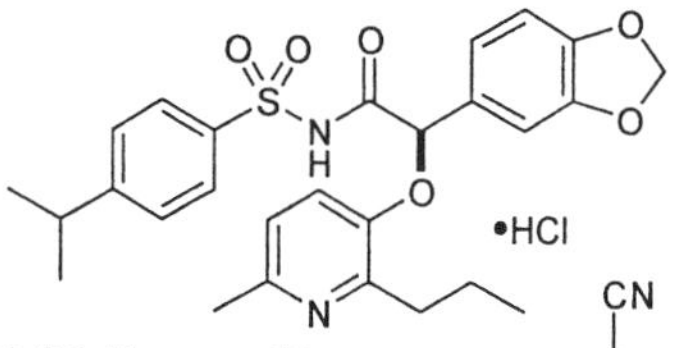   **37** 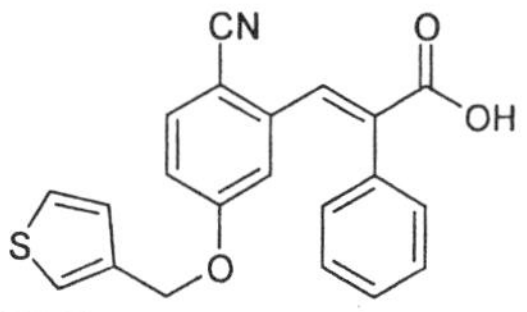   **38**
          39 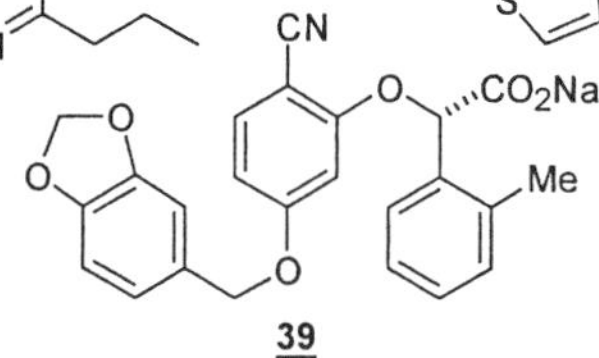

SB 234551, **34**, a new class of pyrazole acrylic acid-based ET_A receptor selective antagonist, was disclosed (71). It has K_i values of 0.13 and 500 nM for ET_A and ET_B receptor, respectively. It also displayed high functional activities in antagonizing the ET-1 induced contraction in isolated rat aorta and isolated human pulmonary artery with K_B values of 1.9 and 1.0 nM, respectively. Compound **34** (0.1-1.0 mg/kg i.v.) dose-dependently inhibited the pressor response to exogenous ET-1 in conscious rats. It has an oral bioavailability of 30% and a plasma half-live of 125 min in rats. Similarly, minor modifications of compound **34** yielded SB 247083, **35**, another ET_A selective antagonist with very comparable *in vitro* activity profiles (K_i = 0.41 and 467 nM, and K_B 3.5 and 340 nM, respectively for ET_A and ET_B) (72). ATZ1993, **36**, a benz[r]indazole-3-carboxylic acid-based ET antagonist, was reported to have pK_i

values of 8.69 and 7.20 for ET_A and ET_B receptors, respectively (73). Administrated orally (30 mg/kg) for 7 weeks, compound **36** demonstrated inhibition of intimal hyperplasia after balloon denudation of the rabbit carotid artery. PABSA, **37**, an acylsulfonamide-based nonpeptide ET antagonist, similar to L-749,329 was recently disclosed and characterized pharmacologically (74). Compound **37** has K_is of 0.11 and 25 nM, and functional K_Bs of 0.46 and 94 nM in isolated rabbit vessels for ET_A and ET_B receptors, respectively. Compound **37** (10-100 mg/kg, p.o.) also has sustained (≥24 h) hypotensive effect in DOCA-salt hypertensive rats. Analysis of the SAR established in the earlier 4-aryloxy-4-arylbutyric acid series led to the identification of a stilbene acid-based ET_A selective antagonist, RPR111723, **38**, which has an IC_{50} of 80 nM for the ET_A receptor and a pK_B of 6.5 in the functional assay, measured on rat aortic strips (75). Further exploration of all the lead structures through superimposition studies within the receptor model yielded RPR118031A, **39**, which exhibited an IC_{50} (ET_A) of 6 nM, a pK_B of 8.1, a bioavailability of 84%, and *in vivo* potency at 75 μmol/kg p.o. in rats (76).

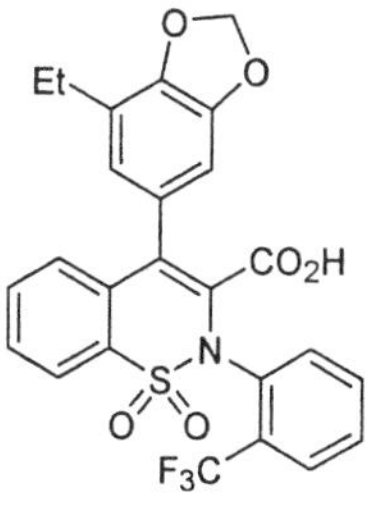

40

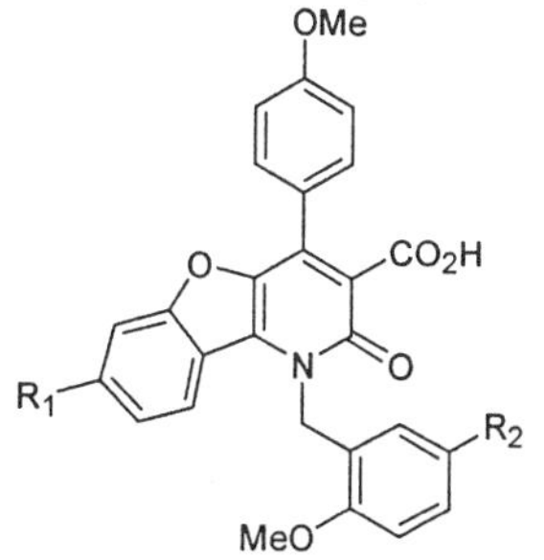

41 R_1 = H, R_2 = OMe
42 R_1 = EtNSO$_2$Me, R_2 = H

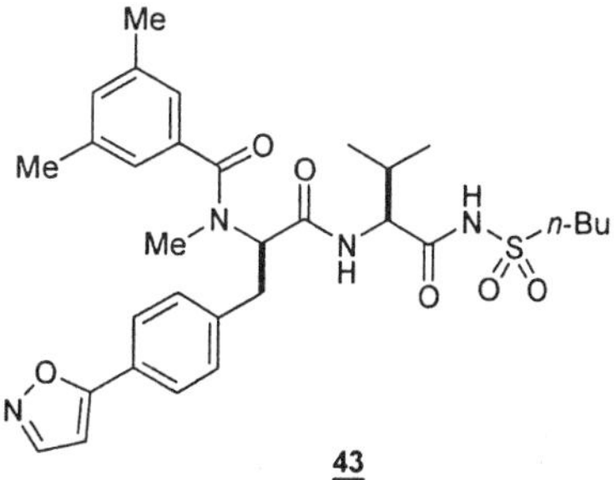

43

A novel series of substituted benzothiazine-1,1-dioxide-3-carboxylic acids was identified from a screening and optimized to yield PD 164800 (77). This series of compound was further developed into a highly ET_A selective antagonist PD 180988 (CI1034), **40**, with IC_{50}s of 0.46 nM and 2,200 nM for ET_A and ET_B receptor, respectively. Compound **40** selectively reversed the pulmonary vasoconstrictor response to hypoxia in a lamb model with minimal systemic effect when given at 30 μg/kg bolus followed by a 50 μg/kg/hr infusion (78). Recently, a series of non-selective antagonists based on benzofuro[3,2-b]pyridine carboxylic acid as pharmacophore, was described in the literature (63). Variations at the core structure led to more potent non-selective and ET_B selective inhibitors, respectively. Compound **41**, has IC_{50} of 21 nM for ET_A receptor and 41 nM for ET_B receptor, respectively. In contrast, additional substituent on the benzofuran led to a ET_B selective antagonist **42**, which has an IC_{50} of 3.6 nM for the ET_B receptor, and greater than 10 μM for the ET_A receptor (79). Dipeptide IRL 3630, **43**, a mixed ET_A/ET_B antagonist with K_i of 1.5 nM and 1.2 nM for ET_A and ET_B receptors, respectively, was developed through structural modification of IRL 2500 (80), an ET_B selective antagonist. Compound **43** was found to be stable on incubation with human and rat plasmas (81).

Conclusion — Recently, tremendous progress has been made in the field of endothelin antagonism. Potent and selective ECE inhibitors are now available, so are those inhibiting NEP and ACE concomitantly. A large number of highly potent, orally available, balanced or selective, long or short acting ET antagonists have been discovered, many of which are currently in clinical trials. Since endothelin antagonism can potentially provides therapeutic intervention in a number of diseases, it will be of great interest to see how these compounds are developed to fulfill the therapeutic potential of ET antagonism.

References

1. M. Yanagisawa, H. Kurihara, S. Kimura, Y. Tomobe, M. Kobayashi, Y. Mitsui, Y. Yazaki, K. Goto, T. Masaki, Nature, 332, 411 (1988).
2. A. Inoue, M. Yanagisawa, S. Kimura, Y. Kasuya, T. Miyauchi, K. Goto, T. Masaki, Proc. Natl. Acad. Sci. U.S.A., 86, 2863 (1989).
3. E. R. Levin, New Engl. J. Med., 333, 356 (1995).
4. G. M. Rubanyi, M. A. Polokoff, Pharmacol. Rev., 46, 325 (1994).
5. T. J. Opgenorth, Adv. Pharmacol., 33, 1 (1995).
6. H. Krum, P. Martin, Curr. Opin. Cardiovasc. Pulm. Renal Invest. Drugs, 1, 316 (1999).
7. M. R. MacLean, Pulm. Pharmacol. Ther., 11, 125 (1998).
8. T. Miyauchi, S. Sakai, Respir. Circ., 47, 31 (1999).
9. D. M. Pollock, J. S. Polakowski, C. D. Wegner, T. J. Opgenorth, Ren. Fail., 19, 753 (1997).
10. M. Zimmermann, J. Neurosurg. Sci., 41, 139 (1997).
11. S. E. Burke, N. L. Lubbers, G. D. Gagne, J. L. Wessale, B. D. Dayton, C. D. Wegner, T. J. Opgenorth, J. Cardiovasc. Pharmacol., 30, 33 (1997).
12. J. B. Nelson, K. Chan-Tack, S. P. Hedican, S. R. Magnuson, T. J. Opgenorth, G. S. Bova, J. W. Simons, Cancer Res., 56, 663 (1996).
13. F. E. Strachan, D. J. Webb, Emerging Drugs, 3, 95 (1998).
14. T. Miyauchi, T. Masaki, Annu. Rev. Physiol., 61, 391 (1999).
15. G. A. Gray, B. Battistini, D. J. Webb, Trends Pharmacol. Sci., 21, 38 (2000).
16. X.-M. Cheng, K. Ahn, S. J. Haleen, Annu. Rep. Med. Chem., 32, 61 (1997).
17. S. Telemaque, N. Emoto, D. Dewit, M. Yanagisawa, J. Cardiovasc. Pharmacol., 31, S548 (1998).
18. A. J. Turner, L. J. Murphy, Biochem. Pharmacol., 51, 91 (1996).
19. A. J. Turner, K. Tanzawa, FASEB J., 11, 355 (1997).
20. O. Valdenaire, E. Rohrbacher, M.-G. Mattei, J. Biol. Chem., 270, 29794 (1995).
21. A. Azarani, G. Boileau, P. Crine, Biochem. J., 333, 439 (1998).
22. A. Schweizer, O. Valdenaire, P. Nelbock, U. Deuschle, J. B. Dumas Milne Edwards, J. G. Stumpf, B. M. Loffler. Biochem. J., 328, 817 (1997).
23. A. J. Turner, K. Barnes, A. Schweizer, O. Valdenaire, Trends Pharmacol. Sci., 19, 483 (1998).
24. A. P. Davenport, R. E. Kuc, C. Plumpton, J. W. Mockridge, P. J. Barker, N. S. Huskisson, Histochem. J., 30, 359 (1998).
25. H. Yanagisawa, M. Yanagisawa, R. P. Kapur, J. A. Richardson, S. C. Williams, D. E. Clouthier, D. Dewit, N. Emoto, R. E. Hammer, Development, 125, 825 (1998).
26. A. Y. Jeng, S. De Lombaert, Curr. Pharm. Des., 3, 597 (1997).
27. B. M. Loffler, Curr. Opin. Cardiovasc. Pulm. Renal Invest. Drugs, 1, 352 (1999).
28. Y. Asai, N. Nonaka, S.-I. Suzuki, M. Nishio, K. Takahashi, H. Shima, K. Ohmori, T. Ohnuki, S. Komatsubara, J. Antibiotics, 52, 607 (1999).
29. Y. Tsurumi, H. Ueda, K. Hayashi, S. Takase, M. Nishikawa, S. Kiyoto, M. Okuhara, J. Antibiotics, 48, 1066 (1995).
30. M. A. Massa, W. C. Patt, K. Ahn, A. M. Sisneros, S. B. Herman, A. Doherty, Bioorg. Med. Chem. Lett., 8, 2117 (1998).
31. S. De Lombaert, R. D. Ghai, A. Jeng, A. J. Trapani, R. L. Webb, Biochem. Biophys. Res. Commun., 204, 407 (1994).
32. S. De Lombaert, L. B. Stamford, L. Blanchard, J. Tan, D. Hoyer, C. G. Diefenbacher, D. Wei, E. M. Wallace, M. A. Moskal, P. Savage, A. Y. Jeng, Bioorg. Med. Chem. Lett., 7, 1059 (1997).
33. E. M. Wallace, J. A. Moliterni, M. A. Moskal, A. D. Neubert, N. Marcopulos, L. B. Stamford, A. J. Trapani, P. Savage, M. Chou, A. Y. Jeng, J. Med. Chem., 41, 1513, (1998).
34. S. De Lombaert, L. Blanchard, L. B. Stamford, J. Tan, E. M. Wallace, Y. Satoh, J. Fitt, D. Hoyer, D. Simonsbergen, J. Moliterni, N. Marcopoulos, P. Savage, M. Chou, A. J. Trapani, A. Y. Jeng, J. Med. Chem., 43, 488 (2000).
35. D. Thormahlen, E. Udvary, S. Rozsa, J. G. Rapp, S. Tovofic, paper # 180, Sixth International Conference on Endothelin (ET-6), Oct. 11-13, 1999, Montreal, Canada.
36. G. M. Ksander, P. Savage, A. J. Trapani, J. L. Balwierczak, A. Y. Jeng, J. Cardiovasc. Pharmacol., 31(Suppl. 1), S71 (1998).

37. S. Vemulapalli, M. Chintala, A. Stamford, R. Watkins, P. Chiu, E. Sybertz, A. B. Fawzi, Cardiovasc. Drug Rev., 15, 260 (1997).
38. H. Arai, S. Hori, I. Aramori, H. Ohkubo, S. Nakanishi, Nature, 348, 730 (1990).
39. T. Sakurai, M. Yanagisawa, Y. Takuwa, H. Miyazaki, S. Kimura, K. Goto, T. Masaki, Nature, 348, 732 (1990).
40. T. D. Warner, J. A. Mitchell, G. De Nucci, J. R. Vane, J. Cardiovasc. Pharmacol., 13 (suppl 5), S85 (1989).
41. K. Ishikawa, M. Ihara, K. Noguchi, T. Mase, N. Mino, T. Saeki, T. Fukuroda, T. Fukami, S. Ozaki, T. Nagase, M. Nishikibe, M. Yano, Proc. Natl. Acad. Sci. U.S.A., 91, 4892 (1994).
42. R. Scatena, Curr. Opin. Cardiovasc. Pulm. Renal Invest. Drugs, 1, 375 (1999).
43. J. R. Wu-Wong, Curr. Opin. Cardiovasc. Pulm. Renal Invest. Drugs, 1, 443 (1999).
44. M. Clozel, V. Breu, G. A. Gray, B. Kalina, B. M. Loffler, K. Burri, J. M. Cassal, G. Hirth, M. Muller, W. Neidhart et al: J. Pharmacol. Exp. Ther., 270, 228 (1994).
45. W. Neidhart, V. Breu, K. Burri, M. Clozel, G. Hirth, U. Klinkhammer, T. Giller, H. Ramuz, Bioorg. Med. Chem. Lett., 7, 2223 (1997).
46. S. Roux, V. Breu, T. Giller, W. Neidhart, H, Ramuz, P. Coassolo, J. R. Clozel, M. Clozel, J. Pharmacol. Exp. Ther., 283, 1110 (1997).
47   M. Clozel, H. Ramuz, J.-P. Clozel, V. Breu, P. Hess, B.-M. Loeffler, P. Coassolo, S. Roux, paper # O49, Sixth International Conference on Endothelin (ET-6), Oct. 11-13, 1999, Montreal, Canada.
48. T. Hoshino, R. Yamauchi, K. Kikkawa, H. Yabana, S. Murata, J. Pharmacol. Exp. Ther., 286, 643 (1998).
49. C. Wilson, S.-J. Hunt, E. Tang, N. Wright, E. Kelly, S. Palmer, C. Heys, S. Mellor, R. James, R. Bialecki, J. Pharmacol. Exp. Ther., 290, 1085 (1999).
50. N. Murugesan, Z. Gu, P. D. Stein, S. Bisaha, S. Spergel, R. Girotra, V. G. Lee, J. Lloyd, R. N. Misra, J. Schmidt, A. Mathur, L. Stratton, Y. F. Kelly, E. Bird, T. Waldron, E. C.-K. Liu, R. Zhang, H. Lee, R. Serafino, B. Abboa-Offei, P. Mathers, M. Giancarli, A. A. Seymour, M. L. Webb, S. Moreland, J. C. Barrish, J. T. Hunt, J. Med. Chem., 41, 5198 (1998).
51   N. Murugesan, paper # 021, Sixth International Conference on Endothelin (ET-6), Oct. 11-13, 1999, Montreal, Canada.
52. C. Wu, M. F. Chan, F. Stavros, B. Raju, I. Okun, S. Mong, K. M. Keller, T. Brock, T. P. Kogan, R. A. F. Dixon, J. Med. Chem., 40, 1690 (1997).
53. C. Wu, E. R. Decker, N. Blok, H. Bui, Q. Chen, B. Raju, A. R. Bourgoyne, V. Knowles, R. J. Biediger, R. V. Market, S. Lin, B. Dupre, T. P. Kogan, G. W. Holland, T. A. Brock, R. A. F. Dixon. J. Med. Chem., 42, 4485 (1999).
54   C. Wu, E. R. Decker, N. Blok, H. Bui, V. Knowles, A. Bourgoyne, G. W. Holland, T. A. Brock, R. A. F. Dixon, poster # 022, Sixth International Conference on Endothelin (ET-6), Oct. 11-13, 1999, Montreal, Canada.
55. M. F. Chan, A. Kois, E. J. Verner, B. G. Raju, R. S. Castillo, C. Wu, I. Okun, F. D. Stravros, V. N. Balaji, Bioorg. Med. Chem., 6, 2301 (1998).
56. M. Winn, T. W. von Geldern, T. J. Opgenorth, H.-S. Jae, A. S. Tasker, S. A. Boyd, J. A. Kester, R. Bal, B. Sorensen, J. R. Wu-Wong, W. Chiou, D. B. Dixon, E. I. Novosad, L. Hernandez, K. C. Marsh, J. Med. Chem., 39, 1039 (1996).
57. T. W. von Geldern, A. S. Tasker, B. K. Sorensen, M. Winn, B. G. Szczepankiewicz, D. B. Dixon, W. J. Chiou, L. Wang, J. L. Wessale, A. Adler, K. C. Marsh, B. Nguyen, T. J. Opgenorth, J. Med. Chem., 42, 3668 (1999).
58. G. Liu, N. S. Kozmina, M. Winn, T. W. von Geldern, W. J. Chiou, D. B. Dixon, B. Nguyen, K. C. Marsh, T. J. Opgenorth, J. Med. Chem., 42, 3679 (1999).
59. G. Liu, K. J. Henry, Jr., B. G. Szczepankiewicz, M. Winn, N. S. Kozmina, S. A. Boyd, J. Wasicak, T. W. von Geldern, J. R. Wu-Wong, W. J. Chiou, D. B. Dixon, B. Nguyen, K. C. Marsh, T. J. Opgenorth, J. Med. Chem., 41, 3261 (1998).
60. W. C. Patt, J. J. Edmunds, J. T. Repine, K. A. Berryman, B. R. Reisdorph, C. Lee, M. S. Plummer, A. Shahripour, S. J. Haleen, J. A. Keiser, M. A. Flynn, K. M. Welch, E. E. Reynolds, R. Rubin, B. Tobias, H. Hallak, A. M. Doherty, J. Med. Chem., 40, 1063 (1997).
61. W. C. Patt, X.-M. Cheng, J. T. Repine, C. Lee, B. R. Reisdorph, M. A. Massa, A. M. Doherty, K. M. Welch, J. W. Bryant, M. A. Flynn, D. M. Walker, R. L. Schroeder, S. J. Haleen, J. A. Keiser, J. Med. Chem., 42, 2162 (1999).
62. S. Anzali, W. W. K. R. Mederski, M. Osswald, D. Dorsch, Bioorg. Med. Chem. Lett., 8, 11 (1998).

63. W. W. K. R. Mederski, M. Osswald, D. Dorsch, S. Anzali, M. Christadler, C.-J. Schmitges, C. Wilm, Bioorg. Med. Chem. Lett., 8, 17 (1998).

64. W. W. K. R. Mederski, D. Dorsch, M. Osswald, S. Anzali, M. Christadler, C.-J. Schmitges, P. Schelling, C. Wilm, M. Fluck, Bioorg. Med. Chem. Lett., 8, 1771 (1998).

65. E. H. Ohlstein, P. Nambi, A. Lago, D. W. Hay, G. Beck, K. L. Fong, E. P. Eddy, P. Smith, H. Ellens, J. D. Elliott, J. Pharmacol. Exp. Ther., 276, 609 (1996).

66. M. Nishikibe, H. Ohta, M. Okada, K. Ishikawa, T. Hayama, T. Fukuroda, K. Noguchi, M. Saito, T. Kanoh, S. Ozaki, T. Kamei, K. Hara, D. William, S. Kivlighn, S. Krause, R. Gabel, G. Zingaro, N. Nolan, J. O'Brien, F. Clayton, J. Lynch, D. Pettibone, P. Siegl, J. Pharmacol. Exp. Ther., 289, 1262 (1999).

67. H. Riechers, H.-P. Albrecht, W. Amberg, E. Baumann, H. Bernard, H.-J. Böhm, D. Klinge, A. Kling, S. Müller, M. Raschack, L. Unger, N. Walker, W. Wernet, J. Med. Chem., 39, 2123 (1996).

68. W. Amberg, S. Hergenröder, H. Hillen, R. Jansen, G. Kettschau, A. Ling, D. Klinge, M. Raschack, H. Riechers, L. Unger, J. Med. Chem., 42, 3026 (1999).

69. T. Kikuchi, T. Ohtaki, A. Kawata, T. Imada, T. Asami, Y. Masuda, T. Sugo, K. Kusumoto, K. Kubo, T. Watanabe, M. Wakimasu, M. Fujino, Biochem. Biophys. Res. Commun., 200, 1708, (1994).

70. N. Cho, Y. Nara, M. Harada, T. Sugo, Y. Masuda, A. Abe, K.Kusumoto, Y. Itoh, T. Ohtaki, T. Watanabe, S. Furuya, Chem. Pharm. Bull., 46, 1724 (1998).

71. E. L. Ohlstein, P. Nambi, D. W. P. Hay, M. Gellai, D. P. Brooks, J. Luengo, J.-N. Xiang, J. D. Elliott, J. Pharmacol. Exp. Ther., 286, 650 (1998).

72. S. A. Douglas, P. Nambi, M. Gellai, J. I. Luengo, J.-N. Xiang, D. P. Brooks, R. R. Ruffolo, Jr. J. D. Elliott, E. H. Ohlstein, J. Cardiovasc. Pharmacol., 31(Suppl. 1), S273 (1998).

73. H. Azuma, J. Sato, H. Masuda, M. Goto, S. Tamaoki, A. Sugimoto, H. Hamasaki, H. Yamashita, Jpn. J. Pharmacol., 81, 21 (1999).

74. T. Iwasaki, S. Mihara, T. Shimamura, M. Kawakami, Y. Hayasaki-Kajiwara, N. Naya, M. Fujimoto, M. Nakajima, J. Cardiovasc. Pharmacol., 34, 139 (1999).

75. P. C. Astles, T. J. Brown, F. Halley, C. M. Handscombe, N. V. Harris, C. McCarthy, I. M. McLay, P. Lockey, T. Majid, B. Porter, A. G. Roach, C. Smith, R. Walsh, J. Med. Chem., 41, 2745 (1998).

76. P. C. Astles, T. J. Brown, F. Halley, C. M. Handscombe, N. V. Harris, T. Majid, C. McCarthy, I. M. McLay, A. Morley, B. Porter, A. G. Roach, C. Sargent, C. Smith, R. J. A. Walsh, J. Med. Chem., 43, 900 (2000).

77. K. A. Berryman, J. J. Edmunds, A. M. Bunker, S. Haleen, J. Bryant, K. M. Welch, A. M. Doherty, Bioorg. Med. Chem., 6, 1447 (1998).

78 Y. Coe, S. J. Haleen, K. M. Welch, F. Coceani, paper # 188, Sixth International Conference on Endothelin (ET-6), Oct. 11-13, 1999, Montreal, Canada.

79. W. W. K. R. Mederski, M. Osswald, D. Dorsch, M. Christadler, C.-J. Schmitges, C. Wilm, Bioorg. Med. Chem. Lett., 9, 619 (1999).

80. Th. Früh, H. Saika, L, Svensson, Th. Pitterna, J. Sakaki, T. Okada, Y. Urade, K. Oda, Y. Fujitani, M. Takimoto, T. Yamamura, T. Inui, M. Makatani, M. Takai, I. Umemura, N. Teno, H. Toh, K. Hayakawa, T. Murata, Bioorg. Med. Chem. Lett., 6, 2323 (1996).

81. J. Sakaki, T. Murata, Y. Yuumoto, I. Nakamura, K. Hayakawa, Bioorg. Med. Chem. Lett., 8, 2247 (1998).

Chapter 9. Factor Xa Inhibitors: Recent Advances In Anticoagulant Agents

Bing-Yan Zhu and Robert M. Scarborough

COR Therapeutics, Inc., 256 East Grand Ave., South San Francisco, CA 94080

Introduction - Factor Xa is a trypsin-like serine protease, which plays a pivotal role in the blood coagulation cascade. Factor Xa is at the final convergence point of both intrinsic and extrinsic coagulation pathways. It forms a prothrombinase complex with factor Va, Ca^{2+} and phospholipid. Factor Xa is the active enzyme in the prothrombinase complex that converts inactive prothrombin to active thrombin, which ultimately leads to blood clotting. Factor Xa has emerged as a very attractive target to develop orally active antithrombotic agents to replace warfarin. Potent and specific factor Xa inhibitors (naturally occurring protein inhibitors such as r-TAP and r-ATS, and small molecule inhibitors such as DX-9065a, YM-60828, SN-429, and RPR-120844) are quite efficacious in several animal models of thrombosis. These factor Xa inhibitors only prolong bleeding time or increase blood loss at much higher doses than fully efficacious doses, while warfarin significantly increases bleeding time and blood loss at the effective doses.

Several comprehensive review articles on factor Xa and thrombin inhibitors have already been published in 1999 (1-4). This chapter will focus on the most recent advances reported in 1999 in the design and discovery of novel factor Xa inhibitors as potential orally active anticoagulants. A number of series of novel and structurally simple non-benzamidine factor Xa inhibitors have been discovered which inhibit factor Xa and factor Xa in the prothrombinase complex with low nM potency. These inhibitors do not inhibit any other trypsin-like serine proteinases such as thrombin, trypsin, APC, plasmin, t-PA, urokinase, and kallikrein. Some of the non-benzamidine factor Xa inhibitors show the desired pharmacokinetic properties and have antithrombotic effects in animal thrombosis models as well. Several new molecules such as ZD-4927 and DPC-423 (structures not disclosed) have advanced into clinical trials as oral agents. ZK-807834 (CI-1031) is under clinical investigation as an intravenous agent.

Diamidino and Dibasic Factor Xa Inhibitors - From a drug discovery and development perspective, diamidino based factor Xa inhibitors such as **1** (DX-9065a) and **2** (YM-60828) continue to be important tools for understanding the biological roles of factor Xa (1). These dibasic inhibitors also proved the principle that small molecule factor Xa inhibitors could be promising anticoagulant agents with much better efficacy/safety index relative to that of warfarin (1). Equally important, both inhibitors **1** and **2** also serve as good structural templates to design more potent and specific factor Xa inhibitors with improved pharmacokinetic and pharmacodynamic profiles. The X-ray structure of compound **1** complexed with factor Xa also provided the structural basis for the design of other novel inhibitors of factor Xa (1). Inhibitors **1** and **2** also paved the way to understand and identify the key issues (physicochemical and biochemical properties) required to identify not only *in vitro* potent factor Xa inhibitors but also potent *in vivo* antithrombotic agents. Compound **1** has a K_i = 41 nM and IC_{50} = 70 nM against free factor Xa. Compound **2** inhibits free factor Xa with a K_i = 1.3 nM and factor Xa in the prothrombinase complex with an IC_{50} = 7.7 nM. The antithrombotic effects of compounds **1** and **2** have been extensively studied in several animal thrombosis models (1). Particularly, compound **2** has shown strong antithrombotic effects *in vivo* due to the overall favorable profiles of excellent inhibitory activity against factor Xa and factor Xa in the prothrombinase complex, very good aqueous solubility and the associated low plasma protein binding. It has been correlated in the YM-60828 series that more lipophilic compounds generally show less anticoagulant activities due to decreased aqueous solubility and the associated higher plasma protein binding (1). More importantly, factor Xa inhibitors **1** and **2** do not cause bleeding complications or significant blood loss at the effective doses (1). The major drawback of these diamidino-based factor Xa inhibitors is their low oral bioavailability (1). Interestingly, Compound **1** has a long $t_{1/2}$ (> 20 h) in man, and it has low total body clearance (120 mL/min) and low plasma protein binding (ca. 60%) in human plasma (5).

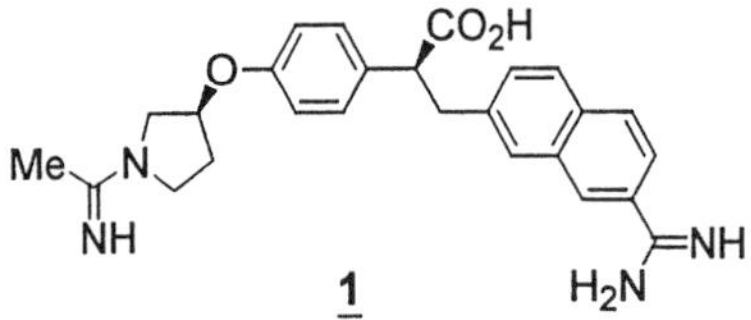

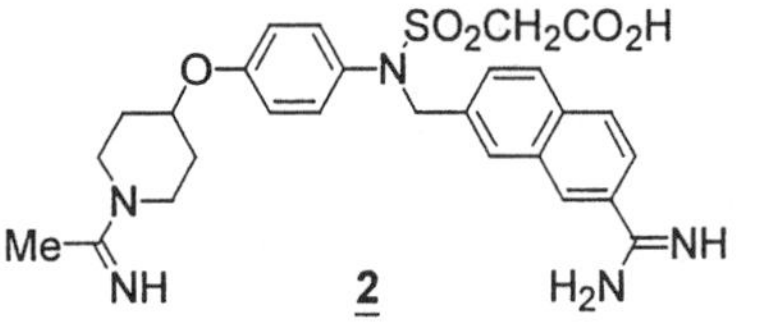

Several series of diamidino factor Xa inhibitors, which share a similar pharmacophore with YM-60828, were reported during 1999. Compound **3** exemplifies the replacement of S4-binding piperidinyloxy unit with hexahydro-1,4-diazepine ring (6). The P1 naphthylamidine can be replaced with fused heterobicyclic-arylamidines as exemplified by compound **4** which has an IC_{50} of 10 nM as a factor Xa inhibitor (7). A biphenylamidine unit was also found to be good replacement of the naphthylamidine in the YM-60828 series as represented by **5** (IC_{50} = 4 nM) (8). Compound **6** incorporates the P1 naphthylamidine, which is linked to the P4 moiety through an amide bond (9).

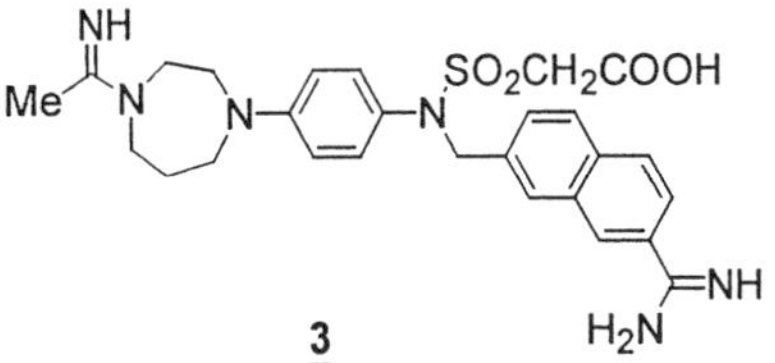

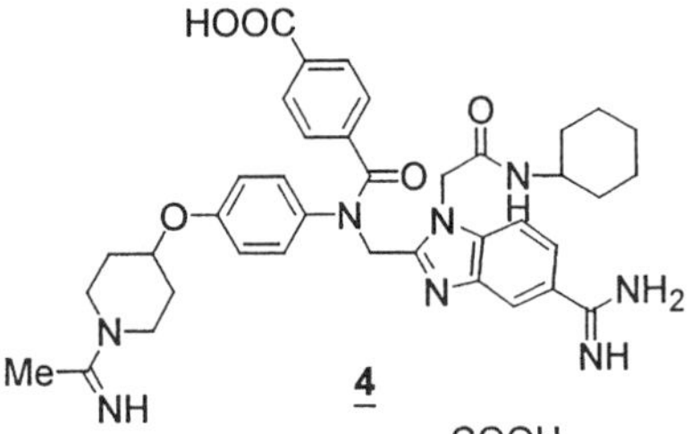

Several series of fused heterobicyclic templates which link the P1 naphthylamidine with P4 methyl iminomethylpiperidinyloxy were reported (10, 11). Compounds **7** - **10** have strong anti-factor Xa inhibitory potency with IC_{50} = 6 nM, 6 nM, 6 nM and 1 nM, respectively (10). Compounds **7** and **9** also showed strong *in vivo* antithrombotic effects in a rabbit venous thrombosis model.

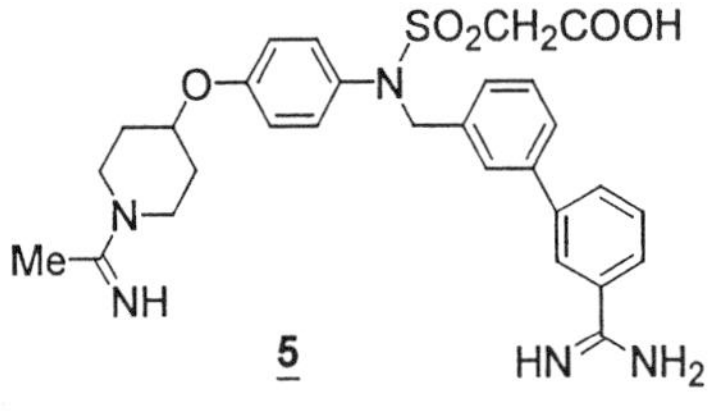

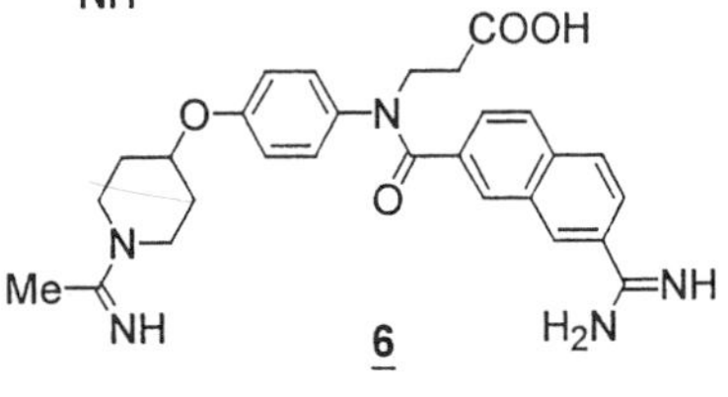

Other dibasic compounds include the pentapeptide **11** (SEL-2711, K_i = 3 nM) and **12** (ZK-807834, CI-1031, K_i = 0.11 nM) (1). Both inhibitors are remarkably potent and specific for factor Xa and factor Xa in the prothrombinase complex, and are highly potent in animal thrombosis models (1). The observed *in vivo* antithrombotic effects are attributed to their superior potency against factor Xa in the prothrombinase complex (IC_{50} = 3 nM and 6 nM for **11** and **12**, respectively), their high aqueous solubility and low plasma protein binding. Interestingly, when compound **11** was crystallized with thrombin, it was found to bind to the active site of thrombin in a retro-binding fashion (12). The benzamidine inserts into the S1 pocket and the *N*-methyl pyridine ring binds in the aryl-binding S4 pocket. A retro-binding orientation of **11** has been proposed to interact with factor Xa (12). ZK-807834 is currently in phase II clinical trial as an intravenous anticoagulant agent. It should be pointed out that the basic P4 *N*-methyl imidazoline ring in the ZK-807834 series affords compounds with much more favorable pharmacokinetic profiles relative to neutral P4 Me2NCO- analogs. This novel P4 *N*-Me-imidazoline-phenyl element might provide an alternative to increase the aqueous solubility while retaining the *in vitro* potency as well as pharmacokinetic properties of these inhibitors. Several benzamidine and non-benzamidine factor Xa inhibitors containing this novel P4 element have been disclosed (discussed later).

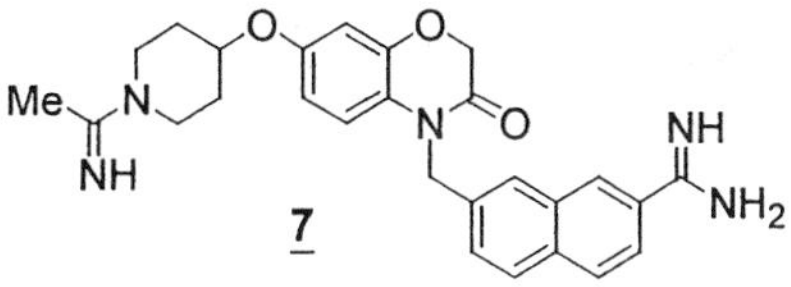

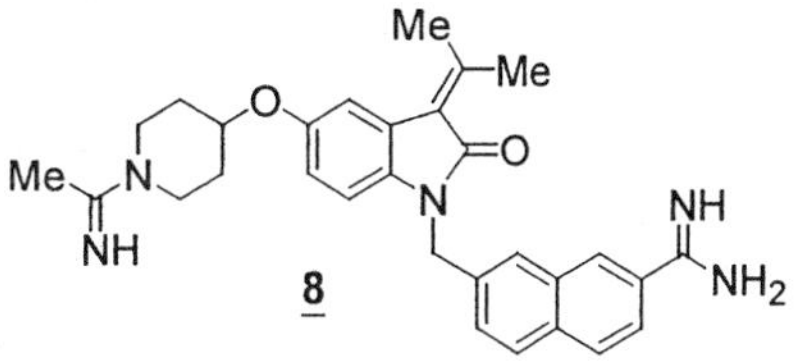

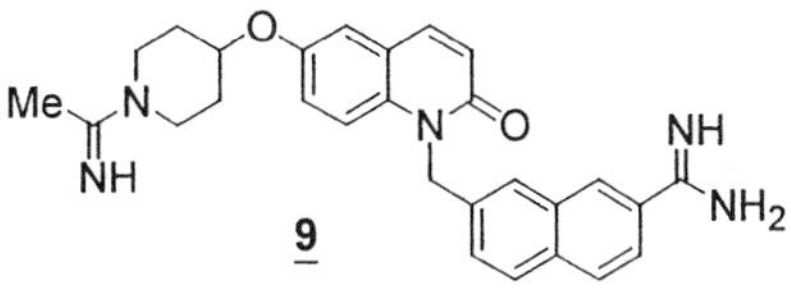

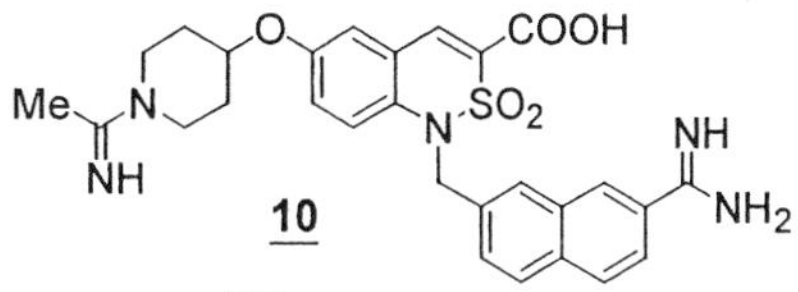

<u>Mono-Benzamidine and Non-Benzamidine Factor Xa Inhibitors</u> - Due to the fact that diamidino based factor Xa inhibitors in general have low oral bioavailability and that the amidino group [-C(=NH)-NH2] is likely one major contributor to the low oral bioavailability due to its high basicity, extensive research has been focused on the discovery of mono-benzamidine based factor Xa inhibitors. Incorporation of a – CO_2H group into mono-benzamidine molecules tends to increase slightly their oral bioavailability and duration of action as documented in factor Xa inhibitor and thrombin inhibitor fields (4). Other goals for the discovery of mono-benzamidine factor Xa inhibitors have been: 1. to utilize amidine pro-drug approaches for promising benzamidine leads to improve oral bioavailability; 2. to incorporate less basic benzamidine surrogates (such as 1-aminoisoquinoline) into potent mono-benzamidine leads; and 3. to utilize the highly potent mono-benzamidine molecules as structural leads to design potent non-benzamidine factor Xa inhibitors. However, it is still quite a challenge to discover mono-benzamidine factor Xa inhibitors with high oral bioavailability (including pro-drugs), long $t_{1/2}$, low clearance, and desired *in vivo* efficacy. Nevertheless, mono-benzamidine factor Xa inhibitors have been shown to display strong *in vivo* antithrombotic efficacy, provided that such inhibitors are highly potent *in vitro*, have relatively high aqueous solubility and low plasma protein binding. Mono-benzamidine factor Xa inhibitors also displayed a much wider safety window following intravenous administration.

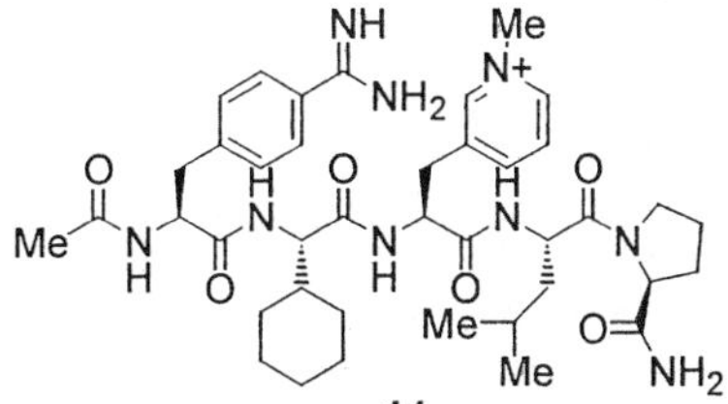

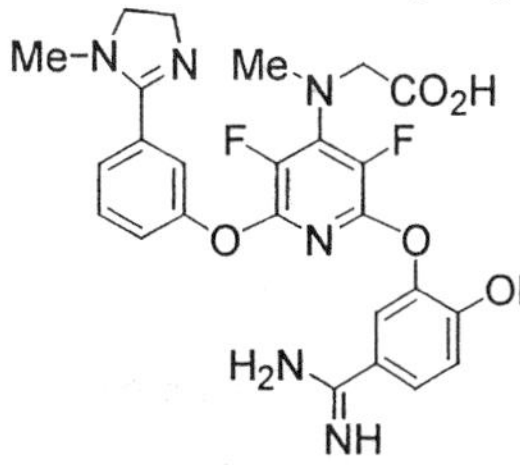

A series of 3-sulfoamido-pyrrolidinones have been found to be a useful template for the presentation of factor Xa S1 and S4-subsites-binding ligands. Optimization of the P1 and P4 moieties has resulted in several relatively potent and selective factor Xa inhibitors (13-15). Compound **13a** (RPR-120844) with a thienylamidine unit as P1 inhibits free factor Xa with a $K_i = 7$ nM. It is also a potent inhibitor of factor Xa bound in the prothrombinase complex with an $IC_{50} = 73$ nM to 141 nM. Compound **13b** (RPR-130492) containing a hydroxyl group in the P1 benzamidine ring has a $K_i = 3$ nM. In general, addition of a hydroxyl group *para* to the amidine group in P1 benzamidine ring increases the *in vitro* potency. A biphenyl or bi-heteroaryl group such as a 3-pyridinylthiophene ring system can substitute for the P4 naphthalene ring, as exemplified by compound **14** (RPR-130737, $K_i = 2$ nM). Compound **13a** has been studied in several thrombosis models (16, 17). In a rabbit venous thrombosis model, it inhibits thrombosis formation by 45%, 60% and 72% with doses of 10, 30, and 100 μg/kg/min (plasma concentrations are about 700 nM, 1 μM and 2 μM, respectively). The *in vivo* IC_{50} value for **13a** is about 100 fold the K_i value *in vitro*. However, these benzamidines have low oral bioavailability in dogs and very short $t_{1/2}$ (compound **13a** has $t_{1/2}$ values of 20 min. and 15 min. in rats and dogs, respectively).

| **13a** | **13b** | **14** |

In order to improve the oral bioavailability and $t_{1/2}$ of the benzamidine factor Xa inhibitors, direct replacement of highly basic benzamidine with less basic surrogates such as 1-aminoisoquinoline has been extensively explored in both the thrombin and factor Xa inhibitor fields. Fortunately, direct replacement of the *meta*-benzamidine ring with a 1-aminoisoquinoline unit only slightly reduced the *in vitro* potency and actually increased specificity for factor Xa within the γ-lactam series. Systematic optimization of the P1 and P4 elements lead to a single-digit nanomolar inhibitor of factor Xa with excellent selectivity in the 3-sulfoamido-pyrrolidinone series (18-20). For example, compound **15** has a $K_i = 180$ nM while the corresponding benzamidine compound has a $K_i = 47$ nM. Addition of an H-bonding amino group to the isoquinoline ring at the 6-position increased the *in vitro* potency by 2-fold, leading to inhibitor **16** with a $K_i = 80$ nM. Optimization of the P4 elements by utilizing fused heterobicyclic rings such as thienopyridine leads to more potent factor Xa inhibitors (**17** - **20**) with improved aqueous solubility. Compound **17** (RPR-208815, $K_i = 22$ nM) is equipotent to its benzamidine counterpart with oral bioavailability of 33% in dogs, compared to negligible oral bioavailability for the similar benzamidine analog. Again, addition of an H-bonding amino group at the 6-position of the isoquinoline ring in compound **17** increased potency, leading to **18** with a $K_i = 6$ nM. Compounds **19** and **20** have K_i values of 140 nM and 167 nM, respectively. These data clearly demonstrated that potent benzamidine based factor Xa inhibitors could be converted into aminoisoquinolines and other heterobicyclic surrogates with only slightly reduced *in vitro* potency, but with significantly increased oral bioavailability and specificity. Nevertheless, *the in vivo* efficacy of these benzamidine surrogate analogs has not been reported and it is not known if these inhibitors have the desired antithrombotic effects. However, the 1-aminoisoquinoline ring system may serve as a general surrogate for the benzamidine ring. Several series of factor Xa inhibitors containing the 1-aminoisoquinoline ring systems as P1 *meta*-benzamidine surrogates have been disclosed and will be discussed later.

15 R = H
16 R = NH$_2$

17 X = N, R$_1$, R$_2$ = H
18 X = N, R$_1$ = H, R$_2$ = NH$_2$
19 X = CH, R$_1$, R$_2$ = H
20 X = CH, R$_1$ = Cl, R$_2$ = H

The X-ray structure of **17** (RPR-208815), complexed with factor Xa has revealed unique binding interactions (21). The 1-amino-isoquinoline ring is inserted into the S1 specificity pocket. The N atom of the isoquinoline ring is engaged in a water mediated H-bond with the acidic side chain of Asp-189 of factor Xa. The 1-NH$_2$ group of the isoquinoline ring forms an H-bond with C=O of Gly-218 of factor Xa rather than a twin-twin interaction with the side chain of Asp-189 of factor Xa. The thienopyridine ring fits well in the S4 pocket. The N atom of the thienopyridine ring is exposed to the solvent.

An extensive search for P1 *meta*-benzamidine surrogates other than 1-aminoisoquinoline in the pyrrolidinone-arylsulfonamide series lead to the discovery of fused (5,6)-aromatic heterobicycles, notablely, azaindoles (pyrrolo-pyridines) as novel P1 elements (22). The aminothienopyridine **21**, a direct analog of **15**, has a K$_i$ = 250 nM. Azaindoles **22** and **23** have K$_i$ = 110 nM and 330 nM, respectively. When the N atom of the P1 pyridine ring was moved to the 5-position (5-azaindole), similar *in vitro* potency was observed (**24**, K$_i$ = 100 nM). When a bi-aryl ring (such as bithienyl) was utilized as P4 element instead of fused bicycles, potency was improved by 10-fold (**25**, K$_i$ = 10 nM). When the N atom of the P1 pyridine was placed in the 6-position (6-azaindole) of the azaindole ring, factor Xa inhibitors with relatively high potency were achieved as exemplified by **26** (RPR-208707, K$_i$ = 18 nM). The X-ray structure of **26** complexed with factor Xa has revealed very unique interactions (21), which may contribute to its high potency. 6-Azaindole ring was found to reside in the S1 pocket but does not form a salt bridge with the side chain of Asp-189, which is typical for benzamidines. Instead, the pyridine N atom forms a water mediated H-bond with the side chain of Asp-189 and the pyrrolo N-H atom is H-bond to the C=O of Gly-218. The thienopyridine ring fits tightly in the S4 pocket with a high degree of aromatic interactions. The P4 pyridine N is facing out of the S4 pocket towards solvent. It can be seen that multiple H-bond interactions are critical for the high affinity towards factor Xa. These non-benzamidine molecules also displayed improved specificity for factor Xa relative to the benzamidine counterparts. Although the azaindoles were discovered as very novel benzamidine surrogates in this series, these derivatives have only moderate *in vitro* potency. Neither pharmacokinetic profiles nor *in vivo* antithrombotic effects have been disclosed for these azaindole derivatives. However, azaindoles as novel P1 elements in the γ-lactam series have been incorporated into more promising piperazinone series (but as P4 elements), which will be discussed later.

A series of P4 bi-phenyl- and related-carboxamide derivatives with *meta*-benzamidine as P1 were discovered as potent and specific inhibitors of factor Xa. Optimization of β-aminoester side chains and P4 elements lead to highly potent benzamidine based factor Xa inhibitors. Systematic SAR shows inhibitors represented by compounds **27** are optimal in terms of inhibitory activity (23). The biphenyl group substituted either by positively charged moieties (CH$_2$NH$_2$, or CH$_2$NMe$_2$) or H-bonding groups such as NO$_2$ and CONH$_2$ are the preferred P4 elements since these P4 elements can have optimal hydrophobic interactions and additional ionic/H-bond interactions with the unique S4 pocket of factor Xa. For example, when the distal P4 phenyl ring is substituted by H-bonding moieties such as NO_2 or $CONH_2$ group, the potency of **27b**, **27c** and **27d** was increased relative to **27a**. Nevertheless, compound **27d** suffered from poor aqueous solubility. Compounds **27e** - **27g** are highly potent *in vitro* with good aqueous solubility. The hydrophobic side chains at the β-position of the β-amino ester were found to increase *in vitro* potency (Me, Et, or Bn groups are preferred). The *in vivo* antithrombotic activity has been evaluated for this series of compounds in a rat arterial thrombosis model (effective at the reducing the size of thrombus mass and increased occlusion time with minimal effects on bleeding times). The X-ray structure of **27e** (RPR-128515) has revealed that the amidine group forms bidentate interactions with the side chain of Asp-189 and additional H-bonds with C=O and NH of Gly-218 of factor Xa (21). The aminomethylbiphenyl group of **27e** was found to be in the S4 pocket. The CH_2NH_2 group in the P4 biphenyl group forms a salt bridge with the Glu-97 side chain of factor Xa.

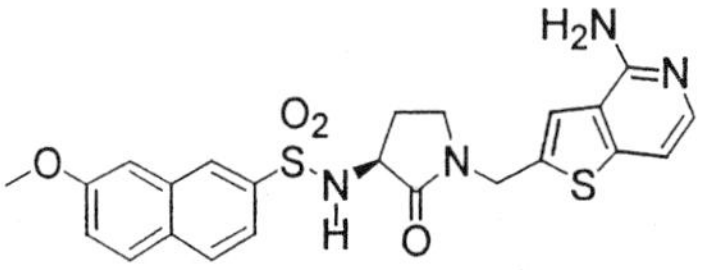

21

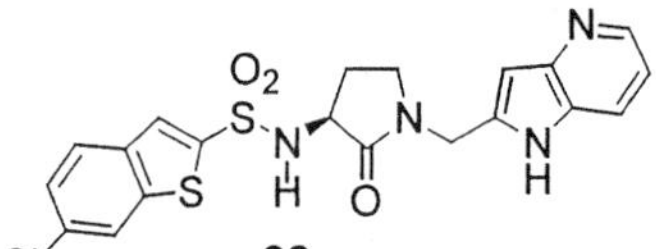

22

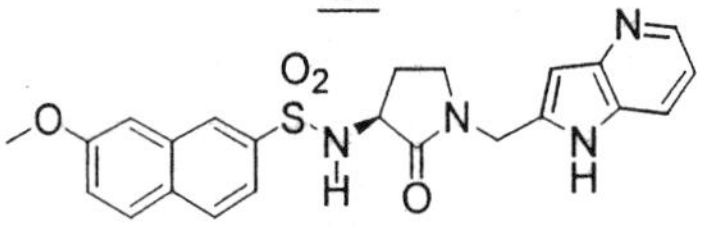

23

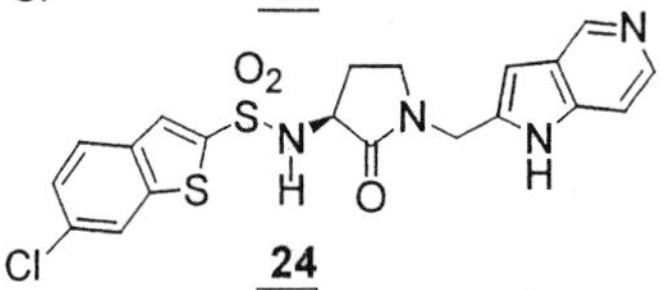

24

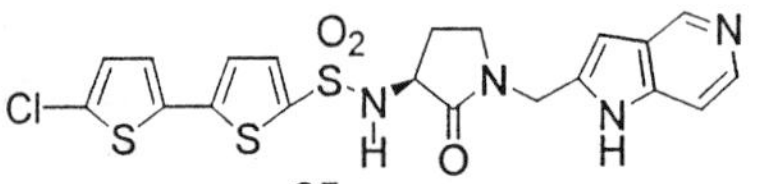

25

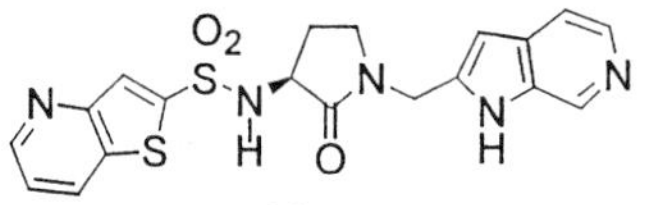

26

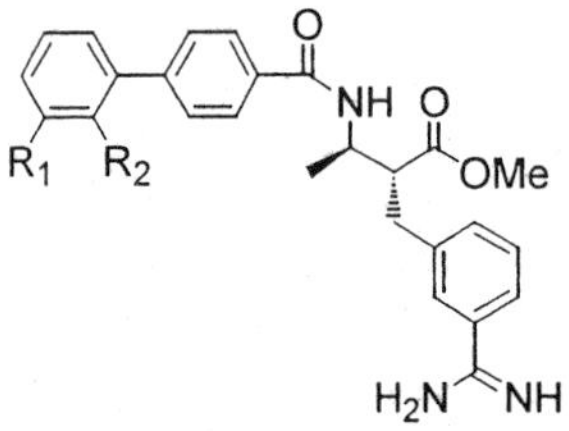

27a, R_1, R_2 = H, K_i = 5.3 nM
27b, R_1 = H, R_2 = NO_2, K_i = 1.1 nM
27c, R_1 = NO_2, R_2 = H, K_i = 1.1 nM
27d, R_1 = $CONH_2$, R_2 = H, K_i = 0.5 nM
27e, R_1 = CH_2NH_2, R_2 = H, K_i = 0.9 nM
27f, R_1 = CH_2NMe_2, R_2 = H, K_i = 1.0 nM
27g, R_1 = $CH_2N^+Me_3$, R_2 = H, K_i = 2.0 nM

Achiral analogs of inhibitors **27** have been developed and sub-nM potency was achieved by combining optimized P4 elements and addition of a hydroxy group to the P1 benzamidine ring (at the position *para* to the amidine group) (24). These analogs, as exemplified by **28** and **29**, are more specific for factor Xa than compounds **27**, probably due to the hydroxy-group effect. Interestingly, compound **29** (K_i = 0.75 nM) is almost 10 times more potent than compound **28** (K_i = 5 nM), indicating that the P4 pyridine N atom might form additional H-bonds within the S4 pocket of factor Xa. These data clearly point out that in addition to the hydrophobic interactions, multiple H-bonds/ionic interactions can occur in the S4 pocket of factor Xa for improving potency and specificity. The novel 4-[4-(2-substituted)pyridinyl]-phenyl P4 element in compound **29** is extensively utilized in the piperazinesulfonamide series, which will be discussed later. The $CONH_2$ group in the distal P4 phenyl ring can be replaced with the SO_2NH_2 group while retaining *in* *vitro* potency. Selected members from this series were found to be effective in animal models of thrombosis upon intravenous dosing.

Replacement of the P1 *meta*-benzamidine with an *N*-amidino-piperidine ring (cyclic guanidine) in the biphenylcarboxamide series was also disclosed in the patent literature (25). Compounds **30** and **31** exemplify the preferred compounds. The cyclized guanidine moieties were also utilized in the thrombin inhibitor field (4). However, biological activities of these compounds have not been disclosed.

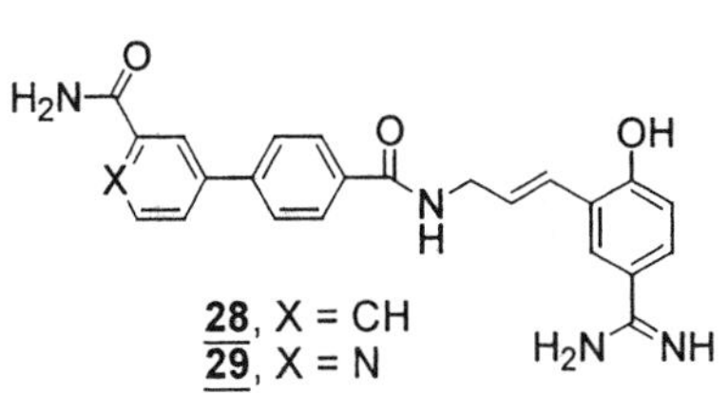

28, X = CH
29, X = N

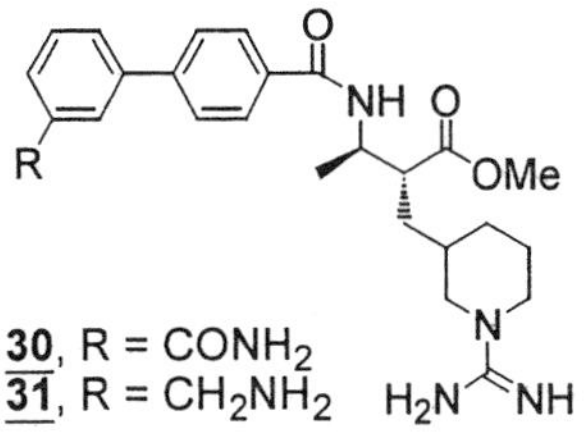

30, R = CONH$_2$
31, R = CH$_2$NH$_2$

A related series of benzamidine derivatives as factor Xa inhibitors with structural similarity to **27** - **29** was disclosed in patent applications in which a P1 3-phenoxyamidine is utilized (1, 26). These molecules, which are structurally simplified relative to **27**- **29**, also inhibit factor Xa with low nM potency. As demonstrated by the patent applications, very diverse P4 elements can be utilized to give strong inhibitory potency against factor Xa. Compounds **32a** through **32f** highlight the diversity of P4 elements, and some *in vitro* SAR. The additional R groups include aromatic and non-aromatic 5 or 6 -membered ring systems (such as phenyl, 4-pyridyl, 1-piperdinyl, and imidazolyl rings), amidines, guanidines, cyclic amidines and cyclic guanidine moieties.

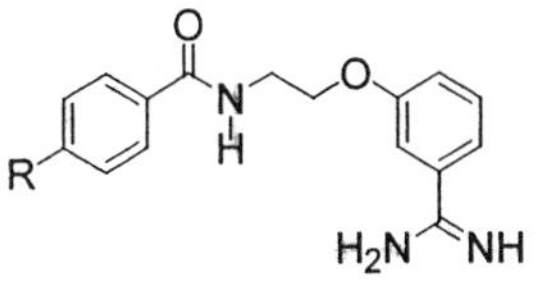

32a, R = NH$_2$-C(=NH)-, IC$_{50}$ = 316 nM
32b, R = Me$_2$N-C(=NH)-, IC$_{50}$ = 79 nM
32c, R = Me$_2$NCO-, IC$_{50}$ = 25 nM

32d, R =   IC$_{50}$ = 19 nM

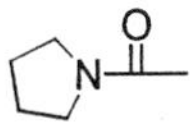

32e, R =   IC$_{50}$ = 251 nM

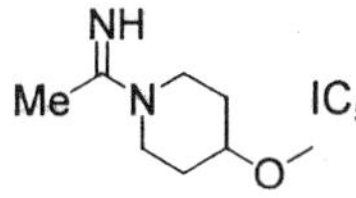

32f, R =   IC$_{50}$ = 50 nM

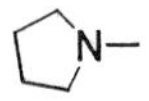

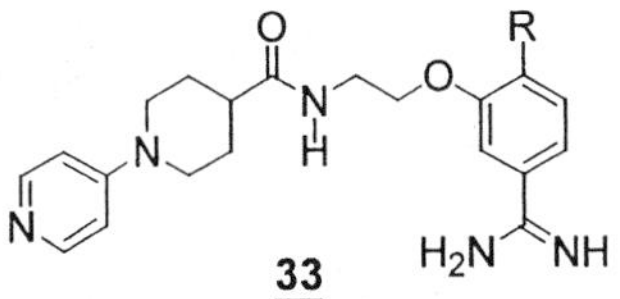

33a, R = H, IC$_{50}$ = 316 nM
33b, R = OH, IC$_{50}$ = 10 nM
33c, R = CH$_2$OH, IC$_{50}$ = 50 nM
33d, R = (CH$_2$)$_3$OH , IC$_{50}$ = 16 nM
33e, R = CH$_2$CH$_2$COOH, IC$_{50}$ = 40 nM
33f, R = CH$_2$CH$_2$COOEt, IC$_{50}$ = 63 nM
33g, R = CH$_2$CH$_2$SO$_3$H, IC$_{50}$ = 63 nM

33

Addition of H-bonding groups to the P1 benzamidine ring (*para* to the amidine group) in compounds **32** increases their *in vitro* potency against factor Xa, as illustrated by compounds **33a** – **33g** (26). It seems that different lengths linking these H-bonding groups to the P1 phenyl ring are well tolerated and varieties of H-bonding groups are accepted as well. These H-bonding groups can form H-bonds with Ser-195, and/or the catalytic triad in the S1 specificity pocket of factor Xa. Combination of the potency-enhancing elements leads to inhibitors **34** and **35** with IC$_{50}$ values of 8 nM and 3 nM, respectively (26). The 1-aminoisoquinoline derivatives **36a**, **36b**, and **37** of the benzamidine leads showed promising *in vitro* potency as factor Xa inhibitors (IC_{50} = 251 nM, 79 nM and 25 nM, respectively) (27).

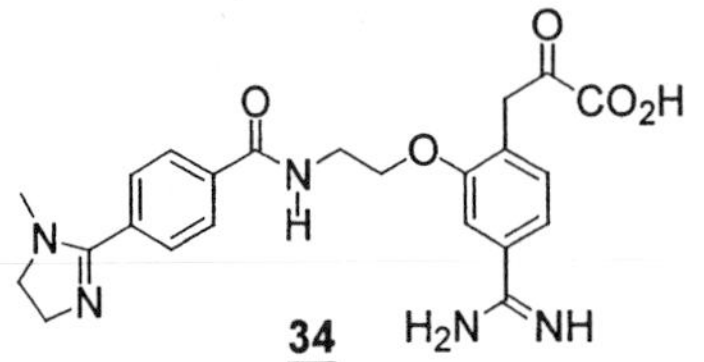

34 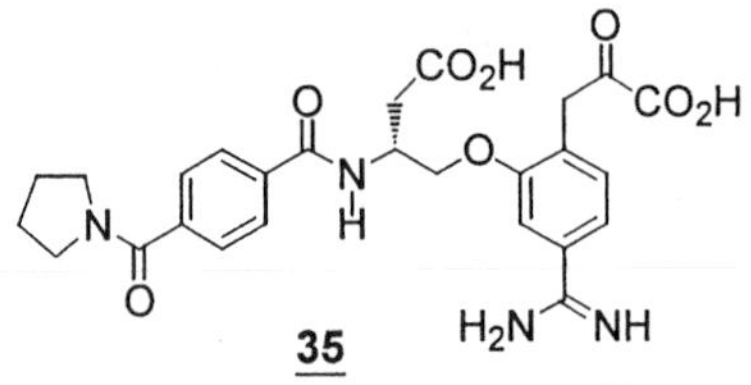   **35**

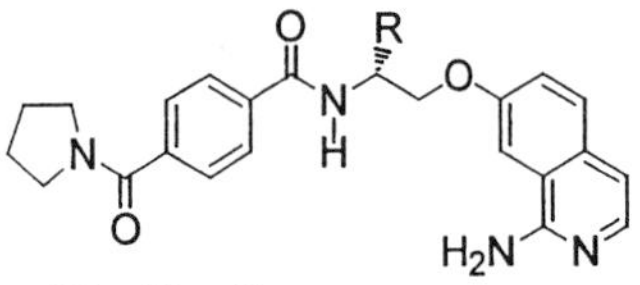

36a, R = H
36b, R= CH_2CH_2COOH **37**

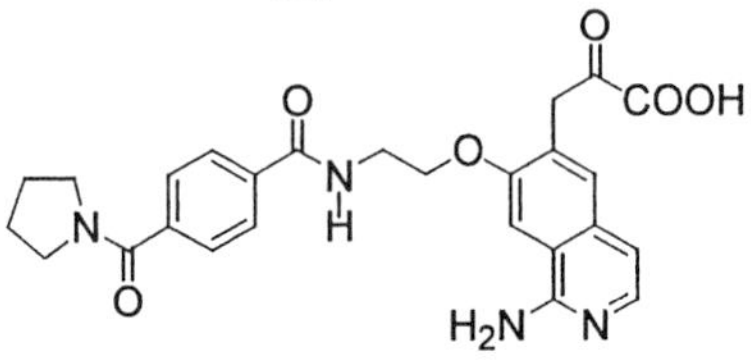

   Biphenylsulfonamides and biphenylsulfones as S4 binding motifs have been extensively utilized to link P1 benzamidine through various linkers (28-33). Compound **38** (SN-429) with a K_i = 10 pM is an excellent example.   The 2-sulfonamide (SO_2NH_2) and sulfone (SO_2Me) in the distal phenyl ring of the P4 biphenyl ring is critical for potency probably due to the H-bond interactions with the side chain of Glu-97 of factor Xa.   The –NHCO- linkage is also critical since *N*-methylation [-N(Me)CO-] of **38** was detrimental and reduced the *in vitro* potency 1000-fold to K_i = 11 nM. When the amide bond (-NHCO-) is reversed to (–CONH-) for **38**, the potency was decreased by 290-fold to K_i = 2.9 nM. The 3-Me group of the pyrazole ring in **38** can be replaced with CF_3 (preferred), H (K_i = 0.16 nM), SO_2Me (K_i = 10 pM) and CH_2NHSO_2Me (K_i = 50 pM). The proximal phenyl ring of the P4 biphenyl group can be substituted with F (preferred) or Cl atom. Additionally, the proximal phenyl ring can be replaced with pyridine or pyrimidine rings as shown by **39a** (K_i = 6 pM) and **39b** (SN-116, K_i = 40 pM).

   As summarized in Table 1, compound **38** is highly potent *in vivo* as an antithrombotic agent in a rabbit AVST model with ID_{50} = 0.023 µM/kg/h (*in vivo* IC_{50} = 23 nM, which is 2000 fold the K_i value *in vitro*). In the same thrombosis model, ID_{50} for DX-9065a and *r*-TAP are 0.4 and 0.009 µM/kg/h, respectively. Although compound **38** is only 10 times less potent by ID_{50} values than *r*-TAP *in vivo*, it is only 17 times more potent than DX-9065a *in vivo* despite being more than 4000 times more potent than DX-9065a *in vitro*. This may be due to higher clearance and/or higher plasma protein binding of SN-429 *in vivo* than DX-9065a. Additionally, relatively high free (unbound) inhibitor concentrations in the plasma may be required to effectively inhibit thrombosis formation due to the explosive nature of coagulation cascade, such that the inhibitor can bind to factor Xa fast enough to block the triggered coagulation cascade.   Even with the low pM *in vitro* potency of the inhibitors to factor Xa, higher concentrations may still be required to offset the diffusion rate. It has been observed that in general, the IC_{50} values *in vivo* are more than 100-fold the corresponding IC_{50} values *in vitro*, excluding diamidino factor Xa inhibitors which have very low plasma protein binding. Compounds **39a** and **39b** are also highly potent *in vivo* as antithrombotic agents with ID_{50} = 0.07 and 0.034 µM/kg/hr, respectively.   Compound **39b** has a longer plasma $t_{1/2}$ and lower clearance than **39a** (Table 1).   A number of aromatic and non-aromatic five membered rings as central templates have also been disclosed. Compounds **40** - **43** are typical examples. As summarized in Table 1, Compounds **40** - **43** are highly potent inhibitors of factor Xa *in vitro* and potent *in vivo* in rabbit AVST thrombosis models. The *in vivo* IC_{50} values (the plasma concentrations of inhibitors which reduce thrombosis weight by 50% in rabbit AVST models) for some of the compounds were also disclosed. The *in vivo* IC_{50} values = 1220 nM, 440 nM and 62 nM for **42a** (SF-303, R = CO_2Me, X, Y = CH), **42b** [(-)-SF-324, X, Y = N, R = CO_2Me], and **42c** [(-)-SK-549, X, Y = CH, R = 1-tetrazole], respectively. For **42d** (SM-084), R = 1-tetrazole, X = N, and Y = CH. However, mono-benzamidine inhibitors such as SN-429 in general suffer from low oral bioavailability (3% in dogs for SN-429) and short plasma $t_{1/2}$ (Table 1), and show only moderate selectivity between factor Xa and other trypsin-like serine proteases.

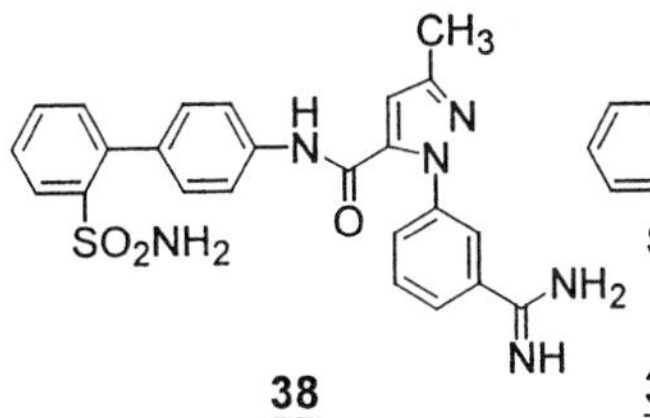

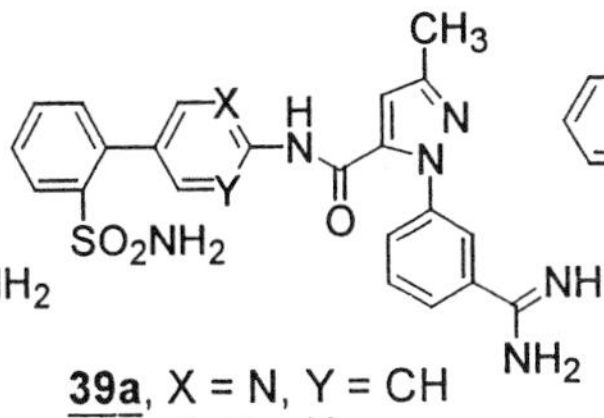

Table 1. Summary of the *In Vitro* and *In Vivo* AVST Data of Compound **38** Series

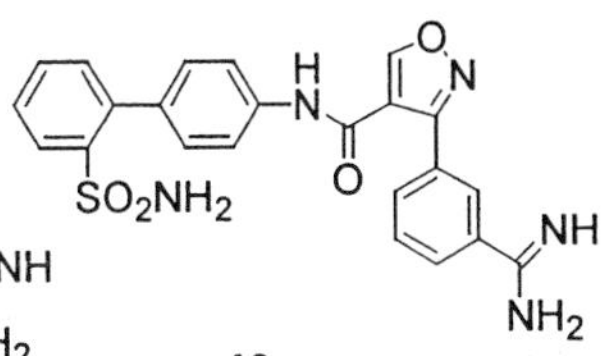

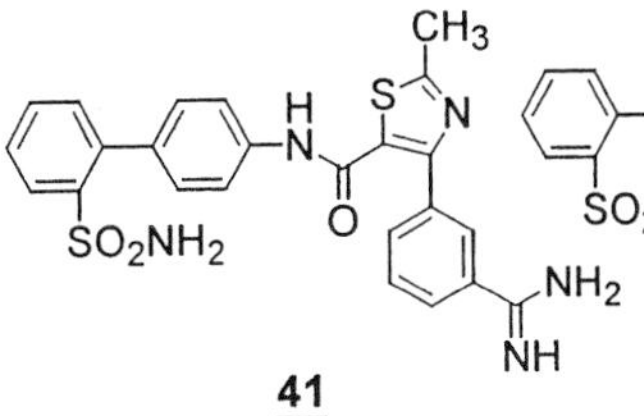

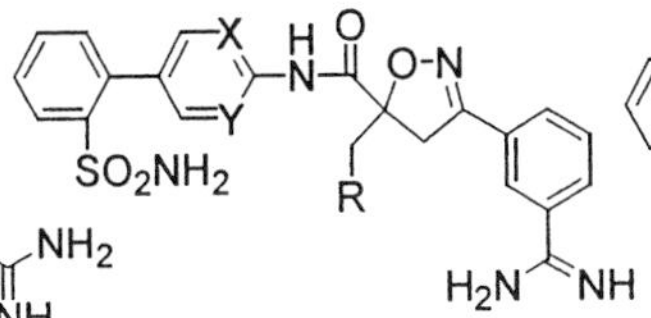

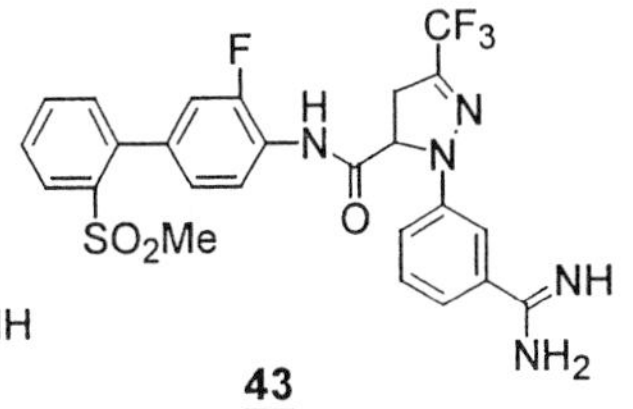

Compound	K_i (nM)	ID_{50} (μM/kg/h)	CL (L/h/kg) in dog	$t_{1/2}$ (h) in dog
38, SN-429	0.010	0.023	0.67	0.82
39a	0.006	0.07	0.76	0.31
39b, SN-116	0.060	0.034	0.36	3.3
40, SA-862	0.150	0.31	1.5	4.3
41, SN-292	0.060	0.5	1.82	2.56
42a, SF-303	6	0.6	NA	NA
42b, (-)-SF-324	2.3	0.15	NA	NA
42c, (-)-SK-549	0.52	0.035	0.7	1.6
42d, (-)-SM-084	0.110	0.032	1.3	1.1
r-TAP	0.18	0.009		
DX-9065a	41	0.4		

As stated earlier, discovery of highly potent mono-benzamidines (K_i in low pM range as in **38** series) has provided lead structures for further development of non-benzamidine factor Xa inhibitors. The SN-429 series is an excellent illustration of this approach. Here, the P1 benzamidine unit in SN-429 series was replaced with its surrogate, 1-aminoisoquinoline, giving **44a** (SQ-311) and **44b** (SQ-315) (34). Both compounds are highly potent factor Xa inhibitors (K_i = 0.33 nM and 0.14 nM for **44a** and **44b**, respectively) and are much more specific for factor Xa than SN-429. Compounds **44a** and **44b** have F = 13% and 66% in dogs with $t_{1/2}$ (PO) = 4.4 hr and 3.4 hr, respectively. Compounds **44a** and **44b** inhibit thrombosis formation in a rabbit AVST model *with in vivo* IC_{50} = 690 nM and 218 nM, respectively. The ID_{50}

value is 1.2 µM/kg/hr for **44b** in the same animal model. The *in vivo* IC_{50} values of **44a** and **44b** are more than 1500 fold the IC_{50} values *in vitro*, possibly due to high plasma protein binding. Additionally, since the parent P1 benzamidine molecules such as SN-429 and its analogs are so potent (K_i in low pM range), the amidino group was simply replaced by a CH_2NH_2 group (30, 35), as exemplified by **45** and **46**. P1 benzylamine group might still form H-bonds with the S1 pocket (Asp-189 or Gly-218) of factor Xa. P1 Benzylamine based thrombin inhibitors show significant improvement in oral bioavailability and duration of action in rats relative to the benzamidines (36). Compound **45** is highly aqueous soluble, and has a K_i = 2.7 nM, F = 13%, and a long $t_{1/2}$ (9.3 h, PO) in dogs (34). Compounds **45**, **46** and related analogs (such as SO_2Me for SO_2NH_2, CF_3 for Me in the pyrazole ring, and fluoro substitution in the proximal P4 phenyl ring) are the most preferred compounds and might represent the most promising leads from this series (30, 35). Compounds **47** - **49** represent other preferred compounds with novel S4 binding elements (35).

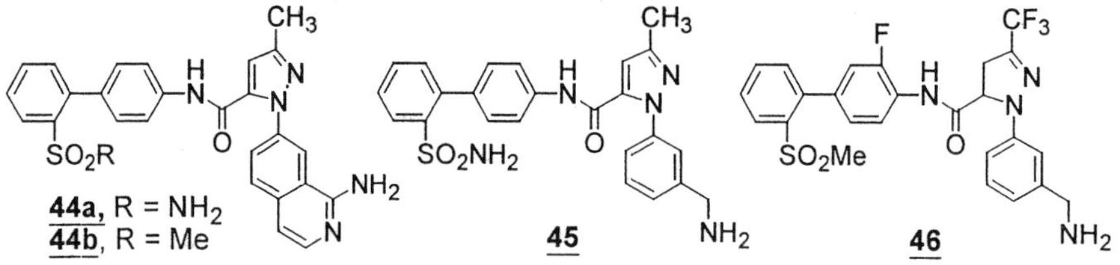

44a, R = NH_2
44b, R = Me

45

46

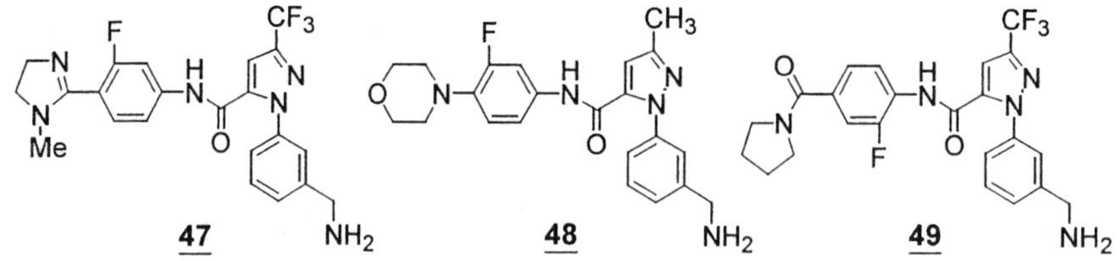

47

48

49

Additionally, the amidino group of SN-429 has been deleted from the parent molecule and replaced by hydrophobic motifs such as OMe, Cl, F, or Br, leading to **50a** which has poor aqueous solubility (37). The OMe group of **50a** might insert into the small hydrophobic hole of factor Xa in the S1 site, which is formed by Tyr-228, Ala-90 and Val-213. The addition of the H-bonding groups such as $CONH_2$ or CH_2NH_2 group to the P1 phenyl ring (*ortho* to the pyrazole ring) leads to **50b**, **50c** and **51** (38).

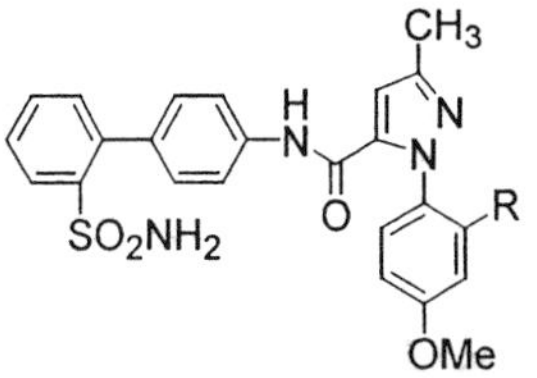

50a, R = H
50b, R = $CONH_2$
50c, R = CH_2NH_2

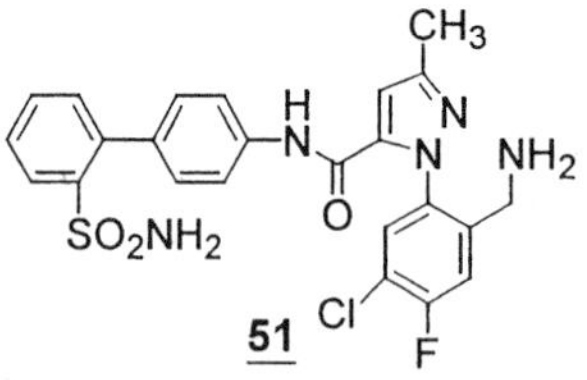

51

Fused heterobicyclic-arylamidines such as indolylamidine as P1 motifs were reported to be potent and specific inhibitors of factor Xa *in vitro* (39). Compounds **52a** (X= CH), **52b** (X= C-F), **52c** (X= C-Cl), **52d** (SE-170, X= C-Br), **52e** (X= C-I), and **52f** (X= N) have K_i values of 2.1 nM, 3.1 nM, 0.7 nM, 0.32 nM, 0.6 nM, and 0.43 nM, respectively. ID_{50} values in the rabbit AVST thrombosis model are 2.0 µM/kg/h, 0.14 µM/kg/h and 0.3 µM/kg/h for **52a**, **52d**, and **52f**, respectively. The *in vivo* IC_{50} = 60 nM for **52d** in the same rabbit AVST thrombosis model. Compounds **53** typify other novel fused heterobicyclic-arylamidines as the P1 elements (40).

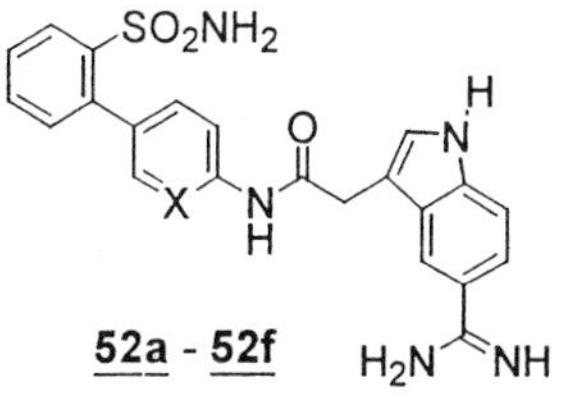

52a - 52f

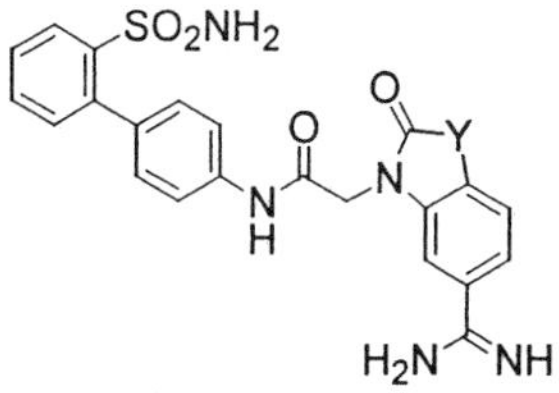

53

Y = O, CH2,
NH, CH2CH2,
CH2O, CH2NH

A novel series of P1 3-amidinoaniline derivatives containing novel P4 elements were reported to be potent factor Xa inhibitors as shown by **54a** (IC$_{50}$ = 4 nM) and **55** (IC$_{50}$ = 4 nM). Compound **54b** containing a carboxylate group is the specifically claimed compound (41).

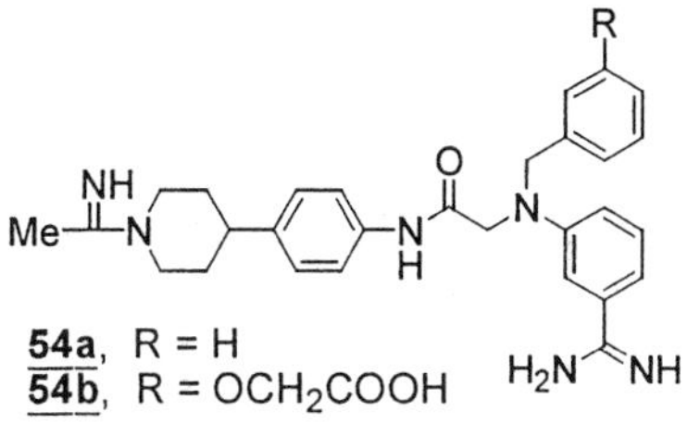

54a, R = H
54b, R = OCH$_2$COOH

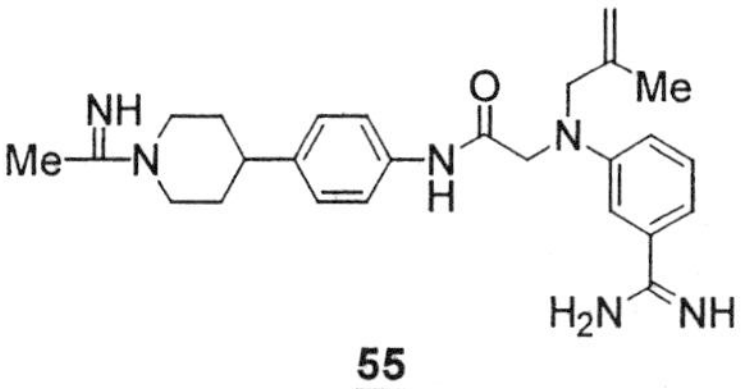

55

A series of indole derivatives, where indole serves as a central template to link P1 benzamidine and P4 elements, were disclosed (42). Compound **56a**, **56b** and **56c** have K$_i$ = 90 nM, 9 nM and 7 nM, respectively. Hydrophobic P4 derivatives **57a**, **57b** and **57c** inhibit factor Xa with K$_i$ = 4.8 nM, 11 nM and 9 nM, respectively.

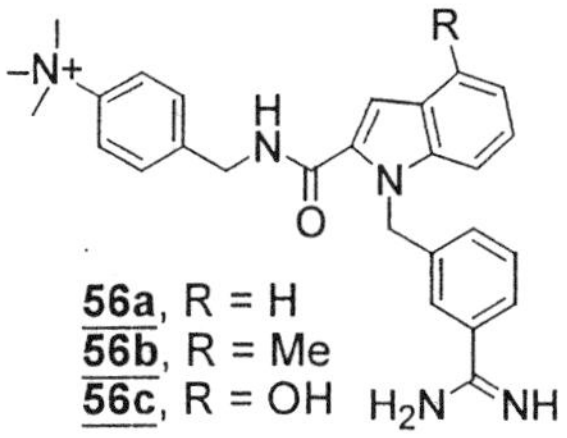

56a, R = H
56b, R = Me
56c, R = OH

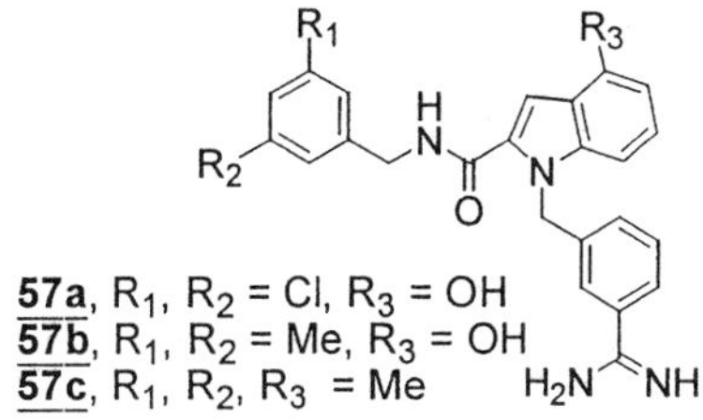

57a, R$_1$, R$_2$ = Cl, R$_3$ = OH
57b, R$_1$, R$_2$ = Me, R$_3$ = OH
57c, R$_1$, R$_2$, R$_3$ = Me

Several series of quinolones, quinoxalinones, benzoxazinones and thiazinones as central templates to link P1 *meta*-benzamidine and a novel P4 element were disclosed as factor Xa inhibitors (43-45). Compounds **58a**, **58b**, **59a** and **59b** have IC$_{50}$ values of 7 nM, <1 nM, 65 nM, and 9 nM, respectively. Compounds **58a** and **59b** also showed antithrombotic effects in a rabbit venous thrombosis model.

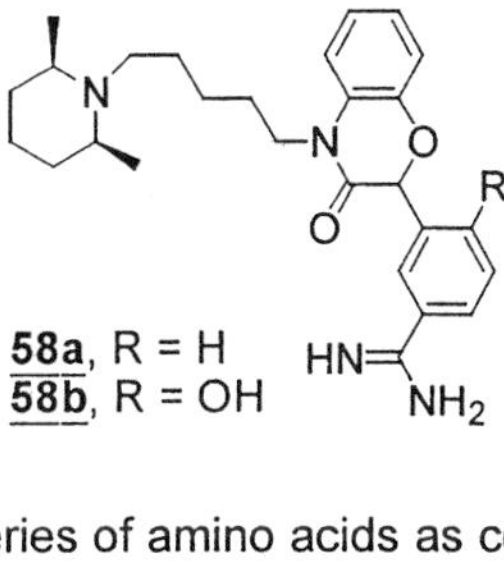

58a, R = H
58b, R = OH

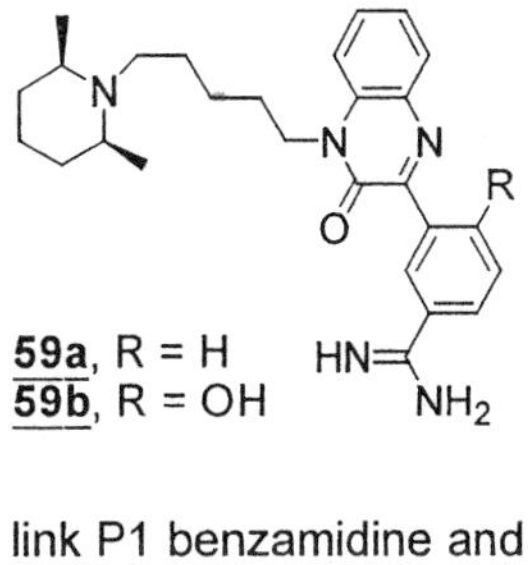

59a, R = H
59b, R = OH

A series of amino acids as central templates to link P1 benzamidine and various P4 elements were disclosed in a recent patent application. Compounds **60** and **61** have K$_i$ values of 15 nM and 14 nM, respectively (46). The 1-amino-isoquinoline derivatives **62** and **63** have K$_i$ values of 8 nM and 70 nM, respectively (47).

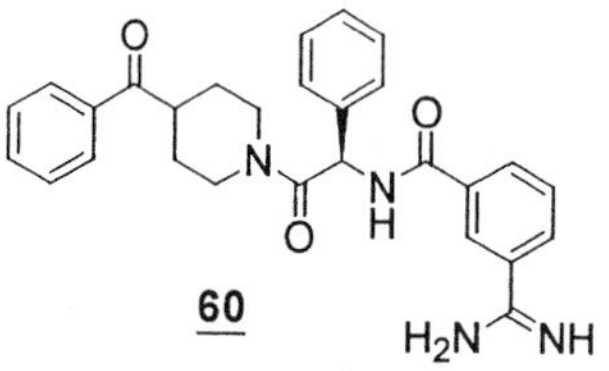

60

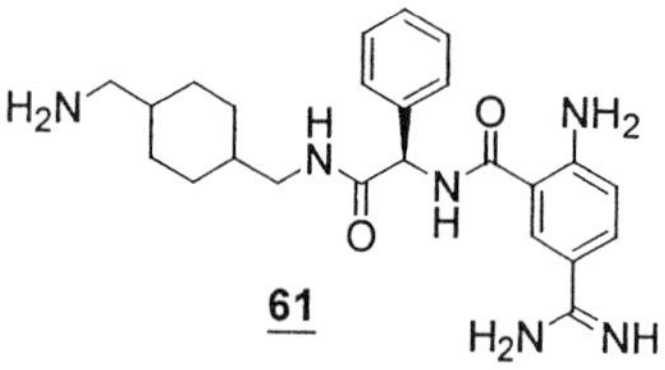

61

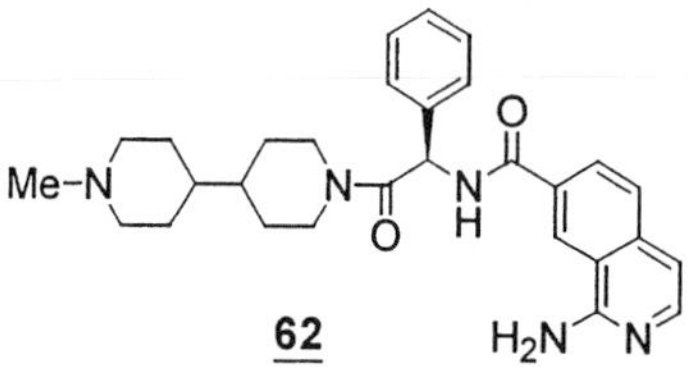

62

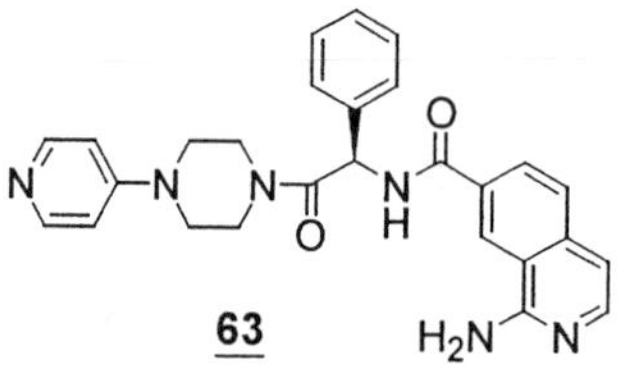

63

A series of amidine-derived 1,2-dibenzamidobenzene inhibitors of factor Xa has been reported (48). Compounds **64**, **65** and **66** have K_i values of 71 nM, 4 nM and 2 nM, respectively. Compound **66** containing a carboxyl group showed *in vivo* antithrombotic efficacy in rabbit arterio-venous thrombosis model with ID_{50} = 70 µg/kg/h. The *meta*-benzamidine is proposed to interact with the S1 pocket of factor Xa, and hydrophobic P4 elements interact with the S4 pocket. A series of non-amidino ortho-phenyldiamide derivatives were also discovered to be very selective inhibitors of factor Xa with moderate potency (49-51). Compounds **67a** and **67b** have K_i = 50 nM and 53 nM, respectively. The amide bond (-NHCO-) linking the P1 element with the central phenyl ring can be reversed to –CONH- with retaining potency (K_i for **68a** and **68b** = 50 nM and 17 nM, respectively). However, the amide bond orientation (-CONH) linking P4 with the central phenyl ring is critical for potency. Fused heterobicycles as P1 elements also showed anti-factor Xa activity (K_i for **69** = 29 nM). However, compounds **68a** – **69** showed poor *in vitro* anticoagulant activity due to high plasma protein binding. In order to improve *in vivo* antithrombotic efficacy, more hydrophilic or charged S4-binding elements were incorporated to increase aqueous solubility and reduce plasma protein binding. For example, non-benzamidine inhibitor **69a** (clog D = 0.4) has a K_i = 10 nM and it doubled the prothrombin time (PT) at 0.58 µM concentration. Compound **69a** showed antithrombotic efficacy in a rabbit arterio-venous thrombosis model with ED_{50} = 1.8 mg/kg/hr (*in vivo* IC_{50} ca. 500 nM) (52).

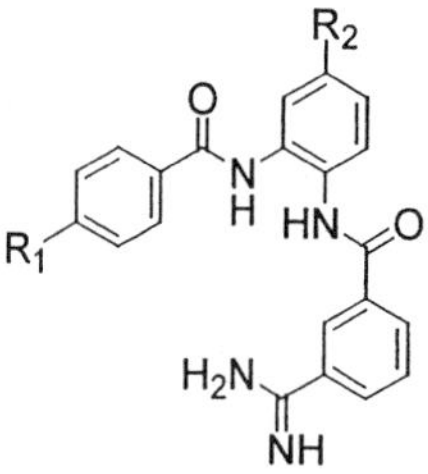

64, R_1 = OMe, R_2 = H
65, R_1 = *t*-Bu, R_2 = H
66, R_1 = *t*-Bu, R_2 = CO_2H

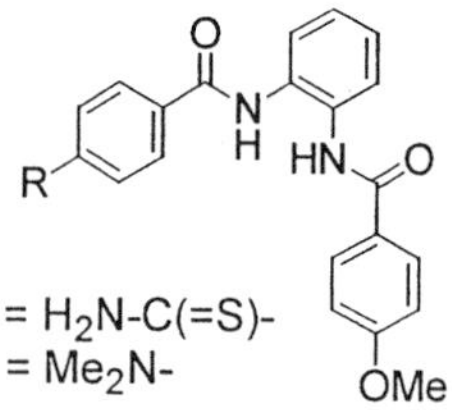

67a, R = H_2N-C(=S)-
67b, R = Me_2N-

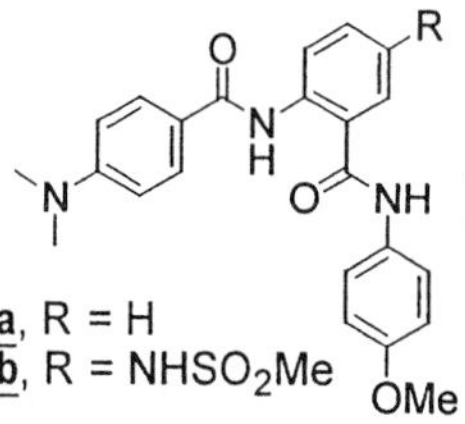

68a, R = H
68b, R = $NHSO_2Me$

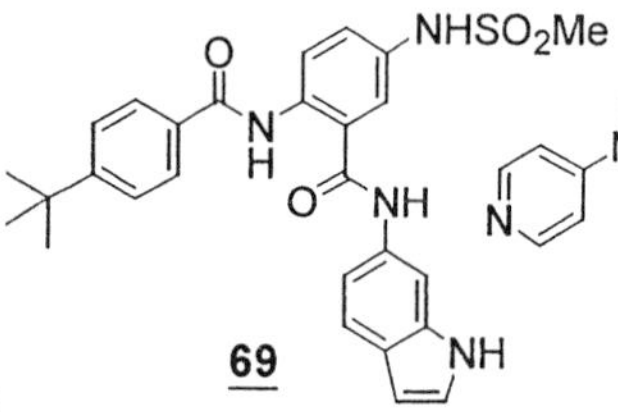

69

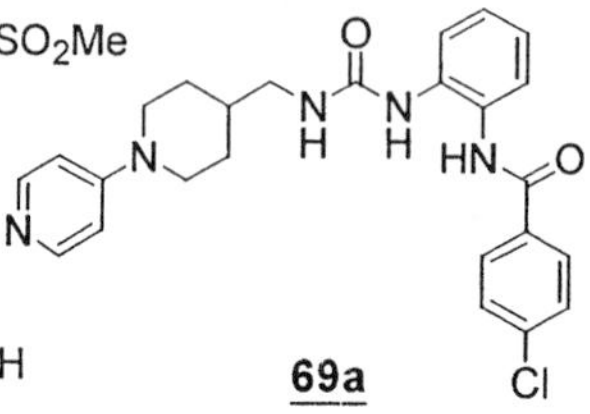

69a

Another series of non-benzamidine factor Xa inhibitors containing heteroarylcarbonylaminobenzamides (ortho-anthranilamides) has been disclosed (53). Compounds **70** and **71** exemplify some of the preferred inhibitors. Extremely diverse P4 motifs have been synthesized from this series, demonstrating the high tolerance of the S4 pocket of factor Xa to various P4 elements. The 5-chloro-2-aminopyridine ring should bind in the S1 pocket of factor Xa. Although no specific biological activities were disclosed, it is stated that the compounds routinely inhibit factor Xa with $K_i < 3$ nM.

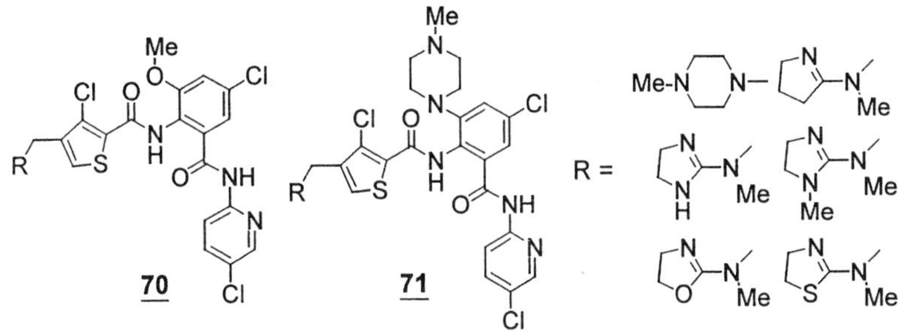

Other Non-Benzamidine Based Factor Xa Inhibitors – It has been thought that the P1 benzamidine group is a requirement for a molecule to have superior anti-factor Xa potency. Although true in many cases, several series of promising P1 non-benzamidine structures were reported in 1999. These P1 non-benzamidine structures have strong anti-factor Xa potency and are remarkably specific for factor Xa *versus* other trypsin-like serine proteinases (trypsin, thrombin, t-PA, urokinase, plasmin, kallekrein, and APC), and are much more specific for factor Xa than P1 benzamidines in general. Due to the increased lipophilicity, non-benzamidine factor Xa inhibitors tend to have improved oral bioavailability than benzamidines. However, due to the increased lipophilicity, some of the non-benzamidine inhibitors have lower aqueous solubility and much higher plasma protein binding. It has been observed that very high plasma protein binding (>99%) has detrimental effects on the *in vivo* antithrombotic efficacy of thrombin inhibitors even with sub-nM *in vitro* inhibitory potency (4). It has also been correlated for factor Xa inhibitors that at the given *in vitro* potency and specificity, increasing lipophilicity generally increases the plasma protein binding which leads to less potent anticoagulants (1).

A series of piperazine-naphthylsulfonamide derivatives have been disclosed in the patent literature (1). Compound **72a** has a claimed IC_{50} value of 3 nM. The piperazine ring can be substituted by acid, ester and amide groups, which retain the *in vitro* potency as exemplified by compound **72b** ($IC_{50} = 2$ nM) and **72c** ($IC_{50} = 4$ nM). Molecular modeling studies by the authors show that arylsulfonamides actually insert into S1 pocket and basic pyridine ring occupies the S4 pocket. The piperazine ring has a chair conformation. This binding mode is much preferred to the "reversed" binding mode. Compound **73** is an open chain version of the piperazinesulfonamides (54).

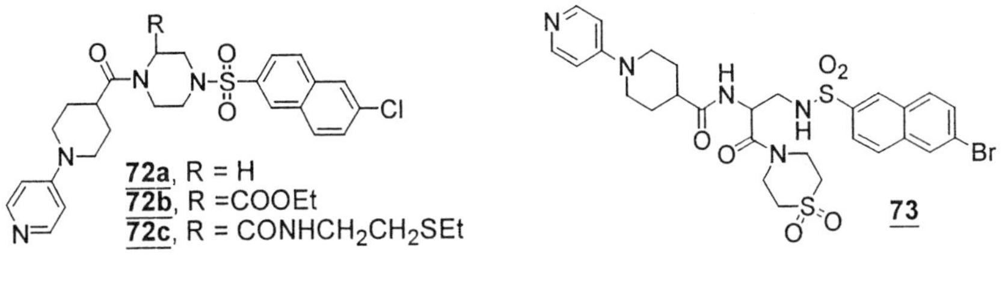

72a, R = H
72b, R =COOEt
72c, R = CONHCH$_2$CH$_2$SEt

73

4-(4-Pyridyl)phenylcarbonylpiperazine-arylsulfonamides were also reported as potent and specific inhibitors of factor Xa (1). Although the free base form of **74** has limited water solubility, a reduced particle size form of the HCl salt of **74** has improved water solubility and increased oral bioavailability (55). Compound **75** exemplifies the replacement of the P4 proximal phenyl ring by a pyridine or pyrimidine ring (56-61). Interestingly, molecular modeling studies by the authors show that the arylsulfonamides insert into the S1 specificity pocket of factor Xa. The $-SO_2-$ group is critical since it could form a H-bond with Ser-195 or Gln-192 of factor Xa and it makes the aryl ring turn 90° into the S1 pocket. The biaryl groups of these inhibitors are proposed to occupy the S4 pocket of factor Xa. The piperazine ring adopts a chair conformation. Although no X-ray structures of piperazine-sulfonamides with factor Xa have been reported, structure-activity data support this hypothesis. Other P1 elements were also disclosed to replace P1 naphthylsulfonamides (56-61). The preferred P1 elements are 2-indolyl, 2-benzimidazolyl, 2-benzofuranyl, 2-benzo[b]thiophene, or 2-pyrrolo[2,3-b]pyridyl sulfonamides substituted by F, Cl or Br. An example is compound **76**, which has an IC_{50} value of 5 nM. 5-Chloroindolyl, 5 or 6-chloro-benzo[b]thiophene elements are the most preferred P1 groups (shown in **77**).

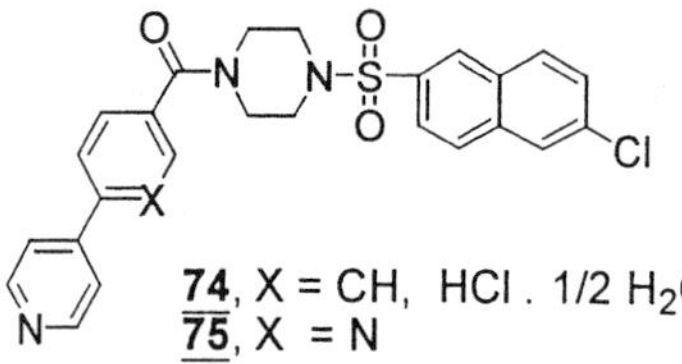

74, X = CH, HCl . 1/2 H_2O
75, X = N

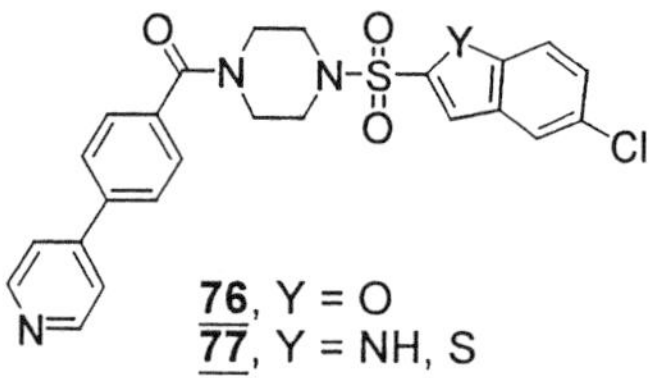

76, Y = O
77, Y = NH, S

A number of pharmaceutical companies have modified lead compound **74** by focusing on improving aqueous solubility and inhibitory potency. Other biaryl P4 elements such as 4-pyridinyl, 3-pyridazinyl, 4-pyrimidinyl, 1-imidazolyl, 4-imidazolyl or thiazolyl substituted by H, CH_2NH_2, NH_2, and NMe_2 in combination with preferred P1 sulfonamides are disclosed in patent applications (56-61). Compounds **78** represented by the combination of P1, P3 and P4 as well as substitutions (R_1 and R_2 are H or varieties of side chains) on the piperazine ring are an illustration of these extensive efforts.

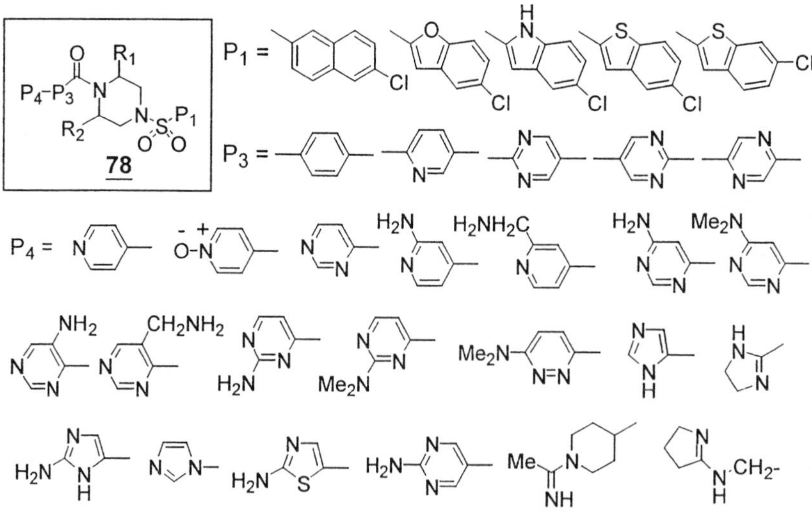

Compounds 78a – 78h illustrate some of the preferred piperazinesulfonamide inhibitors within compounds 78 selected from several patent applications (56-61). It is interesting to note that pyridine *N*-oxide is also a preferred P4 element.

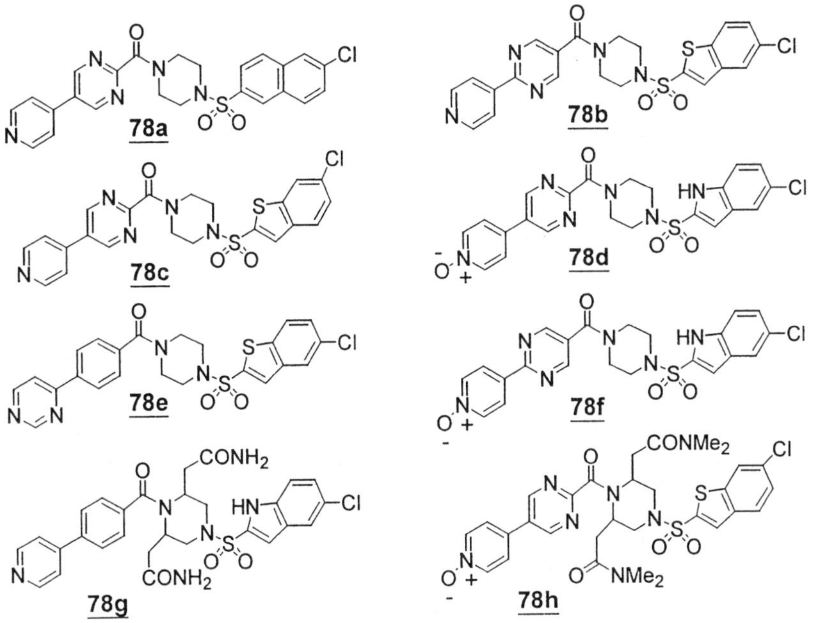

Other S4 binding elements (P4) have also been disclosed within the piperazine-arylsulfonamides series as exemplified by 79 – 82. Compounds 79 and 80 inhibit factor Xa with IC_{50} = 19 nM and 0.7 nM, respectively (1, 61). Replacement of the biaryl P4 within the piperazinesulfonamide series with fused heterobicyclics such as thiazolylpiperidine led to compounds 81a and 81b with IC_{50} of 6.6 nM and 7.8 nM, respectively (60). Deletion of the carboxamide bond in 78c leads to triaryl-like compounds 82 (61).

81a, R_1 = Me, R_2 = H
81b, R_1, R_2 = Me

82, R = H, $CH_2CONHMe$, X = NH, S

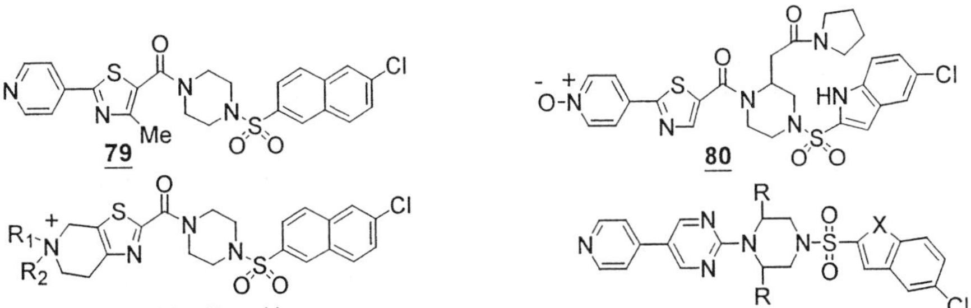

A series of piperazinone derivatives (83a - 83e), which are directly derived from piperazinesulfonamides 72, were reported to be potent inhibitors of factor Xa (62-65). Similar to the piperazine series, the piperazinone ring (next to the P4) can be substituted by various functional groups such as COOH, COOEt, CH_2NMe_2, and CH_2NHAc. Compounds 83b, 83c, 83d, and 83e have IC_{50} = 19 nM, 7.4 nM, 8.5 nM and 7.4 nM, respectively. Piperazinone derivatives 84 - 85, which directly mimic biarylcarbonylpiperazinesulfonamides 74 - 78, have also been disclosed (62-65). An example is compound 85a which has an IC_{50} = 50 nM.

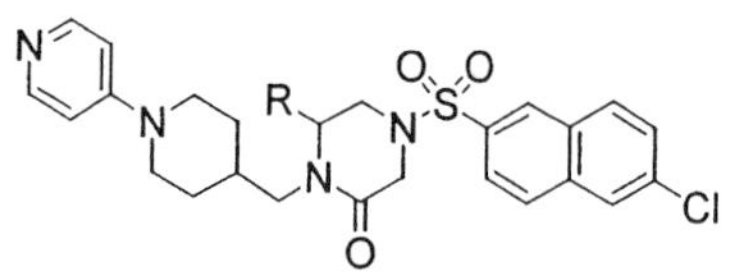

83a, R = H
83b, R = CO$_2$Et
83c, R = COOH
83d, R = CH$_2$NMe$_2$
83e, R = CH$_2$NHAc

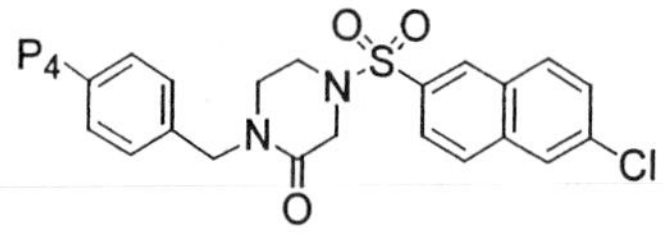

P$_4$ =

84a   **84b**   **84c**

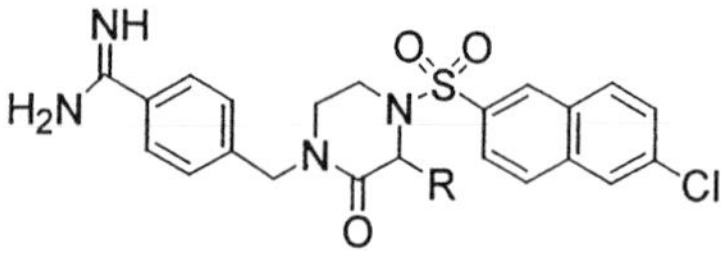

85a, R = H
85b, R = CH$_2$COOH

Another series of piperazinone derivatives with fused aromatic heterobicycles (such as azaindole and 4-amino-quinazoline) as P4 elements were also disclosed. This series of piperazinones are the extension of 3-sulfonamido-pyrrolidinone series (compounds **13** - **26**) from the drug design perspective. Compounds **86a** and **86b** are potent inhibitors of factor Xa (K$_i$ = 1.3 nM and 18 nM, respectively), indicating that the *para*-benzamidine is the preferred P4 element (66). The corresponding *para*-benzamidine mimetic analogs **87a** and **87b** are even more potent (K$_i$ = 1.4 nM and 0.7 nM, respectively). However, the binding orientation of piperazinones **86a** and **86b** with factor Xa is reversed compared to the γ-lactam series. The X-ray structures of benzamidine compounds **86a** and **86b** complexed with factor Xa have revealed that the benzamidine groups actually bind in the S4 pocket of factor Xa (66). The 6-chloro-benzo[b]thiophenesulfonamide interacts with the S1 pocket of factor Xa. The chloro-atom on the benzothiophene ring points into the center of side chain of Tyr-228 of factor Xa and sulfur atom faces to side chain of Ser-195 of factor Xa. The X-ray structures of **86a** and **86b** support the modeling results for the piperazinesulfonamide series. Compounds **88** - **93** are the preferred compounds disclosed (67). For example, compound **88** (RPR-200443) has a K$_i$ value of 4 nM.

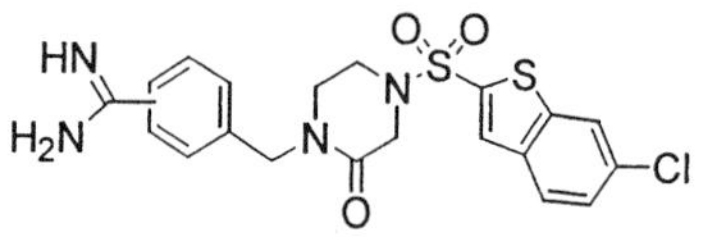

86a, *para*-amidine
86b, *meta*-amidine

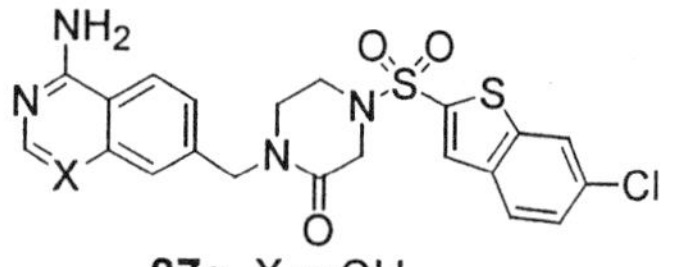

87a, X = CH
87b, X = N

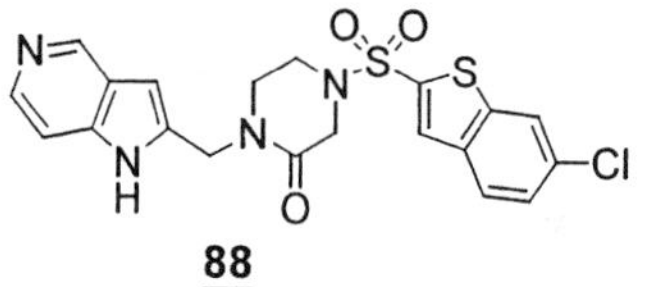

88

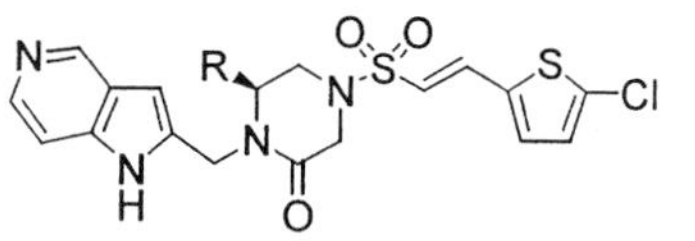

89, R = H, CH$_2$OH

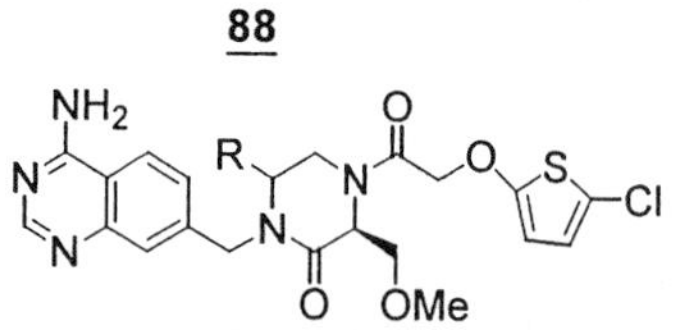

90, R = H, Me

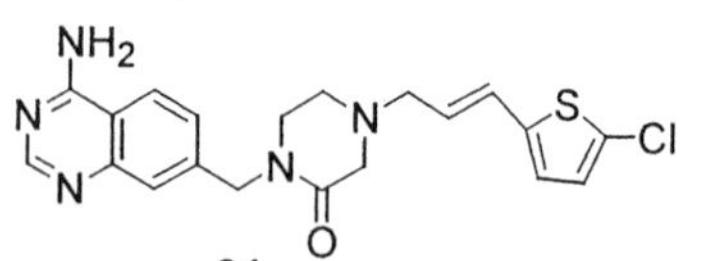

91

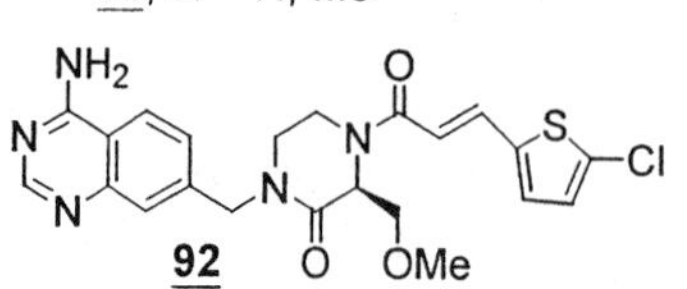

92

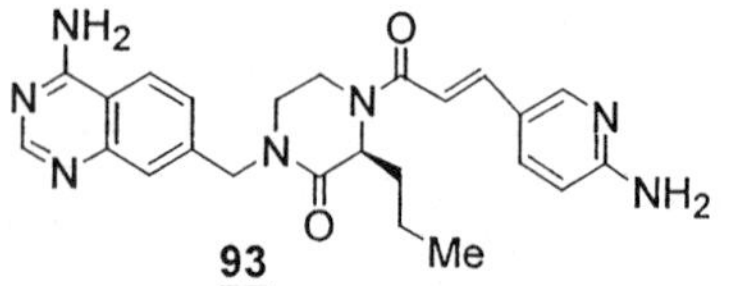

93

Drug Design Perspective and Issues - Potent factor Xa inhibitors generally have L-shaped conformations. P4 and P1 elements must have almost a 90° turn to achieve optimized interactions with S1 and S4 pockets of factor Xa. The S1 pocket of factor Xa is much more sensitive than the S4 pocket to ligand structural variations. Only a limited group of S1 binding elements lead to highly potent factor Xa inhibitors. On the other hand, the S4 pocket is much more tolerant of P4 structural variations. As illustrated in this review, very diverse P4 elements can be utilized to give highly potent inhibitors of factor Xa. The diverse P4 elements provide the opportunities to modulate the physiochemical properties (clog D, aqueous solubility and plasma protein binding) to deliver desired *in vivo* antithrombotic efficacy and desired pharmacokinetic profiles. However, due to the symmetric nature of the S1 and S4 pockets (both S1 and S4 pockets can bind to hydrophobic or positively charged motifs), it is sometimes not obvious how to predict the binding modes (orientations) of specific inhibitors with factor Xa. Additionally, a subtle structure change of a lead structure can reverse the binding orientation. It might also be possible for an inhibitor to have dual binding modes. Thus, within a series it may be difficult to carry out systematic structure-activity relationships if the binding orientation changes due to very subtle structural modifications or an incorrect binding mode is presumed.

Development Issues – Generally, benzamidine based factor Xa inhibitors suffer from low oral bioavailability, and/or high clearance, and/or short $t_{1/2}$ *in vivo*. Additionally, P1 benzamidine factor Xa inhibitors display moderate selectivity. However, benzamidine based factor Xa inhibitors showed very promising *in vivo* antithrombotic efficacy after intravenous administration in animal thrombosis models. The observed *in vivo* antithrombotic effects are attributed to the overall favorable parameters such as excellent *in vitro* potency (low nM to low pM), high aqueous solubility (clog D < 0) and low plasma protein binding (<80%). Some of the non-benzamidine factor Xa inhibitors summarized in this review are very potent (<10 nM) against factor Xa, and are much more specific for factor Xa than P1 benzamidine factor Xa inhibitors. They do not inhibit other trypsin-like serine proteases such as trypsin, thrombin, t-PA, APC, urokinase, and plasmin. Some highly potent and specific non-benzamidine inhibitors of factor Xa also have high oral bioavailability (F) and long plasma $t_{1/2}$. However, due to their increased lipophilicity, some non-benzamidine inhibitors have relatively low aqueous solubility and may have high plasma protein binding (>99%). The undesired low aqueous solubility and associated high plasma protein binding may render them inactive *in vivo* as antithrombotic agents even with low nM *in vitro* inhibitory activity. However, several series of non-benzamidine factor Xa inhibitors have very good aqueous solubility and may have relatively low plasma protein binding. Coupled with their excellent inhibitory activities (low nM to pM range) and specificity for factor Xa, these non-benzamidine inhibitors show promising *in vivo* antithrombotic efficacy. It needs to be emphasized that factor Xa inhibitors, which have high potency (pM range) against factor Xa in the prothrombinase complex, can tolerate higher (98%) plasma protein binding for the desired efficacy. *In vivo* efficacy depends on several factors, including *in vitro* potency against factor Xa in the prothrombinase complex, binding kinetics of factor Xa inhibition, specificity, actual free (unbound) inhibitor concentrations in plasma, and aqueous solubility. However, very high aqueous solubility (especially charged groups) may jeopardize oral bioavailability. Thus, from an *in vivo* efficacy perspective, high aqueous solubility (log D < 0) and low plasma protein binding are preferred. From a pharmacokinetic perspective, moderate lipophilicity and good aqueous solubility (preferably log D in the range of 1-4) are preferred. Therefore, in order to obtain clinically viable oral factor Xa inhibitors, very careful and systematic approaches to optimize *in vitro* potency (preferably K_i < 1 nM), specificity, and balanced physicochemical properties are necessary to achieve both good pharmacokinetic profiles (high oral bioavailability and long $t_{1/2}$) and antithrombotic efficacy.

Conclusions – During the past few years, significant activity has been directed towards the identification of small molecule factor Xa inhibitors as replacements for warfarin. In animal thrombosis models, not only are the small molecule factor Xa inhibitors effective in preventing thrombosis formation, but they also have a much better efficacy/safety window than warfarin. Much progress has been made during the last year to discover both benzamidine and non-benzamidine factor Xa inhibitors. In particular, many classes of non-benzamidine factor Xa inhibitors have been discovered and developed into highly potent, specific and reversible inhibitors of factor Xa. Some non-benzamidine molecules have highly oral bioavailability and other desirable pharmaceutical properties, and are efficacious in animal thrombosis models as well. During next several years, more clinical candidates are expected to enter clinical trials. The first results of these trials should provide valuable insights into the usefulness of factor Xa inhibitors as oral anticoagulants to replace warfarin.

References

1. B.Y. Zhu, and B.M. Scarborough, Curr. Opin. Cardiovascular, Pulmonary and Renal Invest. Drugs, 1, 63 (1999) and references therein.
2. J.M. Fevig and R.R. Wexler, Annu. Rep. Med. Chem., 34, 81 (1999).
3. W.R. Ewing, H.W. Pauls, and A.P. Spada, Drugs Future, 24, 771 (1999).
4. J.B.M. Rewinkel, and A.E.P. Adang, Current Pharm. Design, 5, 1043 (1999).
5. N. Murayama, M. Tanaka, S. Kunitada, H, Yamada, T. Inoue, Y. Terada, M. Fujita, and Y. Ikeda, Clin. Pharmacol. Ther., 66, 258 (1999).
6. H, Koshio, F, Hirayama, T. Ishihara, M. Funatsu, T. Kawasaki, and Y. Matsumoto, WO Patent 9905124 (1999).
7. S. Katoh, K. Yokota, and M. Hayashi, WO Patent 9952895 (1999).
8. W. Huang, P. Zhang, Y. Song, P. Wong, K. Tran, U. Sinha, R.M. Scarborough and B.-Y. Zhu, 219th ACS National Meeting, San Francisco, CA, March 26-30, 2000, Book of Abstracts, ORG-240.
9. F. Hirayama, H. Koshio, T. Ishihara, H, Kaizawa, T. Kawasaki, and Y. Matsumoto, WO Patent 9911617 (1999).
10. B.-Y. Zhu, W. Li, W. Huang, K. Kane-Maguire, C.K. Marlowe, Y. Song, P. Zhang, L. Wang, J. Fan, P. Wong, K. Tran, L. Xing, and R.M. Scarborough, 219th ACS National Meeting, San Francisco, CA, March 26-30 (2000), Book of Abstracts, ORG-237.
11. F. Hirayama, H. Koshio, T. Ishihara, H. Kaizawa, N. Katayama, Y. Taniuchi, and Y. Matsumoto, WO Patent 9937643 (1999).
12. I, Mochalkin, and A. Tulinsky, Acta Crystallogr., Sect. D: Biol. Crystallogr., D55, 785 (1999).
13. M.R. Becker, W.R. Ewing, R.S. Davis, H.W. Pauls, C. Ly, A. Li, H.J. Mason, Y.M. Choi-Sledeski, A.P. Spada, V. Chu, K.D. Brown, D.J. Colussi, R.J. Leadley, R. Bently, J. Bostwick, C.Kasiewski, and S. Morgan, Bioorg. Med. Chem. Lett., 9, 2753 (1999).
14. W.R. Ewing, M.R. Becker, V.E. Manetta, R.S. Davis, H.W. Pauls, H. Mason, Y.M. Choi-Sledeski, D. Green, D. Cha, A.P. Spada, D.L. Cheney, J.S. Mason, S. Maignan, J-P. Guilloteau, K. Brown, D. Colussi, R. Bentley, J. Bostwick, C.J. Kasiewski, S.R. Morgan, R.J. Leadley, C.T. Dunwiddie, M.H. Perrone, and V. Chu, J. Med. Chem., 42, 3557 (1999).
15. Y.M. Choi-Sledeski, D.G. McGarry, D.M. Green, H.J. Mason, M.R. Becker, R.S. Davis, W.R. Ewing, W.P. Dankulich, V.E. Manetta, R.L. Morris, A.P. Spada, D.L. Cheney, K.D. Brown, D.J. Colussi, V. Chu, C.L. Heran, S.R. Morgan, R.G. Bentley, R.J. Leadley, S. Maignan, J-P. Guilloteau, C.T. Dunwiddie, and H.W. Pauls, J. Med. Chem., 42, 3572 (1999).
16. J.S. Bostwick, R. Bentley, S. Morgan, K. Brown, V. Chu, W.R. Ewing, A.P. Spada, H. Pauls, M.H. Perrone, C.T. Dunwiddie, R.J. Leadley, Thromb. Haemostasis, 81, 157 (1999).
17. R.J. Leadley, S.R. Morgan, R. Bentley, J.S. Bostwick, C.J. Kasiewski, C. Heran, V. Chu, K. Brown, P. Moxey, W.R. Ewing, H. Pauls, A.P. Spada, M.H. Perrone, and C.T. Dunwiddie, J. Cardiovasc. Pharmacol., 34, 791 (1999).
18. Y.M. Choi-Sledeski, H.W. Pauls, M.R. Becker, W.R. Ewing, A.P. Spada, D.M. Green, R. Davis, C. Ly, C.J. Gardiner, Y. Gong, J. Jiang, A. Li, H.J. Mason, G. Liang, V.

Chu, K. Brown, D. Colussi, S.R. Morgan, and R.J. Leadley, 217th ACS National Meeting, Anaheim, Calif., March 21-25 (1999), Book of Abstracts, MEDI-080.

19. Y.M. Choi-Sledeski, M.R. Becker, D.M. Green, R. Davis, W.R. Ewing, H.J. Mason, C. Ly, A. Spada, G. Liang, D. Cheney, J. Barton, V. Chu, K. Brown, D. Colussi, R. Bentley, R. Leadley, C. Dunwiddie, and H.W. Pauls, Bioorg. Med. Chem. Lett., 9, 2539 (1999).

20. W.R. Ewing, M.R. Becker, A. Li, J.Z. Jiang, R.S. Davis, Y.M. Choi-Sledeski, H.W. Pauls, A. Parameswaran, A.P. Spada, V. Chu, D.J. Colussi, K.A. Brown, and R.J. Leadley, 218th ACS National Meeting, New Orleans, Aug. 22-26 (1999), Book of Abstracts, MEDI-028.

21. S. Maignan, J.P. Guilloteau, V. Mikol, Y.M. Choi-Sledeski, M.R. Becker, S.I. Klein, W.R. Ewing, H.W. Pauls, and A.P. Spada, 218th ACS National Meeting, New Orleans, Aug. 22-26 (1999), Book of Abstracts, MEDI-020.

22. M.R. Becker, Y.M. Choi-Sledeski, H.W. Pauls, Y. Gong, W.R. Ewing, R.S. Davis, V. Chu, K. Brown, D. Colussi, R.J. Leadley, R. Bentley, C.J. Kasiewski, S.R. Morgan, V. Mikol, S. Maignan, and J.-P. Guilloteau, 218th ACS National Meeting, New Orleans, Aug. 22-26 (1999), Book of Abstracts, MEDI-033.

23. M. Czekaj, S.I. Klein, C.J. Gardner, K.R. Guertin, A.L. Zulli, H. Pauls, A.P. Spada, V. Chu, K. Brown, D. Colussi, R.J. Leadley, C.T. Dunwiddie, S.R. Morgan, C.L. Heran, M.H. Perrone, S. Maignan, and J.-P. Guilloteau, 218th ACS National Meeting, New Orleans, Aug. 22-26 (1999), Book of Abstracts, MEDI-032.

24. Y, Gong, H.W. Pauls, A.P. Spada, M. Czekaj, G. Liang, V. Chu, D.J. Colussi, K.D. Brown, and J. Gao, Bioorg. Med. Chem. Lett., 10, 217 (2000).

25. S.I. Klein, and K.R. Guertin, WO Patent 9948870 (1999).

26. T. Nakagawa, K. Sagi, K. Yoshida, Y. Fukuda, M. Shoji, S. Takehana, T. Kayahara, and A. Takahara, WO Patent 9964392 (1999).

27. T. Nakagawa, S. Makino, K. Sagi, M. Takayanagi, T. Kayahara, and S. Takehana, WO Patent 9947503 (1999).

28. D.J.P. Pinto, M.J. Orwat, S. Wang, E. Amparo, J.R. Pruitt, K.A. Rossi, R.S. Alexander, J.M. Fevig, J. Cacciola, P.Y.S. Lam, R.M. Knabb, P.C. Wong, and R.R. Wexler, Book of Abstracts, 217th ACS National Meeting, Anaheim, Calif., March 21-25 (1999), MEDI-006.

29. J.M. Fevig, J. Buriak, J. Pinto, Q. Han, S. Wang, R.S. Alexander, K.A. Rossi, R.M. Knabb, P.C. Wong, M.R. Wright, and R.R. Wexler, 217th ACS National Meeting, Anaheim, Calif., March 21-25 (1999), Book of Abstracts, MEDI-087.

30. D.J.P. Pinto, WO Patent 9950255 (1999).

31. M.L. Quan, A.Y. Liauw, C.D. Ellis, J.R. Pruitt, D.J. Carini, L.L. Bostrom, P.P. Huang, K. Harrison, R.M. Knabb, M.J. Thoolen, P.C. Wong, and R.R. Wexler, J. Med. Chem., 42, 2752 (1999).

32. M.L. Quan, C.D. Ellis, A.Y. Liauw, R.S. Alexander, R.M. Knabb, G. Lam, M.R. Wright, P.C. Wong, and R.R. Wexler, J. Med. Chem., 42, 2760 (1999).

33. P.C. Wong, M.L. Quan, E.J. Crain, C.A. Watson, R.R. Wexler, and R.M. Knabb, J. Pharmacol. Exp. Ther., 292, 351 (2000).

34. P.Y.S. Lam, R. Li, C.G. Clark, D.J. Pinto, R.S. Alexander, K.A. Rossi, R.M. Knabb, P.C. Wong, B.J. Aungst, S.A. Bai, M.R. Wright, and R.R. Wexler. 219th ACS National Meeting, San Francisco, CA, March 26-30 (2000), Book of Abstracts, MEDI-160.

35. D.J.P. Pinto, J.R. Pruitt, J. Cacciola, J.M. Fevig, Q. Han, M.J. Orwat, M.L. Quan, and K.A. Rossi, U.S. Patent 6,020,357 (2000).

36. K. Lee, W.H. Jung, C. W. Park, C.Y. Hong, I.C. Kim, S. Kim, Y.S. Oh, O.W. Kwon, S.H Lee, H.D. Park, S.W. Kim, Y.H. Lee, and Y.J. Yoo, Bioorg. Med. Chem. Lett., 8, 2563 (1998).

37. R.A. Galemmo, C. Dominguez, J.M. Fevig, Q. Han, P.Y-S. Lam, D.J.P. Pinto, J.R. Pruitt, and M.L. Quan, U.S. Patent 5,998,424 (1999).

38. R.A. Galemmo, D.J.P. Pinto, L.L. Bostrom, and K.A. Rossi, WO Patent 9932454 (1999).

39. Q. Han, D. Dominguez, P.F.W. Stutten, D.E. Duffy, J.M. Park, R.A. Galemmo, K.A. Rossi, R.A. Alexander, P.C. Wong, R.M. Knabb, and R.R. Wexler, 218th ACS National Meeting, New Orleans, LA, August 22-26 (1999), Book of Abstracts, MEDI-030.

40. Q. Han, C. Dominguez, E.C. Amparo, J.M. Park, M.L. Quan, and K.A. Rossi, WO Patent 9912903 (1999).

41. S. Akahane, M. Uchida, H. Isawa, N. Kikuchi, T. Ozawa, H. Kobayashi, Y. Kai, and K. Akahane, WO Patent 9910316 (1999).

42. E. Defossa, U. Heinelt, O. Klingler, G. Zoller, F. Al-Obeidi, A. Walser, P. Wildgoose, and H. Matter, WO Patent 9933800 (1999).

43. D.A. Dudley, and J.J. Edmunds, WO Patent 9950254 (1999).

44. K.A. Berryman, D.M. Downing, D.A. Dudley, J.J. Edmunds, L.S. Narasimhan, and S.T. Rapundalo, WO Patent 9950257 (1999).

45. D.A. Dudley, and J.J. Edmunds, WO Patent 9950263 (1999).

46. J.W. Liebeschuetz, W.A. Wylie, B. Waszkowycz, C.W. Murray, A.D. Rimmer, P.M. Welsh, S.D. Jones, J.M.E. Roscoe, S.C. Young, and P.J. Morgan, WO Patent 9911658 (1999).

47. J.W. Liebeschuetz, W.A. Wylie, B. Waszkowycz, C.W. Murray, A.D. Rimmer, P.M. Welsh; S.D. Jones, J.M.E. Roscoe, S.C. Young, P.J. Morgan, N.P. Camp, and A.P.A. Crew, WO Patent 9911657 (1999).

48. M.R. Wiley, L.C. Weir, S. Briggs, N.A. Bryan, J. Buben, C. Campbell, N.Y. Chirgadze, R.C. Conrad, T.J. Craft, J.V. Ficorilli, J.B. Franciskovich, L.L. Froelich, D.S. Gifford-Moore, T. Goodson, D.K. Herron, V.J. Klimkowski, K.D. Kurz, J.A. Kyle, J.J. Masters, A.M. Ratz, G. Milot, R.T. Shuman, T. Smith, G.F. Smith, A.L. Tebbe, J.M. Tinsley, R.D. Towner, A. Wilson, and Y.K. Yee, J. Med. Chem., 43, 883 (2000).

49. D.K. Herron, T. Goodson, M.R. Wiley, L.C. Weir, J.A. Kyle, Y.K. Yee, A.L. Tebbe, J.M. Tinsley, D. Mendel, J.J. Masters, J.B. Franciskovich, J.S. Sawyer, D.Q. Beight, A.M. Ratz, G. Milot, S.E. Hall, V.J. Klimkowski, J.H. Wikel, B.J. Eastwood, R.D. Towner, D.S. Gifford-Moore, T.J. Craft, and G.F. Smith, J. Med. Chem., 43, 859 (2000).

50. Y.K. Yee, A.L. Tebbe, J.H. Linebarger, D.W. Beight, T.J. Craft, D. Gifford-Moore, T. Goodson, D.K. Herron, V.J. Klimkowski, J.A. Kyle, J.S. Sawyer, G.F. Smith, J.M. Tinsley, R.D. Towner, L. Weir, and M.R. Wiley, J. Med. Chem., 43, 873 (2000).

51. F. Grams, R. Kucznierz, H. Leinert, K. Stegmeier, and W. Von Der Saal, WO Patent 9942439 (1999).

52. J.J. Masters, J.B. Franciskovich, J.M. Tinsley, T.J. Craft, D.W. Beight, J. Ficorfilli, L.L. Froelich, D.S. Gifford-Moore, L.A. Hay, D.K. Herron, V.J. Klimkowski, K.D. Kurz, D. Mendel, A.M. Ratz, J.S. Sawyer, G.F. Smith, T. Smith, A.L. Tebbe, R.D. Towner, L.C. Weir, M.R. Wiley, A. Wilson, and Y.K. Yee, 219th ACS National Meeting, San Francisco, CA, March 26-30 (2000), Book of Abstracts, MEDI-187.

53. D.O. Arnaiz, Y-L. Chou, R.E. Karanjawala, M.J. Kochanny, W. Lee, A.M. Liang, M.M. Morrissey, G. B. Phillips, K.L. Sacchi, S.T. Sakata, K.J. Shaw, R.M. Snider, S.C. Wu, B. Ye, Z. Zhao, and B.D. Griedel, WO Patent 9932477 (1999).

54. J. Preston, and A. Stocker, WO 9909027 (1999).

55. M. Ashford, and R. James, WO Patent 9957112 (1999).

56. R. James, T, Nowak, and P. Warner, WO Patent 9906371 (1999).

57. P.W.R. Caulkett, R. James, S.E. Pearson, A.M. Slater, and R.P. Walker, WO Patent 9957113 (1999).

58. T. Nowak, J. Preston, J.W. Rayner, M.J. Smithers, and A. Stocker, WO Patent 9957099 (1999).

59. D. Dorsch, H. Juraszyk, H. Wurziger, S. Bernotat-Danielowski, and G. Melzer, WO Patent 9916751 (1999).

60. S. Kobayashi, S. Komoriya, M. Ito, T. Nagata, A. Mochizuki, N. Haginoya, T. Nagahara, and H. Horino, WO Patent 9916747 (1999).

61. S. Kobayashi, S. Komoriya, N. Haginoya, M. Suzuki, T. Yoshino, T. Nagahara, T. Nagata, H. Horino, M. Ito, and A. Mochizuki, WO Patent 0009480 (2000).

62. D. Dorsch, H. Juraszyk, H. Wurziger, J. Gante, W. Mederski, H-P. Buchstaller, S. Anzali, S. Bernotat-Danielowski, and G. Melzer, WO Patent 9931092 (1999).

63. H. Nishida, Y. Hosaka, Y. Miyazaki, T. Matsusue, and T. Mukaihira, WO Patent 9933805 (1999).

64. D. Dorsch, H. Juraszyk, W. Mederski, H. Wurziger, J. Gante, H-P. Buchstaller, S. Bernotat-danielowski, and G. Melzer, DE Patent 19835950 (1999).

65. H. Tawada, F. Itoh, H. Banno, and Z. Terashita, WO Patent, 9940075 (1999).

66. W.R. Ewing, M.R. Becker, A. Li, R.S. Davis, J.Z. Jiang, A. Zulli, Y.M. Choi-Sledeski, H.W. Pauls, M.R. Myers, A.P. Spada, S. Maignan, J-P. Guilloteau, V. Mikol, K.D. Brown, D. Colussi, V. Chu, R.J. Leadley, D. Cheney, and J. Mason, 219th ACS National Meeting, San Francisco, CA, March 26-30 (2000), Book of Abstracts, MEDI-161.

67. W.R. Ewing, M.R. Becker, Y.M. Choi-Sledeski, H.W. Pauls, W. He, S.M. Condon, R.S. Davis, B.A. Hanney, A.P. Spada, C.J. Burns, J.Z. Jiang, A. Li, M.R. Myers, W.F. Lau, and G.B. Poli, WO Patent 9937304 (1999).

Chapter 10. Antiplatelet Therapies

Joanne M. Smallheer, Richard E. Olson and Ruth R. Wexler
DuPont Pharmaceuticals Company
P.O. Box 80500, Wilmington, DE 19880

<u>Introduction</u> – Platelet activation and subsequent platelet aggregation play a major role in the pathogenesis of thromboembolic diseases such as unstable angina, myocardial infarction, transient ischemic attacks, stroke and peripheral artery disease. Together these diseases comprise the most common cause of mortality in western society. The medical need for more efficacious antiplatelet agents and the growing understanding of platelets in vascular injury have catalyzed an extensive evaluation of novel approaches to control platelet function. An ideal antiplatelet drug should block platelet-dependent thrombus formation reversibly and specifically, regardless of the nature of the agonist and without interfering with normal hemostasis and/or wound healing. Preferably, an antiplatelet agent should also have a rapid onset of action and be available in an oral dosage form suitable for both short-term and long-term dosing. This chapter will highlight recent advances in the development of GPIIb/IIIa receptor antagonists, ADP-receptor antagonists and thrombin receptor antagonists.

GP IIb/IIIa ANTAGONISTS

Platelet activation causes a morphological change in platelets rendering GPIIb/IIIa receptors competent to bind fibrinogen. The binding of fibrinogen to the activated form of GPIIb/IIIa is the final step leading to platelet aggregation. Therefore, regardless of the activating mechanism, the blockade of the GPIIb/IIIa receptor can prevent the formation and/or propagation of a platelet thrombus. The binding of human fibrinogen to platelets is believed to be mediated through two arginine-glycine-aspartic acid (RGD) sequences at residues 95-97 and 572-574 on the fibrinogen α-chain and a dodecapeptide at residues 401-411 (HHLGGAKQAGDV) on the carboxy terminus of the γ-chain (1). From the standpoint of drug design, most of the non-peptide antagonists reported in the literature are designed to be RGD peptidomimetics which incorporate various conformational constraints to maintain the proper orientation and distance of 15-17 Å (16-18 atoms) between the basic nitrogen of the arginine mimic and the carboxyl oxygen of the aspartic acid mimic (1,2).

<u>IV IIb/IIIa Antagonists</u> – There are currently three iv GPIIb/IIIa agents on the market. ReoPro™ (c7E3, abciximab), a chimeric monoclonal antibody which binds to the GPIIb/IIIa receptor, has continued to gain clinical acceptance as an adjunct to percutaneous transluminal coronary angioplasty (PTCA) since its launch in 1995 (3,4). A one year follow-up to the EPILOG trial showed sustained reduction in the primary composite endpoint of death, myocardial infarction or urgent revascularization. The prolonged reduction in rates of mortality is attributed to the blockade of early ischemic events since the direct antiplatelet effect of abciximab persists for no longer than 2-3 weeks after treatment in the acute setting (5). A RGD-containing cyclic peptide, eptifibatide (Integrilin™, <u>1</u>), was approved in 1998 based on the results of the PURSUIT trial which evaluated eptifibatide in unstable angina (UA) or non-Q-wave myocardial infarction (MI) patients (6). The dosing regimen in the PURSUIT trial was based on *ex vivo* platelet aggregation using PPACK as the anticoagulant and maintaining the Ca^{++} concentration close to physiological levels to give a more accurate assessment of platelet activity than was obtained in the previous IMPACT II trial, which used citrated blood (7,8). The infusion rate was 3-4 fold higher than for IMPACT II and was estimated to produce a minimum of 80% inhibition of *ex vivo* ADP-activated platelet aggregation (6). Tirofiban (MK-383,

Aggrastat™, **2**), the first non-peptide IIb/IIIa antagonist to be marketed, was approved for patient-use in 1998 on the basis of the positive results of the PRISM and PRISM-PLUS trials which evaluated this compound in patients with unstable angina (9-11).

Subsequent clinical trials with these agents continue to support their expanded use for acute treatment of patients undergoing PTCA and in combination with thrombolytics for the treatment of MI in the emergency room setting (12). The EPISTENT and ESPRIT trials demonstrated highly improved outcomes for patients undergoing angioplasty who were treated with abciximab or eptifibatide, respectively, in combination with coronary stenting (13,14). In the TIMI 14 and SPEED trials, a combination of abciximab with fibrinolytics was superior to fibrinolytic therapy alone (15,16). In addition, YM-337, the Fab fragment of a humanized monoclonal antibody, and two non-peptide compounds, lamifiban, **3**, and fradafiban, **4**, have been evaluated clinically as iv agents (17-22). Recent data from the PARAGON B trial, which compared lamifiban added to aspirin and heparin vs. aspirin and heparin therapy in UA and non-Q-wave MI patients, however, failed to show a significant difference in the primary clinical endpoint (23).

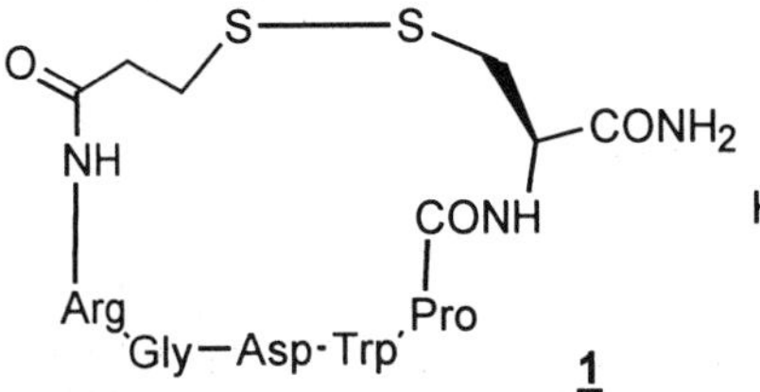

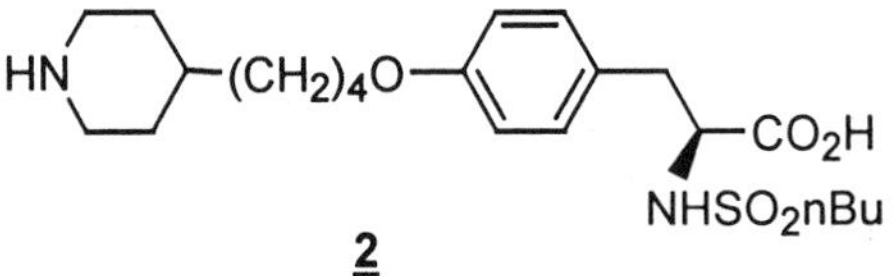

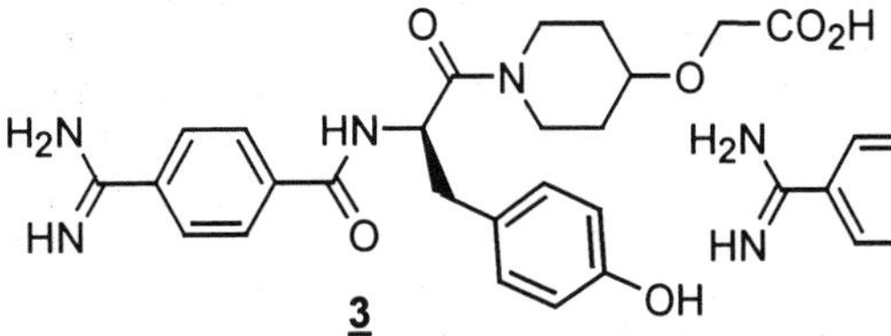

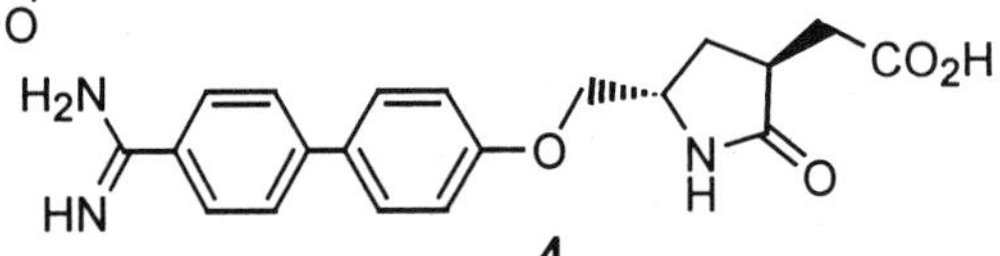

Oral IIb/IIIa Antagonists – While early clinical trials with several oral GPIIb/IIIa agents produced promising data supporting the use of these compounds to prevent reoccurrence of ischemic events, the results of several large Phase III trials have been disappointing and have led to the discontinuation of development of several of the early clinical compounds (2). Development of xemilofiban (SC-54684, **5**) was stopped as a result of the lack of efficacy in the EXCITE trial which evaluated long-term treatment with xemilofiban in 7200 patients undergoing angioplasty with or without stent placement. Patients were randomized to receive xemilofiban for 30-90 min before PTCA and then 10 or 20 mg tid for six months or ticlopidine for 2-4 wk followed by placebo. All patients received aspirin and heparin. At six months after PTCA, endpoint events occurred in 13.9% of patients in the 10 mg tid xemilofiban arm, 12.7% of patients in the 20 mg tid xemilofiban arm and 13.5% of patients on placebo (24-27). The incidence of periprocedural myocardial infarction was reduced from 4.2% with placebo to 3.3% at the 10 mg dose and 2.8% at the 20 mg dose for the first 2 days after angioplasty. However, a subsequent slight but nonsignificant increase in the rate of MI was also observed in patients on low dose xemilofiban, such that the early protective effect was not maintained at six months. The higher dose of xemilofiban also led to an increased incidence of bleeding complications (25,26). The lack of efficacy in this trial may be partly due to the fact that the dosing of xemilofiban was not started early enough to achieve 80% inhibition of platelet aggregation at the time of PTCA (28,29).

Enrollment in the OPUS-TIMI 16 trial, a study of long-term orbofiban (SC-57099B **6**) therapy was suspended when interim evaluation indicated an increase in early mortality at 30 days in orbofiban treated patients (30). Patients who had completed 30 days of treatment were continued in the trial (26). Patients were randomized to receive either placebo, orbofiban 50 mg bid for 30 days then 30 mg bid or orbofiban 50 mg bid in addition to aspirin. Mortality was increased by 0.8% in the 50 mg bid arm and by 1.5% in the 50 mg/30 mg bid arm compared to placebo (26) The increased mortality rate was confined to the subset of patients with reduced creatinine clearance (26). In a study of 71 patients dosed at 30, 40 and 50 mg orbofiban bid, it was found that while platelet aggregation was inhibited in a dose-dependent fashion, platelet adhesion decreased with the medium and high doses, but increased with the low dose indicating that low drug levels and low receptor inhibition enhanced platelet adhesion (31). When platelet activation was measured in a subset of patients randomized to receive 50 mg bid orbofiban or placebo in the OPUS-TIMI 16 trial, a significant increase in platelet activation, as measured by a higher expression of P-selectin, was observed at day 5-7 in patients receiving orbofiban (32). In a further substudy of patients from the OPUS-TIMI 16 trial, it was found that the dosage of orbofiban used in the trial did not reduce the increased platelet reactivity, as measured by fibrinogen binding in response to ADP and P-selectin expression, at 14 days after development of an acute coronary syndrome, although at 4 months a decrease in fibrinogen binding was observed (33). Increased platelet reactivity was observed in patients treated with orbofiban, which further suggests that in low doses, orbofiban acts as a partial agonist of GPIIb/IIIa (34).

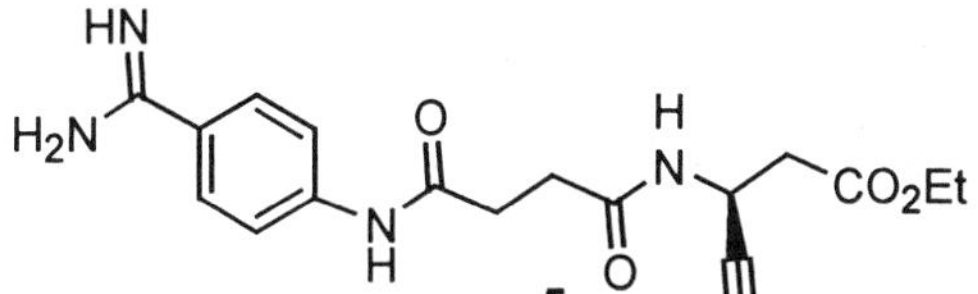

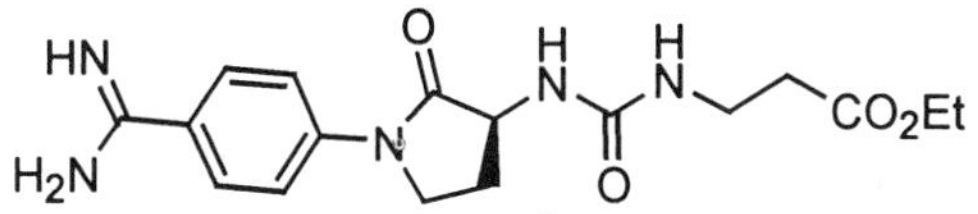

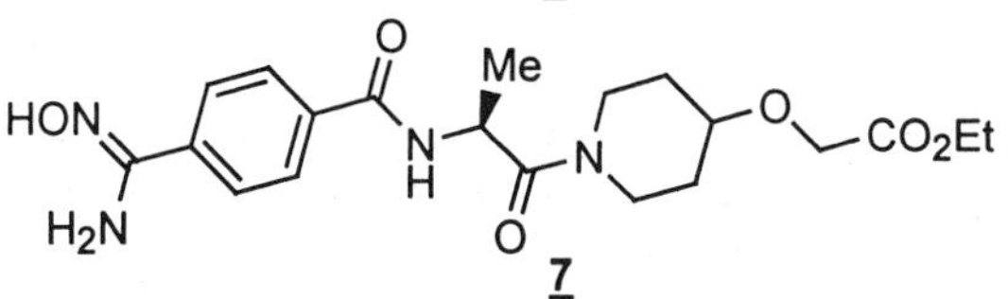

Sibrafiban (RO48-3657, **7**) was evaluated in the SYMPHONY trial (9000 patients) designed to assess high and low dose sibrafiban vs. aspirin in patients after an acute coronary syndrome. The high dose of sibrafiban was selected based on previous data from the TIMI 12 trial to avoid bleeding and the low dose was chosen to maintain 25% inhibition of ADP-induced platelet aggregation at trough (20,35). In the SYMPHONY trial, there was no significant difference between the sibrafiban treated patients vs. those receiving aspirin with respect to the 90-day incidence of a composite of mortality, myocardial infarction and severe recurrent ischemia. Patients in the high dose sibrafiban arm had a higher percentage of bleeding events (28). Based on these results, a second trial, SYMPHONY 2, which included a treatment arm combining low dose sibrafiban and aspirin was halted prematurely. The combination of sibrafiban with aspirin also failed to show any increased efficacy. A statistically significant increase in mortality (2.4% vs. 1.3%) and MI (6.9 vs. 5.3%) was observed with high dose sibrafiban (36)

A meta-analysis of the EXCITE, OPUS and SYMPHONY trials shows a statistically significant increase in death and the combined endpoint of death or MI independent of the dose of drug or the use of aspirin (37). A number of mechanisms for the observed increase in mortality have been proposed. These include consideration of whether GPIIb/IIIa antagonist interaction with the target receptor or non-target sites has negative consequences. Following an acute coronary event, patients can remain in a prothrombotic state for several months, which argues for prolonged treatment with antiplatelet agents (38,39). Heightened platelet activation as assessed by increased P-selectin expression (40) and fibrinogen binding has been reported for abciximab and fradafiban (41,42). Such prothrombotic actions have been observed in orbofiban treated patients as noted above (32,34) and might be especially relevant in periods of lower antiaggregation activity associated with trough drug levels. An alternate mechanism has been put forth based on the observation that xemilofiban and orbofiban cause apoptosis and procaspase-3 activation in rat cardiomyocytes mediated by an RGD recognition site on caspase-3. The effect of orbofiban on apoptosis was more pronounced in hypoxic cells, a situation potentially representative of ischemic myocardium (43). The relevance of these observations to the clinical setting remains to be elucidated.

Two other early compounds to reach clinical trials, Klerval (RPR-109891, structure unpublished) and ZD-2486, **8**, were discontinued after Phase II trials (21).

8

Despite the difficulties encountered by the first wave of oral agents, a number of oral GPIIb/IIIa receptor antagonists remain in clinical trials with several undergoing evaluation in once daily dosing paradigms. In addition, new preclinical compounds continue to appear as a result of ongoing efforts to identify a clinically useful oral agent. Lefradafiban (BIBU-104, **9**), a double prodrug form of fradafiban, **4**, is currently in Phase III trials in Europe for the treatment of unstable angina (22). High rates of bleeding were observed with this compound in the Phase II FROST trial, and the highest dose was stopped prematurely due to excess bleeding (26). Lotrafiban (SB-214587, **10**) entered Phase III trials in March of 1999 (44). The BRAVO trial will assess the efficacy of bid treatment with lotrafiban vs. placebo in acute coronary syndrome with a primary endpoint of a composite outcome of death, MI, stroke and need for vascular surgery at 2 yrs. All patients in this trial will also receive aspirin (45).

Roxifiban (DMP754) is a methyl ester prodrug of XV459 (**11**, R = H), a potent inhibitor of ADP-induced human platelet rich plasma (PRP) aggregation with an IC_{50} of 19 ± 3.8 nM (46). It has a terminal half-life in dogs of 12 h and suitable pharmacokinetics for once daily dosing (47,48). Like abciximab, **11** binds with high affinity to both activated and unactivated human platelets (49) and has a long dissociation half-life (7 min as compared to seconds for xemilofiban, orbofiban or sibrafiban) (49-52). This may translate into reduced potential for prothrombotic effects since the drug has a longer receptor residency and may therefore present less opportunity for fibrinogen binding to drug-activated platelets (42,52). The combination of tight binding and slow off-rate results in a lower peak to trough ratio than has previously been achieved with other oral agents. Measurement of platelet aggregation in human blood is not affected by Ca^{++} concentration and is identical when measured in either citrated or heparinized blood (53). Roxifban has completed Phase II trials (ROCKET 1 and ROCKET 2).

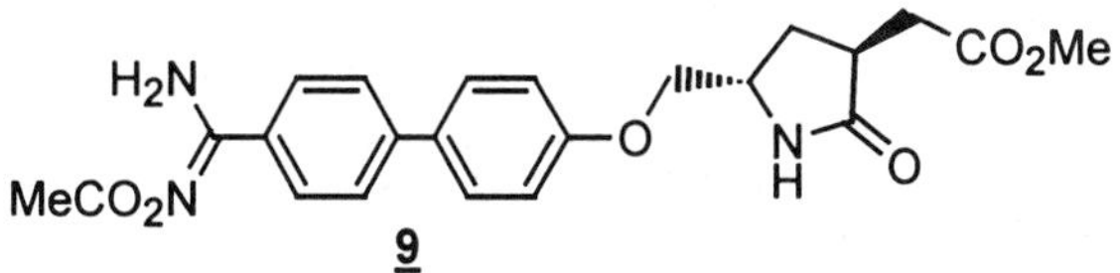

9

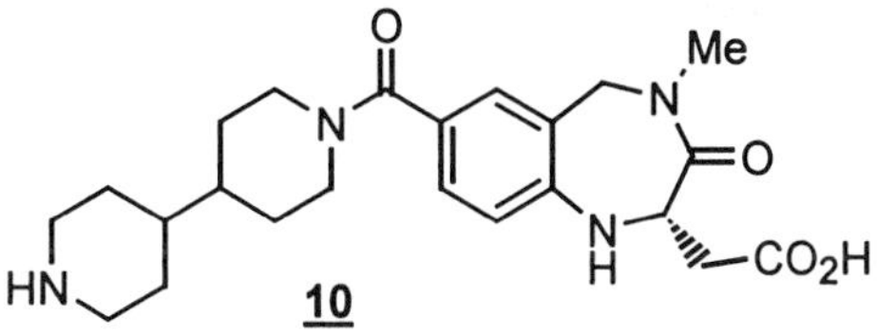

10

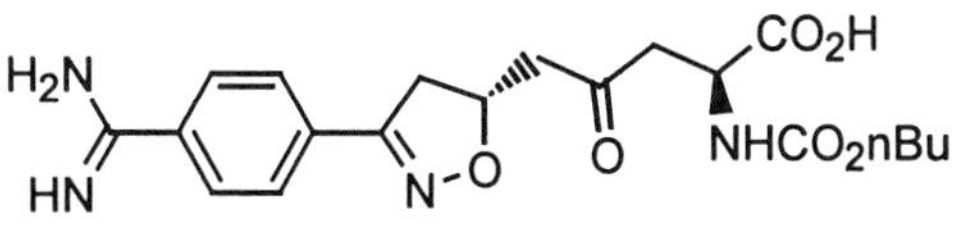

11

Elarofiban, **12a**, (RWJ-53308, R = 3-pyridyl) which also recently completed Phase II trials, resulted from the optimization of RWJ-50042, **12b**, (R = H) an early prototype compound which was designed to mimic the gamma chain of fibrinogen (54). In Phase I trials, a 1 mg/kg dose had a t½ of 25.5 h and inhibited ADP-induced platelet aggregation by 51% after 2 h (55). A similar profile to abciximab was observed in the effect of calcium concentration on measured inhibition of platelet aggregation (56). Preliminary results of Phase I and II studies with cromafiban (CT50352, structure unpublished) have also been reported. In healthy volunteers a terminal half-life of 20 h and a Cmax at 2 h which corresponded with the maximum inhibition of *ex vivo* platelet aggregation was observed (57). Another compound with a long half-life which has reached advanced clinical trials is gantofiban (EMD-122347, **13a**, R = Et, R^1 = CO_2Me) which is rapidly absorbed after oral administration and converted into its active metabolite, EMD-132338, **13b** (R = R^1 = H). In Phase I clinical trials a dose-independent terminal half-life of 21.2±6.0 h was observed after single oral doses with reproducible pharmacokinetics (58).

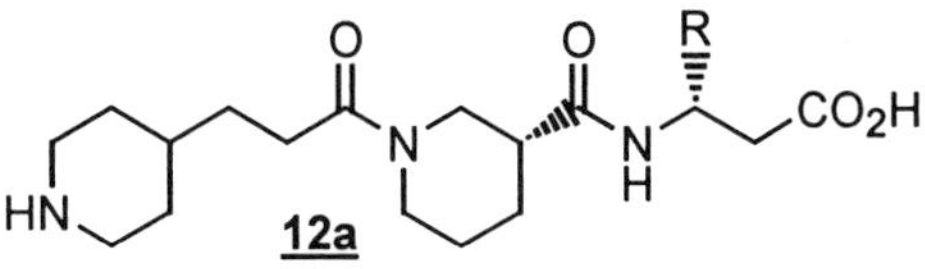

12a
12b

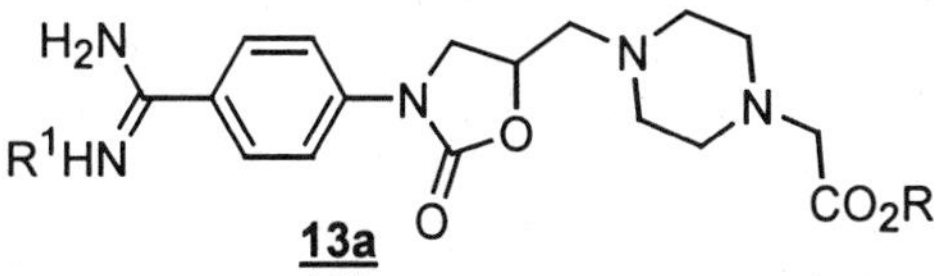

13a
13b

SR121787, **14,** is a diester, N-ethylcarbamoyl prodrug of SR121566 (IC$_{50}$ = 46 nM). A 2 mg/kg oral dose in baboons resulted in rapid inhibition of *ex vivo* platelet aggregation (15 min) and significant inhibition (43%) was still detected 24 h post dose (51). No bleeding events or drop in platelet counts were observed at doses which gave maximal inhibition of platelet aggregation. SR121787 was active in the rabbit arterial shunt model of thrombosis with antithrombotic activity correlating with *ex vivo* platelet aggregation. (59). A combination of low doses of SR121787 and either heparin or an indirect inhibitor of factor Xa, SR90107/ORG31540 resulted in a synergistic effect on antithrombotic efficacy in the rabbit carotid artery lesion model. (60). While the combination of SR121787 and heparin or high dose SR121787 alone resulted in an increased bleeding risk in this model, the combination of SR-121787 with an antithrombin dependent factor Xa inhibitor did not (60).

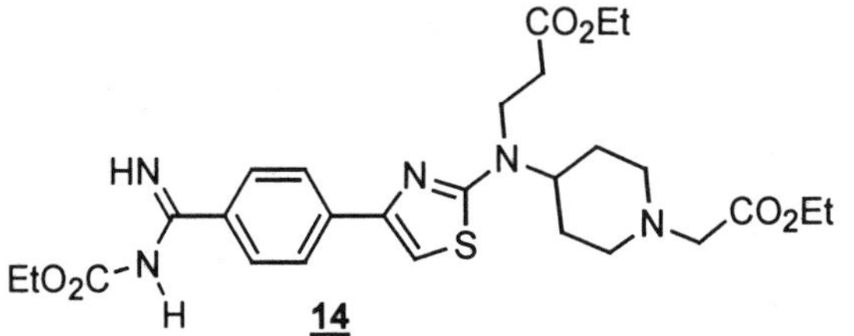

ME3230, (EF5289, **15a**, R = tBuOCH$_2$OCO, R^1 = Et), a prodrug of EF5154 (**15b**, R = R1 = H, IC$_{50}$ = 47 nM) was selected for development based upon its good permeability across CaCO-2 membranes, metabolic stability in human liver and plasma, and oral bioavailability (42.8% in dogs and 24% in rhesus monkeys) (61). A 1 mg/kg dose in dogs resulted in complete inhibition of platelet aggregation in 1-2 h with 50% inhibition at 8 h which corresponds to tha 3.7 h half life (62). Another ketopiperazine containing compound, TAK-029, **16**, (IC$_{50}$ = 30 nM) when dosed orally in guinea pigs at 10 mg/kg produced 50% inhibition of *ex vivo* platelet aggregation 24 h after dosing (63).

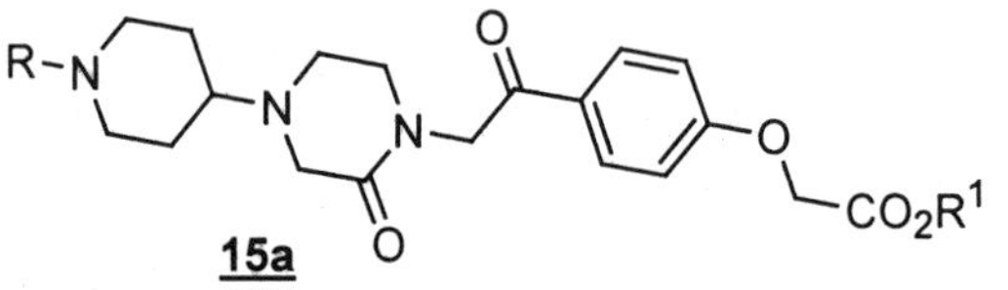

Ester prodrug, MS-180, **17a**, (R = Et) is converted *in vivo* into its active metabolite **17b** (R = H, IC$_{50}$ = 35 nM). Oral administration of **17a** in dogs resulted in dose dependent *ex vivo* inhibition of platelet aggregation with maximum inhibition at 2-4 h post dose (64). Intraduodenal administration of a 1 mg/kg dose in a guinea pig thrombosis model delayed or substantially prevented occlusive thrombus formation. An 80% inhibition of *ex vivo* platelet aggregation was achieved with only a slight effect on the bleeding time (65). AR0510, the ethyl ester prodrug of **18** (IC$_{50}$ = 18 nM), which incorporates a *trans*-cyclohexane moiety as a conformational restraint, inhibited *ex vivo* platelet aggregation by 80% for 6 h after a 10 mg/kg oral dose in dogs (66).

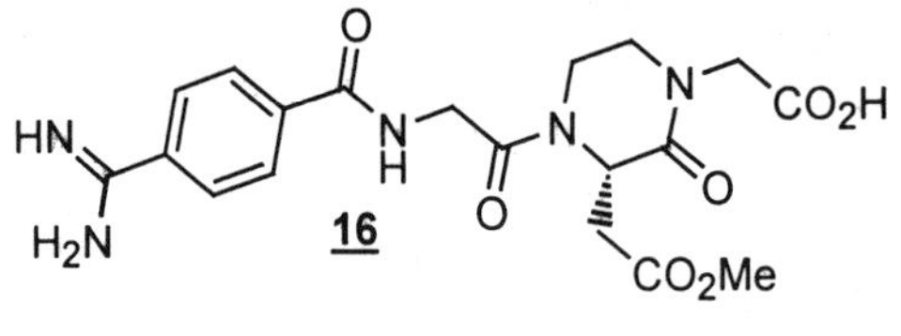

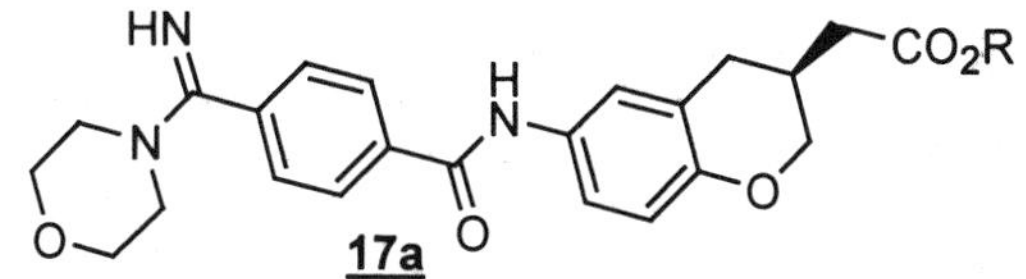

17a
17b

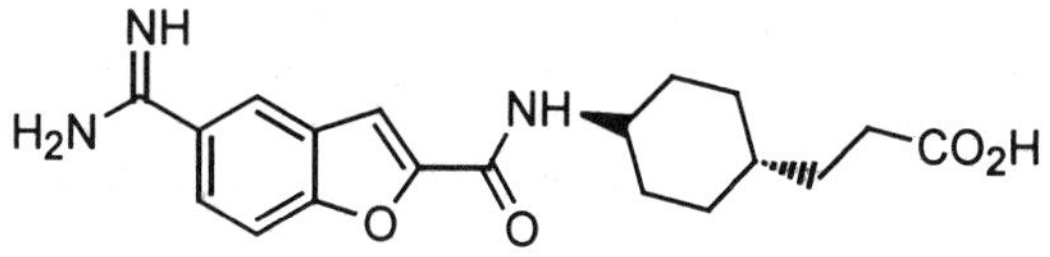

18

Hydantoin S1197, **19**, is orally active when administered as its ethyl ester prodrug S5740 (67). The pharmacokinetics of S1197 after single oral doses of the prodrug in the form of HR1740 (acetate salt) and HMR1794 (maleate salt) were linear and displayed a terminal half-life of 16.8±8.5 h. Significant inhibition of platelet aggregation was observed at 24h at doses of 150 and 200 mg HMR1794 (68). SM-20302, **20**, (1mg/kg, po) significantly prolonged the time to occlusion in an arterial thrombosis model in guinea pigs (69). In the electrolytic injury model of coronary artery thomobosis in dogs, it was found that doses of SM-20302 which produced 57-59% inhibition of platelet aggregation for <24 h led to long-term (5days) reduction in arterial thrombosis (69,70). The *in vivo* antithrombotic efficacy of SM-20302 correlates with its *ex vivo* platelet inhibition in heparinized blood but not with that in citrated blood (69-71).

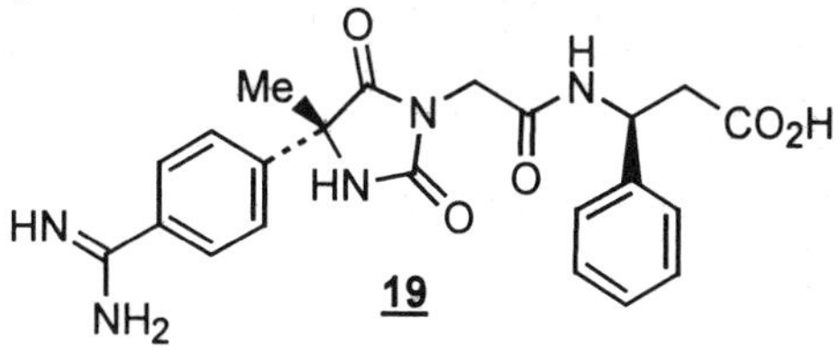

19

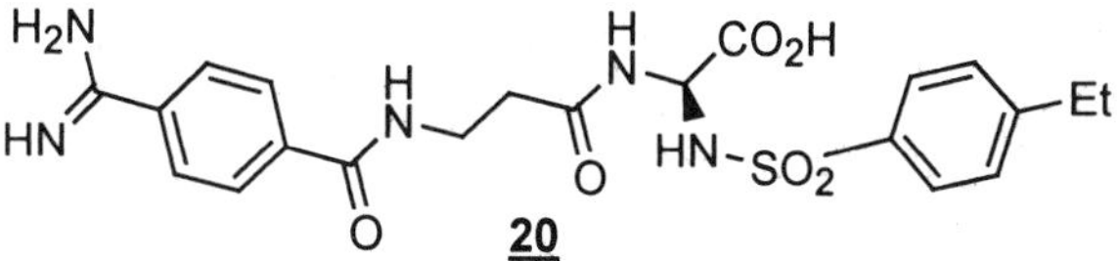

20

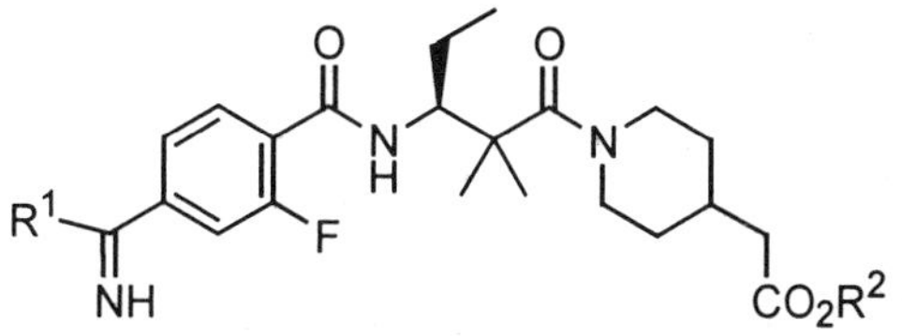

21a $R^1 = NH_2$, $R^2 = H$ **21b** $R^1 = NH_2$, $R^2 = Et$

21c $R^1 =$ 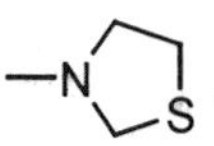   , $R^2 = Et$

NSL-96173, **21a**, identified from an SAR study of a series of compounds that incorporate a trisubstituted β-amino acid as a conformational restriction, had limited bioavailability even as its ethyl ester prodrug **21b**. Further modification by introduction of a thiazolidine at one of the amidine nitrogens provided NSL-96184, **21c**, which had improved C_{max} and prolonged half-life after oral administration without loss of potency. The free acid of NSL-96184 inhibited platelet aggregation in human PRP with an IC_{50} of 45 nM. Oral administration of a 10 mg/kg dose in guinea pigs resulted in a quick onset ($\leq$ 30 min) but relatively short duration (t½ of 1.8 h) possibly due to oxidative metabolism of the sulfur atom in the thiazolidine ring (72).

A potent thiazole compound, UR-12947, **22**, (IC_{50} = 3.5 nM), when dosed po in dogs at 0.1-1 mg/kg, resulted in 80-90% inhibition of *ex vivo* platelet aggregation with peak inhibition at 2 h with 40% inhibition maintained out to 24 h (73). A series of compounds incorporating 6,6-bicyclic scaffolds such as isoquinoline, tetralin, tetralone, and benzopyran have been described. Tetralone **23** from this series is reported to be a 60 nM inhibitor with 23% oral bioavailability in rats (74).

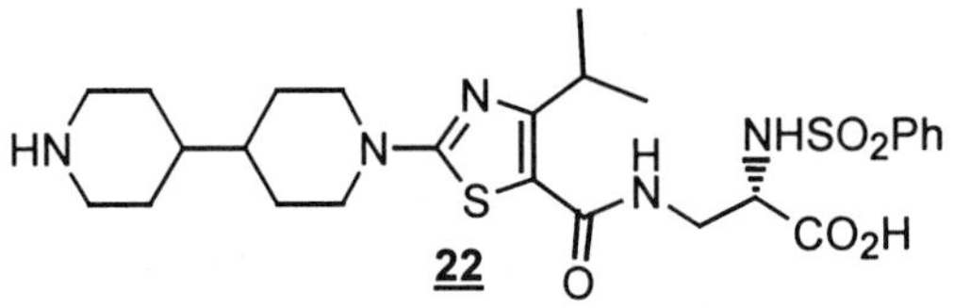

22

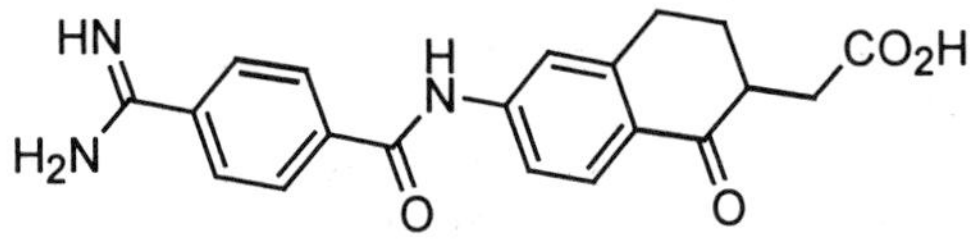

23

An extensive SAR study around the alpha carbamate substituent of roxifiban (**11**) resulted in the discovery that aryl and heteroarylsulfonamide analogs showed prolonged duration of *ex vivo* platelet aggregation after oral dosing in dogs, particularly when an ortho substitutent was present on the aryl group (75). DMP802, **24**, as the mesylate salt, was found to have the best combination of solubility and duration of action in this series. Analogs of roxifiban with an isoxazolidine (**25**) or isoxazole (**26**) scaffold are also potent antagonists, with IC_{50} = 28 and 50 nM, respectively (76,77).

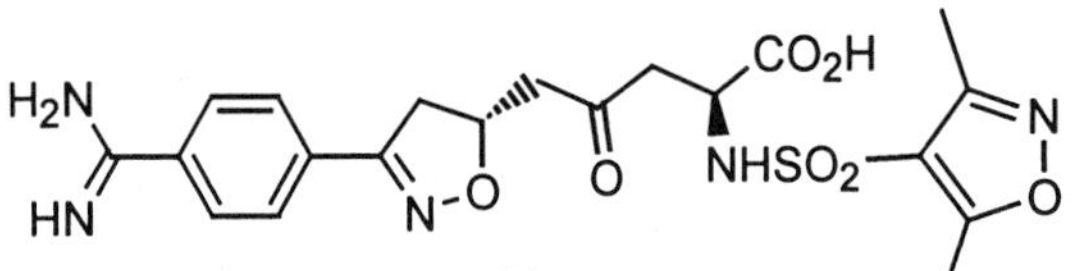

24

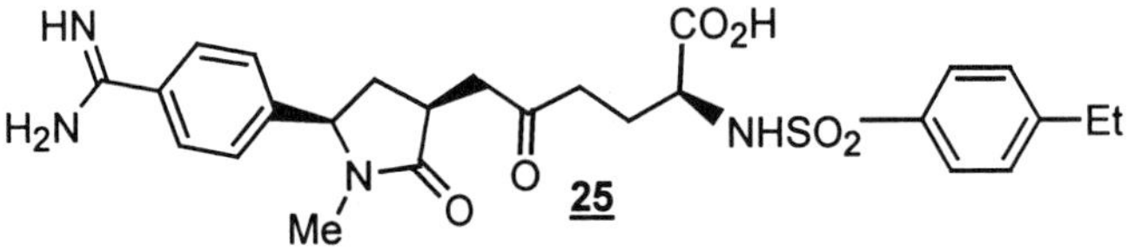

25

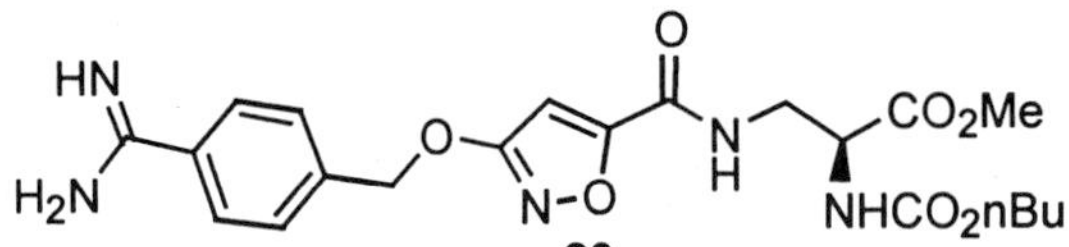

26

Thienothiophene dervative **27** is an 8 nM inhibitor of platelet aggregation and has high affinity for both resting and activated platelets (78). Compound **28** (L-750,034 (IC_{50} = 17 nM) was designed to minimize the distance between the arginine and aspartic acid mimics. It showed good oral activity in dogs with a single 2 mg/kg dose resulting in inhibition of *ex vivo* platelet aggregation by >40% for 4 h (79). UR-3216 (structure undisclosed) when dosed po in monkeys maintained an *ex vivo* antiaggregatory effect for >20 h (80,81). A series of nitroaryl based inhibitors with varying integrin selectivity led to compounds such as **29** which are selective IIb/IIIa antagonists (82). Tricyclic compound, **30**, which is an extension of a previously reported benzodiazepinedione analog, had an IC_{50} of 54 nM (83).

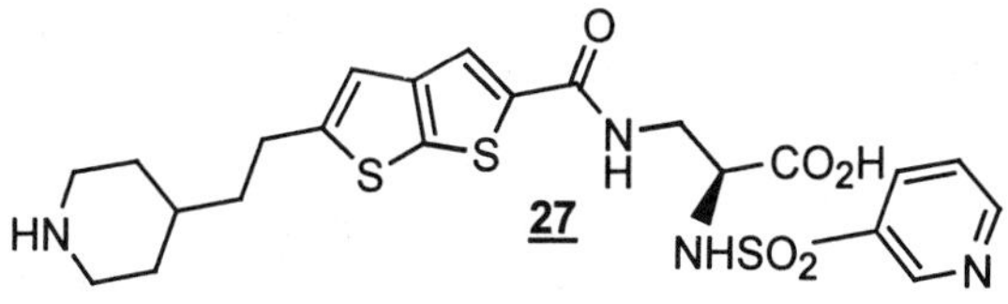

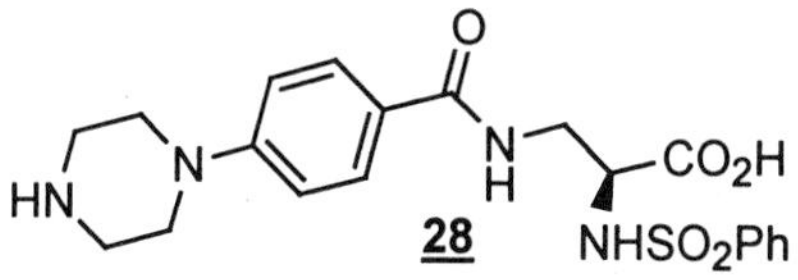

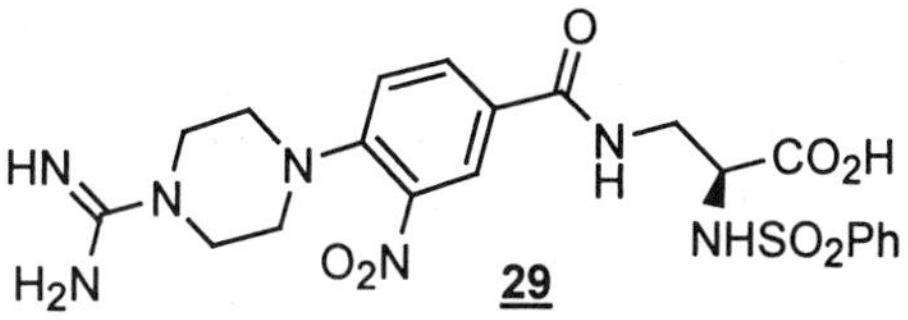

Acetylenic compound **31** was also potent with an IC_{50} value of 37 nM (84). Triazolopyridine **32**, and piperazone, **33**, (CRL-42789) are novel structures that have recently appeared in the patent literature (85,86). While not an exhaustive list, these compounds provide an insight into the diversity of structural classes and the continuing interest in identifying a safe, efficacious and orally bioavailable GPIIb/IIIa agent.

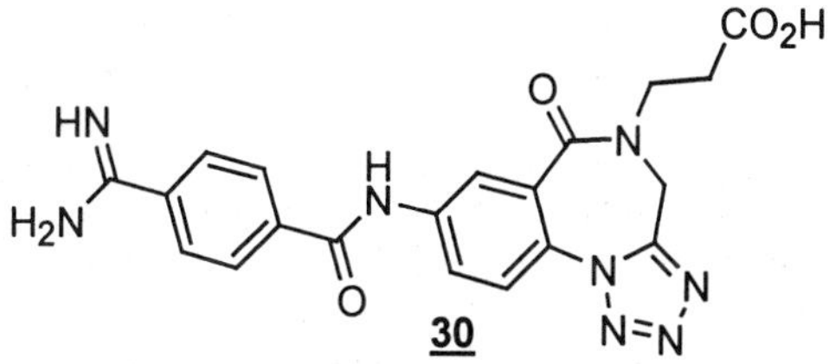

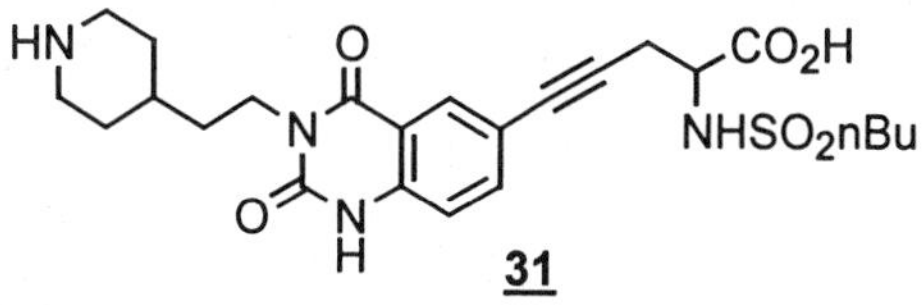

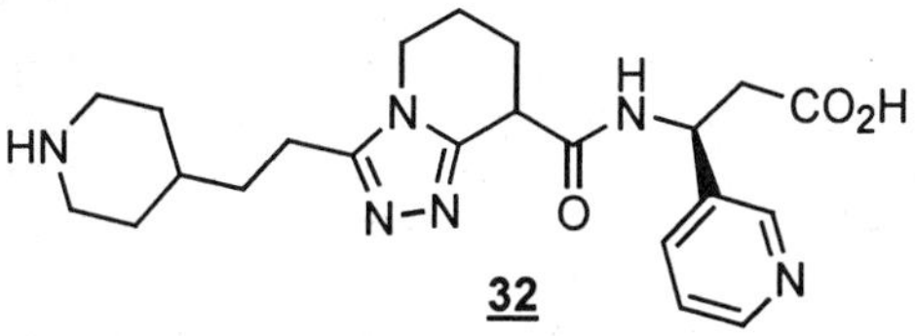

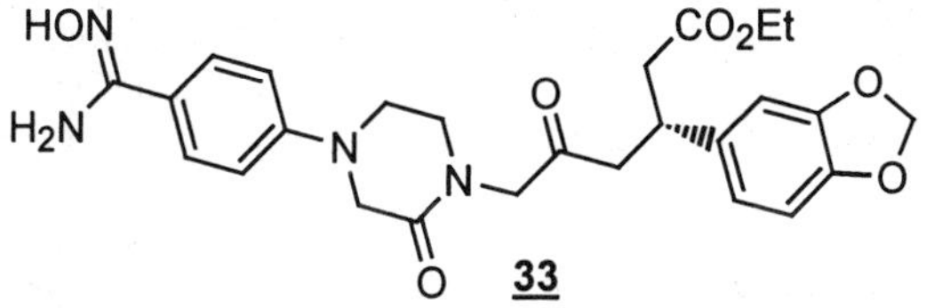

ISSUES FACING THE USE OF ORAL GPIIB/IIIA ANTAGONISTS:

The principal challenge to the prophylactic use of oral agents is the demonstration of a clear therapeutic benefit. Issues of patient selection, optimal dose and pharmacological profile for an oral agent, appropriate platelet monitoring, and safety considerations relevant to long-term treatment remain unresolved.

Drug-platelet binding profiles – The platelet binding profile of GPIIb/IIIa antagonists appears to be a major determinant of their pharmacokinetic and pharmacodynamic profiles. Agents under current study vary in their GPIIb/IIIa affinity and in their selectivity for the activated and unactivated platelet receptor (87). Significant affinity for unactivated GPIIb/IIIa is a property shared by certain high affinity GPIIb/IIIa antagonists. Among the intravenous agents, abciximab is characterized by approximately equal affinity for activated and unactivated platelets (88), while tirofiban and eptifibatide have higher affinity for activated over unactivated platelets (88). The platelet dissociation half-life of abciximab is longer (3-4 hours vs. 11 seconds for tirofiban) (88), and its biological half-life, as measured by restoration of ADP-induced platelet aggregation, is significantly longer (12 hours vs. 2-4 hours for 50% return of platelet function), a potential benefit for improved efficacy (89). Although direct clinical comparisons have not been conducted, inter-trial analyses of similar patient groups for each of these agents have suggested a superiority of outcomes for abciximab (89). The oral compounds also vary in their platelet binding profiles. The active metabolites of orbofiban, **6**, and roxifiban, **11**, demonstrated Kd values for activated human platelets of 113 and 2.85 nM, respectively, and 663 and 3.87 nM for resting platelets, perhaps accounting for the differing dosing regimens (49,90,91). The clinical ramifications of the differing binding profiles need further investigation. The larger reservoir of free drug in circulation for lower affinity agents may enable a more robust antithrombotic response in the presence of activated platelets, which can express new GPIIb/IIIa receptors through agonist-stimulated α-granule release (88). On the other hand, the prolonged platelet occupancy for high affinity agents may allow less frequent dosing, with concomitant decreases in peak-trough drug level variation and a lower incidence of bleeding events.

Patient selection – Improved understanding of risk factors and diagnostic methods might identify patients likely to suffer severe cardiac events in the 1-6 month period following a primary ischemic episode (92-94). Recent attention has focused on identifying biochemical markers capable of detecting earlier myocardial injury, even in the absence of cardiac enzyme elevations (95,96). Two specific markers of myocardial necrosis, troponin T and troponin I, have attracted attention as potential aids in treatment decisions. In ACS patients treated with the parenteral GPIIb/IIIa antagonists abciximab and tirofiban, retrospective analyses showed that elevated levels of troponin T and troponin I identified subgroups more likely to benefit from GPIIb/IIIa blockade (97,98). In a phase II trial (FROST) of the oral GPIIb/IIIa antagonist, lefradafiban, **13**, higher troponin levels predicted a greater reduction in cardiac events (99). The prospective diagnostic value for troponin levels has not been demonstrated in a Phase III trial. C-Reactive protein, a marker for systemic inflammation, continues to receive attention as a potential predictor of cardiovascular disease although no studies have been undertaken in conjunction with GPIIb/IIIa antagonists (100-103). In a possible application of pharmacogenetics to ischemic disease, several groups have found an association of a polymorphism known as PlA2 (L33P) on GPIIIa with an increased risk of myocardial infarction (104,105) Analyses of GPIIb/IIIa treatment groups on the basis of GPIIIa polymorphisms have not been reported.

Dose selection – efficacy vs. bleeding – The proper balance of antithrombotic efficacy and maintenance of hemostasis in the context of long term prevention of ischemic events is not known. The limited efficacy observed for oral agents to date suggests that further reductions in platelet activity may be necessary, although whether this applies to peak or trough antithrombotic effects is unclear. In trials of parenteral GPIIb/IIIa antagonists in the treatment of acute coronary syndromes, high levels of platelet blockade (> 80% IPA) were found to be necessary for efficacy (106). The maintenance of such high levels of platelet inhibition, if needed, may prove difficult in the chronic setting. It may be especially difficult to maintain the trough levels needed for efficacy without having excessive peak levels resulting in bleeding unless a compound has a long half-life. Since the highest dose of sibrafiban in SYMPHONY was set with a view towards bleeding events observed in the TIMI 12 trial (35), and higher doses may thus be problematic, optimization of oral GPIIb/IIIa therapy may need to incorporate the minimization of peak-trough ratios associated with longer duration (q.d.) agents (28).

In order to set appropriate doses for GPIIb/IIIa antagonists, numerous means of assessing antiplatelet activity have been used as surrogates of antithrombotic efficacy (88,107,108). The methods most frequently employed in clinical trials have used turbidometric aggregometry. An exaggeration of drug antiplatelet effect in citrated (calcium poor) PRP may have led to an overestimation of the antiaggregation effect of eptifibatide in the first efficacy trial of this agent, and contributed to the modest efficacy observed (8,109). A whole blood assay capable of improved discrimination at therapeutic levels of inhibition was recently used to assess the effect of lefradafiban (110). Receptor occupancy has attracted attention as another surrogate for antithrombotic efficacy (88). A method once used extensively as a surrogate for clinical bleeding, template bleeding times, has been shown to be of little value as a predictor of hemorrhagic events in trials of GPIIb/IIIa antagonists (107).

Thrombocytopenia – Thrombocytopenia, defined as a drop in platelet count below 100-149 x 10-9/L (mild) to <20 x 10-9/L (profound), has been observed after administration of parenteral and oral GPIIb/IIIa antagonists (111-113). The risk among the parenteral agents appears to be higher with abciximab (111), although one analysis found similar incident rates with tirofiban (114). The etiology of the platelet loss has been postulated to be immune mediated, possibly involving neoepitopes induced by the drug-receptor interaction. Drug-dependent antibodies (DDab) have been associated with thrombocytopenia observed after experimental GPIIb/IIIa antagonists were given to primates (115). Drug-induced thrombocytopenia (DIT) was observed in 0.44% of patients treated with xemilofiban and DDab were identified in patient sera (116). The incidence of thrombocytopenia in the first 30 days of the OPUS-TIMI 16 trial was 0.7% for orbifiban patients vs. 0.1% on placebo. Post-30 days, the rate of incidence was 0.1% for both groups, suggesting a need for early monitoring of platelet counts (117). Methods for diagnosing patients at risk of developing DIT in response to GPIIb/IIIa antagonists have been disclosed, but no reports on the use of such an assay to prescreen patients has been reported to date (118,119). In the SYMPHONY trial, thrombocytopenia was infrequent, and did not differ among the treatment groups (28). Fradafiban reportedly does not induce platelet GPIIb/IIIa ligand-induced binding sites in patients and low incidences of thrombocytopenia were also reported for lefradafiban in Phase I and II trials (22). Non-immune mediated thrombocytopenia involving activation and aggregation of platelets following abciximab administration has been reported (120). The clinical management of confirmed thrombocytopenia centers on removal of known causative factors and minimization of bleeding risk (111,121). In the majority of cases, normalization of platelet counts has followed cessation of drug. In cases of profound thrombocytopenia, platelet transfusions have been necessary (111).

ADP RECEPTOR ANTAGONISTS

Adenosine diphosphate (ADP) plays a key role in hemostasis and thrombosis by directly stimulating platelet aggregation. Stimulation of platelets by ADP leads to rapid calcium influx, mobilization of calcium from intracellular stores, activation of phospholipase C, thromboxane A2 production and inhibition of adenylyl cyclase. ADP also causes platelets to undergo a shape change and activate fibrinogen receptors which leads to platelet aggregation. Initially, all the physiological and intracellular signaling events triggered by ADP in platelets were attributed to the P2T receptor (P2 receptors on thrombocytes). More recently, due to the availability of selective P2 receptor ligands, the platelet P2 receptor has been resolved into three P2 receptor subtypes, the Ca^{2+} ion-gated P2X1 receptor and two G-protein coupled receptors (P2Y1, and P2YAC) (122-127). The ionotropic P2X1 receptor has been shown to mediate the rapid influx of calcium. Activation of the P2Y1 receptor is essential when ADP and collagen are the principal agents involved in platelet aggregation and is required for platelet shape change. P2Y1-null mice are resistant to thromboembolism induced by intravenous injection of ADP or collagen and adrenaline but have no spontaneous bleeding tendency (126). The P2YAC receptor, as yet uncloned, is coupled to the inhibition of adenylyl cyclase.

The thienopyridine, clopidogrel, **34**, a follow-on analog of ticlopidine, **35**, is an effective antiplatelet drug, useful for the treatment of ischemic cerebrovascular, cardiac and peripheral artery diseases (128). Clopidogrel has antithrombotic activity several times greater than ticlopidine and aspirin, and has a favorable safety and tolerability profile. Data from the CLASSICS trial, in which 1020 patients undergoing stent inplantation were randomized to a combination of 75 mg clopidogrel for four weeks (with or without a 800 mg loading dose) and aspirin or ticlopidine and aspirin, demonstrated that clopidogrel was equivalent in efficacy with a more benign side effect profile (129). Clopidogrel is an irreversible inhibitor of ADP-induced platelet aggregation and ADP-induced inhibition of C-AMP accumulation in platelets. Clopidogrel's anti-aggregating activity is due to a short-acting metabolite generated in the liver by a cytochrome P_{450}-dependent pathway. Up until recently, clopidogrel's mechanism of action was not well understood, although it was known to inhibit ADP-induced platelet aggregation. The data from a study in healthy volunteers receiving 75 mg of clopidogrel daily for seven days indicates that clopidogrel impairs the ADP receptor coupled to Gi/adenylyl cyclase (P2YAC) but does not affect platelet ADP receptors coupled to cation influx (P2X1) or calcium mobilization (P2Y1) (130). CS-747, **36**, another thienopyridine in clincal development, is more potent and longer-lasting than clopidogrel (131).

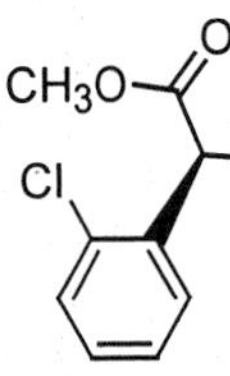

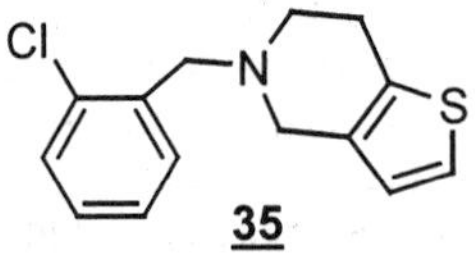

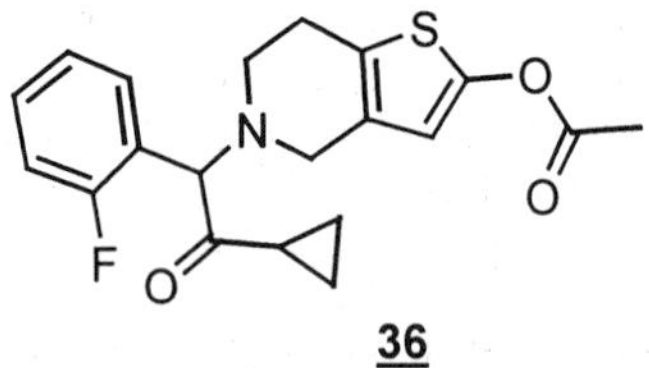

AR-C69931MX, **37**, an ATP-analog which is currently in Phase II trials, is being developed as an intravenous agent for the treatment of unstable angina and for use in the setting of PTCA (132,133). It is a potent and selective P2T (recently redefined as P2YAC) receptor antagonist which was designed to have a rapid metabolic clearance (t½ is 2.6 minutes in humans). This compound has superseded AR-C67085MX, **38**, which also reached Phase II evaluation. AR-C69931MX and AR-C67085MX have IC_{50}'s against ADP-induced aggregation of 0.4 nM and 2.5 nM, respectively (133).

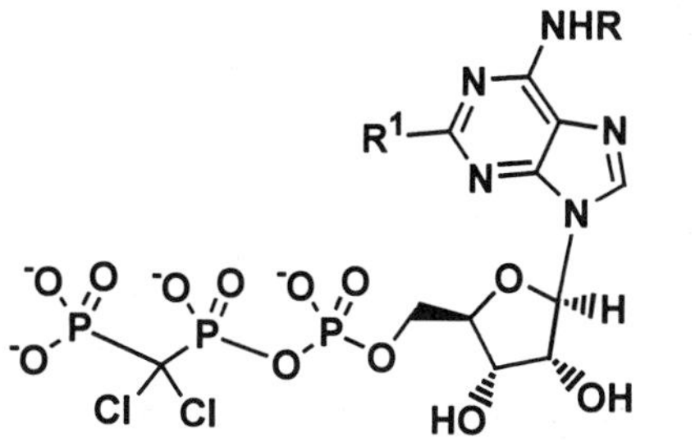

37, R = (CH$_2$)$_2$SMe, R^1 = S(CH$_2$)$_2$CF$_3$
38, R = H, R^1 = SPr

Comparison of the effect on bleeding time of AR-C69931MX compared to lamifiban in anesthetized dogs suggests that the risk of persistent hemorrhage will be significantly less for AR-C69931MX. This may prove to be a potential advantage for P2YAC antagonists over GPIIa/IIIa antagonists. In Phase I studies, AR-C69931MX demonstrated 100% inhibition of platelet aggregation with only a 1.5-fold increase in bleeding time which returned to normal 30 minutes after discontinuation of intravenous infusion. Data from the first clinical study with AR-C69931MX in patients with acute coronary syndrome at either 2 µg/kg/min or 4 µg/kg/min intravenously in addition to aspirin and heparin showed that it is a potent, short acting platelet ADP-receptor antagonist which is well tolerated as adjunctive therapy and affords stable inhibition of ADP-induced platelet aggregation with concomitant extension of bleeding time (134). Like clopidogrel, AR-C69931MX and related ATP analogs inhibit ADP-induced platelet aggregation via the adenylyl cyclase pathway, but do not affect calcium mobilization or shape change. Thus these compounds act on the P2YAC or P2CYC receptor without affecting the P2Y1 receptor (124,127,128,135).

THROMBIN RECEPTOR ANTAGONISTS AND AGONISTS

Since thrombin is the most potent activator of platelets, understanding the process by which thrombin activates platelets remains an important goal toward understanding hemostasis and thrombosis and may result in novel antiplatelet therapies (136,137). Thrombin signaling is mediated by a family of G-protein coupled protease-activated receptors (PARs) which are activated by thrombin cleavage of an N-terminal domain exposing a new NH$_2$-terminus which acts as a tethered ligand. It has been shown that PAR1 and more recently PAR4 mediate most, if not all, of the thrombin signaling in human platelets (137-139). The amino acid sequences of the tethered activating ligands for the human receptors are SFLLRN for PAR1 and GYPGQV for PAR4. At low concentrations of thrombin (1 nM), inhibition of PAR1 but not PAR4, markedly attenuated platelet aggregation, suggesting that inhibition of PAR1 alone may be sufficient to exert an antithrombotic effect (140). At high thrombin concentrations (30 nM), neither PAR1 nor PAR4 blockade alone significantly inhibited platelet aggregation, however, simultaneous inhibition of both PAR1 and PAR4 blocked platelet activation and aggregation, hence it may be necessary to inhibit both PAR1 and PAR4 to prevent thrombosis (136,140).

Naturally-occuring Thrombin Receptor Antagonists – Eryloside F, (**39**) an isolate from an extract of the marine sponge *Erylus formosus*, has been shown to be a potent thrombin receptor antagonist. Eryloside F was approximately 6-fold more potent against SFLLRN than against U-46619 (a stable thromboxane A2 mimetic)-induced platelet aggregation with IC$_{50}$ values of 0.3 and 1.7 µM respectively (141).

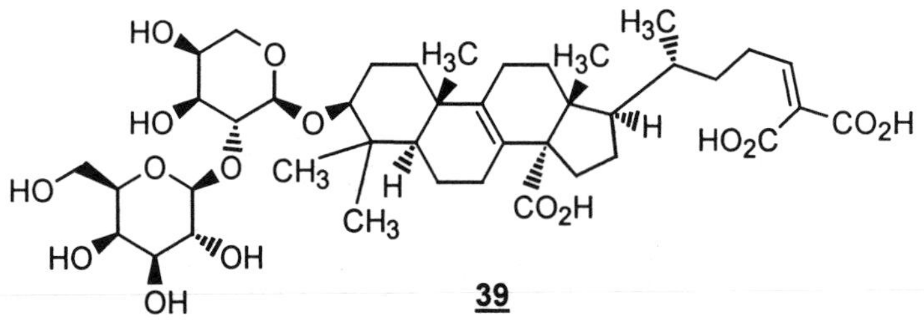

39

<u>Peptide and peptidomimetic agonists and antagonists</u> – Analoging the tethered-ligand sequence of PAR1 has led to the development of potent PAR1 peptide antagonists such as BMS-197525 (N-*trans*-cinnamoyl-*p*-fluoroPhe-*p*-guanidinoPhe-Leu-Arg-NH$_2$) which blocked SFLLRNP-NH$_2$ induced platelet aggregation with an IC$_{50}$ of 0.20 µM and bound to PAR1 with an IC$_{50}$ of 0.008 µM (142-144). Macrocyclic hexapeptide analogs of SFLLRN were shown to be less potent in inducing platelet aggregation relative to SFLLRN-NH$_2$ and did not act as antagonists of α-thrombin, with the most potent macrocycle being **40** (IC$_{50}$ = 24 µM) (145).

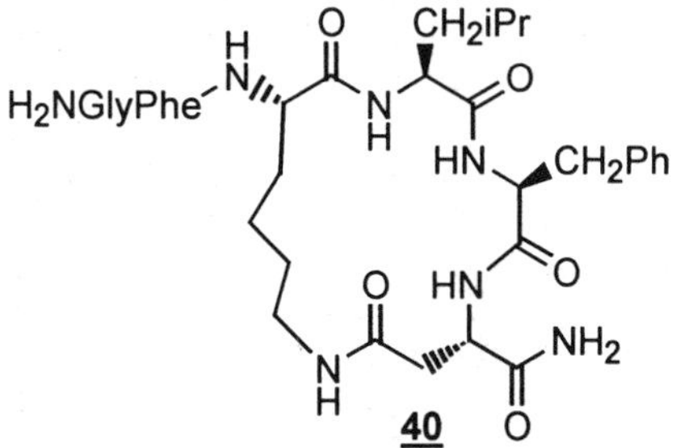

40

A series of azole-based carboxamides designed to be peptidomimetic analogs of SFLLRN, in which oxazole or thiazole is employed at positions 2/3 as a dipeptide mimetic, has been reported. Oxazole **41** binds to PAR1 with an IC$_{50}$ of 1.6 µM and had IC$_{50}$ values of 25 µM and 6.6 µM against α-thrombin and SFLLRN-NH$_2$ induced platelet aggregation, respectively (146). RWJ-56110, **42**, which employs a 6-aminoindole template designed based on distance parameters taken from models of SFLLRN-NH$_2$ generated computationally, is a potent, selective PAR1 antagonist with an IC$_{50}$ of 0.44 µM vs. PAR1 binding. Compound **42** inhibits platelet aggregation induced by α–thrombin (IC$_{50}$ = 0.34 µM) and SFLLRN-NH$_2$ (IC$_{50}$'s = 0.16 µM). It has no effect on PAR2, PAR3, or PAR4 (147).

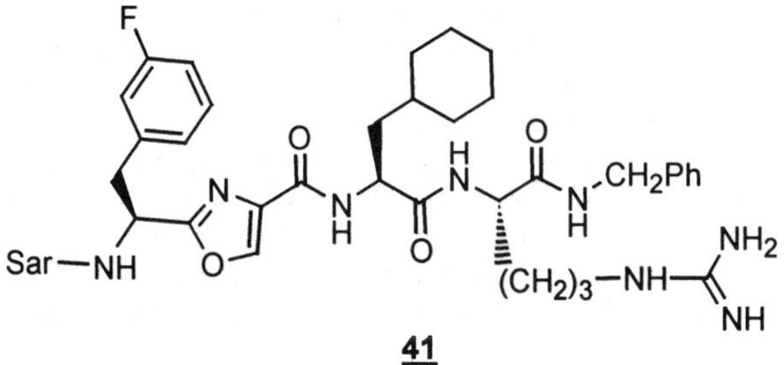

41

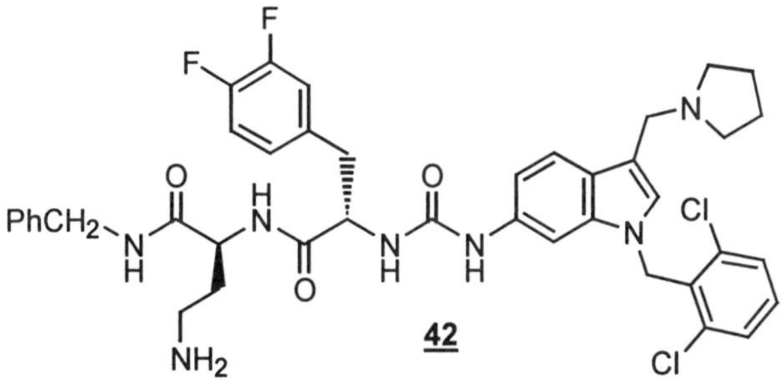

42

Peptidomimetic agonists of the thrombin receptor have also been reported. A series of heterocycle-peptide hybrids containing a CHa-Arg-Phe motif or closely related tripeptide and an N-terminal aminotriazole (e.g., **43**) was found to be nearly as potent as SFLLRN-NH$_2$ in inducing platelet aggregation (148). Replacement of the aminotriazole with other substituted heterocycles also resulted in similar agonist potency (IC$_{50}$ of 0.5-1 µM compared to 0.28 µM for SFLLRN-NH$_2$). An arylethenoyl cap on the N-terminus resulted in analogs with mixed agonist-antagonist activity (148).

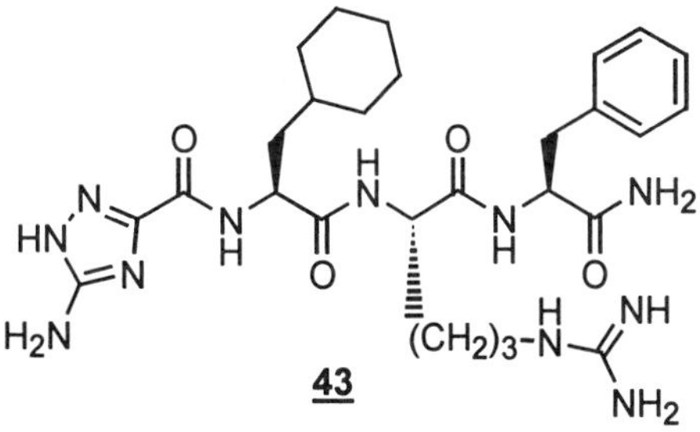

43

Non-peptide thrombin receptor antagonists – Two series of non-peptide PAR1 antagonists, represented by thiazolidinone, FR171113, **44**, and pyrroloquinazolines, **45** and **46**, recently emerged. FR171113 inhibits thrombin-induced platelet aggregation and SFLLRN-NH$_2$ induced platelet aggregation with an IC$_{50}$ = 290 nM and 150 nM, respectively (149). FR1711113 did not inhibit platelet aggregation induced by arachidonic acid, U-46619, platelet activating factor, adrenaline and calcium ionophore A23187 in human PRP suggesting that this compound is a specific inhibitor of thrombin receptor-mediated platelet aggregation. Compounds **45** and **46** bind to the thrombin receptor with IC$_{50}$'s of 52 nM and 56 nM, respectively. These two derivatives were also shown to transiently inhibit platelet aggregation induced by 0.1 µM thrombin (150).

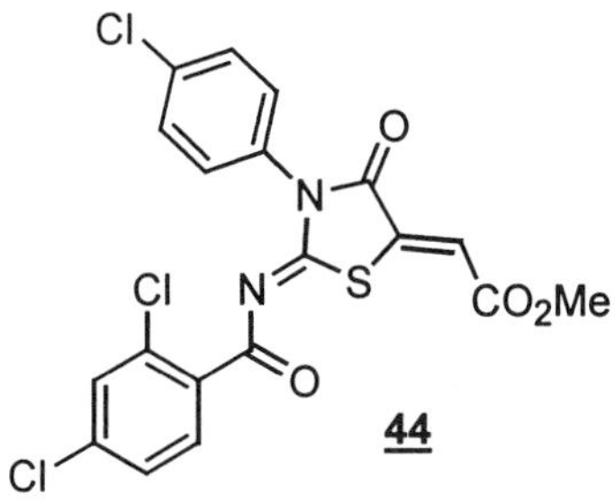

44

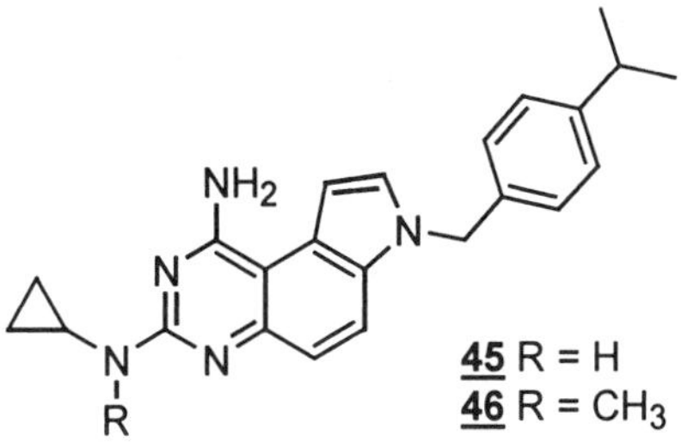

45 R = H
46 R = CH$_3$

CONCLUSION

Extensive clinical experience with aspirin, ticlopidine and clopidogrel has demonstrated the significant impact of antiplatelet therapy on reducing mortality and non-fatal events in patients with acute coronary syndrome. More recently, short-term intravenous administration of GPIIb/IIIa antagonists to ACS patients has resulted in further improvements in patient outcomes. Since platelets remain in an activated state for weeks or perhaps months after an acute coronary event, the allure of long-term antiplatelet therapy continues to fuel the search for a suitable oral agent. The evaluation of the newer oral GPIIb/IIIa antagonists with longer half-lives and different platelet binding affinity will determine if chronic treatment with such an agent is feasible. Meanwhile, the investigation of other mechanisms of platelet aggregation, shows promise of yielding additional compounds for clinical investigation.

References

1.   E.J. Topol, T.V. Byzova and E.F. Plow, Lancet, 353, 227 (1999).
2.   M. Verstraete, Circulation, 101, e76 (2000).
3.   A.A.J. Adgey, Eur. Heart J., 19, D10 (1998).
4.   G. Manoharan, S.J. Maynard and A.A.J. Adgey, Exp. Opin. Invest. Drugs, 8, 555 (1999).
5.   A.M. Lincoff, J.E. Tcheng, R.M. Califf, D.J. Kereiakes, T.A. Kelly, G.C. Timmis, N.S. Kleiman, J.E. Booth, C. Balog, C.F. Cabot, K.M. Anderson, H.F. Weisman and E.J. Topol, Circulation, 99, 1951 (1999).
6.   R.M. Scarborough, Am. Heart J., 138, 1093 (1999).
7.   R.A. Harrington, Amer. J. Cardiology, 80, 34B (1997).
8.   D.R. Phillips, W. Teng, A. Arfsten, L. Nannizzi-Alaimo, M.M. White, C. Longhurst, S.J. Shattil, A. Randolph, J.A. Jakubowski, L.K. Jennings and R.M. Scarborough, Circulation, 96, 1488 (1997).
9.   J.J. Cook, B. Bednar, J.J. Lynch, Jr., R.J. Gould, M.S. Egbertson, W. Halczenko, M.E. Duggan, G.D. Hartman, M.W. Lo, G.M. Murphy, L.I. Deckelbaum, F.L. Sax and E. Barr, Cardiovas. Drug Rev., 17, 199 (2000).
10.   H. White, N. Engl. J. Med., 338, 1498 (1998).
11.   P. Theroux, N. Engl. J. Med., 338, 1488 (1998).
12.   M. Ferrario, A. Repetto, S. Lucreziotti and D. Ardissino, Am. Heart J., 138, S121 (1999).
13.   E.J. Topol, D.B. Mark, A.M. Lincoff, E. Cohen, J. Burton, N. Kleiman, D. Talley, S. Sapp, J. Booth, C.F. Cabot, K.M. Anderson and R.M. Califf, Lancet, 354, 2019 (1999).
14.   Scrip, 2525, 21 (2000).
15.   E.M. Antman, R.P. Giugliano, C.M. Gibson, C.H. McCabe, P. Coussement, N.S. Kleiman, A. Vahanian, A.A.J. Adgey, I. Menown, H.-J. Rupprecht, R. Van der Wieken, J. Ducas, J. Scherer, K. Anderson, F. Van de Werf and E. Braunwald, Circulation, 99, 2720 (1999).
16.   R.M. Califf, Am. Heart J., 139, S33 (2000).
17.   S. Harder, U. Klinkhardt and H.K. Breddin, Clin. Pharm. Ther., 65, 179 (1999).
18.   S. Harder, C.M. Kirchmaier, H.J. Krzywanek, D. Westrup, J.-W. Bae and H.K. Breddin, Circulation, 100, 1175 (1999).
19.   S. Harder, U. Klinkhardt, K. Krzywanek, C.M. Kirchmaier, D. Westrup and H.K. Breddin, Annals of Hematology, 78, A58 (1999).
20.   M. Dooley and K.L. Goa, Drugs, 57, 225 (1999).
21.   A. Cases, X. Rabasseda and J. Castañer, Drugs Future, 24, 261 (1999).
22.   R.C. Carroll, Curr. Opin. Cardiovasc. Pulm. Renal Invest. Drugs, 1, 104 (1999).
23.   Scrip, 2523, 22 (2000).
24.   P.W. Serruys, H. Darius, C. Moris, W. O' Neill, M. Knudtson, R. Anders, J.H. Kleiman, K. Barker, G.A. van Es and J.O. Symons, Eur. Heart J., 20, 529 (1999).
25.   W.W. O'Neill, P.W. Serruys, M. Knudtson, G.C. Timmis, C. van der Zwann, J.H. Kleiman, J.C. Alexander and R.J. Anders, Circulation, 100, I187 (1999).
26.   J.H. Alexander, S. Al-Khatib, W. Cantor, A.B. Greenbaum, M.P. Hudson, D.F. Kong and M. Shah, Am. Heart J., 138, 175 (1999).
27.   W.W. O'Neill, P. Serruys, M. Knudtson, G.-A. Van Es, G.C. Timmis, C. van der Zwaan, J. Kleiman, J. Gong, E.B. Roecher, R. Drieling, J. Alexander and R. Anders, N. Engl. J. Med., 342, 1316 (2000).
28.   N.K. Newby, Lancet, 355, 337 (2000).

29. C. Heeschen and C.W. Hamm, Lancet, 355, 330 (2000).
30. M. Merlos, P.A. Leeson and J. Castañer, Drugs of the Future, 23, 1190 (1998).
31. P. Yue, F. Mony, V. David, M. Ghitescu and P. Theroux, Circulation, 100, I620 (1999).
32. M. Casey, C. Fornari, G.E. Bozovich, M.L. Iglesias Varela, B. Mautner and C.P. Cannon, Circulation, 100, I681 (1999).
33. M.B. Holmes, B.E. Sobel and D.J. Schneider, Circulation, 100, I431 (1999).
34. M.B. Holmes, B.E. Sobel, C.P. Cannon and D.J. Schneider, Am. J. Cardiol., 85, 491 (2000).
35. L.K. Newby, Am. Heart J., 138, 210 (1999).
36. Scrip, 2524, 22 (2000).
37. D.P. Chew, D.L. Bhatt and E.J. Topol, J. Amer. Coll. Cardiol., 35, 393 (2000).
38. M.T. Roe and D.J. Moliterno, J. Thromb. Thrombolysis, 7, 247 (1999).
39. K.A. Ault, C.P. Cannon, J. Mitchell, J. McCahan, R.P. Tracy, W.F. Novotny, J.D. Reimann and E. Braunwald, J. Amer. Coll. Cardiol., 33, 634 (1999).
40. D.J. Schneider, D.J. Taatjes and B.E. Sobel, Cardiovasc. Res., 45, 437 (2000).
41. K. Peter, M. Schwarz, J. Ylanne, B. Kohler, M. Moser, T. Nordt, P. Salbach, W. Kuber and C. Bode, Blood, 92, 3240 (1998).
42. K. Peter, M. Schwarz, T. Nordt, B. Kohler, J. Ruef, W. Kuebler and C. Bode, Ann. of Hematology, 78, A21 (1999).
43. S.R. Adderley and D.J. Fitzgerald, J. Biol. Chem., 275, 5760 (2000).
44. R.C. Carroll, Curr. Opin. Cardiovasc. Pulm. Renal Invest. Drugs, 1, 131 (1999).
45. P. Jong and A. Langer, Exp. Opin. Invest. Drugs, 8, 1453 (1999).
46. C.B. Xue and S.A. Mousa, Drugs of the Future, 23, 707 (1998).
47. C.B. Xue, J. Wityak, T.M. Sielecki, D.J. Pinto, D.G. Batt, G.A. Cain, M. Sworin, A.L. Rockwell, J.J. Roderick, S. Wang, M.J. Orwat, W.E. Frietze, L.L. Bostrom, J. Liu, C.A. Higley, F.W. Rankin, A.E. Tobin, G. Emmett, G.K. Lalka, J.Y. Sze, S.V. Di Meo, S.A. Mousa, M.J. Thoolen, A.L. Racanelli, E.A. Hausner, T.M. Reilly, W.F. DeGrado, R.R. Wexler and R.E. Olson, J. Med. Chem., 40, 2064 (1997).
48. W.F. Ebling, D.M. Kornhauser, S. Ma, V.A. Cain, J.S. Oliver and J. Pieniaszek, H. J., J. Clin. Pharmacol., 39, 983 (1999).
49. S.A. Mousa, J.M. Bozarth, W. Lorelli, M.S. Forsythe, M.J.M.C. Thoolen, A.M. Slee, T.M. Reilly and P.A. Friedman, J. Pharm. Exp. Ther., 286, 1277 (1998).
50. S.A. Mousa, R. Kapil and D.-X. Mu, Arterioscler. Thromb. Vasc. Biol., 19, 2535 (1999).
51. S.A. Mousa, M. Forsythe, J. Bozarth, A. Youssef, J. Wityak, R. Olson and T. Sielecki, J. Cardiovasc. Pharmacol., 32, 736 (1998).
52. C.P. Cannon, Curr. Opin. Cardiovasc. Pulm. Renal Invest. Drugs, 2, 114 (2000).
53. M. Forsythe and S.A. Mousa, Thromb. Haemost., 82, A837 (1999).
54. W.J. Hoekstra, B.E. Maryanoff, B.P. Damiano, P. Andrade-Gordon, J.H. Cohen, M.J. Costanzo, B.J. Haertlein, L.R. Hecker, B.L. Hulshizer, J.A. Kauffman, P. Keane, D.F. McComsey, J.A. Mitchell, L. Scott, R.D. Shah and S.C. Yabut, J. Med. Chem., 42, 5254 (1999).
55. A. Van Hecken, M. Depre, K. Wynants, J. Arnout, D. Doose, R. Abels, E. Vercammen, D. Gibson and P.J. de Schepper, Arch. Pharmacol., 358, R463 (1998).
56. S.C. Kapoor, B. Lages, C.K.N. Li, T. Hoffmann and F. Catella-Lawson, conf abst. 41st Annual Mtg of the Amer. Soc. Hematology, New Orleans, LA 12/3-7/99, Poster 90 (1999).
57. D.D. Gretler, M.M. Kitt, J.L. Lambing, G.L. Park, T. Mant and R.M. Scarborough, Clin. Pharmacol. Ther, 63, 210 (1998).
58. B. Meibohm, R. Neugebauer, K.U. Buhring, M. Schulte and A. Kovar, J. Clin. Pharmacol., 39, 978 (1999).
59. P. Savi, A. Badorc, A. Lale, M.F. Bordes, J. Bornia, C. Labouret, A. Bernat, P. de Cointet, P. Hoffmann, J.P. Maffrand and J.M. Herbert, Thromb. Haemost., 80, 469 (1998).
60. A. Bernat, P. Savi, A. Lale, P. Hoffmann and J.M. Herbert, J. Cardiovasc. Pharmacol., 33, 573 (1999).
61. K. Ota, K. Katano, T. Miura, K. Kobayashi, T. Imai and T. Ando, conf. abst. AFMC/AIMECS '99 Intl. Med. Chem. Symp., Poster P38 (1999).
62. S. Ohuchi, H. Iida, M. Nagasawa, T. Sugano, K. Kataha and M. Ishikawa, conf. abst. XVIIth Congress of the Intl. Soc. Thromb. Haemost., , Poster 2706 (1999).
63. H. Sugihara, H. Fukushi, T. Miyawaki, Y. Imai, Z. Terashita, M. Kawamura, Y. Fujisawa and S. Kita, J. Med. Chem., 41, 489 (1998).
64. K. Okumura, T. Shimazaki, Y. Aoki, H. Yamashita, E. Tanaka, S. Banba, K. Yazawa, K. Kibayashi and H. Banno, J. Med. Chem., 41, 4036 (1998).

65. H. Banno, H. Kawazura, T. Yutaka, N. Sakuma, T. Kitamori, J. Hosoya, K. Kibayashi, H. Yamashita, K. Umemura and M. Nakashima, Eur. J. Pharmacol., 367, 275 (1999).

66. S. Ono, T. Yoshida, K. Maeda, K. Kosaka, Y. Inoue, T. Imada, C. Fukaya and N. Nakamura, Chem. Pharm. Bull., 47, 1694 (1999).

67. H.U. Stilz, W. Guba, B. Jablonka, M. Just, O. Klingler, W. Konig, V. Wehner and G. Zoller, Lett. Pept. Sci., 5, 215 (1998).

68. D. Trenk, E. Stengele, R. Adler, M. Just, D. Brockmeier, M. Seibert-Grafe and E. Jaehnchen, Clin. Pharm. Ther., 65, 150 (1999).

69. S. Horisawa, M. Kaneko, Y. Ikeda, Y. Ueki and T. Sakurama, Thromb. Res., 94, 227 (1999).

70. J. Huang, S.S. Rebello, L.A. Rosenberg, M. Kaneko, T. Sakurama and B.R. Lucchesi, Eur. J. Pharmacol., 366, 203 (1999).

71. S.S. Rebello, J.B. Huang, K. Saito and B.R. Lucchesi, Arterioscler. Thromb. Vasc. Biol., 18, 954 (1998).

72. Y. Hayashi, J. Katada, T. Harada, A. Tachiki, K. Iijima, Y. Takiguchi, M. Muramatsu, H. Miyazaki, T. Asari, T. Okazaki, Y. Sato, E. Yasuda, M. Yano, I. Uno and I. Ojima, J. Med. Chem., 41, 2345 (1998).

73. E. Carceller, L.A. Gomez, M. Merlos, J. Garcia-Rafanell and J. Forn, 216th ACS Meeting Abstracts, Medi 079, (1998).

74. M.J. Fisher, A.E. Arfstan, U. Giese, B.P. Gunn, C.S. Harms, V. Khau, M.D. Kinnick, T.D. Lindstrom, M.J. Martinelli, H.J. Mest, M. Mohr, J. Morin, J.M., J.T. Mullaney, A. Nunes, M. Paal, A. Rapp, G. Ruehter, K.J. Ruterbories, D.J. Sall, R.M. Scarborough, T. Schotten, B. Sommer, W. Stenzel, R.D. Towner, S.L. Um, B.G. Utterback, R.T. Vasileff, S. Voeelkers, V.L. Wyss and J.A. Jakubowski, J. Med. Chem., 42, 4875 (1999).

75. R.E. Olson, T.M. Sielecki, J. Wityak, D.J. Pinto, D.G. Batt, W.E. Frietze, J. Liu, A.E. Tobin, M.J. Orwat, S.V. Di Meo, G.C. Houghton, G.K. Lalka, S.A. Mousa, A.L. Racanelli, E.A. Hausner, R.P. Kapil, S.R. Rabel, M.J. Thoolen, T.M. Reilly, P.S. Anderson and R.R. Wexler, J. Med. Chem, 42, 1178 (1999).

76. P.N. Confalone, F. Jin and S.A. Mousa, Bioorg. Med. Chem. Lett., 9, 55 (1999).

77. C.-B. Xue, J. Roderick, S. Mousa, R.E. Olson and W.F. DeGrado, Bioorg. Med. Chem. Lett, 8, 3499 (1998).

78. M.S. Egbertson, J.J. Cook, B. Bednar, J.D. Prugh, R.A. Bednar, S.L. Gaul, R.J. Gould, G.D. Hartman, C.F. Homnick, M.A. Holahan, L.A. Libby, J. Lynch, J.J., R.J. Lynch, G.R. Sitko, M.T. Stranieri and L.M. Vassallo, J. Med. Chem., 42, 2409 (1999).

79. G.D. Hartman, M.E. Duggan, W.F. Hoffman, R.J. Meissner, J.J. Perkins, A.E. Zartman, A.M. Naylor-Olsen, J.J. Cook, J.D. Glass, R.J. Lynch, G. Zhang and R.J. Gould, Bioorg. Med. Chem. Lett., 9, 863 (1999).

80. Y. Aga, K. Baba, T. Nakanishi, J. Kita, H. Ueno, M. Tanaka, T. Motoyama and Y. Kuroki, Jpn. J. Pharm., 79, 192P (1999).

81. K. Baba, M. Doi, T. Motoyama, K. Takata, M. Tanaka, Y. Kuroki, A. Matsuda and H. Ueno, Jpn. J. Pharmacol., 79, 192P (1999).

82. K.C. Nicolaou, J.I. Trujillo, B. Jandeleit, K. Chibale, M. Rosenfeld, B. Diefenbach, D.A. Cheresh and S.L. Goodman, Bioorg. Med. Chem., 6, 1185 (1998).

83. K.D. Robarge, M.S. Dina, T.C. Somers, A. Lee, T.E. Rawson, A.G. Olivero, M.H. Tischler, R.R. Webb, II, K.J. Weese, I. Aliagas and B.K. Blackburn, Bioorg. Med. Chem., 6, 2345 (1998).

84. N.J. Liverton, D.J. Armstrong, D.A. Claremon, D.C. Remy, J.J. Baldwin, R.J. Lynch, G. Zhang and R.J. Gould, Bioorg. Med. Chem. Lett., 8, 483 (1998).

85. W.J. Hoekstra, E.C. Lawson and B.E. Maryanoff, PCT application WO 0006570-A1, (2000).

86. C. Yue, M. Henry, T. Giboulot and B. Lesur, French patent application FR 2781221-A1, (2000).

87. R.A. Bednar, S.L. Gaul, T.G. Hamill, M.S. Egbertson, J.A. Shafer, G.D. Hartman, R.J. Gould and B. Bednar, J. Pharmacol. Exp. Ther., 285, 1317 (1998).

88. R.M. Scarborough, N.S. Kleiman and D.R. Phillips, Circulation, 100, 437 (1999).

89. J.-F. Tanguay, Eur. Heart J. Suppl., 1, E27 (1999).

90. S.A. Mousa, J.M. Bozarth and M. Forsythe, Blood, 94, 222a (1999).

91. S.A. Mousa, S. Khurana and M.A. Forsythe, Arterioscler. Thromb. Vasc. Biol., 20, 1162 (2000).

92. J.A. Ambrose and G. Dangas, Arch. Intern. Med., 160, 25 (2000).

93. S.L. Chierchia, Eur. Heart J. Suppl., 1, N2 (1999).

94. E. Braunwald, R.M. Califf, C.P. Cannon, K.A.A. Fox, V. Fuster, B. Gibler, R.a. Harrington, S.B. King III, N.S. Kleiman, P. Theroux, E.J. Topol, F. Van de Werf, H.D. White and J.T. Willerson, Am. J. Med., 108, 41 (2000).
95. F. Hartmann, M. Kampmann, N. Frey, M. Muller-Bardorff and H.A. Katus, Eur. Heart J., 19, N2 (1998).
96. R.H. Christenson and H.M.E. Azzazy, Clin. Chem., 44, 1855 (1998).
97. C. Heeschen, C.W. Hamm, B. Goldmann, A. Deu, L. Langenbrink and H.D. White, Lancet, 354, 1757 (1999).
98. C.W. Hamm, C. Heeschen, B. Goldmann, A. Vahanian, J. Adgey, C.M. Miguel, W. Rutsch, J. Berger, J. Kootstra and M.L. Simoons, N. Engl. J. Med., 340, 1623 (1999).
99. M. Akkerhuis, C. van der Zwaan, R.G. Wilcox, K.L. Neuhaus, A. Vahanian, J.L. Boland, C.W. Hamm, T. Baardman, J. Hoffmann, J.W. Deckers and M.L. Simoons, Circulation, 100, I292 (1999).
100. J. Danesh, JAMA, 282, 2169 (1999).
101. M. Visser, L.M. Bouter, G. McQuillan, M.H. Wener and T.B. Harris, JAMA, 282, 2131 (1999).
102. W. Koenig, M. Sund, M. Frohlich, H.G. Fischer, H. Lowel, A. Doring, W.L. Hutchinson and M.B. Pepys, Circulation, 99, 237 (1999).
103. J. Danesh, R. Collins, P. Appleby and R. Peto, JAMA, 279, 1477 (1998).
104. J.L. Anderson, G.J. King, T.L. Bair, S.P. Elmer, J.B. Muhlestein, J. Habashi and J.F. Carlquist, J. Amer. Coll. Cardiol., 33, 727 (1999).
105. P.J. Goldschmidt-Clermont, C.M. Roos and G.E. Cooke, J. Thromb. Thrombol., 8, 89 (1999).
106. D.A. Vorchheimer and V. Fuster, Circulation, 97, 312 (1998).
107. N.S. Kleiman, Am. Heart J., 138, S263 (1999).
108. B.S. Coller, Blood Coagulation And Fibrinolysis, 10, 0957 (1999).
109. S.S. Rebello, J. Huang, J.D. Faul and B.R. Lucchesi, J. Thromb. Thrombol., 9, 23 (2000).
110. R.F. Storey, J.A. May, R.G. Wilcox and S. Heptinstall, Thromb. Haemost., 82, 1307 (1999).
111. M. Madan and S.D. Berkowitz, Am. Heart J., 138, S317 (1999).
112. J.C. Blankenship, F.J. Menapace and S.L. Demko, J. Am. Coll. Cardiol., 33, 255A (1999).
113. J.E. Tcheng, Am. Heart J., 139, S38 (2000).
114. R.P. Guigliano and R.R. Hyatt, J. Amer. Coll. Card., 31, 185A (1998).
115. B. Bednar, J.J. Cook, M.A. Holahan, M.E. Cunningham, P.A. Jumes, R.A. Bednar, G.D. Hartman and R.J. Gould, Blood, 94, 587 (1999).
116. J.A. Brassard, R.A. Cooper, S.R. Kupfer, W.J. Konocsar, B.R. Curtis, M.D. Kane and R.H. Aster, Blood, 94, 647a (1999).
117. S.A. Coulter, C.P. Cannon, R.A. Cooper, C.H. McCabe, R. Aster, A. Charlesworth, A.M. Skene and E. Braunwald, J. Amer. Coll. Cardiol., 35, 393 (2000).
118. B. Bednar, D.M. Bollag and R.J. Gould, PCT patent application WO9919463-A1 published 4/22/99.
119. D.A. Seiffert, J.T. Billheimer, L.A. Breth, T.C. Burn, I.B. Dicker, H.J. George, J.M. Hollis, G.F. Hollis, J.E. Kochie and K.T. O'Neil, PCT patent application WO9938014-A-1, published 7/29/99.
120. K. Peter, A. Straub, B. Kohler, M. Volkmann, M. Schwarz, W. Kubler and C. Bode, Amer. J. Cardiol., 84, 519 (1999).
121. J. Llevadot, S.A. Coulter and R.P. Guigliano, J. Thromb. Thrombol., 9, 175 (2000).
122. S.P. Kunapuli and J.L. Daniel, Biochem. J., 336, 513 (1998).
123. S.P. Kunapuli, Platelets, 9, 343 (1998).
124. J.L. Daniel, C. Dangelmaier, J. Jin, B. Ashby, J.B. Smith and S.P. Kunapuli, J. Biol. Chem., 273, 2024 (1998).
125. L.F. Brass, J. Clin. Invest., 104, 1663 (1999).
126. C. Leon, B. Hechler, M. Freund, A. Eckly, C. Vial, P. Ohlmann, A. Dierich, M. LeMeur, J.-P. Cazenave and C. Gachet, J. Clin. Invest., 104, 1731 (1999).
127. J.-E. Fabre, M. Nguyen, A. Latour, J.A. Keifer, L.P. Audoly, T.M. Coffman and B.H. Koller, Nature Medicine, 5, 1199 (1999).
128. J.-M. Herbert, P. Savi and J.-P. Maffrand, Eur. Heart J., 1, A31 (1999).
129. F.H. Jafary and C.D. Kimmelstiel, J. Thromb. Thrombol., 9, 157 (2000).
130. J. Geiger, J. Brich, P. Honig-Liedl, M. Eigenthaler, P. Schanzenbacher, J.M. Herbert and U. Walter, Arterioscler. Thromb. Vasc. Biol., 19, 2007 (1999).

131. A. Sugidachi, F. Asai, T. Ogawa, T. Inoue and H. Koike, Brit. J. Pharmacol., 129, 1439 (2000). Structure obtained from IDDB Online Database, Current Drugs Ltd., 1999.
132. S.C. Chattaraj, Curr. Opin. Cardiovasc. Pulm. Renal Invest. Drugs, 1, 600 (1999).
133. A.H. Ingall, J. Dixon, A. Bailey, M.E. Coombs, D. Cox, J.I. McInally, S.F. Hunt, N.D. Kindon, B.J. Teobald, P.A. Willis, R.G. Humphries, P. Leff, J.A. Clegg, J.A. Smith and W. Tomlinson, J. Med. Chem., 42, 213 (1999).
134. R.F. Storey, K.G. Oldroyd and R.G. Wilcox, Circulation, 100, I710 (1999).
135. B. Hechler, C. Leon, C. Vial, P. Vigne, C. Frelin, J.-P. Cazenave and C. Gachet, Blood, 92, 152 (1998).
136. S.R. Coughlin, Thromb. Haemost., 82, 353 (1999).
137. T.-K. Vu, D.T. Hung, V.I. Wheaton and S.R. Coughlin, Cell, 64, 1057 (1991).
138. M.L. Kahn, Y.-W. Zheng, W. Huang, V. Bigornia, D. Zeng, S. Moff, R.V.J. Farese, C. Tam and S.R. Coughlin, Nature, 394, 690 (1998).
139. W.F. Xu, H. Andersen, T.E. Whitmore, S.R. Presnell, D.P. Yee, A. Ching, T. Gilbert, E.W. Davie and D.C. Foster, Proc. Natl. Acad. Sci. USA, 95, 6642 (1998).
140. M.L. Kahn, M. Nakanishi-Matsui, M.J. Shapiro, H. Ishihara and S.R. Coughlin, J. Clin. Invest., 103, 879 (1999).
141. P. Stead, S. Hiscox, P.S. Robinson, N.B. Pike, P.J. Sidebottom, A.D. Roberts, N.L. Taylor, A.E. Wright, S.A. Pomponi and D. Langley, Bioorg. Med. Chem. Lett., 10, 661 (2000).
142. M.D. Hollenberg In "Bioactive Peptides in Drug Discovery and Design: Medical Aspects"; Matsoukas, J., Mavromoustakos, T., Eds.; IOS Press 1999, p 265.
143. M.S. Bernatowicz, C.E. Klimas, K.S. Hartl, M. Peluso, N.J. Allegretto and S.M. Seiler, J. Med. Chem., 39, 4879 (1996).
144. T. Fujita, M. Nakajima, Y. Inoue, T. Nose and Y. Shimohigashi, Bioorg. Med. Chem. Lett., 9, 1351 (1999).
145. D.F. McComsey, L.R. Hecker, P. Andrade-Gordon, M.F. Addo and B.E. Maryanoff, Bioorg. Med. Chem. Lett., 9, 255 (1999).
146. W.J. Hoekstra, B.L. Hulshizer, D.F. McComsey, P. Andrade-Gordon, J.A. Kauffman, M.F. Addo, D. Oksenberg, R.M. Scarborough and B.E. Maryanoff, Bioorg. Med. Chem. Lett., 8, 1649 (1998).
147. P. Andrade-Gordon, B.E. Maryanoff, C.K. Derian, H.-C. Zhang, M.F. Addo, A.L. Darrow, A.J. Eckardt, W.J. Hoekstra, D.F. McComsey, D. Oksenberg, E.E. Reynolds, R.J. Santulli, R.M. Scarborough, C.E. Smith and K.B. White, Proc. Natl. Acad. Sci. USA, 96, 12257 (1999).
148. D.F. McComsey, M.J. Hawkins, P. Andrade-Gordon, M.F. Addo, D. Oksenberg and B.E. Maryanoff, Bioorg. Med. Chem. Lett., 9, 1423 (1999).
149. Y. Kato, Y. Kita, M. Nishio, Y. Hirasawa, K. Ito, T. Yamanaka, Y. Motoyama and J. Seki, Eur. J. Pharmacol., 384, 197 (1999).
150. H.-S. Ahn, L. Arik, G. Boykow, D.A. Burnett, M.A. Caplen, M. Czarniecki, M.S. Domalski, C. Foster, M. Manna, A.W. Stamford and Y. Wu, Bioorg. Med. Chem. Lett., 9, 2073 (1999).

Section III. CANCER AND INFECTIOUS DISEASES

Editor: Jacob J. Plattner, Chiron Corporation
Emeryville, CA 94608

Chapter 11. Anti-angiogenesis as a Therapeutic Strategy for Cancer

Wendy J. Fantl and Steven Rosenberg
Chiron Corporation
4560 Horton Street, Emeryville, CA 94608

Introduction - Angiogenesis, the growth of new blood vessels from pre-existing vasculature, plays a critical role in a large spectrum of diseases including, most notably, cancer (1-4). Inhibition of angiogenesis or dissolution of existing tumor vasculature as a therapeutic strategy for cancer is attractive for at least three reasons (2,5,6). First, since all solid tumors and some leukemias are dependent on a blood supply for their growth and maintenance, targeting the tumor vasculature can be considered as a general strategy for all solid tumors regardless of their genotype. Second, the problem of drug resistance associated with cytotoxic drugs targeting genetically unstable tumor cells is not anticipated to be a problem when targeting genetically stable endothelial cells. Third, in normal adults only 0.01% of endothelial cells are engaged in cell division, whereas in tumors this is increased by several orders of magnitude. Survival of a tumor requires continuous replenishment of oxygen and nutrients. To fulfill this function, tumor cells activate the quiescent vasculature to undergo neovascularisation via an "angiogenic switch" which alters the local balance of positive and negative angiogenic regulators (5,7). This field was last reviewed in this series in 1997 (8).

BIOLOGICAL RATIONALE

Based on these concepts, anti-angiogenic strategies are to inhibit positive regulators or to induce or use negative regulators themselves as therapeutic agents. However, given the intricate network of factors discovered, the situation is more complex and alternative strategies for therapeutic intervention include the tumor endothelial cell itself or endothelial cell interactions with the extracellular matrix. This review will focus on the most recent reports on the use of proteins (antibodies, protein fragments, interferons) and small molecules that inhibit growth and maintenance of tumor vascularization.

PHYSIOLOGICAL AND PATHOLOGICAL ANGIOGENESIS

There are many mechanistic similarities between physiological and pathological angiogenesis, particularly in the pathways regulated by the vascular endothelial growth factors (VEGFs), and fibroblast growth factors (FGFs) (9 - 13). Angiopoietins 1 and 2 (Ang1 and 2) acting through the endothelial cell-specific Tie receptor tyrosine kinases stimulate vascular remodeling. The Ephrin receptor tyrosine kinases are involved in distinguishing between arteries and veins (9,14-16). Angiogenesis is also dependent on endothelial cell migration through the extracellular matrix by means of focalized proteolysis of specific matrix components. In the case of tumor angiogenesis, the matrix metalloproteinases (MMPs) and components of the urokinase plasminogen activator (uPA) pathway are attractive targets for inhibiting angiogenesis (17,18-20). A recent study showed that plasminogen activator inhibitor-1 (PAI-1) is required for efficient tumor angiogenesis, consistent with the highly significant prognostic impact of PAI-1 levels in human breast cancer (19,21). Maintenance of vascular structure is dependent upon the adhesive properties of endothelial cells mediated by integrins and cadherins (20,22).

Many of these molecules are being evaluated as potential targets that could either be pro- or anti-angiogenic. Before discussing ongoing studies, some emerging biological concepts will be outlined which bear on the efficacy of anti-angiogenic therapies.

That expansion of an avascular tumor mass beyond 1mm is dependent upon the development of a new vasculature is well-documented (1,7,23). However, recent studies have described situations in which tumors co-opt host blood vessels to form an initial tumor mass that is well-vascularized (15,24). This is followed by regression of the tumor vasculature which results in apoptosis of the tumor. However, cells at the tumor boundary secrete factors that stimulate angiogenesis, and in so doing, rescue the tumor. It appears that Ang1 and VEGF are involved in these processes. Further, recent evidence has described a functional population of circulating endothelial precursor cells that are able to initiate neovascularisation (25). A second consideration for a clinical anti-angiogenic strategy relates to a multistage pancreatic islet carcinogenesis model in mice which showed that the efficacy of an anti-angiogenic drug was dependent on the carcinogenic stage at which it was administered (26).

Antibody Therapies - Three antibody therapies will be described targeting different mechanistic aspects of angiogenesis. As described above, angiogenesis depends on specific molecular interactions between endothelial cells and the extracellular matrix (20). Integrins are heterodimeric transmembrane proteins that mediate cell adhesion to the extracellular matrix, which triggers signal transduction pathways that regulate cell survival, proliferation and migration (22). There is inherent specificity built into the integrin signaling system since there are 15 α and 8 β subunits which can pair in limited combinations and interact with a variety of extracellular matrix proteins. Further, differences in the composition of extracellular matrix proteins seen in different tissues may change the angiogenic properties of endothelial cells. Of relevance to this review is the high expression of the $\alpha_v\beta_3$ heterodimer seen on tumor endothelial cells. Antibodies and small peptides against the $\alpha_v\beta_3$ heterodimer are being used as an anti-angiogenic strategy, including Vitaxin, a humanized version of the LM609 monoclonal antibody against $\alpha_v\beta_3$ is in Phase II clinical trials (4, 22).

Another strategy to inhibit angiogenesis is to reduce the amount of an angiogenic promoter (discussed above), for example VEGF. A recent study showed that VEGF also has a role in tumor vessel maintenance (27). A humanized monoclonal antibody to VEGF is in phase II trials for a variety of cancers (28). Complementing these approaches are preclinical trials which have shown that fibrosarcoma cells transfected with soluble extracellular domain of the VEGF receptor 1 (Flt) show decreased growth after implantation into animals (29).

The third possibility for inhibiting angiogenesis and possibly disrupting preformed vessels is the use of a vascular targeting agent against specific endothelial cell markers. An example is a monoclonal antibody directed against the endoglin glycoprotein now in preclinical studies. Endoglin is a component of the transforming growth factor beta receptor system, and is essential for developmental angiogenesis (30,31). It is highly expressed in endothelial cells and was recently shown to be elevated in the angiogenic endothelium at the tumor edge (32).

Protein Fragments - Since the discovery of thrombospondin and its fragments as endogenous angiogenic inhibitors (33,34), many other endogenous proteolytically produced protein fragments have been described as potent inhibitors of angiogenesis. Two important criteria need to be considered for expression of these protein fragments for therapeutic use. Both are related to protein production in microorganisms and could influence the interpretation of both *in vitro* and preclinical studies. It is essential that the purification protocol for any protein from bacteria for anti-angiogenic studies must include stringent conditions for endotoxin removal. Endotoxins are polysaccharide chains covalently bound to lipid A located on the surface of Gram-negative bacteria (35,36) and can stick to hydrophobic and charged protein surfaces. Of note, is the exquisite sensitivity in terms of cell cytotoxicity that endothelial cells have toward endotoxin in the range of 0.01-10µg/ml (37). Further, given the heterogeneous nature of endotoxin, different strains of bacteria produce endotoxins of substantially different potencies, a factor to be taken into account when selecting a bacterial host. The second consideration for protein expression from any cell type is to have the correct folding and disulfide bond formation (38).

In some clinical and experimental situations a primary tumor inhibits growth of its metastases. It has been postulated that the tumor itself may be the source of an anti-angiogenic factor(s) which have half-lives long enough to inhibit tumor growth at distal sites (39,40). One of the first anti-angiogenic factors to be purified from a murine tumor was angiostatin and its sequence revealed that this protein is a 38 kDa internal fragment of plasminogen, where the parent molecule is devoid of any inhibitory activity (40). Subsequently, endostatin was isolated and shown to be a 20 kDa fragment of collagen XVIII (41). Both angiostatin and endostatin were shown to inhibit endothelial cell migration and proliferation stimulated by FGF and VEGF, to stimulate endothelial cell apoptosis and to inhibit tumor growth in a wide variety of preclinical models (42-45). Encouraging results in other preclinical models showed that angiostatin may synergize with radiation in reducing tumor volume (46). At present, endostatin is in Phase 1 clinical trials at three clinical centers in the US.

Since the descriptions of angiostatin and endostatin, other anti-angiogenic protein fragments have been discovered. Interestingly, kringle 5 of plasminogen, which lies outside angiostatin seems to be a more potent angiogenic inhibitor *in vitro* (47-48). Fragments derived from other proteins include; vasostatin (N-terminal domain of calreticulin) (49,50), canstatin (a 24KDa domain derived from the α2 chain of type IV collagen), restin (a fragment of collagen XV), PR39 (31), the METH proteins (for metalloprotease and thrombospondin domains) and derived fragments (51,52), PEX(C-terminal fragment of matrix metalloprotease 2) (51) and fragments derived from serpins that are involved in the coagulation pathway (53). The biology of these proteins is not understood particularly in terms of the enzymes responsible for their generation. An alternate way in which these fragments could be therapeutically useful is to increase their production *in vivo*. For example, the combination of uPA with captopril (which acts as a sulfhydryl donor) generated angiostatin *in vitro* (54)

Modulation of FGF Levels by Interferons - FGF1 and 2 are potent stimulators of angiogenesis, are expressed in many tumors and their expression levels often correlate with vascularity (13). Even though FGF is a more potent endothelial cell mitogen than VEGF *in vitro*, the current consensus is that VEGF is the more prominent regulator of *in vivo* angiogenesis. Reports have described a reduction in FGF mRNA and protein levels mediated by interferons (IFN) α and β (55). In a transgenic model of pancreatic cancer, IFN treatment severely decreased the rate of tumor growth with a concomitant decrease in vessel density (56). There are numerous clinical trials in progress to test the efficacy of interferons especially in combination therapies.

SYNTHETIC AND SMALL MOLECULAR WEIGHT INHIBITORS

Protease Inhibitors - There has been significant progress in this area over the last few years, as inhibitors of the matrix metalloproteinases (MMPIs) have moved further into clinical development, and a number of other proteases involved in neovascularization, particularly the urokinase-type plasminogen activator (uPA) system, have been investigated. Both these areas have been reviewed recently in this series, including MMPIs in this volume (57,58). It should be noted that although these protease inhibitors are anti-angiogenic in some assays, they work by inhibition of extracellular matrix remodelling, which modulates the behavior of several cell types involved in tumor progression, including transformed cells, inflammatory cells, as well as endothelial cells. In addition, recent data in wound healing studies with knock-out mice and inhibitors suggest that there is functional overlap between the MMP and uPA systems, *in vivo* (59).

Matrix Metalloproteinase Inhibitors – The molecules which are in the latter stages of clinical development are BB-2516 (Marimastat) **1** and AG-3340 (Prinomastat) **2**, the former is a broad spectrum inhibitor, while the latter has reduced inhibitory activity towards MMP-1(60).

A third compound which was in Phase III clinical trials, BAY12-9566 (Tanomastat) **3**, has recently been withdrawn, due to a higher rate of disease progression in the treated group in a small cell lung cancer trial (18). Recent data from the Phase III trial of **1** in gastric cancer showed that although the trial did not meet its primary endpoint of a survival benefit, there was a trend towards efficacy,

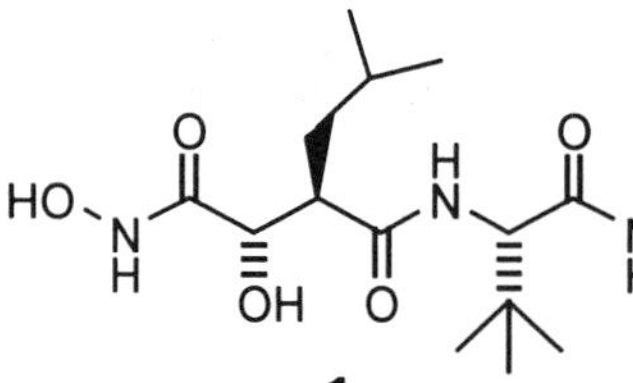

1 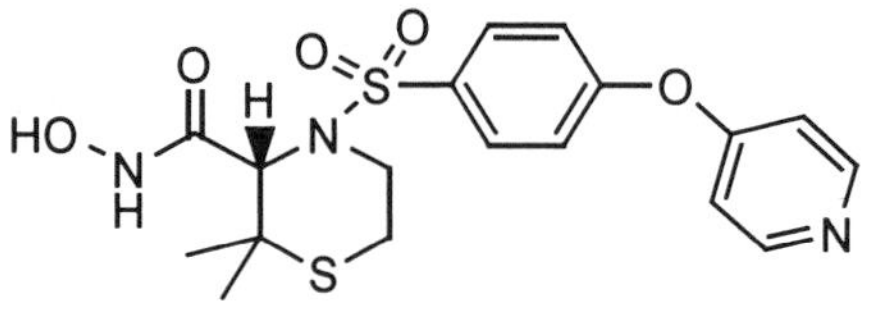   **2**

especially in patients without metastases at the time of study entry. All of these molecules show a similar dose limiting toxicity of moderate to severe arthralgia, which is of unknown cause (61). The recent observation of a number of severe phenotypes in MT-MMP-1 knock-out mice suggests

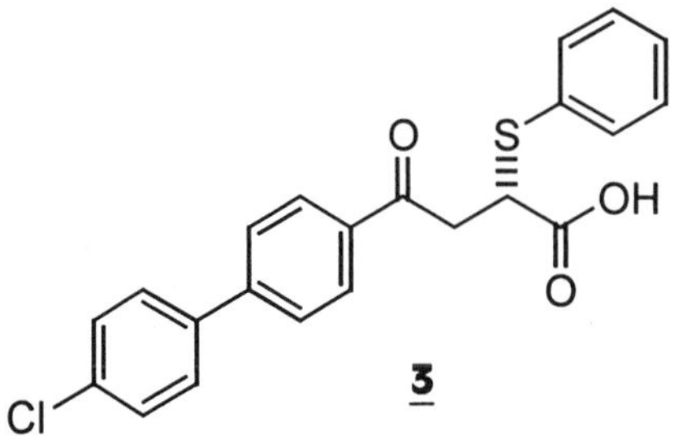

3 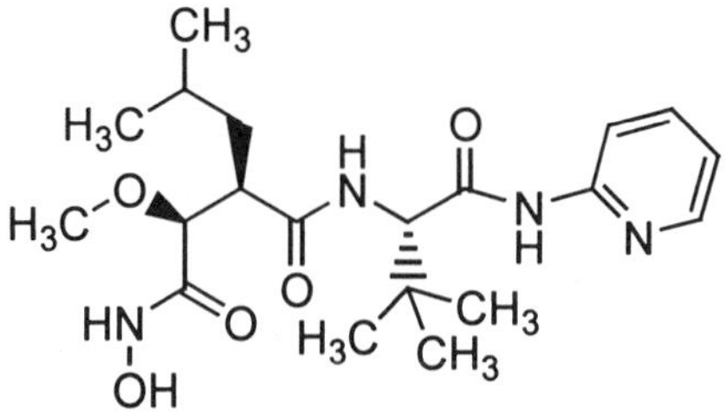   **4**

that inhibition of this enzyme could be responsible for the side-effects seen in humans (62). A series of new molecules have been disclosed in the patent literature over the last year, including BBI-3644 (Solimastat) **4** a second generation MMPI, which has significantly reduced arthralgia in animal models as compared to **1** (63). The challenge in the MMPI area will be to design a clinical trial which yields a robust effect, while managing the side-effect profile.

Urokinase Inhibitors – Recent work has shown that urokinase is an *in vivo* angiogenic factor in the rat corneal pocket assay (64). The mechanism of this effect is likely to be release of pro-angiogenic factors from the extracellular matrix, especially bFGF (65). Two new classes of inhibitors of this enzyme have

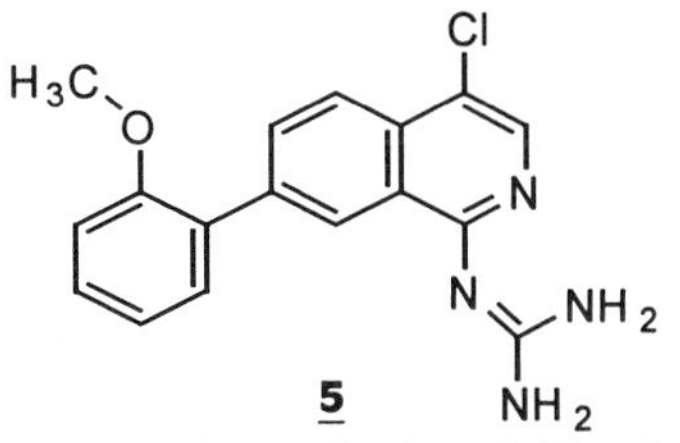

recently been disclosed. The first are guanidine isoquinolines, such as **5**, which has *in vitro* potency in the mid-nM range and greater than 5-fold selectivity versus plasmin and >10 fold versus tissue-type plasminogen activator (66). Another series of potent inhibitors of trypsin-like serine proteases include compounds such as **6**, which show selective inhibition of urokinase (67).

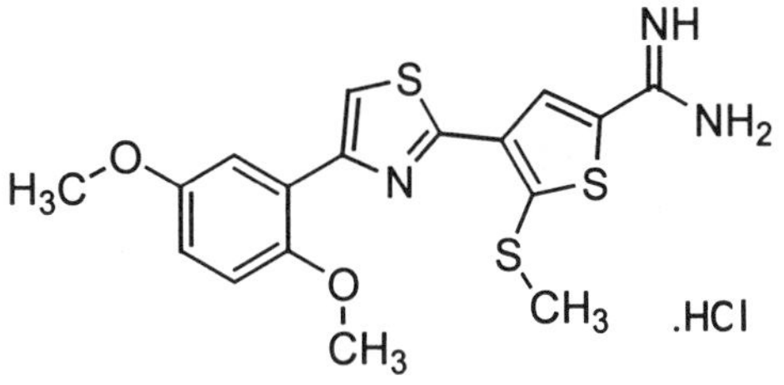

Tyrosine Kinase Inhibitors - Several receptor tyrosine kinases have been implicated in neovascularization, including the VEGFRs (KDR/flk-1) and flt), the FGFRs, and those for the angiopoietins (TIE-1 and TIE-2) (14,68). Inhibitors of the intracellular tyrosine kinase activities of these receptors have been the focus of medicinal chemistry efforts in this area. As has been seen for other receptor tyrosine kinases, it is possible to identify small molecule, ATP competitive, inhibitors which show significant selectivity for specific receptors (69,70). A major question is whether broad spectrum inhibition is preferable as opposed to selective inhibitors of KDR/flk-1. Thus, for example the compound furthest in clinical development, SU5416, **7**, is a selective KDR inhibitor, with an IC_{50} of 20 nM (71).

In contrast, SU6668 (another indolinone, structure not disclosed) is a potent and broad-spectrum inhibitor of VEGF, FGF, and platelet-derived growth factor (PDGF) receptor tyrosine kinases (IC$_{50}$ = 0.02, 1.3 and 0.06 uM), respectively. SU6668 showed increased endothelial cell and tumor cell apoptosis over SU5416 in a colorectal liver metastasis model, suggesting increased efficacy with the broader

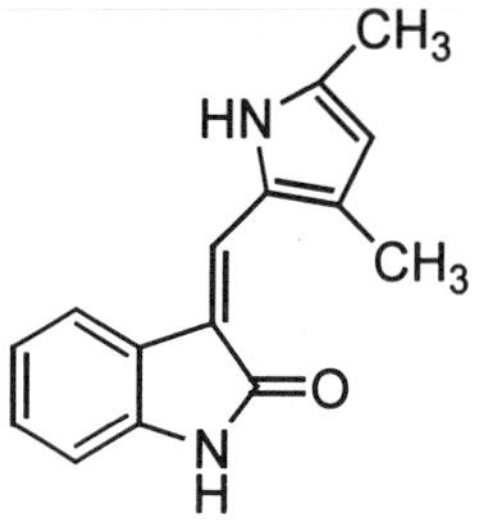

7 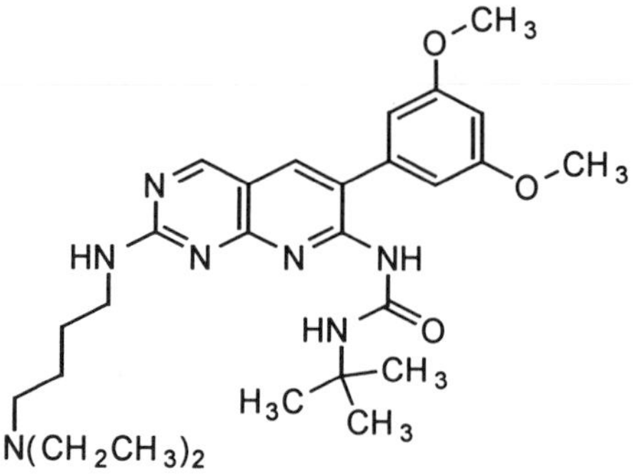   **8**

inhibitory range (72). The three-dimensional structures of two members of this series, SU5402 and SU4984, have been determined in complex with the FGF receptor kinase domain, directly demonstrating that these molecules bind as the adenine ring system, making some key hydrogen bonds which are also made by ATP (73).

Another molecule in the pyrido[2,3d] pyrimidine class which is a potent inhibitor of both FGFR and VEGFR kinase activities is PD173074, **8**, which blocks neovascularization of the mouse cornea stimulated by VEGF and FGF (74). This compound has also been co-crystallized with the kinase domain of FGFR-1, and shows a high degree of surface complementarity to the ATP binding site of this receptor (74).

The anilinoquinazolines and related heterocycles, exemplified by ZD-4190, **9**, show selectivity for VEGFR kinases (75). This compound has significant anti-tumor activity in a number of animal models after chronic oral administration (76). A very recent report describes the X-ray structures of two members of this series

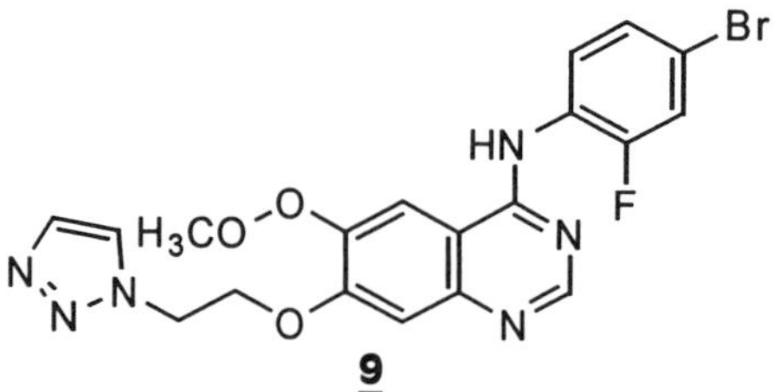

9

bound to related kinases (77). The three dimensional structure of a derivative of the kinase domain of KDR, has only recently been determined by X-ray crystallography (78).

Integrin Antagonists – The role of integrins in angiogenesis has been recently reviewed (22). At the time of the last review in this series on angiogenesis, only peptidic inhibitors of the angiogenic integrins $\alpha_v\beta_3$ and $\alpha_v\beta_5$ were described (8). A key goal has been to obtain selective $\alpha_v\beta_3$ and $\alpha_v\beta_5$ antagonists, with minimal activity against GPIIbIIIa. Such molecules may be anti-angiogenic agents, and given the expression of $\alpha_v\beta_3$ on osteoclasts and smooth muscle cells, a role in treatment of osteoporosis and restenosis, as well (79). It appears that the small molecules now identified are RGD mimetics, where the spacing and pKa of functional groups is crucial for potency and selectivity. Compounds with significant

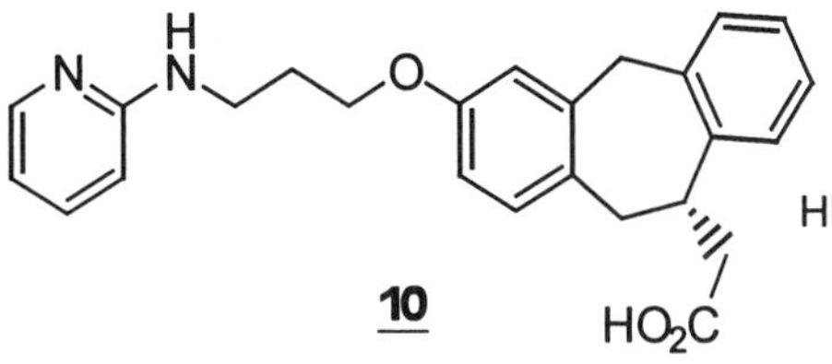

oral bioavailability such as SB265123 **10**, are selective inhibitors of $\alpha_v\beta_3$ and $\alpha_v\beta_5$ with IC$_{50}$s in the low nM range, and have shown inhibition of bone resorption in two rat models (80). A pro-drug strategy has given SG545, **11**, which shows significant activity in Matrigel angiogenesis and human colorectal xenograft models (81). In comparative studies this molecule was more effective than other known angiogenesis inhibitors. Another molecule with demonstrated *in vivo* activity is SC-68448, **12**, which shows about 100 fold selectivity versus GPIIbIIIa, and reduced angiogenesis in both bFGF induced corneal neovascularization and a Leydig cell tumor model (82). Other series with potent *in vitro* activity include the indazoles, **13** and isoxazolines **14**. The former shows higher potency against $\alpha_v\beta_3$ (IC$_{50}$= 2.3 nM) but only 10 fold selectivity (83), whereas the latter has an IC$_{50}$ of 34 nM against $\alpha_v\beta_3$, but is more than 1000 fold selective (84).

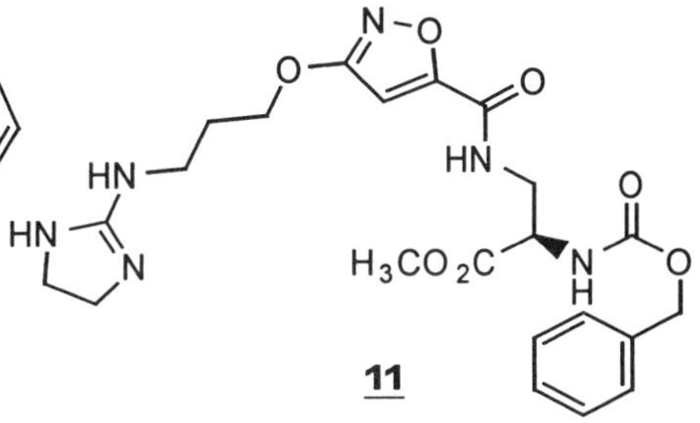

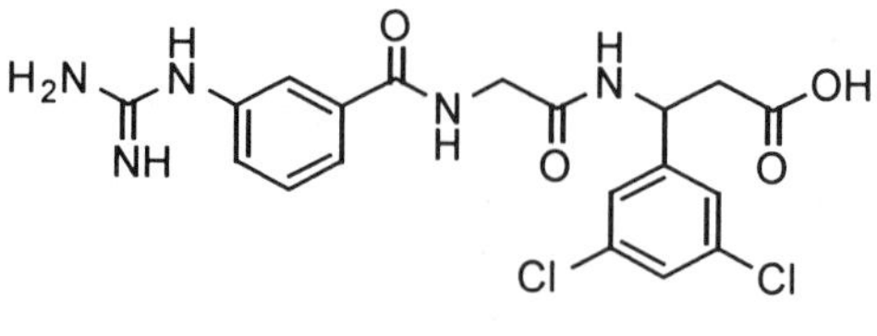

12

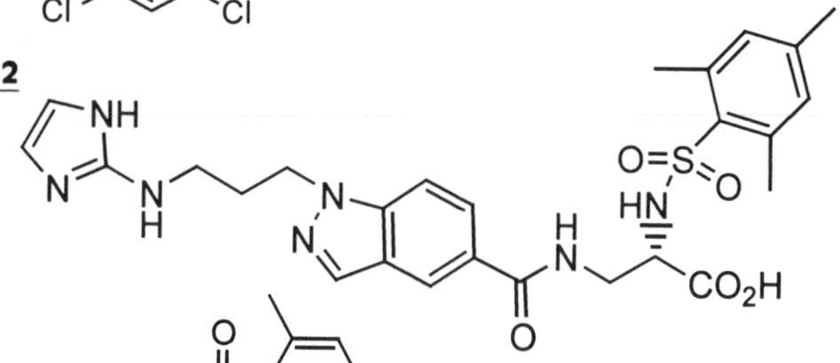

13

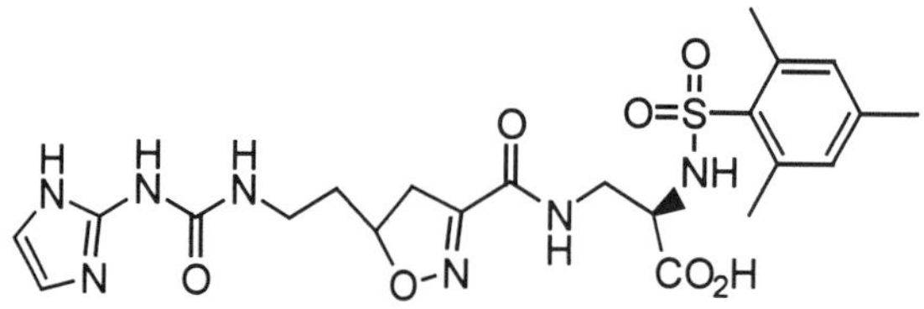

14

<u>Fumagillin/TNP470 and Derivatives</u> – The fumagillin derivative, TNP-470, **15**, is in clinical development for a variety of solid tumors. The molecular target by which this molecule inhibits endothelial cell proliferation was elucidated in an elegant study

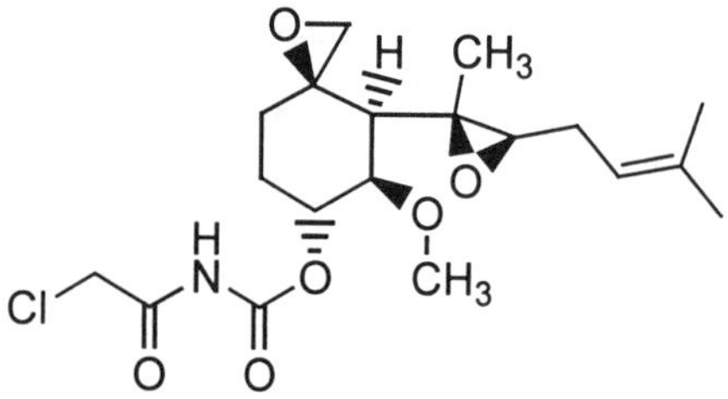

15

where a derivative of fumagillin was biotinylated, and the covalently modified target then identified as methionine aminopeptidase-2 (MAP-2), by mass spectrometry (85). A concurrent study showed that **15** also binds to MAP-2, and that a series of analogs show parallel potency in endothelial cell proliferation and MAP-2 enzyme inhibition assays (86). Subsequent structural studies have defined the molecular details of the covalent complex between this class of compounds and human MAP-2 (87). Despite these advances, the mechanism by which MAP-2 is required for endothelial cell proliferation still needs to be elucidated.

PRESENT ISSUES AND FUTURE PROSPECTS

The development of protein fragments, in particular, as anti-angiogenic therapeutics has been the focus of great attention. Issues for these molecules still include reproducibility of *in vivo* results and production of reproducible lots of material for clinical trials. The recently initiated clinical trials of endostatin may provide the data on whether this approach will live up to its preclinical publicity. Generally, the key issues facing the anti-angiogenic approach to cancer therapy is the definition of clinical settings in which single agents or anti-angiogenic agents in combination with cytotoxics meet clinical endpoints of reduced disease progression or increased survival. Considerations from basic science and early clinical studies suggest that the minimal residual disease setting is the most fertile ground for such studies.   A key part of discovering such clinical trial formats is the definition of pharmacodynamic markers of biochemical efficacy for a given target or mechanism. This will enable the rational identification of dosing regimens and disease settings which are most likely to yield positive results.

Future prospects for these therapeutic modalities will likely include combination therapies of anti-angiogenic agents which work by complementary mechanisms, such as MMPIs and tyrosine kinase inhibitors.   In addition, new combinations of cytotoxics, hormonal therapies, and anti-angiogenics are likely to be the preferred treatments for cancer in the future.

References

1.   J. Folkman, Cancer Res., 46, 467, (1986).
2.   L.M. Ellis, and I.J. Fidler, Eur. J. Cancer, 32A, 2451, (1996).
3.   G. Christofori, Angiogenesis, 2, 21, (1998).
4.   E. Keshet, and S.A. Ben-Sasson, J. Clin. Invest., 104, 1497, (1999).
5.   A.J. Hayes, L.Y. Li, and M.E. Lippman, BMJ, 318, 853, (1999).
6.   V. Brower, Nat. Biotechnol., 17, 963, (1999).
7.   D. Hanahan, and J. Folkman, Cell, 86, 353, (1996).
8.   D. Powell, J. Skotnicki, and J. Upeslacis. (1997) in Ann. Rep. Med. Chem., Vol. 32, pp. 161, Academic Press
9.   P. Carmeliet, and D. Collen, Kidney Int, 53, 1519, (1998).
10. T. Veikkola, M. Karkkainen, L. Claesson-Welsh, and K. Alitalo, Cancer Res., 60, 203, (2000).
11. T. Veikkola, and K. Alitalo, Semin. Cancer Biol., 9, 211, (1999).
12. J. Rak, and R.S. Kerbel, Nat. Med., 3, 1083, (1997).
13. I. Vlodavsky, and G. Christofori. (1999) in Fibroblast Growth Factors in Tumor Progression and Angiogenesis (B.A. Teicher, Ed.), Humana Press.
14. N.W. Gale, and G.D. Yancopoulos, Genes and Development, 13, 1055, (1999).
15. J. Holash, S.J. Wiegand, and G.D. Yancopoulos, Oncogene, 18, 5356, (1999).
16. G.D. Yancopoulos, M. Klagsbrun, and J. Folkman, Cell, 93, 661, (1998).
17. T.H. Vu, J.M. Shipley, G. Bergers, J.E. Berger, J.A. Helms, D. Hanahan, S.D. Shapiro, R.M. Senior, and Z. Werb, Cell, 93, 411, (1998).
18. W.D. Klohs, and J.M. Hamby, Curr. Opin. Biotechnol., 10, 544, (1999).
19. K. Bajou, A. Noel, R.D. Gerard, V. Masson, N. Brunner, C. Holst-Hansen, M. Skobe, N.E. Fusenig, P. Carmeliet, D. Collen, and J.M. Foidart, Nat. Med., 4, 923, (1998).
20. A.E. Aplin, A. Howe, S.K. Alahari, and R.L. Juliano, Pharmacol. Rev., 50, 197, (1998).
21. J.A. Foekens, H.A. Peters, M.P. Look, H. Portengen, M. Schmitt, M.D. Kramer, N. Brunner, F. Janicke, M.E. Meijer-van Gelder, S.C. Henzen-Logmans, W.L.J. van Putten, and J.G.M. Klijn, Cancer Res., 60, 636, (2000).
22. B.P. Eliceiri, and D.A. Cheresh, J. Clin. Invest., 103, 1227, (1999).
23. J. Folkman, Nat. Med., 1, 27, (1995).
24. J. Holash, P.C. Maisonpierre, D. Compton, P. Boland, C.R. Alexander, D. Zagzag, G.D. Yancopoulos, and S.J. Wiegand, Science, 284, 1994, (1999).
25. M. Peichev, A.J. Naiyer, D. Pereira, Z. Zhu, W.J. Lane, M. Williams, M.C. Oz, D.J. Hicklin, L. Witte, M.A. Moore, and S. Rafii, Blood, 95, 952, (2000).
26. G. Bergers, K. Javaherian, K.M. Lo, J. Folkman, and D. Hanahan, Science, 284, 808, (1999).
27. R.K. Jain, N. Safabakhsh, A. Sckell, Y. Chen, P. Jiang, L. Benjamin, F. Yuan, and E. Keshet, Proc. Natl Acad. Sci. U S A, 95, 10820, (1998).
28. N. Ferrara and K. Alitalo, Nat. Med., 5, 1359, (1999)
29. R.L. Kendall, and K.A. Thomas, Proc. Natl. Acad. Sci. U S A, 90, 10705, (1993).
30. D.Y. Li, L.K. Sorensen, B.S. Brooke, L.D. Urness, E.C. Davis, D.G. Taylor, B.B. Boak, and D.P. Wendel, Science, 284, 1534, (1999).
31. J. Li, M. Post, R. Volk, Y. Gao, M. Li, C. Metais, K. Sato, J. Tsai, W. Aird, R.D. Rosenberg, T.G. Hampton, F. Sellke, P. Carmeliet, and M. Simons, Nat. Med., 6, 49, (2000).
32. D.W. Miller, W. Graulich, B. Karges, S. Stahl, M. Ernst, A. Ramaswamy, H.H. Sedlacek, R. Muller, and J. Adamkiewicz, Int. J. Cancer, 81, 568, (1999).
33. F. Rastinejad, P.J. Polverini, and N.P. Bouck, Cell, 56, 345, (1989).

34. D.J. Good, P.J. Polverini, F. Rastinejad, M.M. Le Beau, R.S. Lemons, W.A. Frazier, and N.P. Bouck, Proc. Natl. Acad. Sci. U S A, 87, 6624, (1990).

35. E.T. Rietschel, T. Kirakae, F. Ulrich Schade, U. Mamat, G. Schmidt, H. Loppnow, A.J. Ulmer, U. Zahringer, U. Seydel, F. Di Padova, M. Schreier, and H. Brade, The FASEB J., 8, 217, (1994).

36. E.T. Rietschel, H. Brade, O. Holst, L. Brade, S. Muller-Loennies, U. Mamat, U. Zahringer, F. Beckmann, U. Seydel, K. Brandenburg, A.J. Ulmer, T. Mattern, H. Heine, J. Schletter, H. Loppnow, U. Schonbeck, H.D. Flad, S. Hauschildt, U.F. Schade, F. Di Padova, S. Kusumoto, and R.R. Schumann, Curr. Top. Microbiol. Immunol., 216, 39, (1996).

37. J.M. Harlan, L.A. Harker, G.E. Striker, and L.J. Weaver, Thrombosis Res., 29, 15, (1983).

38. P.H. Bessette, F. Aslund, J. Beckwith, and G. Georgiou, Proc. Natl. Acad. Sci. U S A, 96, 13703, (1999).

39. J. Folkman, Eur. J. Cancer, 32A, 2534, (1996).

40. M.S. O'Reilly, L. Holmgren, Y. Shing, C. Chen, R.A. Rosenthal, M. Moses, W.S. Lane, Y. Cao, E.H. Sage, and J. Folkman, Cell, 79, 315, (1994).

41. M.S. O'Reilly, T. Boehm, Y. Shing, N. Fukai, G. Vasios, W.S. Lane, E. Flynn, J.R. Birkhead, B.R. Olsen, and J. Folkman, Cell, 88, 277, (1997).

42. M. Dhanabal, R. Ramchandran, M.J. Waterman, H. Lu, B. Knebelmann, M. Segal, and V.P. Sukhatme, J. Biol. Chem., 274, 11721, (1999).

43. L. Claesson-Welsh, M. Welsh, N. Ito, B. Anand-Apte, S. Soker, B. Zetter, M. O'Reilly, and J. Folkman, Proc. Natl. Acad. Sci. U S A, 95, 5579, (1998).

44. M. Dhanabal, R. Volk, R. Ramchandran, M. Simons, and V.P. Sukhatme, Biochem. Biophys. Res. Commun., 258, 345, (1999).

45. K.L.B. Sim, Angiogenesis, 2, 37, (1998).

46. H.J. Mauceri, N.N. Hanna, M.A. Beckett, D.H. Gorski, M.J. Staba, K.A. Stellato, K. Bigelow, R. Heimann, S. Gately, M. Dhanabal, G.A. Soff, V.P. Sukhatme, D.W. Kufe, and R.R. Weichselbaum, Nature, 394, 287, (1998).

47. H. Lu, M. Dhanabal, R. Volk, M.J. Waterman, R. Ramchandran, B. Knebelmann, M. Segal, and V.P. Sukhatme, Biochem. Biophys. Res. Commun., 258, 668, (1999).

48. Y. Cao, A. Chen, S.S.A. An, R.W. Ji, D. Davidson, and M. Llinas, J. Biol. Chem., 272, 22924, (1997).

49. S.E. Pike, L. Yao, K.D. Jones, B. Cherney, E. Appella, K. Sakaguchi, H. Nakhasi, J. Teruya-Feldstein, P. Wirth, G. Gupta, and G. Tosato, J. Exp. Med., 188, 2349, (1998).

50. S.E. Pike, L. Yao, J. Setsuda, K.D. Jones, B. Cherney, E. Appella, K. Sakaguchi, H. Nakhasi, C.D. Atreya, J. Teruya-Feldstein, P. Wirth, G. Gupta, and G. Tosato, Blood, 94, 2461, (1999).

51. G.D. Kamphaus, P.C. Colorado, D.J. Panka, H. Hopfer, R. Ramchandran, A. Torre, Y. Maeshima, J.W. Mier, V.P. Sukhatme, and R. Kalluri, J. Biol. Chem., 275, 1209, (2000).

52. F. Vazquez, G. Hastings, M.A. Ortega, T.F. Lane, S. Oikemus, M. Lombardo, and M.L. Iruela-Arispe, J. Biol. Chem., 274, 23349, (1999).

53. T. Browder, J. Folkman, and S. Pirie-Shepherd, J. Biol. Chem., 275, 1521, (2000).

54. S. Gately, P. Twardowski, M.S. Stack, D.L. Cundiff, D. Grella, F.J. Castellino, J. Enghild, H.C. Kwaan, F. Lee, R.A. Kramer, O. Volpert, N. Bouck, and G.A. Soff, Proc. Natl. Acad. Sci. U S A, 94, 10868, (1997).

55. R.K. Singh, M. Gutman, C.D. Bucana, R. Sanchez, N. Llansa, and I.J. Fidler, Proc. Natl. Acad. Sci. U S A, 92, 4562, (1995).

56. S. Parangi, M. O'Reilly, G. Christofori, L. Holmgren, J. Grosfeld, J. Folkman, and D. Hanahan, Proc. Natl. Acad. Sci. U S A, 93, 2002, (1996).

57. J.W. Skiles, L.G. Monovich, and A.Y. Jeng, in Ann. Rep. Med. Chem., (A.M. Doherty, ed) Vol. 35, Academic Press (2000).

58. S. Rosenberg. in Ann. Rep. Med. Chem., (A.M. Doherty, ed) Vol. 34, Academic Press (1999), pp. 121-128.

59. L.R. Lund, J. Romer, T.H. Bugge, B.S. Nielsen, T.L. Fandsen, J.L. Degen, R.W. Stephens, and K. Dano, EMBO J., 18, 4645, (1999).

60. A.R. Nelson, B. Fingleton, M.L. Rothenberg, and L.M. Matrisian, J. Clin. Onc., 5, 1135 (2000)

61. J.W. Hutchinson, G.M. Tierney, S.L. Parsons, and T.R. Davis, Br. J. Bone and Joint Surgery, 80, 907, (1998).

62. K. Holmbeck, P. Bianco, J. Caterina, S. Yamada, M. Kromer, S.A. Kuznetsov, M. Mankani, P.G. Robey, A.R. Poole, I. Pidoux, J.R. Ward, and H. Birkedal-Hansen, Cell, 99, 81, (1999).

63. R.P. Beckett, M. Whittaker, A. Miller, and M.F. Martin. (1999) World Patent 9925693

64. G. Fibbi, R. Caldini, M. Chevanne, M. Pucci, N. Schiavone, L. Morbidelli, A. Parenti, H.J. Granger, M. Del Rosso, and M. Ziche, Lab. Invest., 78, 1109, (1998).
65. D. Ribatti, D. Leali, A. Vacca, R. Giulani, A. Gualandris, L. Roveali, M.I. Nolli, and M. Presta, J. Cell Science, 112, 13, (1999).
66. C.G. Barber, P.V. Fish, and R.P. Dickinson. World Patent 9920608 (1999).
67. C.R. Illig, N.L. Subasinghe, J.B. Hoffman, K.J. Wilson, and M.J. Rudolph. World Patent. 9940088 (1999).
68. R. Friesel, and T. Maciag, Thrombosis and Haem., 82, 748 - 754, (1999).
69. A. Levitzki, Pharm.Ther., 82, 231, (1999).
70. J.M. Hamby, and H.D. Showalter, Pharm.Ther., 82, 169, (1999).
71. T.A. Fong, L.K. Shawver, L. Sun, C. Tang, H. App, T.J. Powell, Y.H. Kim, R. Schreck, X. Wang, W. Risau, A. Ullrich, K.P. Hirth, and G. McMahon, Cancer Res., 59, 99, (1999).
72. R.M. Shaheen, D.W. Davis, W. Liu, B.K. Zebrowski, M.R. Wilson, C.D. Bucana, D.J. McConkey, G. McMahon, and L.M. Ellis, Cancer Res., 59, 5412, (1999).
73. M. Mohammadi, G. McMahon, L. Sun, C. Tang, P. Hirth, B.K. Yeh, S.R. Hubbard, and J. Schlessinger, Science, 276, 955, (1997).
74. M. Mohammadi, S. Froum, J.M. Hamby, M.C. Schroeder, R.L. Panek, G.H. Lu, A.V. Eliseenkova, D. Green, J. Schlesinger, and S.R. Hubbard, EMBO J., 17, 5896, (1998).
75. L.F. Hennequin, A.P. Thomas, C. Johnstone, E.S.E. Stokes, P.A. Ple, J.-J.M. Lohmann, D.J. Ogilvie, M. Dukes, S.R. Wedge, J.O. Curwen, J. Kendrew, and C. Lambert-van der Brempt, J. Med. Chem., 42, 5369, (1999).
76. D.J. Ogilvie, S.R. Wedge, M. Dukes, J. Kendrew, J.O. Curwen, A.P. Thomas, L.F. Hennequin, P. Ple, E.S.E. Stockes, C. Johnstone, P. Wadsworth, G.H.P. Richmond, and B. Curry. in *Annual Meeting American Association for Cancer Research* (1999) p. 69.
77. L. Shewchuk, A. Hassell, B. Wisely, W. Rocque, W. Holmes, J. Veal, and L.F. Kuyper, J. Med. Chem., 43, 133, (2000).
78. M.A. McTigue, J.A. Wickersham, C. Pinko, R.E. Showalter, C.V. Parast, A. Russell-Tempczyk, M.R. Gehring, B. Mroczkowski, C.C. Kan, J.E. Villafranca, and K. Appelt, Structure Fold and Design, 7, 319, (1999).
79. M.A. Horton, Int. J. Biochem. Cell Biol., 29, 721, (1997).
80. M.W. Lark, G.B. Sroup, S.M. Hwang, I.E. James, D.J. Rieman, F.H. Drake, J.N. Bradbeer, A. Mathur, K.F. Erhard, K.A. Newlander, S.T. Ross, K.L. Salyers, B.R. Smith, W.H. Miller, W.F. Huffman, and M. Gowen, J. Pharm.Exp. Ther., 291, 612, (1999).
81. J.S. Kerr, R.S. Wexler, S.A. Mousa, C.S. Robinson, E.J. Wexler, S. Mohamed, M.E. Voss, J.J. Devenny, P.M. Czerniak, A. Gudzelak, and A.M. Slee, Anticancer Res., 19, 959, (1999).
82. C.P. Carron, D.M. Meyer, J.A. Pegg, V.W. Engleman, M.A. Nickols, S.L. Settle, W.F. Westlin, P.G. Ruminski, and G.A. Nickols, Cancer Res., 58, 1930, (1998).
83. D.G. Batt, J.J. Petraitis, G.C. Houghton, D.P. Modi, G.A. Cain, M.H. Corjay, S.A. Mousa, P.J. Bouchard, M.S. Forsythe, P.P. Harlow, F. Barbera, S.M. Spitz, R.R. Wexler, and P.K. Jadhav, J. Med. Chem., 43, 41, (2000).
84. W.J. Pitts, J. Wityak, J.M. Smallheer, A.E. Tobin, J.W. Jetter, J.S. Buynitsky, P.P. Harlow, K.A. Solomon, M.H. Corjay, S.A. Mousa, R.R. Wexler, and P.K. Jadhav, J. Med. Chem., 43, 27, (2000).
85. N. Sin, L. Meng, M.Q. Wang, J.J. Wen, W.G. Bornmann, and C.M. Crews, Proc Natl Acad Sci USA, 94, 6099, (1997).
86. E.C. Griffith, Z. Su, B.E. Turk, S. Chen, Y.H. Chang, Z. Wu, K. Biemann, and J.O. Liu, Chem. Biol., 4, 461, (1997).
87. S. Liu, Widom, J., Kemp, C.W., Crews, C.M., Clardy, J., Science, 282, 1324, (1998).

Chapter 12.   Progress in the Oxazolidinone Antibacterials

Robert C. Gadwood and Dean A. Shinabarger
Pharmacia & Upjohn
Kalamazoo, MI 49007

Introduction - The oxazolidinones are a new class of antibacterial agents with activity against a broad spectrum of gram-positive pathogens including staphylococci, streptococci, and enterococci.   Because they are totally synthetic and mechanistically novel, the oxazolidinones lack cross-resistance to every clinically significant resistance mechanism tested to date (1,2).   Consequently, these antibacterial agents are effective against drug resistant gram-positive bacteria including methicillin resistant *S. aureus* (MRSA), vancomycin resistant enterococci (VRE) and penicillin resistant *S. pneumoniae* (PRSP).   The incidence of drug resistance in gram-positive bacteria is growing rapidly and has become a significant public health threat (3-6).

The oxazolidinone structure is relatively simple and allows for diverse synthetic modification. Because of this, and because of the importance of the oxazolidinones as new antibacterials, many reports have appeared describing new oxazolidinone analogues. This review will primarily cover new oxazolidinone analogues that have appeared in the last two years. The early chemistry and biology of the oxazolidinone antibacterials has been previously reviewed (7-12).

Linezolid - A number of reviews summarizing the discovery and antibacterial activity of linezolid (**1**) have appeared (13-16). Linezolid has completed clinical trials and is currently awaiting approval for use as an antibacterial agent. Results from Phase II and Phase III clinical trials showed that linezolid treatment afforded high cure rates in cases of complicated and uncomplicated skin and soft tissue infections, community and clinically acquired pneumonia, VRE infected patients, and MRSA infection (17).

Several new *in vitro* studies with linezolid have appeared. Linezolid has potent activity against anaerobic clinical isolates including those from bite wound infections and from clinically significant infections in South Africa (18-21). The potent *in vitro* activity of linezolid against staphylococci (22-27), *S. pneumoniae* (28-30), and enterococci (31-33) has been confirmed in several new studies. These studies also included multiply drug-resistant isolates. Linezolid was tested against the fastidious gram-negative respiratory pathogens *B. pertussis* and *B. parapertussis* and was found to have good activity against the former but weak activity against the latter (34).   Linezolid has weak *in vitro* activity against *Legionella* spp. and *Brucella melitensis* (35-36).

Linezolid was effective in an *in vivo* murine model of *Mycobacterium tuberculosis*, although the close analogue PNU-100480 (**2**) had slightly better activity (37,38).   In a rat intra-abdominal abscess model employing *E. faecalis* and *E. faecium*, linezolid was found to have only modest activity (39). Since linezolid had potent in vitro activity against the strains employed in this model, it was suggested that pharmacokinetic factors were responsible.

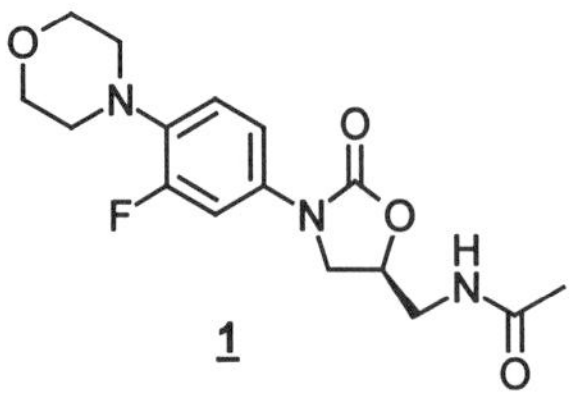

1 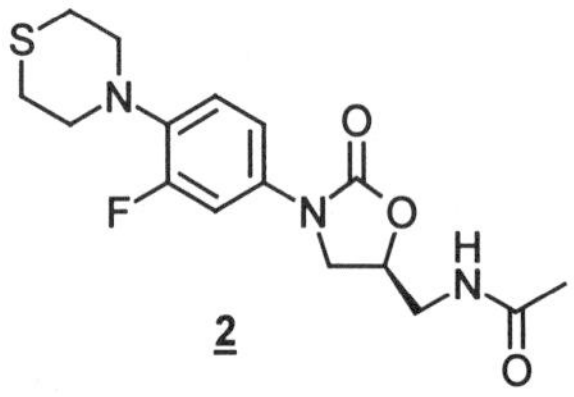   **2**

Mechanism of Action - The oxazolidinones inhibit protein synthesis in actively growing bacteria. Mechanism of action studies support oxazolidinone binding to the bacterial 50S ribosomal subunit and inhibition of formation of the 70S ribosomal initiation complex (40). Protein synthesis inhibition was first observed in *Bacillus subtilus* cultures and later demonstrated in *acrAB* or *tolC* mutants of *Escherichia coli*, indicating that the high MIC observed for oxazolidinones against gram-negative bacteria was likely due to drug efflux from the interior of the cell (41-42). Cell-free studies showed that linezolid and eperezolid (**3**) are potent inhibitors of the bacterial transcription/translation system (43). Further studies with cell-free extracts of *E. coli* revealed that a four-fold increase in the amount of mRNA added to the assay decreased the IC_{50} for linezolid three-fold (43). Extracts prepared from *E. coli* cells grown in the presence of an oxazolidinone could not translate MS2 RNA (44).

Initiation complex formation involving mRNA, 70S ribosomes, and fMet-tRNA was inhibited by 50% in the presence of linezolid (45). Initiation complex formation with 30S ribosomes was inhibited as well, suggesting that linezolid recognizes a similar binding site formed when fMet-tRNA binds to ribosomes. However, high IC_{50} values indicate that additional factors might be needed in order to form the complete oxazolidinone-binding site on the ribosome (40). Several studies have failed to demonstrate inhibition of elongation, termination, or N-formylmethionyl-tRNA biosynthesis by the oxazolidinones (41,43). In the only report examining binding of an oxazolidinone to the ribosome, it was determined that labeled eperezolid was bound to 50S but not 30S subunits (46). An oxazolidinone photoaffinity probe (**4**) was found to covalently bind to both the 16S and 23S rRNA (47). On 16S rRNA, an oxazolidinone footprint was found at A864 in the central domain, whereas with 23S rRNA, the modified residues were U2113, A2114, U2118, A2119, and C2153, all of which are found in the peptidyl transferase domain V. The crosslinking of oxazolidinones to 23S rRNA is particularly intriguing, especially since the domain V point mutations located at G2447U and G2576U can independently confer resistance to this class of compounds (48,49).

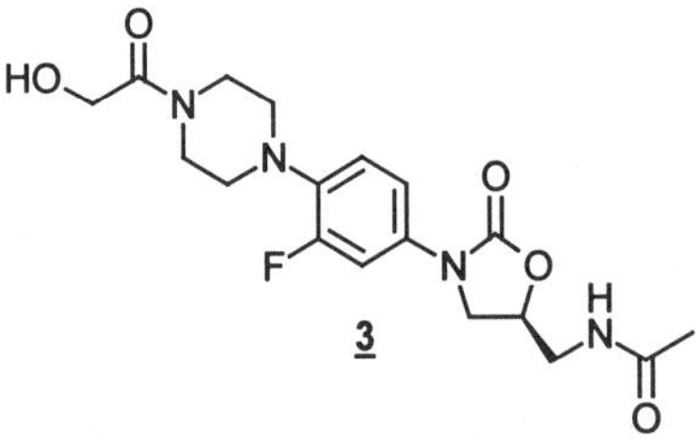

3 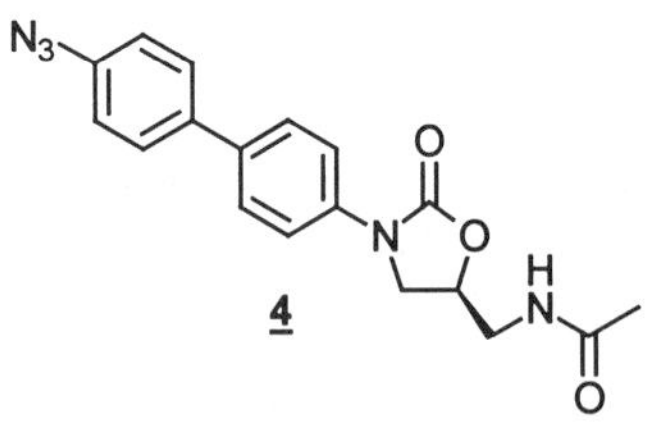   **4**

NEW OXAZOLIDINONE ANALOGUES

A-ring modifications - The antibacterial oxazolidinone pharmacophore consists of an oxazolidinone ring (the A-ring) with an aromatic ring (the B-ring) attached at nitrogen. Frequently (as is the case with linezolid) a third ring (the C-ring) is also present as a substituent on the B-ring. The oxazolidinone ring is always substituted at C5 with a functionalized methylene, most commonly an acetamidomethylene. Relatively few successful modifications of the A-ring have been reported (5), although an isoxazoline (**5**) is an acceptable oxazolidinone bioisostere (50,51). An oxazinone analogue of linezolid (**6**) has been reported, but its antibacterial activity was not described (52).

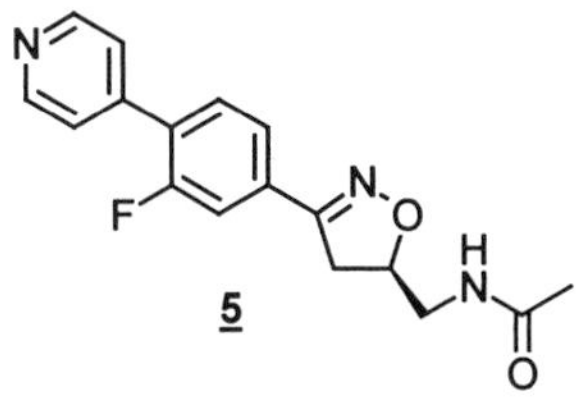

5 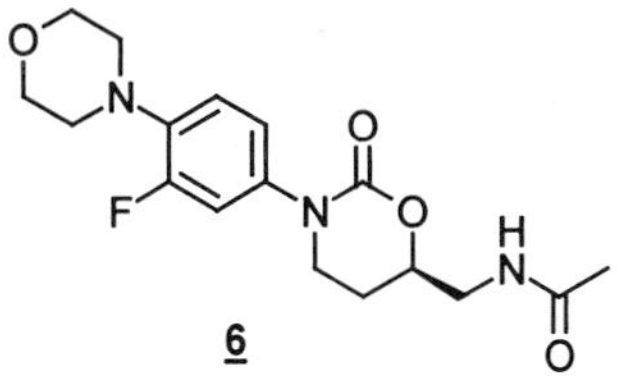   **6**

A series of tricyclic analogues having a methylene bridge between the A-ring and the B-ring have been reported, most recently in chiral form (53,54). The *trans*-fused ring system **7** is active, but the *cis*-fused ring system **8** is inactive. The direct linezolid analogue **9** is only weakly active. The 3-pyridyl analogue **10** is the most active compound in this series (53).

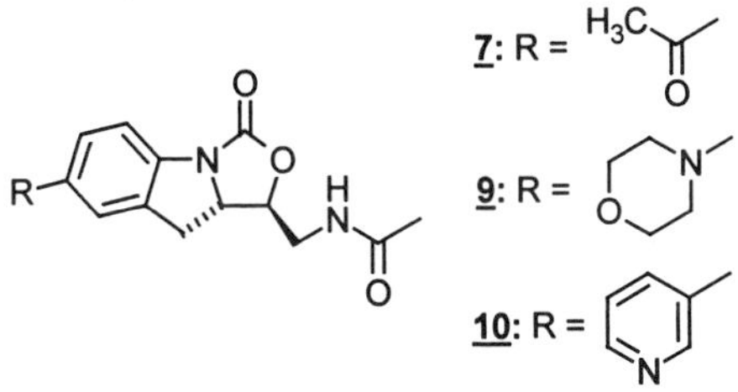

7: R =

9: R =

10: R =

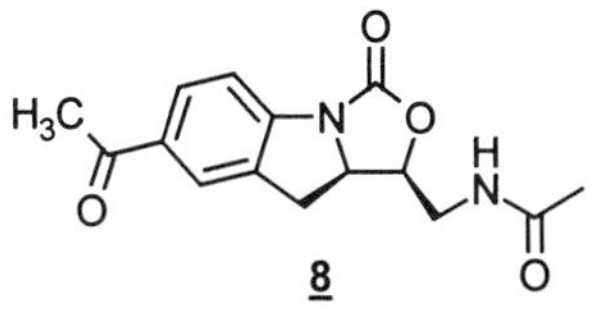

8

A variety of modifications of the acetamide side chain are tolerated.   A study with benzthiazolones (**11**) showed that good antimicrobial activity was retained when R was small alkyl or alkoxy (55,56).   Some of these modified analogues have improved pharmacokinetic properties (57).   Thioamide and thiourea side chains have also been described and in some cases offer an improvement in antibacterial activity (58-60). The thioamide **12** is four-fold more potent than the corresponding carboxamide, with MIC values against staphyloccoci, streptococci, and enterococci ranging from 0.12 to 0.5 μg/mL (61,62). In addition, **12** shows good activity against *H. influenzae* (MIC values of 0.5 to 1 μg/mL) (61). The more radically modified aryl ether analogues **13** and **14** are also active (63,64).

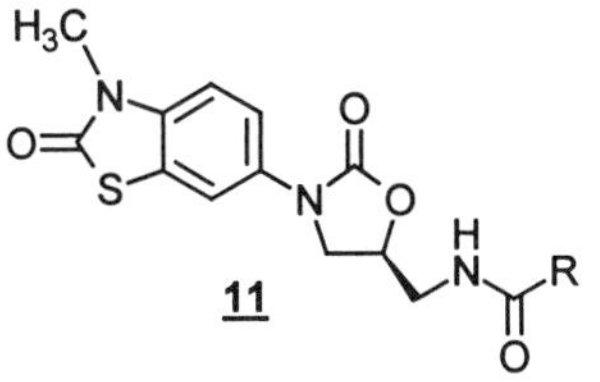

11 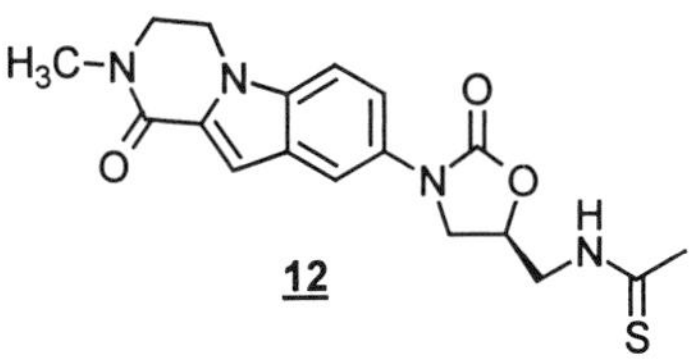   **12**

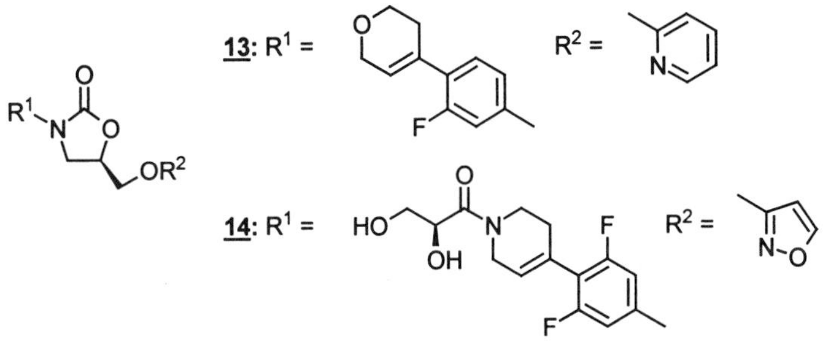

13: R¹ =   R² =

14: R¹ =   HO   R² =

<u>B-ring modifications</u> - Heteroaromatics can be employed as B-ring bioisosteres. Pyridyl and thiophenyl analogues (**15** and **16**, respectively) retain antibacterial activity against *S. aureus* (65-69). Both **15** and **16** have *in vivo* activity in mice upon subcutaneous administration (70). The pyridothienyl oxazolidinone **17** and the thiazolopyridine **18** both have modest activity against staphyloccoci (71,72). The related benzofuran **19** has good gram-positive antibacterial activity (73).

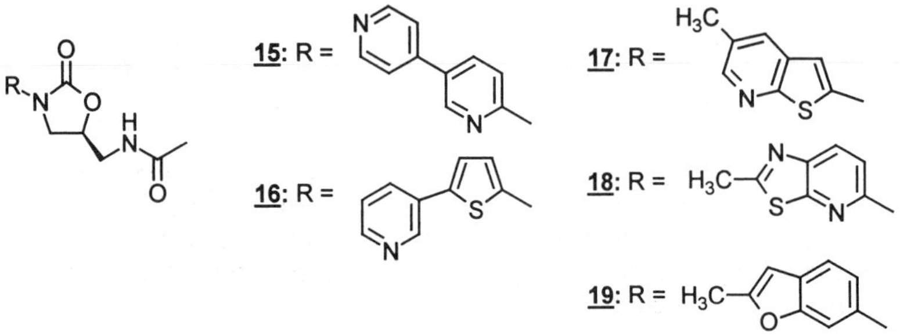

15: R =

16: R =

17: R =

18: R = H₃C-

19: R = H₃C-

Oxazolidinones that incorporate a fused B-ring system containing a lactam group have good antibacterial activity. The benzoxazinone **20** has *in vitro* activity comparable to linezolid and better i.v. pharmacokinetic parameters (74,75). Other fused B-ring examples include 2-quinolone, benzoxazolone, and oxazolopyridine (**21**, **22**, and **23**, respectively) (76-78). The previously described benzthiazolones **11** are also in this family (55,56).

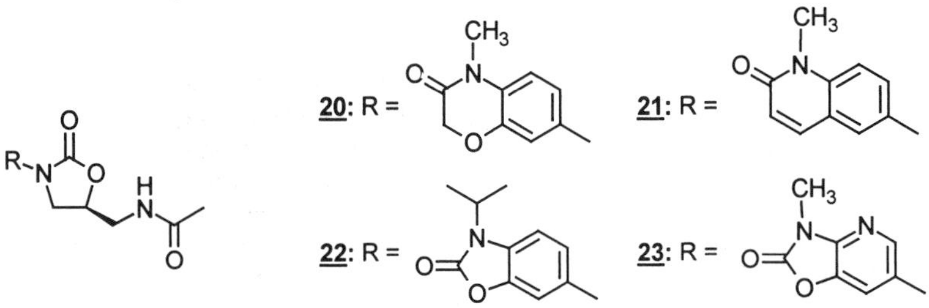

20: R =

21: R =

22: R =

23: R =

<u>C-ring modifications</u> - A great deal of flexibility exists in the heterocyclic C-ring SAR of the oxazolidinones and several bicyclic C-ring variations have appeared. The bicyclo[3.1.1]azaheptanone analogue **24** has good activity, as do the dihydropyrrolo[1.2-c]oxazole and the imidazo[2.1-b]thiazoloyl analogues (**25** and **26**, respectively) (79-82). Compound **26** is particularly active, with MIC values of 0.03 to 0.5 μg/mL against gram-positive organisms. This is also one of the few examples of an active oxazolidinone with a carbonyl linkage between the B and C-rings.

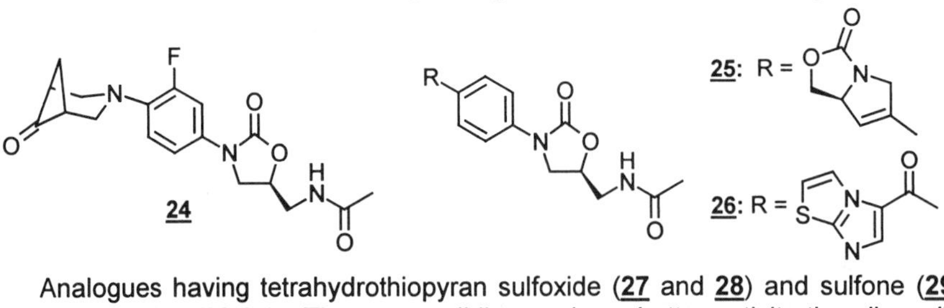

24

25: R =

26: R =

Analogues having tetrahydrothiopyran sulfoxide (**27** and **28**) and sulfone (**29**) C-rings are active (83,84). These oxazolidinones have better activity than linezolid against fastidious gram-negative organisms, with MIC values of 2 - 4 μg/mL against *H. influenzae* and *M. catarrhalis*. Both **27** and **28** have oral activity in mice comparable to linezolid. The *cis*-sulfoxide **27** is converted to the *trans*-isomer **28** *in vivo* (85). Phenyloxazolidinones having dihydrothiopyranyl (**30**), dihydropyranyl (**31**), and tetrahydropyridinyl (**32**) C-rings are also potent antibacterials (86-88).

Analogues with piperazinone (**33**) and diazepinone (**34**) C-rings also retain activity (89,90).

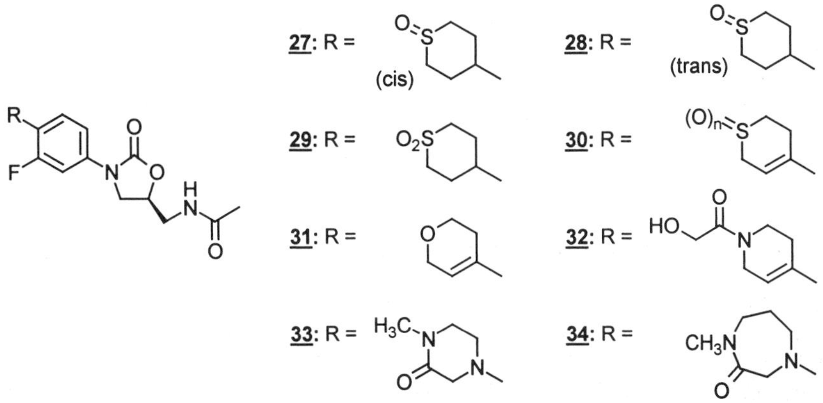

Oxazolidinone analogues with heteroaromatic C-rings often have potent activity against fastidious gram-negative organisms in addition to gram-positive pathogens. Examples include the cyanopyrrole **35** and the cyanopyrazole **36**, which have MIC values of <0.125 to 0.5 µg/mL against gram-positive organisms and 1 to 4 µg/mL against *H. influenzae* and *M. catarrahlis* (91). These compounds also have oral activity in mice. Analogues with imidazole, triazole, and tetrazole C-rings also have good activity (92,93). The 5-cyanothiazole and 5-cyanothiophene analogues (**37** and **38**, respectively) are even more potent against both gram-positive (MIC values <0.125 µg/mL) and fastidious gram-negative organisms (MIC values of 0.25 to 2 µg/mL) and are orally active (94,95). Similar activity is reported for the cyanoethylthiadiazole **39** (96,97). The pyrazoles **40** and **41** are active, but less potent than other members of this group (98). Other examples of the heteroaromatic C-ring family include the isoxazole **42**, the isomeric thiazole **43**, and the substituted imidazoles **44** and **45** (95,99,100).

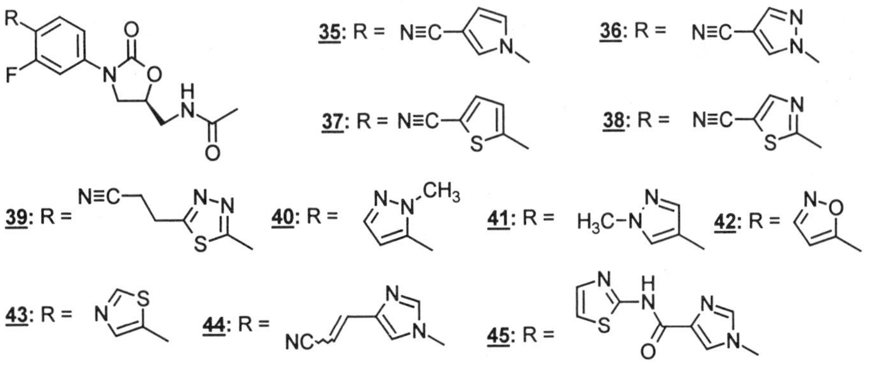

A number of examples have been reported wherein the C-ring and the B-ring are bridged by one or two atoms to form a rigid tricyclic system. The bridged bis-aryl oxazolidinone **46** has potent *in vitro* activity against gram-positive bacteria (MIC values of 0.125 to 1 µg/mL) and also has activity against *H. influenzae* (MIC value of 4 µg/mL) (101,102). However, **46** is inactive *in vivo* in mice after subcutaneous administration. The imidazobenzoxazinyl system **47** is another example with good potency against *S. aureus* (103). The pyrazinoindolyl system **48** is more potent *in vitro* than linezolid, but has poor oral activity due to unfavorable pharmacokinetics

(61,62).   The pyrazinobenzoxazinyl oxazolidinone **49** is only modestly active (104,105).

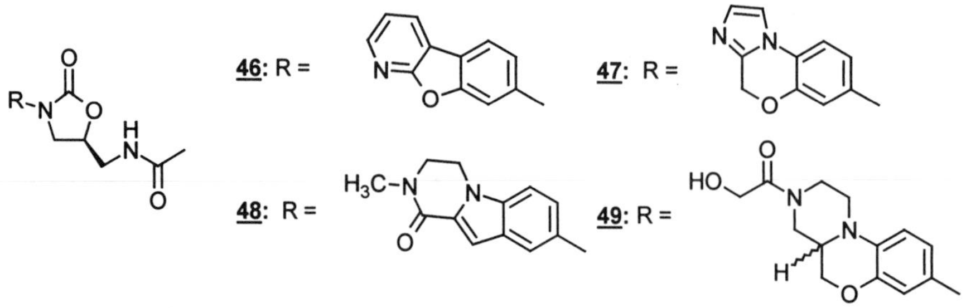

46: R =   **47**: R =

48: R =   **49**: R =

D-rings - The oxazolidinone pharmacophore also tolerates another ring (D-ring) attached to a piperazinyl C-ring. Oxazolidinones with 1,3-thiazole, 1,3,4-thiadiazole, and 1,2,4-thiadiazole rings attached to a piperazine C-ring (**50**, **51**, and **52**, respectively) show potent antibacterial activity against staphylococci and streptococci (106,107).   Similarly, the D-ring can be a pyridine (**53**) or a pyrimidine (**54** and **55**) (108-111). Each of these has good in vitro activity against gram-positive organisms and oral activity in the *in vivo* mouse model. Isoxazoyl piperazines (**56**) have enhanced activity (112). A QSAR analysis of these compounds suggests that steric factors are the most important determinant of activity (113).

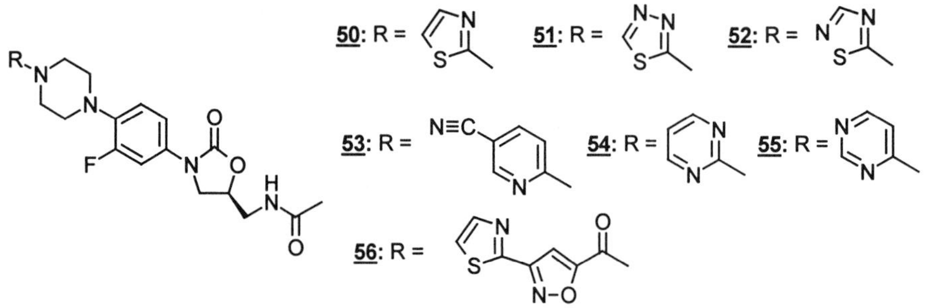

50: R =   **51**: R =   **52**: R =

53: R =   **54**: R =   **55**: R =

56: R =

Miscellaneous - The cephalosporin-oxazolidinone hybrid **57** has broad spectrum *in vitro* activity against both gram-positive and gram-negative organisms (114). Methods for making oxazolidinones using combinatorial chemistry techniques have been reported (115).   The pyridylamide **58** has oral activity in mice (PD_{50} = 7.5 mg/kg for p.o. administration).

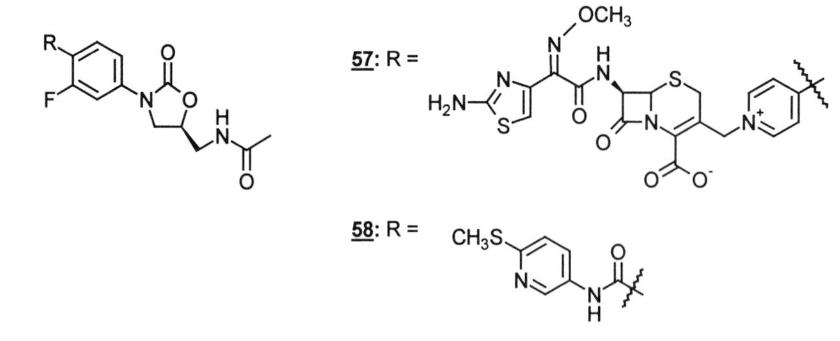

57: R =

58: R =

Conclusion - Clinical data with linezolid illustrate the potential for oxazolidinones to be effective weapons against gram-positive bacteria in humans. The importance of their activity against drug resistant strains of gram-positive pathogens is highlighted by a backdrop of escalating bacterial resistance. New analogues have been found with increased gram-positive potency and also activity against fastidious gram-negative organisms. The relative ease with which the oxazolidinones can be prepared has allowed the preparation of a diverse array of structural types that populate a rich SAR.

References

1. S. J. Brickner, D. K. Hutchinson, M. R. Barbachyn, P. R. Manninen, D. A. Ulanowicz, S. A. Garmon, K. C. Grega, S. K. Hendges, D. S. Toops, C. W. Ford, and G. E. Zurenko, Antimicrob. Agents Chemother., 39, 673 (1996).
2. M. Fines, and R. Leclercq, 39th Interscience Conference on Antimicrobial Agents and Chemotherapy, San Francisco, CA, USA (1999). Abstract C-847.
3. W. Witte, J. Antimicrob. Chemother., 44, Suppl. A, 1 (1999).
4. H. Labischinski, K. Ehlert and B. Wieland, Exp. Opin. Invest. Drugs, 7, 1245 (1998).
5. S. J. Brickner, Chem. Ind., 131 (1997).
6. D. T. W. Chu, J. Plattner and L. Katz, J. Med. Chem., 39, 3853 (1996).
7. B. Riedl and R. Endermann, Exp. Opin. Ther. Patents, 9 625 (1999).
8. M. R. Barbachyn, S. J. Brickner, R. C. Gadwood, S. A. Garmon, K. C. Grega, D. K. Hutchinson, K. Munesada, R. J. Reischer, M. Taniguchi, L. M. Thomasco, D. S. Toops, H. Yamada, C. W. Ford and G. E. Zurenko, Adv. Exp. Med. Biol., 456 (Resolving the Antibiotic Paradox), 219 (1998).
9. S. J. Brickner, Curr. Pharm. Des., 2, 175 (1996).
10. M. Müller and K.-L. Schimz, Cell. Mol. Life Sci., 56, 280 (1999).
11. C. S. Kim and S. J. Chang, Korean J. Med. Chem., 7, 145 (1997).
12. M. R. Barbachyn, S. J. Brickner, G. J. Cleek, R. C. Gadwood, K. C. Grega, S. K. Hendges, D. K. Hutchinson, P. R. Manninen, K. Munesada, R. C. Thomas, L. M. Thomasco, D. S. Toops and D. A. Ulanowicz, Spec. Publ. - R. Soc. Chem., 198, 15 (1997).
13. C. W. Ford, J. C. Hamel, D. Stapert, J. K. Moerman, D. K. Hutchinson, M. R. Barbachyn and G. E. Zurenko, Trends Microbiol., 5, 196 (1997).
14. L. D. Dresser and M. J. Rybak, Pharmacother., 18, 456 (1998).
15. G. E. Zurenko, C. W. Ford, D. K. Hutchinson, S. J. Brickner and M. R. Barbachyn, Exp. Opin. Invest. Drugs, 6, 151 (1997).
16. J. Lizondo, X. Rabasseda and J. Casteñer, Drugs Future, 21, 1116 (1996).
17. L. E. Lawrence, M. J. Pucci, M. B. Frosco and J. F. Barrett, Exp. Opin. Invest. Drugs, 8, 2201 (1999).
18. B. Yagi, G. E. Zurenko, Anaerobe, 3, 301 (1997).
19. C. Edlund, H. Oh and C. E. Nord, Clin. Microbiol. Infect., 5, 51 (1999).
20. E. J. C. Goldstein, D. M. Citron and C. V. Merriam, Antimicrob. Agents Chemother., 43, 1469 (1999).
21. M. M. Lubbe, P. L. Botha and L. J. Chalkley, Eur. J. Clin. Microbiol. Infect. Dis., 18, 46 (1999).
22. J. H. Jorgensen, M. L. McElmeel and C. W. Trippy, Antimicrob. Agents Chemother., 41, 465 (1997).
23. R. Wise, J. M. Andrews, F. J. Boswell and J. P. Ashby, J. Antimicrob. Chemother., 42, 721 (1998).
24. M. J. Rybak, D. M. Cappelletty, T. Moldovan, J. R. Aeschlimann and G. W. Kaatz, Antimicrob. Agents Chemother., 42, 721 (1998).
25. M. E. Jones, M. R. Visser, M. Klootwijk, P. Jeisig, J. Verhoef and F.-J. Schmitz, Antimicrob. Agents Chemother., 43, 421 (1999).
26. F.-J. Schmitz, A. Krey, R. Geisel, J. Verhoef, H.-P. Heinz and A. C. Fluit, Eur. J. Clin. Microbio. Infect. Dis., 18, 528 (1999).
27. C. von Eiff and G. Peters, J. Antimicrob. Chemother., 43, 569 (1999).
28. J. A. Kearney, K. Barbadora, E. O. Mason, E. R. Wald and M. Green, Int. J. Antimicrob. Agents, 12, 141 (1999).
29. R. Patel, M. S. Rouse, K. E. Piper and J. M. Steckelberg, Diagn. Microbiol. Infect. Dis., 34, 1119 (1999).
30. O. Mansor, J. Pawlak and L. Saravolatz, J. Antimicrob. Chemother., 43, 31 (1999).
31. G. D. Bostic, M. B. Perri, L. A. Thai and M. J. Zervos, Diagn. Microbiol. Infect. Dis., 30, 109 (1998).

32. M. C. Struwig, P. L. Botha and L. J. Chalkley, Antimicrob. Agents Chemother., 42, 2752 (1998).
33. G. A. Noskin, F. Siddiqui, V. Stosor, D. Jacek and L. R Peterson, Antimicrob. Agents Chemother., 43, 2059 (1999).
34. J. E. Hoppe, J. Chemother., 11, 220 (1999).
35. T. Schülin, C. B. Wennersten, M. J. Ferraro, R. C. Moellering, Jr. and G. M. Eliopoulos, Antimicrob. Agents Chemotherap., 42, 1520 (1998).
36. I. Trujillano-Martin, E. Garcia-Sanchez, M. J. Fresnadillo, J. E. Garcia-Sanchez, J. A. Garcia-Rodriguez and I. M. Martinez, Int. J. Antimicrob. Agents, 12 185 (1999).
37. M. R. Barbachyn, S. J. Brickner and D. K. Hutchinson, U. S. Patent 5,880,118 (1999).
38. M. H. Cynamon, S. P. Klemens, C. A. Sharpe and S. Chase, Antimicrob. Agents Chemother., 43, 1189 (1999).
39. T. Schulin, C. Thauvin-Eliopoulos, R. C. Moellering, Jr. and G. M. Eliopoulos, Antimicrob. Agents Chemother., 43, 2873 (1999).
40. D. Shinabarger, Exp. Opin. Invest. Drugs, 8, 1195 (1999).
41. D. C. Eustice, P. A. Feldman, and A. M. Slee, Biochem. Biophys. Res. Comm., 150, 965 (1988).
42. J. M. Buysse, W. F. Demyan, D. S. Dunyak, D. Stapert, J. C. Hamel, and C. W. Ford. 36th Interscience Conference on Antimicrobial Agents and Chemotherapy, New Orleans, LA, USA (1996). Abstract C-42.
43. D. L. Shinabarger, K. R. Marotti, R. W. Murray, A. H. Lin, E. P. Melchior, S. M. Swaney, D. S. Dunyak, W. F. Demyan, and J. M. Buysse, Antimicrob. Agents Chemother., 41, 2132 (1997).
44. D. C. Eustice, P. A. Feldman, I. Zajac, and A. M. Slee, Biochem. Antimicrob. Agents Chemother., 32, 1218 (1988).
45. S. M. Swaney, H. Aoki, M. C. Ganoza, and D. L. Shinabarger, Antimicrob. Agents Chemother., 42, 3251 (1998).
46. A. H. Lin, R. W. Murray, T. J. Vidmar, and K. R. Marroti, Antimicrob. Agents Chemother., 41, 2127, (1997).
47. N. B. Mattasova, M. V. Rodnina, R. Endermann, H. P. Kroll, U. Pleiss, H. Wild, and W. Wintermeyer, RNA., 7, 939 (1999).
48. S. M. Swaney, D. L. Shinabarger, R. D. Schaadt, J. H. Bock, J. L. Slightom, and G. E. Zurenko, 38th Interscience Conference on Antimicrobial Agents and Chemotherapy, San Diego, California, USA (1998). Abstract C-104.
49. G. E. Zurenko, W. M. Todd, B. Hafkin, B. Meyers, C. Kauffman, J. Bock, J. Slightom, and D. Shinabarger, 39th Interscience Conference on Antimicrobial Agents and Chemotherapy, SanFrancisco, California, USA (1999). Abstract C-848.
50. M. R. Barbachyn, G. J. Cleek, L. A. Dolak, S. A. Garmon, J. Morris, E. P. Seest, R. C. Thomas, W. Watt, D. G. Wishka, C. W. Ford and G. E. Zurenko, 39th Interscience Conference on Antimicrobial Agents and Chemotherapy, San Francisco, California, USA (1999). Abstract F-572.
51. M. R. Barbachyn, G. J. Cleek and R. C. Thomas, U. S. Patent 5,990,136 (1999).
52. B. B. Lohray, S. Baskaran, B. Y. Reddy and K. S. Rao, Tetrahedron Lett., 39, 6555 (1998).
53. D. M. Gleave, S. J. Brickner, P. R. Manninen, D. A. Allwine, K. D. Lovasz, D. C. Rohrer, J. A. Tucker, G. E. Zurenko and C. W. Ford, Bioorg. Med. Chem. Lett., 8, 1231 (1998).
54. M. D. Gleave and S. J. Brickner, J. Org. Chem., 61, 6470 (1996).
55. D. Häbich, S. Bartel, R. Endermann, W. Guarnieri, M. Härter, H.-P. Kroll, S. Raddatz, B. Riedl, U. Rosentreter, M. Ruppelt, A. Stolle and H. Wild, 38th Interscience Conference on Antimicrobial Agents and Chemotherapy, San Diego, California, USA (1998). Abstract F-130.
56. A. Stolle, D. Häbich, S. Bartel, B. Riedl, M. Ruppelt, H. Wild, R. Endermann, K.-D. Bremm, H.-P. Kroll, H. Labischinski, K. Schaller and H.-O. Werling, EP 738 726 (1996).
57. R. Endermann, S. Bartel, W. Guarnieri, D. Häbich, M. Härter, H.-P. Kroll, S. Raddatz, B. Riedl, U. Rosentreter, M. Ruppelt, A. Stolle and H. Wild, 38th Interscience Conference on Antimicrobial Agents and Chemotherapy, San Diego, California, USA (1998). Abstract F-129.
58. B. Riedl, D. Häbich, A. Stolle, M. Ruppelt, S. Bartel, W. Guarnieri, R. Endermann and H.-P. Kroll, U. S. Patent 5,792,765 (1998).
59. J. B. Hester, E. G. Nidy, S. C. Perricone and T. J. Poel, WO 9854161.
60. T. Yoshida, R. Tokuyama and Y. Tomita, WO 9912914 (1999).

61. S. Bartel, R. Endermann, W. Guarnieri, D. Häbich, M. Härter, K. Henniger, 39th Interscience Conference on Antimicrobial Agents and Chemotherapy, San Francisco, California, USA (1999). Abstract F-565.
62. S. Bartel, W. Guarnieri, D. Häbich, S. Raddatz, B. Ridl, U. Rosentreter, M. Ruppelt, A. Stolle, H. Wild, R. Endermann and H-P. Kroll, WO 9937652.
63. M. B. Gravestock, WO 9964417 (1999).
64. M. B. Gravestock, WO 9964416 (1999).
65. S. Bartel, R. Endermann, W. Guarnieri, D. Häbich, H.-P. Kroll and S. Raddatz, 37th Interscience Conference on Antimicrobial Agents and Chemotherapy, Toronto, Ontario, Canada (1997). Abstract F18.
66. B. Riedl, D. Häbich, A. Stolle, H. Wild, R. Endermann, K. D. Bremm, H.-P. Kroll, H. Labischinski, K. Schaller and H.-O. Werling, U. S. Patent 5,843,967 (1998).
67. S. Bartel, R. Endermann, W. Guarnieri, D. Häbich, H.-P. Kroll and S. Raddatz, 37th Interscience Conference on Antimicrobial Agents and Chemotherapy, Toronto, Ontario, Canada (1997). Abstract F17.
68. B. Riedl, D. Häbich, A. Stolle, H. Wild, R. Endermann, K. D. Bremm, H.-P. Kroll, H. Labischinski, K. Schaller and H.-O. Werling, U. S. Patent 5,698,574 (1997).
69. B. Riedl, D. Häbich, A. Stolle, M. Ruppelt, S. Bartel, W. Guarnieri, R. Endermann and H.-P. Kroll, EP 789 026 (1997).
70. S. Bartel, R. Endermann, W. Guarnieri, D. Häbich, H.-P. Kroll, S. Raddatz, B. Riedl, U. Rosentreter, M. Ruppelt, A. Stolle and H. Wild, 37th Interscience Conference on Antimicrobial Agents and Chemotherapy, Toronto, Ontario, Canada (1997). Abstract F19.
71. B. Reidl, D. Häbich, A. Stolle, M. Ruppelt, S. Bartel, W. Guarnieri, R. Endermann and H-P. Kroll, U. S. Patent 5,827,857 (1998).
72. A. Stolle, D. Häbich, B. Riedl, M. Ruppelt, S. Bartel, W. Guarnieri, H. Wild, R. Endermann and H-P. Kroll, U. S. Patent 5,869,659 (1999).
73. B. Reidl, D. Häbich, A. Stolle, H. Wild, R. Endermann, K. D. Bremm, H.-P. Kroll, H. Labischinski, K. Schaller and H.-O. Werling, U. S. Patent 5,684,023 (1997).
74. D. Häbich, S. Bartel, R. Endermann, W. Guarnieri, M. Härter and K. Henniger, 39th Interscience Conference on Antimicrobial Agents and Chemotherapy, San Francisco, California, USA (1999). Abstract F-566.
75. S. Bartel, W. Guarnieri, D. Häbich, S. Raddatz, B. Ridl, U. Rosentreter, M. Ruppelt, A. Stolle, H. Wild, R. Endermann and H-P. Kroll, WO 9937641 (1999).
76. D. Häbich, A. Stolle, B. Riedl, M. Ruppelt, S. Bartel, W. Guarnieri, R. Endermann and H-P. Kroll, U. S. Patent 5,861,413 (1999).
77. S. Bartel, R. Endermann, W. Guarnieri, D. Häbich, M. Härter, H.-P. Kroll, S. Raddatz, B. Riedl, U. Rosentreter, M. Ruppelt, A. Stolle and H. Wild, 38th Interscience Conference on Antimicrobial Agents and Chemotherapy, San Diego, California, USA (1998). Abstract F-131.
78. A. Stolle, D. Häbich, B. Riedl, M. Ruppelt, S. Bartel, W. Guarnieri, H. Wild, R. Endermann and H-P. Kroll, U. S. Patent 5,869,659 (1999).
79. Y. H. Yoon, H. S. Kim, K. H. Lee, K. H. Lee, J. A. Kang and Y. H. Lee, U. S. Patent 5,929,083 (1999).
80. M. R. Barbachyn, R. C. Thomas, G. L. Cleek, L. M. Thomasco and R. C. Gadwood, U. S. Patent 5,952,324 (1999).
81. M. B. Gravestock, WO 9910342 (1998).
82. S. D. Mills, WO 9911642 (1998).
83. T. J. Poel, R. C. Thomas, M. R. Barbachyn, C. W. Ford, G. E. Zurenko, W. J. Adams, S. M. Sims, W. Watt and L. A. Dolak, 39th Interscience Conference on Antimicrobial Agents and Chemotherapy, San Francisco, California, USA (1999). Abstract F-568.
84. T. J. Poel, J. P. Martin, M. R. Barbachyn, WO 9929688.
85. J. M. Friis, E. M. Shobe, J. Palandra, K. E. Rousch, R. E. Ouding and W. J. Adams, 39th Interscience Conference on Antimicrobial Agents and Chemotherapy, San Francisco, California, USA (1999). Abstract F-569.
86. T. J. Poel, R. C. Thomas, C. W. Ford and G. E. Zurenko, , 37th Interscience Conference on Antimicrobial Agents and Chemotherapy, Toronto, Ontario, Canada (1997). Abstract F22.
87. D. K. Hutchinson, M. D. Ennis, R. L. Hoffman, R. C. Thomas, T. J. Poel, M. R. Barbachyn, S. J. Brickner and D. J. Anderson, U. S. Patent 5,968,962 (1999).
88. M. Gravestock, U. S. Patent 5,981,528 (1999).
89. M. J. Betts, WO 9727188.
90. J. B. Hester, U. S. Patent 5,998,406 (1999).

91. M. J. Genin, D. K. Hutchinson, D. A. Allwine, J. B. Hester, D. E. Emmert, S. A. Garmon, C. W. Ford, G. E. Zurenko, J. C. Hamel, R. D. Schaadt, D. Stapert, B. H. Yagi, J. M. Friis, E. M. Shobe and W. J. Adams, J. Med. Chem., 41, 5144 (1998).

92. D. K. Hutchinson, M. R. Barbachyn, J. B. Hester, D. E. Emmert, S. A. Garmon, R. J. Reischer, L. S. Stelzer, G. E. Zurenko, C. W. Ford, R. C. Gadwood and M. J. Genin, 38th Interscience Conference on Antimicrobial Agents and Chemotherapy, San Diego, California, USA (1998). Abstract F-137.

93. D. K. Hutchinson, U. S. Patent 5,910,504 (1999).

94. R. C. Gadwood, E. A. Walker, L. M. Thomasco, M. R. Barbachyn, K. C. Grega, M. J. Genin, G. E. Zurenko and C. W. Ford, 38th Interscience Conference on Antimicrobial Agents and Chemotherapy, San Diego, California, USA (1998). Abstract F-139.

95. M. J. Genin, D. A. Allwine, D. J. Anderson, M. R. Barbachyn, R. C. Gadwood, S. A. Garmon, K. C. Grega, J. B. Hester, D. K. Hutchinson, T. M. Judge, C. S. Lee, K. Munesada, T. J. Poel, R. J. Reischer, L. S. Stelzer, R. C. Thomas, L. M. Thomasco, D. S. Toops, A. J. Wolf, C. W. Ford and G. E. Zurenko, 39th Interscience Conference on Antimicrobial Agents and Chemotherapy, San Francisco, California, USA (September 26-29, 1999). Abstract F-570.

96. R. C. Gadwood, L. M. Thomasco, E. A. Weaver, G. E. Zurenko and C. W. Ford, 39th Interscience Conference on Antimicrobial Agents and Chemotherapy, San Francisco, California, USA (September 26-29, 1999). Abstract F-571.

97. R. C. Gadwood, L. M. Thomasco, D. J. Anderson, U. S. Patent 5,977,373 (1999).

98. M. J. Genin, D. A. Allwine, M. R. Barbachyn, K. C. Grega, L. A. Dolak, R. M. Jensen, E. P. Seest, C. W. Ford and G. E. Zurenko, 17th International Congress of Heterocyclic Chemistry, Vienna, Austria (1999). Abstract PO-269.

99. M. J. Betts, M. L. Swain, WO 9928317 (1998).

100. M. J. Betts and D. A. Roberts, WO 9910343 (1998).

101. S. Bartel, R. Endermann, W. Guarnieri, D. Häbich, M. Härter, H.-P. Kroll, S. Raddatz, R. Riedl, U. Rosentreter, M. Ruppelt, A. Stolle and H. Wild, 37th IUPAC Congress, Berlin, Germany (1999). Abstract SYN-2-140.

102. S. Bartel, W. Guarnieri, B. Riedl, D. Häbich, A. Stolle, M. Ruppelt, S. Raddatz, U. Rosentreter, H. Wild, R. Endermann and H-P. Kroll, WO 9903846.

103. S. Raddatz, S. Bartel, W. Guarnieri, U. Rosentreter, M. Ruppelt, H. Wild, R. Endermann, H. Kroll and K. Henninger,WO 9940094 (1998).

104. G. J. Cleek, R. C. Thomas, H. Yamada, L. A. Dolak, E. P. Seest, G. E. Zurenko, R. D. Schaadt, B. Yagi, W. F. Demyan and W. Watt, 214th ACS National Meeting, Las Vegas, NV, USA (September, 1997). Abstract MEDI-150.

105. A. Yamada, G. J. Cleek, D. K. Hutchinson and R. C. Thomas, U. S. Patent 5,922,707 (1999).

106. R. C. Gadwood, B. V. Kamdar, T. J. Poel, L. M. Thomasco, E. A. Walker and D. S. Toops, 26th National Medicinal Chemistry Symposium, Richmond, VA, USA (1998). Abstract B-9.

107. R. C. Gadwood, M. R. Barbachyn, D. S. Toops, H. W. Smith and V. A. Vaillancourt, U. S. Patent 5,736,545 (1998).

108. J. A. Tucker, D. A. Allwine, K. C. Grega, M. R. Barbachyn, J. L. Klock, J. L. Adamski, S. J. Brickner, D. K. Hutchinson, C. W. Ford, G. E. Zurenko, R. A. Conradi, P. S. Burton and R. M. Jensen, J. Med. Chem., 41, 3727 (1998).

109. J. A. Tucker, S. J. Brickner and D. A. Ulanowicz, U. S. Patent 5,719,154 (1998).

110. M. J. Betts, C. J. Darbyshire and C. J. Midgley, WO 9801447 (1997).

111. M. J. Betts, C. J. Darbyshire and C. J. Midgley, WO 9801446 (1997).

112. A. N. Pae, H. Y. Kim, H. J. Joo, B. H. Kim, Y. S. Cho, K. I. Choi, J. H. Choi and H.Y. Koh, Bioorg. Med. Chem. Lett., 9, 2679 (1999).

113. A. N. Pae, S. Y. Kim, H. Y. Kim, H. J. Joo, Y. S. Cho, K. I. Choi, J. H. Choi and H.Y. Koh, Bioorg. Med. Chem. Lett., 9, 2685 (1999).

114. Y. H. Yoon, K. H. Lee, S. B. Song, H. S. Whang, K. H. Lee, J. H. Kim, D. H. Kim, Y. G. Kim, J. A. Kang, Y. H. Lee, WO 9933839 (1998).

115. M. F. Gordeev, Luehr, G. W., D. Patel, Z-J. Ni and E. Gordon, WO9937630 (1999).

Chapter 13. Progress in Macrolide and Ketolide Antibacterials

Daniel T. W. Chu

Chiron Corporation, Emeryville, CA 94608

Introduction - The macolide antibiotic erythromycin was first found in the fermentation products of a strain of *Streptomyces erythreus* by McGuire (1). It has been used for the treatment of various bacterial infections in both out-patient and in-patient settings for more than forty years. Erythromycin derivatives such as clarithromycin and azithromycin are frequently prescribed for the treatment of both upper and lower respiratory tract infections (2,3). In addition to having better activity than their parent erythromycin A against anaerobes, streptococci, *Haemophilius influenzae*, *Legionella* spp., *Branhamella* spp., *Chlamydia* spp., *Mycoplasma* spp. and *Pasteurella multocida*, these derivatives possess improved pharmacokinetic profiles. However, MRSA and penicillin-resistant *S. pneumoniae* are found to become increasingly resistant to these newer macrolides. The emergence of resistant community-acquired as well as nosocomial pathogens to the clinically used antibacterial agents, including macrolides, has resulted in increased activity in macrolide research. This report summarizes the published literature in the past two years on macrolide research activities such as macrolide resistance, search for agents to overcome bacterial resistance and the identification of various non-antibacterial uses of macrolide. Historical information related to various macrolide resistance mechanisms will be included. Three reviews on recent developments in macrolide research have appeared (4-6).

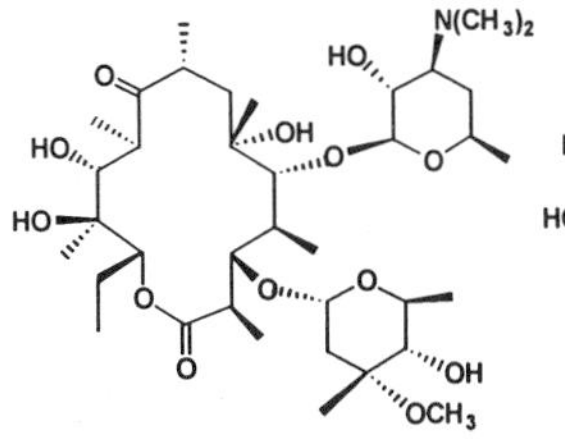

| Erythromycin A | Clarithromycin | Azithromycin |

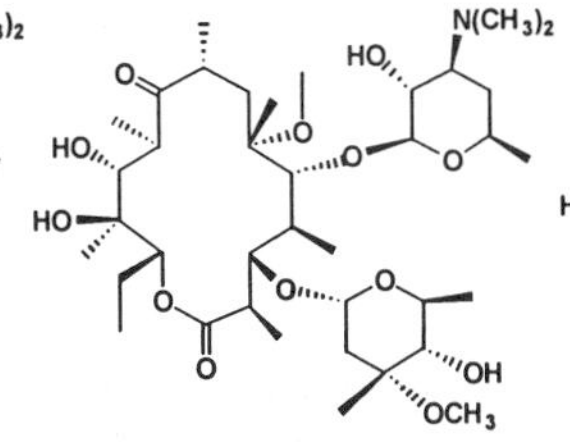

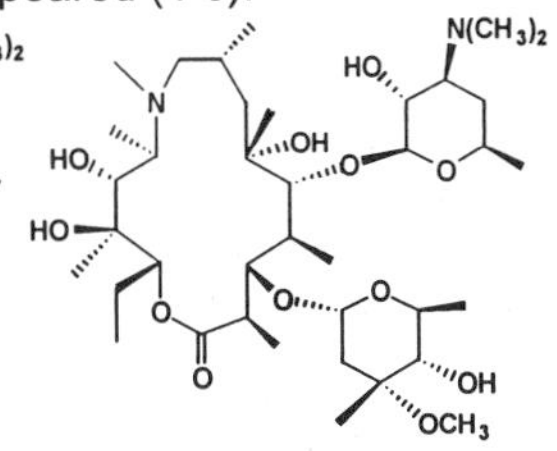

MACROLIDE BACTERIAL RESISTANCE

Macrolides inhibit protein synthesis by interacting with bacterial 23S rRNA making contacts to hairpin 35 in domain II of the rRNA and to the peptidyl transferase loop in domain V (7). This interaction stimulates the dissociation of the peptidyl-tRNA from the ribosomes during the translocation step resulting in chain termination (8). Resistance to macrolides is found to occur by one of the three mechanisms: target-site modification, active efflux and enzymatic inactivation. The mechanism of resistance to macrolides in pathogenic bacteria was summarized and reviewed (9,10). A minireview on new nomenclature for macrolide and macrolide-lincosamide-streptogramin B (MLS$_B$) resistance determinants was published (11).

Historical Prospective - Two types of target-site modification resistance for macrolides were described. The first mechanism is the post-transcriptional modification of the 23S rRNA by adenine-N^{6,6}-dimethyltransferase (MLS$_B$ resistance) and the second is by site-specific mutations in the 23 rRNA gene. The MLS$_B$ resistance is mediated by plasmid or transposon genes called *erm* (erythromycin-resistant methylases) encoding enzymes that catalyze the N^{6,6}-dimethylation of the adenosine-2058 residue of bacterial 23S rRNA (12,13). The *erm* genes can be expressed constitutively or inducibly (14,15). The enzyme induction is effected by exposure of the organism to both 14- and 15 - but not 16-membered macrolides. The N-dimethyltransferases have been found in *Staphylococcus*, *Streptococcus*, *Enterococcus* and *Bacillus* strains. The macrolide-resistant *S. pneumoniae* and *S. pyogenes* are commonly found to have the active efflux genes *mefA*, which encodes a membrane-associated protein and *mefE* (16,17). A different macrolide efflux gene *mreA* was found in a strain of *S. agalactiae* displaying resistance to both 14- and 15- as well as 16-membered macrolide (18). Enzymatic inactivation of the macrolide antibiotic was found to be mediated by phosphorylation, glycosylation, or esterification (19-25). However, in terms of clinical importance, enzymatic inactivation of the macrolide is of no significance. Strains that carry the Erm methylase give the MLS$_B$ phenotype while those streptococci harboring the *mef* genes produce the M phenotype (i.e. macrolide resistant but clindamycin and streptogramin B susceptible). *S. pneumoniae* having rRNA mutation is associated with the ML phenotype and is resistant to 14-,15-, and 16-membered macrolide and lincosamides. MS phenotype in *S. pneumoniae* is found to have several mutations in ribosomal protein L4 while having *msrA* gene in *S. aureus*. Both MS phenotypes are resistant to macrolide and streptogramins.

Resistance Update - A novel *erm* gene, designated *ermTR*, was identified from an erythromycin-resistant clinical *S. pyogenes* isolate with MLS$_B$ inducible resistance (26). Its nucleotide sequence is 82.5% identical to *ermA* found in *S. aureus*. Molecular characterization of seven *Streptococcus mitis* erythromycin-resistant strains showed the presence of either the *mef* or *ermB* gene (27). Although the *ermB* gene has previously been found in viridans group streptococci (28), it represents the first time a *mef* gene was found in this group. Novel genetic arrangement for an ErmB MLS$_B$ resistance determinant was found in the *Clostridium difficile* 630 strain. It contains two copies of an *ermB* gene, separated by a 1.34-kb direct repeat. Both *ermB* genes are flanked by variants of the direct repeat sequence (29). Site-specific mutation in the 23S rRNA gene of *H. pylori* confers two types of resistance to the antibiotics. The A2142G mutation provides high-level resistance to all MLS$_B$ antibiotics while the A2143G mutation gave intermediate level of resistance to clarithromycin and clindamycin, but not streptogramin B (30,31).

A rapid and easy method for the detection of clinically relevant mutation in the 23S rRNA gene of *H. Pylori* was developed. The high sensitivity of the PCR-based reverse hydridization method (PCR-LiPA) provides more accurate data, especially when multiple strains are present (32). Several clinical *Neisseria gonorrhoeae* isolates with reduced susceptibility to azithromycin were found to carry the self-mobile rRNA methylase gene(s) *ermF*, or *ermB* and *ermF* while the oral commensal *Neisseria* spp. were shown to carry one or more rRNA methylase genes, including *ermB*, *ermC*, and/or *ermF* (33). The erm genes are associated with complete conjugative elements. Thus, this represents the first description of complete transposable elements in *Neisseria*. The *mtrCDE*-encoded efflux pump has been suggested to be one of the resistant mechanisms. The *mtrCDE* genes constitute a single transcriptional unit that is negatively regulated by the adjacent *mtrR* gene product. A *N. gonorrhoeae* strain with reduced susceptibility to erythromycin can also be due to the presence of MtrR transcriptional repressor protein. A recent paper reported that the *mtrR* promoter and coding region mutation in *N. gonorrhoeae* clinical isolates from individuals in Uruguay with reduced azithromycin susceptibility occur frequently (34). A few *S. agalactiae* clinical isolates with the M phenotype were found to harbor *mefA* and *mefE* genes (35). The efflux system was also found to be the major resistant mechanism in *Burkholderia pseudomallei* (36). A highly macrolide-resistant *Escherichia coli* clinical isolate BM2506 having macrolide 2'-phosphotransferase II (MPH(2')II) was found to harbor two plasmids, pTZ3721 and pTZ3723. The *mphB* gene is located in these two plasmids and can be transferred to other *E. coli* strains by conjugation or mobilization (37). A glycosyl transferase encoded by *gimA* from *Streptomyces ambofaciens* was characterized and found to be located on downstream of *srmA*, a gene that confers resistance to spiramycin (38). The different macrolide efflux resistant determinants were reviewed (39).

PREVALENCE OF MACROLIDE RESISTANCE

In a susceptibility surveillance study in Spain between 1996-1997 of patients with respiratory tract infection, among 1,113 *S. pneumoniae* isolates, 37% isolates were resistant to macrolides (40). The susceptibility of 302 *S. pneumoniae* central Italian isolates revealed that erythromycin resistance rates increased from 7.1 in 1993 to 32.8% in 1997. Erythromycin resistant isolates mostly carried *ermAM* determinant and a minority (5.8%) carried the *mefA* determinant (41). In a 1996-1997 winter US study, among 1276 *S. pneumoniae* clinical isolates, 23% were resistant to erythromycin, clarithromycin and azithromycin (42). In a Canada surveillance study on prevalence and mechanism of erythromycin-resistant *S. pneumoniae* involving 113 hospital and private laboratories, only 2.9 % of the isolates were erythromycin resistant, with 43.5% showing the MLS$_B$ phenotype (43). In Norway, the prevalence of erythromycin-resistant pneumococci is slightly higher than in Canada (44).

In Finland, 14.5% of *S. pyogenes* isolates were erythromycin-resistant. All of the M phenotype isolates had *mefA* gene which was found to be transferred by conjugation. The MLS$_B$ -resistant isolates had *ermTR* gene with one exception having the *ermB* gene (45). In Spain, 27% of the β-hemolytic group A streptococci isolates isolated in 11 hospital between May 1996 - April 1997 were found to be erythromycin resistant, displaying the M phenotype (46). However, in a Ontario, Canada surveillance study, 2.1% of the group A streptococci was found to erythromycin resistant, consistent with the finding with *S. pneumoniae* that, macrolide resistant streptococci is relatively low (47). The distribution of clinical isolates of erythromycin-resistant *S. pyogenes* in Italian laboratories from 1995 to 1998 was reported and 52.5% of the resistant strains were assigned to M phenotype, 31.0% to iMLS phenotype and 16.5% to cMLS phenotype (48). 12% of group C streptococci clinical isolates in Spain were found to be erythromycin-resistant (46). The resistance mechanism in Lancefield group C and group G streptococci in Finland macrolide resistant isolates were found to be different. About 95% of the group C resistant streptococci had the *mefE* gene whereas 94% of the group G had the *ermTR* methylase gene (49). 55% of *S. oralis* isolates in Taiwan were found to be erythromycin-resistant (50). In France, the erythromycin resistance rate in viridans group streptococci was reported to around 40% (51). Azithromycin-resistant pneumococci obtained by subculturing have no *ermB* or *mefE* genes, contrary to those found in clinical resistant isolated (52). Antimicrobial susceptibility study from the SENTRY antimicrobial surveillance program conducted in United States and Canada found that macrolides are highly active against *Moraxella catarrhalis* isolates with less than 1% resistant, while for the *H. influenzae* isolates, less than 5% strains were found to resistant to macrolide (42,53). The distribution of genes en-coding resistance to MLS antibiotics among staphylococci was reported. Resistance was mainly due to the presence of *ermA* or *ermC* genes detected in 88% of the resistant isolates from 32 French hospitals in 1995. Macrolide resistance due to *msrA* was more prevalent in coagulase-negative staphylococci than in *S. aureus* (54).

AZITHROMYCIN AND CLARITHROMYCIN UPDATE

The median MICs (minimum inhibitory concentrations) against *Mycobacterium-avium-M. intracellulare* complex was found to be 8 μg/ml for azithromycin and ≤2 μg/ml for clarithromycin (55). The MIC value for clarithromycin against 46 strains of *M. ulerans* ranged from 0.125-2 μg/ml at pH 6.6 and from 0.125-0.5 μg/ml at pH 7.4 (56). The MIC$_{90}$ value for clarithromycin tested against *Actinobacillus actinomycetem-comitans* is ≤ 2 μg/ml (57). The susceptibility of *Coxiella burnetii*, the etiologic agent for Q fever, to clarithromycin was found to have an MIC value ranging from 2-4 μg/ml (58). The pH effect of carbon dioxide on the susceptibility testing of azithromycin and clarithromycin against 178 clinical isolates from the lower respiratory tract of patients with chronic obstructive pulmonary disease were studied. The MICs measured in air alone were lower than those measured in 5% carbon dioxide. However, testing of isolates in 5% carbon dioxide on pH-adjusted medium (pH 8.4) reduced the loss of activity (59). Molecular investigation of the postantibiotic effects of clarithromycin on *S. aureus* cells indicated the reduction of the number of 50S ribosomal subunits to 13% of the untreated control while the 30S subunit formation was not affected (60).

A simple assay using infected J774 cells for the quantitation of activities of clarithromycin and azithromycin against intracellular *Legionella pneumophila* has been developed (61). This MTT assay system permits comparative and quantitative evaluation of intracellular activities of macrolides and processing a large number of samples efficiently. A model of continuous *Chlamydia pneumoniae* infections *in vitro* closely resembling to the actual events *in vivo* was developed (62). HEp-2 cells inoculated with CM-1 and TW-183 strains have been persistently infected for a periods of over 1.5 and 2 years, respectively. Using this model, azithromycin was found to reduce but not to completely eliminate the organism. This observation is in agreement with the failure of antibiotic therapy against *C. pneumoniae* infection in humans. A liquid chromatographic method using a cyanopropyl column with electrochemical detection for the determination of macrolides was reported (63).

A study of the pharmacodynamics of clarithromycin and azithromycin on their efficacy against pulmonary *H. influenzae* infection in rats suggested that duration of therapy should be considered as a key parameter in the evaluation of macrolide efficacy (64). The serum area under the curve when the drug concentration in serum is plotted against time was found to be the best predictor of the *in vivo* efficacy of macrolides in a mouse thigh model (65). The effects on interaction of macrolides and other chemical substances were reported. Using a chemostat continuous-culture system against slow-growing *H. pylori*, combination of amoxicillin and clarithromycin was found to be bactericidal at pHs 6.5 and 7.0 (66). The combination of azithromycin-lansoprazole was found to be synergic on 60% of *H. pylori* strains (67). The combination of azithromycin and rifampin produced good activity and higher rates of eradication of *C. pneumoniae* from lung tissues than azithromycin alone in experimental mouse pneumonitis (68). Atovaquone was found to interact pharmacokinetically with azithromycin in human immunodeficiency virus-infected children. At steady-state, the values of azithromycin's area under the concentration-time curve from 0 to 24 hour and maximum concentration in serum were consistently lower than azithromycin administered alone (69). Administration of grapefruit juice increased the time to peak concentration of both clarithromycin and 14-hydroxylclarithromycin, but did not affect other pharmacokinetic parameters (70). Oral administration of cimetidine, however, prolonged clarithromycin absorption (71). Although zafirlukast is a known inhibitor of CYP3A4, it does not appear to exert a clinically significant pharmacokinetic effect on azithromycin, clarithromycin or 14-hydroxy-clarithromycin (72).

Several clinical studies on macrolides have been reported. Due to their safety profile, macrolides represent one of the few potential therapeutic options for pregnant women and children infected with *Orientia tsutsugamushi*, the etiologic agent of scrub typhus. Azithromycin administered to two pregnant women infected with drug-resistant strains of *O. tsutsugamushi* in northern Thailand rapidly abated the symptoms, signs and fever in these patients (73). A comparative trial between azithromycin and cipro-floxacin for treatment of uncomplicated typhoid fever in Egypt on 123 adults infected with or without multidrug resistance strains of *Salmonella typhi* showed that both azithromycin and ciprofloxacin were similarly effective, clinically and bacteriologically (74). A prospective, open-label, randomized study of azithromycin in acute otitis media in children was conducted. Bacteriologic failure after 3 to 4 days of treatment occurred in high proportion (53%) of culture-positive patients, especially in those infected with *H. influenzae*. It was suggested that the current susceptibility breakpoint for azithromycin for *H. influenzae* should be considerably lowered for acute otitis media caused by *H. influenzae* (75). The pharmacokinetic study on serum and leukocyte exposure following oral azithromycin, given over 3- or 5- day in healthy subjects indicated that the exposures of serum and both types of WBCs were similar with both regimens and that azithromycin can be administered over either 3 or 5 days (76). The intracellular disposition of clarithromycin and azithromycin in AIDS patients requiring *Mycobacterium avium* complex (MAC) prophylaxis was investigated. Both drugs displayed sustained intracellular concentrations in mononuclear and polymorphonuclear leukocytes exceeding their MICs for MAC for the entire dosing period (77). A multicenter, randomized, dose-ranging study on azithromycin for treating disseminated MAC patients with AIDS produced symptomatic improvement on patients with both 600 and 1,200 mg daily dose regimens (78).

NOVEL MACROLIDES TO OVERCOME BACTERIAL RESISTANCE

Macrolide derivatives having activity against erythromycin resistant *S. pyogenes* were first reported in 1989 (79). Several 11,12 carbamate clarithromycin analogs and several 11,12-carbonate erythromycin analogs with additional modifications at the 4" position of the cladinose were found to be active against both inducible and constitutive-resistant *S. pyogenes* . Based on this initial discovery, recent chemical modifications of erythromycin have generated several new classes of macrolide derivatives having potent activity against erythromycin resistant bacteria.

Ketolides - Ketolides are the 3-descladinosyl-3-oxo-11,12-cyclic carbamate analogs of erythromycin or clarithromycin. They were found to be active against penicillin-resistant and erythromycin-resistant *S. pneumoniae*. These derivatives do not induce MLS$_B$ resistance in staphylococci and streptococci (80). Three ketolides undergoing clinical development reported so far are the HMR 3004, HMR 3647 and ABT-773. All three compounds were found to be highly active against respiratory pathogens including erythromycin-resistant strains (81, 82).

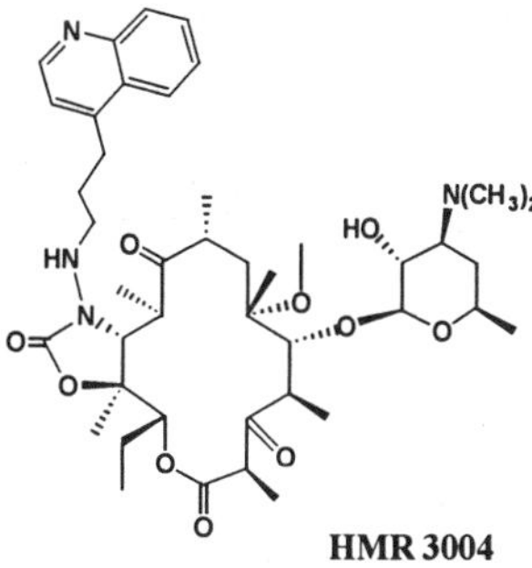

HMR 3004 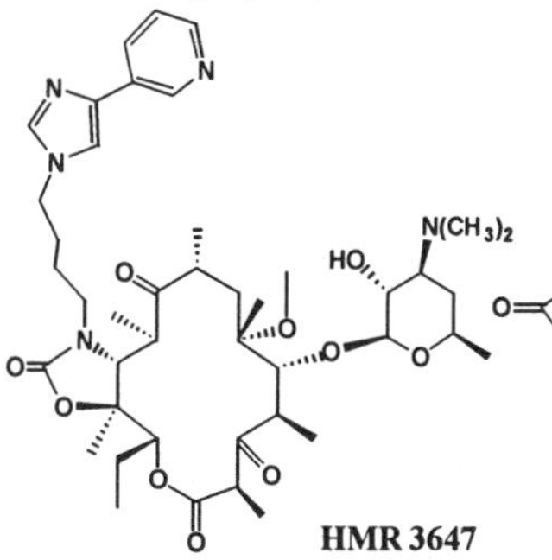   HMR 3647 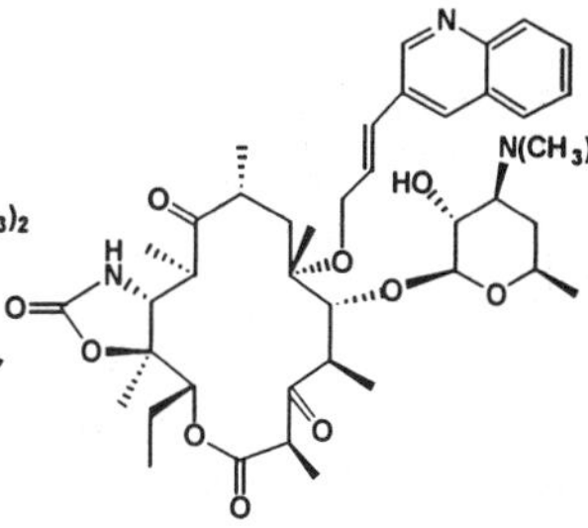   ABT-773

The ketolide HMR 3004 was found to be very active when tested against penicillin- and erythromycin-resistant pneumococci (MIC$_{90}$ = 0.25 µg/ml) (83,84). Tested against 379 anaerobes, it had a MIC$_{50}$ of 1.0 µg/ml (85). Tested against 500 Gram-positive organisms, including multiply resistant enterococci, streptococci and staphylococci, HMR 3004 had a 100% susceptibility at a concentration of ≤ 1 µg/ml (81,86). It was found to be strongly and rapidly accumulated by polymorphonuclear leukocytes (87) and very active against respiratory infections in animal study (88). This compound was active against beta-lactamase-producing *H. influenzae* in a murine model of experimental pneumonia and was found to be more active than azithromycin, ciprofloxacin, clarithromycin, erythromycin and pristinamycin (89).

A closely related ketolide, HMR 3647 (telithromycin), was prepared and is under-going extensive clinical development (90). A new drug application (NDA) to the U.S. Food and Drug Administration (FDA) for telithromycin was filed on March 6, 2000. IHMR 3647 was found to be as active as HMR 3004 against anaerobes (91). When tested against 419 human anaerobic isolates, its MIC$_{90}$ values (in µg/ml) against *Peptostreptococcus* species, *Bacteroides fragilis, Clostridium perfringens, C. difficile, Prevotella bivia, P. orisbuccae, Bilophila wadsworthia, Veillonella* species and

Fusobacterium ulcerans were ≤ 0.06, 2, 0.06, 1, 0.5, 1, 2, 4 and >16, respectively (92). The activity of telithromycin against anaerobic bacteria, compared to those of eight antibacterial agents by time-kill methodology was examined. HMR 3647 had the lowest MICs, especially against non-*B. fragilis* group with MIC ≤ 2 µg/ml (93). The MIC_{90} values for telithromycin against penicillin- and erythromycin-susceptible pneumococci was 0.03 µg/ml and against penicillin- and erythromycin-resistant pnemococci was 0.25 µg/ml (94). A recent study gave the MIC_{90} value of 0.5 µg/ml for *S. pneumoniae* (81).

HMR 3647 was active (MIC_{90} values ranged from ≤0.015-2 µg/ml) against most of aerobic and facultative non-spore-forming Gram-positive bacilli tested with the exception of *Corynebacterium striatum, coryneform* CDC group 12 and *Oerskovia* species (95). Another study, however, reported high activity for HMR 3647 against *Corynebacterium* species except *C. jeikeium* and *C. urealyticum* with MIC_{90} values ranging from 0.06-0.25 µg/ml (96). It was generally more active than other marketed macrolide antibiotics against almost all the aerobic and fastidious facultative isolates with MIC_{90} value of 1 µg/ml (97). Against *H. influenzae*, it is as active as azithromycin having MIC_{90} value of 4 µg/ml (98). When tested against *Moraxella catarrhalis*, *Neisseria meningitidis* and *N. gonorrhoeae*, HMR 3647 has MIC_{90} values of 0.12, 0.12 and 0.25 µg/ml, respectively (99). It was found to have high activity against *Chlamydia pneumoniae*, *Mycoplasma pneumoniae*, *E. faecalis*, *Bordetella pertussis*, *Toxoplasma gondii*, Lactobacillus, *Leuconostoc*, and *Pediococcus* species as well as rapidly growing mycobacteria (100-106).

In a murine model of experimental pneumonia, HMR 3647 was effective against β-lactamase-producing *H. influenzae* (89). It was found to be effective for the treatment of *Legionella pneumophia* in a guinea pigs pneumonia experimental model (107). The interaction between tetithromycin and human polymorphonuclear neutrophils (PMNs) was studied (108). HMR 3647 is trapped in PMNs, where it is concentrated up to 300 times. It is poorly released by these cells with 80% of the compound still remain in cell after 2 hours in fresh medium. In PMNs and NB4 cells, more than 75% and 63%, respectively, of the molecules were found to accumulate in the azurophil granule fraction (109). The pharmacodynamic properties of HMR 3647 demonstrated by time-kill kinetics and postantibiotic effects on enterococci and *B. fragilis* were found to be similar to those obtained with macrolides (110). Additional studies such as *in vitro* activities against various organisms, pharmacodynamic properties, post-antibiotic effect and bactericidal activity, *in vitro* selection of resistance, resistance phenotype and uptake by myelomonocytic cell lines of HMR 3647 were presented in the 39th Interscience Conference on Antimicrobial Agents and Chemotherapy (ICAAC) (111).

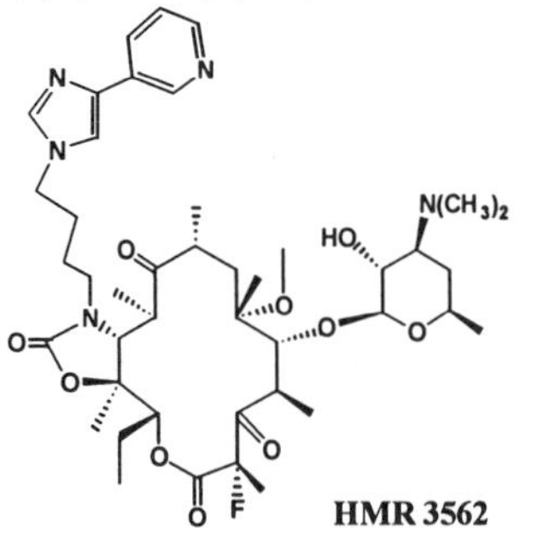

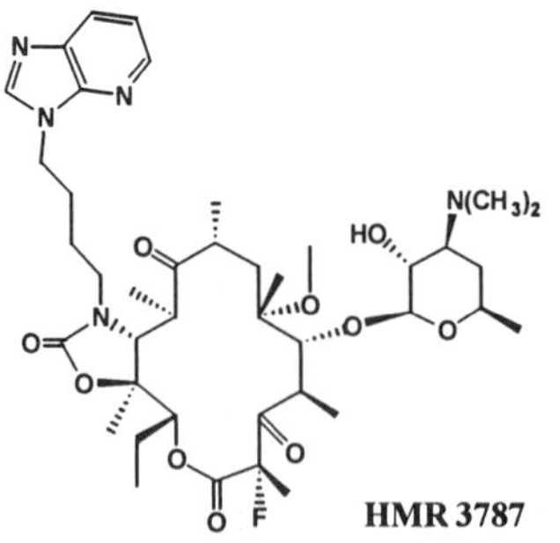

Two 2-fluoro ketolides, HMR 3562 and HMR 3787 were found to have potent *in vitro* antibacterial activity against inducibly resistant *S. aureus* and *S. pneumoniae* as well as constitutively resistant *S. pneumoniae*. They possess better activity against *H. influenzae* compared to their corresponding non-2-fluorinated parent (112). Both HMR 3562 and HMR 3787 display high therapeutic efficacy in mice infected by different common respiratory pathogens, such as multidrug resistant pneumococci and *H. influenzae* (113).

A-197579   A-201316

A novel series of 6-O-substituted ketolides was prepared and reported to possess excellent activity against inducibly resistant *S. aureus* and *S. pneumoniae* as well as constitutively resistant *S. pneumoniae* (114). Analogs with aryl group at the 6-O-position with the propenyl spacer in combination with a 11,12-cyclic carbamate and a 3-keto group exhibited the best activity. In this study, the unsubstituted 11,12-carbamate analogs are more active than both the carbazate or tricyclic analogs (ABT-773 vs. A-201316 vs. A-197579). The MIC_{90} value in µg/ml for ABT-773 against penicillin-susceptible and -resistant *S. pneumoniae* was ≤ 0.06 and 0.03, respectively (115). Against 1529 isolates of *H. influenzae*, ABT-773 has a MIC_{90} of 4 µg/ml identical to the azithromycin value. ABT-773 is very active when tested against *M. catarrhalis* having a MIC_{90} value of 0.06 µg/ml. The activity of ABT-773 was compared to the activity of HMR 3647 against over 500 Gram-positive clinical isolates including 298 *S. pneumoniae*, 97 *S. aureus* and 120 *S. pyogenes*. Macrolide resistant isolates with ribosomal methylase ErmAM (erm) or macrolide efflux (mef) were tested. ABT-773 had superior activity against macrolide resistant *S. pneumoniae*: MIC_{90} value in µg/ml; erm 0.015 vs. 0.12 for HMR 3647, mef 0.12 vs. 1 µg/ml for HMR 3647. ABT-773 also had superior activity against macrolide resistant *S. pyogenes*: MIC_{90} value in µg/ml: erm 0.5 vs. > 8 for HMR 3647, mef 0.12 vs. 1 µg/ml for HMR 3647 (116). The high activity of ABT-773 against *S. pneumoniae* is probably due to several factors. ABT-773 rapidly accumulates in both macrolide sensitive and resistant *S. pneumoniae* (117). The accumulation to MLS phenotype of *S. pneumoniae* suggests that ABT-733 possesses some affinity to methylated ribosome. The inactivation of a macrolide efflux pump (mef) had no effect on ABT-773 accumulation rate. It binds to unmethylated ribosome 10-100 fold tighter than erythromycin and does not induce *erm* (118). ABT-773 is active against *H. pylori* (MIC_{90}=0.25 µg/ml) and *L. pneumophila* (MIC=0.015 µg/ml) (119). It has significant activity against *T. gondii* both *in vitro* and in two murine models of acute toxoplasmosis (120). ABT-773 has favorable pharmacokinetic properties. Plasma elimination half-lives averaged 1.6, 4.5, 3.0 and 5.9 hours after IV dosing in mouse, rat, monkey and dog, respectively. Peak plasma concentrations averaged 1.47, 0.52, 0.56 and 0.84 µg/ml with bioavailability of 49.5, 60.0, 35.8 and 44.1% after oral dosing in the same species. Lung concentration of ABT-773 was >25-fold higher than plasma concentration after oral dosing in rat (121). In an experimental rat lung infections caused by *S. pneumoniae*, ABT-773 was found to be 5-fold more efficacious than azithromycin against *mefE* -bearing strain and has excellent efficacy against *ermAM*-bearing strain while azithromycin was inactive (122). ABT-773 and HMR 3647 were found to be equivalent in efficacy against macrolide susceptible *S. aureus* and *S. pneumoniae* in mouse protection tests and rat pulmonary infections. ABT-773 showed 3 to 16-fold improved efficacy over HMR 3647 against constitutive macrolide resistant *S. pneumoniae* (*ermAM*) in rat pulmonary infection. It also showed improved efficacy over HMR 3647 in rat lung infections caused by *S. pneumoniae* (*mefE*) and *H. influenzae* (123).

Anhydrolides and Acylides - A series of anhydrolides (the 2,3-anhydroerythromycin analogs, exemplified by A-179461) in which a carbon-carbon double bond was introduced at the C2-C3 position and having a sp^2 carbon at C3 position (same position as the 3-keto group in the ketolides) was prepared (124,125). A-179461, having a similar structure to HMR 3004, was found to have fairly similar *in vitro* antibacterial activity as HMR 3004 against Gram-positive organisms. Thus, functionalities other than a ketone at the C3 position can be used as a substitute for the 3-cladinosyl residue to produce active macrolide derivative (126). Acylides, 3-O-acyl-5-O-desosaminyl-erythronolide 11,12-carbamate and 3-O-acyl-5-O-desosaminyl-erythronolide 6,9;11,12-dicarbonate derivatives were reported to have potent in vitro antibacterial activity against gram positive organisms similar to HMR 3647 (127).

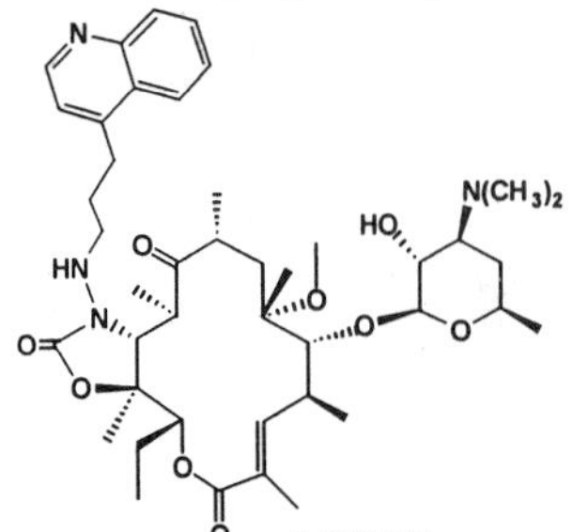

A-179461 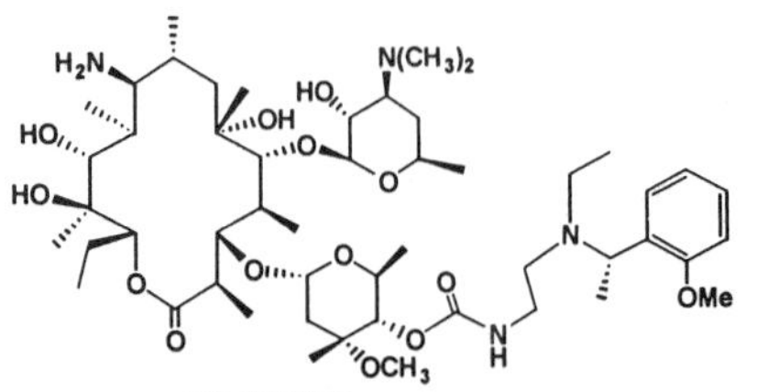   CP-544372

4" Carbamate Macrolides - A novel series of 4" carbamates of 14,15-membered macrolides in which the 4" hydroxyl group is replaced by a carbamate group, have been shown to possess potent *in vitro* activity against Gram-positive and Gram-negative respiratory pathogens, including macrolide-resistant strains of *S. pneumoniae*. CP-544372 possessed good *in vitro* activity comparable to HMR 3647 (MICs in µg/ml against *S. pyogenes mefA*, *S. pneumoniae ermB* and *mefE* were 0.5, 0.16 and 0.08, respectively). It has a higher maximum drug concentration (C_{max}) in the lung compared to serum. In mice given CP-544372 at 1.6-100 mg/kg orally, the serum levels of C_{max} were 0.001-0.66 µg/ml, and AUC was 0.13-5.5 µg*h/ml, and the mean terminal half-life was 6.5 hours. In murine acute pneumonia models of infection induced by macrolide-sensitive and -resistant strains of pneumococci or *H. influenzae*, CP-544372 was shown to be orally active (128).

Conclusion - Novel macrolides having C4" carbamate functional groups and/ or ketolides have been prepared by chemical modification and were found to have potent activities against macrolide-resistant strains.

References

1. M.A. Sanda and G.L. Mandell, in L.S. Goodman and A. GilmanEds, The Pharmacological Basis of Therapeutics, 8th ed. Macmillian Publishing Co.,New York, N.Y. (1985).
2. S. Morimoto, Y. Takahashi, Y. Watanabe and S. Omura, J. Antibiot. 37, 343 (1984).
3. G. Bright, A. Nage, J. Bordner, K. Desai, J. Dibrino et al, J. antibiot. 41, 1029 (1988).
4. D.T.W. Chu, Cur.Opin. Microbiology, 2,467 (1999).
5. M. Fines and R. Leclercq, Cur. Opin. Anti-infective Invest. Drugs, 1 (4), 443 (1999).
6. A. Bryskier, Exp. Opin. Invest. Drugs 8 (8), 1171 (1999).
7. L.H. Hansen, P. Mauvais and S. Douthwiaite, Mol. Microbiol. 31, 623 (1999).
8. A. Brisson-Noel, P. Trieu-Cuot P. Courvalin, J. Antimicrob. Chemother. 22,(s. B), 13 (1988).
9. Key concepts in understanding macrolide reisistance, program & abstrats, pp12-13, ICMASK&O5, Serville, Spain (2000).
10. Y. Nakajima, J. Infect. Chemother. 5, 61 (1999).
11. M.C. Roberts, J. Sutcliffe, P. Courvalin, L.B. Jensen, J. Rood and H. Seppala, Antimicrob. Agents Chemother. 35, 1273 (1991).
12. Q. Leclercq and P. Courvalin, Antimicrob. Agents Chemother. 35, 1267 (1991).
13. Q. Leclercq and P. Courvalin, Antimicrob. Agents Chemother. 35, 1273 (1991).
14. D. Dubnau, Crit. Rev. Biochem. 16, 103 (1986).
15. B. Weisblum, Br. Med. Bull, 40, 47 (1984).
16. J. Clancy, J. Petitpas, F. Dib-Haji, W. Tuan, M. Cronan, V. Kamath, J. Bergeron and J.A. Retsema, Mol. Microbiol. 22, 867 (1996).

17. A. Tait-Kamradt, J. Clancy, M.F. Cronan and F. Dib-Haji, Antimicrob. Agents Chemother. 41, 2251 (1997).
18. J. Clancy, F. Dib-Haji, J. Petitpas, and W. Yuan, Antimicrob. Agents Chemother. 41, 2719 (1997).
19. N. Noguchi, A. Emura, H. Matsuyama, K. O'Hara, M. Sasatsu and M. Kono, Antimicrob. Agents Chemother. 39, 2359 (1995).
20. N. Noguchi, J. Katayama and K. O'Hara, Microbiol. Lett. 144, 197 (1996).
21. E. Cundliffe, Antimicrob. Agents Chemother. 36, 348 (1992).
22. M.S. Kuo, D.G. Chirby, A.D. Argoudelis, J.I. Cialdelia, J.H. Coats and V.P. Marshall, Antimicrob. Agents Chemother. 33, 2089 (1989).
23. J. Sassaki, K. Mizoue, S. Morimoto and S. Omura, J. antibiot. (Tokyo) 49, 1110 (1996).
24. M. Arthur, D. Autissier and P. Courvalin, Nucleic Acids Res. 14, 4987 (1986).
25. H. Qunissi and P. Courvalin, Gene, 35, 271 (1985).
26. H. Seppala, M. Skumik, H. Soini, M.C. Roberts and P. Huovinen, Antimicrob. Agents Chemother. 42, 257 (1998).
27. S.M. Poutanen, J. de Azavedo, B.M. Willey, D.E. Low and K.S. MacDonald, Antimicrob. Agents Chemother. 43, 1505 (1999).
28. D. Clermont and T. Horaud, Antimicrob. Agents Chemother. 34, 1685 (1990).
29. K.A. Farrow, N. Lyras and J.I. Rood, Antimicrob. Agents Chemother. 44, 411 (2000).
30. G.E. Wang and D. Taylor, Antimicrob. Agents Chemother. 42, 1682749 (1998).
31. Y.J. Debets-Ossenkopp, A.B. Brinkman, E.J. Kuipers, G. Vandenbroucke and J.G. Kusters, Antimicrob. Agents Chemother. 42, 1952 (1998).
32. L-J. van Doorn, Y.J. Debets-Ossenkopp, A. Marais, R. Sanna, F. Mergaud, J.G. Kusters and W.G.Y. Quint, Antimicrob. Agents Chemother. 43, 1779 (1999).
33. M.C. Roberts, W.O. Chung, D. Roe, M. Xia, C. Marquez, G. Borthagaray, W.L. Whittington and K.K. Holmes, Antimicrob. Agents Chemother. 43, 1367 (1999).
34. L. Zarantonelli, G. Borthagaray, E-H. Lee and W.M. Shafer, Antimicrob. Agents Chemother. 43, 2468 (1999).
35. C. Arpin, H. Daube, F. Tessier, C. Quentin, Antimicrob. Agents Chemother. 43, 944 (1999).
36. R.A. Moore, D. DeShazer, S. Reckseidler, A. Weissman and D. E. Woods, Antimicrob. Agents Chemother. 43, 465 (1999).
37. J. Katayama, H. Okada, K. O'Hara, and N. Noguchi, Biol. Pharm. Bull. 21, 326 (1998).
38. A. Gourmelen, M.H. Blondeler-Touault, J.L. Pernodet, Antimicrob. Agents Chemother. 42, 2612 (1998).
39. J. Sutcliff, Cur. Opin. Anti-infective Invest. Drugs, 1(4), 403 (1999).
40. F. Baquero, J.A. Garcia-Rodriguez, J.G. de Lomas, L. Aguilar and The Spanish Surveillance Group for Respiratory Pathogens, Antimicrob. Agents Chemother. 43, 357 (1999).
41. P. Oster, A. Zanchi, S. Cresti, M. Lattanzi, F. Montagnani, C. Cellesi and G.M. Rossolini, Antimicrob. Agents Chemother. 43, 2510 (1999).
42. C. Thornsberry, P.T. Ogilvie, H.P. Holley, Jr and D.F. Sahm, Antimicrob. Agents Chemother. 43, 2612 (1999).
43. N.J. Johnston, J.C. Azavedo, J.D. Kellner and D.E. Low, Antimicrob. Agents Chemother. 42, 2425 (1998).
44. T. Bergan, P. Gaustad, E.A. Hoiby, B.P. Berdal, G. Furuberg, J. Baann and T. Tonjum, Intern. J. Antimicrob. Agents, 10, 77 (1999).
45. J. Kataja, P. Huovinen, M. Skurnik, The Finnish Study Group for Antimicrobial Resistance, and H. Seppala, Antimicrob. Agents Chemother. 43, 48 (1999).
46. F. Baquero, J.A. Garcia-Rodriguez, J.G. de Lomas, L. Aguilar and The Spanish Surveillance Group for Respiratory Pathogens, Antimicrob. Agents Chemother. 43, 178 (1999).
47. J.C.S. de Azavedo, R.H. Yeung, D.J. Bast, C.L. Duncan, S.B. Borgia and D.E. Low, Antimicrob. Agents Chemother. 43, 2144 (1999).
48. E. Giovanetti, M.P. Montanari, M. mingoia and P. E. Varaldo, Antimicrob. Agents Chemother. 43, 1935 (1999).
49. J. Kataja, H. Seppala, M. Skurnik, H. Sarkkinen and P. Huovenen, Antimicrob. Agents Chemother. 42, 1439 (1998).
50. L. Teng, P. Hsueh, Y. Chen, S. Ho and K. Luh, J. Antimicrob. Chemother. 41, 621 (1998).
51. C. Arpin, M-H Canron, J. Maugein and C. Quentin, Antimicrob. Agents Chemother. 43, 2335 (1999).
52. G.A. Pankuch, S.E. Jueneman, T.A. Davies, M.R. Jacob and P.C. Appelbaum, Antimicrob. Agents Chemother. 42, 2914 (1998).
53. G.V. Doern, R.N. Jones, M.A. Pfaller, K. Kugler and The SENTRY participants group, Antimicrob. Agents Chemother. 43, 385 (1999).
54. G. Lina, A. Quaglia, M-E. Reverdy, R. Leclercq, F. Vandenesch and J. Etienne, Antimicrob. Agents Chemother. 43, 1062 (1999).
55. L. Steele-Moore, K. Stark , W.J. Holloway, Antimicrob. Agents Chemother. 43, 1530 (1999).

56. F. Portaeis, H. Traore, K. De Ridder and W.M. Meyers, Antimicrob. Agents Chemother. 42, 2070 (1998).
57. R. Piccolomini, G. Catamo and G. Di Bonaventura, Antimicrob. Agents Chemother. 42, 3000 (1998).
58. A. Gikas, J. Spridaki, A. Psaroulaki, D. Kofterithis and Y. Tselentis, Antimicrob. Agents Chemother. 42, 2747 (1998).
59. M. M. Johnson, S.L. Hill and L.J. Piddock, Antimicrob. Agents Chemother. 43, 1862 (1999).
60. W. S. Champney and C.L. Tober, Antimicrob. Agents Chemother. 43, 1324 (1999).
61. F. Higa, N. Kusano, M. Tateyama, T. Shinzato, N. Arakaki, K. Kawakami and A. Sato, Antimicrob. Agents Chemother. 42, 1392 (1998).
62. A. Kutlin, P. Roblin and M. Hammerschlag, Antimicrob. Agents Chemother. 43, 2268 (1999).
63. F. Kees, S. Spangler and M.Wellenholer, J. Chromatography A, 812, 287 (1998).
64. J.D. Alder, P.J. Ewing, A.M. Nilius, M. Mitten, A. Tovcimak, A. Oleksijew, K. Jarvis, L. Paige and K.T. Tanaka, Antimicrob. Agents Chemother. 42, 2385 (1998).
65. C. Carbon, *Clin. Infectious Dis.* 27, 28 (1998).
66. I.J. Hassan, R.M. Stack, J. Greenman and M.R. Millar, Antimicrob. Agents Chemother. 43, 1387 (1999).
67. T. Malizia, M. Tejada, F. Marchetti, P. Favini, G. Pizzarelli, M. Campa and S. Senesi, J. Antimicro Chemother 41(suppl B) , 29 (1998).
68. K. Wolf and R. Malinverni, Antimicrob. Agents Chemother. 43, 1491 (1999).
69. L.Y. Ngo, R. Yogev, W.M. Dankner, W.T. Hughes, S. Burchett, J. Xu, B. Sadler and J.D. Unadkat for the ACTG 254 team, Antimicrob. Agents Chemother. 43, 1516 (1999).
70. K.L. Cheng, A.N. Nafziger, C.A. Peloquin and G.W. Amsden, Antimicrob. Agents Chemother. 42, 2385 (1998).
71. G.W. Amsden, K.T. Cheng, C.A> Oeloquin and A.N. Nafziger, Antimicrob. Agents Chemother. 42, 1578 (1998).
72. K.W. Garvey, C.A. Peloquin, P.G. Godo, A.N. Nafziger and G.W. Amsden, Antimicrob. Agents Chemother. 43, 1152 (1999).
73. G. Watt, P. Kantipong, K. Jongsakul, P. Watcharapichat and D. Phulsuksombati, Antimicrob. Agents Chemother. 43, 2817 (1999).
74. N.I. Girgins, T. Butler, R.W. Frenck, Y. Sultan, F.M. Brown, D. Tribble and R. Khakhria, Antimicrob. Agents Chemother. 43, 1441 (1999).
75. R. Dagan, E. Leibovitz, D.M. Fliss, A. Leiberman, M.R. Jacobs, W. Crag and P. Yagupsky, Antimicrob. Agents Chemother. 44, 43 (2000).
76. G.W. Amsden, A.N. Nafziger and G. Foulds, Antimicrob. Agents Chemother. 43, 163 (1999).
77. K.Q. Bui, J. McNabb, C. Li, C.H. Nightingale and D.P. Nicolau, Antimicrob. Agents Chemother. 43, 2302 (1999).
78. S.L. Koletar, A.J. Berry, M.H. Cynamon, J. Jacobson, J.S. Currier, R.R. MacGregor, M.W. Dunne and D.J. Williams, Antimicrob. Agents Chemother. 43, 2869 (1999).
79. P. Fernandes, W. Baker, L. Freiberg, D. Hardy and E. McDonald, Antimicrob. Agents Chemother. 33, 78 (1989).
80. A. Bonnefoy, A.M. Girard, C. Agouridas and J.F. Chantot, J. Antimicrob. Chemother. 40, 85 (1997).
81. K. Malathum, T.M. Coque, K.V. Singh and B.E. Murray, Antimicrob. Agents Chemother. 43, 930 (1999).
82. A.B. Brueggemann, G.V. Doern, H.K. Huynh, E.M. Wingert and P.R. Rhomberg, Antimicrob. Agents Chemother. 44, 447 (2000).
83. L.M. Ednie, S.K. Spangler, M.R. Jacobs and P.C. Appelbaum, Antimicrob. Agents Chemother. 41, 1033 (1997).
84. C. Jamjian, D.J. Biedenbach and R.N. Jones, Antimicrob. Agents Chemother. 41, 454 (1997).
85. L.M. Ednie, S.K. Spangler, M.R. Jacobs and P.C. Appelbaum, Antimicrob. Agents Chemother. 41, 1037 (1997).
86. T. Schulin, C.B. Wennersten, R.C. Moellering, Jr. and G.M. Eliopoulos, Antimicrob. Agents Chemother. 41, 1196 (1997).
87. D. Vazifeh, H. Abdelghaffar and M. Labro, Antimicrob. Agents Chemother. 41, 2099 (1997).
88. C. agouridas, A. Bonnefoy and J.F. Chantot, Antimicrob. Agents Chemother. 41, 2149 (1997).
89. K.E. Piper, M.S. Rouse, J.M. Steckelberg, W.R. Wilson and R. Patel, Antimicrob. Agents Chemother. 43, 708 (1999).
90. A. Denis, C. Agouridas, J.M. Auger, Y. Benedetti, A. Bonnefoy, F. Bretin, J.F. Chantot, A. Dussarat, C. Fromentin, S.G. D'Ambrieres, S. Lachaud, P. Laurin, O. Le Martret, V. Loyau, N. Tessot, J.M. Pejac, and S. Perron, Bioorg. Med. Chem Lett. 9, 3075 (1999).

91. L. Ednie, M. Jacobs and P.C. Appelbaum, Antimicrob. Agents Chemother. 41, 2019 (1997).
92. E.J.C. Goldstein, D.M. Citron, C.V. Merriam, Y. Warren and K. Tyrrell, Antimicrob. Agents Chemother. 43, 2801 (1999).
93. K.L. Credito, L.M. Ednie, M.R. Jacobs and P.C. Appelbaum, Antimicrob. Agents Chemother. 43, 2027 (1999).
94. G.A. Pankuch, M.A. Vissalli, M.R. Jacobs and P.C. Appelbaum, Antimicrob. Agents Chemother. 42, 624 (1998).
95. F. Soriano, R. Fernandez-Roblas, R. Calvo and G. Garcia-Calvo, Antimicrob. Agents Chemother. 42, 1028 (1998).
96. L. Martinez, A. Pascual, A.I. Suarez and E. J. Perea, Antimicrob. Agents Chemother. 42, 3290 (1998).
97. E.J.C. Goldstein, D.M. Citron, S.H. Gerardo, M. Hudspeth and V. Merriam, Antimicrob. Agents Chemother. 42, 1127 (1998).
98. A.L. Barry, P.C. Fuchs and S.D. Brown, Antimicrob. Agents Chemother. 42, 2138 (1998).
99. J.A. Saez-Nieto and J.A. Vazquez, Antimicrob. Agents Chemother. 43, 983 (1999).
100. P.M. Robin and M.R. Hammerschlag, Antimicrob. Agents Chemother. 42, 1515 (1998).
101. C.M. Bebear, H. Rnaudin, M.D. Aydin, J.F. Chantot and C. Bebear, J. Antimicrob. Chemother. 39, 669 (1997).
102. D.B. Hoellman, G. Lin, M.R. Jacobs and P.C. Appelbaum, Antimicrob. Agents Chemother. 43, 166 (1999).
103. J.E. Hoppe and A. Bryskier, Antimicrob. Agents Chemother. 42, 965 (1998).
104. F.G. Araujo, etal, Antimicrob. Agents Chemother. 41, 2137 (1997).
105. M. Zarazaga, Y. Saenz, A. Portillo, C. Tenorio, F. Ruiz-Larrea, R. Del Campo, F. Baquero and C. Torres, Antimicrob. Agents Chemother. 43, 3039 (1999).
106. R. Fernandez-Roblas, J. Esteban, F. Cabria, J.C. Lopez, M.S. Jimenez and F. Soriano, Antimicrob. Agents Chemother. 44, 181 (2000).
107. P.H. Edelstein and M.A. Edelstein, Antimicrob. Agents Chemother. 43, 90 (1999).
108. D. Vazifeh, A. Preira, A. Bryskier and M.T. Labro, Antimicrob. Agents Chemother. 42, 1944 (1998).
109. C.Miossec-Bartoll, L. Pilatre, P. Peyron, E. N'Diaye, V. Collart-Dutilleul, I. Maridonneau-Parini and A. Diu-Hercend, Antimicrob. Agents Chemother. 43, 2457 (1999).
110. H.J. Boswell, J.M. Andrews and R. Wise, J. Antimicrob. Chemother. 41, 1651 (1998).
111. 39th ICAAC, San Francisco (1999). Paper Nos. 10, 540, 1227, 1240, 1241, 1242, 1244, 1247, 1248, 1249, 1929.
112. 39th ICAAC, San Francisco (1999). Paper Nos. 2153, 2154.
113. 39th ICAAC, San Francisco (1999). Paper No. 2156.
114. Y.S. Or, R.F. Clark, S. Wang, D.T.W. Chu, A.M. Nilius, R.K. Flamm, M. Mitten, P. Ewing, J. Alder and Z. Ma, J. Med. Chem. 43, 1045 (2000).
115. A.B. Brueggemann, G.V. Doern, H.K. Huynh, E.M. Wingert and P.R. Rhomberg, Antimicrob. Agents Chemother. 44, 447 (2000).
116. 39th ICAAC, San Francisco (1999). Paper Nos. 2136, 2138, 2139, 2140.
117. 39th ICAAC, San Francisco (1999). Paper No. 2137.
118. 39th ICAAC, San Francisco (1999). Paper Nos. 2134, 2135.
119. 39th ICAAC, San Francisco (1999). Paper Nos. 2145, 2146.
120. 39th ICAAC, San Francisco (1999). Paper No. 2147.
121. 39th ICAAC, San Francisco (1999). Paper No. 2148
122. 39th ICAAC, San Francisco (1999). Paper No. 2151.
123. 39th ICAAC, San Francisco (1999). Paper No. 2150.
124. R. Elliott, D. Pireh, G. Griesgraber, A. Nilius, P.I. Ewing, M.H. Bui, P.M. Raney, R.K. Flamm, K. Kim and R.F. Henry, D. Chu, J.J. Plattner and Y.S. Or, J. Med. Chem. 41, 1651 (1998).
125. G. Griesgraber, M.J. Kramer, R. Elliott, A.M. Nilius, P.I. Ewing, P.M. Raney, M.H. Bui, R.K. Flamm, D.T. Chu, J.J. Plattner and Y.S. Or, J. Med. Chem. 41, 1660 (1998).
126. R.L. Elliott, D. Pireh, A.M. Nilius, P.M. Johnson, R.K. Flamm, D.T.W. Chu, J.J. Plattner and Y.S. Or, Biorog. Med. Chem. Lett 7, 641 (1997).
127. 39th ICAAC, San Francisco (1999). Paper Nos. 2159, 2160.
128. 38th ICAAC, San Diego (1998). Abst. nos. F120, F121, F122.

Chapter 14. Progress with Antifungal Agents and Approaches to Combat Fungal Resistance

William J. Watkins and Thomas E. Renau
Microcide Pharmaceuticals, 850 Maude Ave., Mountain View, CA 94043

Introduction - The rising incidence of serious fungal infections, especially in the immunocompromised, attests to the need for more effective therapies. All current agents have some serious liabilities: inadequate spectrum, limited dosage forms, narrow therapeutic window and rapid emergence of resistance. Several excellent reviews of the incidence and treatment of systemic mycoses have recently appeared (1,2). Thus the search for new agents, particularly those with a novel mode of action, continues unabated (3). Two chapters in this series relating to this field have been published in the last three years (4, 5). In this chapter, we review publications since 1998 pertaining to new agents under development for the treatment of systemic fungal infections, and detail those mechanisms of resistance to established agents that have been demonstrated to be of clinical relevance, together with approaches to combat them.

NEW AGENTS UNDER DEVELOPMENT

Azoles - The publication of an excellent review of current and emerging azole antifungal agents in January 1999 (2) serves as a timely summary of progress up to early 1998, and the material herein covers important advances since then. The most advanced clinical candidates are conveniently compared in a recent edition of Current Opinion in Anti-Infective Investigational Drugs (6).

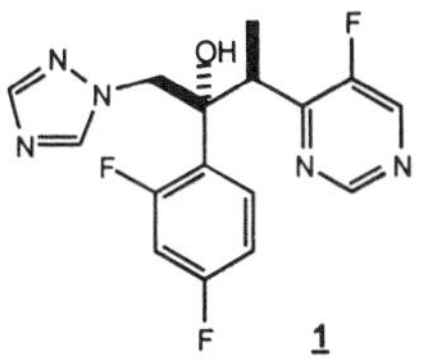

Voriconazole (**1**), previously known as UK-109,496 (Pfizer), is in Phase III clinical trials. Unlike fluconazole, this derivative has potent activity against a wide variety of fungi, including all the clinically important pathogens. Several publications substantiating this have appeared with regard to newer or more rare pathogens (7-10), and comparison with other azoles (11-14). Voriconazole is clearly more potent than itraconazole against *Aspergillus* spp. (12) and is comparable to posaconazole (previously known as SCH 56592) and BMS-207,147 in its activity against *C. albicans*; however, it appears slightly more potent against *C. glabrata* (14). In general, *Candida* spp. that are less susceptible to fluconazole also exhibit higher MICs to voriconazole and other azoles (15). Despite this, the voriconazole MIC_{90} for 1,300 bloodstream isolates was 0.5µg/ml (14). Laboratory strains resistant to voriconazole have been identified, and their susceptibilities to other agents compared (16).

The *in vitro* activity of posaconazole (**2**) appears similar to that of voriconazole and BMS-207,147 (8,14,17). Currently in Phase III clinical trials, compelling evidence of the efficacy of posaconazole in a variety of animal models of infection, including those caused by rarer pathogens, continues to appear (18-23). Following synthesis of radio-labelled drug (24), the elimination of the compound in the rat has been studied (25). Significant enterohepatic recirculation was seen; the primary routes of elimination are *via* the bile and feces.

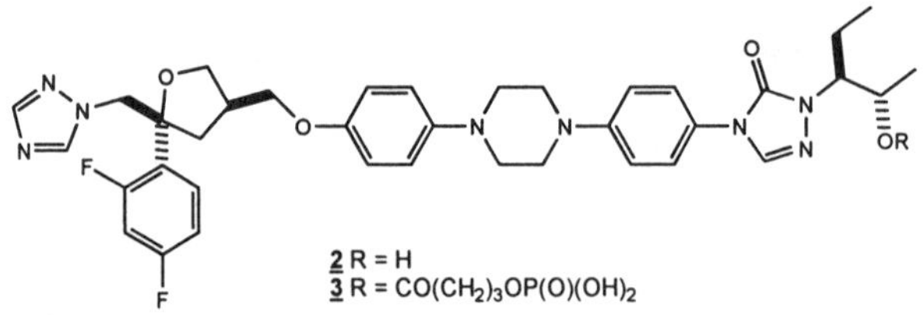

2 R = H
3 R = CO(CH$_2$)$_3$OP(O)(OH)$_2$

Posaconazole exhibits high oral bioavailability, but its low solubility precludes convenient formulation for intravenous use. A program aimed at identifying a more soluble pro-drug (26, 27) has identified SCH 59884 (**3**) as a promising candidate. This agent is inactive *in vitro*, but is de-phosphorylated *in vivo* to produce the active 4-hydroxybutyrate ester of SCH 56592. This, in turn, is hydrolyzed to the parent compound in human serum.

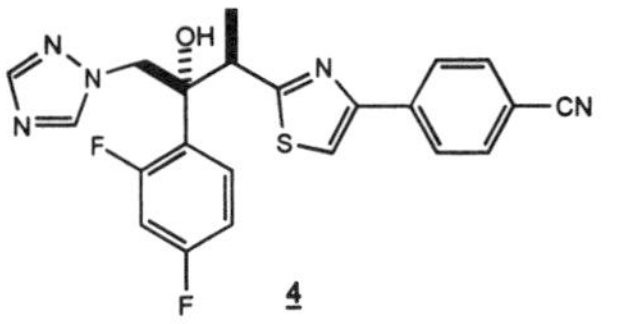

BMS-207,147 (**4**) has *in vitro* activity that is broadly similar in spectrum and potency to that of voriconazole and posaconazole (8,28-30). *In vivo*, the compound exhibits good oral bioavailability and a long serum half-life in both rats and dogs. The activity of the agent in a rabbit model of invasive aspergillosis has been demonstrated (31).

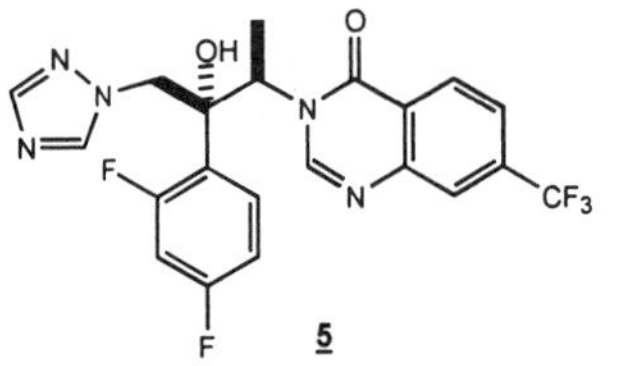

An azole antifungal discovery program has continued (32,33) at J. Uriach & Cia (Spain). Following careful analysis of serum half-lives in different animal species, the program culminated in the discovery of a new variant in UR-9825 (**5**). This compound is reported to have activity *in vitro* roughly similar to both voriconazole and BMS-207,147. Activity in candidiasis models in the rat and rabbit was shown to be comparable to fluconazole, and the compound was also shown to be effective in a rat model of disseminated aspergillosis. Upon oral administration to rats at 100 mg/kg bid or 250 mg/kg qd for 28 days, no overt signs of toxicity were observed. UR-9825 is currently in Phase I clinical trials.

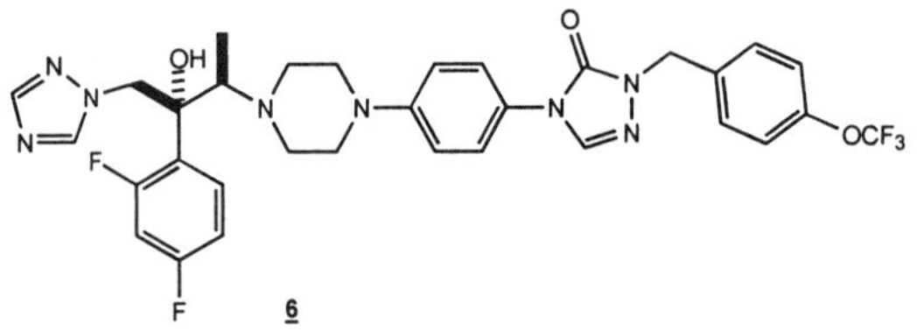

The 38th ICAAC in San Diego saw the first publications relating to Syn2869 (**6**), a novel broad-spectrum agent discovered in a collaboration between Synphar Laboratories (Edmonton, Canada) and Taiho Pharmaceutical Co. (Tokushima, Japan). The compound contains the piperazine-phenyl-triazolone side chain common to itraconazole and posaconazole, and displays similar potency and spectrum to the latter (34,35). The synthesis and SAR of the compound and its relatives has been reported (36), and activity superior to itraconazole has been demonstrated in animal models of *C. albicans, C. glabrata* and *C. neoformans* (37-39). Pharmacokinetics in mice and rabbits have been evaluated: the oral bioavailability of Syn2869 in mice is 60%, and higher tissue/serum ratios than those seen for itraconazole were claimed to contribute to the greater efficacy of the compound in a model of pulmonary invasive aspergillosis. The activity of the compound against less common mold pathogens has also been assessed (40).

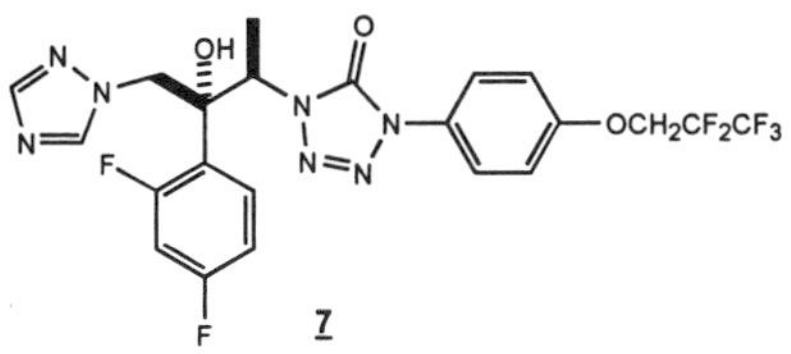

The synthesis and SAR leading to the discovery of the Takeda azole TAK-187 (**7**) has been published (41,42).

Echinocandins/pneumocandins - Echinocandins and the related pneumocandins are natural products discovered in the 1970s that act as non-competitive inhibitors of (1,3)-β-D-glucan synthase, an enzyme complex that forms glucan polymers in the fungal cell wall (43). There are three water-soluble derivatives of the echinocandins and pneumocandins that are in late stage clinical development, as well as several preclinical agents (44, 45). The most recent data regarding these agents is described below.

LY 303366 (**8**) is a pentyloxyterphenyl side chain derivative of echinocandin B discovered at Eli Lilly and recently licensed for parenteral use to Versicor. The agent is undergoing Phase II clinical trials (46). Recent studies have shown that the MIC's of LY 303366 against *Candida* spp. range from 0.08-5.12 µg/mL and that similar activity is obtained *in vitro* against *Aspergillus* spp. (47). Supporting these *in vitro* findings, several studies describing the potent activity of LY 303366 in animal models of disseminated candidiasis and pulmonary aspergillosis have been reported (48, 49). LY 303366 is also active in animal models of esophageal candidiasis and aspergillosis that are resistant to fluconazole and itraconazole, respectively (50, 51). The pharmacokinetics of the compound in healthy and HIV-infected volunteers after single-dose administration has been reported (52). The pharmacokinetics of the compound in healthy and HIV-infected volunteers after single-dose administration has been reported.

LY 303366 was radioiodinated and used as a probe in microsomal preparations of *C. albicans* containing glucan synthase activity (53). Two proteins of 40 and 18 kDa were identified using the photoaffinity agent, and analysis of the 40 kDa fragment revealed it to be a protein not previously described as being involved in glucan synthesis or in the mechanism of action of the echinocandins.

Recent SAR studies have focused principally on i) improving the water solubility of the compounds by introducing phosphonate and phosphate ester prodrugs on the phenolic hydroxy group (54), and ii) addressing the instability of LY 303366 under strongly basic conditions by incorporating nitrogen-containing ethers at the hemiaminal hydroxy group (55).

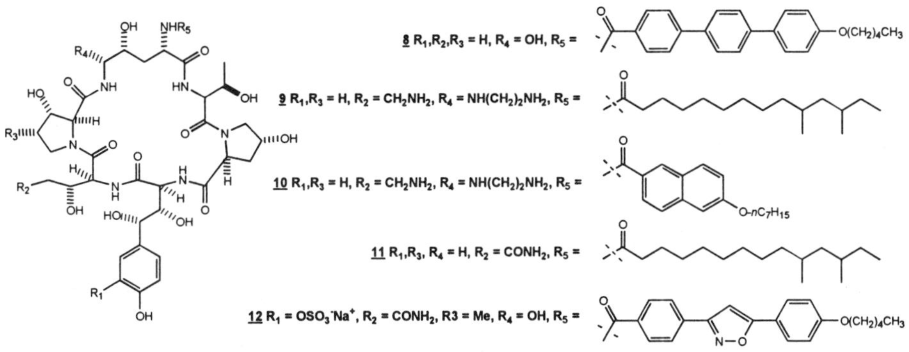

8 R_1,R_2,R_3 = H, R_4 = OH, R_5 =

9 R_1,R_3 = H, R_2 = CH_2NH_2, R_4 = $NH(CH_2)_2NH_2$, R_5 =

10 R_1,R_3 = H, R_2 = CH_2NH_2, R_4 = $NH(CH_2)_2NH_2$, R_5 =

11 R_1,R_3,R_4 = H, R_2 = $CONH_2$, R_5 =

12 R_1 = $OSO_3^-Na^+$, R_2 = $CONH_2$, $R3$ = Me, R_4 = OH, R_5 =

MK-0991 (**9**), previously classified as L-743872 (56), is undergoing Phase III trials and, like LY 303366, has shown *in vitro* and *in vivo* activity against *Candida* (including azole-resistant *Candida*), *Aspergillus* and other fungi. Recent *in vitro* data with MK-0991 has shown it to be highly effective against fluconazole -susceptible and -resistant *Candida* spp., with MIC's ranging from ≤ 0.19 to 0.78 µg/mL (57). The compound was fungicidal at concentrations ≤ 1.5 µg/mL for 73% of the 50 yeast isolates tested. In a recent study using 400 blood stream isolates of *Candida* spp. from 30 medical centers, MK-0991 and LY 303366 had MIC_{90}'s of 0.25 and 1 µg/mL, respectively (58). All of the isolates from this study for which fluconzaole and itraconzole had elevated MIC's (≥ 64 µg/mL and ≥ 1µg/mL, respectively) were inhibited by ≤ 0.5 µg/mL of both agents.

Using the nucleus of MK-0991 as a template, a regio-, chemo- and stereoselective semisynthesis of the echinocandin analog **10** has been developed from the cyclic hexapeptide **11** in four steps in 83% overall yield (59).

FK-463 (**12**) (60), like LY 303366 and MK-0991, has potent *in vitro* activity against a variety of *Candida* (MIC range ≤ 0.004-2 µg/mL) and *Aspergillus* (≤ 0.004-0.03 µg/mL) species (61,62). The compound is also active in a number of animal efficacy models (63, 64).

FK-463 was identified as part of a program to modify the lipophilic acyl side chain of the echinocandins (65). Thus, the target compound was prepared by condensation of an active ester of 4-(5-(4-pentoxyphenyl)isoxazol-3-yl)benzoic acid with the cyclic peptide nucleus obtained by enzymatic deacylation of the natural product.

The compound has favorable pharmacokinetics and was well tolerated in a single dose Phase I study in healthy volunteers (66). In a Phase II study in an AIDS population, FK-463 was effective in improving or clearing the clinical signs and symptoms of esophageal candidiasis at 12.5, 25 and 50 mg once daily for up to 21 days (67). In addition, once daily dosing for 14-21 days revealed no safety-related concerns.

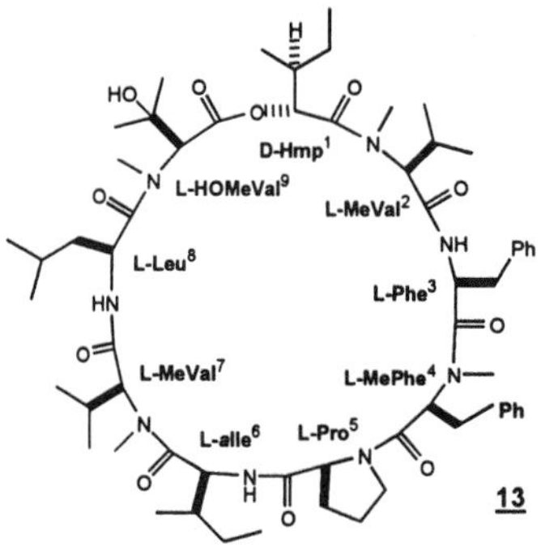

Aureobasidins - Aureobasidin A (**13**), a cyclic depsipeptide produced by *Aureobasidium pullulan*, inhibits inositol phosphorylceramide synthase (IPC synthase), an enzyme essential and unique in fungal sphingolipid biosynthesis (68). Using a fluorometric assay, it has been shown that **13** acts as a tight-binding, non-competitive inhibitor (K_i 0.55 nM) with respect to an analog of the substrate of IPC synthase (69).

A structural feature unique to the aureobasidins is the N-methylation of four of seven amide bonds. **13** has been examined using X-ray crystallography, and associated modelling studies imply that the high degree of N-methylation contributes to the relative stability of a unique, arrowhead-like conformation that may be associated with the biological activity (70).

The syntheses of **13** and several related cyclopeptide derivatives have been reported (71). Aeurobasidin derivatives with modifications at positions 6, 7 or 8 were prepared as part of a study to elaborate the SAR of the natural product. While analogs having L-glutamic acid at positions 6 or 8 showed weak activity, esterification of the γ-carboxyl group with benzyl or shorter alkyl (C_4-C_6) alcohols significantly enhanced the potency (72). Introduction of a longer C_{14} alkyl chain resulted in total loss of antifungal activity.

Pradimicins and benanomycins – These are dihydrobenzonaphthacene quinones conjugated with a D-amino acid and a disaccharide side-chain (73). They bind to cell wall mannoproteins in a calcium-dependent manner that causes disruption of the plasma membrane and leakage of intracellular potassium. Spectroscopic studies on the interaction of BMS181184 (**14**), a water-soluble pradimicin derivative, suggest that two molecules of **14** bind one Ca^{2+} ion, and each compound binds two mannosyl residues (74).

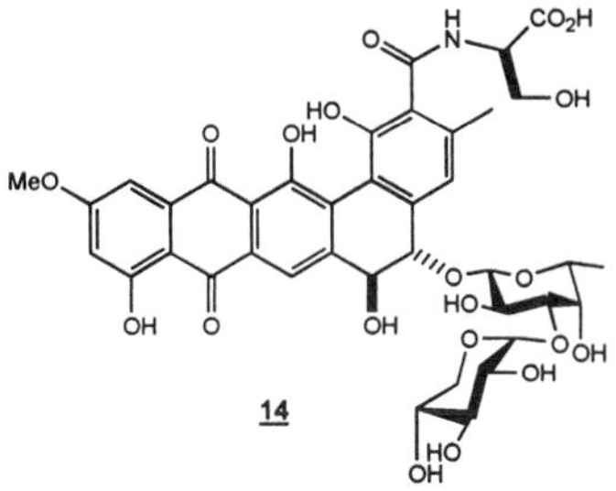

14 possesses activity towards *Aspergillus* spp. *in vitro* but is less potent than itraconazole or amphotericin B (75). In a model of invasive pulmonary aspergillosis in persistently neutropenic rabbits, daily doses of 50 and 150 mg/kg of **14** were as effective as amphotericin B at 1 mg/kg/day (76).

The total synthesis of pradimicinone, the common aglycon of the pradimicin-benanomicin antibiotics, has been achieved (77).

Polyoxins and nikkomycins – These are naturally-occurring nucleoside peptide antibiotics that inhibit chitin synthase, an enzyme that catalyzes the polymerization of N-acetylglucosamine, a major component of the fungal cell wall. A comprehensive review of synthetic efforts and subsequent biological studies on these agents has recently been reported (78).

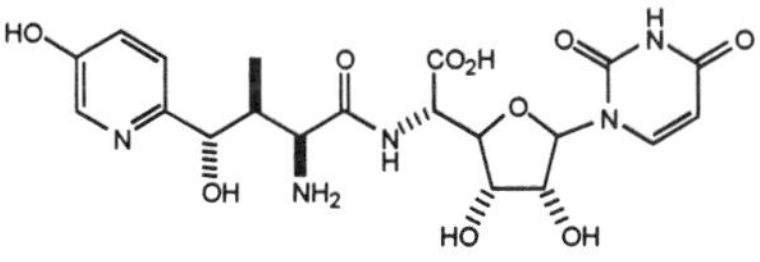

Nikkomycin Z (**15**) (79), the most advanced of these agents, has demonstrated additive and synergistic interactions with either fluconazole or itraconazole against *C. albicans* and *C. neoformans in vitro* (80). Marked synergism was also observed between nikkomycin Z and itraconazole against *A. fumigatus*. Nikkomycin Z is active against the less common endemic mycoses such as histoplasmosis, where pronounced synergistic interactions with fluconazole have been observed both *in vitro* and *in vivo* (81).

Sordarins - Both fungal and mammalian cells require two proteins, elongation factors 1 (EF1) and 2 (EF2), for ribosomal translocation during protein synthesis. A family of selective EF2 inhibitors, derived from the tetracyclic diterpene glycoside natural product sordarin (**16**), has been identified that demonstrates activity *in vitro* against a wide range of pathogenic fungi including *Candida* spp., *Cryptococcus neoformans*, and *P. carinii* (82-84). *In vivo*, sordarin derivatives have shown efficacy against systemic infections in mice caused by fluconazole-sensitive and -resistant *Candida albicans*, with ED_{50}'s ranging from 10-25 mg/kg (85). The toxicological properties of the new sordarin derivatives have been evaluated in several *in vitro* and *in vivo* preclinical studies (86, 87). Overall, the compounds have demonstrated no evidence of genotoxicity in the Ames test, are not clastogenic in cultured human lymphocytes and are well tolerated in rats and dogs.

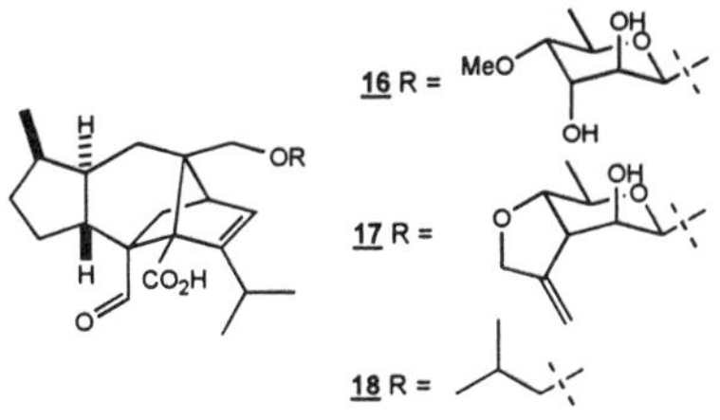

Whereas modification of the sugar unit of **16** affords the tetrahydrofuran derivative GM 237254 (**17**) and structurally related analogs (83), researchers at Merck have reported the preparation and evaluation of a variety of alkyl-substituted derivatives such as L- 793,422 (**18**) (88). Compounds of this type clearly demonstrate that a certain degree of lipophilicity on the side-chain is important for optimal antifungal acitivity. Biological studies have confirmed that L-793,422 and GM 237354 share the same mode of action (89).

An enantiospecific synthesis of the monocyclic core of sordarin has been achieved via the conversion of (+)-3,9-dibromocamphor into a 1,1,2,2,5-penta-substituted cyclopentane bearing all of the key functionalities present in the natural product (90).

CLINICALLY-IMPORTANT MECHANISMS OF ANTIFUNGAL RESISTANCE

Clinical failure in the treatment of systemic fungal infections may be due to a variety of factors, only one of which is the emergence of resistance of the pathogenic species. Characterization of the emergence of resistance is therefore not a straightforward matter.   Nonetheless, laboratory and epidemiologal data have confirmed that resistance, particularly to fluconazole, is spreading rapidly and has become a major problem in late-stage AIDS patients (91). The development of such resistance is commonly due to multiple mechanisms (92-94). Two recent reviews addressing this topic have appeared (95, 96).

In principle, a fungal pathogen can acquire resistance by adopting one or more of the following general mechanisms: antibiotic modification; target modification; functional compensation for the consequences of target inhibition; and alterations in intracellular drug accumulation. The first of these, although common in other classes of antibiotics, has not been observed for azole antifungals.

Target modification is clearly a common contributor to clinical resistance to azole therapy, and has been implicated directly for *C. neoformans* (97), *C. albicans* (98-100), *C. glabrata* (101), and by inference for other *Candida* spp. (102).   Different mutations have been documented, making the rational design of new agents less prone to this resistance mechanism difficult.   Target modification has also been implicated as a mode of resistance to 5-fluorocytosine (103), the sordarins (84) and the aureobasidins (104); efflux has also been implicated for the latter (105).

Functional compensation for the effects of an antifungal agent has been documented in two areas. Fungal resistance to the polyenes *in vitro* is associated with a marked decrease in ergosterol content; however, clinical resistance to Amphotericin B (AmB) in common fungal pathogens remains rare. For example, all 193 *Candida* spp. isolates obtained over an 8-year period (1990-1997) at a hospital in Madrid, Spain were susceptible to amphotericin B, whereas 19% were resistant to fluconazole (106). Azole resistance occurs when compensatory changes exist in other enzymes in the ergosterol biosynthesis pathway, notably inactivation of Δ-5,6-desaturase. This leads to the build-up of non-toxic 14α-methylated steroids in fungal membranes. The change in membrane composition can lead to cross-resistance to polyenes (107, 108).

Probably the most prevalent cause of resistance to azole therapy is caused by reduction in intracellular drug concentrations due to active efflux. In *C. albicans*, two types of efflux pump have been shown to be clinically relevant: a member of the Major Facilitator superfamily known as MDR1 (or BEN), and ATP-binding cassette (ABC)-type transporters CDR1 (109) and CDR2 (110). Fluconazole is a substrate for all three of these pumps, whereas other azoles are affected only by CDR1/2. Homologs of the ABC-type transporters have been shown to confer resistance in clinical isolates of *C. glabrata* (CgCDR1/2) (98,111) and *A. fumigatus* (ADR1) (112). Tools for the convenient characterization of such resistance mechanisms in clinical isolates are becoming available (113).

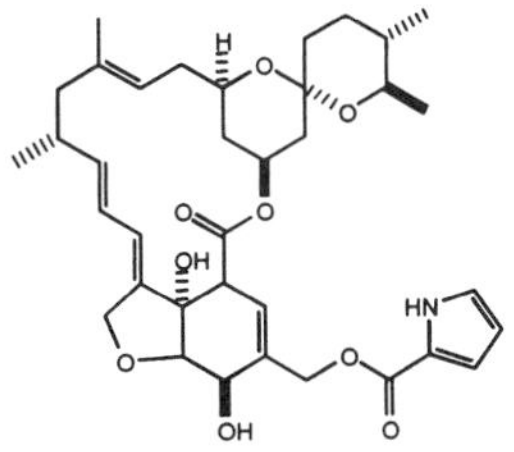

19

The attractiveness of efflux pumps in pathogenic fungi as targets for inhibition is therefore readily apparent. The first reports of inhibitors of ABC-type pumps in *C. albicans* and *C. glabrata* have recently appeared (114). The agents lack antifungal activity, and were characterized by their ability to increase intrinsic susceptibility to known pump substrates (azoles, terbinafine, rhodamine 6G), but not to agents not subject to efflux (amphotericin B). In a fluorescence assay, the compounds were shown to increase intracellular accumulation of rhodamine 6G. Such compounds can reverse CDR-mediated azole resistance in *C. albicans* (64 to 128-fold reduction in MIC of fluconazole or posaconazole) and reduce intrinsic resistance in *C. glabrata* (8-16-fold reduction in MIC). A representative of the class, milbemycin α-9 (MC-510,027, **19**), was shown to dramatically reduce the MIC_{90} of a broad panel of clinical isolates of *Candida* (115).

An agent from a different series, MC-005,172 (structure not published), was used in a pharmacodynamic assessment of the potential of efflux pump inhibitors to enhance the activity of fluconazole *in vivo* (116). By correlation of fungal kidney burden with AUC/MIC, 100mg/kg of the agent was shown to be capable of reducing the effective MIC of fluconazole from 128 to 5μg/ml.

CONCLUSION

Several new azoles are poised to have an impact on the treatment of mycoses, and the echinocandins continue to show promise as a novel class of antifungal agents, particularly for parenteral use. Other classes continue to be explored, as the demand for alternative therapies remains high.

Clinical failure to antifungal therapy due to resistance to existing agents is spreading rapidly, and is often multifactorial. The use of inhibitors of efflux pumps shows promise as a novel approach that may restore the potency of established compounds.

References

1.   P.L. Fidel, J.A. Vazquez and J.D. Sobel, Clin. Microbiol. Rev., 12, 80 (1999).
2.   D.J. Sheehan, C.A. Hitchcock and C.M. Sibley, Clin. Microbiol. Rev., 12, 40 (1999).
3.   N.H. Georgopapadakou, Current Opin. Microbio., 1, 547, 1998.
4.   P.A. Lartey and C.M. Moehle in "Ann. Rep. Med. Chem.", J.A. Bristol, Ed., Academic Press, NY, 1997, pp. 151.
5.   J.M. Balkovec in "Ann. Rep. Med. Chem.", J.A. Bristol, Ed., Academic Press, NY, 1998, pp. 173.
6.   D.T.W. Chu, J.J. Plattner and E. De Clercq, eds, Curr. Opin. Anti-Infect. Inv. Drugs, 1, (1999).
7.   G.M. Gonzalez, R. Tijerina, D.A. Sutton and M.G. Rinaldi, 39th ICAAC (1999), Abs. No. 1511.
8.   M.A. Pfaller, S.A. Messer, S. Gee, S. Joly, C. Pujol, D.J. Sullivan, D. C. Coleman, D. C. and D.R. Soll, J. Clin. Microbiol., 37, 870 (1999).
9.   M. Cuenca-Estrella, B. Ruiz-Diez, J.V. Martinez-Suarez, A. Monzon and J.L. Rodriguez-Tudela, J. Antimicrob. Chemother., 43, 149 (1999).
10.   M. A Ghannoum, I. Okogbule-Wonodi, N. Bhat and H. Sanati, J. Chemother., 11, 34 (1999).
11.   G. Quindos, S. Bernal, M. Chavez, M.J. Gutierrez, M.M.E. Strella and A. Valverde, 38th ICAAC (1998), Abs. No. J-6.
12.   F. Marco, M.A. Pfaller, S.A. Messer and R.N. Jones, Med. Mycol., 36, 433 (1998).
13.   M.H. Nguyen and C.Y. Yu, Antimicrob. Agents Chemother., 42, 471 (1998).
14.   M.A. Pfaller, S.A. Messer, R.J. Hollis, R.N. Jones, G.V. Doern, M.E. Brandt and R.A. Hajjeh, Antimicrob. Agents Chemother., 42, 3242 (1998).

15.   C.J. Clancy, A. Fothergill, M.G.Rinaldi and M.H.Nguyen, 39th ICAAC (1999), Abs. No. 1518.

16.   E.K. Manavathu, J.L. Cutright, P.H. Chandrasekar and O.C. Abraham, Abs. Gen. Meeting Amer. Soc. Microb., 99, 14 (1999).

17.   S.T. Yildiran, M.A. Saracli, A.W. Fothergill and M.G. Rinaldi, 39th ICAAC (1999), Abs. No. 1519.

18.   M. Lozano-Chiu, S. Arikan, V.L. Paetznick, E.J. Anaissie, D. Loebenberg and J.H. Rex, Antimicrob. Agents Chemother., 43, 589 (1999).

19.   L. Najvar, J.R. Graybill, R. Bocanegra and H. Al-Abdely, 38th ICAAC (1998), Abs. No. J-68.

20.   P. Melby, J.R. Graybill and H. Al-Abdely, 38th ICAAC (1998), Abs. No. B-58.

21.   L. Najvar, J.R. Graybill, R. Bocanegra and H. Al-Abdely, 38th ICAAC (1998), Abs. No. J-69.

22.   J.A. Urbina, G. Payares, L.M. Contreras, A. Liendo, C. Sanoja, J. Molina, M. Piras, R. Piras, N. Perez, P. Wincker, and D. Loebenberg, Antimicrob. Agents Chemother., 42, 1771 (1998).

23.   J. Molina, O. Martins-Filho, Z. Brener, A.J. Romanha, D. Loebenberg and J.A. Urbina, Antimicrob. Agents Chemother., 44, 150 (2000).

24.   C.V. Magatti, D. Hesk, M.J. Lauzon, S.S. Saluja and X. Wang, J. Label. Comp. Radiopharm., 41, 731 (1998).

25.   P. Krieter, J. Achanfuo-Yeboah, M. Shea, M. Thonoor, M. Cayen and J. Patrick, 39th ICAAC (1999), Abs. No. 1199.

26.   A.A. Nomeir, P. Kumari, S. Gupta, D. Loebenberg, A. Cacciapuoti, R. Hare, C.C. Lin and M.N. Cayen, 39th ICAAC (1999), Abs. No. 1934.

27.   D. Loebenberg, F. Menzel Jr., E. Corcoran, C. Mendrick, K. Raynor, A.F. Cacciapuoti and R.S. Hare, 39th ICAAC (1999), Abs. No. 1933.

28.   J.C. Fung-Tomc, E. Huczko, B. Minassian and D.P. Bonner, Antimicrob. Agents Chemother., 42, 313 (1998).

29.   D.J. Diekema, M.A. Pfaller, S.A. Messer, A. Houston, R.J. Hollis, G.V. Doern, R.N. Jones and the Sentry Participants Group, Antimicrob. Agents Chemother., 43, 2236 (1999).

30.   C.B. Moore, C.M. Walls and D.W. Denning, 39th ICAAC (1999), Abs. No. 1931.

31.   K. Shock, S. Marino, V. Andriole and T. Baumgartner, 38th ICAAC (1998), Abs. No. J-54.

32.   J. Bartroli, E. Turmo, M. Alguero, E. Boncompte, M.L. Vericat, L. Conte, J. Ramis, M. Merlos, J. Garcia-Rafanell and J. Forn, J. Med. Chem., 41, 1855 (1998).

33.   J. Bartroli, E. Turmo, M. Alguero, E. Boncompte, M.L. Vericat, L. Conte, J. Ramis, M. Merlos, J. Garcia-Rafanell and J. Forn, J. Med. Chem., 41, 1869 (1998).

34.   A.P. Gibb and H. Van Den Elzen, 38th ICAAC (1998), Abs. No. F-147.

35.   S.M. Salama, A. Gandhi, H. Atwal, J. Simon, J. Khan, R.G. Micetich and M. Daneshtalab, 38th ICAAC (1998), Abs. No. F-150.

36.   P. Spevak, I. Sidhu, G. Samari, S.M. Salama, D.Q. Nguyen, R.G. Micetich, J. Khan, G. Kasitu, C. Ha, Y. Bathini, M.D. Abel, T. Furukawa and N. Unemi, 38th ICAAC (1998), Abs. No. F-148.

37.   N. Unemi, T. Uji, H. Saito, K. Nishida, F. Higashitani, T. Furukawa and H. Yamaguchi, 38th ICAAC (1998), Abs. No. F-149.

38.   S.M. Salama, H. Atwal, A. Gandhi, J. Khan, H. Montaseri, M. Poglod, R.G. Micetich and M. Daneshtalab, 38th ICAAC (1998), Abs. No. F-151.

39.   J. M. Khan, H. Montaseri, M. Poglod, H.Z.Bu, S. Salama, R.G. Micetich and M. Daneshtalab, 38th ICAAC (1998), Abs. No. F-152.

40.   E.M. Johnson, A. Szekely and D.W. Warnock, Antimicrob. Agents Chemother., 43, 1260 (1999).

41.   T. Kitazaki, A. Tasaka, N. Tamura, Y. Matsushita, H. Hosono, R. Hayashi, K. Okonogi and K. Itoh, Chem. Pharm. Bull., 47, 351 (1999).

42.   T. Kitazaki, A. Tasaka, H. Hosono, Y. Matsushita, and K. Itoh, Chem. Pharm. Bull., 47, 360 (1999).

43.   D.W. Denning, J. Antimicrob. Chemother., 40, 611, (1997).

44.   S. Hawser, M. Borgonovi, A. Markus and D. Isert, J. Antibiot, 52, 305 (1999).

45.   A.M. Nilius, P.M. Raney, D.M. Hensey, W. Weibo, Q. Li and R.K. Flamm, 39th ICAAC (1999), Abs. No. 1938.

46.   S. Hawser, Curr. Opin. Anti-Infect. Invest. Drugs, 1, 353 (1999).

47.   K.L. Oakley, C.B. Moore and D.W. Denning, Antimicrob. Agents Chemother., 42, 2726 (1998).

48.   R. Petraitiene, V. Petraitis, A.H. Groll, M. Candelario, T. Sein, A. Bell, C.A. Lyman, C.L. McMillian, J. Bacher and T.J. Walsh, Antimicrob. Agents Chemother., 43, 2148 (1999).

49.   V. Petraitis, R. Petraitiene, A.H. Groll, A. Bell, D.P. Callender, T. Sein, R.L. Schaufele, C.L. McMillian, J. Bacher and T.J. Walsh, Antimicrob. Agents Chemother., 42, 2898 (1998).

50.   V. Petraitis, R. Petraitiene, M. Canderlario, A. Fiel-Ridley, A. Groll, T. Sein, R.L. Schaufele, J. Bacher and T.J. Walsh, 38th ICAAC (1998), Abs. No. J-72.

51.   L.K. Najvar, R. Bocanegra, S.E. Sanche and J.R. Graybill, 39th ICAAC (1999), Abs. No. 2002.

52.   L. Ni, B. Smith, B. Hathcer, M. Goldman, C. McMillian, M. Turik, I. Rajman, L.J. Wheat and V.S. Watkins, 38th ICAAC (1998), Abs. No. J-134.

53.   J.A. Radding, S.A. Heidler and W.W. Turner, Antimicrob. Agents Chemother., 42, 1187 (1998).

54.   M.J. Rodriquez, V. Vasudevan, J.A. Jamison, P.S. Borromeo and W.W. Turner, Bioorg. Med. Chem. Lett., 9, 1863 (1999).

55.   J.A. Jamison, L.M. LaGrandeur, M.J. Rodriquez, W.W. Turner and D.J. Zeckner, J. Antibiot., 51, 239 (1998).

56.   A.H. Groll and T. J. Walsh, IDrugs, 2, 1201 (1999).

57.   F. Barchiesi, A.M. Schimizzi, A.W. Fothergill, G. Scalise and M.G. Rinaldi, Eur. J. Clin. Microbiol. Infect. Dis. 18, 302 (1999).

58.   F. Marco, M.A. Pfaller, S.A. Messer and R.N. Jones, Diagn. Microbiol. Infect. Dis. 32, 33 (1998).

59.   M. Journet, D. Cai, L.M. DiMichele, D.L. Hughes, R.D. Larsen, T.R. Verhoeven, P.J. Reider, J. Org. Chem. 64, 2411 (1999).

60.   R.A. Fromtling and J. Castaner, Drugs Future, 23, 1273 (1998).

61.   K. Maki, Y. Morishita, Y. Iguchi, E. Watabe, K. Otomo, N. Teratani, Y. Watanabe, F. Ikeda, S. Tawara, T. Goto, M. Tomishima, H. Ohki, A. Yamada, K. Kawabata, H. Takasugi, H. Tanaka, K. Sakane, F. Matsumoto and S. Kuwahara, 38th ICAAC (1998), Abs. No. F-141.

62.   S. Tawara, F. Ikeda, K. Maki, Y. Morishita, K. Otomo, N. Teratani, T. Goto, M. Tomishima, H. Ohki, A. Yamada, K. Kawabata, H. Takasugi, K. Sakane, H. Tanaka, F. Matsumoto and S. Kuwahara, Antimicrob. Agents Chemother., 44, 57 (2000).

63.   Y. Wakai, S. Matsumoto, K. Maki, E. Watabe, K. Otomo, T. Nakai, K. Hatano, Y. Watanabe, F. Ikeda, S. Tawara, T. Goto, F. Matsumoto and S. Kuwahara. 38th ICAAC (1998), Abs. No. F-143.

64.   S. Matsunoto, Y. Wakai, K. Maki, E. Watabe, T. Ushitani, K. Otomo, T. Nakai, Y. Watanabe, F. Ikeda, S. Tawara, T. Goto, F. Matsumoto and S. Kuwahara, 38th ICAAC (1998), Abs. No. F-142.

65.   M. Tomishima, H. Ohki, A. Yanada, H. Takasugi, K. Maki, S. Tawara and H. Tanaka, J. Antibiot., 52, 674 (1999).

66.   J. Azuma, I. Yamamoto, M. Ogura, T. Mukai, H. Suematsu, H. Kageyama, K. Nakahara, K. Yoshida and T. Takaya, 38th ICAAC (1998), Abs. No. F-146.

67.   K. Pettengell, J. Mynhardt, T. Kluyts and P. Soni, 39th ICAAC (1999), Abs. No. 1421.

68.   K. Takesako. K. Ikai, F. Haruna, M. Endo, K. Shimanaka, E. Sono, T. Nakamura and I. Kato, J. Antibiot., 44, 919 (1991).

69.   W. Zhong, D.J. Murphy and N.H. Georgopapadakou, FEBS Lett., 463, 241 (1999).

70.   Y. In, T. Ishida and K. Takesako, J. Pept. Res., 53, 492 (1999).

71.   U. Schmidt, A. Schumacher, J. Mittendorf and B. Riedl, J. Pept. Res., 52, 143 (1998).

72.   T. Kurome, T. Inoue, K. Takesako and I. Kato, J. Antibiot., 51, 359 (1998).

73.   J.C. Fung-Tomc, B. Minassian, E. Huczko, B. Kolek, D.P. Bonner and R.E. Kessler, Antimicrob. Agents Chemother., 39, 295 (1995).

74.   K. Fujikawa, Y. Tsukamoto, T. Oki and Y.C. Lee, Glycobiology, 8, 407 (1998).

75.   K.L. Oakley, C.B. Moore and D.W. Denning, Int. J. Antimicrob. Agents, 12, 267 (1999).

76.   C.E. Gonzalez, A.H. Groll, N. Giri, D. Shetty, I. Al-Mohsen, T. Sein, E. Feuerstein, J. Bacher, S. Piscitelli and T.J. Walsh, Antimicrob. Agents Chemother., 42, 2399 (1998).

77.   M. Kitamura, K. Ohmori, T. Kawase and K. Suzuki, Angew. Chem. Int. Ed., 38, 1229 (1999).

78.   D. Zhang and M.J. Miller, Curr. Pharm. Des., 5, 73 (1999).

79.   N.H. Georgopapadakou, Curr. Opin. Anti-Infect. Drugs, 1, 346 (1999).

80.   R.K. Li and M.G. Rinaldi, Antimicrob. Agents Chemother., 43, 1401 (1999).

81.   J.R. Graybill, L.K. Najvar, R. Bocanegra, R.F. Hector and M.F. Luther, Antimicrob. Agents Chemother., 42, 2371 (1998).

82.   D. Gargallo-Viola, Curr. Opin. Anti-Infect. Invest. Drugs, 1, 297 (1999).

83.   E. Herreros, C.M. Martinez, M.J. Almela, M.S. Marriott, F.G. de las Heras and D. Gargallo-Viola, Antimicrob. Agents Chemother., 42, 2863 (1998).
84.   L. Capa, A. Mendoza, J.L. Lavandera, F.G. de las Heras and J.F. Garcia-Bustos, Antimicrob. Agents Chemother., 42, 2694 (1998).
85.   A. Martinez, E. Jiminez, P. Aviles, J. Caballero, F.G. de las Heras and D. Gargallo-Viola, 39th ICAAC (1999), Abs. No. 294.
86.   E. Herreros, M.J. Almela, S. Lozano, C.M. Martinez and D. Gargallo-Viola, 39th ICAAC (1999), Abs. No. 158.
87.   D.G. Gatehouse, T.C. Williams, A.T. Sullivan, G.H. Apperley, S.P. Close, S.R. Nesfield, A. Martinez and D. Gargallo-Viola, 38th ICAAC (1998), Abs. No. J-75.
88.   B. Tse, J.M. Balkovec, C.M. Blazey, M-J. Hsu, J. Nielsen, D. Schmatz, Bioorg. Med. Chem. Lett., 8, 2269 (1998).
89.   M.C. Justice, M-J. Hsu, B. Tse, T. ku, J. Balkovec, D. Schmatz and J. Nielsen, J. Biol. Chem., 273, 3148 (1998).
90.   J.C. Cuevas and J.L. Martos, Tetrahedron Lett., 39, 8553 (1998).
91.   T. White, 39th ICAAC (1999), Abs. No. 1126.
92.   R. Franz, S.L. Kelly, D.C. Lamb, D.E. Kelly, M. Ruhnke and J. Morschhaeuser, Antimicrob. Agents Chemother., 42, 3065 (1998).
93.   J.L. Lopez-Ribot, R.K. McAtee, L.N. Lee, W.R. Kirkpatrick, T.C.White, D. Sanglard and T.F. Patterson, Antimicrob. Agents Chemother., 42, 2932 (1998).
94.   S. Perea, R.A. Cantu, R.K. McAtee, W.R. Kirkpatrick, T.F. Patterson and J.L. Lopez-Ribot, 39th ICAAC (1999), Abs. No. 296.
95.   D.A. Stevens and K. Holmberg, Curr. Opin. Anti-Infect. Inv. Drugs, 1, 306 (1999).
96.   P. Marichal, Curr. Opin. Anti-Infect. Inv. Drugs, 1, 318 (1999).
97.   K. Venkateswarlu, M. Taylor, N.J. Manning, M.G. Rinaldi and S.L. Kelly, Antimicrob. Agents Chemother., 41, 748 (1997).
98.   D. Sanglard, F. Ischer, L. Koymans and J. Bille, Antimicrob. Agents Chemother., 42, 241 (1998).
99.   K. Asai, N. Tsuchimori, K. Okonogi, J.R. Perfect, O. Gotoh and Y. Yoshida, Antimicrob. Agents Chemother., 43, 1163 (1999).
100.   D.C. Lamb, D.E. Kelly, T.C. White and S.L. Kelly, Antimicrob. Agents Chemother., 44, 63 (2000).
101.   D. Sanglard, F. Ischer, D.C. Calabrese, P.A. Majcherczyk and J. Bille, Antimicrob. Agents Chemother., 43, 2753 (1999).
102.   T. Fukuoka, Y. Fu and S.G. Filler, 38th ICAAC (1998), Abs. No. J-90.
103.   H. Vanden Bosche, P. Marichal and F.C. Odds, Trends Microbiol., 2, 393 (1994).
104.   M. Kuroda, T. Hashida-Okado, R. Yasumoto, K. Gomi, I. Kato and K. Takesako, Mol. Gen. Genet., 261, 290 (1999).
105.   A. Ogawa, T. Hashika-Okado, M. Endo, H. Yoshioka, T. Tsuruo, K. Takesako, I. Kato, Antimicrob. Agents Chemother., 42, 755 (1998).
106.   L. Torres, L. Alcala, T. Pelaez, M. Diaz, M. Marin, M. Rodriquez-Creizems and E. Bouza, 38th ICAAC (1998), Abs. No. J-28.
107.   S.L. Kelly, D.C. Lamb, D.E. Kelly, J. Loeffler and H. Einsele, Lancet, 348, 1523 (1996).
108.   F.S. Nolte, T. Parkinson, D.J. Falconer, S. Dix, J. Williams, C. Gilmore, R. Geller and J.R. Wingard, Antimicrob. Agents Chemother., 41, 196 (1997).
109.   G.D. Albertson, M. Nimi, R.D. Cannon and H.F. Jenkinson, Antimicrob. Agents Chemother., 40, 2835 (1996).
110.   D. Sanglard, F. Ischer, M. Monod and J. Bille, Microbiology, 143 (Pt 2), 405 (1997).
111.   D. Sanglard, F. Ischer and J. Bille, 38th ICAAC (1998), Abs. No. C-148.
112.   J.W. Slaven, M.J. Anderson, D. Sanglard, G.K. Dixon, J. Bille, I.S. Roberts and D.W. Denning, 39th ICAAC (1999), Abs. No. 447.
113.   S. Maesaki, P. Marichal, H.Vanden Bossche, D. Sanglard and S. Kohno, J. Antimicrob. Chemother., 44, 27 (1999).
114.   O. Lomovskaya, M. Warren, A. Mistry, A. Staley, J. Galazzo, H. Fuernkranz, M. Lee, G. Miller and D. Sanglard, 39th ICAAC (1999), Abs. No. 1269.
115.   S.Chamberland, J. Blais, D.P. Cotter, M.K. Hoang, J. Galazzo, A. Staley, M. Lee and G.H. Miller, 39th ICAAC (1999), Abs. No. 1270.
116.   K. Sorensen, E. Corcoran, S. Chen, D. Clark, V. Tembe, O. Lomovskaya and M. Dudley, 39th ICAAC (1999), Abs. No. 1271.

Chapter 15. Matrix Metalloproteinase Inhibitors for Treatment of Cancer

Jerry W. Skiles, Lauren G. Monovich, and Arco Y. Jeng
Novartis Institute for Biomedical Research, 556 Morris Ave.,
Summit, NJ 07901

<u>Introduction</u> Cancer is the second most common cause of death in the advanced countries; approximately one in five persons will die of this disease. It imposes great cost on society and individuals *via* premature disability, mortality and high treatment costs. Despite advances in the diagnosis and management of the disease and billions of dollars spent in research, only modest improvements in cure and survival rate have been realized. The primary treatment approach has relied upon cytotoxic strategies to limit tumor growth and metastasis. However, cytotoxic drugs and radiation therapy often lead to unacceptable side effects. Although the exact mechanisms responsible for the formation of the tumors and the onset of metastasis are not fully understood, the critical event signaling the initiation of the metastatic cascade in tumor invasion is thought to be the interaction of the tumor with the basement membrane. The matrix metalloproteinases (MMPs) are documented to be involved in the proteolysis of the basement membrane and other extracellular matrix (ECM) components, and they appear to play an essential role in angiogenesis, tumor growth, and metastasis. Therefore, in recent years inhibitors of the MMPs have been proposed as a possible new means of controlling tumor growth and metastasis while exhibiting a low toxicity profile compared to existing therapies. In this chapter, the biochemistry of MMPs and the rationale for treatment of cancer with MMP inhibitors are briefly reviewed. Major emphasis is on recently published, potent MMP inhibitors and their pharmacological properties. The results of clinical trials of MMP inhibitors are briefly summarized.

<u>Overview of the MMPs</u> Several excellent reviews on the design of MMP inhibitors have appeared (1-16). The MMPs are a family of zinc-containing, calcium-dependent enzymes involved in tissue remodelling and degradation of the ECM proteins, angiogenesis, and cell motility (17-20). Currently, 20 human MMPs are known (Table 1). The MMPs belong to the matrixin family, and they may be subdivided into five classes according to their substrate specificity, primary structures, and their cellular localization. MMP-23 is the most recently discovered MMP to be cloned and characterized (21). The enzymes are expressed as inactive zymogens which are activated by serine proteases, e.g., furin and plasmin, and other MMPs. The zymogens are excreted by a variety of connective tissue and pro-inflammatory cells, including fibroblasts, osteoblasts, endothelial cells, macrophages, neutrophils, and lymphocytes. The MMPs generally consist of four distinct domains: an N-terminal pro-domain, a catalytic domain, a hinge region, and a C-terminal hemopexin-like domain. With the exception of MMP-7 and MMP-23, all human MMPs contain a conserved hemopexin-like domain. This domain is important for macromolecular substrate recognition as well as interaction with tissue inhibitors of metalloproteinases (TIMPs). The membrane-type MMPs (MT-MMPs) contain an additional transmembrane domain that anchors them in the cell surface. The activity of the MMPs is normally regulated through the presence of endogenous inhibitors such as the TIMPs. An imbalance between the MMPs and their natural inhibitors (i.e., TIMPs) is believed to be a contributing factor in the manifestation of numerous disease states involving the degradation of ECM components: osteoarthritis, rheumatoid arthritis, angiogenesis, cancer, pulmonary emphysema, corneal ulceration, atherosclerotic plaque rupturing, aortic aneurysms, and periodontal disease.

Table 1. The Human MMP Family: Classes, Numbers and Names

Collagenases		MT-MMP (membrane type)	
MMP-1	Collagenase-1	MMP-14	MT1-MMP
MMP-8	Neutrophil Collagenase	MMP-15	MT2-MMP
MMP-13	Collagenase-3	MMP-16	MT3-MMP
MMP-18	Collagenase-4	MMP-17	MT4-MMP
Gelatinases		MMP-24	MT5-MMP
MMP-2	Gelatinase-A (72 kDa)	Other Enzymes	
MMP-9	Gelatinase-B (92 kDa)	MMP-12	Macrophage metalloelastase
Stromelysins		MMP-19	Unnamed
MMP-3	Stromelysin-1	MMP-20	Enamelysin
MMP-10	Stromelysin-2	MMP-23	Same as MMP-21 and MMP-22
MMP-11	Stromelysin-3		
MMP-7	Matrilysin (Pump-1)		

Rationale for treatment of cancer with MMP inhibitors - Numerous publications have convincingly documented the involvement of MMPs in various cancers. For example, enhanced expression of collagenase and/or gelatinase mRNAs have been found in benign and malignant ovarian tumors, colon cancer, and oral squamous cell carcinomas, etc. (22-24). In breast cancers, increased expressions of most MMPs have been observed (25). Furthermore, increases in the production of collagenase and/or gelatinase in malignant tissues have been demonstrated by immunohistochemical staining (22, 26) and in brain tumors by both Western blot analysis and gelatin zymography (27). Moreover, the activities of MMPs in bladder cancer correlate with tumor grade and invasion (28).

It has also been shown that the MMPs facilitate tumor growth, local invasion, and metastases (29). Initiation of cellular invasion requires attachment of the cells to a basement membrane followed by proteolysis of the basement membrane by the MMPs and a migration through these lesions. Following invasion, cell proliferation and continued invasion results in production of metastatic foci. However, the growth in the size of solid tumors is limited without the formation of new blood vessels to supply nutrients and oxygen. MMPs have been demonstrated to promote angiogenesis. The new blood vessel formation allows expansion of tumor foci in three dimensions and an enhanced metastatic behavior that correlates directly with the degree of vascularization of the primary tumor. Recently, reduced angiogenesis and tumor progression have been demonstrated in mice deficient in MMP-2 (30). These results strongly suggest that MMP inhibitors may prove to be clinically relevant agents for blocking tumor invasion and metastasis and, hence, for the treatment of cancers.

SMALL MOLECULE INHIBITORS OF MMPs

Several excellent reviews on the design of small molecule MMP inhibitors have appeared (1-7). In general, peptidomimetics that incorporate a zinc binding group and P and/or P' side-chains to interact with the enzyme subsites have been the most common structural features of inhibitors. Widely utilized zinc binding moieties include: hydroxamic acids, carboxylic acids, thiols, phosphonates, and "novel" chelators.

Hydroxamic acids - Owing to the high affinity of hydroxamates for zinc, hydroxamic acids are the most populated class of MMP inhibitors (1, 2). The surge of new structural information and analysis of enzyme-substrate complexes is ongoing (19). Structural analysis of bound succinate-based hydroxamic acid inhibitors, such as batimastat (BB-94, **1**) which is currently in Phase II clinical trials for cancer, reveals that the P_1 α-thienylthiomethylene substituent and the P_2' phenylalanine side-chain that occur in **1** are directed toward solvent and thus make no direct contact with the enzyme. The lysine-derived succinate macrocycle such as **2** also exhibits a broad spectrum MMP inhibitory activity (1, 2, 31). The ring size and ring substitution of macrocycles such as **2** affect both potency and selectivity (32). For example, substitution for the P_2' and/or P_3' groups commonly held by aromatic amino acid-derived amides, such as in batimastat (**1**), are tolerated in succinate-based hydroxamic acid inhibitors **3**, **4** and **5**. Aryl and heteroaryl ketones, such as the indole **3**, maintain potency. The inhibitor **3** has an IC_{50} of 1.2 nM for MMP-3 inhibition (33). Several structurally related compounds of **3** are reported to have improved oral bioavailability (1). Other examples of succinate based hydroxamates include Trocade (Ro 32-3555, **36**, Table 2), which is in Phase III clinical trials for rheumatoid arthritis (34, 35), and the piperidine **4**. The inhibitor **4** selectively inhibits MMP-1 with an IC_{50} of 6 nM (1, 36). α-Aminohydroxamate **5** (IC_{50} = 6.5 nM for MMP-1) shows promising bioavailability, 50 % in rats and 70 % in cynomolgus monkeys (37).

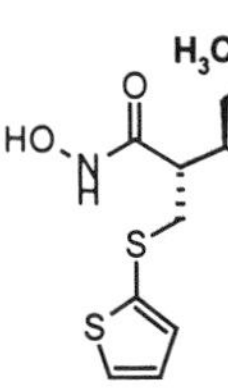

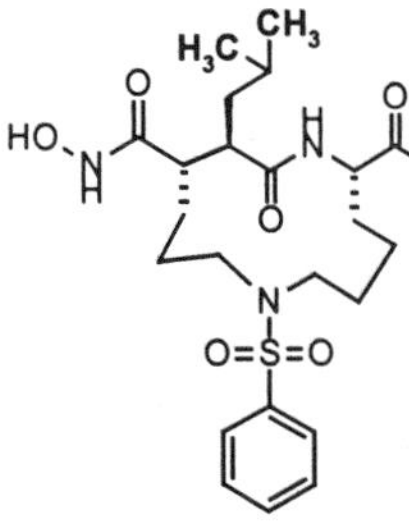

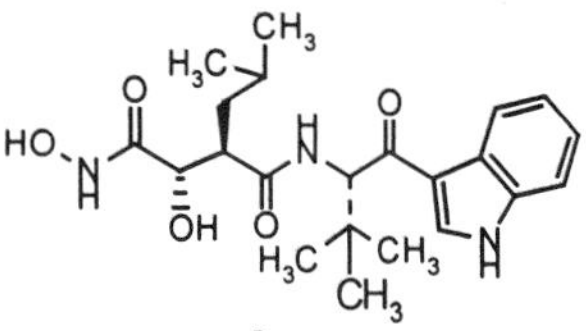

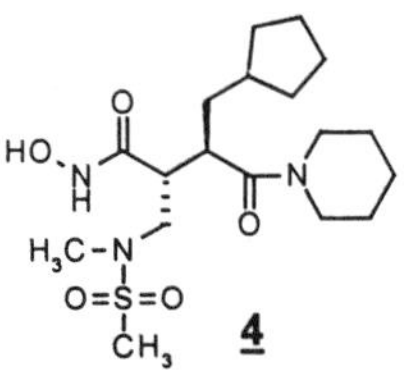

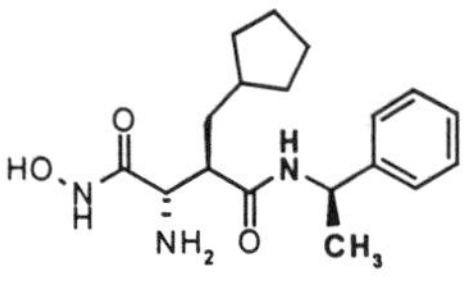

Sulfonamide hydroxamate inhibitors such as **7**, **8**, and **9** capitalize on the CGS 27023A (**6**) scaffold by capturing the α-side-chain in a ring. The SAR of **6** as well as related analogs have been published (38, 39). In addition, the bound conformation of **6** complexed with MMP-3 as well as the full solution NMR structure of the **6**: MMP-3 complex have recently appeared (40, 41). The synthesis of cyclic sulfonamide-based hydroxamic acid inhibitors, including sulfonylated piperidines, tetrahydroisoquinolines, diazepines, thiazines, and others, continues to be fruitful. For instance, benzodiazepines, exemplified by **7**, possess a broad MMP inhibitory activity (IC_{50} = 20 nM for MMP-1, 1 nM for MMP-9 and MMP-13, and 13 nM for tumor necrosis factor-α converting-enzyme (TACE)) (42). Thiazepine **8** closely resembles the thiazine AG-3340 (**30**, Table 2) and other series of acyclic penicillamine-based hydroxamate inhibitors (43, 44). A proline-based scaffold, such as **9** (IC_{50} = 10 nM for MMP-1, 3 nM for MMP-3, and < 0.5 nM for MMP-13), has been reported to produce potent, broad spectrum MMP inhibitors when an sp^2-hybridized carbon is present in the pyrrolidine ring (45).

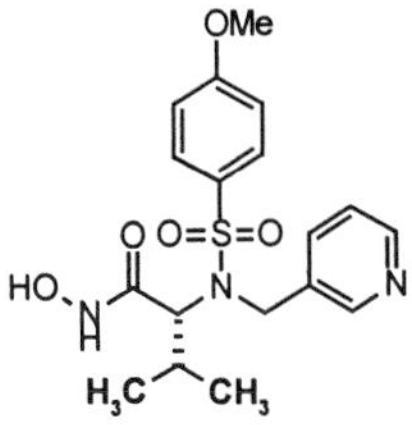

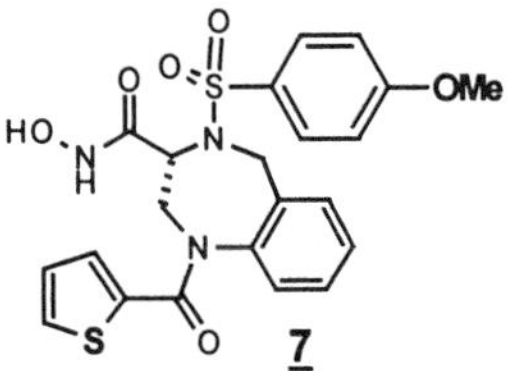

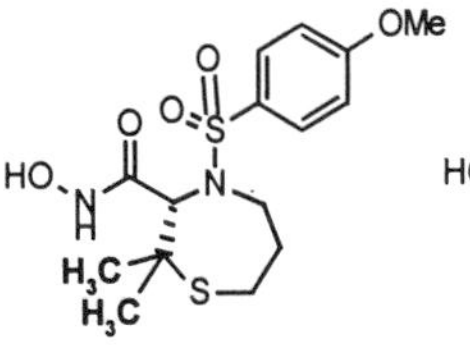

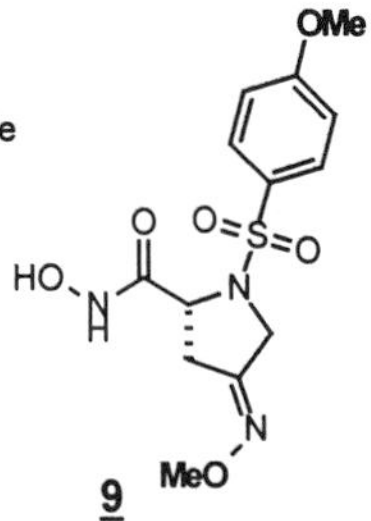

Substituted aromatics such as sulfonamide **10** (IC$_{50}$ = 24 nM for MMP-9) which is active in an *ex-vivo* enzyme inhibition assay (46), 2, 3-disubstituted thiophene **11** having an IC$_{50}$ of 19 nM for MMP-1 inhibition, and the orally active pyrazolyl-pyrimidine **12**, represent alternative permutations of the same strategy (47-49).

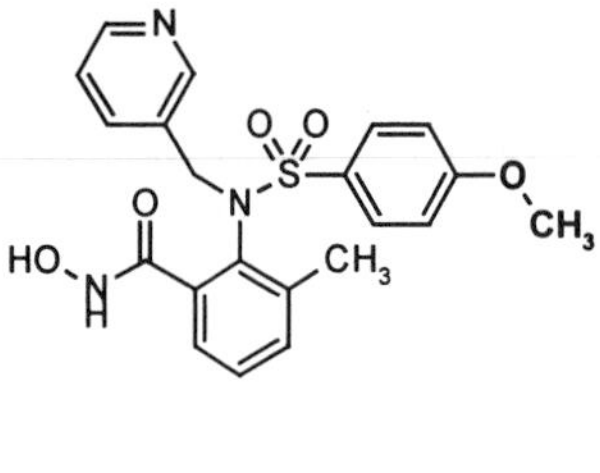

10 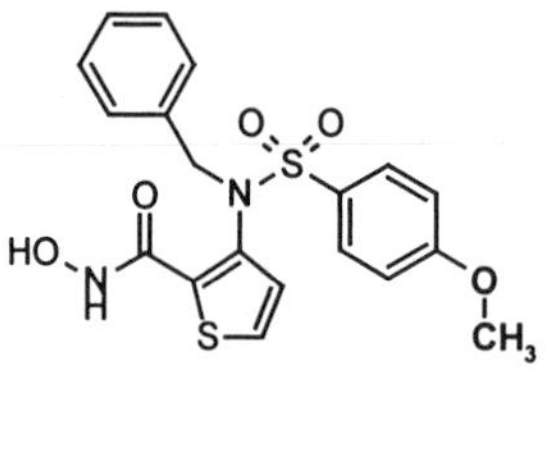   **11** 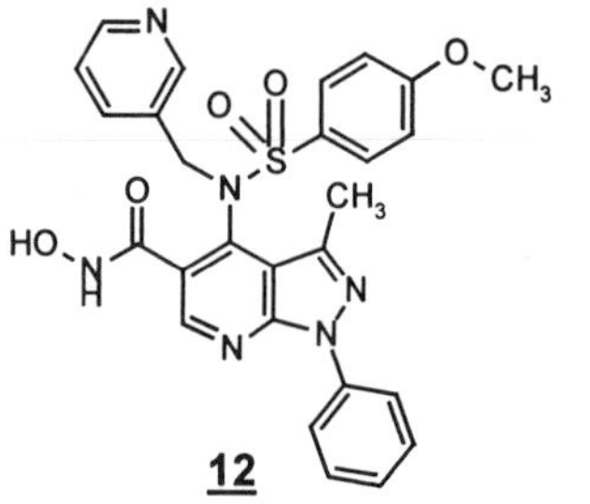   **12**

Other modifications of the template **6** have the picolyl moiety replaced with other groups in order to improve water solubility and/or potency. Structure **13** builds (50) on earlier efforts bearing *N*-ethylmorpholino groups (51, 52). (*R*)-Phosphinamide **14** provides a good bioisosteric replacement for the sulfone of **6** and has comparable inhibitory activity toward various MMPs (53). X-ray crystallographic analysis (53) places the lone phosphinamide oxygen of **14** within hydrogen bonding distance to both the Ala[165] and Leu[164] residues of MMP-3 and agrees with previous reports (40, 41) that attribute hydrogen bonding to the pair of sulfone oxygens in CGS 27023A (**6**). Alternatively, the sulfonamide nitrogen may be substituted with a carbon atom, as in sulfone **15**. Racemic sulfone **15** has been reported to inhibit phosphodiesterase type 4 (PDE4) with an IC$_{50}$ of 30 nM in addition to inhibiting MMP-1, -2, and -3, with K$_i$ values of 2,000, 60, and 600 nM, respectively (54).

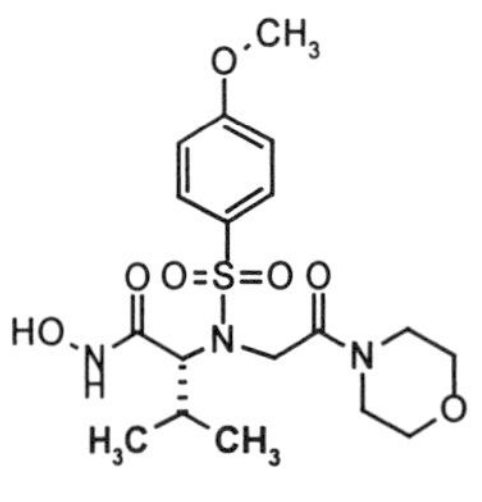

13 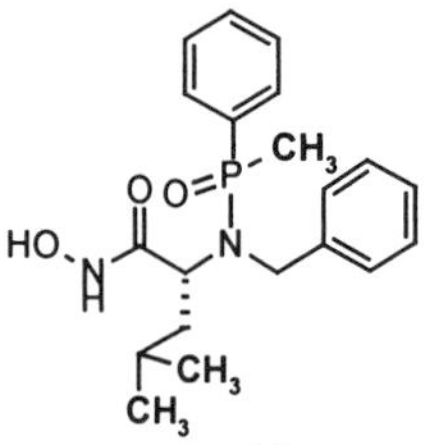   **14** 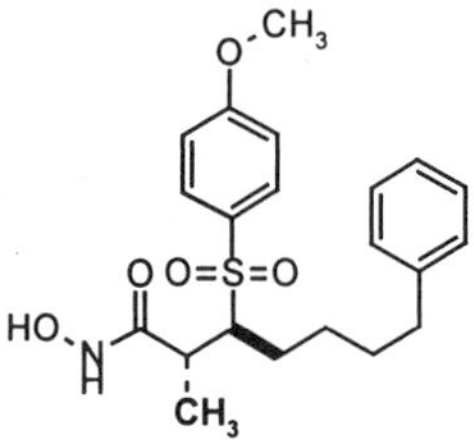   **15**

Reverse hydroxamic acids - Reverse hydroxamic acids are less popular due to either a perceived or actual *in vivo* instability. However, compound **16** (GW-3333) has IC$_{50}$ values of 20 and 42 nM for MMP-3 and TACE inhibition, respectively (55). This compound is orally active in rats having a bioavailability of 25 - 50 % with a t(1/2) of 2-4 h. Inhibitor **16** has also been shown to be orally active in a TNF-dependent peptidoglycan-polysaccharide reactivation arthritis model (55). A simpler and less peptidic structure, such as racemic formylhydroxylamine **17**, inhibits MMP-3 with an IC$_{50}$ of 9.1 nM (56).

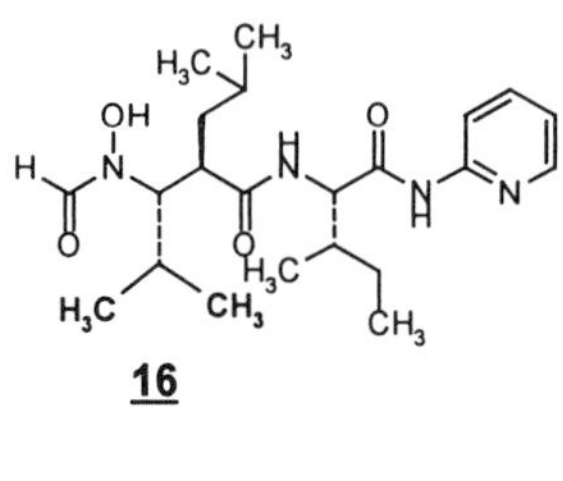

16 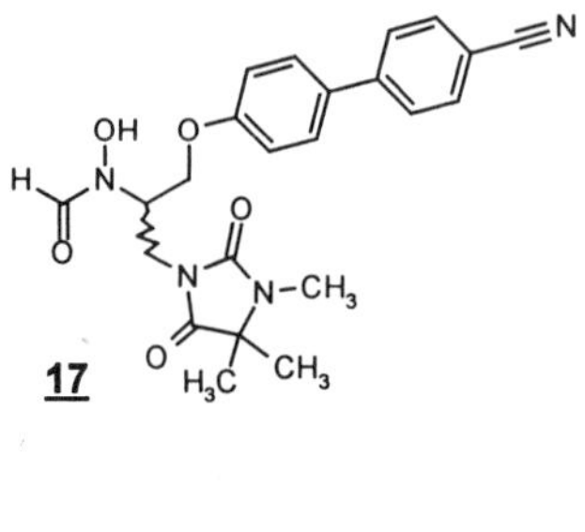   **17**

<u>Carboxylic acids</u> - As a result of the weaker zinc-binding ability of the carboxylic acid relative to the hydroxamic acid, the potency of the carboxylates correlates most closely with the interaction between long, lipophilic groups and the P_1' pocket (15). Detailed molecular modeling of acid <u>18</u>, which is a potent MMP-2 and -9 inhibitor, yields some insight (57). Like hydroxamic acids, the carboxylate MMP inhibitors originate from both succinate- and non-peptidic (sulfonamide) scaffolds. The orally active succinate-derived carboxylate <u>19</u> (AG-3433) (IC_{50} of 0.9 nM for MMP-2, 19 nM for MMP-3, 4545 nM for MMP-7, and 3.3 nM for MMP-13) contains a geminally-substituted lactone, replacing the more common P_2/P_3 structures (58). Currently <u>19</u> (AG-3433) is in Phase I clinical trials (Table 2) for cancer. The macrocycle strategy linking P_1' groups to P_2 has been revisited on the carboxylate MMP inhibitors, producing structures such as <u>20</u> (59). New biaryl ketones, like <u>21</u> (IC_{50} of 48 nM for MMP-3), have appeared (60).

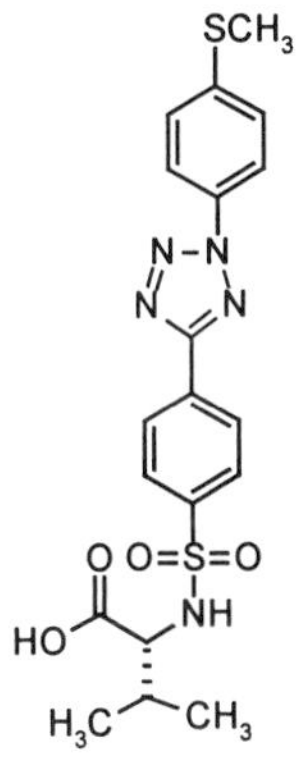

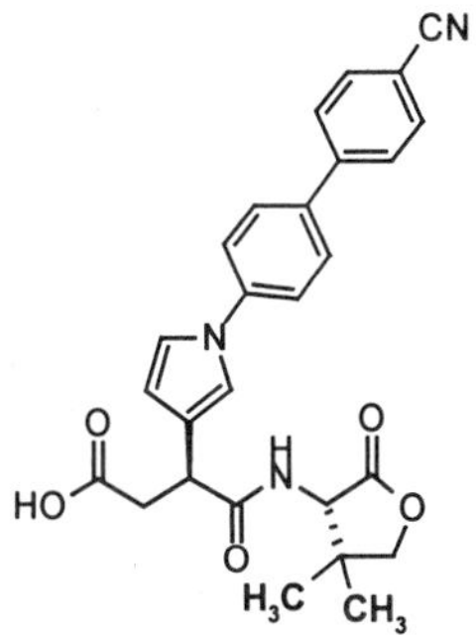

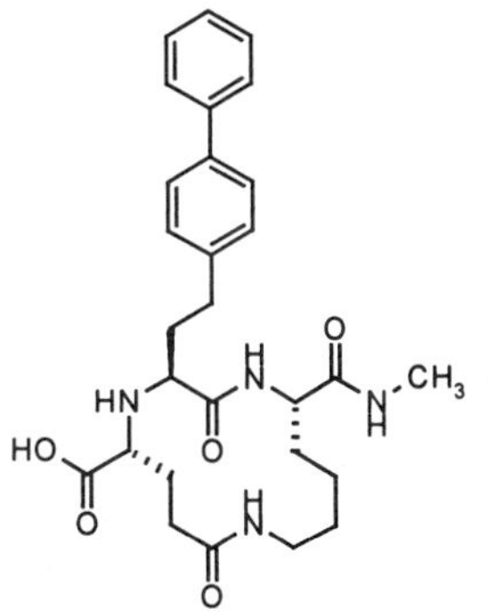

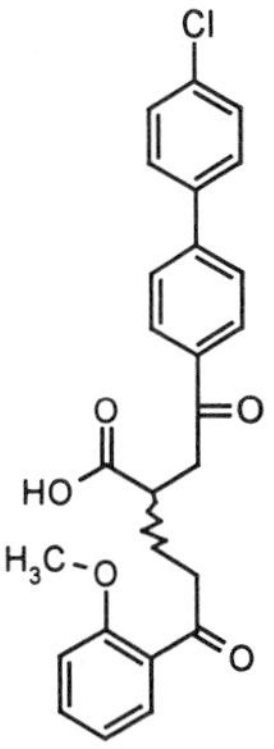

<u>18</u>   <u>19</u>   <u>20</u>   <u>21</u>

<u>Thiols</u> - The thiol MMP-inhibitors have been extensively reviewed (1, 2). Synthesis of β-hydroxythiol inhibitors, such as <u>22</u>, has recently appeared (61). Thiol <u>23</u> and the conformationally constrained *trans* diastereomer <u>24</u> (racemate) exhibit similar potencies toward MMP-13 with respective IC_{50} values of 0.5 and 0.4 nM (62, 63). The acyclic thiol <u>23</u> (IC_{50} =1500 nM) is a better inhibitor of MMP-1 than is <u>24</u> (IC_{50} >10,000 nM).

<u>22</u>   <u>23</u>   <u>24</u>

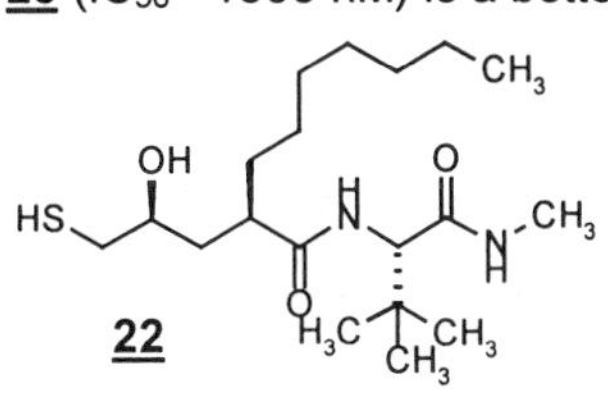

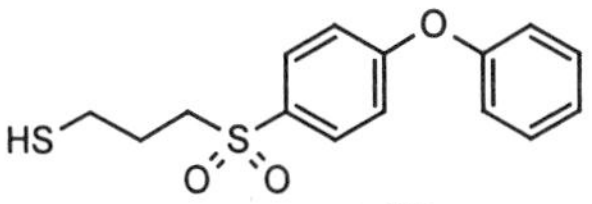

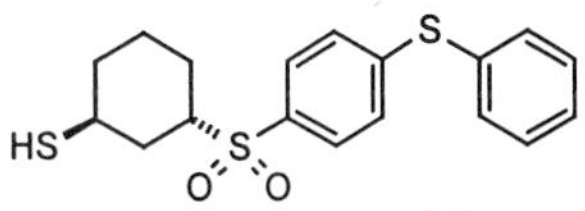

A combinatorial chemistry approach to synthesize diketopiperazines has identified thiol <u>25</u> with improved MMP-1 (IC_{50} = 47 nM) selectivity versus MMP-3 and -9 having IC_{50} values of >40000 and 1200 nM, respectively (64).

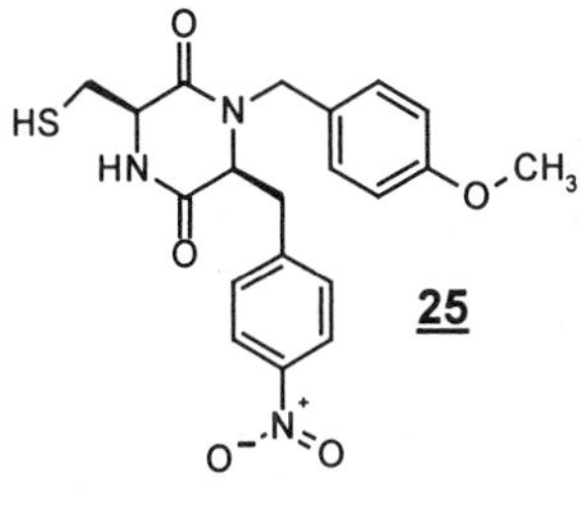

<u>25</u>

Phosphorus-based compounds - X-ray crystallographic evidence places the lipophilic 4-benzylphenyl moiety of phosphinate **26** in the S_2 pocket of MMP-1 (IC_{50} = 60 nM). However, compounds of this series have poor absorption and very low oral activity probably due to charge (65).

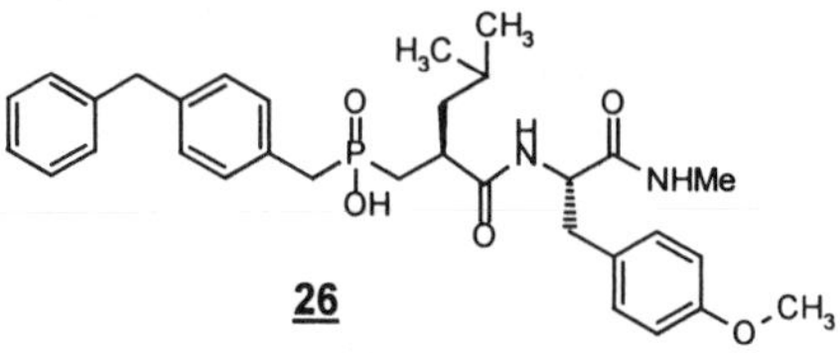

26

Novel or other zinc-binding groups - The X-ray crystal structure of thiadiazole **27** bound to MMP-3 (K_i = 18 nM) shows that the catalytic zinc coordinates with the thiocarbonyl, and the thiadiazole NH engages in potential hydrogen bonding to the catalytic Glu^{202} residue of MMP-3. More surprisingly, the remainder of the inhibitor binds to the unprimed pockets of the enzyme ("left-handed" binding) and partially accounts for the selectivity of this series for stromelysin (66). The pentafluoro-substitution present in the most potent members of the series is also noteworthy and believed to facilitate π-π stacking interactions with Tyr^{155} of MMP-3, a residue that does not occur in MMP-1.

Pyrimidine-2, 4, 6-triones, as exemplified by **28** (IC_{50} of 12 nM for MMP-8), represents a novel new class of MMP inhibitors (67). Besides their high *in vitro* efficacy, the optimized compounds have excellent bioavailability. In murine Lewis lung carcinomas, this series has high *in vivo* activity in blocking primary tumor and metastasis. The gelatinase inhibitor **28** shows comparable *in vivo* activity to the non-specific first generation compounds such as batimastat (**1**) and AG-3340 (**30**). These compounds do not show the sedative effects of known barbiturates.

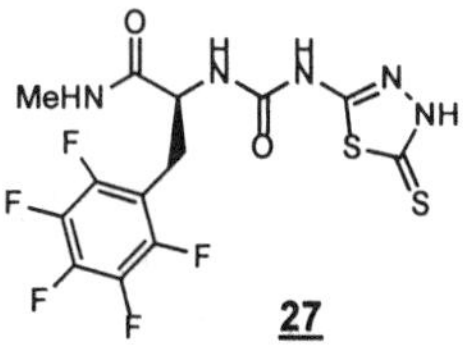

27 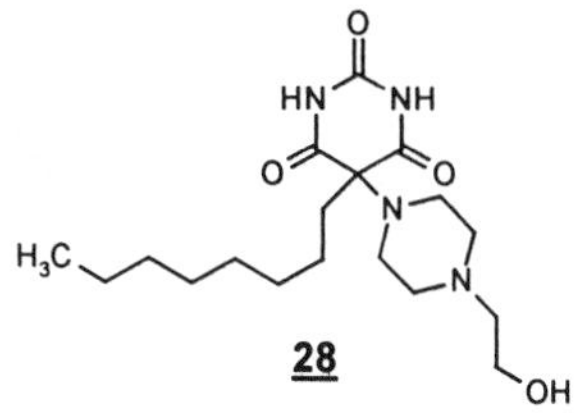   **28**

MMP INHIBITORS IN CLINICAL TRIALS FOR CANCER

Marimastat (BB-2516, **29**) is an orally active MMP inhibitor that is in Phase III clinical trials for various cancers. It was the first orally administered MMP inhibitor in pivotal clinical trials. Marimastat (**29**) is the subject of ten Phase III investigations world-wide. In August 1999, the results from a study of 145 patients with inoperable gastric cancer showed that the primary endpoint of the trial (improved survival) was not met. However, a survival benefit was seen at certain study groups, especially in long-term follow-up patients. The study enrolled 369 patients receiving 10 mg of **29** or placebo.

In a recent study of 239 patients with advanced pancreatic cancer, no difference was found between patients receiving **29** plus gemcitabine and those given gemcitabine alone (68). The primary endpoint was survival difference between patients receiving the combination of the two drugs and those given gemcitabine alone. At the end point, 90 % mortality was recorded in one of the treatment groups. The results showed that the survival curves were almost identical, and there was virtually no difference between the two groups at any time point. The median survival was 164 days in the gemcitabine alone group and 165.5 days in the combination group. In addition, there were no significant benefits in secondary endpoints or in safety or quality of life between the two groups. As in the previous trial, there was some indication of benefit with **29** in patients with less extensive disease.

A third study with **29** is focused on patients with the least aggressive cancers. Two of the four trials due to be reported in 2000 involve patients with small-cell lung cancer. These patients have already responded to chemotherapy, so their tumor burden has been reduced to some extent. The other two studies are for patients with glioblastoma and ovarian cancer, both of which are aggressive cancers with high tumor burdens.

Prinomastat (AG-3340, **30**) is being developed for the treatment of cancer and age-related macular degeneration. In May of 1999, **30** entered Phase III trials for lung and prostate cancers. The recommended dose for these trials was 5 to 25 mg, b.i.d. Following the demonstration of an enhanced efficacy of chemotherapy when supplemented with **30**, pilot combination studies and double-blinded, placebo-controlled Phase III trials in 700 patients are in progress for the treatment of non-small cell lung cancer (NSCLC) or advanced hormone-refractory prostate cancer (69). A Phase III clinical trial of **30** in combination with chemotherapy in patients with advanced NSCLC has been initiated. This trial is designed to evaluate the safety and efficacy of **30** as part of first-line therapy in combination with chemotherapy. This trial advances the ongoing clinical development of **30** in combination with the anticancer drugs paclitaxel/carboplatin for the treatment of advanced NSCLC and with mitoxantrone/prednisone for the treatment of hormone-refractory prostate cancer. The primary objective of this study is to compare time of overall survival between patients receiving **30** or placebo in combination with gemcitabine and cisplatin (70).

Prinomastat (**30**) is well tolerated at doses of up to 100 mg, **30** b.i.d., p.o., are generally well tolerated. In most studies, no dose-limiting toxicities occur during the first four weeks of treatment. Musculoskeletal and joint-related events can occur after four weeks of treatment with higher doses. These events include joint stiffness, swollen joints and, in a few patients, some limits on the mobility of certain joints. The side effects typically begin in the shoulders, knees or hands, with involvement of additional joints in a dose- and time-dependent manner. These side effects are reversible and can be managed by treatment rests for two to four weeks and subsequent dose reduction. In some studies, fatigue, believed to be attributable to prinomastat administration, has been observed.

The carboxylate BAY 12-9566 (**32**) was being developed as a potential treatment for preventing metastasis in several types of neoplasm as well as for osteoarthritis. In September 1999, all trials for all indications with **32** were halted following recommendation from an independent safety monitoring board that the Phase III NSCLC trial be stopped. The NSCLC trial was terminated following the discovery that **32** performed worse than placebo. The compound had also reached Phase III clinical trials in pancreatic cancer, Phase II trials in several other types of cancer and in osteoarthritis. This inhibitor was also being developed the for the potential treatment of rheumatoid arthritis, for which early Phase II studies had been initiated.

The thiol compound D-2163 (BMS-275,291, **34**) is an inhibitor of a broad range of MMPs known to be associated with the growth and spread of tumors, but does not inhibit metalloproteinase-mediated shedding events, which may be involved in the side effects associated with the first generation MMP inhibitors (71). In preclinical studies **34** did not exhibit any deleterious effects on tendons or joints. The inhibitor **34** demonstrates ten-fold increase in systemic exposure for a given dose when compared with **24** (qv), as well as having a superior selectivity profile. No histopathological effects were observed in a marmoset model of joint pain and tendinitis. In a single-dose Phase I study (25 mg to 900 mg), **34** demonstrated excellent blood levels. Phase II patient studies conducted by BMS started in the first quarter of 1999. A similar compound, D-1927 (**35**), is being developed for inflammation.

Several other MMP inhibitors are also in various stages of clinical development for cancer and arthritis (Table 2). As of March 31, 2000 Trocade (Ro 32-3555, **36**), which was in Phase III clinical trials for arthritis, has been withdrawn from all future clinical studies.

Table 2. MMP Inhibitors in Clinical Trials for Cancer and Arthritis

Structure	Comments	Structure	Comments
1	British Biotech Batimastat BB-94 Cancer Phase II	**33**	Agouron AG-3433 Cancer Phase I
29	British Biotech Matimastat BB-2516 Cancer Phase III	**34**	Chiroscience D-2163 Cancer Phase I
30	Agouron Prinomastat AG-3340 Cancer Phase III	**35**	Chiroscience BMS D-1927 BMS-275291 Inflammation Phase II
31	Roche Biosciences RS-130,830 Arthritis Phase II	**36**	Roche Trocade Ro 32-3555 Arthritis Phase III Withdrawn
32	Bayer Bay 12-9566 Cancer Phase III Arthritis Phase II Withdrawn		

References

1. M. Whittaker, C.D. Floyd, P. Brown, and A.J.H. Gearing, Chem. Rev., 99, 2735 (1999).
2. M.R. Michaelides and M.L. Curtin, Curr. Pharm. Design, 5, 787 (1999).
3. A. Zask, J.I. Levin, L.M. Killar, and J.S. Skotnicki, Curr. Pharm. Des., 2, 624 (1996).
4. R.P. Beckett and M. Whittaker, Exp. Opin. Ther. Patents, 8, 259 (1998).
5. W.K. Hagmann, M.W. Lark, and J.W. Becker, in "Ann. Rep. Med. Chem." Vol. 31, J.A. Bristol, Ed., Academic Press, New York, 1996, p. 231.
6. J.B. Summers and S.K. Davidsen, in "Ann. Rep. Med. Chem." Vol. 33, J.A. Bristol, Ed., Academic Press, New York, 1998, p. 131.
7. R.P. Beckett, A.H. Davidson, A.H. Drummond, P. Huxley, and M. Whittaker, Drug Dis. Today, 1, 16 (1996).
8. A.H. Davidson, A.H. Drummond, W.A. Galloway, and M. Whittaker, Chem. Ind., 7, 258 (1997).
9. A.D. White, T.M.A. Bocan, P.A. Boxer, J.T. Peterson, and D. Schrier, Curr. Pharm. Des., 3, 45 (1997).
10. "Annals of the New York Academy of Sciences", Vol 878, R.A. Greenwald, S. Zucker, and L.M. Golub, Eds., The New Academy of Sciences, New York, NY, 1999.
11. P.D. Brown and M. Whittaker, in "Antiangiogenic Agents Cancer Ther.", 205 (1999).
12. L.L. Johnson, R. Dyer, and D.J. Hupe, Curr. Opin. Chem. Biol., 2, 466 (1998).

13. N. Borkakoti, Curr. Opin. Drug Discov. Develop., 2, 449 (1999).
14. N. Borkakoti, Progress Biophys. Mol. Biol., 70, 73 (1998).
15. R.E. Babine and S.L. Bender, Chem. Rev., 97, 1359 (1997).
16. D.E. Levy and A.M. Ezrin, Emerging Drugs, 2, 205 (1997).
17. J. F. Woessner, FASEB J., 5, 2145 (1991).
18. K. Sugita, Invest. Drugs, 2, 327 (1999).
19. J.M. Ray and W.G. Stetler-Stevenson, Exp. Opin. Invest. Drugs, 5, 323 (1996).
20. H.T. Zhang and A.L. Harris, Exp. Opin. Invest. Drugs, 7, 1629 (1998).
21. G.Velasco, A.M. Pendas, A. Fueyo, V. Knauper, G. Murphy, and C. Lopez-Ortin, J. Biol. Chem., 274, 4570 (1999).
22. H. Autio-Harmainen, T. Karttunen, T. Hurskainen, M. Höyhtyä, A. Kauppila, and K. Tryggvason, Lab. Invest., 69, 312 (1993).
23. C. Pyke, E. Ralfkiær, K. Tryggvason, and K. Dano, Am. J. Pathol., 142. 359 (1993).
24. S.T. Gray, R.J. Wilkins, and K. Yun, Am. J. Pathol., 141, 301 (1992).
25. K.J. Heppner, L.M. Matrisian, R.A. Jensen, and W.H. Rodgers, Am. J. Pathol., 149, 273 (1996).
26. C. Monteagudo, M.J. Merino, J. San-Juan, L.A. Liotta, and W.G. Stetler-Stevenson, Am. J. Pathol., 136, 585 (1990).
27. J.S. Rao, P.A. Steck, S. Mohanam, W.G. Stetler-Stevenson, L.A. Liotta, and R.Sawaya, Cancer Res., 53, 2208 (1993).
28. B. Davies, J. Waxman, H. Wasan, P. Abel, G. Williams, T. Krausz, D. Neal, D. Thomas, A. Hanby, and F. Balkwill, Cancer Res., 53, 5365 (1993).
29. W.C. Powell, J. D. Knox, M. Navre, T.M. Grogan, J. Kittelson, R.B. Nagle, and G.T. Bowden, Cancer Res., 53, 417 (1993).
30. T. Itoh, M. Tanioka, H. Yoshida, T. Yoshioka, H. Nishimoto, and S. Itohara, Cancer Res., 58, 1048 (1998).
31. R.J. Cherney, L. Wang, D.T. Meyer, C.-B. Xue, E.C. Arner, R.A. Copeland, M.B. Covington, K.D. Hardman, Z.R. Wasserman, B.D. Jaffee, and C.P. Decicco, Bioorg. Med. Chem. Lett., 9, 1279 (1999).
32. J.J.-W. Duan, L. Chen, C.-B. Xue, Z.R. Wasserman, K.D. Hardman, M.B. Covington, R.R. Copeland, E.C. Arner, and C.P. Decicco, Bioorg. Med. Chem. Lett., 9, 1453 (1999).
33. S.K. Davidsen, A.S. Florjancic, G.S. Sheppard, J.R. Giesier, L. Xu, Y. Guo, M.L. Curtin, M.R. Michaelides, C.K. Wada, and J.H. Holms, U.S. Patent 5,985,911 (1999).
34. M.J. Broadhurst, P.A. Borwn, G. Lawton, N. Ballantyne, N, Borkakoti, K.M.K. Bottomley, M.I. Cooper, A.J. Eatherton, I.R. Kilford, P.J. Malsher, J.S. Nixon, E.J. Lewis, B.M. Sutton, and K. Wilson, Bioorg. Med. Chem. Lett., 7, 2299 (1997).
35. K.M. Bottomley, W.H. Johnson, and D.S. Walter, J. Enz. Inhib., 13, 79, (1998).
36. R.P. Beckett, F.M. Martin, A. Miller, R.S. Todd, and M. Whittaker, PCT Patent Application, WO9817655 (1998); Chem Abstr. 128, 308398 (1999).
37. M. Alpegiani, P. Bissolino, F. Absate, E. Perrone, R. Corigli, and D. Jabes, PCT Patent Application, WO9902510 (1999); Chem Abstr., 130, 139360 (1999).
38. L.J. MacPherson, E.K. Bayburt, M.P. Capparelli, B.J. Carroll, R. Goldstein, M.R. Justice, L. Zhu, S. Hu, R.A. Melton, L. Fryer, R.L. Goldberg, J.R. Doughty, S. Spirito, V. Blancuzzi, D. Wilson, E. O'Byrne, V. Ganu, and D. Parker, J. Med. Chem., 40, 2525 (1997).
39. A.Y. Jeng, M. Chou, and D.T. Parker, Bioorg. Med. Chem. Lett., 8, 897 (1998).
40. N.C. Gonnella, Y.-C. Li, and X. Zhang, Bioorg. Med. Chem., 5, 2193 (1997).
41. Y.-C. Li, X. Zhang, R. Melton, V. Ganu, and N.C. Gonnella, Biochemistry, 37, 14048 (1998).
42. J.D. Albright, E.G. Delos Santos, and X. Du, PCT Patent Application, WO9937625 (1999)
43. S.L. Bender, 214th National Meeting of the American Chemical Society, Las Vegas, NV, Sep. 7-11, 1997, MEDI 108.
44. S.L. Bender and A.A. Melwyn, U.S. Patent 5,985,900 (1999).
45. M. Cheng, B. De, N.G. Almstead, S. Pikul, M.E. Dowty, C.R. Dietsch, C.M. Dunaway, F. Gu, L.C. Hsieh, M.J. Janusz, Y.O. Taiwo, and M.G. Natchus, J. Med. Chem., 42, 5426 (1999).
46. J.I. Levin, M.T. Du, A.M. Venkatesan, F.C. Nelson, A. Zask, and Y. Gu, PCT Patent Application, WO9816503 (1998).
47. J.I. Levin and F.C. Nelson, U.S. Patent 5,962,481 (1999).
48. A.M. Venkatesan, G.T. Grosu, J.M. Davis, J.L. Baker, and J.I. Levin, PCT Patent Application, WO9942436 (1999).
49. J.I. Levin, A. Zask, J.D. Albright, and X. Du, PCT Patent Application, WO9918076 (1999).
50. R.P. Robinson and J.P. Rizzi, U.S. Patent 5,994,351 (1999).
51. R.M. Heintz, D.P. Getman, J.J. McDonald, G.A. DeCrescenzo, S.C. Howard, and S.Z. Abbas, PCT Patent Application, WO9839329 (1998).
52. D.P. Getman, D.P. Becker, T.E. Barta, C.I. Villamil, S.L. Hockerman, L.J. Bedell, J.N. Freskos, R.M. Heintz, and J.J. McDonald, PCT Patent Application, WO9839313 (1998).

53. S. Pikul, D.K.L. McDow, N.G. Almstead, B. De, M.G. Natchus, M.V. Anastasio, S.J. McPhail, C.E. Snider, Y.O. Taiwo, L. Chen, C.M. Dunaway, F. Gu, and G.E. Mieling, J. Med. Chem., $\underline{42}$, 87 (1999).

54. R.D. Groneberg, C.J. Burns, M.M. Morrissette, J.W. Ullrich, R.L. Morris, S. Darnbrough, S.W. Djuric, S.M. Condon, G.M. McGeehan, R. Labaudiniere, K. Neuenschwander, A.C. Scotese, and J.A. Kline, J. Med. Chem. $\underline{42}$, 541 (1999).

55. T.J. Carty, L.L. LoPresti-Morrow, P.G. Mitchell, P.A. McNiff, and K.F. McClure, Inflamm. Res., $\underline{48}$, 229 (1999).

56. M.L. Curtin, S.K. Davidsen, J.F. Delaria, A.S. Floriancic, J. Giesler, G. Gong, Y. Guo, H.R. Heyman, J.H. Holms, M.R. Michaelides, D.H. Steinman, C.K. Wada, and L. Xu, PCT Patent Application, WO9906361 (1999).

57. R. Kiyama, Y. Tamura, F. Watanabe, H. Tsuzuki, M. Ohtani, and M. Yodo, J. Med. Chem., $\underline{42}$, 1723 (1999).

58. J.G. Deal, S.L. Bender, W.K.M. Chong, R.K. Duvadie, A.M. Caldwell, L. Li, M.A. McTigue, J.A. Wickersham, K. Appelt, D.R. Shalinsky, R.D. Daniels, C.R. McDermott, J. Brekken, S.A. Margosiak, R.A. Kumpf, M.A. Abreo, B.J. Burke, J.A. Register, E.F. Dagostino, D.L. Vanderpool, and O. Santos, 217th National Meeting of the Americamn Chemical Society, Anaheim, CA, March 21-25, 1999, MEDI 197.

59. R.J. Cherney, L. Wang, D.T. Meyer, C.B. Xue, Z.R. Wasserman, K.D. Hardman, P.K. Welch, M.B. Covington, R.A. Copeland, E.C. Arner, W.F. DeGrado, and C.P. Decicco, J. Med. Chem. $\underline{41}$, 1749 (1999).

60. W.J. Scott, M.A. Popp, and D.S. Hartsough, U.S. Patent 5,932,763 (1999).

61. K. Paulvannan and T. Chen, Synlett, $\underline{9}$, 1371 (1999).

62. J.N. Freskos, B.V. Mischke, G.A. DeCrescenzo, R. Heintz, D.P. Getman, S.C. Howard, N.N. Kishore, J.J. McDonald, G.E. Munie, S. Rangwala, C.A. Swearingen, C. Voliva, and D.J. Welsch, Bioorg. Med. Chem. Lett., $\underline{9}$, 943 (1999).

63. J.N. Freskos, J.J. McDonald, B.V. Mischke, P.B. Mullins, H.-S. Shieh, R.A. Stegeman, and A.M. Stevens, Bioorg. Med. Chem. Lett., $\underline{9}$, 1757 (1999).

64. A.K. Szardenings, V. Antonenko, D.A. Campbell, N. DeFrancisco, S. Ida, L. Shi, N. Sharkov, D. Tien, Y. Wang, and M. Navre, J. Med. Chem., $\underline{42}$, 1348 (1999).

65. L.A. Reiter, J.P. Rizzi, J. Pandit, M.J. Lasut, S.M. McGahee, V.D. Parikh, J.F. Blake, D.E. Danley, E.R. Laird, A. Lopez-Anaya, L.L. Lopresti-Morrow, M.N. Mansour, G.J. Martinelli, P.G. Mitchell, B.S. Owens, T.A. Pauly, L.M. Reeves, G.K. Schulte, and S.A. Yocum, Bioorg. Med. Chem. Lett., $\underline{9}$, 127 (1999).

66. E.J. Jacobsen, M.A. Mitchell, S.K. Henges, K.L. Belonga, L.L. Skaletzky, L.S. Stelzer, T.J. Lindberg, E.L. Fritzen, H.J. Schostarez, T.J. O'Sullivan, L.L. Maggiora, C.W. Stuchly, A.L. Laborde, M.F. Kubicek, R.A. Poorman, J.M. Beck, H.R. Millar, G.L. Petzold, P.S. Scott, S.E. Truesdell, T.L. Wallace, J.W. Wilks, C. Fisher, L.V. Goodman, P.S. Kaytes, S.R. Ledbetter, E.A. Powers, G. Vogeli, J.E. Mott, C.M. Trepod, D.J. Staples, E.T. Baldwin and B.C. Finzel, J. Med. Chem., $\underline{42}$, 1525 (1999).

67. A. Oliva, G. De Cillis, F. Grams, G. Zimmermann, E. Menta, and H.-W. Krell, PCT Patent Application, WO 9858925 (1998).

68. SCRIP, Jan. 26 (2000).

69. R. Ogden, International Congress of the Metastasis Society, San Diego (1998).

70. D.R. Shalinsky, J. Brekken, H. Zou, L.A.Bloom, C.D. McDermott, S. Zook, N.M. Varki, and K. Appelt, Clinical Cancer Res., $\underline{57}$, 1905 (1999).

71. J.G. Montana, Lecture at the Society for Medicines Research meeting on Trends in Medicinal Chemistry, London, Dec. 3 (1998).

Chapter 16. Recent Developments in Antiretroviral Therapies

Tomas Cihlar and Norbert Bischofberger
Gilead Sciences, Inc.
Foster City, CA 94404

Introduction – Clinical use of novel potent inhibitors of human immunodeficiency virus (HIV) in the last few years marked a new era in the treatment of HIV infection and for the first time in the history of the disease reverted the AIDS-related death rate. To date, fourteen antiretroviral agents belonging to three distinct classes (nucleoside and non-nucleoside reverse transcriptase inhibitors, and protease inhibitors) have been licensed in the U.S. Combination therapy of at least three potent drugs became the standard of care. The treatments, however, suffer from limitations such as resistance development, frequent cross-resistance within classes, long-term toxicity, difficult dosing regimens, and adverse drug-drug interactions, which all may lead to long-term treatment failure. This chapter reviews the most recent progress in the discovery and development of new antiretroviral agents. Because of an enormous effort occurring both in the academia and pharmaceutical industry, this field is evolving more rapidly than ever.

NUCLEOSIDE REVERSE TRANSCRIPTASE INHIBITORS (NRTIs)

Currently, six nucleosides are licensed in the U.S. for the treatment of HIV infection: Zidovudine (AZT), zalcitabine (ddC), didanosine (ddI), stavudine (d4T), lamivudine (3TC) and abacavir. Abacavir is the most recently licensed nucleoside (1). The anti-HIV activity, resistance profile, metabolism, enzymology, and cytotoxicity was published (2). Abacavir shows a unique intracellular activation; it is converted to the monophosphate and subsequently 6-deaminated to guanosine nucleotide analog (3). Clinical results of abacavir in NRTI-experienced patients were recently presented (4).

Tenofovir DF (<u>1</u>) and FTC (emtricitabine, <u>2</u>) are currently in Phase III clinical evaluation. <u>1</u> is an orally bioavailable prodrug of tenofovir (PMPA) (5). It shows activity against NRTI-resistant strains of HIV-1 (6). In treatment-experienced patients, <u>1</u> showed a dose-dependent antiviral effect which was sustained for 48 weeks without any sign of dose related toxicity (7). <u>2</u> is a fluorinated analog of 3TC. It has shown potent activity with once daily administration (8).

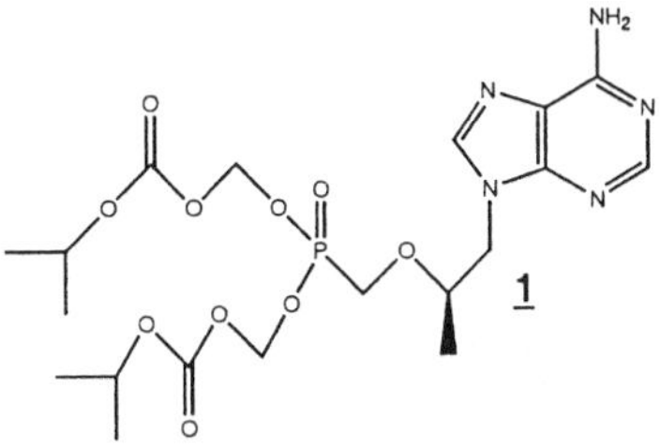

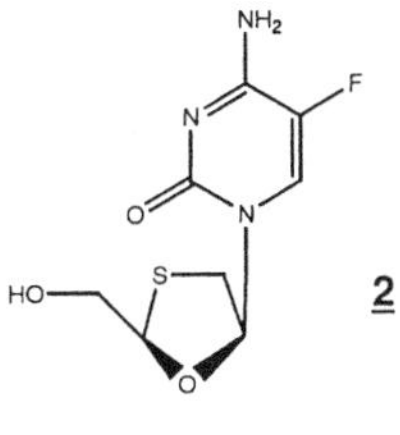

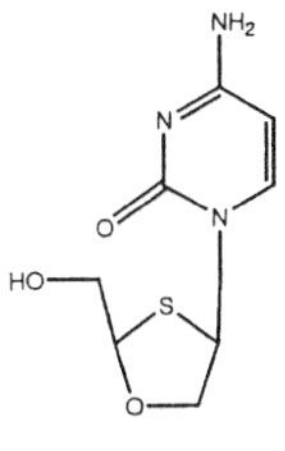

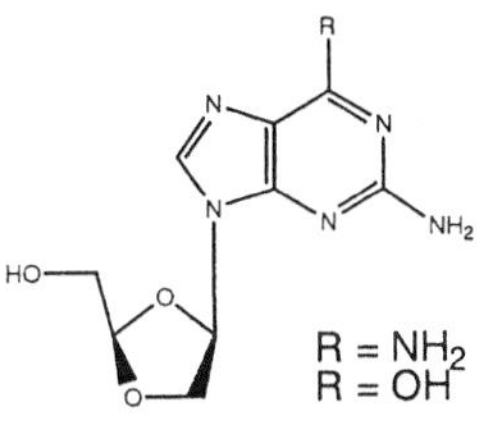

R = NH₂ **4**
R = OH **5**

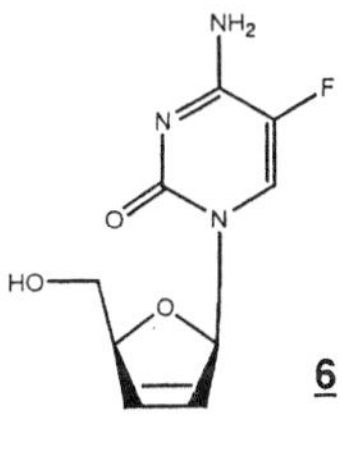

dOTC (**3**), 2'-β-fluoro-2',3'-dideoxyadenosine (F-ddA), and DAPD (**4**) have been evaluated in Phase I/II clinical studies. Compound **3** is a racemic mixture and its anti-HIV activity, intracellular metabolism, and pharmacokinetics (PK) have been published (9). The rationale for utilizing the racemate is the favorable cross-resistance profile of the two enantiomers. In selection experiments, the (+)enantiomer resulted in rapid evolution of the M184 mutation (also selected by 3TC and FTC). This mutant HIV, however, remains sensitive to the (-)enantiomer (10). In clinical studies, **3** showed potent antiretroviral activity (11). The development of **3** was recently discontinued due to toxicities observed in chronic toxicology studies. F-ddA is a chemically stable analog of ddl. In a Phase I/II study, F-ddA showed oral bioavailability of 65-75% and evidence of anti-HIV activity (12). F-ddA was also found to have anti-HIV activity in NRTI-experienced patients (13). The clinical development of FddA has been discontinued due to toxicities. DXG (**5**) and its prodrug **4** have potent anti-HIV activity *in vitro* (14). 3TC- and ddl-resistant viruses remain sensitive to **4** and **5** (15). Compound **5** also retains its activity against HIV-1 strains with multiple NRTI- and NNRTI-associated mutations, including Q151M and 68/69 insertion (16). Upon oral administration to humans, **4** is rapidly absorbed and converted to **5**. The terminal $t_{1/2}$ of **5** was ~7 hrs (17). Compound **4** showed good tolerability and evidence of antiviral activity in a Phase I/II 14-day monotherapy study (18).

Compound **6** (D4FC) exhibits antiviral activity against both hepatitis B virus (HBV) and HIV. Its oral bioavailability in monkeys averages ~40% and it shows good CNS penetration (19). Resistance development *in vitro* occurs slowly and results in uncommon mutations (20).

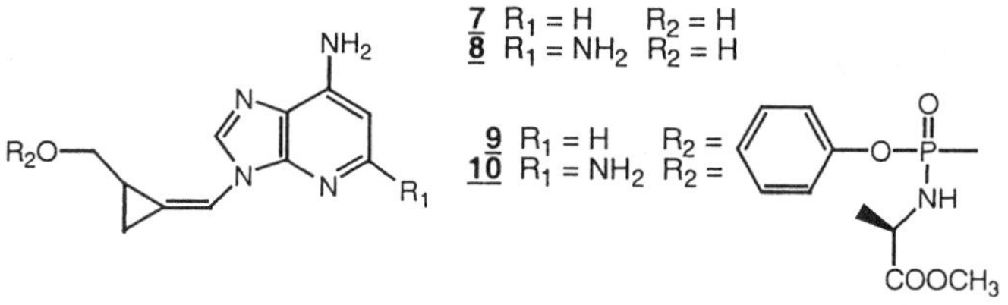

7 $R_1 = H$ $R_2 = H$
8 $R_1 = NH_2$ $R_2 = H$

9 $R_1 = H$ $R_2 =$
10 $R_1 = NH_2$ $R_2 =$

The methylene cyclopropane nucleosides **7** and **8** have shown anti-HIV activity *in vitro*. Conversion of the nucleosides to their phosphoramidate prodrugs **9** and **10** greatly increased their potency (21). In the *in vitro* selection experiments, resistance to **9** and **10** occurred due to the selection of M184I mutation in RT (22).

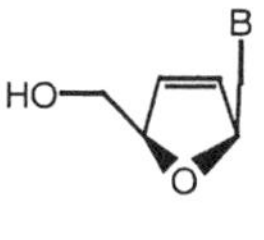

B = Cytosine   **11**
B = 5-Fluorocytosine **12**

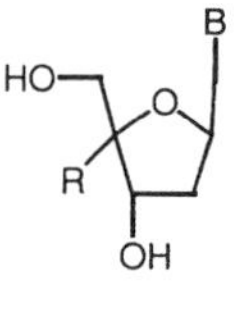

B = Thymine
R = ethyl   **13**
=ethenyl   **14**
=ethynyl   **15**

A number of fluorinated L-nucleosides were synthesized and evaluated (23, 24). The cytosine analogs **11** and **12** (L-FD4C) showed good *in vitro* anti-HIV and HBV activity with no significant cytotoxicity. A variety of 4'-branched thymidines have been described. Compounds **13**, **14**, and **15** were found to have anti-HIV and anti-HSV activity with no significant toxicity (24a).

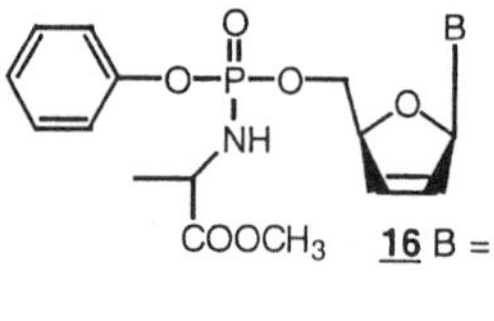

16 B = Thymine 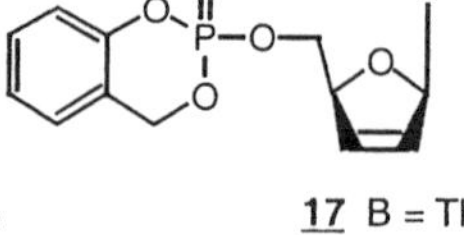   **17** B = Thymine

In order to bypass the first phosphorylation step required in the activation of nucleosides, a number of phosphate and phosphoramidate prodrugs are being evaluated. Compound **16** was found to be significantly more active *in vitro* than the parent d4T and delivered intact d4T-monophosphate intracellularly (25, 26). Similarly, tri-ester **17** exhibited anti-HIV activity and was active in thymidine kinase deficient cells (27).

NON-NUCLEOSIDE REVERSE TRANSCRIPTASE INHIBITORS (NNRTIs)

Two reviews on NNRTIs have recently been published (28, 29). The currently approved NNRTIs are nevirapine, delavirdine, and efavirenz. Emirivine (coactinon, MKC-442, **18**), GW-420867X (**19**), calanolide-A (**20**) and AG-1549 (S-1153, **21**) are currently being evaluated in clinical studies.

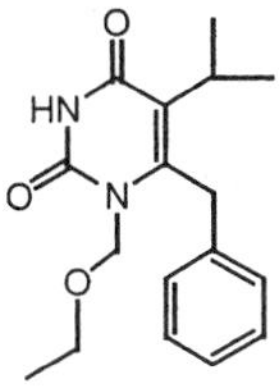

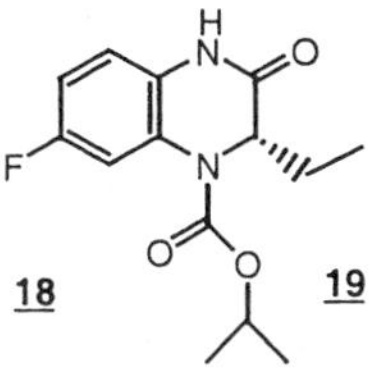

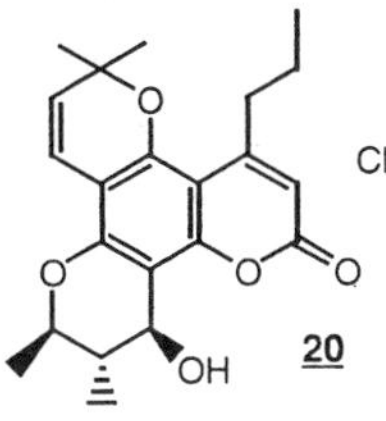

Compound **18** showed linear pharmacokinetics in rats and monkeys and was orally absorbed. In chronic toxicity studies, **18** caused reversible vacuolation of kidney tubular epithelial cells (30). In clinical studies, **18** in combination with d4T and ddI showed rapid and significant suppression of HIV RNA (31). Genotypic analysis indicated that the primary mutation selected by **18** is K103N (32). Compound **19** was selected from a series of quinoxalines based on its potency, which was retained in the presence of serum and its delayed resistance development *in vitro* (33, 34). Short-term Phase I/II studies showed that **19** was well tolerated with evidence of anti-HIV activity (35). Another quinoxaline NNRTI (HBY-097, **22**) (36) was previously evaluated in Phase I/II clinical studies due to its unique *in vitro* resistance profile. It was found that *in vitro,* **22** selected for mutations at positions 179 and 190, whereas *in vivo* the K103N mutation occurred (37, 38). Compound **22** was also found to alter the pharmacokinetics of indinavir (39). Calanolide-A (**20**) is an NNRTI with *in vitro* anti HIV activity in the 30–80 nM range (40). Preliminary results from a Phase I/II study were recently reported (41). Compound **20** showed a dose-dependent antiviral effect and a long half-life (20h), and was well tolerated. A number of analogs of **20** were evaluated for their *in vitro* antiviral HIV activity and resistance development (42). AG-1549 (**21**) is a potent NNRTI with a unique resistance profile and has shown good tolerability and antiviral activity in short-term Phase I/II study (43, 44). Similarly, efavirenz analogs with improved resistance profiles were reported (45). In particular, the 6-chloro (DMP961) and the 5,6-difluoroquinazoline (DMP963) analogs have activity against K103N mutant HIV and are currently being evaluated in Phase I clinical studies (46, 47).

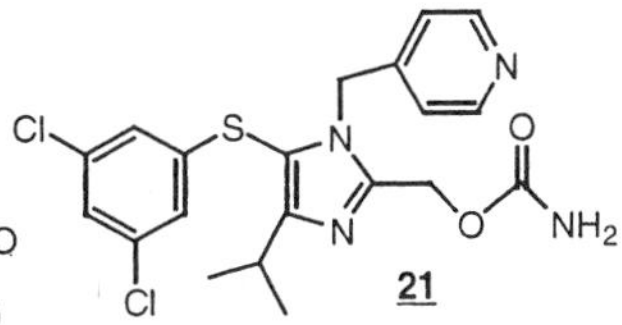

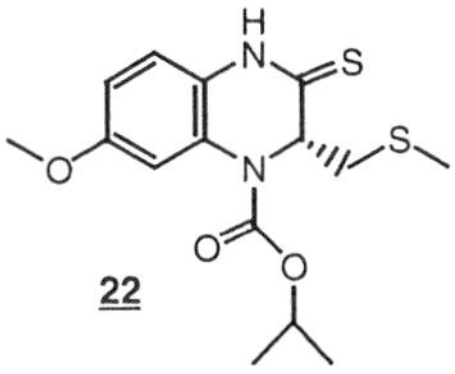

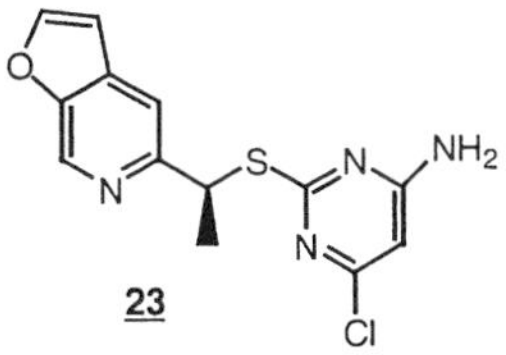

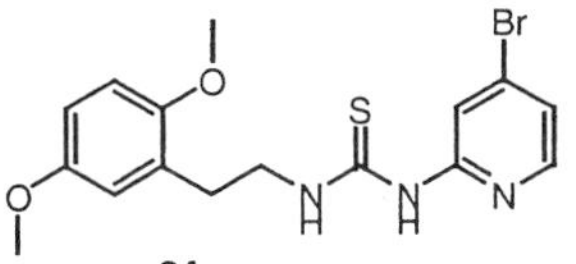

A third generation NNRTI PNU-142721 (**23**) shows good potency and activity against HIV isolates containing multiple NNRTI mutations (48, 49). A number of publications have appeared on phenethylthiazolylthiourea (PETT) derivatives, and bromopyridyl thioureas (50-53). Compound **24** showed potent anti-HIV activity *in vitro* and a good pharmacokinetic profile (54-56). Other NNRTIs described are thiadiazole derivatives (57), alkenyldiarylmethanes (58, 59), pyrrolobenzoxazepinones (60), dihydroalkoxybenzyloxopryimidines (61), thiadiazines (62), pyridoindoles (63) and imidazoles (64). Recently identified SJ-3366 shows activity against both HIV-1 and HIV-2 indicating a potential dual mechanism of action (NNRTI and inhibitor of virus attachment) (65).

HIV PROTEASE INHIBITORS (PIs)

Several reviews summarize recent progress in the discovery and development of PIs (66,67). Currently, five drugs of this class are clinically used for the treatment of HIV infection with amprenavir being the most recently approved PI (68). Cross-resistance, drug-drug interactions due to inhibition of cytochrome P450 (CYP450), and dosing regimens with high pill burden are among the most frequently mentioned limitations of the available Pis (69, 70). Although the resistance profile of amprenavir appears to be unique to some extent (71), it still acts as an inhibitor of CYP450 (72). A phosphate prodrug of amprenavir with improved solubility, which may reduce the pill burden, is under preclinical evaluation (73).

Compound **25** (ABT-378) is in Phase III evaluation (74). Its rational design,

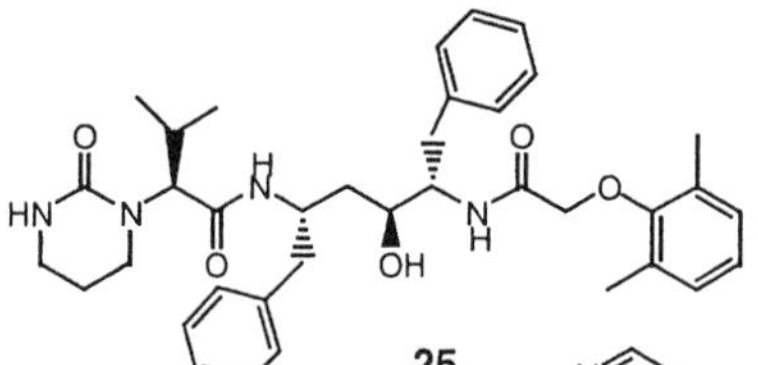

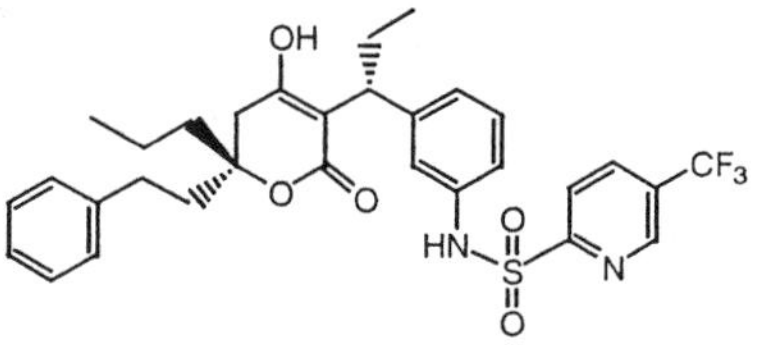

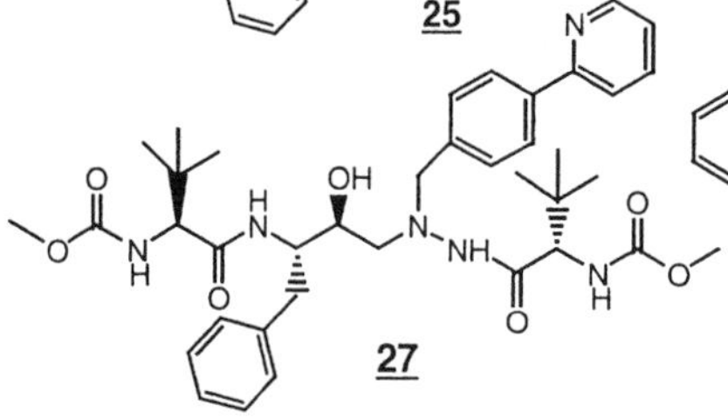

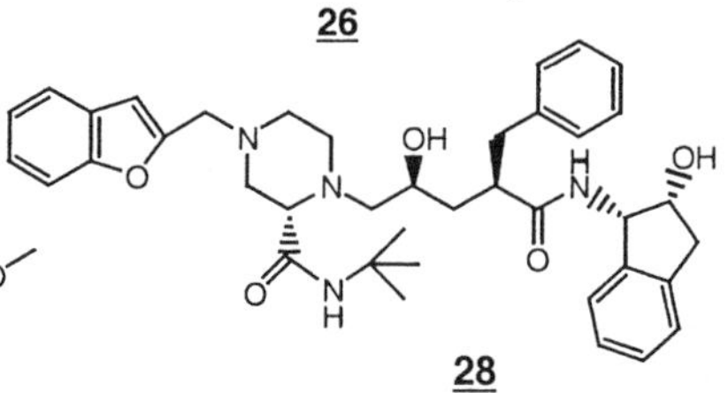

which relied on novel side substitutions in the backbone of ritonavir, resulted in 5- to 10-fold increased potency and limited protein binding compared to ritonavir (75). **25** is co-formulated with ritonavir to inhibit its metabolism by CYP450 (76). This favorable drug-drug interaction results in trough plasma levels of **25** exceeding its EC$_{90}$ by >30-fold (74). Tipranavir (PNU-140690, **26**), a non-peptidic PI with 5,6-dihydropyran-2-one sulfonamide scaffold, is in Phase II clinical studies (77). Similar to **25**, **26** is also efficiently metabolized by CYP450 (78) and its co-administration with ritonavir is clinically investigated. Compound **26** retains full activity against a broad range of PI-resistant HIV variants , presumably because of the flexibility of dihydropyrone ring, which allows to accommodate various amino acid changes in resistant PR (79). Aza-dipeptide analogue **27** (BMS-232632, CGP73547) (80) appears to have a prolonged half-life *in vivo* and is currently being developed as the first once-daily PI (81). Short-term monotherapy with **27** in treatment-naïve HIV patients showed similar efficacy to nelfinavir (82). L-756, 423 (**28**) is being explored in Phase I as a novel potentially once-daily PI. Although it is structurally closely related to indinavir, its hydrophobic benzofuran group interacts more efficiently with the alternate S1-S3 binding site of HIV PR (83). Preliminary data with **28**/indinavir combination (MK-944A) demonstrated a short-term *in vivo* efficacy comparable to the other PIs (84).

PIs containing the cyclic urea scaffold have been broadly explored and initially yielded two C2-symmetrical inhibitors (DMP-323 and DMP-450) (85). However, these compounds suffered from variable pharmacokinetics, moderate potency, and poor resistance profile. Modifications of their P1/P1' and P2/P2' residues resulted in identification of many compounds with excellent *in vitro* potency, e.g. compound **29** (K_i = 18 pM; EC_{90} = 2 nM) (86). Further optimization yielded two nonsymmetrical cyclic ureas, DMP-850 and DMP-851 (**30** and **31**, respectively) containing 3-aminoindazole P2 residues (87). **30** and **31** showed *in vitro* antiviral potency comparable to that of the approved PIs (EC_{90} = 62 and 56 nM, respectively) and were selected as clinical candidates based on their favorable PK in dogs (prolonged $T_{1/2}$ and bioavailability of 60%) (87). Some 6-membered ring cyclic ureas showed excellent potency in cell culture, but limited bioavailability (88).

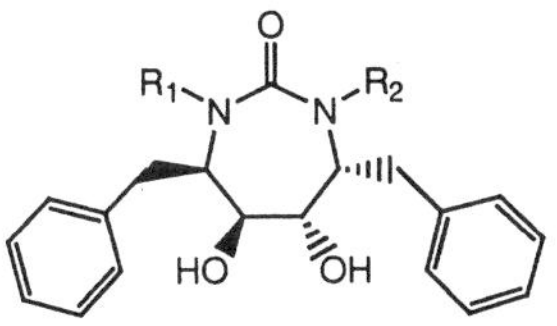

29: $R_1 = R_2 =$

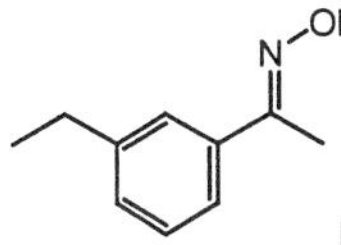

30: R_1 = Benzyl $R_2 =$

31: R_1 = n-Butyl $R_2 =$

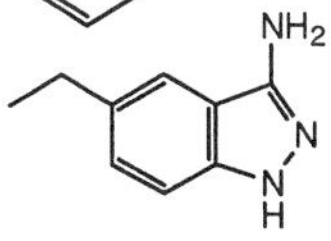

In addition to **26**, **32** (PD-178390) has been identified as a promising lead non-peptidic PI in the series of 5,6-dihydropyran-2-ones (89). It is slightly less potent than **26** (EC_{50} = 200 nM vs. 65 nM), presumably because of the lack of its interaction with the S3 binding pocket of HIV protease and shows bioavailability of 42% in dogs (89). Similar to **26** (79), **32** appears not to share the resistance profile with the other PIs and the virus resistant to **32** did not emerge even during the prolonged *in vitro* selection (90).

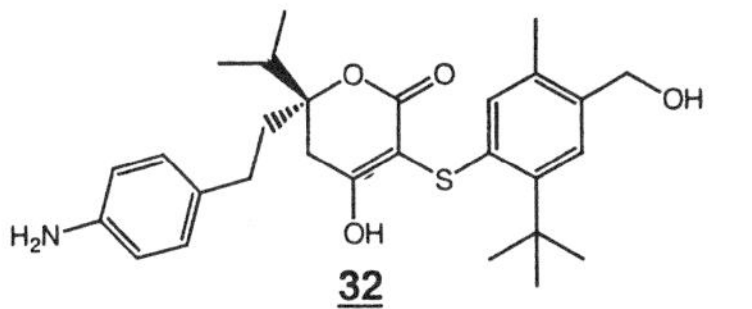

32

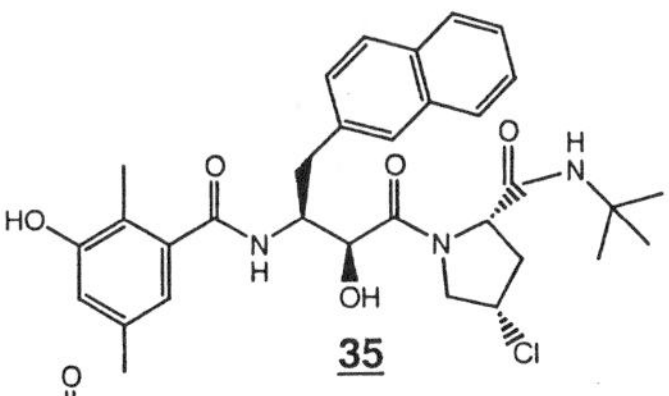

35

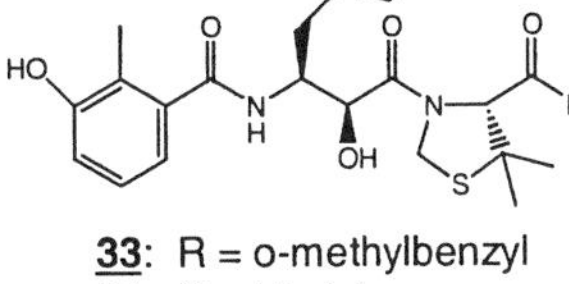

33: R = o-methylbenzyl
34: R = t-butyl

Dipeptide PIs containing 2-hydroxy-3-amino-4-arylbutanoic acid in their scaffold showed promising preclinical characteristics. Compound **33** (JE-2147; AG-1776) and **34** (JE-533; KNI-577) containing the allophenylnorstatine moiety exhibited potent *in vitro* anti-HIV activity with EC_{50} = 15-45 nM (91). **33** appears to fully retain its susceptibility against a variety of HIV strains resistant to multiple approved PIs and exhibits good oral bioavailability and PK profile in two animal species (91-92). *In vitro* selection of a resistant virus in the presence of **33** was significantly delayed compared to **34** (91). Compound **35** is another example from this PI series with a potent *in vitro* antiviral activity (IC_{90} = 27 nM) and favorable PK characteristics (93). Several compounds from the piperazine hydroxyethylamine series substituted with 2-isopropyl thiazolyl P3 ligand (same as in ritonavir), e.g. **36** (A-160621), showed markedly improved potency, reduced plasma protein binding, and less cross-resistance than ritonavir (94). Co-administration of ritonavir increases the bioavailability of **36** in rats by > 50-fold (94). Modifications of amprenavir have also been explored. The replacement of P2 tetrahydrofuran ligand either with a bis-tetrahydrofuran (**37**) or spirocyclic ether (**38**) in conjunction with minor modifications of P2' improved the *in vitro* antiviral potency of amprenavir by 6- and 10-fold, respectively (95,96).

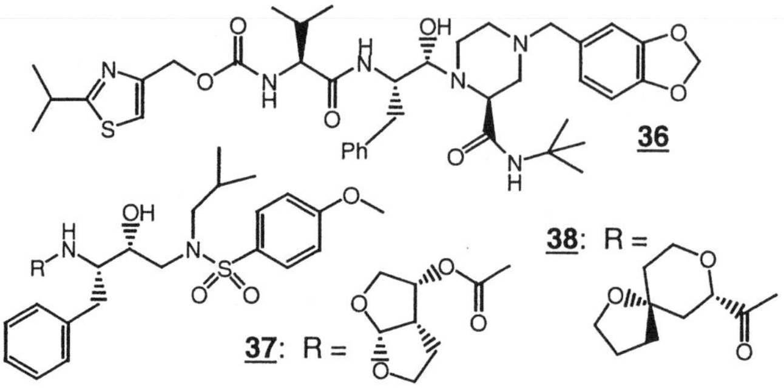

HIV ENTRY INHIBITORS

HIV entry is a complex process involving several specific membrane protein interactions. Initially, viral glycoprotein gp120 mediates the virus attachment via its binding to at least two host membrane receptors, CD4 and the chemokine coreceptor. This interaction induces a conformational change in the viral fusion protein gp41 leading to a fusion of viral envelope with the host cell membrane (97). In addition to gp120-chemokine receptor interaction, the fusion activity of gp41 is currently being explored as an attractive novel target for antiretroviral therapy. At least one agent from each class is undergoing clinical evaluation (98).

Chemokine receptor binders – The majority of HIV-1 isolates rely exclusively on the CCR5 coreceptor for its entry (M-tropic or R5 strains). However, in the later stages of the disease, more pathogenic variants utilizing CXCR4 coreceptor in addition to CCR5 (dual-tropic or R5X4 strains) or CXCR4 only (T-tropic or X4 strains) emerge in approximately 40% of infected individuals (99). Bicyclam compound **39** (AMD-3100) was the first identified CXCR4-specific inhibitor interfering with the replication of X4, but not R5 viruses with EC_{50} = 1-10 nM (100, 101). **39** also showed activity in SCID-hu PBMC mouse model and is currently in Phase II clinical evaluation as an injectable agent due to its limited oral bioavailability (102, 103).

In addition, several positively charged peptides/peptoids capable of blocking CXCR4 coreceptor have been described. While T140 (14-mer peptide) and ALX40-4C (modified nona-D-arginine) are selective inhibitors of X4 strains with EC_{50} values of 3.5 nM and 1 - 10 µM, respectively (104, 105, 106), CGP64222 (9-mer peptoid), which has initially been identified as an inhibitor of Tat-TAR interaction (107) (see below) shows antiviral effect also against R5 viruses (108). Compound **40** (TAK-779) is a small molecule specifically inhibiting the R5 isolates (EC_{50} = 1.6 to 3.7 nM). It exhibits high-affinity binding to CCR5 coreceptor (K_d = 0.45 nM) and a favorable *in vitro* therapeutic index of 40, 000 (109). Preclinical PK studies in several animal species revealed a long half-life particularly in lymph nodes, but only limited oral bioavailability (110). Mutagenesis of CCR5 suggested an interaction of **40** with a small binding pocket, which may be formed by transmembrane domains of the receptor (111). Distamycin analogue NSC-651016 is a unique HIV coreceptor inhibitor exhibiting activity both against R5 and X4 HIV-1 strains at low µM concentration (112). At least part of the activity, which was also observed in the *in* *vivo* murine model, appears to be due to the down-regulation of coreceptor expression (113).

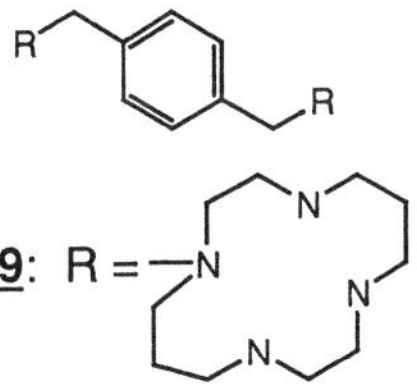

39: R = (structure)   **40**

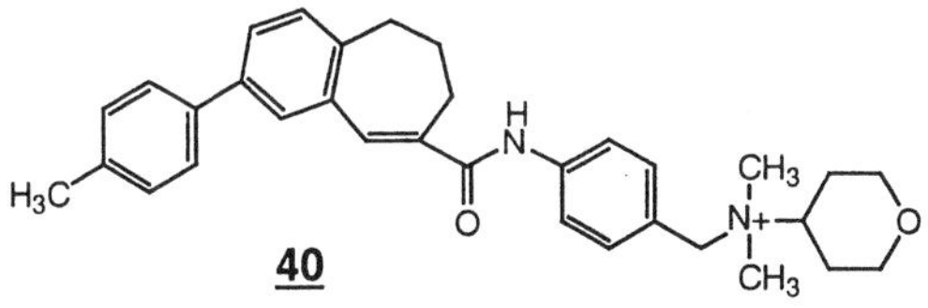

Inhibitors of gp41 fusion activity - Fusion of viral envelope with host plasma membrane is mediated by gp41, a transmembrane subunit of HIV-1 envelope glycoprotein complex (97). Pentafuside (T-20, DP-178), a 36-mer peptide derived from the C-terminal heptad repeat of gp41, is a potent inhibitor of HIV-1 clinical isolates with EC_{50} = 1-2 µg/ml (114). Pentafuside apparently inhibits the formation of fusion-competent conformation of gp41 by interfering with the interaction between its C- and N-terminal heptad repeat (115). Short-term intravenous monotherapy with T-20 showed significant antiviral effect in HIV-infected individuals (116). T-20 is currently undergoing Phase II clinical evaluation (117). Peptide T-1249 (39-mer) is a more potent 2nd generation fusion inhibitor without cross-resistance to T-20 (118). Recently, a hydrophobic cavity on the surface of gp41 coiled-coil core structure, which is presumably important for refolding of gp41 into its fusion-active conformation, has been proposed as a new target for design of anti-HIV agents (119). Modified D-peptides (120) and small molecules (121), which presumably interact with this cavity and display *in vitro* anti-HIV activity, represent a first step towards rational design of novel HIV fusion inhibitors. Triterpene derivative RPR103611 is also active against HIV-1 (EC_{50} = 0.04-0.75 µM) by blocking the virus entry (122). A recent study suggests that the loop between N- and C-terminal heptad repeat of gp41 may be the potential target (123).

HIV INTEGRASE (IN) INHIBITORS

Integration of proviral cDNA into the host genome is an essential step in HIV replication cycle making IN an attractive antiviral target (124). A number of IN inhibitors have been identified in past years (125); however, many of them appear to target the initial step in the integration process, i.e. assembly of preintegration complex. Since these complexes in infected cells are relatively stable and are rather long-lived, an effort has been made to design inhibitors interfering specifically with the second step of integration, i.e. strand transfer. Two closely related types of small molecules showing this characteristic have recently been identified. Diketo acids **41** and **42** inhibit strand transfer catalyzed by recombinant IN with IC_{50} < 0.1 µM. Both compounds also showed *in vitro* antiviral activity with EC_{50} = 1-2 µM (126). Mutations conferring resistance to **41** and **42** mapped to the close vicinity of conserved catalytic residues in IN enzyme demonstrating their specific mechanism of activity. X-ray crystallography studies with a tetrazol derivative **43**, a related inhibitor of DNA strand transfer, revealed the inhibitor localization centrally in the active site of IN enzyme in a close proximity of the acidic catalytic residues (127). These discoveries may initiate a new rational approach towards the development of clinically effective IN inhibitors.

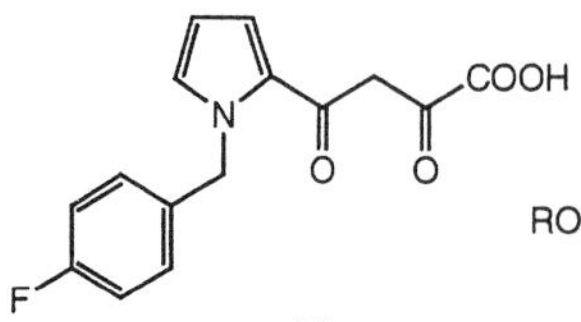

41 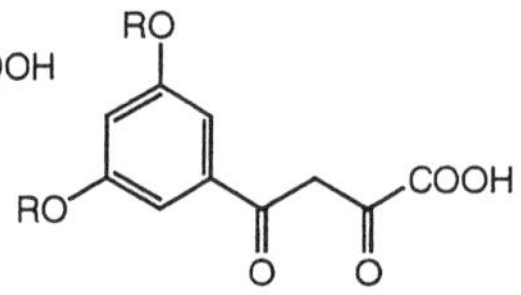   **42:** R = benzyl 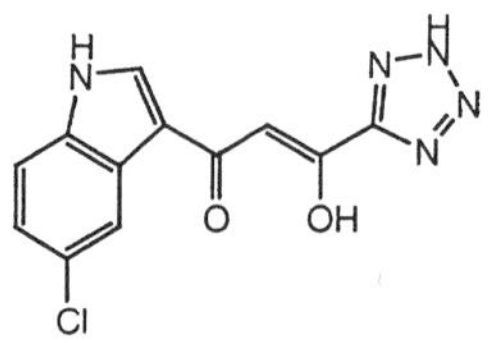   **43**

INHIBITORS OF HIV TRANSCRIPTION

Transcription of the integrated HIV genome into a polycistronic mRNA is transactivated by the binding of viral Tat accessory protein to TAR, a 60-nucleotide stem loop located in 5'-long terminal repeat (5'-LTR) of each HIV transcript (128). Basic peptides such as CGP64222 or Tat10-biotin (107, 129) interfere with *in vitro* Tat/TAR binding and are also active in cellular transactivation assays at low μM concentration. Small molecule CGP40336 inhibits *in vitro* Tat/TAR binding with IC_{50} = 22 nM (130). With exception of CGP64222, which also binds CXCR4 receptor (108), the antiviral effects of the other agents have been less well-characterized. EM2487, small-molecule Tat/TAR inhibitor isolated from *Streptomyces* species, inhibits *in vitro* acute HIV-1 infection with IC_{50} = 270 nM (131). Oxoquinoline derivative K12 and bistriazoloacridone analogue temacrazine , which both inhibit HIV-1 transcription via Tat/TAR-independent mechanism, show potent *in vitro* antiviral effect with EC_{50} = 60-400 and 1-10 nM, respectively (132, 133).

CONCLUSION

During the last 2 to 3 years, the primary goal of antiretroviral therapy shifted to efficient and durable suppression of viral replication. The ultimate goal of the future is viral eradication in infected patients. Considering recent advances in the understanding of replication dynamics and viral reservoirs, it is becoming clear that achieving this goal solely through antiretroviral therapy is very challenging and will require novel highly potent agents with excellent long-term safety profiles and simple dosing regimens. Treatment of patients failing antiretroviral therapy represents another difficult challenge. To succeed, new inhibitors with unique resistance profiles as well as new distinct classes of inhibitors will be needed.

References

1. R. Foster, D. Faulds, Drugs, 55, 729 (1998).
2. S. Daluge, S. Good, M. Faletto, W. Miller, M. St Clair, L. Boone, M. Tisdale, N. Parry, J. Reardon, R. Dornsife, D. Averett, T. Krenitsky, Antimicrob. Agents Chemother., 41, 1082 (1997).
3. M. Faletto, W. Miller, E. Garvey, M. St Clair, S. Daluage, S. Good, Antimicrob. Agents Chemother., 41, 1099 (1997).
4. K. Squires, S. Hammer, V. Degruttola, M. Fischl, D. Bettendorf, L. Demeter, G. Morese, 7th Conference on Retroviruses and Opportunistic Inf., Abstr. 529, (2000).
5. B. Robbins, R. Srinivas, C. Kim, N. Bischofberger, A. Fridland, Antimicrob. Agents Chemother., 42, 612 (1998).
6. R. Srinivas, A. Fridland, Antimicrob. Agents Chemother., 42, 1484 (1998).
7. R. Schooley, R. Meyers, P. Ruane, G. Beall, H. Lampiris, I. McGowan, 39th ICAAC, Abstr. LB-19, (1999).
8. J. Molina, F. Ferchal, C. Rancinan, F. Raffi, 7th Conference on Retroviruses and Opportunistic Infections, Abstr. 518 (2000).
9. J.-M. De Muys, H. Gourdeau, N. Nguyen-Ba, D. Taylor, P. Ahmed, T. Mansour, C. Locas, N. Richard, M. Wainberg, R. Rando, Antimicrob. Agents Chemother., 43, 1835 (1999).
10. N. Richard, H. Salomon, M. Oliveira, R. Rando, M. Wainberg, Nucleosides Nucleotides, 18, 773 (1999).

11. R. Wood, B. Trope, R. Van Leeuwen, D. Martin, L. Proulx, 39th ICAAC, Abstr. 503, (1999).
12. L. Wells, J. Lietzau, R. Little, J. Kelley, H. Ford, J. Pluda, D. Kohler, L. Gillim, H. Mitsuya, R. Yarchoan, 5th Conference on Retroviruses and Opportunistic Infections, Abstr. 651, (1998).
13. L. Little, L. Serchuch, J. Lietzau, M. Edgerly, J. Pluda, J. Kelley, H. Ford, M. Kavlick, H. Mitsuya, R. Yarchoan, 6th Conference on Retroviruses and Opportunistic Inf., Abstr. 380, (1999).
14. Z. Gu, M. Wainberg, P. Nguyen-Ba, L. L'Heureux, J. de Muys, R. Rando, Nucleosides Nucleotides, $\underline{18}$, 891 (1999).
15. Z. Gu, M. Wainberg, N. Nguyen-Ba, L. L'Heureux, J.-M. De Muys, T. Bowlin, R. Rando, Antimicrob. Agents Chemother., $\underline{43}$, 2376 (1999).
16. K. Borroto-Esoda, J. Mewshaw, D. Wakefield, B. Hooper, L. Trost, B. McCreedy, 39th ICAAC, Abstr. 924, (1999).
17. L. Wang, J. Bigley, R. St. Claire, N. Sista, F. Rousseau, 7th Conference on Retroviruses and Opportunistic Infections, Abstr. 103, (2000).
18. D. Richman, H. Kessler, J. Eron, M. Thompson, F. Raff, J. Jacobson, J. Harris, B. McCreedy, J. Bigley, F. Rousseau, 7th Conference on Retroviruses and Opportunistic Inf., Abstr. 668, (2000).
19. L. Ma, S. Hurwitz, J. Zhi, J. Mcatee, D. Liotta, H. McClure, R. Schinazi, Antimicrob. Agents Chemother., $\underline{43}$, 381 (1999).
20. J. Hammond, R. Schinazi, S. Schlueter-Wirtz, J. Mellors, 6th Conference on Retroviruses and Opportunistic Inf., Abstr. 597, (1999).
21. H. Uchida, E. Kodama, K. Yoshimura, Y. Maeda, P. Kosalaraksa, V. Maroun, Y.-L. Qui, J. Zemlicka, H. Mitsya, Antimicrob. Agents Chemother., $\underline{43}$, 1487 (1999).
22. K. Yoshimura, R. Feldman, E. Kodama, M. Kavlick, Y.-L. Qui, J. Zemlicka, H. Mitsuya, Antimicrob. Agents Chemother., $\underline{43}$, 2479 (1999).
23. K. Lee, Y. Choi, J. Hong, R. Schinazi, C. Chu, Nucleosides Nucleotides, $\underline{18}$, 537 (1999).
24. K. Lee, Y. Choi, E. Gullen, S. Schlueter-Wirtz, R. Schinazi, Y. Cheng, C. Chu, J. Med. Chem., $\underline{42}$, 1320 (1999).
24a. I. Sugimoto, S. Shuto, S. Mori, S. Shigeta, A. Matsuda, Bioorg. Med. Chem. Lett., $\underline{8}$, 385 (1999).
25. A. Siddiqui, C. McGuigan, C. Ballatore, F. Zuccotto, I. Gilbert, E. De Clercq, J. Balzarini, J. Med. Chem., $\underline{42}$, 4122 (1999).
26. D. Saboulard, L. Naesens, D. Cahard, A. Salgado, R. Pathirana, S. Velazquez, C. McGuigan, E. De Clercq, J. Balzarini, Mol. Pharmacol., $\underline{56}$, 693 (1999).
27. C. Meier, M. Lorey, E. De Clercq, J. Balzarini, J. Med. Chem., $\underline{23}$, 1417 (1998).
28. O. Pedersen, E. Pedersen, Antiviral Chem. Chemother., $\underline{10}$, 285 (1999).
29. E. De Clercq, Farmaco, $\underline{54}$, 26 (1999).
30. G. Szczech, P. Furman, G. Painter, D. Barry, K. Borroto-Esoda, T. Grizzle, M. Blum, J.-P. Sommadossi, R. Endoh, T. Niwa, M. Yamamoto, C. Moxham, Antimicrob. Agents Chemother., $\underline{44}$, 123 (2000).
31. F. Raffi, K. Arasteh, V. Gathiram, R. Wood, J. McIntyre, M. Zeiler, M. Van Den Berg, B. Bell, J. Quinn, D. Miralles, C. Moxham, F. Rousseau, 7th Conference on Retroviruses and Opportunistic Inf., Abstr. 514, (2000).
32. K. Borroto-Esoda, J. Harris, C. Klish, L. Mewshaw, L. Rimsky-Clarke, B. McCreedy, 75th Conference on Retroviruses and Opportunistic Inf., Abstr. 751, (2000).
33. J.-P. Kleim, V. Burt, M. Maguire, P. Mutch, R. Hazen, M. St Clair, 6th Conference on Retroviruses and Opportunistic Inf., Abstr. 599, (1998).
34. J.-P. Kleim, V. Burt, M. Maguire, R. Ferris, R. Hazen, G. Roberts, M. St Clair, 6th Conference on Retroviruses and Opportunistic Inf., Abstr. 600, (1998).
35. K. Arastech, M. Muller, R. Wood, L. Cass, K. Moore, N. Dallow, A. Jones, V. Burt, J. Kleim, W. Prince, 39th ICAAC, Abstr. 504, (1999).
36. J.-P. Kleim, M. Rosner, I. Winkler, A. Paessens, R. Kirsch, Y. Hsiou, E. Arnold, G. Reiss, Proc. Natl. Acad. Sci., $\underline{93}$, 34 (1996).
37. H. Rubsamen-Waigmann, E. Huguenel, A. Shah, A. Paessens, J. Ruoff, H. Von Briesen, A. Immelmann, U. Dietrich, M. Wainberg, Antiviral Res., $\underline{42}$, 15 (1999).
38. J. Kleim, M. Winters, A. Dunkler, J. Suarez, G. Riess, I. Winkler, J. Balzarini, D. Oette, T. Merigan, J. Infecf. Dis., $\underline{179}$, 709 (1999).
39. S. Hayashi, D. Jayesekera, A. Jaewardene, A. Shah, L. Thevanayagam, F. Aweeka, J. Clin. Pharmacol., $\underline{39}$, 1085 (1999).
40. T. Jenta, D. Soejarto, R. Buckheit, E. Arnold, K. Schweikart, J. Covey, 6th Conference on Retroviruses and Opportunistic Inf., Abstr. 602, (1998).

41. R. Sherer, B. Dutta, R. Anderson, J. Laudette-Aboulhab, A. Kamarulzaman, D. D'Amico, N. Paton, M. Abdullah, R. Pollard, T. Cooley, M. Flavin, Z. Xu, 7th Conference on Retroviruses and Opportunistic Inf., Abstr. 508, (1999).

42. R. Buckheit, E. White, V. Fliakas-Boltz, J. Russell, T. Stup, T. Kinjerski, M. Osterling, A. Weigand, J. Bader, Antimicrob. Agents Chemother., 43, 1827 (1999).

43. T. Fujiwara, A. Sato, M. El-Farrash, S. Miki, K. Abe, Y. Isaka, M. Kodama, Y. Wu, L. Chen, H. Harada, H. Sugimoto, M. Hatanaka, Y. Hinuma, Antimicrob. Agents Chemother., 42, 1340 (1998).

44. J. Hernandez, L. Amador, M. Amantea, H. Chao, P. Hawley, L. Paradiso, 7th Conference on Retroviruses and Opportunistic Inf., Abstr. 669, (1999).

45. M. Patel, R. McHugh, B. Cordova, R. Klabe, S. Erickson-Viitanene, G. Tainor, S. Ko, Bioorg. Med. Chem. Lett., 9, 3221 (1999).

46. J. Corbett, S. Ko, J. Rodgers, S. Jeffrey, L. Bacheler, R. Klabe, S. Diamond, C.-M. Lai, S. Rabel, J. Saye, S. Adams, G. Trainor, P. Anderson, S. Erickson-Viitanene, Antimicrob. Agents Chemother., 43, 2893 (1999).

47. S. Erickson-Viitanene, J. Corbett, S. Ko, J. Rogers, G. Trainor, R. Parsons, S. Diamond, D. Christ, C.-M. Lai, S. Jeffrey, S. Garber, C. Reid, L. Bacheler, R. Klabe, S. Rabel, J. Saye, S. Adams, 6th Conference on Retroviruses and Opportunistic Inf., Abstr. 13, (1998).

48. D. Wishka, D. Graber, L. Kopta, R. Olmsted, J. Friis, J. Hosley, W. Adams, E. Seest, T. Castle, L. Dolak, B. Keiser, Y. Yagi, A. Jeganathan, S. Schlachter, M. Murphy, G. Cleek, R. Nugent, S. Poppe, S. Swaney, F. Han, W. Watt, W. Whitel, T. Poel, R. Thomas, J. Morris, J. Med. Chem., 41, 1357 (1998).

49. L. Wathen, W. Freimuth, T. Sharp, K. Ruzicka, 6th Conference on Retroviruses and Opportunistic Inf., Abstr. 111, (1999).

50. C. Mao, R. Vig, T. Venkatachalam, E. Sudbeck, F. Uckun, Bioorg. Med. Chem. Lett., 8, 2213 (1998).

51. R. Vig, C. Mao, T. Venkatachalam, L. Tuel-Ahlgren, E. Sudbeck, F. Uckun, Bioorg. Med. Chem., 9, 1789 (1998).

52. M. Hogberg, C. Sahlberg, P. Engelhardt, R. Noreen, J. Kangasmetsa, N. Johansson, B. Oberg, L. Vrang, H. Zhang, B. Sahlberg, T. Unge, S. Lovgren, K. Fridborg, K. Backbro, J. Med. Chem., 42, 4150 (1999).

53. C. Mao, E. Sudbeck, T. Venkatachalam, F. Uckun, Bioorg. Med. Chem. Lett., 9, 1593 (1999).

54. C. Mao, E. Sudbeck, T. Venkatachalam, F. Uckun, Antiviral Chem. Chemother., 10, 233 (1999).

55. F. Uckun, C. Mao, S. Pendergrass, D. Maher, D. Zhu, L. Tuel-Ahlgren, T. Venkatachalam, Bioorg. Med. Chem. Lett., 9, 2721 (1999).

56. C. Chen, F. Uckun, Pharm. Res., 16, 1226 (1999).

57. M. Fujiwara, E. Kodama, M. Okamoto, K. Tokuhisa, T. Ide, Y. Hanasaki, K. Katsuura, H. Takayama, N. Aimi, H. Mitsuya, S. Shigeta, K. Konno, T. Yokota, M. Baba, Antiviral Chem. Chemother., 10, 315 (1999).

58. M. Cushamn, A. Casimiro-Garcia, K. Williamson, W. Rice, Bioorg. Med. Chem. Lett., 8, 195 (1998).

59. A. Casimiro-Garcia, M. Micklatcher, J. Turpin, T. Stup, K. Watson, R. Buckheit, M. Cushman, J. Med. Chem., 42, 4861 (1999).

60. G. Campiani, E. Morelli, M. Fabbrini, V. Nacci, G. Greco, E. Novellino, A. Ramunno, G. Maga, S. Spadari, G. Caliendo, A. Bergamini, E. Faggioli, I. Uccella, F. Bolacchi, S. Marini, M. Coletta, A. Nacca, S. Caccia, J. Med. Chem., 42, 4462 (1999).

61. A. Mai, M. Artico, G. Sbardella, S. Massa, E. Novellino, G. Greco, A. Loi, E. Tramontano, M. Marongiu, P. La Colla, J. Med. Chem., 42, 619 (1999).

62. M. Witvrouw, C. Arranz, C. Pannecouque, R. De Clercq, H. Jonckheere, J.-C. Schmidt, A.-M. Vandamme, J. Diaz, T. Ingate, J. Desmyter, R. Esnouf, L. Van Meervelt, S. Vega, J. Balzarini, E. De Clercq, Antimicrob. Agents Chemother., 42, 618 (1998).

63. D. Taylor, P. Ahmed, P. Chambers, A. Tyms, J. Bedard, J. Duchaine, G. Falardeau, J. Lavallee, W. Brown, R. Rando, T. Bowlin, Antiviral Chem. Chemother., 10, 79 (1999).

64. R. Silvestri, M. Artico, S. Massa, T. Marceddu, F. De Montis, P. La Colla, Bioorg. Med. Chem. Lett., 10, 253 (2000).

65. C. Lackman-Smith, k. Watson, A. Wiegand, J. Turpin, R. Buckheit, S. Chung, E. Cho, 39th ICAAC, Abstr. 925, (1999).

66. A. Molla, G. Grannenman, E. Sun, D. Kempf, Antiviral Res., 39, 1 (1998).

67. A. Wlodawer, J. Vondrasek, Annu. Rev. Biophys. Biomol. Struct., 27, 249 (1998).

68. J. Adkins, D. Faulds, Drugs, 55, 837 (1998).

69. D. Boden, M. Markowitz, Antimicrob. Agents Chemother., 42, 2775 (1998).

70. D. Kaul, S. Cinti, P. Carver, P. Kazanjian, Pharmacotherapy, 19, 281 (1999).
71. E. Race, E. Dam, V. Orby, S. Paulous, F. Clavel, AIDS, 13, 2061 (1999).
72. C. Decker, L. Laitinen, G. Bridson, S. Raybuck, R. Tung, P. Chaturvedi, J. Pharm. Sciences, 87, 803 (1998).
73. A. Spaltenstein, C. Baker, Y. Gray-Nunez, I. Kaldor, W. Kazmierski, D. Reynolds, R. Tung, P. Wheelan, E. Furfine, 7th Conference on Retroviruses and Opportunistic Infections, Abstr. 505, (2000).
74. R. Gulick, M. King, S. Brus, K. Real, R. Murphy, C. Hicks, J. Eron, J. Thommes, M. Thompson, C. White, C. Benson, M. Albrecht, H. Kessler, A. Hsu, R. Bertz, D. Kempf, E. Sun, A. Japour, 7th Conference on Retrovirus and Opportunistic Infections, Abstr. 515, (2000).
75. H. Sham, D. Kempf, A. Molla, K. Marsh, G. Kumar, C.-M. Chen, W. Kati, K. Stewart, R. Lal, A. Hsu, D. Betebenner, M. Korneyeva, S. Vasavononda, E. McDonald, A. Saldivar, N. Wideburg, X. Chen, P. Niu, C. Park, V. Jayanti, B. Grabowski, G. Granneman, E. Sun, A. Japour, J. Leonard, J. Plattner, D. Norbeck, Antimicrob. Agents Chemother., 42, 3218 (1998).
76. G. Kumar, J. Dykstra, E. Roberts, V. Jayanti, D. Hickman, J. Uchic, Y. Yao, B. Surber, S. Thomas, G. Granneman, Drug Metab. Dispos., 27, 902 (1999).
77. S. Thaisrivongs, J. Strohbach, Biopolymers, 51, 51 (1999).
78. Y. Wang, C. Daenzer, R. Wood, D. Nickens, M. Borin, W. Freimuth, J. Morales, S. Green, M. Mucci, W. Decian, 7th Conference on Retrovirus and Opportunistic Infections, Abstr. 673, (2000).
79. N. Back, A. van Wijk, D. Remmerswaal, M. van Monfort, M. Nijhuis, R. Schuuramn, C. Boucher, AIDS, 14, 101 (2000).
80. G. Bold, A. Fassler, H. Capraro, R. Cozens, T. Klimkait, J. Lazdins, J. Mestan, B. Poncioni, J. Rosel, D. Stover, M. Tintelnot-Blomley, F. Acemoglu, W. Beck, E. Boss, M. Eschbach, T. Hurlimann, E. Masso, S. Roussel, K. Ucci-Stoll, D. Wyss, M. Lang, J. Med. Chem., 27, 3387 (1998).
81. E. O'Mara, J. Smith, S. Olsen, T. Tanner, A. Schuster, S. Kaul, 6th Conference on Retrovirus and Opportunistic Infections, Abstr. 604, (1999).
82. I. Sanne, P. Piliero, R. Wood, T. Kelleher, A. Cross, A. Mongillo, S. Schnittman, 7th Conference on Retroviruses and Opportunistic Infections, Abstr. 672, (2000).
83. S. Munshi, Z. Chen, Y. Yan, Y. Li, D. Olsen, H. Schock, B. Galvin, B. Dorsey, L. Kuo, Acta. Cryst., D56, 381 (2000).
84. R. Campo, G. Acuna, L. Noriega, M. Makurath, M. Nessly, L. Gilde, R. Leavitt, 7th Conference on Retroviruses and Opportunistic Infections, Abstr. LB8, (2000).
85. G. De Lucca, P. Jadhav, R. Waltermire, B. Aungst, S. Erickson-Vitanen, P. Lam, Pharm. Biotechnol., 11, 257 (1998).
86. Q. Han, C.-H. Chang, R. Li, Y. Ru, P. Jadhav, P. Lam, J. Med. Chem., 41, 2019 (1998).
87. J. Rodger, P. Lam, B. Johnson, H. Wang, S. Ko, S. Seitz, G. Rainor, P. Anderson, R. Klabe, L. Bacheler, B. Cordova, S. Garber, C. Reid, M. Wright, C.-H. Chang, S. Erickson-Viitanene, Chem. & Bio., 5, 597 (1998).
88. G. De Lucca, J. Liang, I. De Lucca, J. Med. Chem., 42, 135 (1999).
89. J. Prasad, F. Boyer, J. Domagala, E. Ellsworth, C. Gajda, H. Hamilton, S. Hagen, L. Markoski, B. Steinbaugh, B. Tait, C. Humbley, E. Lunney, A. Pavlovsky, J. Rubin, D. Ferguson, N. Graham, T. Holler, D. Hupe, C. Nouhan, P. Tummino, A. Urumov, E. Zeikus, G. Zeikus, S. Gracheck, J. Sauders, S. VanderRoest, J. Brodfuehrer, K. Iyer, M. Sinz, S. Gulnik, J. Erickson, Bioorg. Med. Chem., 7, 2775 (1999).
90. L. Sharmeen, A. Heldsinger, B. Neorr, T. McQuade, S. Cracheck, S. Vanderroest, J. Saunders, 5th Conference on Retrovirus and Opportunistic Infections, Abstr. 637, (1998).
91. K. Yoshimura, R. Kato, K. Yusa, M. Kavlick, C. Maroun, A. Nguyen, T. Mimoto, T. Ueno, M. Shintani, J. Falloon, H. Masur, H. Hayashi, J. Erickson, H. Mitsuya, Proc. Natl. Acad. Sci., 96, 8675 (1999).
92. T. Mimoto, R. Kato, H. Takaku, S. Nojima, K. Terashima, S. Misawa, T. Fukazawa, T. Ueno, H. Sato, M. Shintani, Y. Kiso, H. Hayashi, J. Med. Chem., 42, 1789 (1999).
93. E. Takashiro, I. Hayakawa, T. Nitta, A. Kasuya, S. Miyamoto, Y. Ozawa, R. Yagi, I. Yamamoto, T. Shibayama, A. Nakagawa, Y. Yabe, Bioorg. & Med. Chem., 7, 2063 (1999).
94. X. Chen, D. Kempf, H. Sham, B. Green, A. Molla, M. Korneyeva, S. Vasavononda, N. Wideburg, A. Saldivar, K. Marsh, E. McDonald, D. Norbeck, Bioorg. Med. Chem. Lett., 8, 3531 (1998).
95. A. Ghosh, J. Kincaid, W. Cho, D. Walters, K. Krishnan, K. Hussain, Y. Koo, H. Cho, C. Rudall, L. Holland, J. Buthod, Bioorg. Med. Chem. Lett., 8, 687 (1998).

96. A. Ghosh, K. Krishnan, D. Walters, W. Cho, H. Cho, Y. Koo, J. Trevino, L. Holland, J. Buthod, Bioorg. Med. Chem. Lett., 8, 979 (1998).
97. D. Chan, P. Kim, Cell, 93, 681 (1998).
98. E. De Clercq, Drugs R. D., 2, 321 (1999).
99. E. Berger, P. Murphy, J. Farber, Annu. Rev. Immunol., 17, 657 (1999).
100. G. Donzella, D. Schols, S. Lin, J. Este, K. Nagashima, P. Maddon, G. Allaway, T. Sakmar, G. Henson, E. De Clercq, J. Moore, Nature Med., 4, 72 (1998).
101. D. Schols, J. Este, E. De Clercq, Antiviral Res., 35, 147 (1997).
102. D. Schols, E. Fonteyn, G. Bridger, G. Henson, E. De Clercq, 39th ICAAC, Abstr. 915, (1999).
103. C. Hendrix, C. Flexner, R. MacFarland, C. Giandomenico, Schweitzer, G. Henson, 6th Conference on Retroviruses and Opportunistic Infections, Abstr. 610, (1999).
104. Y. Xu, H. Tamamura, R. Arakaki, H. Nakashima, X. Zhang, N. Fujii, T. Uchiyama, T. Hattori, AIDS Res. Hum. Retroviruses, 15, 419 (1999).
105. H. Tamamura, Y. Xu, T. Hattori, X. Zhang, R. Arakaki, K. Kanbara, A. Omagari, A. Otaka, T. Ibuka, N. Yamamoto, H. Nakashima, N. Fujii, Biochem. Biophys. Res. Commun., 253, 877 (1998).
106. B. Doranz, K. Grovir-Ferbas, M. Sharron, S.-H. Mao, M. Goetz, E. Daar, R. Doms, W. O'Brien, J. Exp. Med., 186, 1395 (1997).
107. F. Hamy, E. Felder, G. Heizmann, J. Lazdins, F. Aboul-Ela, G. Varani, J. Karn, T. Klimkait, Proc. Natl. Acad. Sci., 94, 3548 (1997).
108. D. Daelemans, D. Schols, M. Witvrouw, C. Pannecouque, S. Hatse, S. Van Dooren, F. Hamy, T. Klimkait, E. De Clercq, A.-M. Vandamme, Mol. Pharmacol., 57, 116 (2000).
109. M. Baba, O. Nishimura, N. Kanzaki, M. Okamoto, H. Sawada, Y. Iizawa, M. Shiraishi, Y. Aramaki, K. Okonogi, Y. Ogawa, K. Meguro, M. Fujino, Proc. Natl. Acad. Sci., 96, 5698 (1999).
110. M. Baba, M. Okamoto, M. Ino, K. Oonishi, Y. Iizawa, K. Okonogi, O. Nishimura, Y. Ogawa, K. Meguro, M. Jujino, 39th ICAAC, Abstr. 914, (1999).
111. T. Dragic, S. Lin, A. Trkola, D. Thompson, E. Cormier, F. Kajumo, W. Ying, S. Smith, T. Sakmar, 7th Conference on Retrovirus and Opportunistic Infections, Abstr. 498, (2000).
112. O. Howard, J. Oppenheim, M. Hollingshead, J. Covey, J. Bigelow, J. McCormack, J. Buckheit, RW, D. Clanton, J. Turpin, W. Rice, J. Med. Chem., 41, 2184 (1998).
113. O. Howard, T. Korte, N. Tarasova, M. Grimm, J. Turpin, W. Rice, C. Michejda, R. Blumenthal, J. Oppenheim, J. Leukocyte Biol., 64, 6 (1998).
114. C. Wild, D. Shugars, T. Greenwell, C. McDanal, T. Matthews, Proc. Natl. Acad. Sci., 91, 9770 (1994).
115. L. Rimsky, D. Shugars, T. Matthews, J. Virol., 72, 986 (1998).
116. J. Kilby, S. Hopkins, T. Venetta, B. DiMassimo, G. Cloud, J. Lee, L. Alldredge, E. Hunter, D. Lambert, D. Bolognesi, T. Matthews, M. Johnson, M. Nowak, G. Shaw, M. Saag, Nature Med., 4, 1302 (1998).
117. J. Lalezari, J. Eron, M. Carlson, R. Arduino, J. Goodgame, C. Cohen, M. Kilby, E. Nelson, A. Dusek, P. Sista, S. Hopkins, 39th ICAAC, Abstr. LB-18, (1999).
118. D.M. Lambert, 3rd Internatl.l Workshop on HIV Drug Res., Abstr. 10 (1999).
119. D. Chan, C. Chutkowski, P. Kim, Proc. Natl. Acad. Sci., 95, 15613 (1998).
120. D. Eckert, V. Malashkevich, L. Hong, P. Carr, P. Kim, Cell, 89, 103 (1999).
121. A. Debnath, L. Ragigan, S. Jiang, J. Med. Chem., 42, 3203 (1999).
122. J.-F. Mayaux, A. Bousseau, R. Pauwels, T. Huet, Y. Henin, N. Dereu, M. Evers, F. Soler, C. Poujade, E. De Clercq, J.-B. Le Pecq, Proc. Natl. Acad. Sci., 91, 3564 (1994).
123. B. Labrosse, C. Treboute, M. Alizon, J. Virol., 74, 2142 (2000).
124. P. Hindmarsh, J. Leis, Microbiol. Mol. Biol. Rev., 63, 836 (1999).
125. Y. Pommier, N. Neamati, Adv. Virus Res., 52, 427 (1999).
126. D. Hazuda, P. Felock, M. Witmer, A. Wolfe, K. Stillmock, J. Grobler, A. Espeseth, L. Gabryelski, W. Schleif, C. Blau, M. Miller, Science, 287, 646 (2000).
127. Y. Goldgur, R. Craigie, G. Cohen, T. Fujiwara, T. Yoshinaga, T. Fujishita, H. Sugimoto, T. Endo, H. Murai, D. Davies, Proc. Natl. Acad. Sci., 96, 13040 (1999).
128. B. Cullen, Cell, 93, 685 (1998).
129. I. Choudhury, J. Wang, A. Rabson, S. Stein, S. Pooyan, S. Stein, M. Leibowitz, J. AIDS & Hum. Retrovirol., 17, 104 (1998).
130. F. Hamy, V. Brondani, A. Florsheimer, W. Stark, M. Blommers, T. Klimkait, Biochemistry, 37, 5086 (1998).
131. M. Baba, M. Okamoto, H. Takeuchi, 12th ICAR, Abstr. 28, (1999).
132. M. Baba, M. Okamoto, M. Makino, Y. Kimura, T. Ikeuchi, T. Sakaguchi, T. Okamoto, Antimicrob. Agents Chemother., 41, 1250 (1997).

133. J. Turpin, R. Buckheit Jr., D. Derse, M. Hollingshead, K. Williamson, C. Palamone, M. Osterling, S. Hill, L. Graham, C. Schaefer, M. Bu, M. Huang, W. Cholody, C. Michejda, W. Rice, Antimicrob. Agents Chemother., <u>42</u>, 487 (1998).

SECTION IV. IMMUNOLOGY, ENDOCRINOLOGY AND METABOLIC DISEASES

Editor: William K. Hagmann, Merck Research Laboratories, Rahway, NJ 07065

Chapter 17. Chemokines: Targets for Novel Therapeutics

Bharat K. Trivedi[1], Joseph E. Low[1], Kenneth Carson[2] and Gregory J. LaRosa[2]
[1]Parke-Davis Pharmaceutical Research, Division of Warner-Lambert Company,
Departments of Medicinal Chemistry and Molecular Biology, Ann Arbor, MI 48105
[2]Millennium Pharmaceuticals Inc., 75 Sydney Street, Cambridge, MA 02139

Introduction – The trafficking of pro-inflammatory leukocytes is regulated by a complex, multi-step process involving cell-cell adhesion, protein interactions between leukocytes and vascular endothelial cells, chemoattractant factors and their receptors on the surface of leukocytes, and cell movement (1,2). Transplant rejection, and the induction and maintenance of inflammatory, autoimmune, and allergic diseases, is dependent upon the migration of leukocytes from the circulation to extravascular sites. During the past decade, significant advances have been made in our understanding of this process and the factors that play critical roles in the regulation of leukocyte migration. In particular, the discovery and study of the chemokine (for *chemo*attractant cyto*kine*) superfamily has provided a clearer understanding of some of the mechanisms by which the migration of leukocytes is controlled, during normal immune function as well as in pathology (3-5). The human repertoire of chemokines comprises a family of approximately 50 small proteins (70-120 amino acids), that share a high degree of homology in sequence, tertiary structure, and function (6-8). Briefly, the family is comprised of four sub-classes based on the pattern of conserved cysteine residues, this classification also tends to parallel the shared cell type activities. The CXC, CC, C, or CX_3C-chemokines differ in the spacing or presence of the first two cysteine residues. The CXC chemokines tend to act predominantly on neutrophils and lymphocytes, while the CC chemokines act on a variety of leukocytes, but generally not neutrophils. The C and CX3C classes, to date have only one member each, lymphotactin and fractalkine/neurotactin, respectively, and tend to act on lymphocytes and monocytes (9-11).

The activity of chemokines is mediated by cell surface receptors that comprise a subfamily of the G protein-coupled receptor (GPCR) superfamily. To date there are eleven receptors for CC chemokines (12-18), five for the CXCs, and one each for the CX_3C (19) and C (20) chemokines. Following engagement of their receptors, which primarily couple to $G\alpha i$, chemokines activate several signal transduction pathways that induce a variety of functional responses including cellular chemotaxis, granule release, superoxide production, integrin upregulation, cellular differentiation, kinase activation, and proliferation.

Chemokines have generally been characterized as being proinflammatory factors (21-23). Recently, some chemokines have been found to influence normal leukocyte trafficking and immune function, promote or inhibit angiogenesis, and may play a role in the growth and metastasis of cancer cells (24-28). Validation of the chemokine systems as new therapeutic targets for inflammation has advanced prodigiously. The development of a plethora of chemokine and chemokine receptor reagents, such as, blocking monoclonal antibodies, modified antagonistic chemokines, and genetically modified mice has allowed correlations to be made between chemokines and the induction and/or maintenance of disease. Many chemokines are also upregulated in various pathologies, including, rheumatoid arthritis (RA), multiple sclerosis (MS) (29-34), atherosclerosis (35), asthma (36, 37), chronic obstructive pulmonary disorder (COPD) (38), and allergic disease (39). Taken together, this information has provided a very strong rationale for the implementation of chemokine antagonist programs. In the remainder of this review we will focus on the progress made towards identifying low molecular weight antagonists for CCR1, one of the receptors for MIP-1α and RANTES, CXCR2, one of the IL-8 receptors, and CCR2, the MCP-1 receptor (CCR2).

CCR1 Receptor Antagonists - CCR1 (40) is predominantly expressed on monocyte/macrophages, basophils and activated memory T cells. CCR1 interacts with a number of CC-chemokines, but MIP-1α and RANTES are the most studied. The role for CCR1 in human disease is still not clear. However, the presence of CCR1 bearing monocyte/macrophages, and tissue expression of MIP-1α and RANTES, have been associated with RA (41, 42), MS (30, 32, 43), and psoriasis (44-46). Additionally, genetic disruption of CCR1 in mice allows class II mismatched cardiac allografts to survive for extensive lengths of time (47).

Several different series of low molecular weight antagonists have been disclosed (48). For example, compound **1** was shown to have an affinity of 1 nM (vs MIP-1α on human CCR1). No activity was found against twenty-eight other GPCR's (at 10 µM). Dog pharmacokinetics (PK) was presented. The salt had a bioavailability of 70%, and a $T_{1/2} = 3$ hr. MS is the primary indication for this compound. In a rat heart transplant rejection model, compound **1** (rat IC_{50} = 100 nM) was given at 50 mg/kg sc TID. Cyclosporine A was given at a low constant dose (2.5 mg/kg po) or high dose (10 mg/kg po for 4 days). Transplanted hearts were then monitored daily. In this experiment, only the compound **1** and high-dose CsA groups showed increased survival time. In a rabbit kidney transplant rejection model, pellets of compound **1** (rabbit receptor K_i = 50 nM) were introduced at 20 mg/kg sc on day –3. After transplant, survival and kidney function were monitored. Compound **1** treated animals had a significant increase in survival time (from 10 days to 16 days). In a rat experimental autoimmune encephalomyelitis (EAE) model, animals were dosed at 5, 20, or 50 mg/kg sc. Significant effect was seen at 20 and 50 mg/kg sc, with delayed onset of disease at the highest dose.

Compound **2** was recently reported to be a CCR1 receptor antagonist with a K_i of 50 nM (49). Quaternary ammonium salts of **2** were reported to be more potent as CCR1 antagonists (K_i = 5 nM for methyl iodide salt). Variations in linker length, heterocyclic ring size, and halogen substitution pattern were made, most resulting in less active compounds. Compound **2** was reported to be the optimal inhibitor for this series. Constraint of the benzhydryl system into 7-membered heterocycles was performed, but reportedly led to no increase in potency. No biological data has been reported for this series of compounds.

The class of compounds exemplified by **3** was also recently disclosed (50). A variety of ring sizes in the central ring and the left-hand ring have been claimed. No preclinical data has been reported on this series. The class of compounds denoted by **4** was recently disclosed (51-57). For this series, X is described as C-H or N, while Y and Z can alternately be H, CH_2, N, O, or S. A wide range of substituents was claimed for R_{40}.

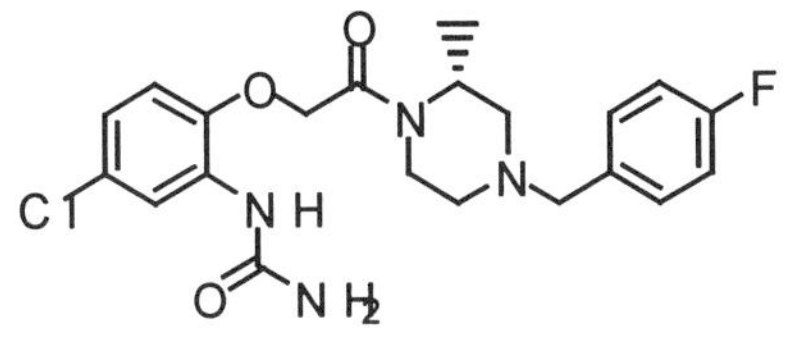

1

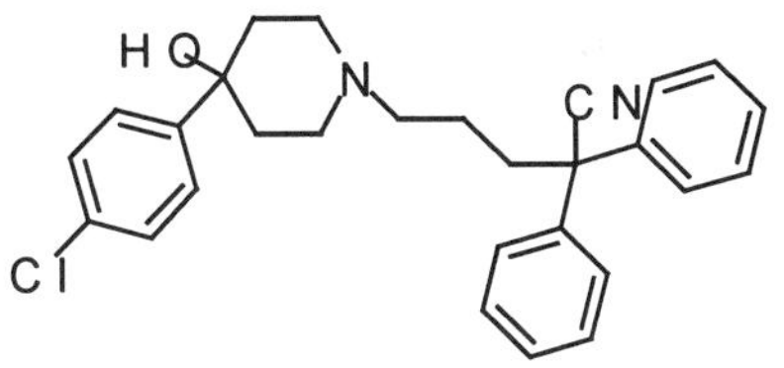

2

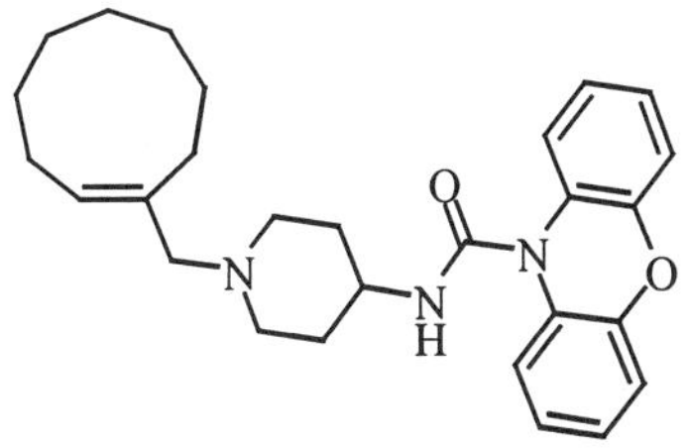

3

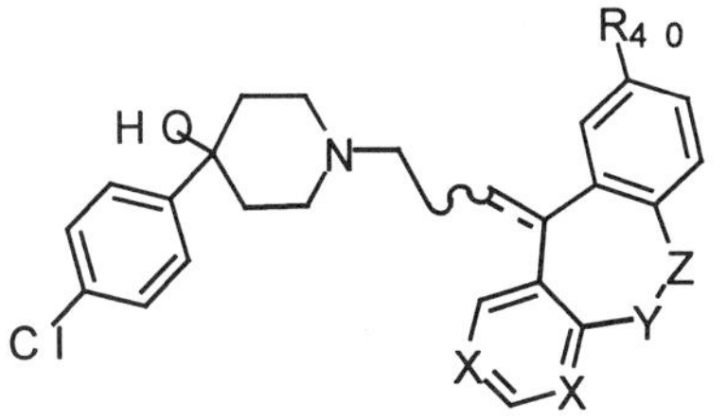

4

CXCR2 Receptor Antagonists - Interleukin-8 (IL-8) is a potent proinflammatory CXC chemokine that binds with equal affinity to two different distinct GPCRs(58,59). CXCR1 (IL-8RA) serves as a receptor for IL-8 and granulcyte chemotactic protein-2 (GCP-2) whereas CXCR2 (IL-8RB) binds IL-8 and GCP-2 plus a number of closely related chemokines containing the Glu-Leu-Arg (ELR) amino acid sequence. This includes growth related oncogene-α (GRO–α), GRO–β, GRO–γ, neutrophil activating peptide-2 (NAP-2), and epithelial cell-derived neutrophil activating protein-78 (ENA-78).

IL-8 receptors are present on a number of cell types including neutrophils (60), CD4$^+$, CD8$^+$ peripheral blood T lymphocytes, eosinophils, and basophils (61-63). In neutrophils, receptor occupation results in intracellular calcium flux, shape change, and a number of pro-inflammatory responses (64-65). This includes respiratory burst, degranulation, generation of biolipids, shedding of L-selectin, up-regulation of CD11b/CD18 adhesion molecule, cellular adhesion, and chemotaxis (65-71).

The exact role of IL-8 in human disease is yet to be determined. IL-8 is found in psoriatic lesions and synovial fluids of patients with RA, osteoarthritis, and gout (72-75). It is also found in elevated levels in bronchoalveolar lavage fluids (BALF) in patients with acute respiratory distress syndrome (ARDS) and COPD (76-77). ELR-positive chemokines such as IL-8 is found in certain types of tumors and have been linked to promotion of tumor growth and metastasis (78). Expression of CXCR2 has been detected in the intima of human atherosclerotic lesion. Atherosclerosis-susceptible, LDL receptor-deficient mice repopulated with marrow cells lacking CXCR2 had reduced intimal accumulation of macrophages in atherosclerotic lesions (79). Anti-IL-8 antibody blockade studies has demonstrated significant protective effects in a number of animal disease models. This includes rabbit models of cerebal ischemia-reprefusion injury, ARDS-like injury, and acute immune complex-induced glomerulonephritis (80-83).

Receptor binding studies conducted with SB 225002 **5** has demonstrated that it is a specific antagonist of the CXCR2 receptor (84). SB 225002 inhibited IL-8 binding to CHO-CXCR2 transfectant membranes with an IC$_{50}$ of 22 nM. GRO-α-stimulated calcium flux in human neutrophil was inhibited with an IC$_{50}$ of 30 nM. IL-8 or GRO-α-stimulated calcium flux in cells (differentiated HL60 or 2AsubE CXCR2 transfectants) expressing predominately CXCR2 receptors were inhibited by SB 225002 with IC_{50}'s of 8-40 nM. Similarly, rabbit neutrophil chemotaxis induced by IL-8 and GRO–α was inhibited with IC_{50}'s of 30 and 70 nM, respectively. Human neutrophil adhesion to activated human umbilical cord vein endothelial cells (HUVEC) under shear-force in the presence of IL-8 was inhibited with an IC_{50} of ~20 nM. In binding studies, the selectivity of SB 225002 against other GPCRs was demonstrated by greater than 150-fold selectivity against CXCR1 and four other GPCRs (fMLP, LTB_4, LTD_4, or C5a). Similarly, SB 225002 did not inhibit IL-8 mediated calcium flux in human neutrophils (IC_{50}>10µM), presumably because this cell expresses both CXCR1 and CXCR2 receptors. Calcium flux in human neutropils stimulated with LTB_4 or RANTES or monocytes stimulated with MIP-1α or MCP-1, was not inhibited by SB 225002. Similarly, IL-8- or LTD_4-mediated calcium flux in RBL-2H3 cells stably transfected with CXCR1 was inhibited with IC_{50} >10 µM. No selectivity information against non-immune related GPCRs has been published for this compound.

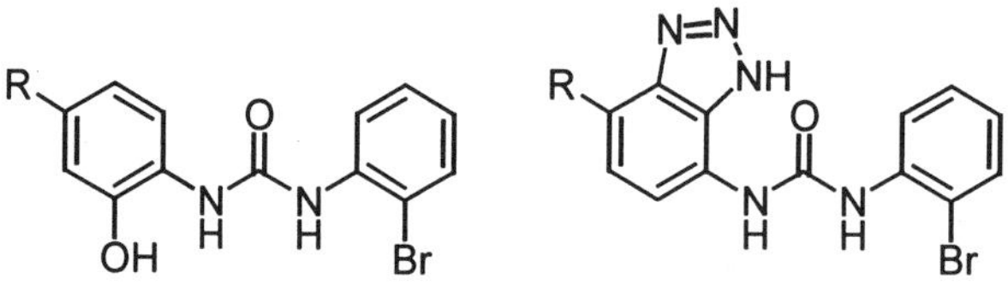

5 R = NO$_2$
6 R = CN

7 R = CN
8 R = Cl

SB 225002 **5** was also tested in a rabbit *in-vivo* model of IL-8- or fMLP-induced neutrophil margination. Intravenous administration of IL-8 or fMLP results in a rapid margination of neutrophils and reduction of whole blood neutrophil count. Co-administration of **5** with the chemokines inhibited IL-8 but not fMLP effects in a dose-dependent manner with significant inhibition observed with a 2.78 or 5.5 µg/kg/min dose.

More recently, an extension of the SAR of **5** in which the –OH group of compound **5** or **6** was replaced in an attempt to improve overall PK profile for this class of compound was disclosed (85). Through systematic SAR evaluation, several analogs were identified in which a triazolr ring was appended onto the 2,3- position of the phenyl ring. Thus, compound **7** was shown to have the binding affinity of 17 nM whereas the corresponding Cl analog was less active with an IC_{50} value of 90 nM. Compound **7** inhibited chemotaxis with an IC_{50} value of 10 nM. This compound was shown to be efficacious *in-vivo*. In the rabbit model of LPS-induced neutrophilia, **7** was active with an ED_{50} value of 10 mg/kg. Interestingly, this compound was evaluated in the LDL receptor deficient mice at 20 and 100 mg/kg/day. Although it was not active at the low dose, at the high dose of 100 mg/kg/day, compound **7** reduced atherosclerotic lesions by 40%. Compound **7** showed significant oral bioavailability both in rat (71%) and in rabbit (25%).

Recently, a series of 2-(alkylaminoalkyl)amino-3-aryl-6,7-dichloroquinoxalines (**9** and **10**) are claimed as selective antagonists of IL-8 binding and neutrophil chemotaxis with IC_{50}'s of 80-400 nM (86,87).

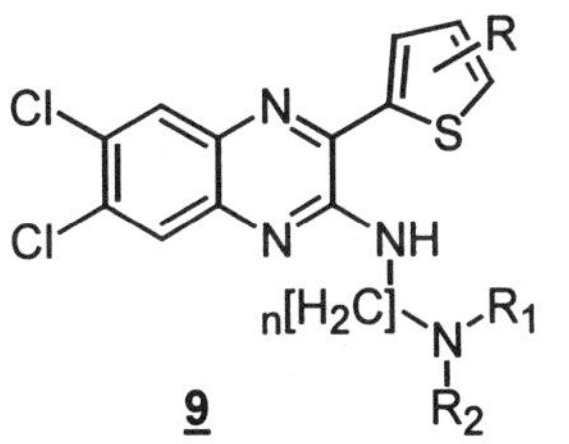

CCR2 Receptor Antagonists - CCR2 is expressed on monocyte/macrophges, basophils, and activated, memory T cells (88). CCR2 binds to several CC-chemokines of the MCP family, including MCP-1, MCP-2, MCP-3, and MCP-4. However, MCP-1 is the highest affinity and most efficacious ligand. CCR2 bearing monocyte/macrophages or T cells, and the expression of MCP-1, have been associated with RA (89, 90), MS (30, 32, 43), and atherosclerosis (91-94). "Proof-of-concept" studies using either MCP-1 or CCR2 blocking reagents, or genetically modified mice, have been reported over the last few years. Blockade of MCP-1/CCR2 by either neutralizing antibodies, or an MCP-1 mutant that acts as an antagonist of CCR2, prevents inflammatory cell infiltration and joint swelling in rodent models of arthritis (95, 96).

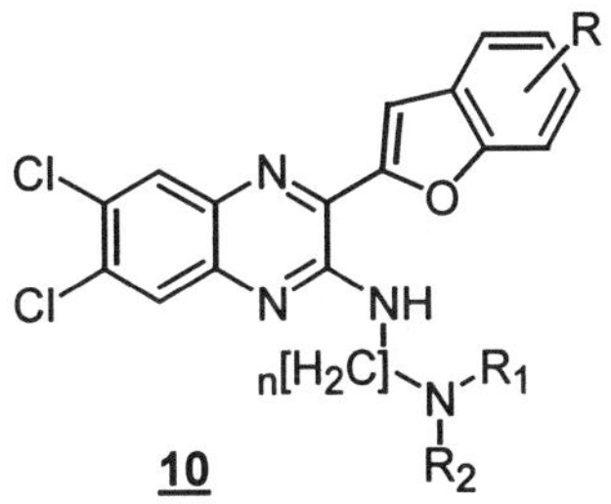

MCP-1 has been shown to be actively involved in recruiting monocytes in atherosclerotic plaques, and is thought to play a pivotal role in the development of atherosclerosis (97). MCP-1 deficient mice crossed with LDL receptor-deficient mice showed significant reduction in atherosclerosis (98), whereas CCR2 receptor deficient mice crossed with apoE knock out mice also showed decreased lesion formation (99). To further substantiate the role of MCP-1, a recent study over expressing MCP-1 in leukocytes, predominantly in macrophages, led to enhanced progression of atherosclerosis (100). This observation was attributed to significant macrophage accumulation in vessel wall. Attempts have been made to identify agents that will modulate expression of MCP-1 at the arterial wall. Trapidil administered at 60 mg/kg/day sc in a cholesterol-fed rabbit model subjected to balloon injury, resulted in a 75% reduction in MCP-1 expression and macrophage accumulation in the arterial wall (101). A recent paper suggested that circulating MCP-1 levels in humans are increased with age (102). This analysis was carried out in 405 healthy individuals with the intent to correlate presence and extent of atherosclerosis. However, no such a correlation was observed. Several review articles have recently appeared summarizing the role of MCP-1 in recruiting monocyte-macrophages in the animal models of atherosclerosis (103-105). Within the family of chemokines, MCP-1 is emerging as one of the most important targets for the discovery of novel therapeutic agents for the treatment of this disease.

Recently, a number of templates have been identified in the literature which bind to the CCR2 receptor. Compound **11** is a potent antagonist to MCP-1 receptor with an IC_{50} value of 33 nM (106). Although it was shown to be a good antagonist in the chemotaxis assay, no other biological data was reported. A series of benzimidazole derivatives **12** (107, 108) and 5-aryl-2, 4-pentadienamide derivatives **13** and **14** (109), were recently shown to be antagonists with sub-micromolar binding affinity for the receptor with similar potency in a chemotaxis assay. Detailed SAR on pentadienamides was recently reported (109), and preliminary SAR studies suggested no significant improvement in the binding affinity of the initial screening lead. Indole 2-carboxylic acid derivatives **15** have recently been disclosed (110, 111) as MCP-1 receptor ligands with activity in both chemotaxis and Ca^{2+} flux assays.

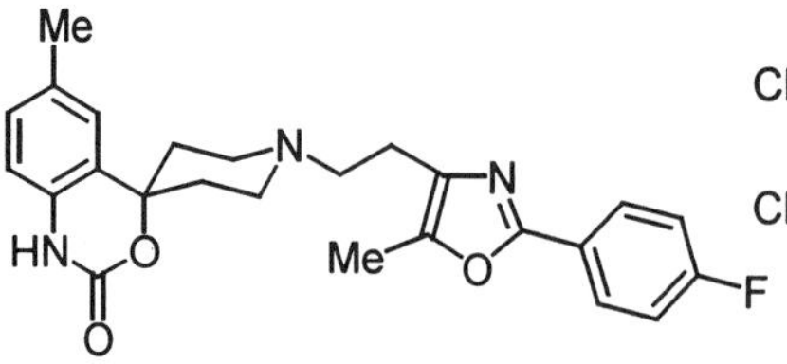

11

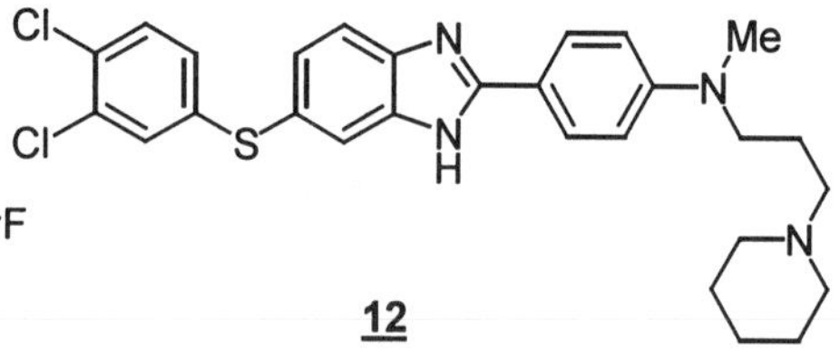

12

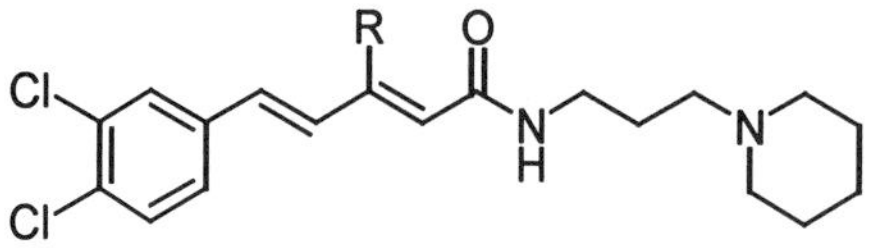

13 R = H
14 R = Me

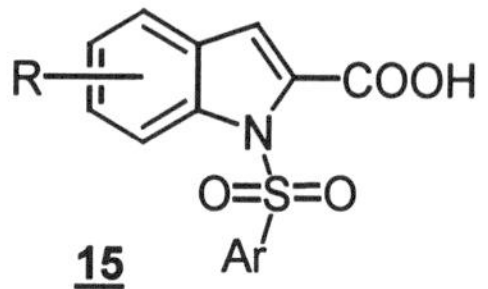

15

A unique series of bicyclic anilide derivatives as potent ligands to the CCR-5 receptor with a well-developed SAR has been disclosed (112-116). An additional patent application revealing an identical genus and list of compounds for the CCR2 receptor has been disclosed (117). Compound **16** (TAK 779) has been identified as the most potent ligand to CCR-5 receptor. Although the binding data for the CCR2 receptor is not disclosed, these compounds may also have excellent affinity for the CCR2 receptor based on the close homology between the CCR5 and CCR2 receptors.

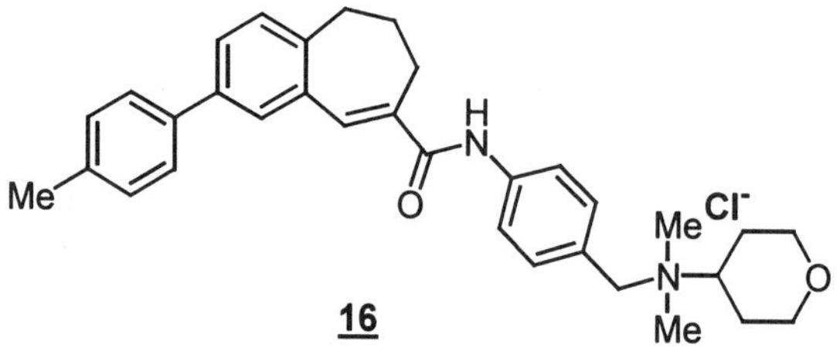

16

A series of modified glycine derivatives with excellent receptor binding affinity has recently been disclosed (118, 119). Compound **17** was identified as one of the preferred structure within this class of compounds. Additional SAR information was presented wherin the R_2 group was modified to an amine group and various R_1 groups, such as NO_2, Cl etc were incorporated in the molecule. All of these analogs were shown to have excellent binding affinity to the CCR2 receptor. The templates identified are unique, and may provide interesting tools to further investigate the role of MCP-1 in a variety of inflammatory disease models including atherosclerosis.

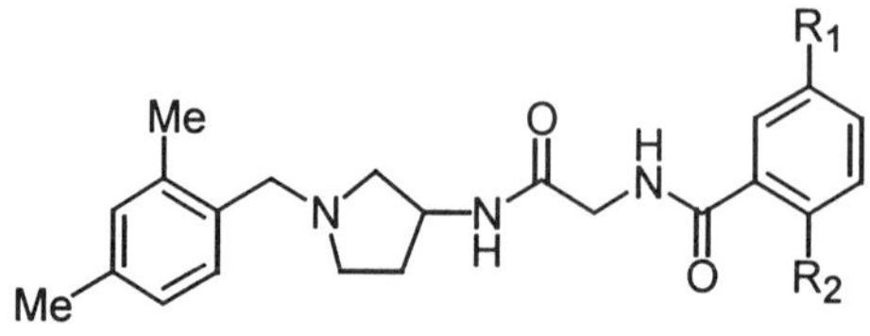

17 $R_1 = CF_3$, $R_2 = H$

Conclusions - The field of chemokines and their receptors is relatively new but is maturing and growing rapidly. Clearly there is an extensive body of evidence showing a correlation between the presence of some chemokines and various disease states. However, there is only a limited evidence for antagonism of a chemokine receptor *in-vivo*, and resolution of a disease, and there is no efficacy data in humans. Nevertheless, there is now a growing body of patent literature describing the discovery and characterization of small molecule, chemokine receptor antagonists. It is encouraging news that this class of GPCR can be antagonized by a low molecular weight organic molecule. Although GPCRs have traditionally been one of the most favored of drug discovery biological targets, the chemokine GPCRs bind a less traditional ligand, a small folded protein, that presumably interacts largely with the receptor extracellular domain(s), rather than in the more traditional transmembrane region binding pocket. So it is hopeful that additional successes in the discovery of pharmaceutically attractive chemokine receptor antagonists will be forthcoming.

References

1. T.A. Springer, Cell, 76, 301 (1994).
2. E.C. Butcher, Adv.Exp.Med.Biol, 323, 181 (1992).
3. J.J. Campbell, E.F. Foxman, and E.C. Butcher, Eur.J.Immunol., 27, 2571 (1997).
4. J.J. Campbell, J. Hedrick, A. Zlotnik, M.A. Siani, D.A. Thompson, and E.C. Butcher, Science (Washington, D. C.), 279, 381 (1998).
5. E.C. Butcher and L.J. Picker, Science, 272, 60 (1996).
6. G.M. Clore and A.M. Gronenborn, FASEB-J, 9, 57 (1995).
7. M. Baggiolini, B. Dewald, and B. Moser, Annu.Rev.Immunol., 15, 675 (1997).
8. M. Baggiolini, Nature (London), 392, 565 (1998).
9. G.S. Kelner, J. Kennedy, K.B. Bacon, S. Kleyensteuber, D.A. Largaespada, N.A. Jenkins, N.G. Copeland, J.F. Bazan, K.W. Moore, T.J. Schall, and A. Zlotnick, Science, 266, 1395 (1994).
10. J.F. Bazan, K.B. Bacon, G. Hardiman, W. Wang, K. Soo, D. Rossi, D.R. Greaves, A. Zlotnik, and T.J. Schall, Nature, 385, 640 (1997).
11. Y. Pan, C. Lloyd, H. Zhou, S. Dolich, J. Deeds, J.A. Gonzalo, J. Vath, M. Gosselin, J. Ma, B. Dussault, E. Woolf, G. Alperin, J. Culpepper, J.C. Gutierrez-Ramos, and D. Gearing, Nature, 387, 611 (1997).
12. O. Yoshie, Immunol.Front., 8, 147 (1998).
13. V.L. Schweickart, A. Epp, C.J. Raport, and P.W. Gray, J.Biol.Chem., 275, 9550 (2000).
14. B. Homey, W. Wang, H. Soto, M.E. Buchanan, A. Wiesenborn, D. Catron, A. Muller, T.K. McClanahan, M.-C. Dieu-Nosjean, R. Orozco, T. Ruzicka, P. Lehmann, E. Oldham, and A. Zlotnik, J.Immunol., 164, 3465, (2000).
15. D.I. Jarmin, M. Rits, D. Bota, N.P. Gerard, G.J. Graham, I. Clark-Lewis, and C. Gerard, J.Immunol., 164, 3460 (2000).
16. B.-S. Youn, C.H. Kim, F.O. Smith, and H.E. Broxmeyer, Blood, 94, 2533 (1999).
17. Zaballos, J. Gutierrez, R. Varona, C. Ardavin, and G. Marquez, J.Immunol., 162, 5671 (1999).
18. B.A. Zabel, W.W. Agace, J.J. Campbell, H.M. Heath, D. Parent, A.I. Roberts, E.C. Ebert, N. Kassam, S. Qin, M. Zovko, G.J. LaRosa, L.L. Yang, D. Soler, E.C. Butcher, P.D. Ponath, C.M. Parker, and D.P. Andrew, J.Exp.Med., 190, 1241 (1999).
19. T. Imai, K. Hieshima, C. Haskell, M. Baba, M. Nagira, M. Nishimura, M. Kakizaki, S. Takagi, H. Nomiyama, T.J. Schall, and O. Yoshie, Cell (Cambridge, Mass.), 91, 521 (1997).
20. T. Yoshida, T. Imai, M. Kakizaki, M. Nishimura, S. Takagi, and O. Yoshie, J.Biol.Chem., 273, 16551 (1998).
21. T.J. Schall and K.B. Bacon, Curr.Opin.Immunol., 6, 865 (1994).
22. K.B. Bacon and T.J. Schall, Int.Arch.Allergy.Immunol., 109, 97 (1996).
23. O.M.Z. Howard, J.J. Oppenheim, and J.M. Wang, J.Clin.Immunol., 19, 280 (1999).
24. S. Jung and D.R. Littman, Curr.Opin.Immunol., 11, 319 (1999).
25. R.M. Strieter, P.J. Polverini, and S.L. Kunkel, J.Biol.Chem., 270, 27348 (1995).
26. N. Fujisawa, S. Hayashi, and E.J. Miller, Melanoma Res., 9, 105 (1999).
27. G. Galffy, K.A. Mohammed, P.A. Dowling, N. Nasreen, M.J. Ward, and V.B. Antony, Cancer Res., 59, 367 (1999).

28. Wuyts, C. Govaerts, S. Struyf, J.-P. Lenaerts, W. Put, R. Conings, P. Proost, and J. Van Damme, Eur.J.Biochem., 260, 421 (1999).

29. L.F. Eng, R.S. Ghirnikar, and Y.L. Lee, Neurochem Res, 21, 511 (1996).

30. R. Godiska, D. Chantry, G.N. Dietsch, and P.W. Gray, J.Neuroimmunol., 58, 167 (1995).

31. P. Van Der Voorn, J. Tekstra, R.H.J. Beelen, C.P. Tensen, P. Van Der Valk, and C.J.A. De Groot, Am.J.Pathol., 154, 45 (1999).

32. T.L. Sorensen, M. Tani, J. Jensen, V. Pierce, C. Lucchinetti, V.A. Folcik, S. Qin, J. Rottman, F. Sellebjerg, R.M. Strieter, J.L. Frederiksen, and R.M. Ransohoff, J.Clin.Invest., 103, 807 (1999).

33. K.E. Balashov, J.B. Rottman, H.L. Weiner, and W.W. Hancock,   Proc.Natl.Acad.Sci. U.S.A., 96, 6873 (1999).

34. R.M. Ransohoff. Chemokines and central nervous system inflammation. in Inflammatory Cells Mediators CNS Dis. 1999: Harwood, Amsterdam, Neth.

35. T.J. Reape and P.H.E. Groot, Atherosclerosis, 147, 213 (1999).

36. J.-A. Gonzalo, C.M. Lloyd, D. Wen, J.P. Albar, T.N.C. Wells, A. Proudfoot, C. Martinez-A., M. Dorf, T. Bjerke, A.J. Coyle, and J.-C. Gutierrez-Ramos, J.Exp.Med., 188, 157 (1998).

37. L.M. Teran, Clin.Exp.Allergy, 29, 287 (1999).

38. R.E. Nocker, D.F. Schoonbrood, E.A. van de Graaf, C.E. Hack, R. Lutter, H.M. Jansen, and T.A. Out, Int.Arch.Allergy Immunol., 109, 183 (1996).

39. R. E. Nocker, D.F. Schoonbrood, E.A. van de Graff, C.E. Hack, R. Lutter, H.M. Jansen and T.A. Out. Int.Arch.Allergy Immunol., 109, 183 (1996).

40. K. Neote, D. DiGregorio, J.Y. Mak, R. Horuk, and T.J. Schall, Cell, 72, 415 (1993).

41. A.E. Koch, S.L. Kunkel, L.A. Harlow, D.D. Mazarakis, G.K. Haines, M.D. Burdick, R.M. Pope, and R.M. Strieter, J.Clin.Invest, 93, 921 (1994).

42. M.V. Volin, M.R. Shah, M. Tokuhira, G.K. Haines, III, J.M. Woods, and A.E. Koch, Clin.Immunol.Immunopathol., 89, 44 (1998).

43. W.J. Karpus, N.W. Lukacs, B.L. McRae, R.M. Strieter, S.L. Kunkel, and S.D. Miller, J.Immunol., 155, 5003 (1995).

44. S.P. Raychaudhuri, W.Y. Jiang, E.M. Farber, T.J. Schall, M.R. Ruff, and C.B. Pert, Acta Derm.Venereol., 79, 9 (1999).

45. M. Fukuoka, Y. Ogino, H. Sato, T. Ohta, K. Komoriya, K. Nishioka, and I. Katayama, Br.J.Dermatol., 138, 63 (1998).

46. Y. Hatano, K. Katagiri, and S. Takayasu, Clin.Exp.Immunol., 117, 237 (1999).

47. W. Gao, P.S. Topham, J.A. King, S.T. Smiley, V. Csizmadia, B. Lu, C.J. Gerard, and W.W. Hancock, J.Clin.Invest., 105, 35 (2000).

48. J.G. Bauman, B.O. Buckman, A.F. Ghannam, J.E. Hesselgesser, R. Horuk, I. Islam, M. Liang, K.B. May, S.D. Monahan, M.M. Morissey, H.P. Ng, G.P. Wie, W. Xu and W. Zheng, WO 09856771 (1999).

49. H.P. Ng, K. May, J.G. Baumann, A. Ghannam, I. Islam, M. Liang, R. Horuk, J. Hesselgesser, R.M. Snider, H.D. Perez, M.M. Morrissey, J.Med.Chem., 42, 4680, (1999).

50. Naya, Y. Owada, T. Saeki, K. Ohwaki, Y. Iwasawa, WO 09804554 (1998).

51. J.R. Luly, Y. Nakasato, and E. Ohshima, WO 00014089 (2000).

52. J.R. Luly, Y. Nakasato, and E. Ohshima, WO 00014086 (2000).

53. J.R. Luly, Y. Nakasato, and E. Ohshima, WO 09937651 (1999).

54. J.R. Luly, Y. Nakasato, and E. Ohshima, WO 09937619 (1999).

55. C.F. Schwender, C.R. Mackay, J.C. Pinto, W. Newman, WO 09937617 (1999).

56. C.R. Mackay, P.D. Ponath, WO09814480 (1998).

57. C.F. Schwender, C.R. Mackay, J.C. Pinto, W. Newman, WO 09802151 (1998).

58. W.E. Holmes, J. Lee, W.J. Kuang, G.C. Rice, W.I. Wood, Science 253, 1278 (1991).

59. P.M. Murphy, H.L. Tiffany, Science, 253, 1280 (1991).

60. T. Yoshimura, K. Matsushima, J.J. Oppenheim, E.J. Leonard, J.Immunol., 139, 788 (1987).

61. C.G. Larsen, A.O. Anderson, E. Appella, J.J. Oppenheim, K. Matsushima, Science, 243, 1464, (1989).

62. R.A. Warringa, L. Koenderman, P.T. Kok, J. Kreukniet, P.L. Bruijnzeel, Blood, 77, 2694 (1991).

63. M.v. White, T. Yoshimura, W. Hook, M.A. Kaliner, E.J. Leonard, Immunol.Lett., 22, 151 (1989).

64. Walz, B. Dewald, V. von-Tscharner, M. Baggiolini. J.Exp.Med. 170, 1745 (1989).

65. M. Thelen, P. Peveri, P. Kernen, V. von-Tscharaner, A. Walz, M. Baggiolini., FASEB J. 2, 2702 (1988).

66. P. Peveri, A. Walz, B. Dewald, M. Baggiolini, J.Exp.Med., 167, 1547 (1988).

67. J.M. Schroder. J.Exp.Med., 170, 847 (1989).

68. A.R. Huber, S.L. Kunkel, R.F. Todd, S.J. Weiss, Science, 254, 99 (1991).
69. P.A. Detmers, S.K. Lo, E. Olsen-Egbert, A. Walz, M. Baggiolini, Z.A. Cohn. J.Exp.Med. 171, 1155 (1990).
70. M.A. Gimbrone Jr, M.S. Obin, A.F. Brock, E.A. Luis, P.E. Hass, C.A. Hebert, Y.K. Yip, D.W. Leung, D.G. Lowe, W.J. Kohr, W.C. Darbonne, K.B. Bechtol, J.B. Baker, Science 246, 1601 (1989).
71. E.J. Leonard, T. Yoshimura, S. Tanaka, M. Raffleld. J.Invest.Dermatol., 96, 690 (1991).
72. J.M. Schroder, E. Christopher, J.Invest.Dermatol., 87, 53 (1986).
73. F.M. Brennan, C.O. Zachariae, D. Chantry, C.G. Larsen, M. Turner, R.N. Maini, K. Matsushima, M. Feldman, Eur.J.Immunol., 20, 2141 (1990).
74. J.A. Symons, W.L. Wong, M.A. Palladino, G.W. Duff, Scand.J.Rheumatol., 21, 92 (1992).
75. R. Terkeltaub, C. Zachariae, D. Santoro, J. Martin, P. Peveri, K. Matsushima, Arthritis.Rheum., 34, 894 (1991).
76. E.J. Miller, A.B. Cohen, S. Nagao, D. Griffith, R.J. Maunder, T.R. Martin, J.P. Weiner-Kronish, M. Sticherling, E. Christopher, M.A. Matthay, Am.Rev.Respir.Dis., 146, 427 (1992).
77. V.C. Broaddus, C.A. Hebert, R.V. Vitangcol, J.M. Hoeffel, M.S. Bernstein, A.M. Boylan, Am.Rev.Respir.Dis., 146, 825 (1992).
78. B.B. Moore, D.A. Arenberg, R.M. Strieter, Trends Cardiovas. Med., 8, 51 (1998).
79. W.A. Boisvert, R. Santiago, L.K. Curtiss, R.A. Terkeltaub, J.Clin.Invest., 101, 353 (1998).
80. T. Matsumoto, K. Ikeda, N. Mukaida, A. Harada, Y. Matsumoto, J. Yamashita, K. Matsushima, Lab.Invest., 77, 119 (1997).
81. K. Yokoi, N. Mukaida, A. Harada, Y. Watanabe, K. Matsushima, Lab. Invest., 76, 375 (1997).
82. V.C. Broaddus, A.M. Boylan, J.M. Hoeffel, K.J. Kim, M. Sadick, J.Immunol., 152, 2960 (1994).
83. T. Wada, N. Tomosugi, T. Naito, H. Yokoyama, K. Kobayashi, A. Harada, N. Mukaida, K. Matsushima, J.Exp.Med., 180, 1135 (1994).
84. J.R. White, J.M. Lee, P.R. Young, R.P. Hertzberg, A.J. Jurewicz, M.A. Chaikin, K. Widdowson, J.J. Foley, L.D. Martin, D.E Griswold, H.M. Sarau, J.Biol.Chem., 273, 10095 (1998).
85. M.R. Palovich, K. Widdowson, J.D. Elliott, J.R. White, H.M. Sarau, M.C. Rutledge, J. Bi, K.A. Dede, G.M. Benson, D.E. Griswold, L.D. Martin, H. Nie, D.M. Schmidt and J.J. Foley, 219th ACS Nat'l.Mtg., San Francisco, April 1999, MEDI-338.
86. Exp.Opin.Ther.Patents 10, 121 (2000)
87. K.G. Carson, D.T. Connor, J.J. Li, J.E. Low, J.R. Luly, S.R. Miller, B.D. Roth, B.K. Trivedi, WO 9942463 (1999).
88. I.F. Charo, S.J. Myers, A. Herman, C. Franci, A.J. Connolly, and S.R. Coughlin, Proc.Nat.Acad.Sci.U.S.A., 91, 2752 (1994).
89. P. Loetscher, B. Dewald, M. Baggiolini, and M. Seitz, Cytokine, 6, 162 (1994).
90. A.E. Koch, S.L. Kunkel, L.A. Harlow, B. Johnson, H.L. Evanoff, G.K. Haines, M.D. Burdick, R.M. Pope, and R.M. Strieter, J.Clin.Invest., 90, 772 (1992).
91. X. Yu, S. Dluz, D.T. Graves, L. Zhang, H.N. Antoniades, W. Hollander, S. Prusty, A.J. Valente, C.J. Schwartz, and G.E. Sonenshein, Proc.Nat.Acad.Sci.U.S.A., 89, 6953 (1992).
92. S. Yla Herttuala, B.A. Lipton, M.E. Rosenfeld, T. Sarkioja, T. Yoshimura, E.J. Leonard, J.L. Witztum, and D. Steinberg, Proc.Nat.Acad.Sci.U.S.A., 88, 5252 (1991).
93. A.E. Koch, S.L. Kunkel, W.H. Pearce, M.R. Shah, D. Parikh, H.L. Evanoff, G.K. Haines, M.D. Burdick, and R.M. Strieter, Am.J.Pathol., 142, 1423 (1993).
94. S.J. Wysocki, M.H. Zheng, A. Smith, M.D. Lamawansa, B.J. Iacopetta, T.A. Robertson, J.M. Papadimitriou, A.K. House, and P.E. Norman, J.Cell.Biochem., 62, 303 (1996).
95. D.J. Schrier, R.C. Schimmer, C.M. Flory, D.K.-L. Tung, and P.A. Ward, J. LeukocyteBiol., 63, 359 (1998).
96. J.H. Gong, L.G. Ratkay, J.D. Waterfield, and I. Clark-Lewis, J.Exp.Med., 186, 131 (1997).
97. N.A. Nelken, S.R. Coughlin, D. Gordon, J.N. Wilcox. J..Clin.Invest., 88, 1121 (1991).
98. L, Gu, Y, Okada, S.K. Clinton, G.K. Sukhova, P. Libby, B.J. Rollins, Mol.Cell., 2, 275, (1998).
99. L. Boring, J. Gosling, M. Cleary, I.F. Charo, Nature, 394, 894 (1998).
100. R.J. Aiello, P.K. Bourassa, S. Lindsey, W. Weng, E. Natoli, B.J. Rollins, P.M. Milos. Thromb Vasc Biol., 19, 1518, (1999).
101. M..Poon, J. Cohen, Z. saddiqui, J.T. Fallonn and M.B. Taubman. Lab Invest, 79, 1369, (1999).
102. H. Inadera, K. Egashira, M. Takemoto, Y. Ouchi and K. Matsushima. J of Interferon and Cytokine Research, 19, 1179 (1999).

103. L.Boring, I.F. Charo, B.J. Rollins in "Chemokines in Diseases: Biology and Clinical Research", C.A. Hebert Ed., Humana Press Inc., Totawa, NJ, 1999, p. 53.

104. I.F. Charo in "Chemokine", A. Mantowani Ed., Chem. Immunol., 72, 30, (1999).

105. T.J. Reape and P.H.E. Groot, Atherosclerosis, 147, 213, (1999).

106. J.M. Lapierre, D. Morgans, R. Wilhelm and T.R. Mirzadagean in 26th National Medicinal Chemistry Symposium, Richmond, VA, 1998.

107. S. Glase, D. Connor, et al. 1998, WO 9806703.

108. S.A. Glase, R.J. Booth, D.T. Connor, S.W. Hunt, D.L. Kellner, T.S. Purchase, B.D. Roth, B.K. Trivedi, D. Yoon, G.J. LaRosa, A.L. Reinhart and L.L. Yang in 26th National Medicinal Chemistry Symposium, Richmond, VA, 1998.

109. K.G. Carson, R.E. Glynn, C.F. Schwender, P.C. Unangst, S.R. Miller, B.K. Trivedi, B.D. Roth, S.W. Hunt, A.L. Reinhart, L. Yang and G.J. LaRosa. "Small Molecule CCR2 Antagonists," abstract presented at the Cambridge Health tech Institute Chemokine Symposium, McLean, VA, November, 1999.

110. J.A. Baker, J.G. Kettle and A.W. Faull. 1999, WO 9907351.

111. A.J. Baker, J.G. Kettle and A.W. Faull. 1999, WO 9907678.

112. O. Nishimura, M. Baba, H. Sawada, K. Kuroshima and Y. Aramaki. 1999, WO 9932100.

113. M. Shiraishi, Y. Aramaki, M. Seto, H. Imoto, Y. Nishikawa, N. Kanazaki, M. Okamoto, H. Sawada, O. Nishimura, M. Baba and M. Fujino in 39th ICAAC Meeting, San Francisco, CA, 1999. Abst. 911.

114. Y. Iizawa, H, Miyake, K. Kuroshima, M. Shiraishi, K. Okonogi and M. Fujino in 39th ICAAC Meeting in San Francisco, CA, 1999. Abst. 912.

115. N. Kanazaki, K. Kuroshima, M. Shiraishi, H. Sawada, O. Nishimura and M. Fujino in 39th ICAAC Meeting in San Francisco, CA, 1999. Abst. 913.

116. M. Baba, M. Okamaoto, M. Ino, K. Oonishi, Y. Iizawa, K. Okonogi, O. Nishimura, Y. Ogawa, K. Meguro and M. Fujino in 39th ICAAC Meeting in San Francisco, CA, 1999. Abst. 914.

117. M. Shiraishi, T. Kitayoshi, Y. Aramaki and S. Honda. 1999, WO 9932468.

118. T. Shiota. 1999, WO 9925686.

119. C.M. Tarby, N. Endo, W. Moree, K. Kataioka, M.M. Ramirez-Weinhouse, M. Imai, E, Bradley, J. Saunders, Y. Kato and P. Meyers in 218th National ACS Meeting in New Orleans, 1999. Abst. 82.

Chapter 18. Inosine Monophosphate Dehydrogenase: Consideration of Structure, Kinetics, and Therapeutic Potential

Jeffrey O. Saunders and Scott A. Raybuck
Vertex Pharmaceuticals, Cambridge, MA 02139

Introduction - Inosine 5'-monophosphate dehydrogenase (IMPDH, E.C.1.1.1.205) catalyzes the NAD dependent oxidation of inosine 5'-monophosphate (IMP) to xanthosine 5'-monophosphate (XMP) and is the rate-limiting enzyme of the *de novo* pathway for guanine nucleotide biosynthesis. Guanine nucleotides are essential to the cell for RNA and DNA synthesis, intermediates in signaling pathways, and as energy sources for metabolic processes. Therefore, IMPDH has become an attractive target for immunosuppressive as well as antiviral and antibiotic therapies (1-3).

The elucidation of the *de novo* purine biosynthesis pathway including the role of IMPDH in purine anabolism occurred over 30 years ago. Despite this, IMPDH inhibitors currently on the market were either discovered from natural sources or have evolved as analogs of nucleosides that demonstrated potent activity in cellular screens. Identification of the inhibition of IMPDH as the primary mechanism of action has largely come from a retrospective analysis. Subsequently, the importance of IMPDH as a key intervention point for antiproliferation has been advanced from a number of biological studies using clinically relevant IMPDH inhibitors (4). Over the last few years the X-ray crystal structures of IMPDH from Chinese hamster, human, and bacterial sources have been solved (5,6). The mechanism of IMPDH catalysis has been established through structural and enzymatic studies (7). This additional information is just beginning to be exploited in the design of more potent inhibitors. The scope of this review will focus on recent developments in the field of IMPDH inhibition and the strategies for future inhibitor design, with an emphasis on those compounds currently in advanced stages.

Role of IMPDH and Therapeutic Indications - The concentration of purine nucleotides in the cell is tightly regulated through two biochemical routes: *de novo* synthesis and the salvage pathway (1, 2). In the salvage pathway, existing purines and their nucleosides and nucleotides are recycled. In the *de novo* synthesis of adenine and guanine nucleotides, the purine ring is assembled in 10 steps from the high-energy intermediate, phosphoribosyl pyrophosphate (PRPP), to yield inosine 5'-monophosphate (IMP). IMP is the branch point between adenine and guanine nucleotide biosynthesis (Figure 1). IMP can be converted to AMP via adenylo-succinate (SAMP) through the sequential actions of adenylosuccinate synthase and adenylosuccinate lyase. *De novo* synthesis of guanine nucleotides from IMP occurs through xanthosine 5'-monophosphate. The oxidation of IMP to XMP is the first committed step in this pathway and is catalyzed by inosine 5'-monophosphate dehydrogenase (IMPDH). XMP can then be converted to GMP by GMP synthase. The parallel purine pathways are linked through their cofactor requirements. Synthesis of SAMP utilizes aspartate and GTP as the energy source; the amination of XMP to give GMP requires both glutamine and ATP. Both AMP and GMP can also be converted back to IMP, if necessary, through the enzymes AMP deaminase and GMP reductase, respectively. Purine nucleoside diphosphates are provided through the action of adenylate kinase or guanylate kinase on the analogous nucleotide. Conversion of diphosphates to dADP and dGDP by ribonucleotide reductase gives intermediates that can then be phosphorylated to their triphosphates, generating the precursors of DNA synthesis.

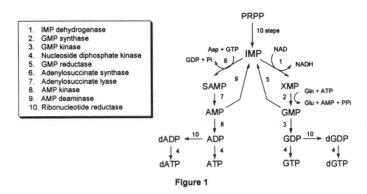

1. IMP dehydrogenase
2. GMP synthase
3. GMP kinase
4. Nucleoside diphosphate kinase
5. GMP reductase
6. Adenylosuccinate synthase
7. Adenylosuccinate lyase
8. AMP kinase
9. AMP deaminase
10. Ribonucleotide reductase

Figure 1

Two human isoforms of IMPDH (type I and II) which share 84% amino acid identity have been identified (8). Expression of mRNA for both isoforms is seen in all tissues examined with type II being at consistently higher levels. Type II is markedly upregulated at the message and protein levels in activated T lymphocytes (9-12). These observations led to the hypothesis that type I is constitutively expressed to maintain a basal level of guanine nucleotides while type II is induced upon proliferation. However, recent reports of upregulation of type I under selective conditions have appeared and may reflect that the two isoforms have distinct roles in cellular metabolism (13). Both human type I and type II proteins are homotetramers of 56 kDa with nearly identical kinetic parameters. Current inhibitors encompassing all type of mechanisms of action do not discriminate between the two isozymes.

The mechanism of IMPDH catalysis has been studied in detail (7). The oxidation of IMP to XMP utilizes NAD as a cofactor and is apparently irreversible. Steady state kinetics of the enzyme from bacterial and mammalian sources had indicated an ordered Bi-Bi kinetic mechanism where IMP binds before NAD and NADH is released before XMP (14, 15). Recent isotope studies now suggest that substrates bind randomly with an ordered release of products (16). A key feature of the IMPDH reaction is the use of an active site cysteine for activation of IMP during enzymatic oxidation (Figure 2). Thiolate attack at C-2 of the inosine ring with subsequent transfer of the C-2 hydride to NAD yields a thioimidate covalently bound to the enzyme. Upon release of NADH, the purine ring remains attached through the cysteine linkage and is at the oxidation level of XMP. This intermediate has been designated E-XMP* (17). The final steps of the reaction are rate limiting for the human enzyme and involve hydrolysis of the enzyme-bound intermediate and release of the last product, XMP (16).

E • IMP	E • IMP • NAD	E-XMP*	E • XMP

Figure 2

The importance of IMPDH and the *de novo* purine biosynthetic pathway in cellular metabolism stems from observations that both mRNA expression and IMPDH activity is significantly enhanced in actively proliferating cell types such as sarcoma and leukemic cells, ovarian tumors, and mitogen activated peripheral blood lymphocytes (9-11). The latter is particularly interesting as B and T lymphocytes appear to depend solely on the *de novo* pathway for guanine nucleotides, making IMPDH an appealing intervention point for immunosuppressive action. In addition, the antiproliferative effects of IMPDH inhibition through the reduction of intracellular guanine nucleotide levels have made this enzyme an attractive drug target for anticancer, antiviral and antibiotic therapies (18).

Structure of IMPDH and Modes of Inhibition - The position of the rate determining step in the catalytic cycle of human IMPDH leads to a significant accumulation of the E-XMP* covalent intermediate and enhances the likelihood for the discovery of small molecules to act as uncompetitive inhibitors, trapping this intermediate before hydrolysis can occur.   This opportunity occurs rarely in other nicotinamide dehydrogenases; the more common mechanism involves direct hydride transfer, which can be partially rate-limiting, and the absence of covalent intermediates. One potent inhibitor is mycophenolic acid, developed and marketed as the morpholino ester prodrug as a treatment for organ rejection. Mycophenolate (MPA) is a potent inhibitor of IMPDH with Ki values in the low nanomolar range for human type II and type I isoforms (19). In 1996, the first X-ray crystal structure of IMPDH from Chinese hamster with MPA bound under turnover conditions was reported (17).   Chinese hamster IMPDH is highly homologous to the human II enzyme, differing in only 6 amino acids of 514.   The global structure of the enzyme contains two domains: a larger domain which is a $(\beta/\alpha)_8$ barrel, and a smaller less ordered domain of unknown function which can be deleted with no obvious effect on catalysis (20). The active site is formed by the loops between the strands of the barrel. The overall structure of the enzyme is similar to other oxido-reductases, such as glycolate oxidase. Two distinct sites for the binding of IMP (substrate) and NAD (cofactor) are observed. One notable difference from other dehydrogenases is that the substrate binding site in IMPDH maps to the cofactor site in other enzymes, implying the NAD site of IMPDH provides a unique opportunity for selective inhibitor design.

The active site of the E-XMP*/MPA complex clearly shows the covalent attachment of Cys 331, which had been identified biochemically, to the C-2 hypoxanthine ring (21). MPA is found to bind in the NAD binding site, stacking against the enzyme-bound purine intermediate. The X-ray crystal structure provides a visualization of the opportunities for additional uncompetitive inhibitors of IMPDH which can act in a similar manner in trapping the E-XMP* intermediate prior to hydrolysis.   Realization of this opportunity poses a significant challenge.   Such compounds must contain suitable aromatic heterocycles that can support both the π-stacking with the purine and maintain the key hydrogen bonds available in the active site.   From a drug design point of view, possible starting points include nicotinamide and NAD mimics, since the first product, NADH, is known to be a weak inhibitor (22).   However, the most attractive non-nucleoside leads will most likely come from *in vitro* screening.

Competitive inhibitors of IMPDH are known and are primarily nucleoside analogs (23). To serve as inhibitors of IMPDH, all such nucleosides must minimally undergo cellular conversion to their mono-phosphorylated forms, largely through the action of adenosine kinase.   Once phosphorylated, potential inhibitors can parallel the metabolic fate of AMP and GMP with the potential for conversion to their di- and triphosphates, as well as serving as substrates for NAD phosphorylase, undergoing conversion to the adenylated forms.   This scenario can lead to multiple inhibitor species whose specific point of action in the cell is difficult to determine. Nevertheless, such compounds have been shown to be efficacious and are currently on the market, e.g. ribavirin and mizoribine. While non-nucleoside analogs are cleaner mechanistically, as the specific examples which follow will illustrate, there is currently no one strategy or compound class for IMPDH inhibition that is superior.

INHIBITORS OF IMPDH - NUCLEOSIDE ANALOGS

Historically, pharmaceutical discovery programs were not directed specifically at the *in vitro* disruption of IMPDH. Rather, cellular assays for evaluating broad spectrum antiviral, antimicrobial and antineoplastic activity were the means through which compounds were discovered and studied. The most prevalent species of inhibitors pursued were the nucleoside analogs, in particular those of guanosine. Despite the elucidation of IMPDH as a primary target of these nucleoside analogs (NAs), and a desirable target for pharmaceutical intervention, the study of NAs has continued to rely on cellular systems where anabolic conversion to activated species takes place. As will be noted later, NAs participate as substrates and/or inhibitors in multiple enzymatic pathways, and a clear understanding of the extent to which this occurs is very difficult to attain. The data, however, clearly show that NAs are not specific inhibitors of IMPDH.

Ribavirin - Ribavirin (1-ß-D-ribofuranosyl-1,2,4-triazole-3-carboxamide, **1**) is a synthetic guanosine analog designed nearly thirty years ago as part of an antiviral program (24). It shows broad-spectrum *in vitro* and *in vivo* activity against a variety of DNA and RNA viruses, including HBV, HCMV, HSV1, HSV2, HVV, Influenza A, Parainfluenza-3, as well as the retrovirus HIV. It was approved in the U.S. in 1986 as Virazole® for use as an agent against respiratory syncytial virus (RSV) and in 1999 as Rebetol® for use in combination with interferon-alpha2b (IFN-α2b) as a treatment for hepatitis C virus (HCV).

Antiviral activity likely derives from one or more of three possible modes of action, each requiring the metabolism of ribavirin to its 5'-mono-, di- or triphosphate. Its ultimate efficacy against any particular virus depends on cellular accessibility and its phosphorylation state in a particular cell type (25). Inhibition by ribavirin triphosphate (RTP, **3**) of viral RNA transcription is well-documented (26-31). This mechanism occurs through the inhibition of RNA polymerase initiation and elongation, resulting in direct inhibition of viral replication. Alternatively, it has been proposed that RTP acts through inhibition of viral mRNA guanylyl transferase, resulting in the disruption of guanine pyrophosphate capping of the 5' end of viral messenger RNA, thus diminishing the efficiency of translation of viral transcripts. Thirdly, ribavirin monophosphate is well characterized as an inhibitor of IMPDH (Ki = 250 nM), providing antiviral effects through depletion of intracellular GTP and dGTP pools (32). The antiviral effect towards many viruses is reversible through the addition of exogenous guanosine, which strongly supports a role for the inhibition of IMPDH (33). In 1999, the X-ray crystal structure of ribavirin 5'-monophosphate (RMP, **2**) bound to a deletion mutant of human IMPDH type II, solved at 1.85Å resolution was reported (5). The structure shows **2** bound in a fashion indistinguishable from that of bound IMP, making use of the same binding interactions with the enzyme.

In clinical trials as a monotherapy for hepatitis C, **1** showed no sustainable effect on serum HCV RNA levels. However, significant sustained decreases in serum transaminase levels and improved liver histology were observed. In combination with IFN-α2b, enhanced virological and biochemical responses were noted in both initial treatment and treatment of relapsers (34). The side effect profile for **1** is limited but includes hemolytic anemia and teratogenicity. Both toxicities have led to restricted use: hemolytic anemia, when accompanied by bone marrow suppression from IFN-α2b, can preclude treatment of patients at risk for heart disease, and women of childbearing age must be treated with caution.

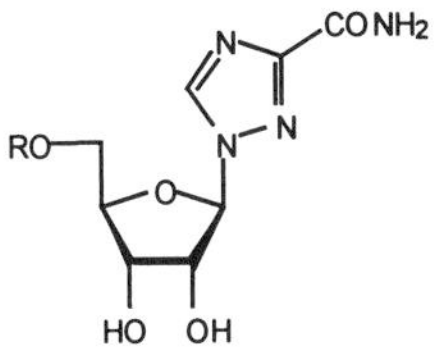

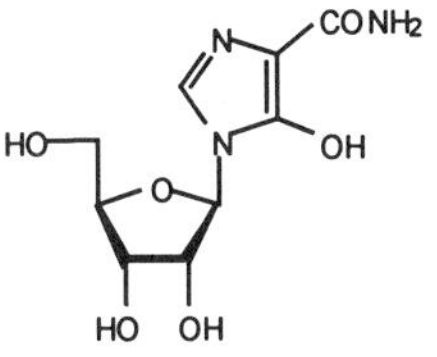

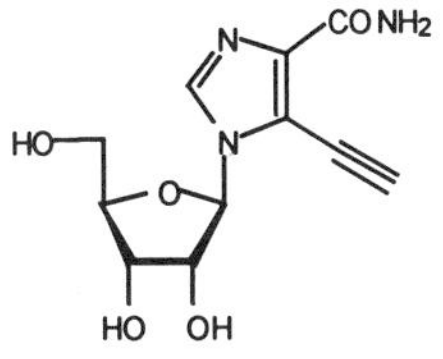

1 R = H
2 R = monophosphate
3 R = triphosphate

4

5

Mizoribine - The natural product Mizoribine (Mzr, Bredinin, **4**), discovered as a fungal metabolite based on its antibiotic activity, has also shown antitumor, antiviral and immunosuppressive activities (35-37). Mzr is an effective competitive inhibitor, following *in vivo* conversion to the 5'-monophosphate (MzrMP), of *de novo* guanine nucleotide synthesis, acting at successive steps on IMPDH and GMP-synthetase, with Ki's of 10nM and 10µM, respectively (37, 38). MzrMP has also been reported to be an inhibitor of DNA polymerases α and β (39). While Mzr is registered and utilized in Japan for treatment of lupus nephritis, nephrotic syndrome, rheumatoid arthritis and for the prevention of renal transplant rejection, its use is accompanied by GI toxicity. Mizoribine has not yet been approved for use in the U.S.

EICAR - Other N- and C-nucleoside analogs have been reported in the literature and thoroughly covered in the recent reviews (23, 36). Of these, EICAR (5-ethynyl-1-ß-D-ribofuranosylimidazole-4-carboxamide, **5**) has had the greatest clinical exposure. This compound is a potent cytostatic agent towards a variety of human solid tumor cells both *in vitro* and *in vivo*, and shows broad spectrum antiviral activity. While **5** appears to act through multiple modes, via several anabolites, the primary mode of action is through the inhibition of IMPDH as the 5'-monophosphate. Designed as a nucleoside analog, **5** is converted to the monophosphate and binds as a competitive inhibitor at the IMP site. The acetylenic moiety, which SAR has shown to be required for inhibitory activity, is believed to experience nucleophilic attack by the active site cystine-331 sulfhydryl group in an irreversible fashion. In cellular assays, in which the accumulation of IMP and the depletion of GTP and dGTP levels can be monitored, EICAR demonstrates an IC_{50} towards IMPDH of 0.80-1.40 µM (40). The antiviral effects of **5** are reversible in the case of many viruses through addition of exogenous guanosine, whereas no reversal appears possible through addition of other nucleosides such as adenosine, inosine, cytidine and uridine. This strongly supports the premise that the primary mode of action through which EICAR exhibits its activity is via the inhibition of IMPDH. Utilizing [^{3}H]-labeled EICAR, the metabolic fate of EICAR in L1210 cells was evaluated. Multiple forms, including the diphosphate, triphosphate and the NAD analog were detected. While analogies can be drawn to other active metabolites of nucleoside analogs, the inhibitory activities of these EICAR-derived species towards RNA-polymerases or guanyltransferases have not been measured.

NAD analogs - The propensity for nucleoside analogs to undergo conversion to their NAD analogs in an *in vivo* setting is well documented. This metabolic fate has been exploited in the case of a number of nucleoside analogs including, but not restricted to, tiazofurin (TR, **6**), selenazofurin (SR, **8**), and benzamide riboside (BR, **9**). In exemplary fashion, TR is initially phosphorylated by adenosine kinase to the 5'-monophosphate, which then participates as a substrate for NAD-phosphorylase giving rise to TAD (**7**). A thorough review of the substantial work to understand the SAR of NAD analogs as antineoplastic agents has recently been published (23).

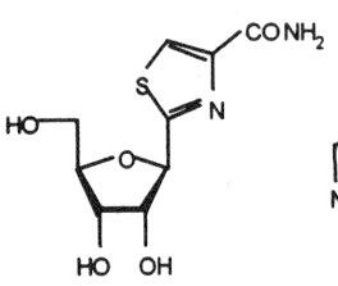

6 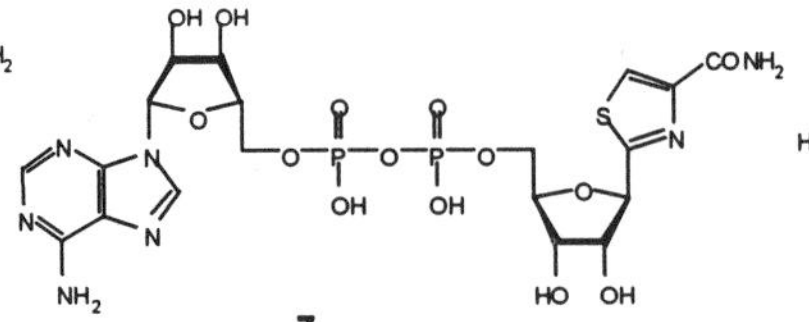   **7** 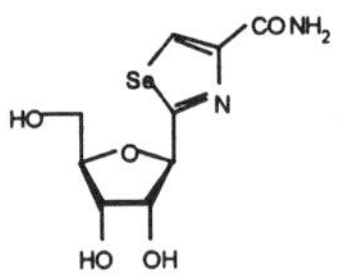   **8** 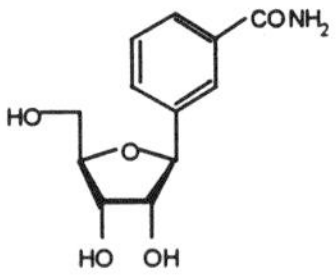   **9**

The X-ray crystal structure of a ternary complex of IMPDH/SAD/6-Cl-IMP, assumed to mimic well the IMPDH/IMP/NAD⁺ enzymatic complex, illustrates the important interactions between each of the components (41). TAD is the sole member of this class to advance into clinical trials and, while showing dose-limiting toxicities in Phase I, demonstrated mixed efficacy in an assortment of antineoplastic Phase II/III clinical trials (42, 43).

INHIBITORS OF IMPDH - NON-NUCLEOSIDE

A second category of IMPDH inhibitors consists of non-nucleosides, which have been found to bind in the NAD site. They are unified by a common mechanism of action, behaving as uncompetitive inhibitors that trap the E-XMP* intermediate before hydrolysis and product release can occur. In that such molecules are not susceptible to the spectrum of *in vivo* modification seen with nucleoside analogs, the opportunity for designing more selective IMPDH inhibitors exists. To date three series have been reported which are based on three very different classes of molecules. However, each class shares structural elements including an aromatic ring system and heteroatoms capable of forming hydrogen bonds that are important for binding in the nicotinamide portion of the NAD binding site. Retrospective structural analysis from two of the series indicates that the respective aromatic systems bind through tight π-stacking with the purine ring of the intermediate. Strategically placed heteroatoms also enter into hydrogen bonding arrangements with polar residues that line the NAD binding site. Maintaining both of these binding components is critical for potency, as corroborated by the very narrow SAR results observed in all three series.

Although a non-nucleoside strategy would appear to benefit greatly from a structure-driven approach, the technical challenges of maintaining an inhibited uncompetitive complex during crystal growth have limited its applicability to date. Once an inhibitor has dissociated from the complex, hydrolysis of the intermediate occurs to give free enzyme and XMP; the inhibitor cannot be driven back on to the correct state of the enzyme thermodynamically. This is typical of weak micromolar inhibitors with fast off-rates that often appear in the early stages of a project where structural insights would be most useful. Thus, to date, the structures of complexes of non-nucloside inhibitors with IMPDH have contained potent uncompetitive inhibitors (Ki < 50 nM).

Mycophenolate Mofetil - Mycophenolate Mofetil (MMF, CellCept™, **11**) is a prodrug of the natural product mycophenolic acid (MPA, **10**), which is a potent, uncompetitive, reversible inhibitor of IMPDH. It was discovered over a century ago and has been extracted from an assortment of *Penicillium* strains (44, 45). MPA binds to human IMPDH type II with a Ki of 7-10 nM and to type I with activities ranging in the literature from 11 nM to 33-37 nM (19, 46). The uncompetitive mechanism involves the trapping of the tetrahedral XMP precursor subsequent to release of NADH. Conditions to trap, crystallize and fully examine this inhibited complex by X-ray crystallography have been reported (17). This information, in conjunction with kinetic and mutagenesis data provide substantial insight into both the enzymatic mechanism of IMPDH and an appreciation for the extremely narrow

SAR generated from the medicinal chemistry effort (47). MPA is found to bind in the NAD binding site. The 6,5 ring system of mycophenolic acid stacks against the thioimidate purine through critical π-interactions. The phenolic hydroxide is observed to hydrogen bond with Thr 333 and Gln 441 of the enzyme, and it has been postulated that this is a mimic for the catalytic water necessary to hydrolyze the enzyme bound intermediate to XMP. All changes to the benzofuranone ring system and any of its substituents lead to a loss of inhibitory activity. The hexenoic tail of MPA is observed in a bent conformation that places the carboxylate within hydrogen bonding distance of the hydroxyl side chain of Ser 276. Modifications of this chain, whether lengthening, shortening, changing the degree of unsaturation, or incorporation of heteroatoms, also lead to significant loss of activity. Only the incorporation of rings in the side-chain, which provide an appropriate conformational restriction, mimicking that seen in the X-ray crystal structure, gives rise to an improved inhibitory activity.

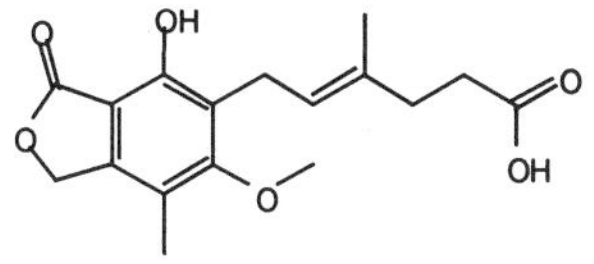

10 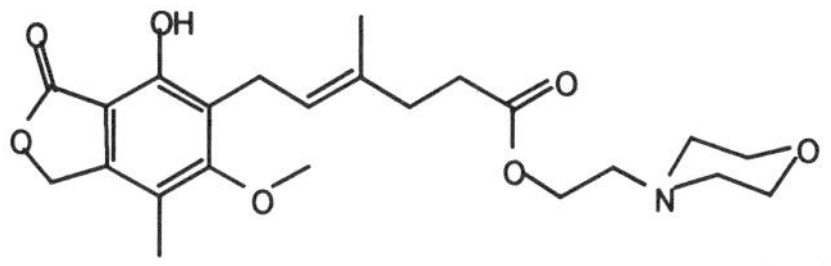   11

Since its last review, **11** has been approved in the U.S. and Europe for the prevention of acute rejection in renal and heart transplantation (48-50). It has also been used successfully to treat small numbers of patients with other immune disorders (51-57). GI toxicity has been well established for **11** and has been linked to its metabolism (58-60). Following administration, the ester prodrug undergoes rapid and complete hydrolysis to MPA, which is in turn extensively glucuronidated (MPAG), a species which does not inhibit IMPDH. The glucuronide undergoes billiary excretion, during which glucuronidases in the bacteria of the lower GI tract give rise to free MPA. Enterohepatic recirculation of MPA is then responsible for an increase in the concentration of this very effective antiproliferative agent in the GI tract, likely resulting in the GI intolerance. More recent reports indicate that effective dosing can be achieved with minimal GI distress (61).

VX-497 - A program based on targeting *in vitro* IMPDH inhibition began with a high-throughput screening effort that yielded **12**, a commercially available, low molecular weight inhibitor with a Ki of 3.5 μM. Refinement of the intrinsic IMPDH inhibitory activity of **12** was driven primarily by biochemical evaluation of the series and secondarily by cellular antiproliferative activity against T and B lymphocytes. This effort led to VX-497 (**13**), a reversible, uncompetitive inhibitor of human IMPDH type II with a Ki of 7-10 nM. VX-497 binds in the same site as mycophenolic acid and inhibits via the same kinetic mechanism. However, as a structurally unique class of IMPDH inhibitor, it does not suffer the same metabolic fate as mycophenolic acid. VX-497 is not glucuronidated nor does it undergo enterohepatic recirculation and thus does not show the same clinical limitations of GI toxicity.

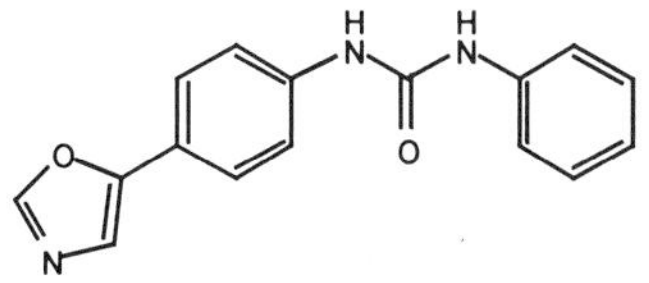

12 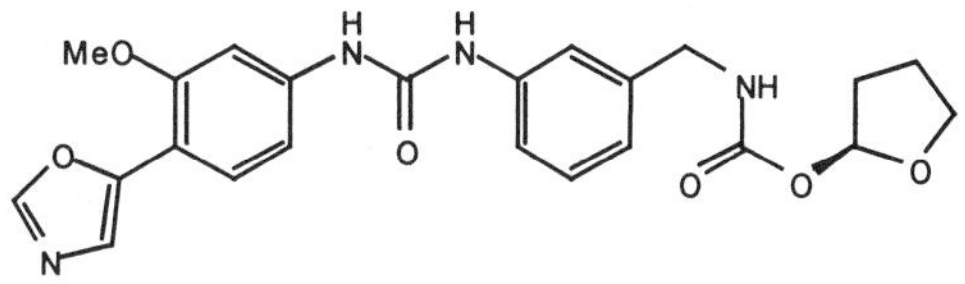   13

Like **10**, an extremely narrow SAR was obtained in this series. As a general rule substitution at only two positions on **12** led to improvements in inhibitory capacity.

On the oxazolylphenyl side of **12** improvements came through substitutions at the position *meta-* to the urea, with methoxy providing nearly an order of magnitude increase in Ki. On the phenyl side of the urea, improvements were again limited to modifications at the position *meta-* to the urea, although a greater range of functionality was accommodated.   Ultimately, the (S)-hydroxy-tetrahydrofuran carbamate of the benzyl amine offered an optimum balance of *in vivo* characteristics. Modifications of the linking urea moiety were detrimental to binding. Structural evaluation of the inhibited IMPDH/VX-497/XMP* complex indicates the importance of π-stacking, hydrogen bonding and other electrostatic interactions. Much of this narrow SAR is readily explained through evaluation of the crystal structure of **13** and related compounds bound to IMPDH. These binding interactions have been described in detail (5).

VX-497 showed considerable success preclinically in a number of models as an immunosuppressive agent.   In a murine skin allograft model, mean survival of vehicle treated animals was 9.9 ± 0.9 days. Graft survival was extended to 13.2 ± 1.3 days in animals treated with 50 mg/kg po **13** and 13.9 ± 1.0 days in animals treated with 85 mg/kg po **13**. In this same study, graft versus host disease (GvHD) developed in the vehicle treated allografted mice. Multiple markers reflective of the presence of GvHD showed significant improvement versus controls when affected mice were treated with 100mg/kg po **13**.   A heterotropic solid organ (heart) transplant study in rats in which well-tolerated, non-toxic doses of 25, 50 and 75 mg/kg po of **13** prolonged graft survival significantly compared to untreated controls, was recently reported (62).

The broad spectrum, *in vitro* antiviral activity of **13** has recently been reported in a study comparing its activity to Ribavirin (**1**) (63). It was demonstrated that **13** exhibits 10 to 100-fold greater potency than **1** against HBV, HCMV, RSV, HSV-1, Parainfluenza 3 virus, EMCV and VEE viral infections in cell culture. Since an *in vitro* HCV replication assay has not been reported, an EMCV replication assay was utilized to compare the additivity of combinations of **13** or **1** with INFα.   Again, **13** was significantly more effective than **1**. VX-497 is presently in Phase II clinical trials for treatment of hepatitis C (64).

Pyridazines - One other discovery program has been described in which an HTS evaluation of 80,000 compounds against IMPDH exposed the single pyridazine hit **14** (65). This compound is an uncompetitive inhibitor of IMPDH with an IC_{50} of 1.93 µM and shows an IC_{50} against the proliferation of EMT6 cells of 39.7 µM. Medicinal chemistry efforts led to three other pyridazine derivatives with improved potency, **15** being the most active at 0.76 µM against IMPDH.

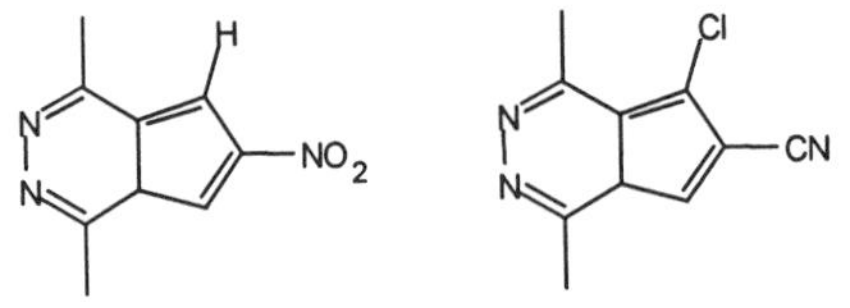

14   **15**

Additionally, *in vitro* and *in vivo* inhibition of guanine nucleotide synthesis was demonstrated.   Compound **15** showed significant immunosuppressive activity, performing better than MPA in the inhibition of delayed type hypersensitivity in mouse, an unexpected result based on the considerably lower intrinsic IMPDH inhibitory activity. No reports of clinical experience with this class of compounds have been reported to date.

<u>Conclusions</u> - IMPDH continues to generate substantial interest as a target for immunosupressive, antiviral, and antimicrobial therapies. Nucleoside analogs have shown promise as competitive inhibitors of this enzyme, but suffer significant anabolic modification. Whether or not the multiple activities of NAs and their anabolites are a desirable attribute is yet to be fully appreciated. Expanded clinical and market exposure of drugs such as Ribavirin will provide further insight. The potential for very selective IMPDH inhibitors continues to progress through investigation of non-nucleoside compounds. The first and most thoroughly evaluated example of this type, mycophenolic acid, was discovered more than a century ago through an antimicrobial screening effort. Subsequently, with the improved understanding of IMPDH and refined tools for uncovering and studying inhibitors of IMPDH, pharmaceutical programs have begun to produce other highly selective non-nucleoside inhibitors.

<h2 style="text-align:center">References</h2>

1.   A.G. Zimmerman, J.J. Gu, J. Laliberte and B.S. Mitchell, Prog. Nucleic Acid Res. Mol. Biol., <u>61</u>, 181 (1998).
2.   G. Weber, H. Nakamura, Y. Natsumeda, T. Szekeres and M. Nagai, Advan. Enzyme Regul., <u>32</u>, 57 (1992).
3.   H.N. Jayaram, D.A. Cooney and M. Grusch, Curr. Med. Chem., <u>6</u>, 561 (1999).
4.   A.C. Allison and E.M. Eugui, Clin. Transplant. Proc., <u>10</u>, 118 (1996).
5.   M.D. Sintchak, E. Nimmesgern, Immunopharmacology, in press (2000).
6.   B.M. Goldstein and T.D. Colby, Curr. Med. Chem. <u>6</u>, 519 (1999).
7.   L. Hedstrom, Curr. Med. Chem., <u>6</u>, 545 (1999).
8.   Y. Natsumeda, S. Ohio, H. Kawasaki, Y. Konno and G. Weber, J. Biol. Chem., <u>265</u>, 5292 (1990).
9.   M. Nagai, Y. Natsumeda, Y. Konno, R. Hoffman, S. Irino and G. Weber, Cancer Res., <u>51</u>, 3886 (1991).
10. M. Nagai, Y. Natsumeda and G. Weber, Cancer Res., <u>52</u>, 258 (1992).
11. Y. Konno, Y. Natsumeda, M. Nagai, Y. Yamaji, S. Ohno, K. Suzuki and G. Weber, J. Biol. Chem., <u>266</u>, 506 (1991).
12. J.J. Gu, J. Spychala and B.S. Mitchell, J. Biol. Chem., <u>272</u>, 4458 (1997).
13. J.S. Dayton, T. Lindsten, C.B. Thompson and B.S. Mitchell, J. Immunol., <u>152</u>, 984 (1994).
14. J. H. Anderson and A.C. Sartorelli, J. Biol. Chem., <u>243</u>, 4762 (1968).
15. B. Xiang, J.C. Taylor and G.D. Markham, J. Biol. Chem., <u>271</u>, 1435 (1996).
16. W. Wang and L. Hedstrom, Biochemistry, <u>36</u>, 8479 (1997).
17. M.D Sintchak, M.A. Fleming, O. Futer, S.A. Raybuck, S.P. Chambers, P.A. Caron, M.A. Murcko and K.P. Wilson, Cell, <u>85</u>, 921 (1996).
18. A.C. Allison and E.M. Eugui, Immunological Revs., <u>136</u>, 258 (1993).
19. P.W. Hager, F.R. Collart, E. Huberman and B.S. Mitchell, Biochem. Pharmacol., <u>49</u>, 1323 (1995).
20. E. Nimmesgern, J. Black, O. Futer, J.R. Fulghum, S.P. Chambers, C.L. Brummel, S.A. Raybuck and M.D. Sintchak, Protein Exp. Purif., <u>17</u>, 282 (1999).
21. J.O. Link and K. Straub, J. Am. Chem. Soc., <u>118</u>, 2091 (1996).
22. A. M. Westley and J. Westley, J. Biol. Chem., <u>271</u>, 5347 (1996).
23. P. Franchetti and M. Grifantini, Curr. Med. Chem., <u>6</u>, 599 (1999).
24. J.T. Witkowski, R.K. Robins, R.W. Sidwell, L.N. Simon, J. Med. Chem., <u>15</u>, 1150 (1972).
25. R.W. Sidwell in "Clinical Application of Actions of Ribavirin," R.A. Smith, V. Knight, J.A.D. Smith, Eds., Academic Press, New York, N.Y., 1984, pp 19.
26. R. Fernandez-Larsson, K. O'Connell, E. Koumas, J.L. Patterson, Antimicrob. Agents Chemother., <u>33</u>, 1668 (1989).
27. P. Toltzis, K. O'Connell, J.L. Patterson, Antimicrob. Agents Chemother., <u>32</u>, 492 (1988).
28. L.F. Cassidy, J.L. Patterson, Antimicrob. Agents Chemother., <u>33</u>, 2009 (1989).
29. J.T. Rankin, S.B. Eppes, J.B. Antczak, W.K. Joklik, Virology, <u>168</u>, 147 (1989).
30. B. Eriksson, B. Helgstrand, N.G. Johansson, A. Larsson, A. Misiorny, J.O. Noren, L. Phillipson, K. Stenberg, S. Stridh, B. Oberg, Antimicrob. Agents Chemother., <u>11</u>, 946 (1977).
31. S.K. Wray, B.E. Gilbert, V. Knight, Antivir. Res., <u>5</u>, 39 (1985).
32. E. DeClercq, Adv. Virus Res., <u>42</u>, 1 (1993).

33.   W. Markland, T.J. McQuaid, J. Jain, A.D. Kwong, Antimicrob. Agents Chemother., in press (2000).
34.   J.G. McHutchison, S.C. Gordon, E.R. Schiff, M.L. Shiffman, W.M. Lee, V.K. Rustgi, Z.D. Goodman, M.H. Ling, S. Cort, J.K. Albrecht, N. Engl. J. Med., 339, 1485 (1998).
35.   Jay R. Luly, Annu. Rep. Med. Chem., 26, 211 (1991).
36.   N. Minakawa, A. Matsuda, Curr. Med. Chem., 6, 575 (1999).
37.   H. Ishikawa, Curr. Med. Chem., 6, 575 (1999).
38.   K.M. Kerr, L. Hedstrom, Biochemistry, 36, 13365 (1997).
39.   T. Horie, Y. Mizushina, M. Takemura, F. Sugawara, A. Matsukage, S. Yoshida, K.
40.   Balzarini, A. Karlsson, L. Wang, C. Bohman, K. Horska, I. Votruba, A. Fridland, A. Van Aerschot, P. Herdewijn, E. DeClercq, J.Biol.Chem., 268, (33), 24591 (1993).
41.   T.D.Colby, K. Vanderveen, M.D. Strickland, G.D. Markham, B.M. Goldstein, Proc. Natl. Acad.Sci., 96, 3531 (1999).
42.   G. Weber, N. Prajda, M. Abonyi, K.Y. Look, G. Tricot, Anticancer Res., 16, 3313 (1996).
43.   G. Tricot, H.N. Jayaram, E. Lapis, Y. Natsumeda, C.R. Nichols, P. Kneebone, N. Heereman, G. Weber, R. Hoffman, Cancer Res., 49, 3696 (1989).
44.   D.B. Gosio, Rivista d'Igiene e Sanita Pubblica 7, (21), 825 (1896).
45.   P.W. Clutterbuck, A.E. Oxford, H. Raistrick, G. Smith, Biochem.J., 26, 1441 (1932).
46.   S.F. Carr, E. Papp, J.C. Wu, Y. Natsumeda, J. Biol. Chem., 268, 27286 (1993).
47.   P.H. Nelson, S. F. Carr, B.H. Devens, E.M. Eugui, F. Franco, C. Gonzalez, R.C. Hawley, D.G. Loughhead, D.J. Milan, E. Papp, J.W. Patterson, S. Rouhafza, E.B. Sjogren, D.B. Smith, R.A. Stephenson, F.X. Talamas, A. Waltos, R.J. Weikert, J.C. Wu, J. Med. Chem., 39, 4181 (1996).
48.   W.H. Parsons, Annu. Rep. Med. Chem., 29, 175 (1994).
49.   H.W. Sollinger, Transplantation Proceedings, 28, 24 (1995).
50.   J. Kobashigawa, L. Miller, D. Renlund, R. Mentzer, E. Alderman, R. Bourge, M. Costanzo, H. Eisen, G. Dureau, R. Ratkovec, M. Hummel, D. Ipe, J. Johnson, A. Keogh, R. Mamelok, D. Mancini, F. Smart, H. Valantine, Transplant, 66, 507 (1998).
51.   D. Glicklich, A. Acharya, Am. J. Kidney Dis., 32, 318 (1998).
52.   W.A. Briggs, M.J. Choi, P.J. Scheel, Am. J. Kidney Dis., 31, 213 (1998).
53.   M.A. Dooley, F.G. Cosio, P.H. Nacham, M.E. Falkenhaim, S.L. Hogan, R.J Falk, J. Am. Soc. Nephrol., 10, 833 (1999).
54.   A. Ho, L. Madger, M. Petri, Arthritis Rheum, 41, S281 (1998).
55.   M.H. Schiff, B. Leischman, Arthritis Rheum, 41, S364 (1998).
56.   R. Nowack, R. Brick, F.J. van der Woulder, Lancet, 359, 774 (1997).
57.   E. Diana, A. Schieppati, G. Remuzzi, Ann. Intern. Med., 130, 422 (1999).
58.   P.A. Keown, Transplantation, 61, (7), 1029 (1996).
59.   R. Pichlmayr. Lancet, 345, 1321 (1995).
60.   H.W. Sollinger, Transplantation, 60(3), 225 (1995).
61.   S.C. Rayhill, H.W. Sollinger in "Transplantation," 147 (1999).
62.   C.J. Decker, A.D. Heiser, T. Faust, G. Ku, S. Moseley, Transplant., 67, (7): S57 (1999).
63.   W. Markland, T.J. McQuaid, J. Jain, A. D. Kwong, Antimicrob. Agents Chemother., 44, 859 (2000).
64.   T. Wright, M. Shiffman, S. Knox, E. Ette, R. Kauffman, J. Alam, Hepatology 30 (4), 408a, 1999.
65.   T.J. Franklin, W.P. Morris, V.N. Jacobs, E.J. Culbert, C.A. Heys, W.H.J. Ward, P.N. Cook, F. Jung, P. Plé, Biochem.Pharmacol., 58, 867 (1999).

Chapter 19. Recent Advances in Therapeutic Approaches to Type 2 Diabetes

John M. Nuss[1] and Allan S. Wagman[2]
Exelixis, Inc.[1], South San Francisco, CA 94080
Chiron Corporation[2], Emeryville, CA 94608

Introduction - Diabetes mellitus is the only non-infectious disease designated as an epidemic by the World Health Organization (1). The prevalence of all types of diabetes is estimated to be 2.3% of the world's population, with the number of diabetics increasing by 4-5% per annum. It is projected that as many as 40-45% of persons aged 65 or older have either Type 2 diabetes or its precursor state, impaired glucose tolerance (IGT). In the US ~10% of the diabetic population suffer from Type 1 diabetes, an autoimmune disease characterized by the loss of pancreatic β-cell function and an absolute deficiency of insulin. The remainder of the diabetic population suffers from Type 2 diabetes or IGT, which, although related to the body's inability to properly respond to insulin, have a more complex etiology (2). Diabetes can be treated by a combination of lifestyle change, dietary change and medication. However, the metabolic disorder underlying diabetes also affects protein and lipid metabolism, leading to serious complications, including peripheral nerve damage, kidney damage, impaired blood circulation, and damage to the retina of the eye. Diabetes is the leading cause of blindness and amputation in western populations, and the direct medical costs alone were estimated to be ~$44bn in the US alone in 1998 (2).

The United Kingdom Prospective Diabetes Study (UKPDS), a long term study of Type 2 diabetics, has shown that rigorous management of blood glucose levels (measured as glycosylated hemoglobin, HbA_{1c}), and blood pressure substantially reduce the incidence of complications (3). The current therapeutic strategies for Type 2 diabetes are very limited, and involve insulin therapy and oral hypoglycemic agents (OHAs) such as sulfonylureas, metformin, and the thiazolidinediones. Combination therapy with one or more of these agents is now a viable option as target blood glucose levels become harder to maintain with monotherapy (3,4). While a wide variety of therapeutic approaches are now being examined for Type 2 diabetes, these can generallly be classified in one of the following categories: 1) insulin or insulin mimetics; 2) agents which effect the secretion of insulin; 3) inhibitors of hepatic glucose production; 4) insulin sensitizers; and 5) agents which inhibit glucose absorption. The current chapter focuses on some of the most interesting, recent (post 1998) medicinal chemistry approaches that have been taken for the treatment of Type 2 diabetes.

INSULIN, INSULIN ANALOGUES OR MODIFIED INSULIN/IMPROVED DELIVERY

Insulin/Modified Insulins - Early or timely initiation of insulin therapy, either as monotherapy or in combination with other agents, can establish good glycemic control in a majority of Type 2 patients, and this mode of therapy does not adversely effect the quality of life of these patients (4,5). Short-acting analogs of insulin such as insulin lispro, an analog of insulin which has favorable pharmacokinetics, and is useful in the control of post-prandial glucose levels, have been reviewed (6).

Improved Delivery Vehicles - Inhalable, oral, intranasal, and transdermal forms of insulin offer considerably more convenience than do the traditional s.c. injectable forms of the drug, along with having more favorable pharmacokinetic characteristics. These alternatives have been extensively reviewed (7,8).

Insulin Mimetics - The natural product L-783,281 **1**, a non-peptidyl fungal metabolite, has been shown to selectively activate the insulin receptor *in vitro*, and oral adminstration of this agent showed a 40% lowering of blood glucose when administered orally for 7 days (9). The mechanism of action of **1** likely involves intracellular activation of the insulin receptor tyrosine kinase (IRTK). A second report of simplified synthetic analogs of these compounds, e.g., **2**, has appeared (10).

1   **2**

ENHANCERS OF INSULIN RELEASE

Sulfonylureas - Sulfonylureas (SU) remain a front line treatment for Type 2 diabetes (11). SUs stimulate insulin release from pancreatic β-cells by a mechanism that involves blocking ATP-sensitive potassium (K+) (KATP) channels. Recently, α–endosulfine, an 18-kD protein, which displaces the binding of sulfonylureas to β-cell membranes, inhibits cloned KATP channels, and stimulates insulin secretion, has been proposed as an endogenous ligand for the SU receptor (12).

Common side effects of SU therapy include hypoglycemia, as their action may occur at times when insulin is not required, and weight gain (11). Prolonged treatment with SU's also exacerbates β-cell exhaustion through over-stimulation of insulin production; SU therapy is not effective when insulin receptor levels decline too far. Newer agents have attempted to address these liabilities. Repaglinide **3** and nateglinide **4** are recently introduced short acting non-sulfonylureas secretagogues which are taken prior to meals to control post-prandial glucose levels and are reported to have fewer hypoglycemic side-effects (13,14). A family of short-acting sulfonylureas, typified by **5**, has recently been reported (15). Administration of **5** at 3 mg/kg po lowered blood glucose levels by > 30%, 10-20%, and < 10% at 30 min, 1 hour, and 2 hours, respectively.   JTT-608 **6** is an insulinotropic agent which stimulates insulin release only in the presence of elevated blood glucose levels, reducing the risk of hypoglycemia (16). In isolated rat β-pancreatic cells, **6** increases insulin secretion by enhancing Ca^{2+} efficacy and increasing Ca^{2+} influx as a result of increased intracellular cAMP levels due to inhibition of phosphodiesterase (PDE) activity.

GLP-1 - The effects of glucagon-like peptide-1 (GLP-1) and analogues on insulin secretion have been recently reviewed (17). N-terminal modified analogs of GLP-1 have been shown to increase the lifetime of this peptide (18). Exendin-4, an GLP-1 mimetic with favorable pharmacokinetic properties, may hold promise as novel therapy to stimulate β-cell growth and differentiation (19). Encapsulated, genetically engineered cells which can secret a mutant, long-lived form of GLP-1 and can be implanted in diabetic patients in the hopes of correcting hyperglycemia have been reported (20). Inhibitors of dipeptidyl peptidase-IV (DPP-IV), which is involved in the majority of GLP-inactivation *in vivo*, can also extend the lifetime of GLP-1 (21). In a detailed biochemical study, pyrrolidine **7**, which is a potent (K_i =11 nM) and specific inhibitor of DPP-IV, has been shown to rapidly inhibit DPP-IV *in vivo* when dosed orally in rats (22). Glucose tolerance was increased by 50% in rats receiving chronic treatment of **7**.

Imidazolines - S-22068 **8** has a potent effect on glucose tolerance and increased significantly insulin secretion in diabetic rats (23). Unlike earlier imidazoline-based secretagogues, **8** does not have a high affinity for α-2 adrenoceptors or the known imidazoline binding sites I$_1$ or I$_2$, but may act through a novel imidazoline-binding site.

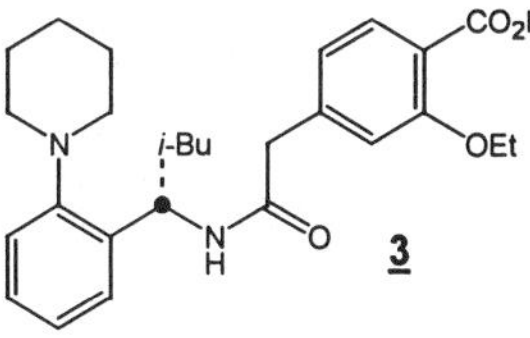

INHIBITORS OF HEPATIC GLUCOSE PRODUCTION

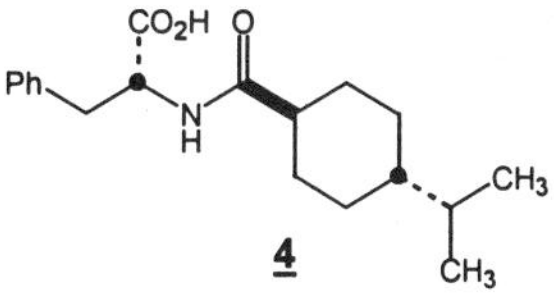

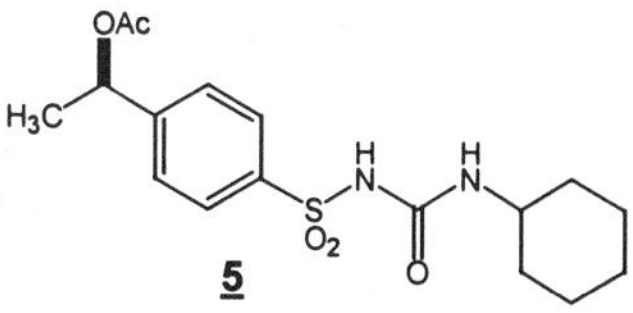

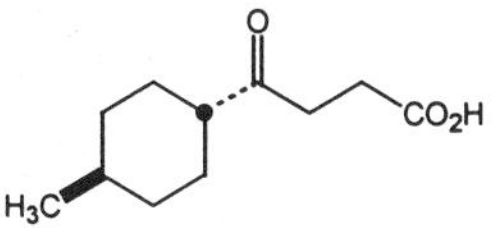

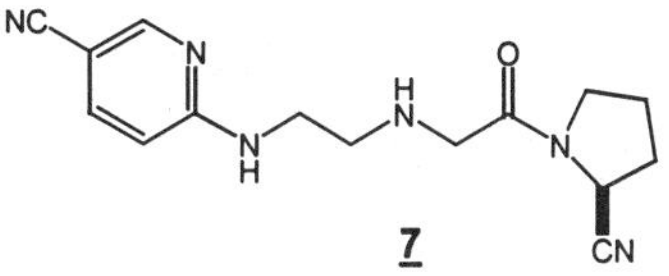

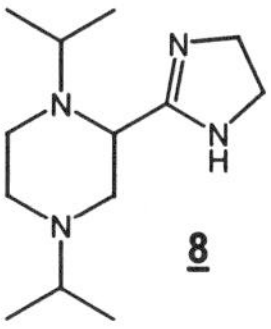

Excessive hepatic glucose production (HGP) is a significant contributor to diabetic hyperglycemia and in Type 2 diabetics, HGP is significantly elevated relative to non-diabetics. Methods for the inhibition of HGP are becoming an important strategy for control of blood glucose levels (24). Several of the novel approaches to the inhibition of HGP by effecting gluconeogenesis or glycogenolysis are discussed in this section.

Biguanides - Metformin **9** remains a primary therapeutic option for Type 2 diabetics (25). Metformin interferes with several processes linked to HGP (gluconeogenesis, glycogenolysis and their regulatory mechanisms), lowering glucose production and resensitizing the liver to insulin. Metformin reduces intestinal absorption of glucose, and increases glucose sensitivity, increasing peripheral glucose uptake and utilization. The precise molecular mechanism of action of metformin continues to be a subject of intense interest and has been reviewed extensively (26). Unlike other OHAs, metformin has been associated with significant weight loss. New studies have described the effectiveness of combination therapy using metformin and other OHAs for the maintenence of blood glucose levels (4,26).

Glucagon Receptor Antagonists - Glucagon receptor antagonists have been the subject of a recent review article (27). Reports describing the discovery and optimization of potent, orally absorbed glucagon receptor antagonists, typified by L-168,049 **10** (K$_b$=25 nM, non-competitive with glucagon), have appeared. Optimization of the potency and selectivity of this series of compounds for glucagon receptor antagonism vs. inhibition of the Ser/Thr p38 (MAPKinase) was described in detail (28,29).

Glycogen Phosphorylase - Reduction of glycogenolysis by the inhibition of glycogen phosphorylase, the enzyme that releases glucose-1-phosphate from glycogen, is another approach to inhibiting HGP. CP-91149 **11** and CP-360626 **12** are potent inhibitors of human liver glycogen phosphorylase and also inhibit glucagon stimulated glycogenolysis in rat and human hepatocytes (30,31). Oral administration of **11** to diabetic mice (25-50 mg/kg) resulted in blood glucose lowering of ~50% (30). Chronic administration of **12** to diabetic ob/ob mice also significantly decreased elevated plasma triglyceride, cholesterol, insulin, and lactate levels (31).

CP-320626 was also found to lower plasma cholesterol in a non-glucose dependent manner by inhibiting cholesterol synthesis *via* direct inhibition of CYP51 (lanosterol 14–α-demethylase) (32).

Pyruvate Dehydrogenase Kinase (PDHK) Inhibitors - PDHK regulates pyruvate dehydrogenase (PDH), the enzyme which catalyzes the decarboxylation of pyruvate to acetyl-CoA.   Activation of PDH increases oxidative glucose metabolism and reduced PDH activity has been linked to reduced glucose utlization in diabetes, as well as enhanced gluconeogenesis and impaired insulin secretion (33). Dichloroacetic acid, an inhibitor of PDHK, among other enzymes, has been shown to lower lactate and glucose levels *in vivo* (34).   Two new classes of potent, specific PDHK inhibitors, represented by **13** and **14**, have been reported (33,35).   **13** has been shown to be orally available and while it significantly reduced lactate in normal, fasted rats, it did not lower blood glucose levels in diabetic animal models (36).

Fructose-1,6-bisphosphatase (FBPase) Inhibition - FBPase is an important enzyme in the glycolytic-gluconeogenic pathway and ensures a unidirectional flux from pyruvate to glucose, and also plays an essential role in the maintenance of normoglycemcia during fasting.   Amide **15** has been been shown to be a potent (IC$_{50}$ = 0.37 µM) inhibitor of this enzyme (37).

Glucose-6-phosphatase (G-6-Pase) Inhibition - G-6-Pase is a multicomponent enzyme, resident in the endoplasmic reticulum (ER), that catalyzes the final step in both gluconeogenesis and glycolysis (38).   Inhibition of glucose-6-phosphate translocase, a component of G-6-Pase which permits entry of glucose-6-phosphate into the ER, is expected to modulate inappropriately high rates of HGP.   **16** was shown to inhibit the multi-subunit G-6-Pase with an IC$_{50}$ of 10 nM (38) by slowing the translocase function, inhibiting glucose output from rat hepatocytes, and significantly lowering circulating plasma glucose concentration in fasted rats and mice.

ENHANCERS OF INSULIN ACTION

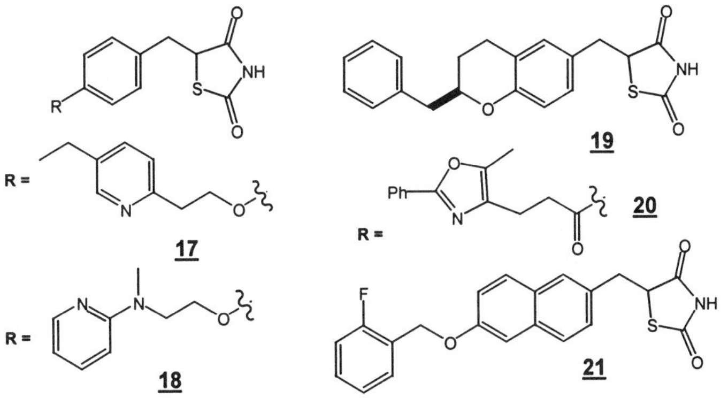

Ligands for Peroxisome-Proliferator Activated Receptor γ (PPARγ) - PPARγ, once only associated with diabetes and obesity, has recently been implicated in the transcriptional control of a wide variety of cellular processes such as the cell cycle, lipid homeostasis, inflammation, carcinogenesis, and immunomodulation (39,40). PPARγ is indicated as a central regulator of the "thrifty gene" response which mediates efficient energy storage (39). PPARγ up-regulates proteins required for the metabolism of glucose and lipids.   Ligands of PPARγ activate the glucose transporter gene resulting in increased production of glucose transporters (e.g., GLUT4) in muscle, liver and fat. PPARγ agonists (or insulin sensitizers) reduce hyperglycemia without increasing the amount of insulin secretion by increasing cellular glucose uptake, reducing cellular glucose production, and increasing insulin sensitivity in resistant tissues.

Thiazolidinediones (TZDs) – The mode of action and clinical benefits of the TZDs have been recently reviewed (40). Evidence is mounting that the PPARγ receptor is the target of the TZDs (41). Idiosyncratic liver dysfunction has been observed in 2% of patients in early troglitazone (Rezulin) trials which has caused market restrictions in Europe and the US (42-45).   Pioglitazone (**17**, Actos) and rosiglitazone (**18**, Avandia) were both launched in the US in 1999. Pioglitazone in doses of 30 mg and 45 mg resulted in reductions in mean HbA_{1c} of 2.3% and 2.6% from baseline in previously untreated patients (46,47).   Rosiglitazone given 4 mg bid results in a 1.5% reduction in HbA_{1c} relative to placebo, and reduces HbA_{1c} 1.2% over metformin alone (48-50). Englitazone **19** may have an effect on gluconeogenesis in Zucker rat livers and has been used to probe the insulin signaling pathway *in vitro* (51,52). Darglitazone **20** has been shown to increase body weight in fatty Zucker rats due to increased feeding (53). MCC-555 (**21**, isaglitazone) prevents or delays the onset of diabetes and has a positive effect on free fatty acids and triglycerides in Zucker rats (54-56).

Other novel TZDs include the phenylsulfone **22** (57), T-174 **23** (58), and KRP-297 **24** (59). Pyrrolidine **25**, when dosed at 30 mg/Kg/day po for 14 days, gives a 73% reduction in blood glucose, with a 68% reduction in triglycerides in *ob/ob* mice (60). This agent did not show significant transactivation of PPARα or PPARγ, while several other congeners in this series which were PPARγ transactivators were poor controllers of blood glucose.

Non-thiazolidinediones (non-TZDs) – There are now a variety of potent non-TZD PPARγ ligands which exhibit antidiabetic activity *in vivo*, including GW-1929 **26** (61,62), GI-262570 **27** (63), GW-409544 (64), JTT-501 **28** (65),and YM-440 (66). GW-1929 is approximately 100-fold more potent in rats than troglitazone in glucose lowering based on serum drug levels (67). The (S)-enantiomer of SB-219994 **29** is considerably more potent and efficacious than its antipode (68). The phenylacetic acid derivative **30** (69) reduces hyperglycemia by 65% at 30 mg/kg/day in Zucker rats. The arylsulfonylamide **31** was obtained from high-throughput screening (70) and the quinoxaline L-764406 **32** has been reported to be a PPARγ partial agonist (71).

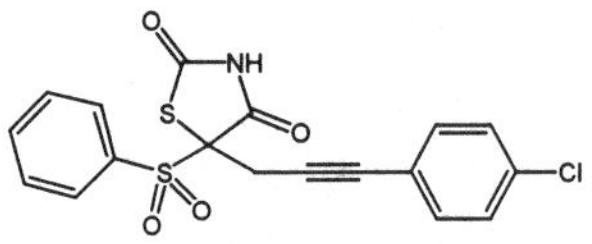

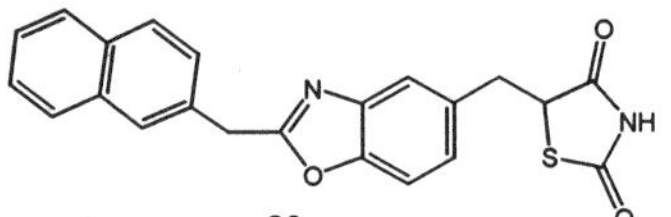

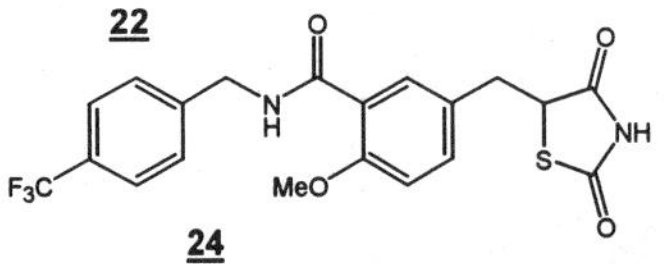

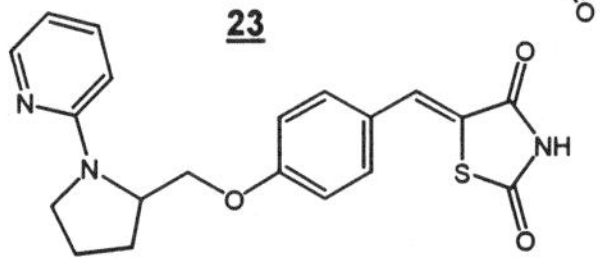

Rexinoids – The retinoid X receptor (RXR) is a ligand activated transcription factor which forms a heterodimeric complex with PPARγ. This complex regulates the expression of genes that control lipid and carbohydrate metabolism as well as the differentiation of adipocytes. 9-cis-Retinoic acid may provide a treatment for NIDDM or delay onset of the disease in at-risk patients (72,73). A series of cyclopentylidene rexinoid analogues **33** are reported to show reduced blood glucose and triglyceride levels (74). A novel series of oxime rexinoids **34** are highly specific activators of the RXR:PPARγ complex (75).

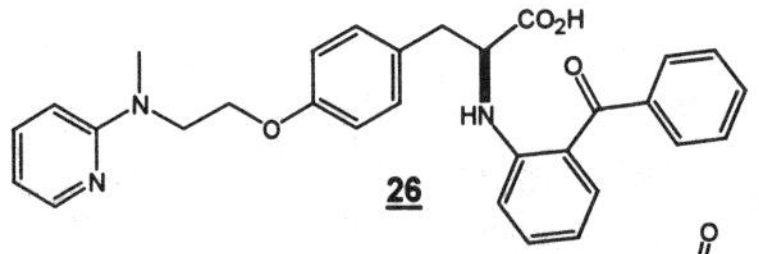

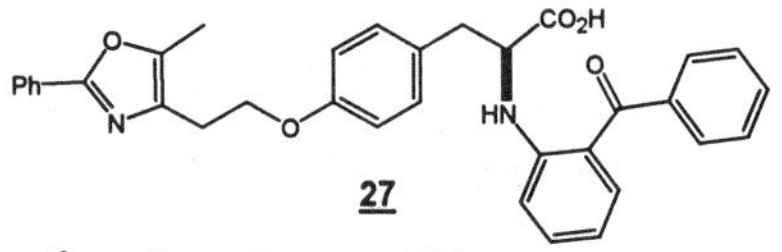

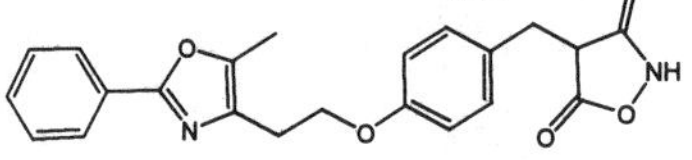

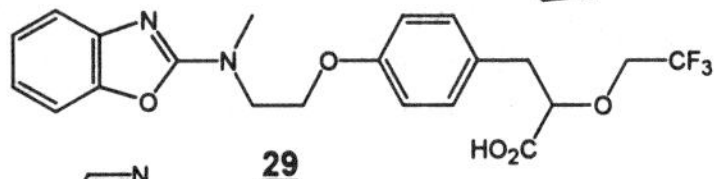

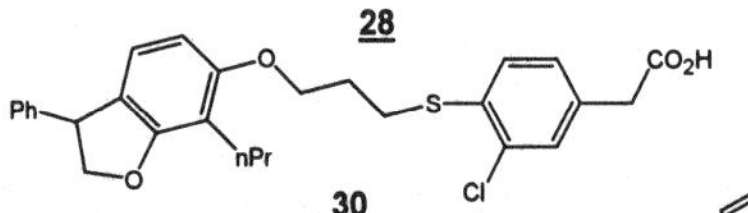

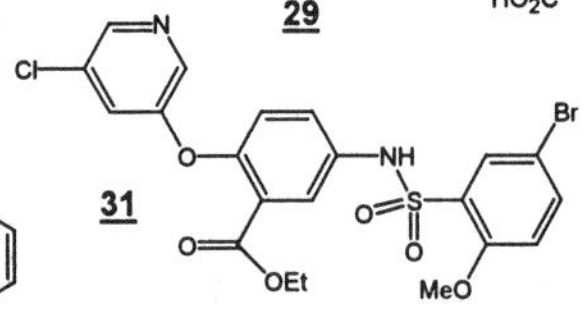

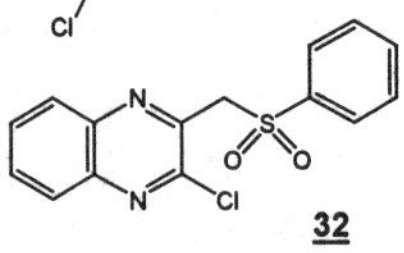

β3 Adrenergic Receptor (β3ARs) Agonists – Agonists of β3ARs increase the release of energy stores as heat in brown adipose tissue, increase lipolysis in white adipose tissue, suppress food consumption, suppress leptin gene expression and serum leptin levels. Uncoupling proteins (UCPs) in the inner mitochondrial membrane generate heat without effecting other energy-consuming processes (76). After treatment with the β3AR agonist CL316,243 **35** (0.2 mg/kg/day), KK-Ay mice showed increased UCP expression in adipose tissue, but not in muscle and heart. Furthermore, the concentration of plasma insulin and free fatty acids were significantly reduced (77). CL316,243 has also been prepared in ester form to improve its bioavailability (78). L-770,644 **36** is a full agonist in rhesus monkeys (0.2 mg/kg) with good *in vivo* properties (79). Other β₃AR agonists have also been reported to have significant antidiabetic effects (80,81).

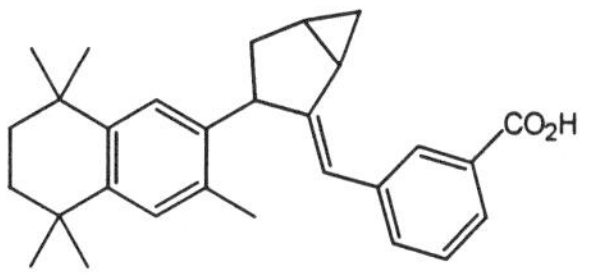

Other Enhancers – The mechanism of insulin sensitization of the benzazine derivative **37** (73% reduction of blood glucose at 50 mg/kg) (82) and azocycloalkane **38** (83) are unknown. Indoles, such as **39**, lower blood glucose significantly in db/db mice, reduce triglycerides, but inhibit phosphodiesterase type 5 (PDE5) (84). Lithium ion, an inhibitor of glycogen synthase kinase-3 (GSK3), a regulatory serine/threonine kinase which inactivates glycogen synthase *via* phosphorylation, stimulates glycogen synthesis activity and other insulin regulated processes resulting in lower blood glucose levels (85). Potent pyrimidine- and pyridine-based inhibitors of human GSK3β, e.g., **40**, have been reported to improve glucose disposal and lower plasma insulin when given orally to *db/db* mice (86).

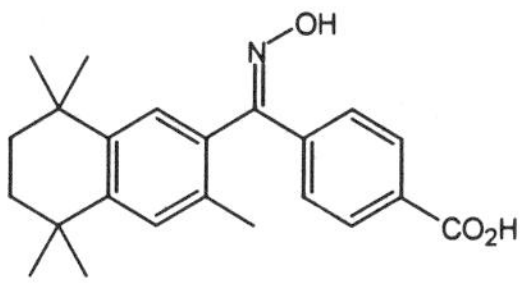

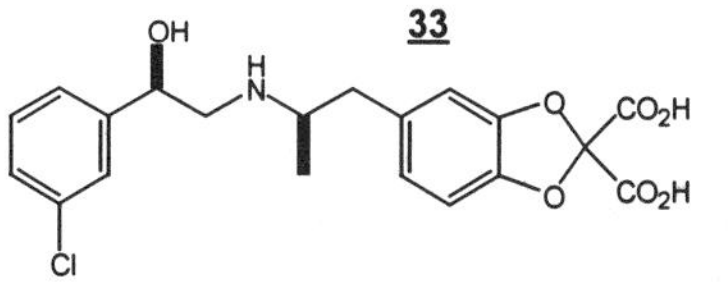

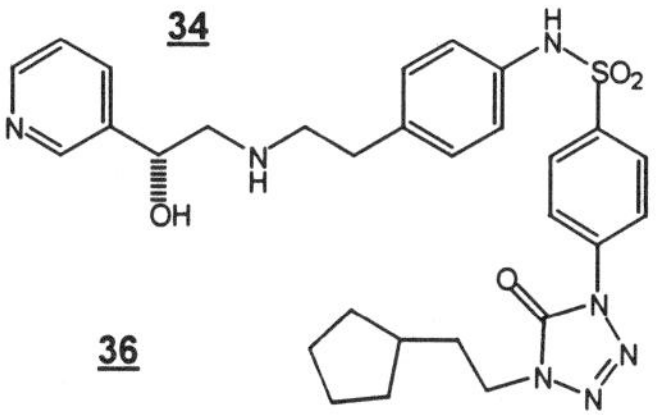

Compounds such as **41**, with an IC₅₀ of ~115 nM for the inhibition of protein tyrosine phosphatase-1 (PTP1B), the phosphatase involved in the inactivation of the insulin receptor following stimulation, have been shown to significantly lower blood glucose and insulin in a diabetic mouse model (87).

INHIBITORS OF GLUCOSE UPTAKE

The pharmacology and clinical use of glycosidase inhibitors such as acarbose, the first member of this class, has been reviewed (88-91). Acarbose delays the process of gastric emptying, moderating postprandial glucose spikes. Several other compounds in this class under development include amylin (pramlintide or the synthetic symlin) (92,93), exendin-4 (AC-2993) (94), and cholecystokinin octapeptide (CCK-8) (95). The Na⁺-glucose cotransporter (SGLT) which is found in the intestines and kidney, actively transports glucose to maintain normal blood glucose concentrations. Inhibitors of SGLT (e.g., T-1095, **42**) are postulated to reduce blood glucose levels (96) by blocking glucose reabsorption and increasing its excretion in urine.

Conclusion - Many new, exciting therapeutic strategies are being explored to treat Type 2 diabetes. The quest for the genetic basis of the disease will likely yield many more new, as yet unknown therapeutic targets as more genomic information is interpreted. Gene therapy approaches employing direct delivery of the insulin gene, either to non-secreting or secreting cells, or through the development of ex-vivo approaches using genetically modified cells, though still in the very early stages, are also under investigation and hold considerable promise in this area (97).

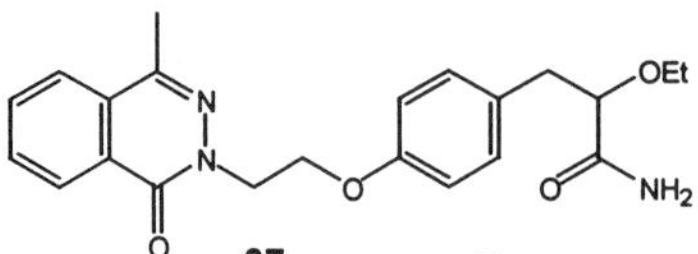

37

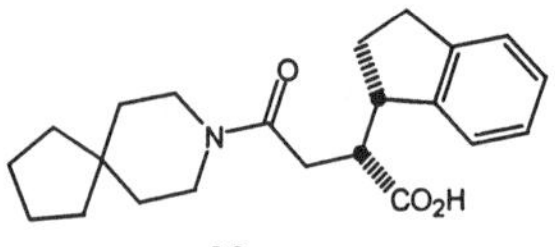

38

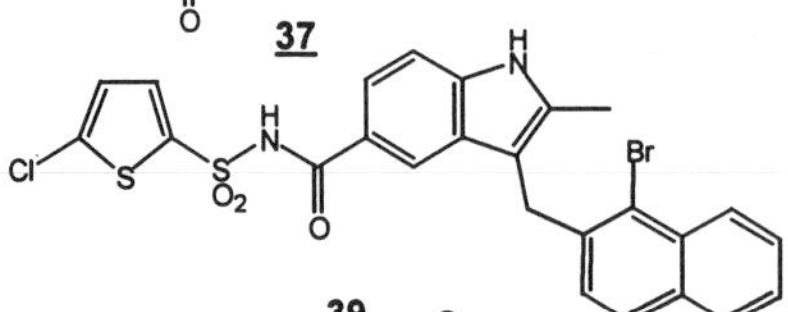

39

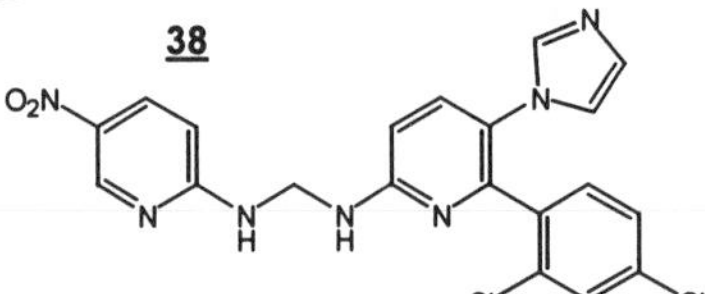

40

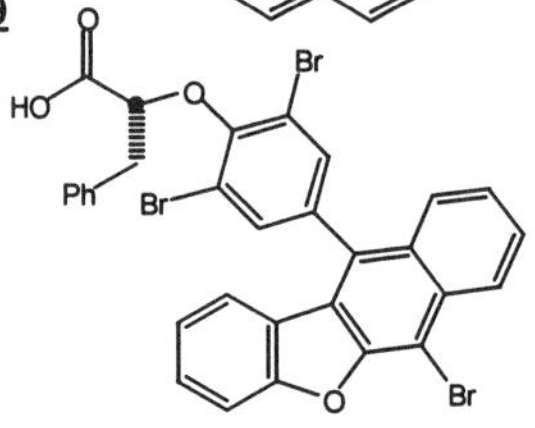

41

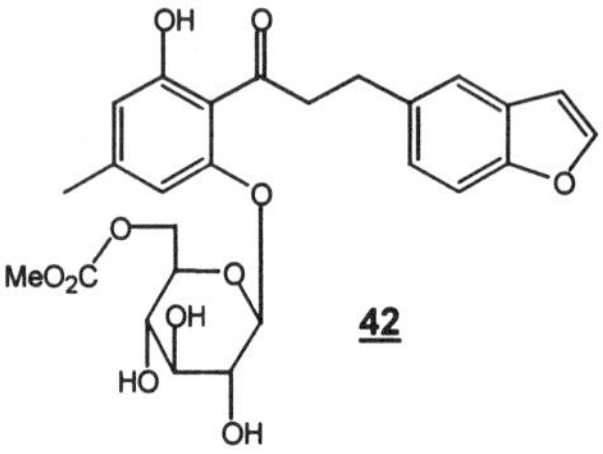

42

References

1. "Prevention of Diabetes Mellitus", World Health Organization Technical Report Series, No. 844 (1994).
2. American Diabetes Association, Diabetes Care, 22 (Suppl. 1), S27 (1999).
3. R.C. Turner, C.A. Cull, V. Frighi, R.R. Holman J. Am. Med. Assoc., 281, 2005 (1999).
4. UK Prospective Diabetes Group, Lancet, 352, 837 (1998).
5. D.R. Matthews, Exp. Clin. Endocrinol. Diabetes, 107(Suppl. 2), S34 (1999).
6. L. Heinemann, J. Diabetes Compl., 13, 105 (1999).
7. J. Brange, A. Volund, Adv. Drug Delivery Rev., 35, 307 (1999).
8. D.J. Chetty, Y.W. Chen, Crit. Rev. Ther. Drug Carrier Syst., 15, 629 (1999).
9. B. Zhang, G. Salituro, D. Szalkowski, et al., Science, 284, 974 (1999).
10. K. Liu, A.B. Jones, H.B. Wood, B. Zhang, PCT Int. Appl. WO 99 51225.
11. M.R. Burge, V. Sood, T.A. Sobhy, A.G. Rassam, D.S. Schade, Diabetes, Obes. Metab. 1, 199 (1999).
12. D. Bataille, L. Heron, A. Virsolvy, K. Peyrollier, A. LeCam, P. Blache Cell. Mol. Life Sci., 56, 78 (2000).
13. W.J. Malaisse, Exp. Opin. Clin. Endocrinol. Diabetes 107(Suppl. 4), S140 (1999).
14. M. Kikuchi, Sogo Rinsho 45, 2765 (1996).
15. H. Akita, Y. Imamura, K. Sanui, K. Kurashima, K. Seri, PCT Int. Appl. WO 99/48864
16. H. Shinkai, H. Ozeki, T. Motomura, T. Ohta, N. Furakawa, and I. Uchida, J.Med.Chem., 41, 5420 (1998).
17. D.J. Drucker, Diabetes, 47, 159 (1998).
18. B. Gallwitz, T. Ropeter, C. Morys-Wortmann, R. Mentlein, W.E. Schmidt, Regul. Pep., 86, 103 (2000).
19. G. Xu, D.A. Stoffers, J.F. Habener, S. Bonner-Weir, Diabetes, 48, 2270 (1999).
20. R. Burcelin, E. Rolland, W. Dolci, S. Germane, V. Carrel, B. Thorens, Ann. NY Acad. Sci., 875, 277 (1999).
21. K. Augustyns, G. Bal, G. Thonus, A. Belyaev, X.M. Zhang, A. Haemers, Curr.. Med. Chem., 6, 311 (1999).
22. T.E. Hughes, M. Mone, M.E. Russell, S.C. Weldon, E.B. Villhauer, Biochemistry, 38, 11597 (1999).
23. G. LeBihan, F. Rondu, A. Pele-Tounian, X. Wang, S. Lidy, E. Touboul, A. Lamouri, G. Dive, J. Huet, B. Pfeiffer, P. Renard and J.J. Godfroid, J.Med.Chem., 42, 1587 (1999).
24. W. Kramer, Exp. Clin. Endocrinol. Diabetes, 107(Suppl. 2), S52 (1999).
25. D. Giugliano, Int. Congr. Ser., 1177(Insulin Resistance, Metabolic Diseases and Diabetic Complications), 231 (1999).
26. N.F. Wiernsparger, Diabetes Metab., 25, 110 (1999).
27. J.N. Livingston, W.R. Schoen, Ann.Rep.Med.Chem., 34, 189 (1999).

28. M.A. Cascieri, G.E. Koch, E. Ber, S.J. Sadowski, E. Ber, S.J. Sadowski, D. Louizides, S.E. deLaszlo, C. Hacker, W.K. Hagmann, M. MacCoss, G.G. Chicchi, and P.P. Vicario, J.Biol.Chem., 274, 8694 (1999).
29. S.E. deLaszlo, C. Hacker, B. Li, D. Kim, M. MacCoss, N. Mantlo, J. Pivnichny, L. Colwell, G.E. Kock, M.A. Cascieri and W.K. Hagmann, Bioorg.Med.Chem.Lett., 9, 641 (1999).
30. D.J. Hoover, S. Lefkowitz-Snow, J.L. Burgess-Henry, W.H. Martin, S.J. Armento, I.A. Stock, R.K. McPherson, P. Genereux, E.M. Gibbs, J.L. Treadway, J.Med.Chem., 41, 2934 (1998).
31. W.H. Martin, D.J. Hoover, S.J. Armento, I.A. Stock, R.K. McPherson, D.E. Danley, R.W. Stevenson, E.J. Barrett, J.L. Treadway, Proc. Nat. Acad. Sci., 95, 1776 (1998).
32. J.L. Treadway, Diabetes, 47 (Suppl. 1): Abst 1115 (1998)
33. T.D. Aicher, R.C. Anderson, G.R. Bebernitz, G.M. Coppola, C.F. Jewell, D.C. Knorr, C. Liu, D.M. Sperbeck, L.J. Brand, R.J. Stohschein, J. Gao, C.C. Vinluan, S.S. Shetty, and W.R. Mann, J.Med.Chem., 42, 2741 (1999).
34. A. Islam, Diabetes, 48 (Suppl. 1), Abst. 2012 (1999)
35. R.J. Butlin, J.N. Burrows, PCT Int. Appl. WO 9962873 (1999)
36. T.D. Aicher, R.C. Anderson, J. Gao, S. Shetty, G.M. Coppola, A. Islam, R.E. Walter, W.R. Mann, J. Med. Chem., 43, 236 (2000).
37. A.M.M. Mjalli, J.C. Mason, K.L. Arienti, K.M. Short, R.D.A. Kimmich, T.K.Jones, PCT Int. Appl. WO 9947549 (1999).
38. J.C. Parker, M.A. VanVolkenburg, A. Maria, C.B. Levy, W.H. Martin, S.H. Burk, Y. Kwon, C. Giragossian, T.G. Gant, P.A. Carpino, R.K. McPherson, P. Vestergaard, J.L. Treadway, Diabetes, 47, 1630 (1998).
39. J. Auwerx, Diabetologia, 42, 1033 (1999).
40. T.M. Wilson, P.J. Brown, D.D. Sternbach, B.R. Henke, J.Med.Chem., 43, 527, (2000).
41. I. Barroso, M. Gumell, S.V.E.F. Crowley, M. Agostinil, J.W. Schwabe, M.A. Soos, G.U. Maslen, T.D.M. Williams, H. Lewis, A.J. Schafer, V.K.K. Chatterlee, S. O'Rahilly, Nature, 402, 880, (1999).
42. Scrip, 2204, 17, (1997).
43. Scrip, 2272, 23, (1997).
44. G.L. Plosker, D. Faulds, Drugs, 57, 409, (1999).
45. M.D. Johnson, L.K. Campbell, R.K. Campbell, Ann.Pharmacother., 32, 337, (1998).
46. The Pink Sheet, July 19 1999, F-D-C Reports, Inc.
47. Y. Yamasaki, R. Kawamori, T. Wasada, A. Sato, Y. Omori, H. Eguchi, M. Tominaga, H. Sasaki, M. Ikeda, M. Kubota, Y. Ishida, T. Hozumi, S. Baba, M. Uehara, M. Shichiri, T. Kaneko, Tohoku J.Exp.Med., 183, 173, (1997).
48. Scrip, 2345, 19, (1999).
49. Press Release, SmithKline Beecham, June 21 (1999).
50. J.A. Balfour, G.L. Plosker, Drugs, 57, 921, (1999).
51. M.D. Adams, P. Raman, R.L. Judd, Biochem.Pharmacol., 55, 1915, (1998).
52. R.W. Stevenson, D.K. Kreutter, K.M. Andrews, P.E. Genereux, E.M. Gibbs, Diabetes, 47, 179, (1998)
53. S.M. Jacinto, Diabetologia, 42(Suppl. 1), Abst. 675, (1999).
54. Scrip, 2335, 28, (1998).
55. L. Pickavance, P. Widdowson, P. King, S. Ishii, H. Tanaka, G. Williams, Br.J.Pharmacol., 125, 767 (1998).
56. R. Upton, P. Widdowson, Ishii, H. Tanaka, G. Williams, Br.J.Pharmacol., 125, 1708 (1998).
57. J. Wrobel, Z. Li, A. Dietrich, M. McCaleb, B. Mihan, J. Sredy, D. Sullivan, J.Med.Chem., 41, 1084, (1998).
58. K. Arakawa, T. Ishihara, M. Aoto, M. Inamasu, A. Saito, K. Ikezawa, Br.J.Pharmacol, 125, 429, (1998).
59. K. Murakami, M. Tsunoda, T. Ide, M. Ohashi, T. Mochizuki, Metab.Clin.Exp., 48, 1450, (1999).
60. B. Lohray, V. Bhushan, A.S. Reddy, B.P. Rao, N.J. Reddy, P. Harikishore, N. Haritha, R.K. Vikramadityan, R. Chakrabarti, R. Rajagopalan, K. Katneni, J.Med.Chem., 42, 2569 (1999).
61. T.M. Willson, Book of Abstracts, 216th ACS National Meeting, Boston, MEDI-381 (1998).
62. K.K. Brown, B.R. Henke, S.G. Blanchard, J.E. Cobb, R. Mook, I. Kaldor, S.A. Kliewer, J.M. Lehmann, J.M. Lenhard, W.W. Harrington, P.J. Novak, W. Faison, J.G. Binz, M.A. Hashim, W.O. Oliver, H.R. Brown, D.J. Parks, K.D. Plunket, W.Q. Tong, J.A. Menius, K. Adkison, S.A. Noble, T.M. Willson, Diabetes, 48, 1415, (1999).
63. B.R. Henke, S.G. Blanchard, M.F. Brackeen, K.K. Brown, J.E. Cobb, J.L. Collins, W.W. Harrington, M.A. Hashim, E.A. Hull-Ryde, I. Kaldor, S.A. Kliewer, D.H. Lake, L.M. Leesnitzer, J.M. Lehmann, J.M. Lenhard, D.J. Parks, K.D. Plunket, T.M. Willson, J.Med.Chem, 41, 5020, (1998).

64. Business Wire, Oct. 28 (1999)
65. H. Maegawa, T. Obata, T. Shibata, T. Fujita, S. Ugi, K. Morino, Y. Nishio, H. Kojima, H. Hidaka, M. Haneda, H. Yasuda, R. Kikkawa, A. Kashigawi, Diabetologia, 42, 151, (1999).
66. T. Suzuki, K. Niigata, T. Takahashi, T. Maruyama, K. Onda, O. Noshiro, Y. Isomura, Book of Abstracts, 215th ACS National Meeting, Dallas, March, MEDI-044, (1998).
67. J.L. Collins, S.G. Blanchard, G.E. Boswell, P.S. Charifson, J.E. Cobb, B.R. Henke, E.A. Hull-Ryde, W.M. Kazmierski, D.H. Lake, L.M. Leesnitzer, J. Lehmann, J.M. Lenhard, L.A. Orband-Miller, Y. Gray-Nunez, D.J. Parks, K.D. Plunkett, W.-Q. Tong, J.Med.Chem., 41, 5037, (1998).
68. P.W. Young, D.R. Buckle, B.C. Cantello, H. Chapman, J.C. Clapham, P.J. Coyle, D. Haigh, R.M. Hindley, J.C. Holder, H. Kallender, A.J. Latter, K.W.M. Lawrie, D. Mossakowska, G.J. Murphy, L. Roxbee Cox, S.A. Smith, J.Pharmacol.Exp.Ther., 284, 751, (1998).
69. C. Santini, J.J. Acton, A.D. Adams, S.D. Aster, G.D. Berger, A. Dadiz, D.W. Graham, D.F. Gratale, W. Han, A.B. Jones, G.F. Patel, K. Sidler, S.P. Sahoo, D. Von Langen, W. Yueh, R.L. Tolman, J.V. Heck, M.D. Liebowitz, J.P. Berger, T.W. Doebber, B. Zhang, D.E. Moller, R.G. Smith, Book of Abstracts, 218th ACS National Meeting, New Orleans, Aug., MEDI-147 (1999).
70. F. De La Brouse-Elwood, J.C. Jaen, L.R. McGee, S.-C. Miao, S.M. Rubenstein, J.-L. Chen, T.D. Cushing, J.A. Flygare, J.B. Houze, P.C. Kearney, WO 9938845, (1999).
71. A. Elbrecht, Y. Chen, A. Adams, J. Berger, P. Griffin, T. Klatt, B. Zhang, J. Menke, G. Zhou, R.G. Smith, D.E. Moller, J.Biol.Chem., 274, 7913, (1999).
72. M.L. Rose, M.A. Paulik, J.M. Lenhard, Expert Opin.Ther.Pat., 9, 1223, (1999).
73. M. Pfahl, W. Lernhardt, A. Fanjol, WO 9842340, (1998).
74. A. Murray, J.B. Hansen, WO 9958486, (1999).
75. S.S.C. Koch, L.J. Dardashti, R.M. Cesario, G.E. Croston, M.F. Boehm, R.A. Heyman, A.M. Nadzan, J.Med.Chem., 42, 742, (1999).
76. H. Yoshitomi, K. Yamazaki, S. Abe, I. Tanaka, Biochem.Biophys.Res.Commun., 253, 85, (1998).
77. M.V. Kumar, R.L. Moore, P.J. Scarpace, Eu.J.Physiology, 438, 681, (1999).
78. F.W. Sum, A. Gilbert, A.M. Venkatesan, K. Lim, V. Wong, M. O'Dell, G. Francisco, Z. Chen, G. Grosu, J. Baker, J. Ellingboe, M. Malamas, I. Gunawan, J. Primeau, E. Largis, K. Steiner, Bioorg.Med.Chem.Lett., 9, 1921, (1999).
79. T.L. Shih, M.R. Candelore, M.A. Cascieri, S.H. Chiu, L.F. Colwell Jr, L. Deng, W.P. Feeney, M.J. Forrest, G.J. Hom, D.E. MacIntyre, R.R. Miller, R.A. Stearns, C.D. Strader, L. Tota, M.J. Wyvratt, M.H. Fisher, A.E. Weber, Bioorg.Med.Chem.Lett., 9, 1251, (1999).
80. T. Kiso, T. Namikawa, T. Tokunaga, K. Sawada, T. Kakita, T. Shogaki, Y. Ohtsubo, Bio.Pharm.Bull. 22, 1073, (1999).
81. R.J. Mathvink, A.M. Barritta, M.R. Candelore, M.A. Cascieri, L. Deng, L. Tota, C. Strader, M.J. Wyvratt, M.H. Fisher, A.E. Weber, Bioorg.Med.Chem.Lett., 9, 1869, (1999).
82. Y. Nagao, F. Konno, J. Kotake, F. Ishii, H. Honda, S. Sato, Eur. Pat. Appl. EP 903343 (1999).
83. M. Bedoya-Zurita, J. Diaz-Martin, M. Daumas, Marc, PCT Int. Appl. WO 9952876 (1999).
84. N. Yamasaki, T. Imoto, T. Oku, H. Kayakiri, O. Onomura, T. Hiramura, PCT Int. Appl. WO 9951574 (1999).
85. I. Hers, J.M. Tavare, R.M. Denton, FEBS Lett., 460, 433, (1999).
86. J.M. Nuss, S.D. Harrison, D.B. Ring, R.S. Boyce, S.P. Brown, D. Goff, K. Johnson, K.B. Pfister, S. Ramurthy, P.A. Renhowe, L. Seely, S. Subramanian, A.S. Wagman, X.A. Zhou, PCT Int. Appl. WO 9965897, (1999).
87. J. Wrobel, J. Sredy, C. Moxham, A. Dietrich, Z. Li, D.R Sawicki, L. Seestaller, L. Wu, A. Katz, D. Sullivan, C. Tio, Z.-Y. Zhang, J.Med.Chem., 42, 3199 (1999).
88. B. Goke, C. Herrmann-Rinke, Diabetes/Metab.Rev., 14(Suppl. 1), S31-S38, (1998).
89. A.J. Scheen, Diabetes Metab., 24, 311, (1998).
90. M. Hanefeld, J.Diabetes Complications, 12, 228, (1998).
91. N. Scorpiglione, M. Belfiglio, F. Carinci, D. Cavaliere, A. DeCurtis, M. Franciosi, E. Mari, M. Sacco, G. Tognoni, A. Nicolucci, Eur.J.Clin.Pharmacol., 55, 239, (1999).
92. O. Schmitz, Diabetes Annu., 12, 269, (1999).
93. W.A. Scherbaum, Exp.Clin.Endocrinol.Diabetes, 106, 97, (1998).
94. BioWorld News, 5, May 17 (2000).
95. I.S. Ebenezer, Meth.Find.Exp.Clin.Pharmacol., 21, 167, (1999).
96. K. Tsujihara, M. Hongu, K. Saito, H. Kawanishi, K. Kuriyama, M. Matsumoto, A. Oku, K. Ueta, M. Tsuda, A. Saito, J.Med.Chem., 42, 5311, (1999).
97. D.J. Freeman, I. Leclerc, G.A. Rutter, Int. J. Mol. Med., 4, 585 (1999).

Chapter 20. Toward the Development of α_{1a} Adrenergic Receptor Antagonists

Mark G. Bock and Michael A. Patane
Merck Research Laboratories, West Point, PA 19486

Introduction - Benign prostatic hyperplasia (BPH) is a pathological disorder in men that develops in response to the action of dihydrotestosterone on the aging prostate gland (1-3). It is a disease that causes voiding difficulties and interferes with the perception of well-being. Although not fully defined, the origin of symptoms in men with BPH appears to result from a combination of two components: static or mechanical constriction of the urethra attributable to increased prostatic mass and spasmodic contractions of smooth muscles under alpha-adrenergic receptor-mediated sympathetic stimulation (dynamic component). The mechanical component of BPH can be counteracted with surgery or androgen-modulating drugs, which reduce the size of the prostate (4). On the other hand, alpha-adrenergic receptor antagonists change the tone of the prostatic capsule (and adenoma), thereby decreasing the pressure in the prostatic part of the urethra and bladder neck without affecting the bladder (5). The latter outcome serves as the theoretical basis for using alpha-adrenergic receptor blockers in the treatment of BPH.

Alpha receptors have been divided into α_1 and α_2 subtypes. Non-selective alpha-adrenergic receptor antagonists were introduced more than two decades ago to alleviate urinary obstruction due to BPH (6,7). Phenoxybenzamine was among the first alpha-adrenergic receptor antagonists used to clinically treat prostatism; however, this and similarly non-selective agents are associated with many adverse effects, particularly cardiovascular responses which manifest as postural hypotension, tachycardia, syncope, and fatigue (8,9). The demonstration that alpha-adrenergic control of prostatic contractions is mediated primarily by α_1-, not α_2-adrenoceptors, was followed by the introduction of newer α_1-selective compounds (10). This second generation of alpha-adrenergic blocking drugs has much less affinity for α_2-type receptors. Examples of marketed compounds which exemplify this profile that have been approved by the FDA for the treatment of BPH include terazosin (Hytrin®) and doxazosin (Cardura®) (11-16); the structurally homologous alfuzosin has been introduced primarily in Europe (17-19). While these inherently selective α_1-adrenoceptor antagonists offer the potential benefit of increased urodynamic effects, this property has not been strictly established in a clinical setting. Indeed, high doses of this class of drugs remain associated with systemic side effects, principally vascular events.

During the past decade, progress within the alpha-adrenergic receptor arena has gained considerable momentum. Among key developments was the confirmation of the existence of heterogeneity among α_1-receptors with current nomenclature recognizing three subtypes (α_{1A}, α_{1B}, and α_{1D}) (20). These subtypes have been identified by molecular cloning (α_{1a}, α_{1b}, and α_{1d}) and have clearly defined native tissue correlates (21). Recently, an additional α_1-receptor subtype (α_{1L}) was identified that has affinity for prazosin and may be present only in the prostate (22). The significance of this receptor is not well understood and it may not be different from the α_{1a} receptor protein, but rather is detected under certain assay conditions (23). A further important result was the recognition that among the α_1-receptor subtypes, the α_{1A}-receptor is chiefly responsible for the contractile tone of the prostatic urethra (24-

27). In the treatment of BPH, blockade of the α_1-adrenoceptors in the urinary outflow tract is needed. Blockade of the α_1-adrenoceptors in the vascular wall, by contrast, is neither needed nor desirable.   This has led to an intensive effort to develop uroselective α_{1A}-subtype adrenergic receptor antagonists with the expectation that such agents would have the potential to offer an improved side-effect profile compared to currently available drugs and thereby achieve an enhanced therapeutic response in BPH patients.

It is our objective in this review to present a synopsis of the advances made toward the development of α_{1A}-subtype selective adrenergic receptor antagonists. In this regard, it must be pointed out that the criteria for 'uroselectivity' have not been strictly defined and the term has been used in several contexts resulting in some confusion. An agent may be 'uroselective' in the sense that it has a preferred affinity for the α_{1A}-receptor in the prostate (pharmacological), or that, in an animal model, it can affect the prostate and urethra without affecting blood pressure (physiological). A clinically meaningful definition of uroselectivity, which can only be made in man, considers desired effects on obstruction and lower urinary tract symptoms relative to adverse effects (28).   In the following chapter, α_{1A}-subtype selective agents are collated according to structural classes and the merits and limitations of key compounds are discussed. Pertinent reviews on this topic have recently appeared (29-31).

Phenoxyethyl amines – Tamsulosin, **1**, is the first α_{1A}-receptor subtype selective antagonist approved by the FDA for the treatment of BPH. Its structure is a distant molecular analog of the endogenous alpha-1 receptor agonist norepinephrine; the absolute stereochemistry adjacent to the biogenic amine is (R). Tamsulosin binds with high affinity for the human α_{1A}- and α_{1D}-receptors (pK_i = 9.7 and 9.8, respectively) with 10-fold lower affinity for the α_{1B}- receptor (32). It is claimed to be uroselective but this proposal has been contested (33). In a Phase I clinical trial, **1** (0.4 mg, multiple dose) administered as a sustained release formulation was well-tolerated with no significant adverse effects (34). Data from two open-label studies with **1** for treating BPH showed that 84% and 90% of patients, respectively, rated treatment efficacy as good, or very good tolerability (35). The clinical experience with **1** suggests that it may provide better tolerability compared to the non-specific α_1-antagonists, terazosin and doxazosin. However, it appears that this profile is not a consequence of its inherent pharmacological selectivity but instead is achieved by optimizing its pharmacokinetic (PK) properties via formulation.

A compound structurally related to tamsulosin is the potent and subtype selective indoline analog KMD-3213, **2**. The trifluoroethyl group in **2** has been inserted presumably to mitigate O-de-alkylation, a primary metabolic pathway for **1** observed in humans (36). As with tamsulosin, the methyl-bearing carbon atom in **2** has the (R)-configuration. KMD-3213 potently binds to the cloned human α_{1a}-receptor with a K_i value of 0.036 nM and it has 583- and 56-fold lower potency at the α_{1b}- and α_{1d}-receptors, respectively (37). Noradrenaline-induced contractions in isolated human prostatic tissue were potently inhibited by **2** with pK_B = 9.45, indicating that it has functional activity in this assay similar to **1** (38). α_{1a}-Antagonist **2** proved superior to **1** in a rat model testing the effects of intraurethral pressure (IUP) response to the agonist phenylephrine; moreover, 24 hours after oral dosing, a dose-dependent inhibition of IUP was found, whereas the effect of **1** disappeared after 18 hours (39). KMD-3213 has been found to exert longer lasting effects on lower urinary tract function than in vascular tissue (40).   This compound is in Phase II in Japan; however, no information regarding the efficacy of **2** in BPH patients has been disclosed (41).

JTH-601, **3**, is a high affinity $\alpha_{1A/1L}$-antagonist (pK_i = 9.4) wherein the achiral linker between the two phenolic nuclei has been truncated relative to **1** and **2** (42). While **3** exhibits only approximately 10-fold selectivity versus the α_{1B}- and α_{1D}-receptors, it is reportedly in early clinical development (43). *In vivo* experiments of **3** support claims of prostatic tissue selectivity (44); **3** is professed to lower IUP and exhibit less pronounced effects on blood pressure than **1** in a phenylephrine-challenge model in rabbits or following ID administration in anaesthetized dogs (45). It is of some pedagogical interest that the principal human metabolite of **3**, JTH-601-G1, **4** is a potent antagonist in human prostate (pA_2 = 8.12) and exhibits α_1-receptor subtype selectivity analogous to the parent compound (46).

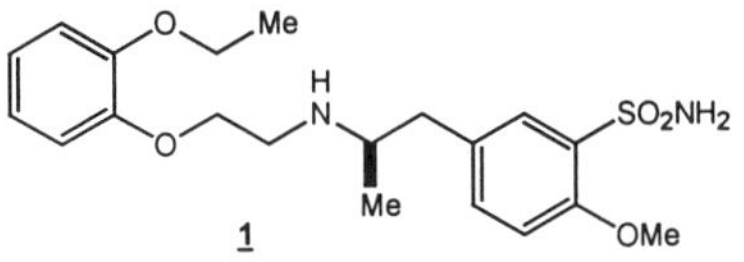

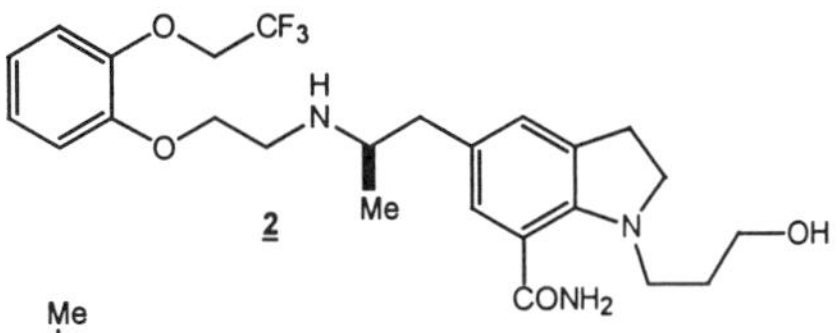

Arylpiperazines – A significant number of N-arylpiperazine α_1-adrenoceptor ligands have been investigated even though there is the potential of compounds containing this structural motif to interact with other seven-transmembrane receptors (47,48). The nicotinamide RS-97078, **5**, was initially identified based on its functional prostatic selectivity using rabbit bladder neck strips and rat thoracic aortic rings (49). Despite its affinity for dopamine (D_2 pK_i 7.4) and serotonin (5-HT$_{1A}$ pK_i 7.6) receptors, **5** was advanced to the clinic where it has been studied in healthy male volunteers following oral administration of single doses ranging from 2.5 to 30 mg (49). Results from these trials have not been disclosed. A second-generation antagonist to emerge is Ro 70-0004 (**6**, RS-100975), a compound which appears to have been derived from earlier work on **5** (50,51). Ro 70-0004 binds to the cloned human α_{1a}-receptor with high affinity (pK_i = 8.9) and selectivity (α_{1b}, 60-fold and α_{1d}, 50-fold). In studies of **6** employing human prostatic tissue (HP) and human renal artery (HRA) under conditions of noradrenaline stimulation, pA_2 values for HP and HRA were 8.6 and 6.9, respectively. This corresponds to a pA_2 ratio (HP/HRA) of 50 and compares favorably with tamsulosin, **1** (HP/HRA = 0.8). *In vivo* studies with **6** support a selective action on lower urinary tissue. Ro 70-0004 was found to be approximately 100-fold less potent than prazosin at producing postural hypotension during head-up tilt in conscious rats (52); similar results were seen in tilt studies using conscious dogs. In anaesthetized mongrel dogs, **6** was 76-fold more potent at inhibiting hypogastric nerve stimulation-induced rises in IUP verses phenylephrine-induced rises in diastolic blood pressure. In the same study, neither prazosin nor tamsulosin, despite the claimed α_{1a}-receptor subtype selectivity for **1**, distinguished between urethral and diastolic blood pressure responses. The observation that ID administration of **6** in dogs results in rapid onset of action (maximal inhibition of IUP responses within 30 minutes) and sustained duration of action (> 4 hours) is noteworthy. Taken together, pre-clinical data indicate that **6** may have advantages over **1**. The assessment of the clinical utility of **6** in the symptomatic treatment of BPH has proceeded to phase II; data from these trials have not been divulged (50).

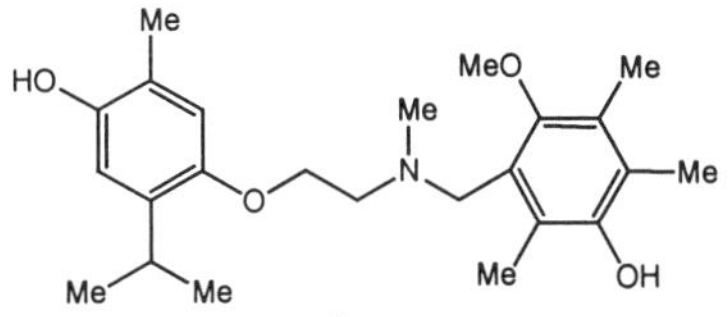

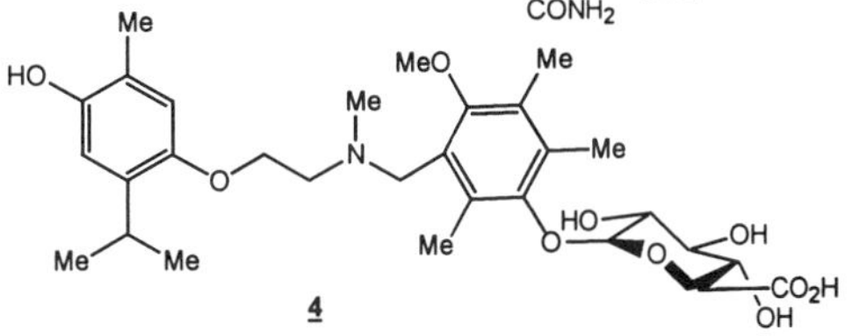

A 4*H*-1-benzopyran-8-carboxamide derivative REC15-2739 (**7**, SB216469) has potent affinity for the human α_{1A}-receptor (pK_i 9.54), with modest selectivity over α_{1B} (pK_i 7.95) and α_{1D} (pK_i 8.87) receptors (53). Studies in anaesthetized dogs indicate that **7** displays some uroselectivity. Under conditions of phenylephrine stimulation, **7** antagonized the increase in IUP with pA_2 = 8.74 compared to blood pressure with pA_2 = 7.51, indicative of 17-fold uroselectivity. It has been observed that **7** is rapidly cleared from plasma as reflected in its short duration of action in conscious dogs relative to terazosin and **1** (54). It is uncertain to what the extent these PK parameters can be extrapolated to a clinical setting; nonetheless, after entering phase II trials, **7** was withdrawn, allegedly because compound efficacy was not supported by clinical data (55).

A recent report discloses the discovery of a new α_1-adrenoceptor antagonist SL 91.0893, **8** (45,56). This arylpiperazine analog exhibits approximately equal affinity for the cloned human α_1-receptor subtypes. Nonetheless, **8** displays dose-dependent effects on IUP in the absence of effects on blood pressure possibly because of its distribution to and accumulation in the prostate. The prostatic concentration of **8** is reported to be 10 times that found in plasma one hour post-dose (PO) and rises to 24 times at 6 hours post-dose.

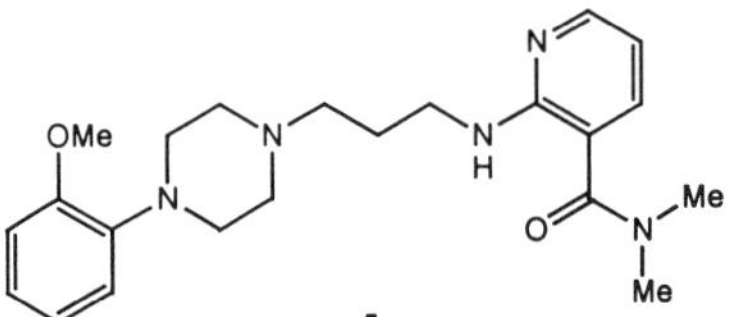

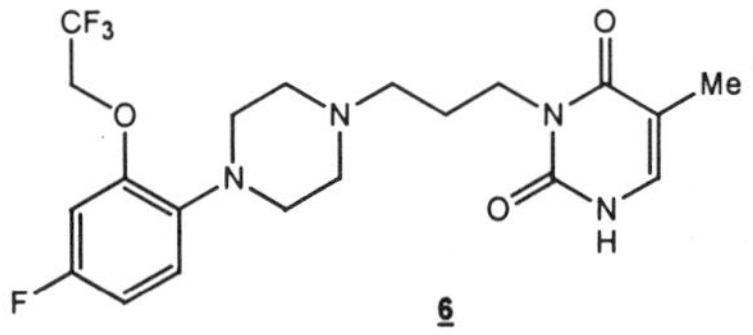

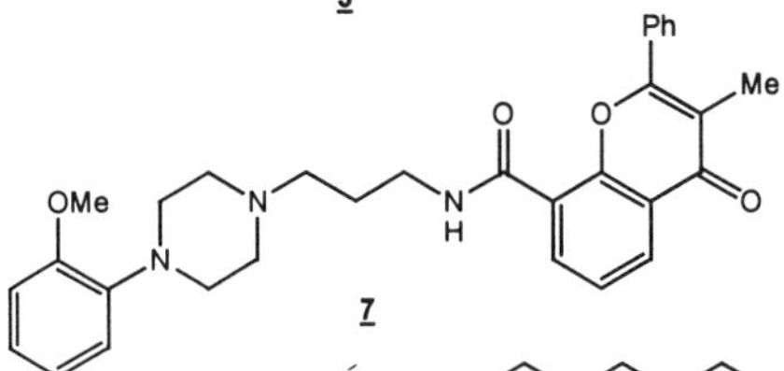

Beginning with screening and literature compounds, a series of hybrid structures that show α_{1A}-receptor selectivity has been generated. Arylpiperazine **9** binds the human α_{1A}-receptor with K_i = 8.7 nM and exhibits 5300- and 42-fold selectivity over α_{1B} and α_{1D}, respectively (57). A dose-dependent reduction in phenylephrine-induced increases in IUP was observed in anaesthetized dogs: 50% and 75% reduction at 30 and 300 μg/kg, respectively. Reductions in phenylephrine-induced increases in mean arterial pressure were less: 25-30% reduction at 30-300 μg/kg (57). More recently, homologs of **9** have been disclosed (58,59). The (*S*)-hydroxy enantiomers **10** and **11** (K_i, α_{1a} = 0.29 nM, 0.33 nM; α_{1b}/α_{1a} >5600, >6000; α_{1d}/α_{1a} 186, 158, respectively) are slightly less potent but more selective at the α_{1a}-receptor than **1**. Both homologs exhibit *in vivo* selectivity comparable to **9**, in spite of their improved *in vitro* profiles. One of these analogs has been selected as a development candidate but the identity of this compound is proprietary (60).

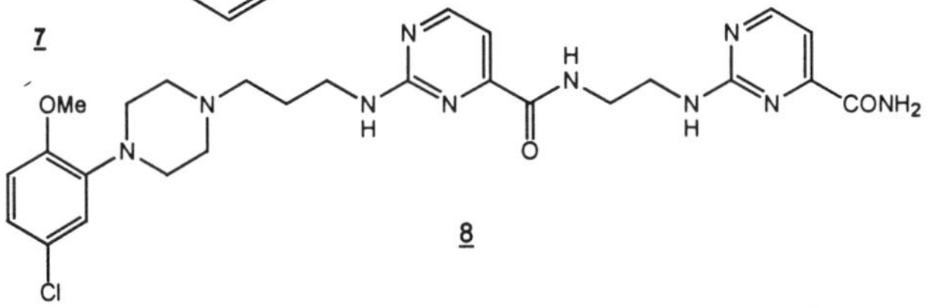

4-Phenyl dihydropyrimidinone **12** is among the most potent arylpiperazine α_{1A}-antagonists reported (61). It displays excellent selectivity versus the other receptor subtypes (K_i, α_{1a}, α_{1b}, α_{1d}: 0.12 nM, 190 nM, 250 nM). Although the genesis of **12** can be traced to the calcium channel blocker niguldipine, **13**, the former displays >1000-fold selectivity for the α_{1a}-receptor; it exhibits a similar selectivity profile for other G-protein coupled receptors like H_1 and the family of serotonin receptors. While **12** shows good tissue selectivity (rat prostate, K_b = 0.28 nM vs. rat aorta, K_b = 550 nM) and functional activity in rodents, its limited PK profile in rat (F = 13%, $t_{1/2}$ = 2 hr) and dog (F = 4%, $t_{1/2}$ = 2.4 hr) will likely preclude it from clinical evaluation.

Arylpiperidines – As a class, arylpiperidine-containing compounds comprise the largest and most diverse collection of α_{1a}-adrenergic antagonists. Among the first α_{1a}-receptor ligands to be identified was the 1,4-dihydropyridine, (S)-(+)-niguldipine **13**, which binds with high affinity to the human α_{1a}-receptor (K_i = 0.2 nM); moreover, it exhibits 340- and 630-fold selectivity in binding to the human α_{1a}-receptor relative to the human α_{1b}- and α_{1d}-receptors (62,63). The Synaptic group effectively eliminated the calcium channel blocking activity of **13** by repositioning the nitro group on the 4-phenyl ring; further ester to amide conversion gave **14** (64). Although **14** has an excellent *in vitro* binding profile (K_i: α_{1a}, 0.18 nM; α_{1b}, 180 nM; α_{1d}, 630 nM; Ca^{2+}, 670 nM), poor PK properties eliminated this compound from development. Additional examples of potent and α_{1a}-receptor subtype selective arylpiperidines with appended 1,4-dihydropyridine and dihydropyrimidine ring systems have been reported (65-67). These closely related structural analogs appear not to overcome the PK liabilities of **14**. Nonetheless, **14** and its analogs served as prototypes for a second generation of arylpiperidines exemplified by the 1,4-dihydropyrimidinone (DHP) **15** (K_i: α_{1a}, 0.2 nM; α_{1b}, 260 nM; α_{1d}, 350 nM) (68). Less lipophilic and potentially more stable to oxidative metabolism than **14**, compound **15** displays incremental improvements in PK properties (rat, F = 19%, $t_{1/2}$ 2 hr; dog F = 26%, $t_{1/2}$ 2.5 hr). Interestingly, although **15** has a relatively short half-life (2 hr) in rats, in an *in situ* rat prostate assay which measures the inhibition response to a phenylephrine-challenge, the duration of action of **15** was in excess of 4 hours. Development of compound **15** was not pursued because *in vitro* and *in vivo* metabolism experiments revealed significant formation of the methyl ester analog of the μ-opioid agonist, *nor*-meperidine. The potent and selective α_{1a}-antagonists **16** (K_i: α_{1a}, 0.7 nM; α_{1b}, 610 nM; α_{1d}, 1280 nM) and **17** (K_i: α_{1a}, 0.4 nM; α_{1b}, 500 nM; α_{1d}, 230 nM) separately address metabolism issues related to **15**; however, their further development is not apparent (69,70).

A recently disclosed arylpiperidine α_{1a}-antagonist is SNAP 7915, **18** (71). The DHP ring in **15** has been truncated to yield a non-racemic oxazolidinone heterocycle which still bears the 3,4-difluorophenyl ring requisite for activity. The preferred configuration of the functional groups at the C-4 and C-5 positions of the oxazolidinone ring is (4S,5S). SNAP 7915 avidly binds to the recombinant human α_{1a}-receptor (0.2 nM) and is >700-fold selective versus α_{1b}- and α_{1d}-receptors. High binding affinity is also observed for rat and dog recombinant α_{1a}-receptors (0.3 and 0.2 nM, respectively) indicating no significant species differences in the binding assays. No remarkable cross-reactivity was exhibited by **18** when screened against a panel of more than 30 G-protein coupled receptors. A dose of 3 µg/kg for **18** versus 16.4 µg/kg for **2** was required to block the effect of phenylephrine-induced increases in IUP in anaesthetized mongrel dogs. In contrast, a dose >300 µg/kg was required for **18** to cause a 15 mm Hg (DBP_{15}) drop in diastolic blood pressure; the DBP_{15} for **2** was 76 µg/kg. Compound **18** displays significant improvements in bioavailability and plasma half-life in rats (25%, 6 hr) and dogs (74%, >12 hr) compared with the foregoing dihydropyridines and dihydropyrimidinones. Future development plans for **18** have not been disclosed.

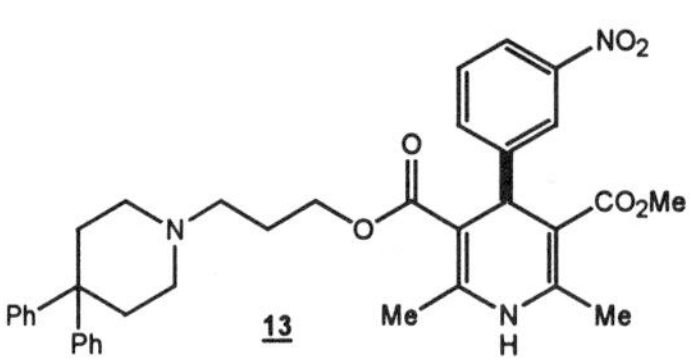

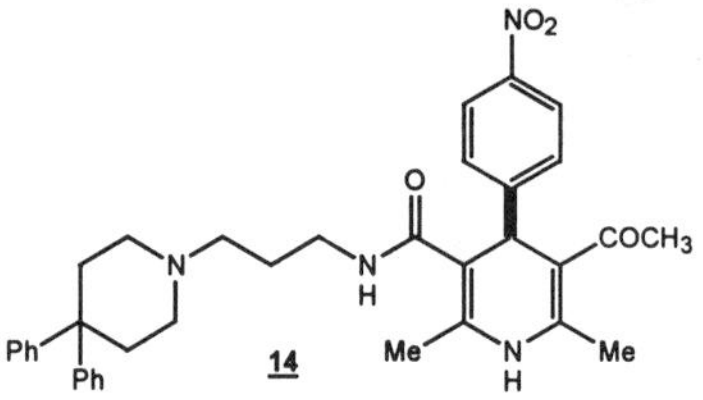

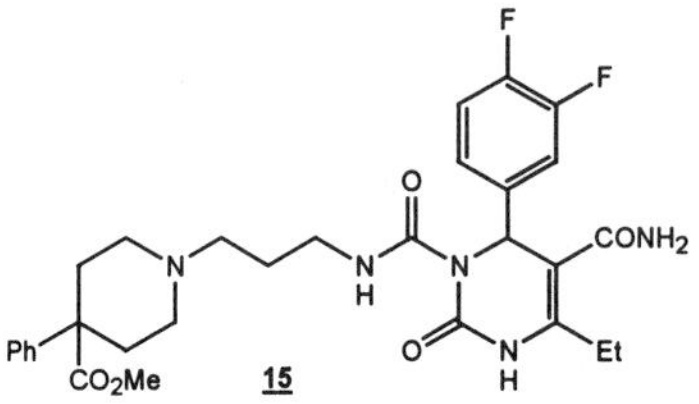

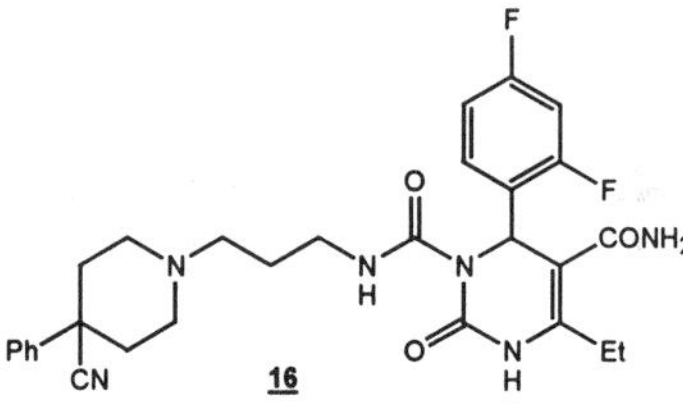

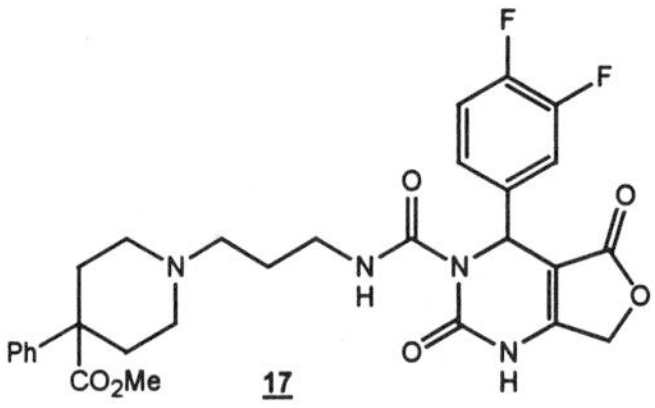

The phenylacetamide **19** contains remnants of the DHP ring system in **16** and is representative of several lower molecular weight analogs that have been derived from this lead (72). Compound **19** has a good receptor binding profile (K_i: α_{1a}, 2.0 nM; α_{1b}, 1400 nM; α_{1d}, 2100 nM) but only marginal PK (rat, F = 8%, $t_{1/2}$ 2 hr; dog, F = 19%, $t_{1/2}$ 2 hr). A stereochemical bias for binding to the human α_{1a}-receptor exists among phenylacetamide analogs with most of the activity residing with the (+)-enantiomer.

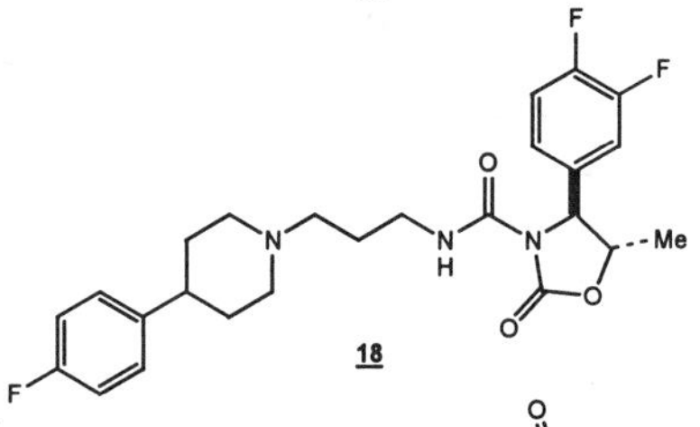

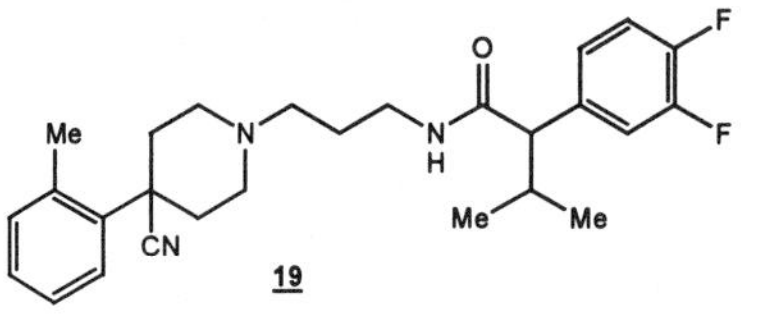

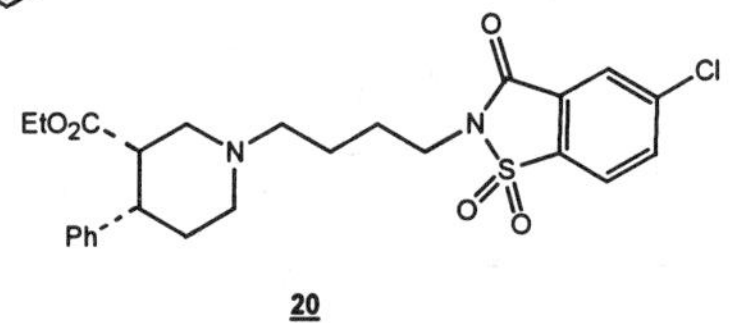

The arylpiperidine **20**, linked by a 4-carbon tether to a saccharin-based nucleus displays 333–940-fold selectivity for binding to the cloned human α_{1a}-receptor over α_{1b} and α_{1d}-receptors, as well as >300-fold selectivity for α_{1a} over at least ten other G-protein coupled receptors (73). Of the four possible diastereomers, the (+)-cis isomer is preferred. PK studies of analogs in this series yielded disappointing results.

<u>Piperidines</u> – Most α_{1a}-receptor antagonists are connected, sometimes only vaguely, by recurring structural themes. According to this criterion, the 4-piperidinyloxazole GG-818, **21** can be related to **1** by virtue of the 4-methoxy-3-aminosulfonylphenethyl fragment. The discovery of **21**, which had briefly entered clinical trials, is the subject of a recent review (74). GG-818 binds human α_{1a}-receptors with sub-nanomolar affinity (pK_i 9.7) and is selective for this receptor over α_{1b} and α_{1d} by 81- and 112-fold. The PK properties of **21** in mongrel dogs are also desirable: F = 93%; $t_{1/2}$ 4hr; low Vdss and CL = 0.9 ml/min/kg. However, **21** inhibits cardiac Ikr with high specificity and it can be assumed that the potential for QTc prolongation in man is the reason why this potent compound was not advanced (75,76).

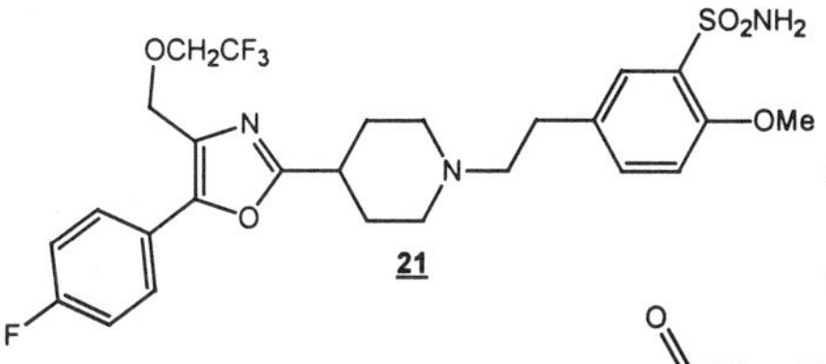

The N-alkyl saccharin-containing piperidine **22** bears a clear relationship to the arylpiperidine **20** (77). It potently binds the human α_{1a}-receptor (K_i = 0.08 nM) with high selectivity (K_i: α_{1b}, α_{1d} = 24, 85 nM, respectively). The effective *in vitro* selectivity of **22** translates to various tissue preparations rich in α_{1a}-receptors. Thus, **22** has more affinity for human (K_i = 0.17 nM) and dog (K_i = 0.24 nM) prostate than aortic tissue (human, dog K_i = 90 and 74 nM, respectively). Functional studies in anaesthetized dogs indicate that **22** is uroselective to the extent that it inhibits phenylephrine-induced increases in IUP (K_b = 5.5 µg/kg), whereas inhibitions in the increase of diastolic blood pressure occurred with K_b = 156 µg/kg. No PK data has been published for this compound.

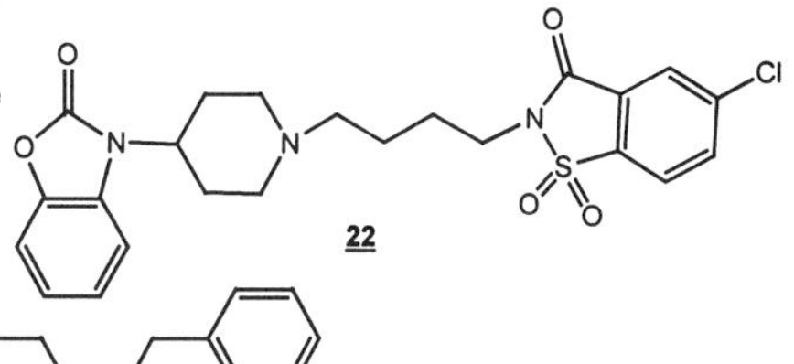

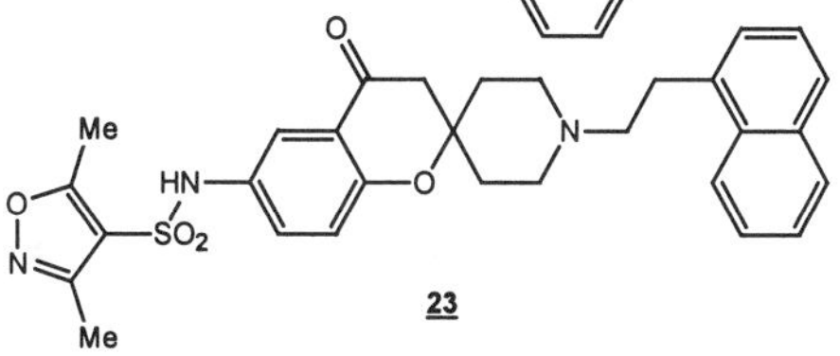

4-Oxospiro[benzopyran-2,4'-piperidines] were originally developed as class III antiarrhythmic agents. In a departure from the SAR precedence established for designing α_{1a}-receptor antagonists, these oxospiro[benzopyran-2,4'-piperidines] were transformed into compounds with affinity and modest selectivity for the human α_{1a}-receptor (78). The optimum analog is **23** (K_i: α_{1a}, 0.69 nM; α_{1b}, 24 nM; α_{1d}, 18 nM); importantly, it shows activity in a rat functional assay and is devoid of antiarrhythmic activity.

<u>Miscellaneous</u> – The excellent binding affinity and selectivity exhibited by arylpiperidine **15** for the α_{1a}-receptor were duplicated with compounds of slightly altered structure. The exchange of the central piperidine ring in **15** with a 4-substituted *cis*-aminocyclohexane ring, as in **24**, has benefits that extend beyond *in vitro* potency (79). Analog **24** displays binding (K_i: α_{1a}, 0.16 nM; α_{1b}, 45 nM; α_{1d}, 120 nM) and functional profiles comparable to **15** but has significantly improved PK properties (rat, F = 23%, $t_{1/2}$ 2.4 hr; dog, F = 43%, $t_{1/2}$ 6.7 hr) (80,81). The current status of **24**, with regard to further development, is unknown.

Following the disclosure of the structure of A-131701, **25**, reports detailing its *in vitro* and *in vivo* pharmacological characteristics appeared (82-84). The compound is only modestly α_{1a}-selective (K_i: α_{1a}, α_{1b}, α_{1d} = 0.22, 6.95, 0.97 nM, respectively), although *in vivo* experiments in conscious dogs reveal that **25** has a prolonged duration of action in measures of prostatic urethral smooth muscle function, with far weaker and more transient effects on cardiovascular α_1-adrenoceptor function. PK analysis of plasma samples from dogs indicates that **25** is 30–50% bioavailable with a half-life of only 0.4–0.8 hr. Somewhat longer half-lives were observed in rat and monkey, with bioavailability values in the 25 to 30% range. Evidence of nonlinear PK was obtained for **25** in dogs and monkeys. Taken together, the *in vitro* and *in vivo* data lend some support to the claims of uroselectivity; the development status of **25** or its homologs is unknown.

24   **25**

<u>Conclusion</u> – The use of selective α_1-adrenoceptor antagonists is an efficacious way to treat BPH (85,86). The ultimate therapeutic benefit of α_1-receptor blockade in patients with BPH depends on the balance between efficacy and tolerability. Presently, there is little compelling evidence that conventional regimens of α_1-antagonists differ with regard to their efficacy in improving lower urinary tract symptoms and obstruction, but there are several indications that α_1-antagonists differ qualitatively with regard to their cardiovascular safety and tolerability (87,88). The continuing evaluation of α_{1a}-subtype adrenoceptor antagonists in clinical settings will test whether or not a higher lever of tolerability (clinical efficacy) can been attained.

References

1.   W.D. Steers and B. Zorn, Disease-a-Month, <u>41</u>(7), 437 (1995).
2.   H. Lepor, J.Urol., <u>141</u>, 1283 (1989).
3.   M: Jonler, M. Riehmann, and R.C. Bruskewitz, Drugs, <u>47</u>, 66 (1994).
4.   J.D. McConnell, J.D. Wilson, F.W. George, J. Geller, F. Pappas, and E. Stoner, J.Clin.Endocrinol.Metab., <u>74</u>, 505 (1992).
5.   L.M. Eri and K.J. Tveter, J.Urol., <u>154</u>, 923 (1995).
6.   H.N. Whitfield, P.T. Doyle, M.E. Mayo, and N. Poopalasingham, Br.J.Urol., <u>47</u>, 823 (1975).
7.   P.F. Boreham, P. Braithwaite, P. Milewski, and H. Pearson, Br.J.Surg., <u>64</u>, 756 (1977).
8.   M. Caine, S. Perlberg, and S. Meretyk, Br.J.Urol., <u>50</u>, 551 (1978).
9.   M. Caine, S. Perlberg, and A. Shapiro, Urology, <u>17</u>, 542, (1981).
10.   J.P. Hieble, M. Caine, and E. Zalaznik, Eur.J.Pharm., <u>107</u>, 111 (1985).
11.   H. Lepor, S. Auerbach, A. Puras-Baez, P. Narayan, M. Soloway, F. Lowe, T. Moon, G. Leifer, and P. Madsen, J.Urol., <u>148</u>, 1467 (1992).
12.   M. K. Brawer, G. Adams, and H. Epstein, Arch.Fam.Med., <u>2</u>, 929 (1993).
13.   J.Y. Gillenwater, R.L. Conn, S.G. Chrysant, J. Roy, M. Gaffney, K. Ice, and N. Dias, J.Urol., <u>154</u>, 110 (1995).
14.   A. Fawzy, K. Braun, G.P. Lewis, M. Gaffney, K. Ice, and N. Dias, J.Urol., <u>154</u>, 105 (1995).
15.   R.A. Janknegt and C.R. Chapple, Eur.Urol., <u>24</u>, 319 (1993).
16.   S.A. Kaplan, K.A. Soldo, and C.A. Olsson, Eur.Urol. <u>28</u>, 223 (1995).
17.   A. Jardin, H. Bensadoun, M.C. Delauche-Cavallier, P. Attali, and BPH-ALF Group, Lancet, <u>337</u>, 1457 (1991).
18.   P. Teillac, M.C. Delauche-Cavallier, P. Attali, Br.J.Urol., <u>70</u>, 58 (1992).
19.   M.I. Wilde, A. Fitton, and D. McTavish, Drugs, <u>45</u>, 410 (1993).
20.   M.C. Michel, B. Kenny, and D.A. Schwinn, Naunyn-Schmiedeberg's Arch.Pharmacol., <u>352</u>, 1, (1995).
21.   J.P. Hieble, D.B. Bylund, D.E. Clarke, D.C. Eikenburg, S.Z. Langer, R.J. Lefkowitz, K.P. Minneman, and R.R. Ruffolo, Jr., Pharmacol.Rev., <u>47</u>, 267 (1995).
22.   I. Muramatsu, M. Oshita, T. Ohmura, S. Kigoshi, H. Akino, M. Gobara, and K. Okada, Br. J.Urol., <u>74</u>, 572 (1994).

23. A.P.D.W. Ford, N.F. Arredondo, D.R. Blue, Jr., D.W. Bonhaus, J. Jasper, M.S. Kava, J. Lesnick, J.R. Pfister, I.A. Shieh, R.L. Vimont, T.J. Williams, J.E. McNeal, T.A. Stamey, and D.E. Clarke, Mol.Pharmacol., 49, 209 (1996).
24. I. Marshall, R.P. Burt, P.O. Anderson, C.R. Chapple, P.M. Greengrass, G.I. Johnson, and M.G. Wyllie, Br.J.Pharmacol., 112, 59 (1992).
25. C. Forray, J.A. Bard, J.M. Wetzel, G. Chiu, E. Shapiro, R. Tang, H. Lepor, P.R. Harting, R.L. Weinshank, T.A. Branchek, and C. Gluchowski, Mol.Pharmacol., 45, 703 (1994).
26. I. Marshall, R.P. Burt, and C.R. Chapple, Br.J.Pharmacol., 115, 781 (1995).
27. J. Tseng-Crank, T. Kost, A. Goetz, S. Hazum, K.M. Roberson, J. Haizlip, N. Godinot, C.N. Robertson, and D. Saussy, Br.J.Pharmacol., 115, 1475 (1995).
28. K.E. Andersson, Eur.Urol, 33(suppl 2), 7 (1998).
29. J.P. Hieble and R.R. Ruffolo, Jr., Pharm.Res., 33, 145 (1996).
30. B. Kenny, S. Ballard, J. Blagg, and D. Fox, J.Med.Chem., 40, 1293 (1997).
31. C. Forray and S.A. Noble, Exp.Opin.Invest.Drugs, 8, 2073 (1999).
32. B.A. Kenny, A.M. Miller, I.J. Williamson, J. O'Connell, D.H. Chalmers, and A.M. Naylor, Br.J.Pharmacol., 118, 871 (1996).
33. D.R. Blue, D.V. Daniels, A.P.D.W. Ford, T.J. Williams, R.E. Eglen, and D.E. Clarke, J.Urol., 157(suppl), 190 (1997).
34. C.R. Chapple, L. Baert, P. Thind, K. Hofner, G.S. Khoe, and A. Spangberg, Eur.Urol., 32, 462 (1997).
35. M.C. Michel, H.U. Bressel, L. Mehlburger, and M. Goepel, Eur.Urol., 34(S. 2), 37 (1998).
36. H. Matsushima, H. Kamimura, Y. Soeishi, T. Watanabe, S. Higuchi, and M. Tsunoo, Drug Met.Disposition, 26, 240 (1998).
37. K. Shibata, R. Foglar, K. Horie, K. Obika, A. Sakamoto, S. Ogawa, and G. Tsujimoto, Mol.Pharmacol., 48, 250 (1995).
38. N. Moriyama, K. Akiyama, S. Murata, J. Taniguchi, N. Ishida, S. Yamazaki, and K. Kawabe, Eur.J.Pharm., 331, 39 (1997).
39. K. Akiyama, M. Hora, S. Tatemichi, N. Masuda, S. Nakamura, R. Yamagishi, and M. Kitazawa, J.Pharm.Exp.Therap., 291, 81 (1999).
40. K. Akiyama, S. Tatemichi, M. Hora, M. Kojima, S. Yamada, T. Ohkura, R. Kimura, and K. Kawabe, Neurourol.Urodyn. 16, 492 (1997).
41. Company Communication, Pharma Projects, G5A, au 134 (1999).
42. I. Muramatsu, M. Takita, F. Suzuki, S. Miyamoto, S. Sakamoto, and T. Ohmura, Eur.J.Pharmacol., 300, 155, (1996).
43. PJP Publication, Ltd, Surrey, UK, Ian Lloyd, ed., G5A, a1254 (1999).
44. S. Yamada, T. Ohkura, R. Kimura, and K. Kawabe, Life Sci., 62, 1585, (1998).
45. J. Sullivan and M. Williams, ID Weekly Highlights, 8, 33 (1998).
46. M. Takahashi, T. Taniguchi, S. Murata, K. Okada, N. Moriyama, S. Yamazaki, and I. Muramatsu, J.Urol, 161, 1350 (1999).
47. A.B. Reitz, D.J. Bennett, P.S. Blum, E.E. Codd, C.A. Maryanoff, M.E. Ortegon, M.J. Renzi, M.K. Scott, R.P. Shank, and J.L. Vaught, J.Med.Chem., 37, 1060 (1994).
48. M.M. Mensonides-Harsema, Y. Liao, H. Tottcher, G.D. Baroszyk, H.E. Greiner, J. Harting, P. de Boer, and K.V. Wikstrom, J.Med.Chem., 43, 432, (2000).
49. T.R. Elworthy, A.P.D.W. Ford, G.W. Bantle, D.J. Morgans, R.S. Ozer, W.S. Palmer, D.B. Repke, M. Romero, L. Sandoval, E.B. Sjogren, F.X. Talamas, A. Vazquez, J. Wu, N.F. Arredondo, D.R. Blue, A. DeSousa, L.M. Gross, M.S. Kava, J. D. Lesnick, R.L. Vimont, T.J. Williams, Q.-M. Zhu, J.R. Pfister, and D.E. Clarke, J.Med.Chem., 40, 2674 (1997).
50. T.J. Williams, D.R. Blue, D.V. Daniels, B. Davis, T. Elworthy, J.R. Gever, M.S. Kava, D. Morgans, F. Padilla, S. Tassa, R.L. Vimont, C.R. Chapple, R. Chess-Williams, R.M. Eglen, D.E. Clarke, and A.P.D.W. Ford, Br.J.Pharmacol., 127, 252 (1999).
51. D.R. Blue, A.P.D.W. Ford, D.J. Morgans, T.J. Williams, and Q.-M. Zhu, Br.J.Pharmacol. 120, 107P (1997).
52. D.R. Blue, A.P.D.W. Ford, D. Morgans, F. Padilla, and D.E. Clarke, Neurourol.Urodyn. 15, 345 (1996).
53. R. Testa, L. Guarneri, C. Taddei, E. Poggesi, P. Angelico, A. Sartani, A. Leonardi, O.N. Gofrit, S. Meretyk, and M. Caine, J.Pharmacol.Exp.Therap., 277, 1237 (1996).
54. A.A. Hancock, M.E. Brune, D.G. Witte, K.C. Marsh, S. Katwala, I. Milicic, L.M. Ireland, D. Crowell, M.D. Meyer and J.F. Kerwin, J.Pharmocol.Exp.Ther. 285, 628 (1998).
55. SCRIP World Pharmaceutical News, 2424, 5, (1998).
56. P. George, B. Marabout, J. Froissant, and J.-P. Merly, EP 0577470 A1, (1994).
57. L. Jolliffe, W. Murray, V. Pulito, A. Reitz, X. Li, L. Mulcahy, C. Maryanoff, and F. Villani, WO 9851298 A1 (1998).
58. G.-H. Kuo, W.V. Murray, C.P. Prouty, WO 9942445 A1 (1999).
59. G.-H. Kuo, W.V. Murray, C.P. Prouty, WO 9942448 A1 (1999).
60. G.-H. Kuo, C. Prouty, W. Murray, V. Pulito, L. Jolliffe, P. Cheung, S. Varga, M. Evangelisto, and J. Wang, 218th ACS National Meeting, New Orleans, Aug. 22-26 (1999), MEDI-225.
61. B. Lagu, D. Tian, D. Nagarathnam, M. R. Marzabadi, W.C. Wong, S.W. Miao, F. Zhang, W. Sun, G. Chiu, J. Fang, C. Forray, R.S.L. Chang, R.W. Ransom, T.B. Chen, S. O'Malley, K. Zhang, K.P. Vyas, and C. Gluchowski, J. Med. Chem., 42, 4794 (1999).
62. R. Boer, A. Grassegger, C. Schudt, and H. Glossmann, Eur.J.Pharmacol.,Mol.Pharmacol.Sect., 172, 131 (1989).

63.   C. Forray, J.A. Bard, J.M. Wetzel, G. Chiu, E. Shapiro, R. Tang, H. Lepor, P.R. Hartig, R. L. Weinshank, T.A. Branchek, and C. Gluchowski, Mol.Pharmacol., 45, 703 (1994).
64.   J.M. Wetzel, S.W. Miao, C. Forray, L.A. Borden, T.A. Branchek, and C. Gluchowski, J.Med.Chem., 38, 1579 (1995).
65.   W.C. Wong, G. Chiu, J.M. Wetzel, M.R. Marzabadi, D. Nagarathnam, D. Wang, J. Fang, S. W. Miao, X. Hong, C. Forray, P.J. Vaysse, T.A. Branchek, C. Gluchowski, R. Tang, and H. Lepor, J.Med.Chem., 41, 2643 (1998).
66.   D. Nagarathnam, J.M. Wetzel, S.W. Miao, M.R. Marzabadi, G. Chiu, W.C. Wong, X. Hong, J. Fang, C. Forray, T.A. Branchek, W.E. Heydorn, R.S.L. Chang, T. Broten, T.W. Schorn, and C. Gluchowski, J.Med.Chem., 41, 5320 (1998).
67.   W.C. Wong, W. Sun, B. Lagu, D. Tian, M.R. Marzabadi, F. Zhang, D. Nagarathnam, S. W. Miao, J.M. Wetzel, J. Peng, C. Forray, R.S.L. Chang, T.B. Chen, R. Ransom, S. O'Malley, T.P. Broten, P. Kling, K.P. Vyas, K. Zhang, and C. Gluchowski, J.Med.Chem., 42, 4804 (1999).
68.   D. Nagarathnam, S.W. Miao, B. Lagu, G. Chiu, J. Fang, T.G. Murali Dhar, J. Zhang, S. Tyagarajan, M.R. Marzabadi, F. Zhang, W.C. Wong, W. Sun, D. Tian, J.M. Wetzel, C. Forray, R.S.L. Chang, T.P. Broten, R.W. Ransom, T.W. Schorn, T.B. Chen, S. O'Malley, P. Kling, K. Schneck, R. Bendesky, C.M. Harrell, K.P. Vyas, and C. Gluchowski, J.Med.Chem., 42, 4764 (1999).
69.   T.G.M. Dhar, D. Nagarathnam, M.R. Marzabadi, B. Lagu, W.C. Wong, G. Chiu, S. Tyagarajan, S.W. Miao, F. Zhang, W. Sun, D. Tian, Q. Shen, J. Zhang, J. M. Wetzel, C. Forray, R.S.L. Chang, T.P. Broten, T.W. Schorn, T.B. Chen, S. O'Malley, R. Ransom, K. Schneck, R. Bendesky, C.M. Harrell, K.P. Vyas, K. Zhang, J. Gilbert, D.J. Pettibone, M.A. Patane, M.G. Bock, R.M. Freidinger, and C. Gluchowski, J.Med.Chem., 42, 4778 (1999).
70.   B. Lagu, D. Tian, G. Chiu, D. Nagarathnam, J. Fang, Q. Shen, C. Forray, R.W. Ransom, R.S.L. Chang, K.P. Vyas, K. Zhang, and C. Gluchowski, Bioorg.Med.Chem.Lett., 10, 175 (2000).
71.   B. Lagu, D. Tian, Y. Jeon, C. Li, D. Nagarathnam, Q. Shen, J.M. Wetzel, C. Forray, R.S.L. Chang, T.P. Broten, R.W. Ransom, T.-B. Chen, S.S. O'Malley, T.W. Schorn, A.D. Rodrigues, K. Kassahun, D.J. Pettibone, R.M. Freidinger, and C. Gluchowski, J.Med.Chem., 43, (2000), in press.
72.   M.A. Patane, R.M. DiPardo, R.C. Newton, R.P. Price, T.P. Broten, R.S.L. Chang, R.W. Ransom, J. Di Salvo, D. Nagarathnam, C. Forray, C. Gluchowski, and M.G. Bock, Bioorg.Med.Chem.Lett., 10, (2000), in press.
73   M.A. Patane, R.M. DiPardo, R.P. Price, R.S.L. Chang, R.W. Ransom, S.S. O'Malley, J. Di Salvo, and M.G. Bock, Bioorg. Med. Chem. Lett., 8, 2495 (1998).
74.   K.K. Adkison, K.A. Halm, J.E. Shaffer, D. Drewry, A.K. Sinhababu, and J. Berman, Pharm.Biotechnol., 11, 423 (1998).
75.   S. Liu, R.L. Rasmusson, S. Wang, D.L. Campbell, J.E. Shaffer, and H.C. Strauss, Biophysical J., 74(2 part 2), A207 (1998).
76.   M.R. Rosen, G.A. Pfeiffer, and A.J. Camm, Eur.Heart J., 19, 1178 (1998).
77.   J.B. Nerenberg, J.M. Erb, W.J. Thompson, H.-Y. Lee, J.P. Guare, P.M. Munson, J.M. Bergman, J.R. Huff, T.P. Broten, R.S.L. Chang, T.B. Chen, S. O'Malley, T.W. Schorn, and A.L. Scott, Bioorg.Med.Chem.Lett., 8, 2467 (1998).
78.   J.B. Nerenberg, J.M. Erb, J.M. Bergman, S. O'Malley, R.S.L. Chang, A.L. Scott, T.P. Broten, and M.G. Bock, Bioorg.Med.Chem.Lett., 9, 291 (1999).
79.   D. Nagarathnam, W.C. Wong, S.W. Miao, B. Lagu, Q. Shen, W. Sun, M.R. Marzabadi, S. Tyagarajan, F. Zhang, C. Forray, M.A. Patane, M.G. Bock, B.E. Evans, K.F. Gilbert, R.S.L. Chang, T.P. Broten, T.W. Schorn, T.B. Chen, S. O'Malley, R.W. Ransom, K. Schneck, R. Bendesky, C.M. Harrell, P. Kling, R. Freidinger, D.J. Pettibone, and C. Gluchowski, Abstract of Papers; 217th American Chemical Society National Meeting, Anaheim, Cal, March 21-25, 1999; MEDI-110.
80.   T.B. Chen, S.S. O'Malley, D. Nagarathnam, S.W. Miao, W.C. Wong, B. Lagu, M.R. Marzabadi, C. Forray, C. Gluchowski, D.J. Pettibone, and R.S.L. Chang, FASEB J., 13, A150.3 (1999).
81.   T. Broten, R. Ransom, A. Scott, T. Schorn, P. Kling, D. Pettibone, P. Siegl, D. Nagarathnam, S. Miao, W.C. Wong, B. Lagu, M.R. Marzabadi, C. Forray, and C. Gluchowski, FASEB J., 13, A150.7 (1999).
82.   M.D. Meyer, R.J. Altenbach, F.Z. Basha, W.A. Carroll, I. Drizin, S.W. Elmore, P.P. Ehrlich, S.A. Lebold, K. Tietje, K.B. Sippy, M.A. Wendt, D.J. Plata, F. Plagge, S.A. Buckner, M.E. Brune, A.A. Hancock, and J.F. Kerwin, Jr., J.Med.Chem., 40, 3141 (1997).
83.   A.A. Hancock, M.E. Brune, D.G. Witte, K.C. Marsh, S. Katwala, I. Milicic, L.M. Ireland, D. Crowell, M.D. Meyer, and J.F. Kerwin, Jr., J.Pharm.Exp.Ther., 285, 628 (1998).
84.   A.A. Hancock, S.A. Buckner, M.E. Brune, S. Katwala, I. Milicic, L.M. Ireland, P.A. Morse, S.M. Knepper, M.D. Meyer, C.R. Chapple, R. Chess-Williams, A.J. Noble, M. Williams, and J.F. Kerwin, Jr., Drug Dev.Res., 44, 140 (1998).
85.   F.M.J. Debruyne and H.G. Van der Poel, Eur.Urol, 36(suppl 1) 54, (1999).
86.   M.G. Wyllie, Eur.Urol, 36(suppl 1) 59, (1999).
87.   J. Sullivan and M. Williams, IDrugs, 1, 652 (1998).
88.   C. de Mey, Eur.Urol, 36(suppl 3) 52, (1999).

Chapter 21. Protein Tyrosine Phosphatase Inhibition

William C. Ripka

Ontogen Corporation, Carlsbad, CA 92009

Introduction - Protein tyrosine phosphatases (PTPs) belong to a growing family of enzymes which are involved in the regulation of a variety of cellular events. The current estimate is that humans have as many as 1000 phosphatase genes (1) and many show high selectivity and specificity. Protein tyrosine phosphorylation regulates cellular processes such as proliferation, differentiation, cell-cell interactions, metabolism, gene transcription, T-cell activation and the antigen-receptor signaling in B cells (2). The state of tyrosine phosphorylation on target proteins is tightly controlled by the concerted actions of protein tyrosine kinases (PTK), enzymes that catalyze the transfer of the γ-phosphate of ATP to the 4-hydroxyl of tyrosyl residues within specific protein/peptide substrates, and phosphatases which remove the phosphate groups. The function of PTP's is rigorously controlled by the localization of these enzymes within the cell and, therefore, intra- and inter-molecular interactions and substrate specificity may well be controlled by factors in addition to the amino acid sequence surrounding the site of dephosphorylation (3). Protein phosphatases (PP's) specifically hydrolyze serine/threonine phosphoesters whereas protein tyrosine phosphatases (PTP's) are phosphotyrosine specific. A sub family of PTP's , the dual specificity phosphatases, are capable of hydrolysis of both phosphotyrosines and phosphoserine/threonines. Although both PP's and PTP's catalyze phosphoester hydrolysis, they utilize completely different structures (<5% homology with PTPs) and a distinctly different catalytic mechanism (4). Nevertheless, the X-ray structure of the dual specificity VHR PTP shows an overall fold similar to both PTP1B and the Yersinia PTP (4). While PTP's are normally considered to down regulate PTK signaling, a number of examples exist where they function as positive mediators. These include SH-PTP2 in insulin and epidermal growth factor (EGF) signaling (5), PTPα in dephosphorylating an inhibitory pTyr residue on src PTK's (6) and cdc25 in removing an inhibitory phosphate on Tyr15 of p34^{cdc2} (7). Additionally, the receptor-linked PTP CD45, is an activator of T-cell receptor mediated signaling (8) and of Ras activation in B-cells stimulated through the antigen receptor. Recent reports that overexpression of certain PTP's occurs in a number of disease states (9-14) has led to considerable interest in PTP inhibitors as therapeutic agents (9). Excellent reviews on the biological function, structural characteristics, and mechanism of catalysis of PTPs and their role in signal transduction have been published (15,16).

Protein Tyrosine Phosphatase General Features - The PTPs are structurally categorized into two major families: receptor-like and non receptor-like. The receptor-like PTPs have highly conserved tandem intracellular catalytic domains, a single membrane spanning region, and a diversity of receptor-like extracellular domains implicated in cell signaling. The cytosolic PTPs contain a single conserved catalytic domain linked to a variety of noncatalytic segments that presumably exert a regulatory or targetting function. Among these noncatalytic segments are Src homology (SH2) domains which are found in the SHP's, a subfamily of intracellular PTPs.

The PTP catalytic domain contains approximately 200-300 amino acids and has an amino acid sequence nearly identical among all PTPs from bacteria to mammals (sequence homology >30%). Unlike the PPs, they do not require metal ions. An important feature of this catalytic domain is a signature motif at the active site, (I/V)HC*XAGXGR(S/T)G. The cysteine in this sequence is the nucleophile required for catalytic activity. Mutants with substitutions of Ser for this Cys retain their ability to bind phosphotyrosine-containing peptides but lack catalytic activity. Although the biological roles of the PTPs and dual specificity phosphatases are distinct, evidence suggests that both classes of enzymes employ a similar mechanism to hydrolyze phosphate monoesters and have in common the characteristic structural motif, HCXXGXXRSCT. Outside the common catalytic domains are various targeting and localization domains which are apparently utilized for controlling and restricting PTP substrate specificity. These include extracellular fibronectin-type or immunoglobulin repeats for the receptor-like PTP's (4). Most receptor-like PTP's contain a second intracellular PTP 'catalytic' domain (D2). It appears the D2 domains of certain types of receptor-like PTPs , e.g. PTPα, PTPε, and LAR, have some intrinsic phosphatase activity while the D2 domains of others, e.g. CD45, PTPγ, have little or none (17). The functional role of this second domain remains to be determined.

Receptor-like PTPs - PTP1B is a cytosolic phosphatase consisting of a single catalytic domain. It is anchored by its C-terminus to the cytoplasmic face of the ER (18). In vitro, it is a nonspecific PTP and dephosphorylates a wide variety of substrates (19). A number of X-ray structures of native and complexed PTP1B exist (20). Although the precise in vivo function of PTP1B has not been established, there is considerable evidence that this enzyme may be involved in the down regulation of insulin signaling by dephosphorylation of specific phosphotyrosine residues on the insulin receptor (9, 21-28). Vanadate, a transition state mimetic and nonspecific competitive inhibitor of PTP1B (9, 29-31) has been shown to exert insulin-like effects in the obese Zucker rat, a genetic model for obesity and type II diabetes (32), and in human clinical trials to be potentially useful in treating certain forms of diabetes (9, 33). The role of PTP1B in diabetes has been more clearly defined as a negative regulator of insulin signaling from results in the PTP1B knockout (KO) mouse (34). This PTP1B-KO mouse showed slightly lower blood glucose levels and 50% lower circulating insulin than normal mice (34). The enhanced insulin sensitivity of the KO was also evident in glucose and insulin tolerance tests. Interestingly, on a high fat diet, the PTP1B-KO mice were resistant to weight gain and remained insulin sensitive, whereas their normal counterparts rapidly gained weight and became insulin resistant. These results demonstrate that PTP1B has a major role in modulating both insulin sensitivity and metabolism and confirms it as a potential therapeutic target in the treatment of type 2 diabetes and obesity. The presence and metabolism of PTP1B has been detected and studied in human skin. Its role in dephosphorylating EGF-R suggests it may play a role in controlling epidermal growth (35). It has been shown that three- to five-fold overexpression of PTP1B suppresses the transformation of cultured mouse fibroblasts induced by the human neu oncogene (36). Statistical analysis has demonstrated a significant association between PTP1B overexpression and breast cancer as well as an association with the overexpression of p185[c-erbB-2], the growth factor receptor protein tyrosine kinase. This kinase is overexpressed in one third of human breast cancer patients and is indicative of a poor prognosis in these patients (37). Finally, increased expression of PTP1B in ovarian cancers that also express PTK's suggests that PTP1B may play a role in growth regulation of such cancers (38).

Other PTPs, in addition to PTP1B, specifically LAR and PTPα, have also been implicated in the insulin-mediated signal transduction pathway (39) and are considered to be negative regulators of insulin action (40-46). In vitro, the recombinant catalytic domain of LAR can efficiently dephosphorylate insulin and EGF receptors (47).

Cdc25 is a member of the dual-specificity family of PPs and is responsible for the activation of the important cell cycle regulatory complex cyclin A/ cdk2 by dephosphorylation of pThr14 and pTyr15 of the cdk2 subunit. (48). Three phosphatases in particular, cdc25-A and –B as well as PTEN, have been implicated in the etiology of tumors. PTEN acts as a tumor suppressor whereas cdc25-A and -B act as oncogenes. Cdc25-A and cdc25-B transform primary mouse embryo fibroblast cells when cotransfected with the Ha-RASG12V oncogene (49). CDK's (cyclin dependent kinases), when phosphorylated on Thr and Tyr residues at their N-terminus, are enzymatically inactive. Removal of the phosphates by cdc25-A and -B activates these kinases leading to stimulation of the cell cycle. Cdc25-A appears to be critical for progression into S phase and cdc25-B promotes progression into mitosis (50).

CD45 is a dual PTP domain, transmembrane protein expressed on all nucleated hematopoietic cells including T-cells and expression of CD45 as well as its catalytic function is required for normal lymphocyte development and antigen receptor signal transduction function (51). Similar to other transmembrane PTPs, CD45 has an amino-terminal extracellular domain, a single transmembrane region, and a cytoplasmic domain that constains tandem PTP domains (52). The second PTP domain appears to be catalytically inactive. Generation of CD45 deficient cells lines has indicated CD45 expression is important for efficient signaling through the T and B cell antigen receptors (53-56). A natural killer cell line deficient in CD45 expression does not efficiently lyse target cells (57). KO mice deficient in CD45 show a pronounced block of thymocyte development and impaired B-cell proliferation demonstrating CD45 is required for antigen-receptor signaling (58). Dephosphorylation of certain protein tyrosine kinases (the src-family PTK's (51, 59) by CD45 leads to up-regulation of their catalytic activity, initiating a cascade of intracellular events leading to T-cell activation (60). CD45 therefore represents a therapeutic target for certain autoimmune and chronic anti-inflammatory diseases characterized by aberrant T-cell activation. Several reviews of CD45 have been published (52, 61-65).

The receptor-like PTPs, RPTPμ, RPTPκ, and RPTPγ, possessing extracellular segments sharing structural similarities with the Ig superfamily of cell adhesion molecules, have been shown to mediate cell aggregation via homphilic interactions (66-68). A role for RPTPμ in regulation cell junctions and cytoskeletal organization is suggested by the finding that RPTPμ is associated with cadherin-catenin complexes at adheren junctions (69).

Nonreceptor-like PTPs- PTPα is broadly expressed and possesses a small, glycosylated extracellular domain, a transmembrane region, and an intracellular component with two catalytic domains (D1,D2) (70, 71). Evidence suggests it is involved in the activation of c-src by dephosphorylation of Tyr-527 (72,73) and the down regulation of insulin receptor signaling (74). VHR is a human dual-specificity phosphatase which preferentially catalyzes the hydrolysis of aromatic phosphatases (75). VHR appears to be significantly more accommodating than PTP1 toward sterically demanding substrates (76). PTPε has been isolated from mouse osteoclast-like cells. It is known that PTP inhibitors can suppress both in vitro osteoclast differentiation and bond resorption suggesting PTP activity may be essential for osteoclast function and could be a useful therapeutic target (77, 78).

Structural information - Detailed structural studies have been carried out for catalytically incompetent PTP1B (C216S) and its complex with DADEpYL (pY=phosphorylated tyrosine) (79,80). The signature motif for PTPs, (H/V)C-X-X-X-X-X-R-(S/T), is found between the β-strand that ends with His-Cys and the α helix which starts with Arg-Ser. This sequence forms the phosphate binding pocket. At the base of the active site cleft is the catalytic cysteine. An Arg points towards the active site to assist phosphate binding and catalysis. About 30-40 residues from the amino terminus of the catalytic Cys is a catalytically important Asp located on a flexible loop (the WPD loop). Binding of the phosphate oxyanion triggers a conformational change which brings a catalytically active Asp, proposed as a general acid/base, into position near the catalytic Cys. The open inactive conformation closes to the active small pocket which binds the phosphate plus a water molecule. The Cys nucleophile attacks the phosphorous atom in the substrate to expell the leaving phenol, resulting in formation of a covalent thiophosphate enzyme intermediate. In a second step, a water molecule attacks the phosphorous atom to yield inorganic phosphate. The WPD loop interactions may offer ways to design inhibitors to discriminate between the various PTPs to achieve selectivity.

Important residues for substrate/inhibitor recognition in PTP1B include Phe182, in the WPD loop and Gln262 which interacts with the phenyl ring of pTyr and defines a portion of the rim of the binding pocket (79). Gln262 is invariant in the PTPs and has been suggested as an important residue for optimal positioning of the nucleophilic water for efficient E-P hydrolysis. Asp48 is a primary determinant of the peptide conformation at the pTyr site forming two H-bonds with the main chain nitrogens of pTyr and the +1 residue.

When the structure of the catalytically inactive mutant C215S of human PTP1B was solved as a complex with a nonpeptidic substrate analog, bis-(paraphosphophenyl)methane (1, BPPM, Km=16 μM), it was found that BPPM binds to PTP1B in two mutually exclusive modes; one in which it occupies the canonical pTyr binding site (active site) and another in which a phosphonophenyl moiety interacts with a set of residues not previously observed to bind aryl phosphates (82). Identification of this second aryl phosphate binding site adjacent to the active site provides a paradigm for the design of tight binding, specific PTP1B inhibitors that can span both sites. (82). There are excellent reviews of the structure and function of PTPs (83-88).

Mechanism and Kinetics - An understanding of the kinetics of the PTPs is an important consideration for the design of inhibitors. Kinetic and trapping experiments show that PTP catalyzed hydrolytic reactions occur via nucleophilic catalysis utilizing the activated cysteine in the active site (89). The reaction is composed of two chemical steps, the formation and breakdown of the phosphoenzyme intermediate. Detailed kinetic analysis of Yersinia PTP substrate hydrolysis has shown the rate to be *independent of the leaving group* and most likely involving the breakdown of the phosphoenzyme intermediate. Thus, the phosphoryl transfer to the enzyme is accompanied by a rapid release of the dephosphorylated product.   Subsequent hydrolysis of the thiol-phosphate intermediate is concomitant with the rapid release of phosphate. Vanadate, with its trigonal bipyramidal geometry, mimics the transition for both chemical steps. The proposed scheme for hydrolysis of the Yersinia PTP is shown in Figures 1 and 2 (90,91).

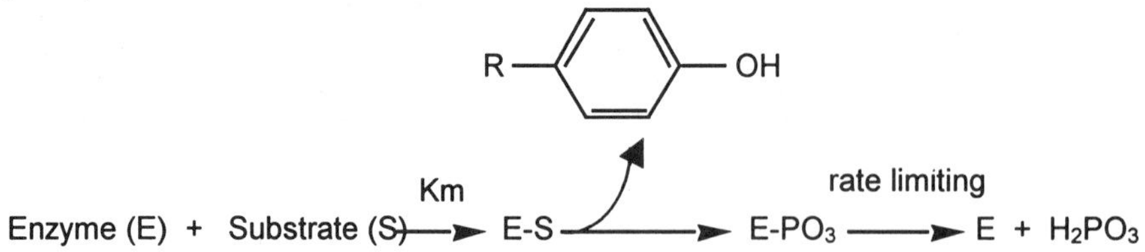

Figure 1.

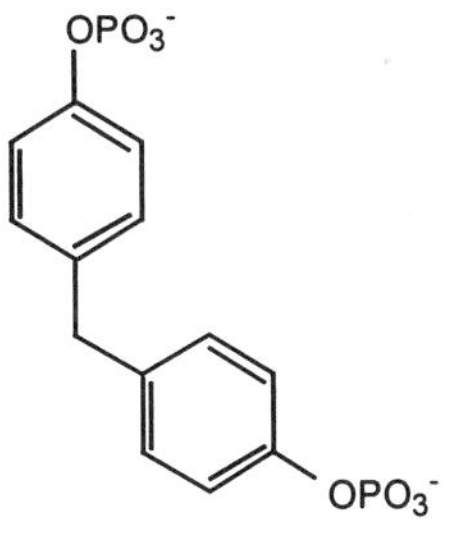

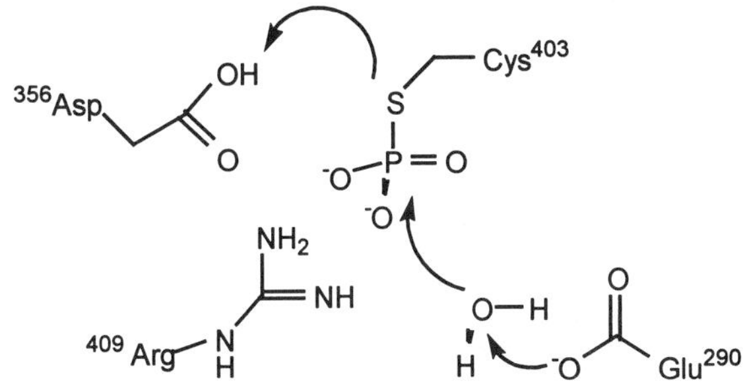

Figure 2.

Although all the PTPs share a similar catalytic domain, their kinetic parameters vary dramatically. The catalytic efficiencies are in the order Yersinia>PTP1B>LAR>yeast PTP1. It has been suggested that the residue following the catalytic Asp181 may be important for enhancing the catalytically competent active site: Yersinia (Gln182), PTP1B(Phe182), LAR(His182), yeast PTP1 (Met182). The kinetic results for PTPs showing decomposition of the Cys phosphate enzyme intermediate as the rate limiting step for overall hydrolysis suggests that the Michaelis constant for substrates will accurately reflect the potencies of potential inhibitors, assuming a satisfactory p-Tyr mimetic. This has turned out to be the case (92).

Phosphotyrosine Mimetics – To generate useful mimetics the pTyr group has been replaced with a number of nonhydrolyzable mimetics such as phosphonates, sulfates, and malonates as well as prodrugs that might address the limited cellular bioavailability due to the poor transport of the ionized phosphonate moiety through cell membranes. Successful mimickry of the pTyr phosphonate group is expected to be highly dependent on both the ionization state and the three-dimensional arrangement of its oxygen functionality. Figure 3 shows the pKa values of a number pTyr mimics based on atom substitutions in the phosphate group (93,94). Molecular modeling showed that the malonate moiety, while 12% larger in volume, nevertheless is accomodated in the catalytic site of PTP1B with a geometry nearly identical to that seen for the phosphonate (95). Malonyl (96, 97) and carboxymethy (98, 99) have been appended to the Tyr oxygen as phosphate mimetics. Modeling (94) and X-ray structures (100) indicate, however, the carboxyl oxygens are extended too far from the phenyl ring pushing the tyrosyl pheny ring out of its normal alignment in the active site. Attachment of the carboxymethyl directly to the pheny ring, however, allows for good overlap of carboxy functionality with the phosphate oxygens while maintaining registry of the phenyl rings (94)..

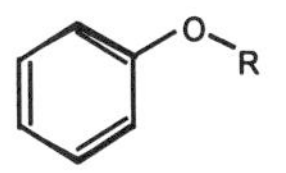

	R= PO$_3$H$_2$	CF$_2$-CO$_2$H	CH$_2$CO$_2$H		pKa
pKa1	2.33 ± .07	1.88 ± .09	3.17		4.08
	R= PO$_3^-$	CF$_2$CO$_2^-$	CH(CO$_2^-$)$_2$	CH(CO$_2^-$)CH$_2$CO$_2^-$	2.03
VDW vol(Å3)	45.8	61.3	74.7	84.4	
VDW radius(Å)	2.44	2.69	2.88	3.00	

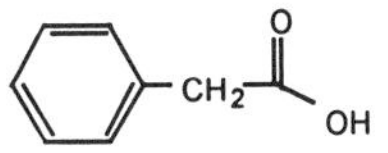

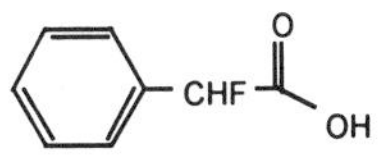

Figure 3

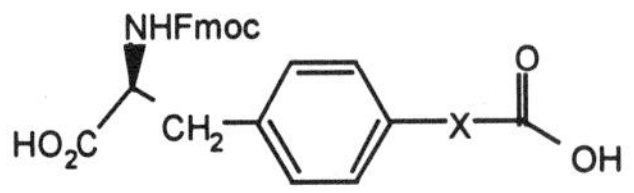

2a X = CH$_2$
2b X = CF$_2$

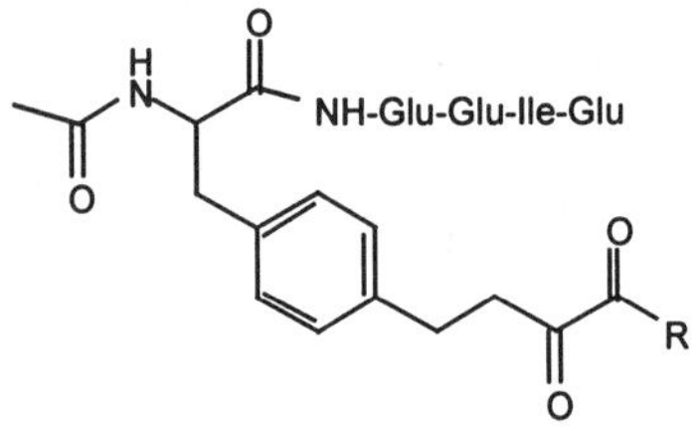

3a R = CH$_3$
3b R = H

The synthesis of the phosphotyrosine mimetics (**2a,b**) have been described (94), and their incorporation into a high affinity p56^{lck} SH2 domain-directed peptide showed them to have good potency. Based on the finding that an α-dicarbonyl moiety successfully served the function of an Arg trap (101), the synthesis of α-dicarbonyl moieties as a substitute for the phosphate group of phosphotyrosine led to compounds **3a** and **3b** (102). This approach circumvents the lability toward phosphatases and, by virture of their uncharged nature, would be expected to have enhanced cell penetration characteristics. In an ELISA, which measured the activity of **3a** and **3b** to inhibit the binding of human pp60^{csrc} SH3-SH2 domain to autophosphorylated EGFR, these compounds were 520 and 370 times less potent, respectively, than the phosphotyrosine analog, Ac-pY-E-E-I-E (IC$_{50}$ $\leq$ 0.5 µM). The linear hexamer peptide sequence, D-A-D-E-pY-L, corresponding to one of the autophosphorylation sites of the EGF receptor (EGF988-993), has been shown to be the minimum peptide for efficient binding and catalysis and has been used successfully for the incorporation of potential p-Tyr mimetics to test their efficacy. pTyr analogues (X) contained within the Ac-D-A-D-E-X-L-NH2 substrate sequence have been prepared and tested as inhibitors of PTP1B. OMT **4** (Ki = 13 µM) and FOMT **5** (Ki = 170 µM) incorporated a malonate equivalent (103). Pmp **7** (Ki = 200 µM) and F2Pmp **8** (Ki = 100 µM) are phosphonate equivalents of the substrate **6** (Km = 60 µM) (104, 105). Salicylate ether **9** (Ki = 3.6 µM) is the most potent of the series.

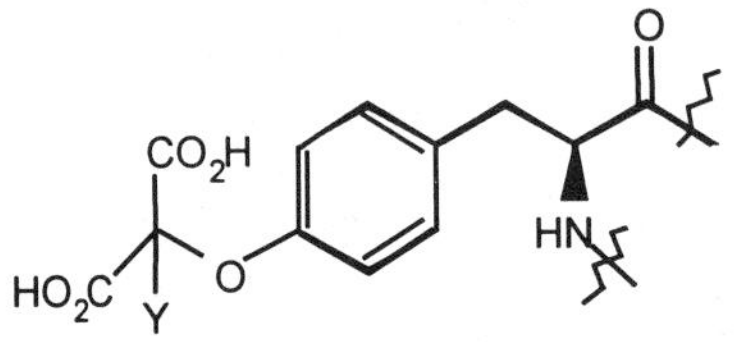

4 Y = H
5 Y = F

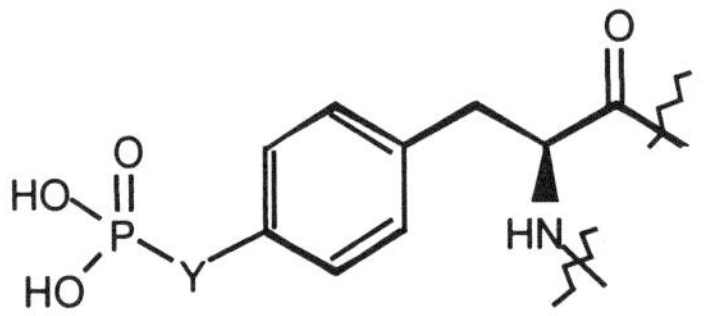

6 Y = O
7 Y = CH$_2$
8 Y = CF$_2$

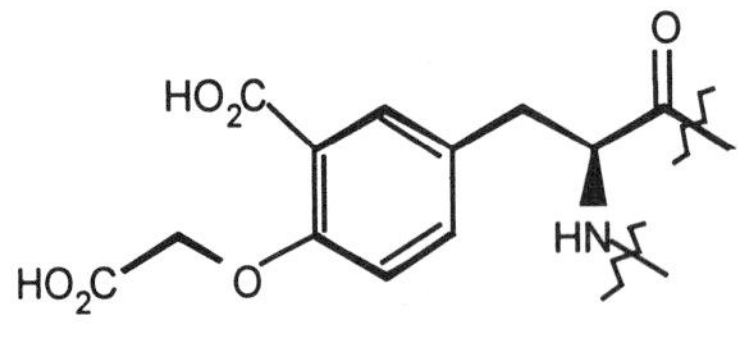

9

Substrate Selectivity – One approach towards PTP inhibitor design utilizes the structure of high affinity peptide substrates. The specific interactions used in substrate binding both inside and outside the pTyr binding pocket can be used as a model for potential novel inhibitors. While primary recognition of PTP substrates occurs through binding pTyr, selectivity undoubtedly arises from the interactions of the amino acid sequence around this residue. These important interactions can be probed by substrate analogs which replace the pTyr with mimetics. Specific consensus substrate peptide motifs for other PTPs could also lay the foundation for structure based rational inhibitor design and identify important regions outside the common active sites that could be targeted for selectivity. Since the identity of the physiological substrates for most PTPs is not known at present, such optimal substrate motifs may also be useful in identifying physiological substrates for PTPs from available sequence data bases (85). In addition to the EGF(988-993) sequence, other potential peptide substrate sequences include KRSpYEEHIP, a peptide modeled after the phosphorylation site (Y316) in the insulin receptor (7, 76). The best peptide-based substrate for VHR has been reported to be DHTGFLpTEpYVATR (k_{cat}= 4.65 s^{-1}, K_m = 140 µM (76). Since the active site substrate specificities of VHR and PTP1 are not identical, even though they both catalyze the hydrolysis of aryl phosphates, it should be possible to develop phosphatase selective inhibitors based on the distinct substrate specificites of these enzymes. Studies to identify the smallest pTyr peptide sequence required by CD45 for substrate recognition used the fyn PTK CD45 substrate as a template and a series of analogs was synthesized (109). From this work the enzyme did not appear to have a strong preference for either the N- or C-terminal peptide sequence of the substrate. The results of kinetic studies also suggested that CD45 does not require extended N- or C-terminal peptide structure, beyond the phosphorylated tyrosine, for efficient substrate turnover. The k_{cat}/K_m values for eleven peptides (from 14-mers to Ac-pTyr-NH$_2$) ranged from 3.5 to 8.0x10^{-3} M^{-1} s^{-1} while individual k_{cat} and K_m constants were within a factor of two of each other (109). The PTP, HPTPβ, is a member of the receptor-linked PTP family with a single cytosolic catalytic domain. From a series of peptide substrates derived from the phosphotyrosyl sites of a variety of target proteins of HPTPβ, an SAR was described (110).

Inhibition - Many of the PTP inhibitors identified to date, such as vanadate, show broad activity against a multitude of phosphatase targets. Since structural features that are important for pTyr recognition in the catalytic site are conserved among different PTPs, it was anticipated that it would be difficult to generate selective inhibitors targeted only at the active site. The experience with substrate analogs, natural product inhibitors, and synthetic inhibitors have shown the importance of exploring structural features outside the active site to achieve selectivity and some success has been achieved.

10   11a R=(CH$_2$)$_2$C$_6$H$_5$ 12   a=double bond, X=O
            11b R=(CH$_2$)$_4$C$_6$H$_5$ 13   a=single bond, X=H$_2$

Natural Products Inhibitors - Sulfircin (10), isolated from a a deep water sponge of the genus Ircinia, was found to be a potent inhibitor of cdc25A (IC$_{50}$ 7.8 µM), PTP1B (IC$_{50}$ 29.8 µM), and VHR (IC$_{\square\square}$ 4.7 µM) (111). A number of analogs (11-13) were synthesized establishing some structure activity relationships. Although the sulfate moiety was found to be critical for inhibitory activity, its stereochemistry did not effect activity. The overall length of the furyl side chain appeared to be important for activity. Acceptable replacments were found for both the furyl sidechain ((CH$_2$)n-C6H5-) as well as the sulfate (malonyl group) (111). Nornuciferene, an aporphine alkaloid, inhibits CD45 with an IC_{50} of 5.3 µM (18). A novel inhibitor of CD45, dephostatin (**14**), was isolated from the culture broth of a strain of Streptomyces (112) and appeared to inhibit CD45 with an IC_{50} of 7.7 µM. Both dephostatin (**14**) and N-methyl-N-nitrosoaniline (**15**) were found to inactivate Yersinia PTP as well. Although the original studies suggested the inhibition was competitive (112), when assayed against Yersinia PTP and PTP1, it was found to inactivate these PTPs in a concentration and time-dependent manner (113).   Kinetic data show the N-nitrosoaniline is the basic functionality for inhibition although the dihydroxy substitution substantially increases the inhibitory potency.   Inactivation was prevented by arsenate indicating modification was at the active site. A specific inhibitor of CD45 and VHR was isolated from microbial metabolites and found to be 3-hexadecanoyl-5-hydroxymethyl-tetronic acid (**16**) (114). The compound inhibited the dephosphorylation activity of CD45 and VHR with IC_{50}'s of 54 and 2.0 µM, respectively.

Dysidiolide (**17**) is a sesterterpene γ-hydroxybutenolide isolated from the marine sponge *Dysidea etheria de Laubenfels* was reported to be a potent inhibitor of cdc25A (115). An X-ray has confirmed the structure of the compound and it was claimed to inhibit the growth of A-549 human lung carcinoma and P388 murine leukemia cell lines with IC_{50}'s of 4.7 and 1.5 µM, respectively. A subsequent report has been unable to confirm the cdc25A inhibitory potency (116).

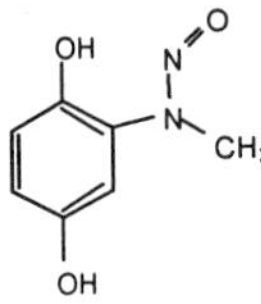

| 14 | 15 | 16 | 17 |

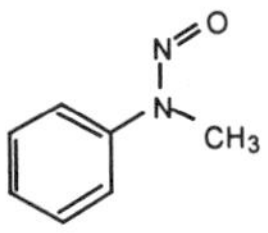

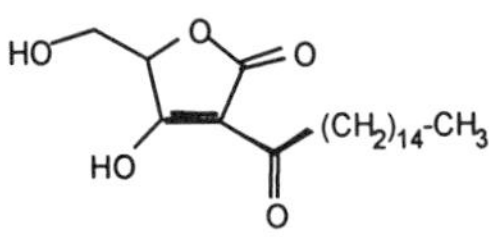

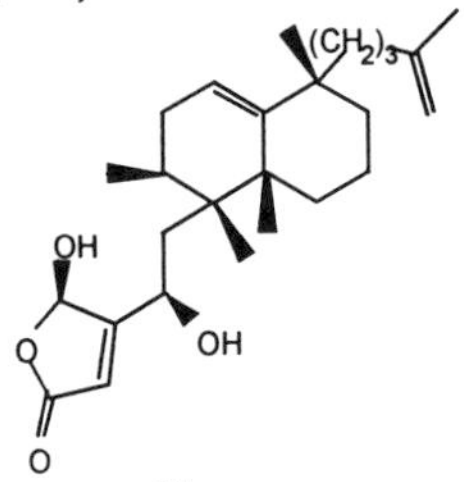

Irreversible Inhibitors - A number of metal containing, reversible inhibitors such as gallium nitrate (30) and vanadate (29,31) have been reported. While they may be mimicking inorganic phosphate and acting as transition state analogs they inhibit many other enzymes and, not unexpectedly, lack selectivity. The properties of the vanadates and their role in PTP inhibition have been reviewed (117,118). Vanadate is a competitive inhibitor for PTP1B with a Ki of 0.38 µM (K_m = 21 µM, K_{cat} 4.85 s^{-1}). It has insulin mimetic activity and has been shown in human clinical trials to be potentially useful in treating insulin and non-insulin dependent diabetes mellitus (119,120).   An orally active insulin mimetic vanadyl complex, bis(pyrrolidine-N-carbodithioato)-oxovanadium (**18**), has been reported (121). In the X-ray structure of the Yersinia PTP and the vanadate oxyanion, which can adopt the pentavalent geometry, the electron density is continuous between the bound anion and the active site Cy403 and the vanadate-cysteine distance is 2.5A , consistent with a covalent bond. The structure shows a distorted trigonal bipyramidal geometry that may mimic the transition state for the thiol-phosphate hydrolysis of the PTP phosphoenzyme intermediate. The conserved Asp356 makes a hydrogen bond of 2.8A to the apical oxygen of vanadate, consistent with its role as a general acid/base. The apical oxygen occupies the position where the phenolic oxygen of the phosphotyrosine would be located in the Michaelis complex and where a water molecule would be during the hydrolysis of the phosphoenzyme intermediate. As a general acid, the aspartic acid donates a proton to the phenolic oxygen to facilitate expulsion of tyrosine during formation of the cysteinylphosphate intermediate and as a general base, aspartate activates a water molecule to facilitate the rate limiting intermediate hydrolysis. Negative charge on the phosphate is stabilized by a bidentate interaction with the conserved Arg and by the backbone amide groups of the conserved active-site loop.

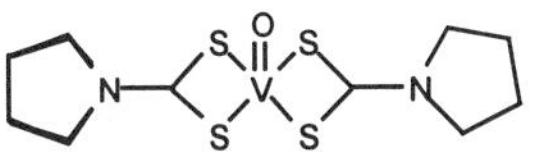

18

Mechanism-based Suicide Inhibitors - Modification of menadione (19) resulted in compounds that showed enhanced inactivation of cdc25A PTP (122).   Flow cytometric analysis showed that a cell cycle delay at the G1/S phase occurred with compound 19 treated cells as compared with control cells.   Compound 19 also inhibited the growth of SK-hep-1 cells via a cell cycle delay due to decreased activity of cdk2 kinase caused by cdc25A phosphatase inactivation (122). Sulfone analogs of napthoquinone (20) were also found to selectively and irreversibly inactivate PTP1B with a K_i of 3.5 μM and an inactivation rate constant of 2.2×10^{-2} sec^{-1} by the mechanism shown (123).

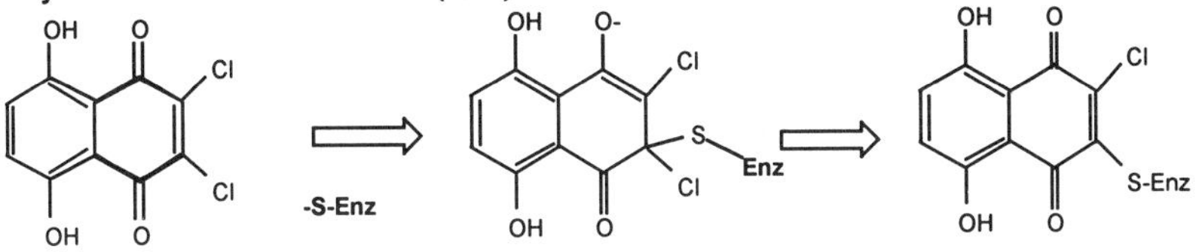

19

Interestingly, compound 20 showed no inactivation up to 40 μM of PTPs cdc25A, cdc25B, cdc25c, LAR, and Yersinia PTP but selectively inactivated PTP1B (123). Compound 21 is a potent inhibitor of PTPs but the probable mechanism of action makes it an unlikely candidate as a therapeutic agent. It was proposed that the cleavage of the phosphate ester leads to rapid elimination of the fluoride ion to form a quinone methide. This methide can then rapidly react with any nearby active site nucleophile to irreversibly inhibit the enzyme (124-126).   Thiol proteases are susceptible to inactivation by affinity reagents containing weakly electrophilic groups that normally are poor participants in SN2 reactions. α-Halobenzylphosphonates (22) have been shown to inactivate Yersinia PTP in a time- and concentration-dependent manner (127). This inactivation is active site directed and irreversible and is much faster than the reaction toward nucleophiles in solution.

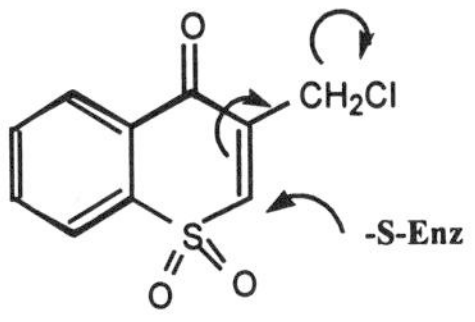

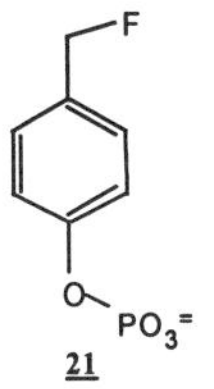

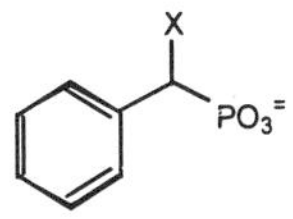

20   21   22 X=F, Cl, Br

Phosphonic Acid Inhibitors - A series of α,α–difluoromethylenephosphonic acid (DFMP) containing non-peptide inhibitors (23) have been synthesized as PTP1B inhibitors (128). None of these were significantly more potent than the parent compound (X=H) with the exception of the m-C6H5 which increased potency by about 17-fold. A second series of bis-DFMP compounds, however, were consistantly more potent (128). The fact that the potency is insensitive to chain length beyond n=2 suggests the second phosphate binding site recently discovered is not involved (82). The increased potency may be related to the strong preference PTP1B has for acidic residues towards the N-terminal from the pTyr.

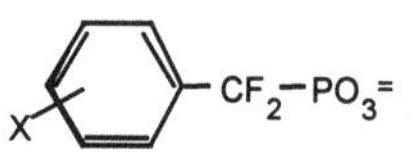

23

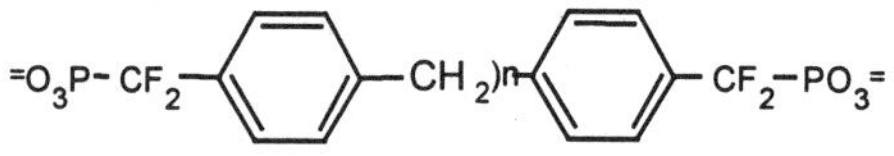

	n
24	1
25	2
26	3
27	4

The X-ray complex of **28** with PTP1B has been reported (129). The structure shows the phosphate held in the catalytic site with extensive hydrophobic interactions withthe napthalene ring more than that possible with a phenyl. The pro-R α-fluorine forms an apparent unconventional hydrogen bond to the amido group of Phe182. Computer modeling suggested that the addition of a hydroxyl to the naphthyl 4-position could allow new H-bonding interactions with Lys120 and Tyr46.

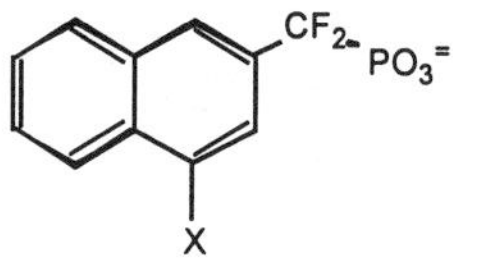

28 X = H
29 X = OH

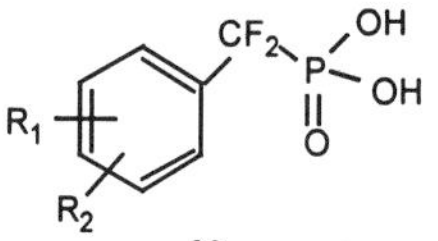

30

The compound (**29**) was two fold more active (Ki = 94 μM) than the parent (Ki = 179 μM). In an effort to achieve the hydrophobic interactions noted for the naphthalenes **28,29** with a different molecular scaffold, analogs based on **30** were synthesized in which the hydrophobic group (R1) might play the same role as the napththe-thalene ring (130). The hydrophilic group (R2) was expected to increase the stability of the bound inhibitor by an H-bonding interaction with Tyr46 and Lys120. The most active compounds in this series possessed a p-ethynyl (IC$_{50}$ = 77.5 μM), a m-ethynyl (IC$_{50}$ = 19 μM), p-phenethylenthynyl-m-N,N-di(sulfonylmethyl)amino (IC$_{50}$ = 88.7 μM), and p-(E)-styryl-m-N,N-di(sulfonylmethyl) (IC$_{50}$= 57.9 μM) (130). The sulfonamide group was introduced for its hydrogen bonding capabilities. The CF$_2$-tetrazole (**32,36**), CF$_2$-sulfonate (**31,35**), and CF$_2$-carboxylate (**33,37**) groups were examined as potential replacements for the difluoromethylene phosphonic acid (DFMP) group (**38**) (131).

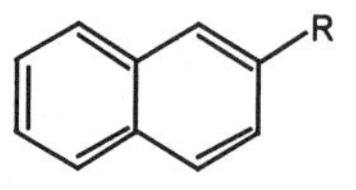

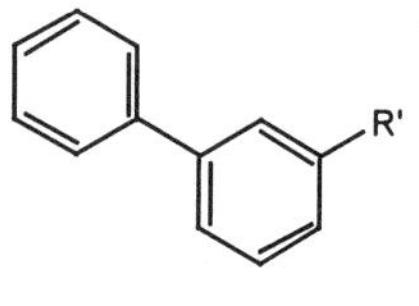

31	CF$_2$SO$_3$H	**35**	CF$_2$SO$_3$H
32	CF$_2$-tetrazole	**36**	CF$_2$-tetrazole
33	CF$_2$-CO$_2$H	**37**	CF$_2$-CO$_2$H
34	CF$_2$-PO$_3$H$_2$	**38**	CF$_2$-PO$_3$H$_2$

A number of aryl containing phosphonates were synthesized and assayed for PTP1B activity (**39-41**) (132). Two compounds **40c** and **41b**, were found to inhibit dephosphorylation of [^{32}P]-insulin receptors by PTP1B with IC$_{50}$ values of 40-50 μM. Of note is the fact that these two compounds also inhibited the serine/threonine phosphatase PP2A with similar potency.

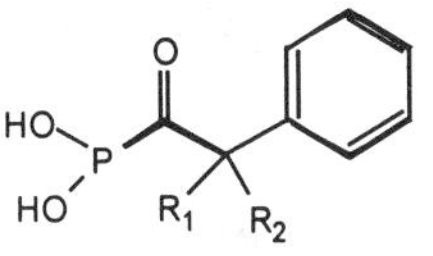

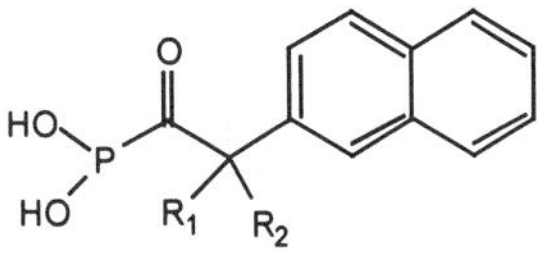

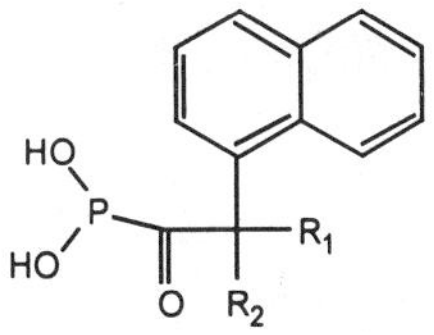

39a R₁ = R₂ = H
39b R₁ = H, R₂ = OH
39c R₁ = H, R₂ = F
39d R₁ = R₂ = F

40a R₁ = R₂ = H
40b R₁ = H, R₂ = OH
40c R₁ = R₂ = F

41a R₁ = H, R₂ = OH
41b R₁ = R₂ = F

Several naphthyl difluoromethylphosphonic acids were designed bearing acidic functionality intended to interact with the PTP1B Arg47, a residue situated just outside the catalytic pocket (133). This residue was shown previously to provide key interactions with acidic residues of phosphotyrosine-containing peptide substrates. These analogs were 7- to 14-fold more potent that the parent **42**. Comparing compounds **42** and **43** shows a 14 fold enhancement by the addition of the second ring of the naphthyl. Incorporation of an acidic group in the naphthyl ring increases activity by 8-fold. Compound **42** also has 14,000 fold less affinity than Ac-Asp-Ala-Asp-Glu-F₂Pmp)-amide (Ki = 0.18 μM)(134). Interestingly, this hexapeptide is only 11 times better than the much smaller compound **44**.

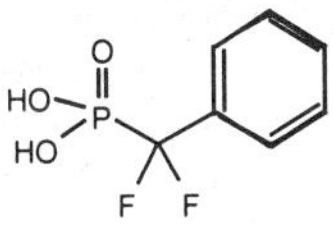

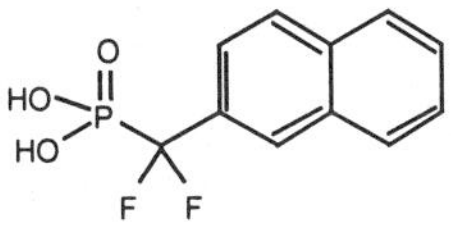

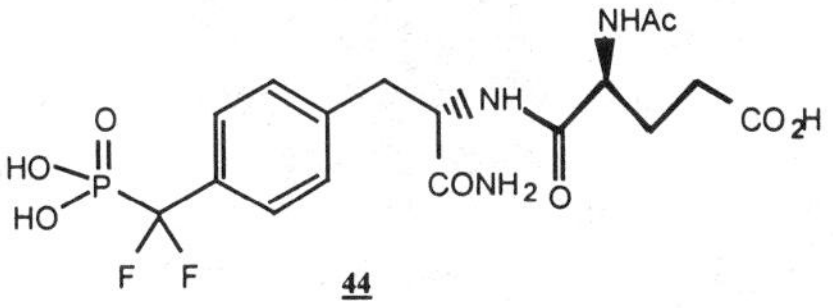

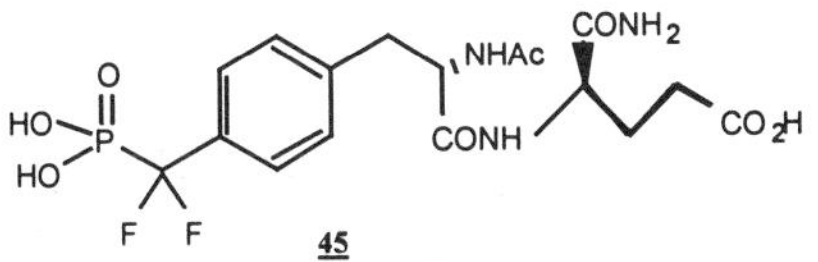

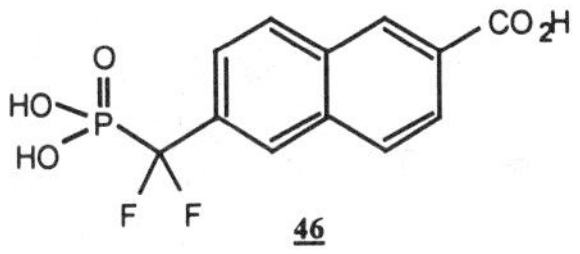

A set of aryl bis-difluorophosphonates were designed to occupy, concurrently the active site and the accessory phosphate site identified in PTP1B (135). In the series the meta-substituted **52** is the most potent PTP inhibitor and 100 times more potent than the parent. Also compound **52** is 40 times more selective for PTP1B than LAR, 200 for PTPα, and 4200 fold than for VHR. The benzylic counter parts of compounds **51** and **52**, namely **54** and **55**, had nearly identical potencies (135). Analogs of compound **58**, e.g **59**, are highly active for PTP1b. The greatest

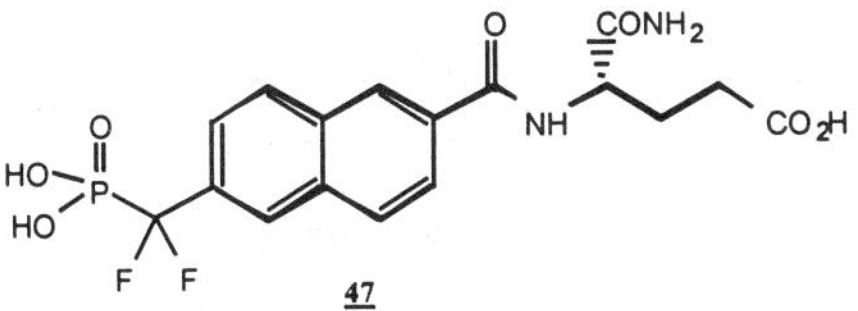

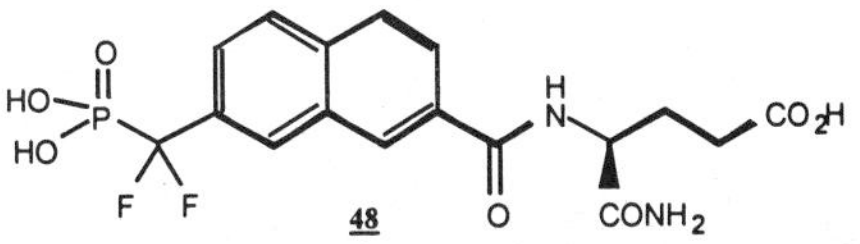

X = COCH₂CH₂-p-C₆H₄-CF₂PO₃H₂

49	X-NH₂
50	X-NH-(CH₂)₂-NH-X
51	X-p-NH-C₆H₄-NH-X
52	X-m-NH-C₆H₄-NH-X
53	X-o-NH-C₆H₄-NH-X
54	X-p-NHCH₂-C₆H₄-CH₂NH-X
55	X-m-NHCH₂-C₆H₄-CH₂NH-X
56	X-N(CH₂CH₂NH-X)₂
57	X-N(CH₂CH₂NH-X)₃

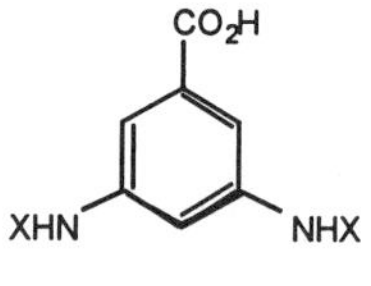

58

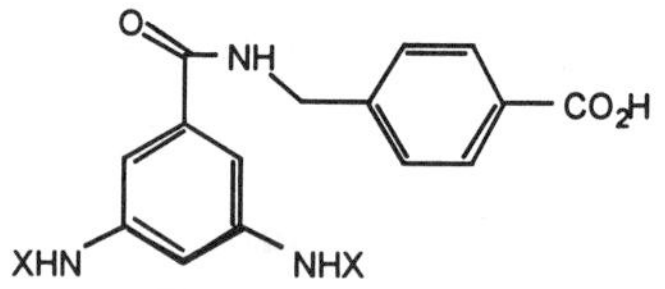

59

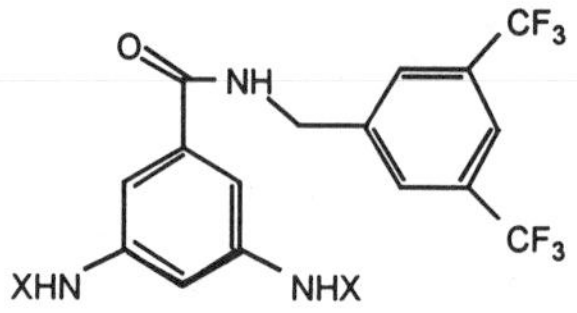

60

selectivity for PTP1B is shown by compounds **59** and **60**. Compound **60** had a Ki of 2.6 μM for PTP1B and is more than 100 fold better than the corresponding values against VHR , LAR, and PTPα. Compound **59** was the most potent PTP1B inhibitor identified in this series. An ether tethered compound, **61**, is a potent inhibitor of LAR with an IC₅₀ of 50 μM (135). The effectiveness of a suitably protected F2Pmp group was evaluated as a prodrug to enhance cellular activity. Compounds **62** , **63**, and **64** were readily taken up into Balbc3T3 cells and provided reasonable intracellular concentrations (136). No uptake was observed for the corresponding phosphate mono- or di-esters. In an effort to identify diesters that would not only show good uptake but also be converted to the active, parent diacids inside the cells, acyloxymethyl esters were investigated. Only phosphonate mono-esters could be obtained with the F2Pmp series even under forcing conditions. The phosphono mono esters (**63,65**), however, were taken up in cells. These analogs were water soluble, permeated cells, and reconverted to the free acid at a useful rate to provide adequate intracellular concentrations of parent di- and tri- acids (136).

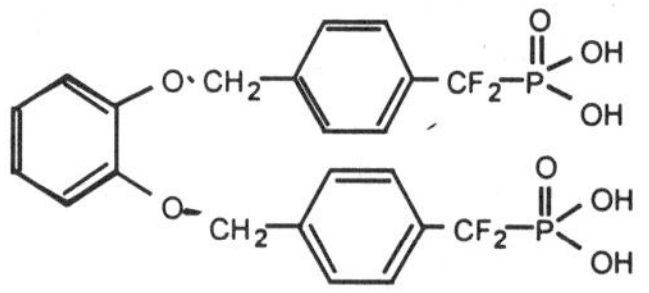

61

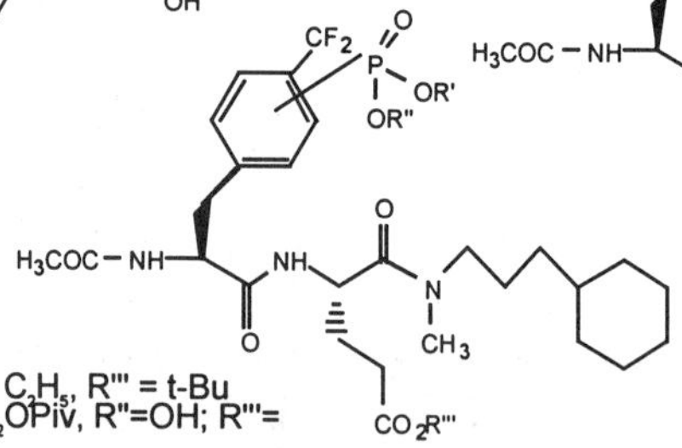

64 R'=R" = C₂H₅
65 R'=CH₂OPiv;

62 R'=R"= C₂H₅, R'" = t-Bu
63 R'= CH₂OPiv, R"=OH; R'"= CO₂R'"

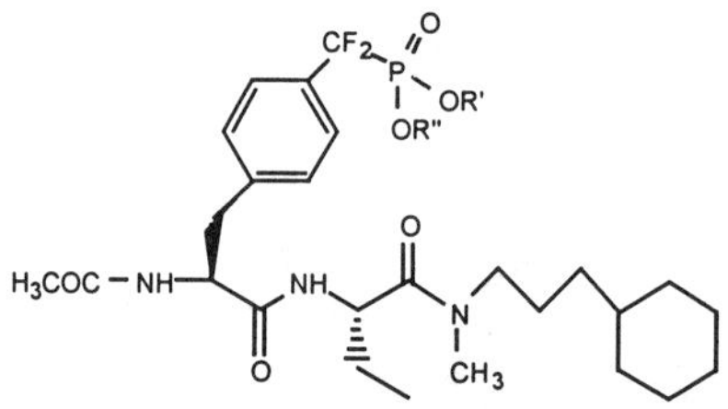

<u>Malonic acid inhibitors</u> - Many pTyr mimetics require highly charged phosphate mimicking functionality for physiological potency. This charged character limits penetration through cell membranes. Carboxy-based pTyr mimetics such as OMT (**4**) potentially allow prodrug protection strategies in which the malonyl carboxy groups can be masked in their ester forms. Once inside the cells the free carboxyls could be liberated by the action of cytoplasmic esterases (137). It has been found, however, that only one ester of the OMT diester containing peptides can be removed enzymatically (138-140). It has been hypothesized that spatial separation of the two carboxyls could potentially provide pTyr mimetics that would be more

readily enzymatically deesterified (141). To be potent mimetics of the phosphate, however, the two carboxyls would need to be sufficiently close that they could have phosphate-mimicking character. Analogs 9 and 66 were based on these considerations (141), When incorporated into the PTP1B consensus peptide, Ac-D-A-D-E-X-L, compounds 67 (Ki = 480 μM) and 68 (Ki= 199 μM) were considerably

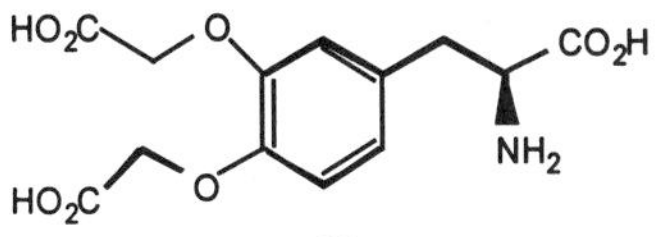

HO2C
O
CO2H
NH2
HO2C
O

66

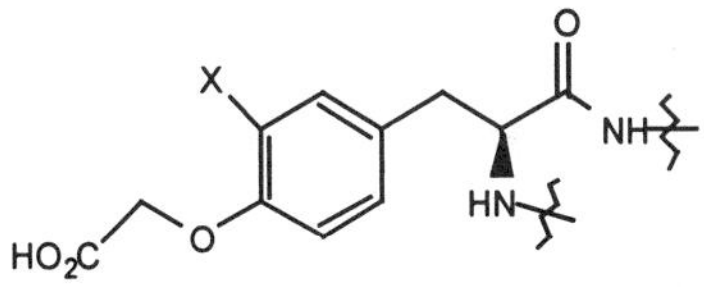

X
O
NH
HN
HO2C
O

67 X = H
68 X = OCH_2CO_2H

less potent than the pTyr substrate (Km = 3.2 μM). Compound 9 (Ki = 3.6 μM), however, had almost equivalent potency suggesting this motif is an excellent mimetic of pTyr.

Three OMT-EGFR containing cyclic peptides were synthesized in an effort to provide a more constrained analog of the potent hexapeptide analogs (142). In a PTP1-based assay the cyclic heptamer (69) was two-fold less potent (Ki= 25.2 μM) relative to the linear parent (4, Ki = 13 μM), while the cyclic octamer (70) showed a five-fold increase in potency (Ki = 2.6 μM). Peptide analog 71, cyclized via a sulfide bridge showed a three fold enhancement in potency (Ki = 0.73 μM). Another recently reported pTyr mimetic, p-malonylphenylalanine (Pmf) (72), when incorporated into high affinity Grb2 SH2 domain-directed scaffolds, shows inhibitory IC_{50}'s as low as 8 nM in Grb2 binding assays (149).

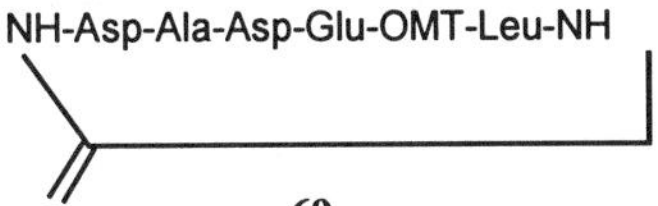

NH-Asp-Ala-Asp-Glu-OMT-Leu-NH
O

69

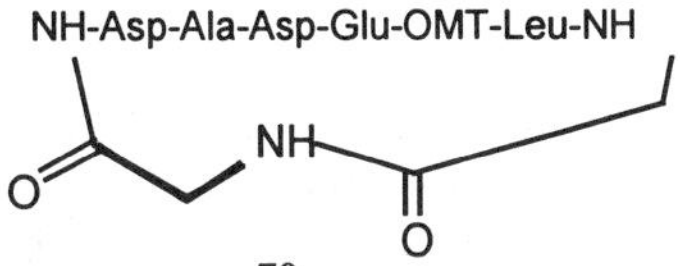

NH-Asp-Ala-Asp-Glu-OMT-Leu-NH
NH
O
O

70

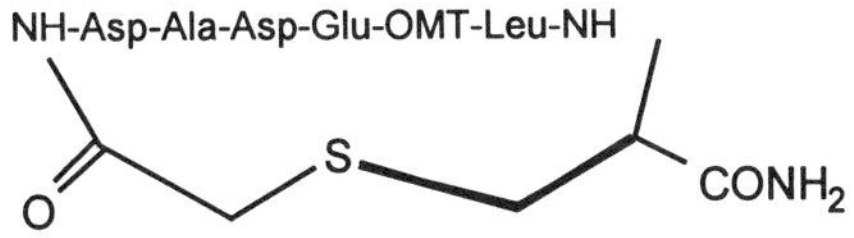

NH-Asp-Ala-Asp-Glu-OMT-Leu-NH
S
O
CONH2

71

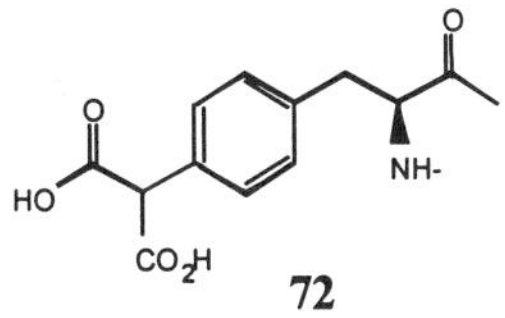

O
O
HO
NH-
CO2H

72

Other Inhibitors - A series of benzbromarone (73) analogs resulted in three potent inhibitors of PTP1B, 74 (IC_{50} = 61 nM), 75 (IC_{50} = 83 nM), and 76 (IC_{50} = 11 nM) (143). It was proposed that the acetic acid residue in 74-76 was binding in the active site as a pTyr mimetic. The α-unsubstituted analogs 77 (IC_{50} = 384 μM) and 78 (IC_{50} = 386 μM) seem not to support this proposal. Additionally, compound 79 (IC_{50} = 145 μM) is only about 2.5 times as active as 77. Nevertheless, compounds 74 and 75 were found to lower glucose levels in the diabetic ob/ob mouse at doses ≤10 mg/kg/day po. Compound 75 showed selectivity against other PTPs of greater than 25 fold.

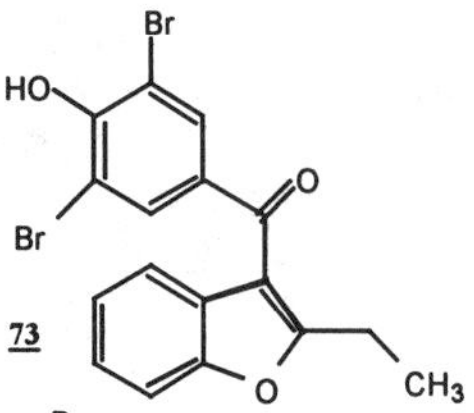

73

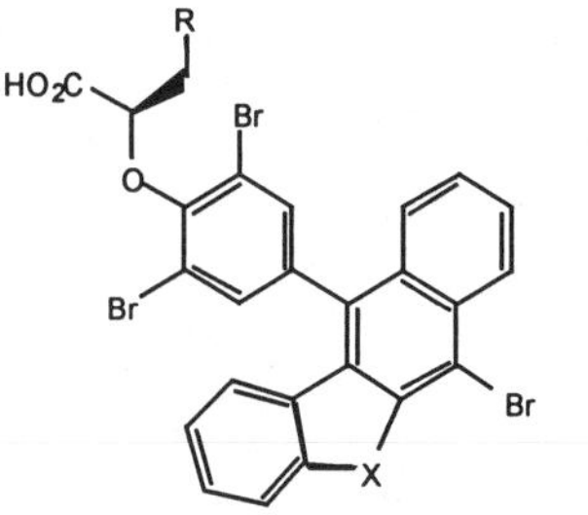

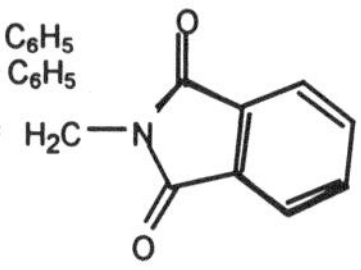

74 X= S, R = C₆H₅
75 X= O, R = C₆H₅
76 X = S, R = H₂C—N

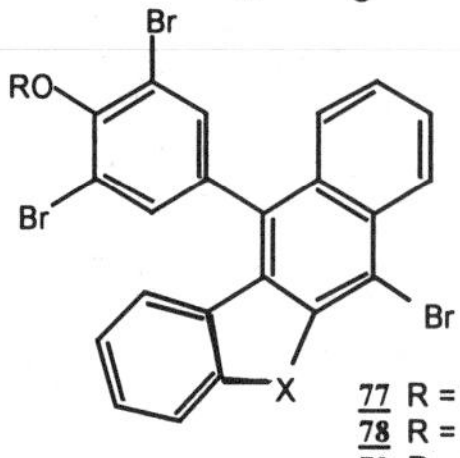

77 R = H
78 R = CH₂CO₂H
79 R = CH₂PO₃H₂

An extensive series of azolidindiones have been reported as PTP1B inhibitors with submicromolar potencies (150). Oxadiazolidinediones **80** and **81** and the corresponding acetic acid analogs, **82** and **83**, had IC$_{50}$'s against PTP1B in the range of 0.12-0.4 µM. Several of these compounds normalized plasma glucose and insulin levels in the ob/ob and db/db diabetic mouse models. A number of benzofuran and benzothiophene biphenyls have been synthesized as PTP1B inhibitors (151). X-ray crystal structures of enzyme-inhibitor complexes show this series to be active site inhibitors. The most potent compounds had IC$_{50}$ values in the 20-50 nM range. A potent and selective inhibitor **84** normalized plasma glucose levels at 25 mg/kg po and 1 mg/kg ip. The X-ray structure of **85** in the PTP1B active site showed the carboxylic acid does not directly interact with R221 but participates in H-bonding with this residue with two bridging water molecules. In the case of compound **86** (IC$_{50}$ = 0.028 µM, PTP1B), even though the salicylic acid occupies a similar space in the active site as the acetic acid-type inhibitors, the orientation is almost opposite to the acid inhibitors.

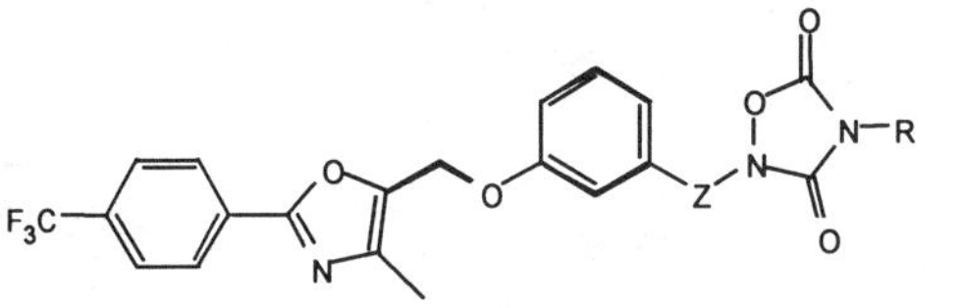

Z

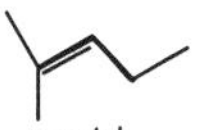

80 R = H

81 R =CH₂CO₂H

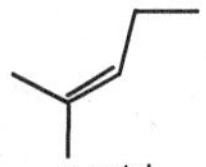

n-octyl

82 R = H

n-octyl   83 R = CH₂CO₂H

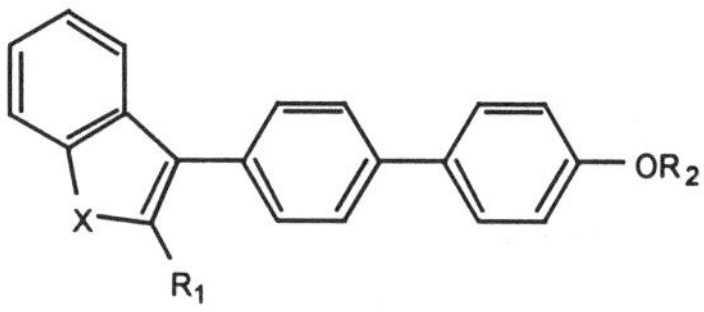

	R_1	R_2	X
84	CH₂C₆H₅	CH(CH₂C₆H₅)CO₂H	O
85	CH₂C₆H₅	CH(CH₂C₆H₅)CO₂H	S
86	C₂H₅	SO₂-(p-CO₂H, m-OH)C₆H₃	S

Enzyme kinetics and X-ray crystallography have identified 2-(oxalylamino)-benzoic acids **87**-**89** as competitive active site inhibitors of PTPs with the oxamic acid moiety serving as the phosphomimetic (152). These compounds are relatively non-selective PTP inhibitors.

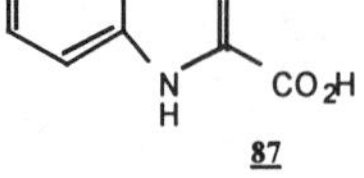

87 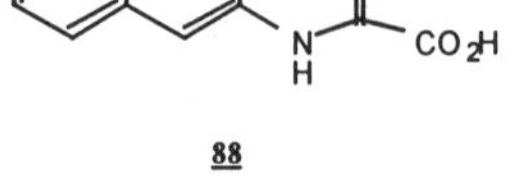   88 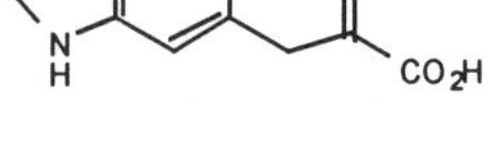   89

A class of nitrothiazoles have been claimed to be potent PTP1B inhibitors: **90** (IC$_{50}$ = 3.9 μM), **91** (IC$_{50}$ = 4.1 μM), **92** (IC$_{50}$ = 13 μM), **93** (IC$_{50}$ = 24 μM) (144).

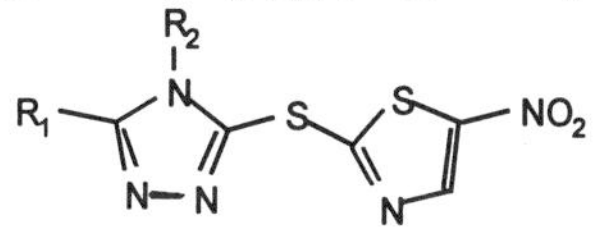

90 R$_1$= 1-ethyl-3-methyl-pyrazole-5-yl, R$_2$ = (CH$_2$)$_3$OCH$_3$
91 R$_1$= t-Bu, R$_2$ = H
92 R$_1$ = OH, R$_2$ = C$_6$H$_5$-
93 R$_1$ = OH, R$_2$ = o-CF$_3$C$_6$H$_4$-

Combinatorial Library Approaches - A four component Ugi reaction was used to prepare a library of dipeptides to explore the binding requirements of the cell cycle phosphatase, cdc25 (145). A variety of phosphate mimetics were incorporated into either the acid or aldehyde components of the Ugi reaction. Typical active compounds from this library are shown with compound **94** (IC$_{50}$ = 16 μM), **95** (IC$_{50}$ = 8.0 μM) and **96** (IC$_{50}$ = 0.7 μM).

Using the DOCK computer program, a method of identifying compounds that are complementary to the ligand binding site of an enzyme with a known 3-dimensional structure, 150,000 compounds in the ACD were screened computationally for good geometric and electrostatic fits to the PTP1B active site (146). This yielded, after further filtering against a variety of criteria, 25 potential lead structures of which seven showed Ki's of <500 μM with five < 100 μM. The most active that were also shown to be competitive inhibtors were **97** (Ki = 39 μM), **98** (Ki = 240 μM), **99** (Ki = 54 μM), **100** (Ki = 61 μM) and **101** (Ki = 510 μM). Compound **100** confirms that the salicylic acid moiety can approximate the binding mode of pTyr and serves as a non-phosphorus containing pTyr mimetic.

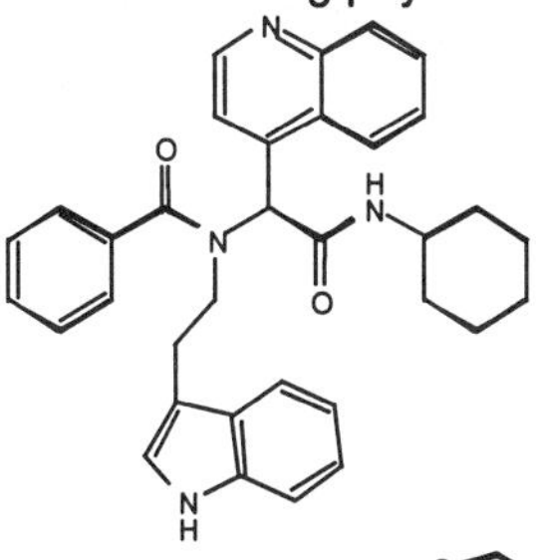

94

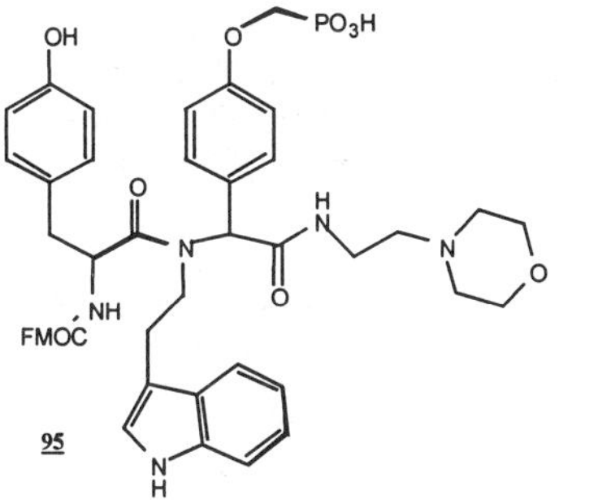

95

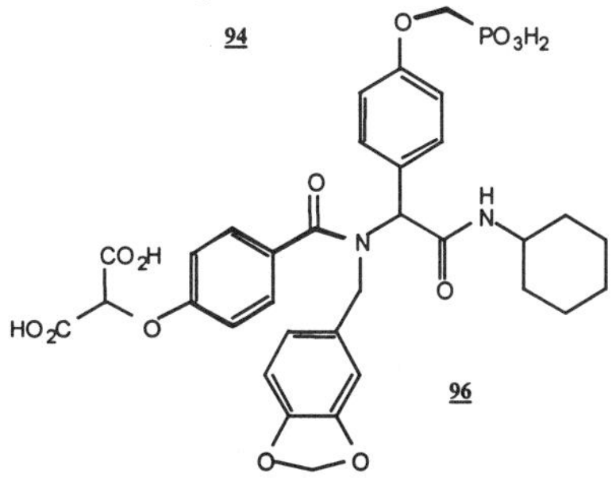

96

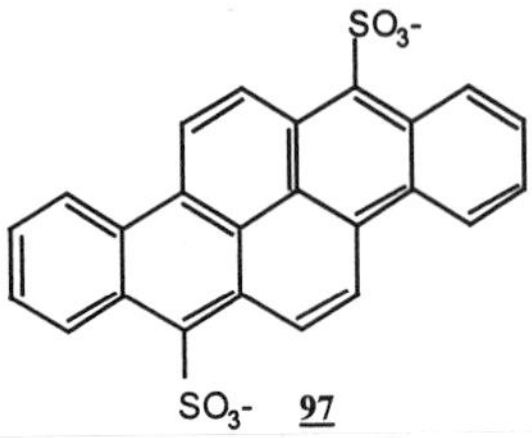

Using a pharmacophore modeled on a natural product inhibitor of phosphothreonine phosphatases, okadaic acid, a library of novel, phosphate free small molecules were synthesized several of which had good PTP activity particularly against PTP1B (147). The most potent compound was **102** with a Ki for PTP1B of 0.85 µM. A radiofrequency tag encoded combinatorial library was used to identify potent tripeptide-substituted cinnamic acid inhibitors of PTP1B. The two most potent inhibitors were **103** (Ki = 490 nM) and **104** (Ki = 79 nM). The inhibitors were competitive with p-nitrophenyl phosphate substrate. Cinnamic acid analogs were also prepared on Rink resin using t-butyl-4-carboxycinnamate in a four component Ugi condensation (148). When screened against the hematopoietic PTP (HePTP), a number of inhibitors were identified with the compound **105** as one of the more active (IC_{50} = 3.9 µM).

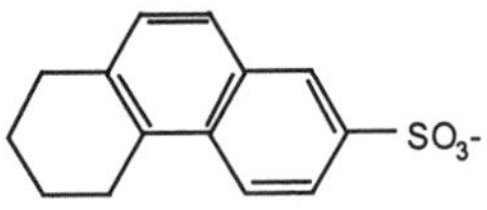

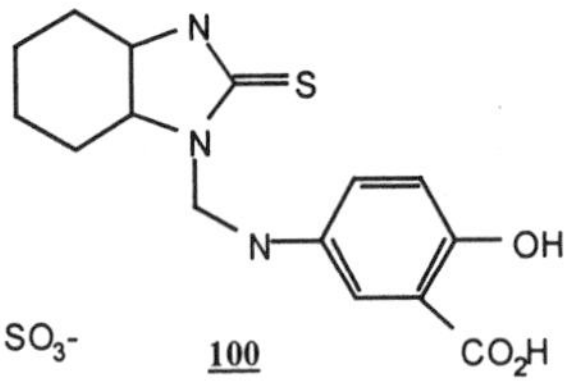

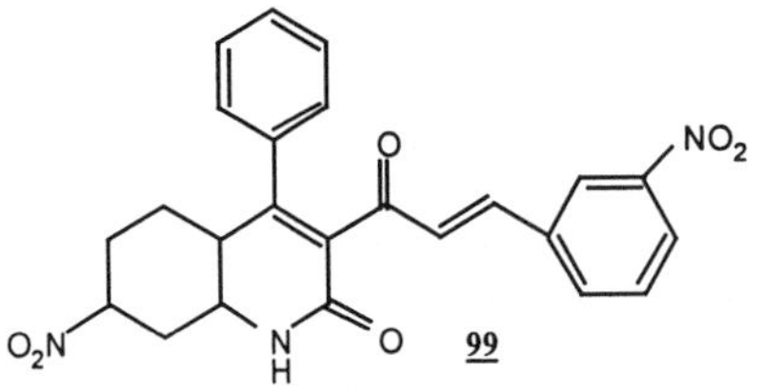

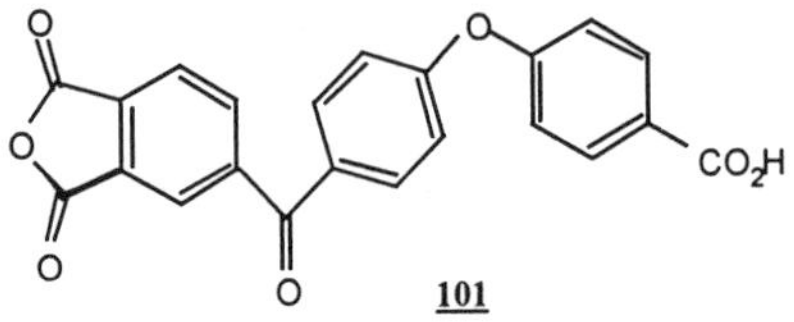

Summary – It has been established in a number of biological studies that the PTP's represent valid therapeutic targets for a number of disease states. Both potent and selective inhibitors have been discovered along with a variety of useful phosphomimetics that may be incorporated into a variety of frameworks. Several of these inhibitors have shown useful *in vivo* activity, particularly in animal models of diabetes.

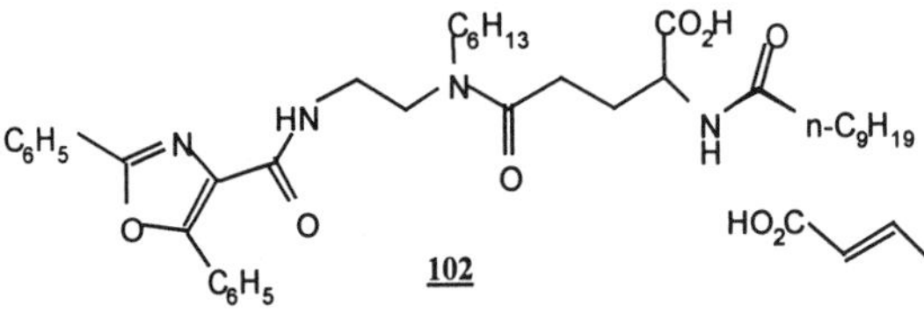

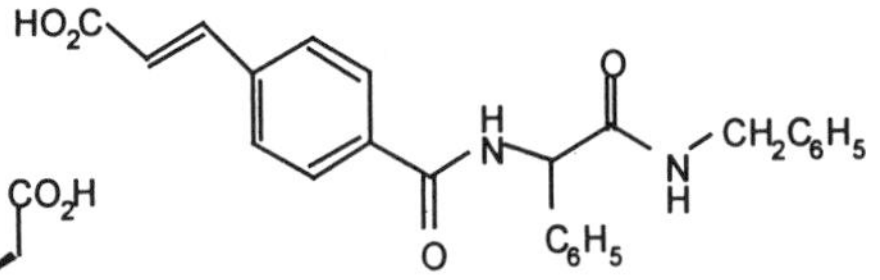

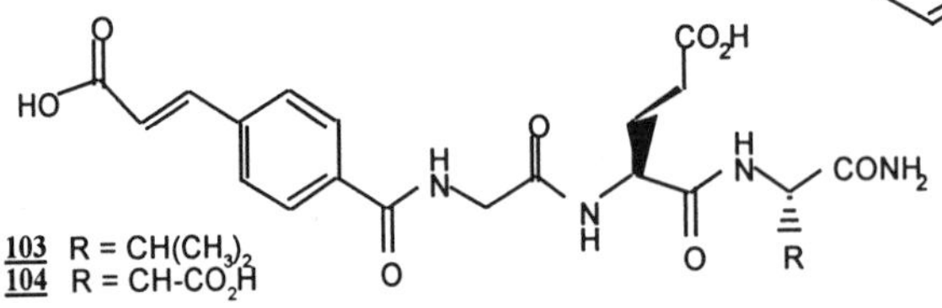

References
1. G. Peters, T. Frimurer, O. Olsen, Biochemistry 37 ,5383 (1998)
2. B. G. Neel, N. K. Tonks, Curr. Opin. Cell Biol., 9, 193 (1997)
3. T. Hunter, Cell, 80, 225 (1995)
4. J. M. Denu, J. A. Stuckey, M. A. Saper, J. E. Dixon, Cell 87, 361 (1996)
5. A. Levitzki, Current Opinion in Cell Biology, 8, 239 (1996)
6. M. Streuli, Current Opinion in Cell Biology, 8, 182 (1996)

7.   J. Montserat, L. Chen, D. S. Lawrence, Z-Y. Zhang, J. Biol. Chem. 271, 7868 (1996)
8.   F. L. Grinnell, Y-C. Lin, US Patent 5,741,777, April 21, 1998
9.   S. Taylor, C. Kotoris, N. Dinaut, Q. Wang, C. Ramachandran, Z. Huang, Bioorg.Med.Chem. 6 1457 (1998)
10. H. Sun, N. K. Tonks, Trends Biochem. Sci. 19, 480 (1994)
11. R.   Ide,   H.   Maegawa,   R.   Kikkawa,   Y.   Shigetta,   A.   Kashiwagi, Biochm.Biophys.Res.Commun. 201, 77 (1994)
12. J. R. Weiner, J. A. Hurteau, B. J. Kerns, R. S. Whitaker, M. R. Conaway, A. Berchuck, R. C. Bast, Am.J.Obstet.Gynecol. 170, 1171 (1994)
13. X. M. Zheng, Y. Wang, C. J. Pallen, Nature 359, 336 (1992)
14. C. J. Pallen, Semin. Cell.Biol. 4, 403 (1993)
15. Z. Zhang, Crit.Rev.Biochem.Mol.Biol. 33 1 (1998)
16. N. K. Tonks, B. G. Neel, Cell, 87, 365 (1996)
17. L. Wu, A. Buist, J. den Hertog, Z-Y. Zhang, J.Biol.Chem. 272,6994 (1997)
18. S. Taylor, C. Kotoris, N. Dinaut, Q. Wang, C. Ramachandran, Z. Huang, Bioorg.Med.Chem. 6 1457 (1998)
19. Z-Y. Zhang, A. B. Walsh, L. Wu, D. J. McNamara, E. M. Dobrusin, W. T. Miller, J.Biol.Chem., 271, 5386 (1996)
20.   G. Peters, T. Frimurer, O. Olsen, Biochemistry 37, 5383 (1998)
21. B. J. Goldstein, Receptor, 2,1 (1993)
22. J. Kusari, K. A. Kenner, K.-I. Suh, D. E. Hill, R. R. Henry, J.Clin.Invest. 293, 1156 (1993)
23. K. A. Kenner, D. E. Hill, J. M. Olefsky, J. Kusari, J. Biol. Chem. 268, 25455 (1993)
24. H. Maegawa, R. Ide, M. Hasegawa, S. Ugi, K. Egawa, M. Iwanishi, R. Kikkawa, Y. Shigeta, A. Kashiwagi, J.Biol.Chem. 270, 7724 (1995)
25. K. A. Kenner, E. Anyanwu, J. M. Olefsky, J. Kusari, J.Biol.Chem. 271, 19810 (1996)
26. F. Ahmad, P.-M. Li, B. J. Goldstein, J.Biol.Chem. 270, 20503 (1995)
27. D. Bandyopadhyay, A. Kusari, K. A. Kenner, f. Liu, J. Chernoff, T. A. Gustafson, J. Kusari, J.Biol.Chem. 272, 1639 (1997)
28. C. Rhamachandran, R. Aebersold, N. K. Tonks, D. A. Pot, Biochemistry 31, 4232 (1992)
29. G. Huyer, S. Liu, J. Kelly, J. Moffat, P. Payette, B. Kennedy, G. Tsaprailis, M. Gresser, C. Ramachandran, J.Biol.Chem 272, 843 (1993)
30. M. Berggren, L. A. Burns, R. T. Abraham, G. Powis, Cancer Res. 53, 1862 (1993)
31. B. Posner, R. Faure, J. N. W. Burgess, A. P. Bevan, D. Lachance, G. Zhang-Sun, I. G. Fantus, J. B. Ng, D. A. Hall, B. S. Lum, A. J. Shaver, J.Biol.Chem. 269, 4596 (1994)
32. S. Pugazhenthi, F. Tanha, B. Dahl, R. L. Khandelwal, Mol.Cell.Biochem.153, 125 (1995)
33. Z. Jia, Biochem. Cell Biol, 75, 17 (1997)
34. M. Elchebly, P. Payette, E. Michaliszyn, W. Cromlish, S. Collins, A. L. Loy, D. Normandin, A. Cheng, J. Himms-Hagen, C-C. Chan, C. Ramachandran, M. J. Gresser, M. L. Tremblay, B. P. Kennedy, Science, 283, 1544 (1999)
35. P. Gunaratne, C. Stoscheck, R. E. Gates, L. Li, L. B. nanney, L. E. King, J.Invest. Dermatol., 103, 701 (1994)
36. S. Brown-Shimer, K. A. Johnson, D. E. Hill, A. M. Bruskin, Cancer Res. 52, 478 (1992)
37. J. R. Wiener, B-J. M. Kerns, E. L. Harvey, M. R. Conaway, J. D. Iglehart, A. Berchuck, R. C. Bast Jr., J.National Cancer Institute, 86, 372 (1994)
38. J. R. Wiener, J. A. Hurteau, B-J. Kerns, R. S. Whitaker, M. R. Conaway, A. Berchuck, R. C. Bast Jr, Am.J.Obstet.Gynecol, 170, 1177 (1994)
39. N. K. Tonks, C. D. Diltz, E. H. Fischer, J.Biol.Chem. 263, 6731 (1988)
40. R.A. Mooney, D.T. Kulas, L.A. Bleyle, J.S. Novak, Biochem.Biophys.Res.Commun. 235, 709 (1997)
41. F. Ahmad, B. J. Goldstein, J.Biol.Chem. 272, 448 (1997)
42. R. Lammers, N. P. Moller, A. Ullrich, FEBS Lett., 404, 37 (1997)
43. B. L. Seely, P. A. Staubs, D. R. Reichart, P. Berhanu, K. L. Milarski, A. R. Saltiel, J. Kusari, J. M. Olefsky, Diabetes 45, 1379 (1996)
44. F.Ahmad, J.L.Azevedo, R.Cortright,G.L.Dohm,B.J.Goldstein, J.Clin.Invest.100,449 (1997)
45. F. Ahmad, P. M. Li, J. Meyerovitch, B. J. Goldstein, J.Biol.Chem.270, 20503 (1995)
46. P. M. Li, W. R. Zhang, B. J. Goldstein, Cell Signal 8, 467 (1996)
47.   B. J. Goldstein, F. Ahmad, W. Ding, P-M. Li, W-R. Zhang, Molecular and Cellular Biochem., 182, 91 (1998)
48. R. E. Cebula, J. L. Blanchard, M. D. Bolsclair, K. Pai, N. J. Bockovich, Bioorg.Med.Chem.Letts, 7, 2015 (1997)
49. K. Galaktionov, A. K. Lee, J. Eckstein, G. Draetta, J. Meckler, M. Loda, D. Beach, Science 269, 1575 (1995)
50.   R. Parsons, Current Opinion in Oncology, 10, 88 (1998)

51. J. B. Bolen, P. A. Thompson, E. Eiseman, I. D. Horak, Adv.Cancer Res., 57, 103 (1991)
52. M. Okumura, M. L. Thomas, Current Opinion in Immunology, 7, 312 (1995)
53. J. T. Pingel, M. L. Thomas, Cell 58, 1055 (1989)
54. G. A. Koretzky, J. Picus, M. L. Thomas, A. Weiss, Nature 346, 66 (1990)
55. C. T. Weaver, J. T. Pingel, J. O. Nelson, M. L. Thomas, Mol.Cell Biol. 11, 4415 (1991)
56. L. B. Justement, K. S. Cambell, N. C. Chien, J. C. Cambier, Science 252, 1839 (1991)
57. G. M. Bell, G. M. Dethloff, J. B. Imboden, J. Immunol. 151, 3646 (1993)
58. K. Kishihara, J. Penninger, V. A. Wallace, T. M. Kundig, K. Kawai, A. Wakeham, E. Timms, K. Pfeffer, P. S. Ohasi, M. L. Thomas, Cell 74, 143 (1993)
59. S. K. Hanks, A. M. Quinn, T. Hunter, Science 241, 42 (1988)
60. J. A. Donovan, G. A. Koretzky, J.Am.Soc.Nephrology 4, 976 (1993)
61. B. G. Neel, Current Opinion in Immunology, 9, 405 (1997)
62. R. Huijsduijnen, Gene 225 , 1 (1998)
63. I. S. Trowbridge, Annu.Rev.Immun, 12, 85 (1994)
64. J. S. Trowbridge, J.Biol.Chem, 266, 23517 (1991)
65. C. A. Janeway, Jr., Ann. Rev. Immunol, 10, 645 (1992)
66. S. M. Brady-Kalnay, A. J. Flint, N. K. Tonks, J. Cell Biol. 122, 961 (1993)
67. J.Y.Sap, Y.P.Jiang, D.Friedlander, M.Grumet, J. Schlessinger, Mol.Cell.Biol.14, 1 (1994)
68. J. Cheng, K. Wu, M.O. Armanini, N. Rourke, D. Dowbenko, L. A. Lasky, J.Biol.Chem. 272, 7264 (1997)
69. S. M. Brady-Kalnay, D. L. Rimm, N. K. Tonks, J.Cell Biol. 130, 977 (1995)
70. N. X. Krueger, M. Streuli, H. Saito, EMBO J. 9, 3241 (1990)
71. J.Sap, P.D'Eustachio, D.Givol, J.Schlessinger, Proc.Natl.Acad.Sci.USA.87, 6112 (1990)
72. J. den Hertog, C. E. G. M. Pals, M. P. Peppelenboach, L. G. Tertoolen, S. W. deLaat, W. Kruijer, EMBO J. 12, 3789 (1993)
73. X. M. Zheng, Y. Wang, C. J. Pallen, Nature, 359, 336 (1992)
74. N. P. H. Moller, K. B. Moller, R. Lammers, A. Kharitoninkov, E. Hoppe, F. C. Wiberg, I. Sures, A. Ullrich, J.Biol.Chem. 270, 23126 (1995)
75. T. Ishibashi, D. P. Bottara, A. Chan, T. Miki, S. Aaronson, Proc.Natl.Acad.Sci.USA 89, 12170 (1992)
76. L. Chen, J. Montserat, D. S. Lawrence, Z-Y. Zhang, Biochemistry, 35, 9349 (1996)
77. (92) A. Schmidt, S. J. Rutledge, N. Endo, E. E. Opas, H. Tanaka, G. Wesolowski, C. T. Leu, Z. Huang, C. Ramachandaran, S. B. Rodan, G. A. Rodan, Proc.Nat.Acad.Sci.USA, 93, 3068 (1996)
78. K. Skorey, H. Ly, J. Kelly, M. Hammond, C. Ramachandran, Z. Huang, M. Gresser, Q. Wang, Journal of Bio Chem., 272, 22473 (1997)
79. K. M. V. Hoffmann, N. K. Tonks, D. Barford, J.Biol.Chem., 272, 27505 (1997)
80. M. Sarmiento, Y. Zhao, S. J. Gordon, Z-Y. Zhang, J. Biol. Chem., 273, 26368 (1994)
81. J. Yang, X. Liang, T. Niu, W. Meng, Z.Zha, G.W. Zhou, J.Biol.Chem. 273, 28199 (1998)
82. Y. A. Puius, Y. Zhao, M. Sullivan, D. S. Lawrence, S. C. Almo, Z-Y. Zhang, Proc.Natl.Acad.Sci USA, 94, 13420 (1997)
83. Z. Zhang, Crit.Rev.Biochem.Mol.Biol. 33, 1 (1998)
84. J. M. Denu, J. A. Stuckey, M. A. Saper, J. E. Dixon, Cell, 87, 361 (1996)
85. Z-Y. Zhang, J. E. Dixon, Advances in Enzymology, 68, 1 (1994)
86. N. K. Tonks, B. G. Neel, Cell, 87, 365 (1996)
87. D. Barford, Z. Jia, N. K. Tonks,Nature Structural Biology, 2, 1043 (1995)
88. Z. Jia, Biochem. Cell Biol, 75, 17 (1997)
89. Z-Y. Zhang, J. E. Dixon, Adv.Enzymology, 68, 1 (1994)
90. Z. Zhang, B. Palfey, L. Wu, Y. Zhao, Biochemistry 34 16389 (1995)
91. J.Denu, D.Lohse, J.Vijayalaksmi, M.Saper, J.Dixon, Proc.Natl.Acad.Sci.93, 2493 (1996)
92. L. Chen, J. Montserat, D. S. Lawrence, Z. Y. Zhang, Biochemistry 35, 9349 (1996)
93. H. Fretz, Tetrahedron 54, 4849 (1998)
94. Z. Yao, T. Gao, J. Voigt, H. Ford, T. Burke Jr., Tet. 55, 2865 (1999)
95. P. Furet, B. Gay, C. Garcia-Echeverria, J. Rahuel, H. Fretz, J. Schoepfer, and G. Caravatti, J.Med.Chem. 40, 3551(1997)
96. T. R. Burke, B. Ye, M. Akamatsu, H. Ford, X. Yan, H. K. Kole, G. Wolf, S. E. Shoelson, P. P. Roller, J.Med.Chem, 39, 1021 (1996)
97. B. Margolis, Prog. Biophys. Mol. Biol. 62, 223 (1994)
98. T. R. Burke, Jr., Z.-J. Yao, M. S. Smyth, B. Ye, Curr. Pharm. Des., 3, 291 (1997)
99. E.A.Lunney, K.S.Para,J.R.Rubin,C.Humblet,J.Fergus, J.Am.Chem.Soc,119,12471 (1997)
100. H. Fretz, P. Furet, J. Schoepfer, C. Garcia-Echeverria, B. Gay, J. Rahuel, G. Caravatti, 15th American Peptide Symposium, Nashville, TN, June 14-19, 1997; Abstract P422
101. S. J. Salamone, F. Jordan, Biochemistry 21, 6383 (1982)

102. M. M. Mehrotra, D. D. Sternbach, M. Rodriguez, P. Churifson, J. Berman, Bioorg. & Med Chem Letts, 6, 1941 (1996)
103. R. P. Roller, L. Wu, Z-Y. Zhang, T. R. Burke Jr., Bioorg.Med.Chem.Lett., 8, 2149 (1998)
104. Z-Y. Zhang, A. M. Thieme-sefler, D. Maclean, D. J. McNamara, E. M. Dobrusin, T. K. Sawyer, J. E. Dixon, Proc. Natl. Acad. Sci USA, 90, 4446 (1993)
105. T.R. Burke Jr., H.K. Kole, P,P. Roller, Biochem.Biophys.Res.Comun, 204, 129 (1994)
106. T.R.Burke, Z-J.Yao, H.Zhao, G.W.A.Milne, L.Wu, Z-Y.Zhang,J.H.Voigt,Tet.4,9981 (1998)
107. I. Marseigne, B. P. Roques, J.Org.Chem. 53, 3621 (1988)
108. T.R. Burke, Jr., M.Smyth, M.Nomizu, A.Otaka, P.P.Roller, J.Org.Chem. 58, 1336 (1993)
109. M. Bobko, H. R. Wolfe, A. Saha, R. E. Dolle, D. K. Fisher, T. J. Higgens, Bioorg.Med.Chem.Lett., 5, 353 (1995)
110. H. Cho, R. Krishnaraj, M. Itoh, E. Kitas, W. Bannwarth, H. Saito, C. T. Walsh, Protein Science 2, 977 (1993)
111. R. E. Cebula, J. L. Blanchard, M. D. Bolsclair, K. Pai, N. J. Bockovich, Bioorg.Med.Chem.Lett., 7, 2015 (1997)
112. M. Imoto, H. Kakeya, T. Sawa, C. Hayashi, M. Hamada, T. Takbucau, K. Umezawa, J.Antibiotics, 46, 1342 (1993)
113 L.Yu, A.McGill, J.Ramirez, P.G.Wang, Z-Y.Zhang, Bioorg.Med.Chem.Lett.,5,1003 (1995)
114. T. Hamaguchi, T. Sudo, H. Osada, FEBS Letts 372, 54 (1995)
115. S. P. Gunasekera, P. J. McCarthy, M. Kelly-Borges, J.Am.Chem.Soc., 118, 8759 (1996)
116. J. L. Blanchard, D. M. Epstein, M. D. Boisclair, J. Rudolph, K. Pai, Bioorg.Med.Chem.Lett., 9, 2537 (1999)
117. S. Mikalsen, O. Kaalhus, J.Biol.Chem. 273, 10036 (1998)
118. G. Huyer, S. Liu, J. Kelly, J. Moffat, P. Payette, B. Kennedy, G. Tsaprailis, M. Gresser, C. Ramachandran, J.Biol.Chem. 272, 843 (1997)
119. A. B. Goldfine, D. C. Simonson, F. Folli, M.-E. Patti, C. R. Kahn, J.Clin.Endocrinol.Metab. 80, 3311 (1995)
120. E. Tsiani, E. Bogdanovic, A. Sorisky, L. Nagy, G. Fantus, Diabetes, 47, 1676 (1998)
121. H. Watanabe, M. Nakai, K. Komazawa, H. Sakurai, J.Med.Chem. 37, 876 (1994)
122. S. Ham, J. Park, S. Lee, W.Kim, K.Kang, K.Choi, Bioorg.Med.Chem.Lett. 8, 2507 (1998)
123. S. W. Ham, J. Park, S-J. Lee, J. S. Yoo, Bioorg.Med.Chem.Lett., 9, 185 (1999)
124. J. K. Meyers, T. S. Widlanski, Science 262, 1451 (1993)
125. Q.P.Wang.U.Dechert,F.Jirik,S.G.Withers,Biochem.Biophys.Res.Commun.200,577 (1994)
126. J. K. Myers, J. D. Cohen, T. S. Widlanski, J.Am.Chem.Soc. 117, 1049 (1995)
127. W. Taylor, Z. Zhang, T. Widlanski, Bioorg.Med.Chem. 4, 1515 (1996)
128. S. Taylor, C. Kotoris, N. Dinaut, Q. Wang, C. Ramachandran, Z. Huang, Bioorg.Med.Chem. 6, 1457 (1998)
129. T. Burke, B. Ye, X. Yan, S. Wang, Z. Jia, L. Chem, Z. Zhang, D. Barford, Biochem. 35, 15989 (1996)
130. T. Yokomatsu, T. Murano, I. Umesue, S. Soeda, H. Shimeno, S. Shibuya, Bioorg.Med.Chem.Lett., 9, 529 (1999)
131. C. Kotoris, M. Chen, S. Taylor, Bioorg.Med.Chem.Lett. 8, 3275 (1998)
132. H. K. Kole, M. S. Smyth, P. L. Russ, T. R. Burke Jr., Biochem J, 311, 1025 (1995)
133. Z. Yao, B. Ye, X. Wu, S.Wang, L.Wu, Z.Zhang, T.Burke, Bioorg.Med.Chem.6,1799 (1998)
134. L. Chen, L. Wu, A. Otaka, M. S. Smyth, P.P. Roller, T. R. Burke, Jr., Biochem.Biophys.Res.Commun. 216, 976 (1995)
135. S. Ham, J. Park, S. Lee, J. Yoo, Bioorg.Med.Chem.Lett. 9, 185 (1999)
136. C. J. Stankovic, N. surendran, E. A. Lunney, M. S. Plummer, K. S. Para, A. Shahripour, J. H. Fergus, J. S. Marks, R. Herrera, S. E. Hubbell, C. Humblet, A. R. Saltiel, B. H. Stewart, T. K. Sawyer, Bioorg.Med.Chem.Lett., 7, 1909 (1997)
137. B. Ye, M. Akamatsu, S. E. Shoelson, G. Wolf, S. Giorgetti-Peraldi, X. J. Yan, P. P. Roller, T. R. Burke, Jr., J.Med.Chem. 38, 4270 (1995)
138. M. Akamatsu, B. Ye, X. J. Yan, H. K. Kole, T. R. Burke, Jr.,, P. P. Roller, 1995 Japan Peptide Symposium Proceedings, 1995, Osaka, Japan
139. F. Bjorkling, J. Boutelje, S. Gatenbeck, K.Hult, T.Norin, P.Szmulik, Tet., 41, 1347 (1985)
140 M. Luyten, S. Muller, B. Herzog, R. Keese, Helv.Chim.Acta 70, 1250 (1987)
141. T. R. Burke, Jr., Z-J. Yao, H. Zhao, G. W. A. Milne, L. Wu, Z-Y. Zhang, J. H. Voigt, Tet. 5, 9981 (1998)
142. M. Akamatsu, P. P. Roller, L. Chen, Z-Y. Zhang, B. Ye, T. R. Burke, Jr., Bioorg.Med.Chem., 5, 157-163 (1997)
143. J. Wrobel, J. sredy, C. Moxham, A. Dietrich, A. Li, D. R. Sawicki, L. Seestaller, L. Wu, A. Katz, D. Sullivan, C. Tio, Z-Y. Zhang, J.Med.Chem. 42, 3199 (1999)

144. G. McMahon, K. P. Hirth, P. Tang, WO 96/40113, Dec. 19, 1996
145. G. Bergnes, C. L. Gilliam, M. D. Boisclair, J. L. Blanchard, K. V. Blake, D. M. Epstein, K. Pai, Bioorg.Med.Chem.Lett. 9, 2849 (1999)
146. M. Sarmiento, L. Wu, Y-F Keng, L. Song, Z. Luo, A. Huang, G-Z. Wu, A. K. Yuan, Z-Y. Zhang, J.Med.Chem. 43, 146 (2000)
147. R. Rice, J. Rusnak, F. Yokokawa, S. Yokokawa, D. Messner, A. Boynton, P. Wipf, J. Lazo, Biochemistry 36, 15965 (1997)
148. X. Cao, E. J. Moran, D. Siev, A. Lio, C. Ohashi, A. M. M. Mjalli, Bioorg.Med.Chem.Lett., 5, 2953(1995)
149. Y. Gao, J. Luo, Z. J. Yao, R. Guo, H. Zou, J. Kelley, J.H. Voight, D. Yang, T.R. Burke, Jr., J.Med.Chem., 43, 911 (2000)
150. M. S. Malamas, J. Sredy, I. Gunawan, B. Mihan, D. R. Sawicki, L. Seestaller, D. Sullivan, B. R. Flam, J.Med.Chem., 43, 995 (2000)
151. M. S. Malamas, J. Sredy, C. Moxham, A. Katz, W. Xu, R. McDevitt, F. O. Adebayo, D. R. Sawicki, L. Seestaller, D. Sullivan, J. R. Taylor, J.Med.Chem., 43, 1293 (2000)
152. H. S. Andersen, L. F. Iverson, C. B. Jeppesen, S. Bramer, K. Norris, H. B. Rasmussen, K. B. Moller, N. P. H. Moller, J.Biol.Chem., 275, 7101 (2000)

Chapter 22. Cholesteryl Ester Transfer Protein as a Potential Therapeutic Target To Improve The HDL To LDL Cholesterol Ratio

James A. Sikorski and Kevin C. Glenn
Pharmacia Discovery Research
St. Louis, MO 63198

Introduction - HMG-CoA reductase inhibitors, or statins, have become first line therapy for lowering plasma cholesterol levels as a way of reducing the incidence of coronary heart disease (CHD). Several prospective clinical trials with statins have shown the importance of lowering low density lipoprotein cholesterol (LDL-C). They showed that lowering LDL-C by 25% to 35% resulted in a 24% to 42% reduction in primary or secondary CHD morbidity and mortality (1). However, 58% to 76% of the statin-treated CHD patients still did not experience therapeutic benefit, suggesting that a significant unmet medical need continues to exist for treating CHD.

The Framingham epidemiological studies demonstrated an inverse relationship between serum high density lipoprotein cholesterol (HDL-C) levels and the incidence of CHD (2). Framingham data showed that within the CHD patient population, 60% had HDL-C <35 mg/dl and LDL-C >130 mg/dl, 25% had HDL-C <35 mg/dl and acceptable LDL-C and triglycerides (TG), and 15% had LDL-C >130 mg/dl with acceptable HDL-C and TG levels. The associated increase in CHD for patients with low HDL-C risk was present in patients both with and without elevated total plasma cholesterol. These data, plus other information, have shown that HDL-C is one of the best predictors of CHD risk in women of all ages and in men after middle-age.

Several studies have shown reduced CHD events with treatments that raised HDL-C. The controlled Helsinki Heart Trial showed that correcting low levels of HDL-C has clinical benefit (3). For each 1 mg/dl rise in HDL-C among the study subjects, the average response was a 3% decrease in the risk of new CHD. In the VA-HIT trial, 2531 men with HDL-C ≤40 mg/dl, LDL-C ≤ 140 mg/dl, TG ≤ 300 mg/dl and CHD were treated for an average of five years with gemfibrozil (4). In the VA-HIT trial, LDL-C was not changed significantly, TG was lowered 31%, HDL-C was raised 6%, and the combined number of CHD events was lowered 24% (a 2-3% decrease in CHD for every 1% rise in HDL-C). By comparison, the prospective statin trials showed only a 1% drop in CHD risk for every 2% decrease in LDL-C (1).

In addition to fibrates, simvastatin is approved to treat low HDL-C levels; however both are only modestly effective on HDL-C. Nicotinic acid treatment is more efficacious, but has troubling side effects. Therefore, a safe and effective drug that raises HDL-C is needed for use alone or with LDL-C lowering agents. Many targets might raise HDL-C including enzymes associated with lipoprotein metabolism [e.g. hepatic triglyceride lipase, lipoprotein lipase, lecithyl cholesterol acyl transferase (LCAT) or phospholipid transfer protein], modulating SR-B1 (the hepatic HDL scavenger receptor), increasing apoAl or other HDL apolipoproteins like apoAl Milano (5), or inhibiting cholesteryl ester transfer protein (CETP).

INHIBITION OF CETP AS A THERAPEUTIC TARGET

CETP mediates the transfer of cholesteryl esters (CE) from HDL to VLDL with a balanced reciprocal exchange of TG. Therefore CETP moves CE from HDL that is known to protect against CHD into proatherogenic VLDL and LDL (6). Indeed early, primarily correlational, reports supported the theory that CETP is atherogenic (7). More recently, data from transgenic mouse studies and detailed evaluation of human genetic epidemiology suggest CETP may also be able to serve a protective role against CHD (6, 8, 9).

CETP may be cardio-protective by its participation in a process termed "reverse cholesterol transport" (RCT), i.e., the transfer of cholesterol from peripheral tissues to the liver via HDL (6). Three pathways are known to exist for HDL delivery of tissue-derived cholesterol including: i) CETP-mediated transfer of CE to TG-rich, apoB-containing lipoproteins that can bind and be cleared by the LDL receptor, ii) SR-B1 selective uptake of HDL CE, and iii) LDL receptor-mediated uptake of apoE-containing HDL particles. The primary recipients of cholesterol from peripheral tissues are pre-β HDL discoidal particles. After esterification of cholesterol by LCAT, CETP transfers CE out of HDL in exchange for TG. HDL-TG is hydrolyzed by hepatic triglyceride lipase to produce discoidal pre-β HDL that can recycle to accept more peripheral tissue cholesterol. In addition, CETP activity enhances LCAT esterification of tissue-derived cholesterol, further contributing to RCT (6). Expression of human CETP in human apoAI transgenic mice increases pre-β HDL particles approximately 2-fold (6). Human apoCIII transgenic mice have increased TG-rich VLDL levels typical of human dyslipidemia (6). Human CETP expression in apoCIII transgenic mice also increases formation of pre-β HDL while reducing HDL-C levels. Yet even though HDL-C levels decrease, the human apoCIII/CETP transgenic mice have reduced fatty streaks compared to apoCIII transgenic mice. A recent, relatively large population-based human study of CETP-deficient subjects also suggests a protective role for CETP (6). Approximately 6% of the 3300 Japanese-American men in the Honolulu Heart Program were heterozygous for CETP deficiency. Heterozygous CETP deficiency was associated with an adjusted 1.7-fold increase in risk of CHD, even though mean plasma CETP levels were 35% lower than normal subjects and HDL-C was ~10% higher.

Observations that support the theory that CETP is pro-atherogenic have been extensively reviewed (6-9) and include the following:

♦   Species variation in CETP expression correlates with their HDL-C levels and susceptibility to dietary cholesterol-induced atherosclerosis.
♦   Expression of human CETP in apoE (-/-) mice (which develop endogenous hypercholesterolemia and atherosclerosis on regular chow diets) have 1.5- to 2-fold higher atherosclerosis than the apoE (-/-) mice alone.
♦   Serum from homozygous CETP-deficient subjects is 1.5- to 2-fold more effective at mediating cellular cholesterol efflux than normal serum.
♦   Metabolic factors that affect CETP activity in turn regulate HDL-C and LDL-C levels and are associated with increased atherogenic risk:
    ◊   alcohol consumption and aerobic exercise elevate HDL-C and reduce plasma CETP protein levels;
    ◊   dietary cholesterol increases CETP activity and mRNA levels;
    ◊   hyperinsulinemia elevates adipose CETP mRNA and reduces HDL-C;
    ◊   elevated CETP levels in nephrotic syndrome correlate with high VLDL and LDL.
♦   In general, inhibition of CETP results in elevated HDL-C and anti-atherogenic properties of HDL include:
    ◊   inhibition of monocyte chemotaxis through endothelial cell monolayers;

◊   inhibition of vascular endothelial cell expression of inflammatory adhesion molecules (e.g. VCAM-1, ICAM-1);

◊   inhibition of LDL aggregation;

◊   inhibition of LDL oxidation;

◊   stimulation of endothelial cell production of PGI2 (inhibiting platelet aggregation);

◊   stimulation of RCT cholesterol efflux from peripheral tissues.

♦ Expression of human CETP in Dahl, salt-sensitive hypertensive rats resulted in elevated cholesterol and triglyceride levels, atherosclerotic lesions, myocardial infarction and decreased survival, mimicking human CHD (10).

Both increased and decreased fatty streaks or atherosclerosis have been described following expression of human CETP depending on the transgenic mouse model being used. Therefore, despite 20 years of research, the real impact of CETP activity on CHD remains unclear. An emerging hypothesis is that a potentially complex dose-relationship exists between CETP levels, HDL-C levels and CHD (6, 11). CETP inhibition may be therapeutically beneficial when treating dyslipidemic patients that have excessive CETP levels that are more than what is needed to promote RCT. The hypothesis suggests that potent CETP inhibitors (>50% inhibition *in vivo*) could reduce CHD risk if they can raise HDL-C levels above 60 mg/dl.

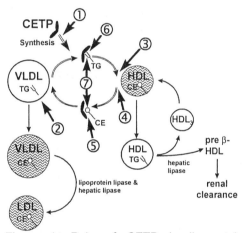

Figure 1: Role of CETP in lipoprotein metabolism: Points where inhibition of CETP could increase plasma HDL-C.

CETP as a Mechanistic Target - CETP is an acidic plasma glycoprotein of molecular weight 74,000. Reduction in CETP activity can be achieved by at least seven different means as shown in Figure 1. Regulating synthesis of CETP (①) is possible since it is well established that dietary cholesterol, hormones (e.g. interleukin-1, insulin, cortico-steroids, and tumor necrosis factor) and lipopolysaccharide (LPS) influence CETP gene expression (6). Numerous studies have shown that CETP biosynthesis and cellular lipid metabolism are tightly linked (6, 11). Recent studies have also shown that repeated administra-tion of CETP antisense oligonucleotides decreased VLDL and LDL-C, raised HDL-C and reduced atherosclerosis in cholesterol-fed rabbits (12, 13). Metabolic variations in donor and acceptor lipoprotein particles, as seen in NIDDM, will also affect CETP activity (②) *in vivo* (6-9). However, neither of these means of regulating CETP activity are relevant to this discussion of CETP inhibitors and so will not be discussed further.

The primary function of CETP inhibitors is to block the balanced exchange of CE and TG between HDL and VLDL. The exact mechanism of CETP-mediated neutral lipid transfer is not fully defined and has been the subject of controversy over the years. The proposed structure of CETP is a sparingly water-soluble cylindrical monomer with a high length to diameter ratio (6, 14). There is likely at least one internal hydrophobic site capable of binding CE, TG or both. This neutral lipid binding pocket may be partially covered by an amphipathic "lid" to further enhance the solvation of these hydrophobic molecules during transport. Lipid transfer kinetic studies now strongly support a mechanism in which CETP shuttles lipids between lipoprotein particles in a ping-pong manner, rather than formation of a ternary complex of CETP with the donor and acceptor lipoproteins (15). As illustrated in Figure 1, at least five points are suggested by the ping-pong model where CETP activity could be inhibited by small molecules: interfering with interface binding (③) of CETP to lipoproteins, blocking desorption (④) of CETP from the surface of lipoproteins, blocking CE (⑤) or TG (⑥) binding to CETP and preventing exchange (⑦) of CETP-bound CE for TG or vice versa.

CETP does not penetrate the lipid surface of lipoproteins but instead undergoes a conformational change upon binding to a lipid surface (11). Apolipoprotein A-I and related synthetic amphipathic peptides are examples of CETP inhibitors that appear to act by altering the interface binding of CETP to lipoproteins (16). LTIP, an endogenous CETP inhibitor now shown to be apo F, also has been shown to prevent or disrupt the association of CETP with lipoproteins (11). LTIP preferentially inhibits lipid transfer involving LDL without affecting CETP-mediated transfer between VLDL and HDL. The monoclonal antibody, TP-2, is an example of a CETP inhibitor that acts by modifying surface desorption of CETP from the lipoprotein particles (6, 7). It has been shown that TP-2 enhances binding of CETP to lipoproteins, thereby reducing the ability of CETP to shuttle neutral lipids between particles. TP-2 also appears to selectively block binding of CE to CETP while not affecting TG binding. By comparison, another monoclonal antibody appears to be able to selectively inhibit TG transfer by CETP, without affecting CE transfer (17).

Proof that inhibition of CETP raises HDL-C while, in some cases also lowering LDL-C, is well established from numerous animal studies (6, 9, 11-13, 18-22). One set of data has come from studies with anti-CETP mouse monoclonal antibodies that bind and inhibit CETP activity (6, 18). Two days after dosing hamsters i.v. with TP-2, a CETP neutralizing mouse monoclonal antibody, endogenous plasma CETP activity was reduced 60%. The TP-2 treatment increased HDL-C 24% while lowering LDL-C + VLDL-C more than 30%. Oral dosing of a small molecule inhibitor of CETP-mediated CE and TG transfer lowered hamster VLDL + LDL-C 22%, lowered total TG 23% and raised HDL-C 10% (19). Peptide vaccines (20), plasmid-based vaccines (21) and intact xenogeneic CETP protein vaccines (22) have each induced endogenous antibodies that bind CETP and neutralize CETP-mediated lipid transfer. These vaccines have been reported to increase HDL-C and result in significantly fewer atherosclerotic aortic lesions in rabbits (21, 22). Also, as mentioned previously, CETP antisense oligonucleotides have been shown to reduce aortic atherosclerosis in rabbits (12). Peptides homologous to the N-termini of apolipoprotein C-I (23) and apolipoprotein C-III (24) have been purified from baboon and pig plasma, respectively, that inhibit CETP. Treatment of rabbits with the C-III peptide raised HDL-C 32% (24).

In vitro assays are the first step in identifying CETP inhibitors. A differential precipitation assay that measures the rate of [^{3}H]CE transfer from HDL to LDL has been used to initially characterize a number of CETP inhibitors including: antibodies (6, 18), peptides (16, 23, 24) apolipoproteins (11, 16) and small molecules (16, 19). A second widely employed assay uses a fluorescent CE analog, NBD-cholesteryl linoleate (NBD-CE) that is self-quenched in a donor phosphatidyl choline emulsion. CETP-mediated transfer of NBD-CE from the emulsion to VLDL allows fluorescence to be measured in VLDL. Several CETP inhibitors that block transfer of NBD-CE were discovered and characterized using this fluorescence-based assay (16, 19).

CHEMISTRY OF SMALL MOLECULE CETP INHIBITORS

As a transfer protein, CETP presents several unique challenges for inhibitor design. Access to potent (nM) inhibitors based on transition-states is unavailable, since there is no chemistry of catalysis. An ideal drug candidate would likely be specific for CETP and not disrupt the integrity of other lipoproteins. While the available biochemical data indicate that human CETP contains independent binding sites for neutral and phospholipids (7), the exact stoichiometry, affinity, and three-dimensional structural details of the protein interactions defining binding for these substrates are currently unknown. Most of the published literature to date summarizes screening hits from academic and industrial labs that have identified novel inhibitor classes that typically exhibit micromolar IC_{50}s for inhibiting CETP-mediated transfer *in vitro*. Many of these have provided the first small molecule probes to more clearly define the biochemical mechanisms available for CETP inactivation. Since 1998, patent filings have started to appear describing new more potent classes (submicromolar IC_{50}s *in vitro*) that are also capable of improving the HDL/LDL cholesterol ratio in animal models following iv or oral dosing. Detailed pharmacological studies of these newer classes have yet to be reported.

Human CETP has a 3- to 8-fold higher intrinsic affinity for CE over TG (14). A number of cholesterol derivatives have been used to analyze the binding require-ments at the CE site. No analogous studies have been reported for the TG site. Short chain CE derivatives, such as cholesteryl bromoacetate **1a**, have been shown to selectively inhibit CE but not TG transfer, thus implying two separate binding domains for neutral lipids in CETP (25). Whereas detailed kinetic analyses demon-strated that the aminosteroid (U-95594, **1b**) inhibited (K_i = 1 μM) the transfer of either TG or CE competitively (26), the vinyl mercurial (U-617, **1c**), selectively inhibited CE, but not TG transfer, and acted as a potent time-dependent inactivator (26). The amino acid residue(s) involved in the inactivation by **1c** have not yet been defined.   Cholesteryl phosphonate esters and 15-ketosterols have also been examined as CETP inhibitors (27,28).

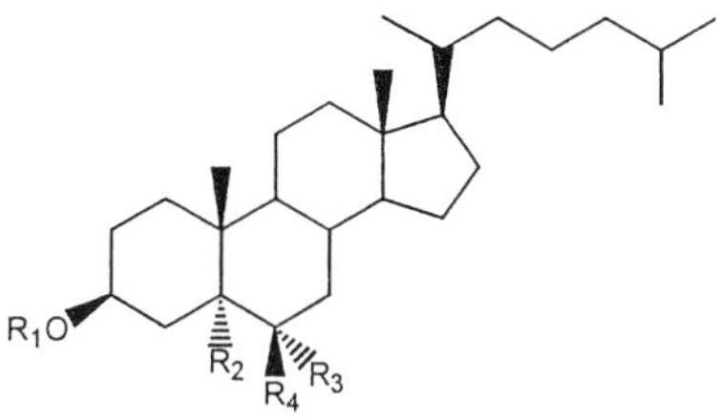

1a R_1 = BrCH$_2$C=O; R_2,R_3,R_4 = H
1b R_1,R_2,R_3 = H; R_4 = NH$_2$
1c R_1 = H; R_2-R_3 = bond; R_4 = HgCl

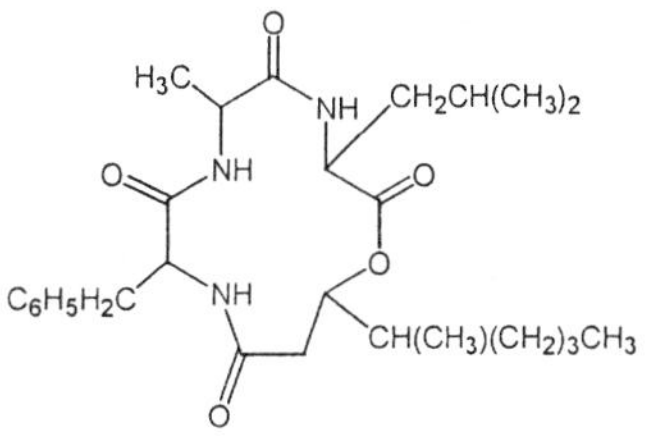

2

Peptidic Inhibitors - Several decapeptides containing the composite motif -Xaa-Arg-Met-Arg-Tyr-Xaa- were identified as CETP inhibitors from a bacteriophage display library (29). One of them, DP1 (NH$_2$-VTWRMWYVPA-CO$_2$H), inhibited CETP-mediated transfer of both CE and TG. Subsequent truncations of DP1 led to a pentapeptide NH$_2$-WRMWY-CO$_2$H (PNU-107368E), that binds directly to CETP and functions as a competitive inhibitor of CETP activity with a K_i of 164 μM. An unusual cyclic depsipeptide (SCH 58149, **2**) CETP inhibitor (IC_{50} = 50 μM) was isolated from a fungal fermentation broth belonging to *Acremonium sp.* (30).

Natural Product Inhibitors - A number of structurally diverse natural products have been identified as CETP inhibitors from both fungal and marine sources. The unusual contiguous quinquacyclopropane derivative (U-106305, **3**) was isolated from fermentation broths of *Streptomyces sp.* and inhibited (IC_{50} = 25 μM) human CETP activity (31). The unusual helical conformation adapted by this inhibitor (32) suggests that CETP can recognize several inhibitor conformational motifs.   The marine natural product Wiedendiol A, **4**, (IC$_{50}$ = 5 μM) competitively blocked CE binding through a direct interaction with the CETP protein (33).

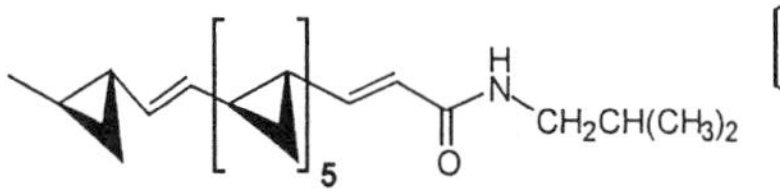

3 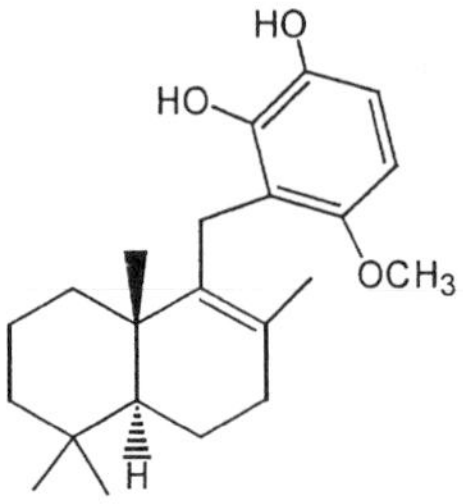   **4** 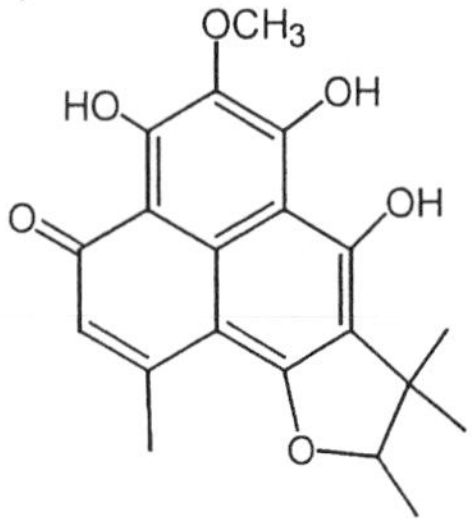   **5**

Cultures of the soil isolate *Penicillium sp.* FO-3657 produced the erabulenol A metabolite, **5**, which inhibited (IC$_{50}$ = 48 μM) human CETP *in vitro* (34). Fungal pigments represented by the azaphilone chaetoviridin B, **6**, (IC$_{50}$ < 6 μM) have been identified as CE transfer inhibitors (35). The related sclerotiorin, **7**, (IC$_{50}$ = 19 μM) has been characterized as a time-dependent irreversible inhibitor that blocked both CE and TG transfer. Following oral dosing of **7** at 10 mg/kg to transgenic mice expressing human CETP and apo A1, inhibition of CETP activity in the plasma was observed *ex vivo* even after 24 hours (35).

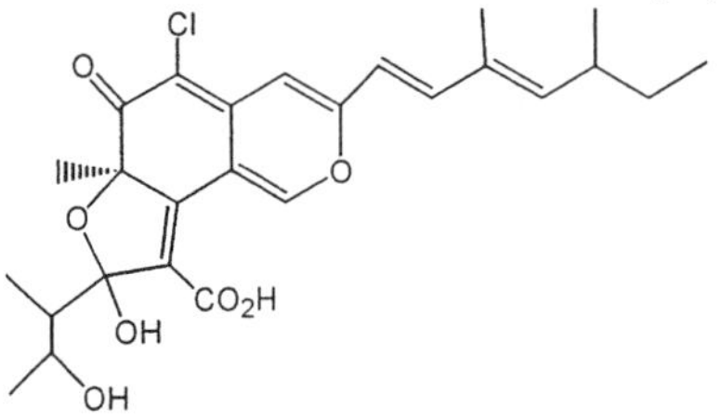

6 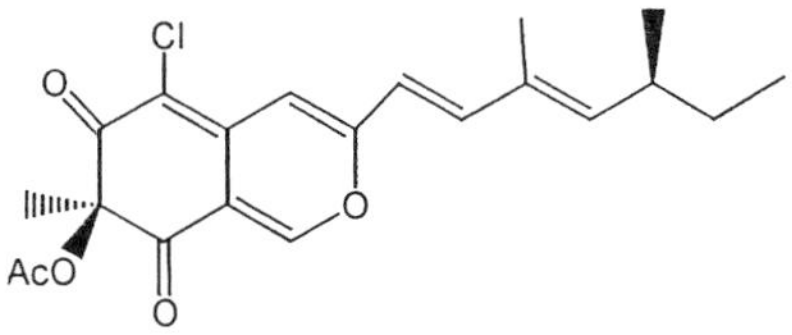   **7**

<u>Sulfur-Containing Inhibitors</u> - Human CETP is monomeric and contains seven cysteine residues (6, 14). This suggests that at least one sulfhydryl residue should be unpaired. Whereas typical cysteine-modifying reagents such as iodoacetamide and *N*-ethylmaleimide are very poor CETP inhibitors (IC$_{50}$ > 1.25 mM), *p*-chloro-mercuriphenylsulfonic acid is an extremely potent (IC$_{50}$ = 0.02 μM), time-dependent inactivator (36). In addition, several hydrophobic disulfides including 2,2'-dithiodi-pyridine, 4,4'-dithiodipyridine and 4,4'-dithiobis(phenylazide) were submicromolar (IC$_{50}$ = 0.5 μM) and time-dependent inhibitors (36). Related disulfides containing more hydrophilic polar or ionizable groups were significantly less active. These results are consistent with CETP's known preference for binding neutral lipids (14).

Several hydrophobic sulfur-containing pyridines **8a-8d** (IC$_{50}$ 0.5-9 μM) and **9a-9b** (IC$_{50}$ 1.5-19 μM) have recently been reported as CETP inhibitors (37). Two of the most potent members of this series are the disulfides **8d** and **9a**. In either case, replacing the reactive disulfide moiety with the more stable thiomethylene ether linkages as in **8c** and **9b** led to significantly weaker inhibitors. Following iv dosing of **9a** to hamsters, >50% inhibition of CETP activity in the plasma could be observed *ex vivo* after 2 and 4 hours (37). Continuous infusion of **9a** at 9.2 mg/kg/day for 8 days to male hamsters maintained on a normal chow diet reportedly produced a 30% reduction in LDL-C and a 26% increase in HDL-C.

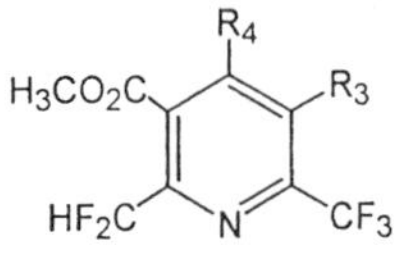

8a   R$_4$ = SH; R$_3$ = CO$_2$CH$_3$ 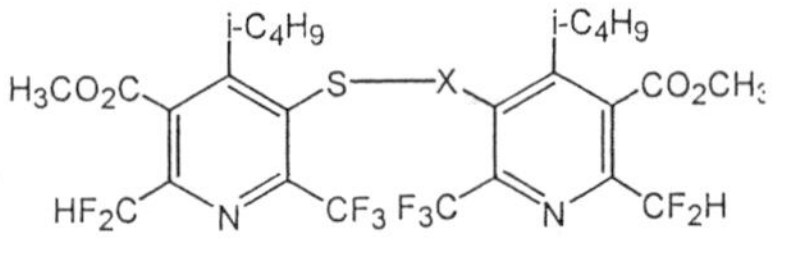   **9a**   X = S

8b $R_4 = i\text{-}C_4H_9$; $R_3 = SH$ **9b** $X = CH_2$
8c $R_4 = i\text{-}C_4H_9$; $R_3 = CH_2S\text{-}t\text{-}C_4H_9$
8d $R_4 = i\text{-}C_4H_9$; $R_3 = S_2\text{-}C_6H_4\text{-}4\text{-}(t\text{-}C_4H_9)$

The 5-thiolo-1,2,4-triazole (PD-140195, **10a**, IC_{50} = 30 µM) selectively inhibited CE but not TG transfer mediated by CETP from rabbit plasma (38). A sensitive *in vitro* fluorescence assay demonstrated an IC_{50} of ~1 µM for **10a** using media from Chinese hamster ovary cells secreting recombinant human CETP. The potency of **10a** was dramatically reduced by the addition of albumin or in whole plasma assays. Following iv infusion of **10a** to anesthetized rabbits at 10 or 20 mg/kg, a transient, concentration-dependent inhibition of CETP activity was observed in the plasma *ex vivo* after 0.5, 1, and 2 hours that returned to control levels after 4 hours (38). Additional analogs, e.g. **10b** (IC_{50} = 2 µM), have recently been described that have greater *in vitro* potency than **10a** under different assay conditions and which retain more activity in human plasma (39). In this series, replacement of the 5-thiol group by hydrogen, as in **10c** (IC_{50} >100 µM), led to a dramatically weaker inhibitor. thus demonstrating the importance of the SH group for potency in this series.

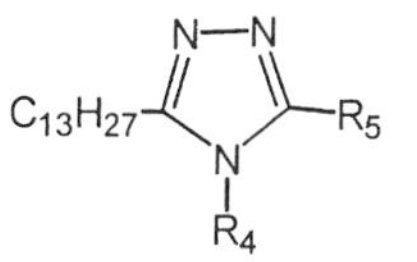

10a $R_4 = C_6H_5$ $R_5 = SH$
10b $R_4 = C_6H_4\text{-}(3\text{-}OCH_3)$; $R_5 = SH$
10c $R_4 = C_6H_4\text{-}(3\text{-}OCH_3)$; $R_5 = H$

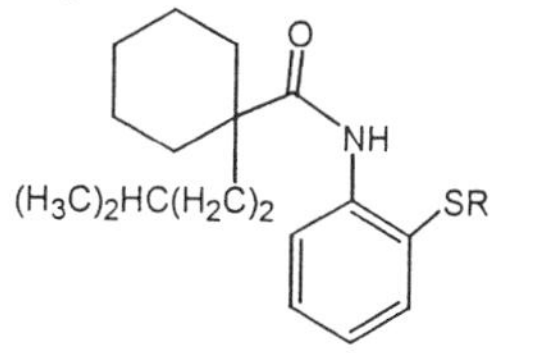

11a $R = H$
11b $R_4 = COCH_3$
11c $R_4 = CO(CH_3)_3$

An extensive series of *ortho*-thioaniline derivatives **11a-c** (IC_{50} = 5-61 µM) has also been reported as CETP inhibitors (40). Single oral doses of **11b** or **11c** at 10, 30, or 100 mg/kg to fasted transgenic mice expressing human CETP, produced a dose-dependent inhibition of CETP activity in the plasma *ex vivo* after 6 hours.

Heterocyclic Inhibitors - The isoflavan (CGS-25159, **12**) inhibited CETP activity in human (IC_{50} = 10 µM) and hamster plasma (IC_{50} = 3 µM). After oral dosing of **12** to hamsters at 10 mg/kg q.d. for four days, 35-60% less CETP activity was observed in the plasma *ex vivo* after 2, 4 and 6 hours (19). Repeated oral dosing of **12** at 10 or 30 mg/kg q.d. for up to 4 weeks to cholesterol-fed hamsters reportedly produced a reduction in total cholesterol (-27% @ 10 mg/kg; -36% @ 30 mg/kg) and a concomitant increase in HDL-C (16% @ 10 mg/kg; 25% @ 30 mg/kg).

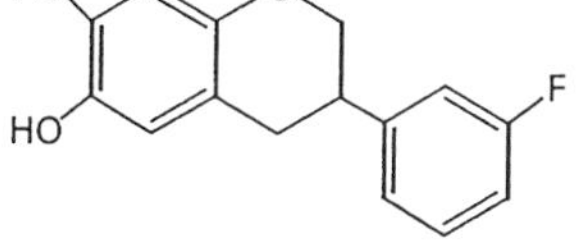

12

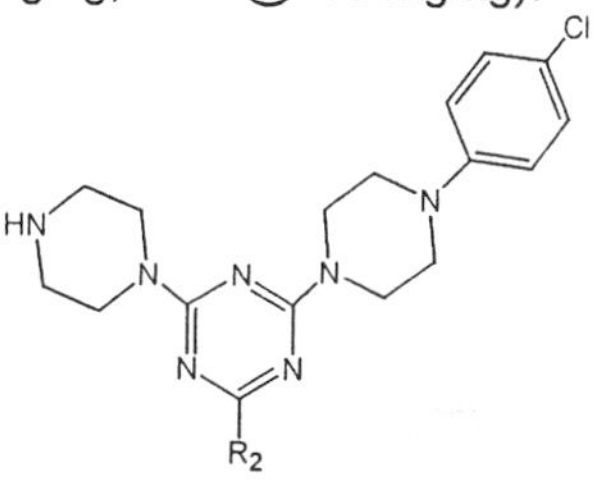

13a $R_2 = C_6H_5CH_2$
13b $R_2 = 2\text{-naphthyl-}CH_2$

Several substituted 1,3,5-triazines represented by **13a** and **13b** (IC_{50} = 9 and 5 µM, resp.) have also been reported as CETP inhibitors (41). Several series of pentasubstituted pyridines (37, 42), benzenes (43), and quinolines (44) lacking sulfur-containing functionalities have also been reported as CETP inhibitors. Some of the most potent members of this class include **14a-d** (IC_{50} = 0.17-0.5 µM) (45). After dosing of racemic **14b** (IC_{50} = 0.17 µM) at 20 mg/kg iv to hamsters, CETP activity in the plasma decreased by 50% *ex vivo* after 2 hours and HDL-C increased by 30% after 24 hours. Oral dosing of a single enantiomer of **14b** to hamsters at 45 mg/kg b.i.d. produced a 20% increase in HDL-C after 24 hours (45).

The related fused thianapthyridinol **15** (IC_{50} = 0.2 µM) had *in vitro* potency comparable to **14b**. However, **15** inhibited 50% of the CETP activity *ex vivo* in hamsters after oral dosing of 10 mg/kg and raised HDL-C by 12% in transgenic mice expressing human CETP after oral dosing at 3 mg/kg b.i.d. for one week (46). The fused indeno[1,2-b]pyridin-5-ol **16** (IC_{50} = 0.06 µM) exhibited greater *in vivo* potency than **14b** even though it had a similar IC_{50} *in vitro*. For example, in hamsters **16** inhibited 64% of the CETP activity *ex vivo* orally at 10 mg/kg and raised HDL-C by 22% in transgenic mice expressing human CETP after dosing in feed at 100 ppm for one week (47).

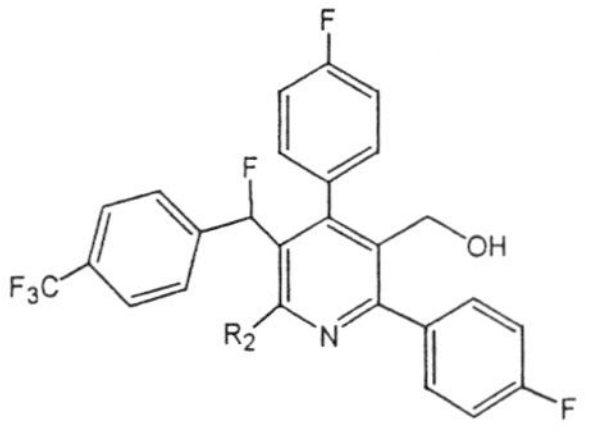

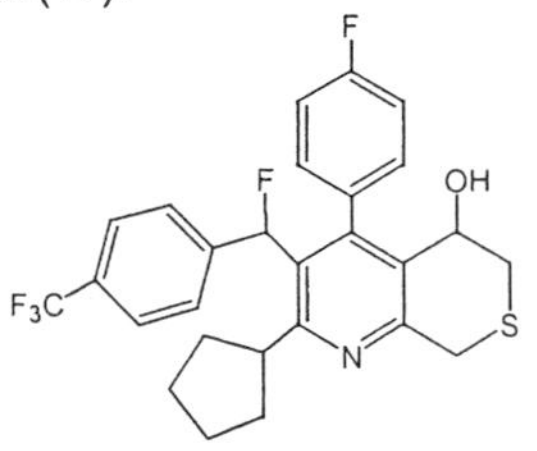

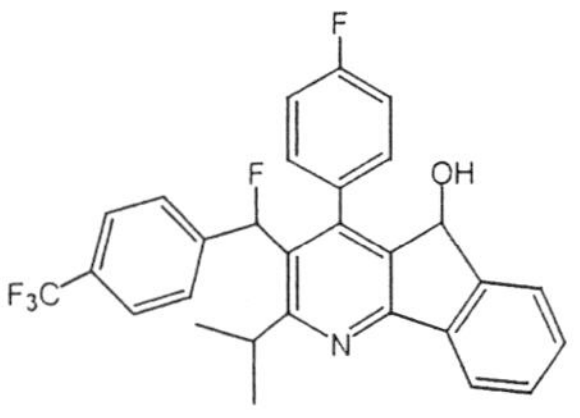

14a R_2 = *i*-C$_3$H$_7$
14b R_2 = cyclopentyl
14c R_2 = cyclohexyl
14d R_2 = cycloheptyl

15

16

Tetrahydroquinoline and tetrahydronaphthalene Inhibitors - Low nanomolar potency was observed in several series of tetrahydroquinolines (48-50), as represented by **17** (IC_{50} = 0.006 µM) (50). A single diastereomer of **17** raised HDL-C by 16% after oral dosing of 3 mg/kg b.i.d. in hamsters. Similarly, **17** also raised HDL-C by 31% and lowered LDL-C by 15% in transgenic mice expressing human CETP after dosing in feed at 400 ppm for one week (50). Several fused tetrahydronaphthalene **18-19** (IC_{50} = 0.003 µM) derivatives have recently been reported as extremely potent CETP inhibitors both *in vitro* and *in vivo* (51, 52). The unusual spirocyclobutyltetrahydronaphthalene **19b** raised HDL-C by 19% after oral dosing of 0.3 mg/kg b.i.d. in hamsters, and **19b** also raised HDL-C by 60% and lowered LDL-C by 18% in transgenic mice expressing human CETP after dosing in feed at 30 ppm for one week (52).

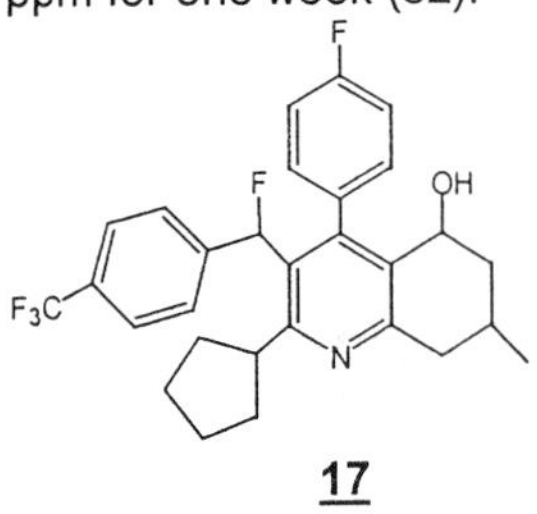

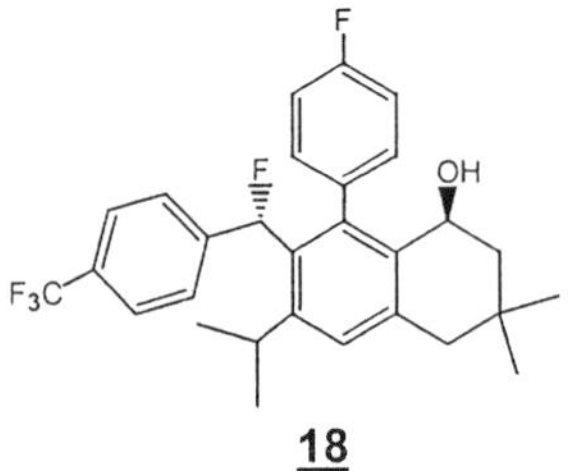

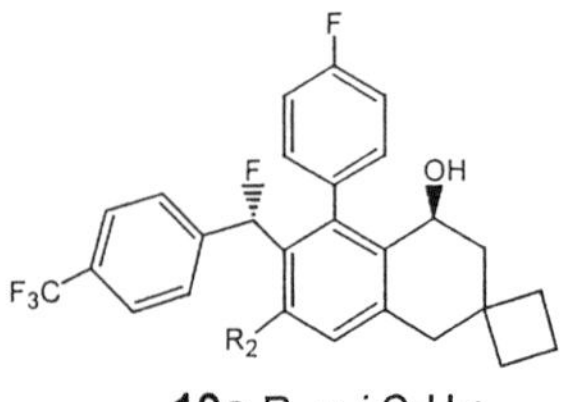

17

18

19a R_2 = *i*-C$_3$H$_7$;
19b R_2 = cyclopentyl

Conclusions - Important advances have been made in the identification of potent *in vitro* inhibitors of CETP over the last few years. Low doses of the first orally bioavailable, non-covalent CETP inhibitors are showing potential to improve the HDL to LDL cholesterol ratio in various animal models. Further development of these lead series may soon allow the first clinical proof of concept in humans to test the hypothesis that inhibition of CETP is another useful therapeutic approach to benefit patients with CHD.

References

1. A.J.J. Wood and R.H. Knopp, N.Engl.J.Med., 34, 498 (1999).
2. W.B. Kannel, Am.Heart J., 110, 1100 (1985).
3. V. Manninen, L. Tenkanen, P. Koskinen, J.K. Huttunen, M. Manttari, O.P. Heinonen and M.H. Frick, Circulation, 85, 37 (1992).
4. H.B. Rubins, S.J. Robins, D. Collins, C.L. Fye, J.W. Anderson, M.B. Elam, F.H. Faas, E. Linares, E. Schafer, G. Schectman, T. Wilt and J. Wittes, N.Engl.J.Med. 34, 410 (1999).
5. K. Garber, Mod.Drug Discovery, 2, 67 (1999).
6. C. Bruce, R.A. Chouinard and A.R. Tall, Annu.Rev.Nutr., 18, 297 (1998).
7. A.R. Tall, J.Lipid. Res., 34, 1255 (1993).
8. L. Laurent, Adv.Vasc.Biol., 5, 217 (1999).
9. P. Moulin, Horm.Res., 45, 238 (1996).
10. V.L. Herrera, S.C. Makrides, H.X. Xie, H. Adari, R.M. Krauss, U.S. Ryan and N. Ruiz-Opazo, Nature Medicine, 5, 1283 (1999).
11. R.E. Morton, Curr.Opin.Lipidol., 10, 321 (1999).
12. M. Sugano, N. Makino, S. Sawada, S. Otsuka, M. Watanabe, H. Okamoto, M. Kamada and A. Mizushima, J.Biol.Chem., 273, 5033 (1998).
13. R. Budzinksi, B. Krist, M. Mark and P. Mueller, DE19731609 (1999).
14. C. Bruce, L.J. Beamer and A.R. Tall, Curr.Opin.Struct. Biol., 8, 426 (1998).
15. D.T. Connolly, J. McIntyre, D. Heuvelman, E. Remsen, R.E. McKinnie, L. Vu, M. Melton, R. Monsell, E. Krul and K. Glenn, Biochem.J., 320, 39 (1996).
16. D.T. Connolly, E.S. Krul, D. Heuvelman and K.C. Glenn, Biochim.Biophys.Acta, 1304, 145 (1996).
17. K. Saito, K-I. Kobori, H. Hashimoto, S. Ito, M. Manabe and S. Yokoyama, J.Lipid Res., 40, 2013 (1999).
18. G.F. Evans, W.R. Bensch, L.D. Apelgren, D. Bailey, R.F. Kaufman, T.F. Bumol and S.H. Zuckerman, J.Lipid Res., 35, 1634 (1994).
19. H.V. Kothari, K.J. Poirier, W.H Lee and Y. Satoh, Atherosclerosis, 128, 59 (1997).
20. P. Needleman and K. Glenn, WO9915655 (1999).
21. L.J. Thomas, WO9741227 (1997).
22. C.W. Rittershaus and L.J. Thomas, WO9920302 (1999).
23. R.S. Kushwaha, H.C. McGill Jr. and P. Kanda, WO9504755 (1995).
24. K-H. Cho, J-Y. Lee, M-S. Choi, J-M. Cho, J-S. Lim and Y.B. Park, Biochim.Biophys.Acta., 1391, 133 (1998).
25. S.J. Busch and J.A.K. Harmony, Lipids, 25, 216 (1990).
26. D.E. Epps, K.A. Greenlee, J.S. Harris, E.W. Thomas, C.K. Castle, J.S. Fisher, R.R. Hozak, C.K. Marschke, G.W. Melchior and F.J. Kezdy, Biochemistry, 34, 12560 (1995).
27. T. Pietzonka, R. Damon, M. Russell and S. Wattanasin, Bioorg.Med.Chem.Lett., 6, 1951 (1996).
28. H.S. Kim, S.H. Oh, D.I. Kim, I.C. Kim, K.H. Cho and Y.B. Park, Bioorg.Med.Chem., 3, 367 (1995).
29. P.D. Bonin, C.A. Bannow, C.W. Smith, H.D. Fischer and L.A. Erickson, J.Pept.Res., 51, 216 (1998).
30. V.R. Hegde, P. Dai, M. Patel, P.R. Das, S. Wang and M.S. Puar, Bioorg.Med.Chem.Lett., 8, 1277 (1998).
31. M.S. Kuo, R.J. Zielinski, J.I.Cialdella, C.K. Marschke, M.J. Dupuis, G.P. Li, D.A. Kloosterman, C.H. Spilman and V.P. Marshall, J.Am.Chem.Soc. 117, 10629 (1995).
32. A.G.M. Barrett, D. Hamprecht, A.J.P. White and D.J. Williams, J.Am.Chem.Soc. 119, 8608 (1997).
33. S. Wang, A. Clemmons, S. Chackalamannil, S.J. Coval, E. Sybertz and R. Burrier, Zhongguo Yaoli Xuebao, 19, 408 (1998).
34. H. Tomoda, N. Tabata, R. Masuma, S.-Y. Si and S. Omura, J.Antibiot., 51, 618 (1998).
35. H. Tomoda, C. Matsushima, N. Tabata, I. Namatame, H. Tanaka, M.J. Bamberger, H. Arai, M. Fukazawa, K. Inoue and S. Omura, J.Antibiot., 52, 160, (1999).
36. D.T. Connolly, D. Heuvelman and K. Glenn, Biochem.Biophys.Res.Comm., 223, 42 (1996).

37. L.F. Lee, K.C. Glenn, D.T. Connolly, D.G. Corley, D.L. Flynn, A. Hamme, S.G. Hegde, M.A. Melton, R.J. Schilling, J.A. Sikorski, N.N. Wall and J.A. Zablocki, WO9941237 (1999).

38. C.L. Bisgaier, A.D. Essenburg, L.L. Minton, R. Homan, C.J. Blankley and A. White, Lipids, 29, 811 (1994).

39. J.A. Sikorski, WO9914204 (1999).

40. H. Shinkai, K. Maeda and H. Okamoto, WO9835937 (1998).

41. Y. Xia, B. Mirzai, S. Chackalamannil, M. Czarniecki, S. Wang, A. Clemmons, H.-S. Ahn and G.C. Boykow, Bioorg.Med.Chem.Lett., 6, 919 (1996).

42. C. Schmeck, A. Brandes, M. Loegers, G. Schmidt, K.-D. Bremm, H. Bischoff, D. Schmidt and J. Schuhmacher, WO9834920 (1998).

43. M. Loegers, A. Brandes, G. Schmidt, K.-D. Bremm, J. Stoltefuss, H. Bischoff and D. Schmidt, WO9915487 (1999).

44. M. Mueller-Gliemann, R. Angerbauer, A. Brandes, M. Loegers, C. Schmeck, G. Schmidt, K.-D. Bremm, H. Bischoff and D. Schmidt, WO9839299 (1998).

45. G. Schmidt, R. Angerbauer, A. Brandes, M. Logers, M. Muller-Gliemann, H. Bischoff, D. Schmidt and S. Wohlfeil, US Patent 5,925,645 (1999).

46. C. Schmeck, M. Mueller-Gliemann, G. Schmidt, A. Brandes, R. Angerbauer, M. Loegers, K.-D. Bremm, H. Bischoff, D. Schmidt and J. Schuhmacher, US Patent 5,932,587 (1999).

47. Brandes, M. Loegers, G. Schmidt, R. Angerbauer, C. Schmeck, K.-D. Bremm, H. Bischoff, D. Schmidt and J. Schuhmacher, EP825185 (1998).

48. J. Stoltefuss, M. Loegers, G. Schmidt, A. Brandes, C. Schmeck, K.-D. Bremm, H. Bischoff and D. Schmidt, WO9914215 (1999).

49. G. Schmidt, J. Stoltefuss, M. Loegers, A. Brandes, C. Schmeck, K.-D. Bremm, H. Bischoff and D. Schmidt, WO9915504 (1999).

50. G. Schmidt, A. Brandes, R. Angerbauer, M. Loegers, M. Mueller-Gliemann, C. Schmeck, K.-D. Bremm, H. Bischoff, D. Schmidt, J. Schuhmacher, H. Giera, H. Paulsen, P. Naab, M. Conrad and J. Stoltefuss, EP818448 (1998).

51. H. Paulsen, S. Antons, A. Brandes, M. Logers, S.N. Muller, P. Naab, C. Schmeck, S. Schneider and J. Stoltefuss, Angew.Chem.,Int.Ed., 38, 3373 (1999).

52. Brandes, M. Logers, J. Stoltefuss, G. Schmidt, K.-D. Bremm, H. Bischoff, D. Schmidt, S. Antons, H. Paulsen, S.N. Muller, P. Naab and C. Schmeck, WO9914174 (1999).

SECTION V. TOPICS IN BIOLOGY

Editor: Janet Allen, Institut de Recherche Jouveinal/Parke-Davis
Fresnes, France

Chapter 23. Pharmacogenomics

Kit Fun Lau and Hakan Sakul
Parke-Davis Laboratory for Molecular Genetics
1501 Harbor Bay Parkway, Alameda, CA 94502

Introduction - Pharmacogenetics can be broadly defined as a field concerned with genetic differences in drug response. Although the field of pharmacogenetics has had limited impact on the pharmaceutical industry until recently, it is almost 50 years old and is well established. In fact, it was more than 100 years ago when investigations of physiological chemists marked the start of events and discoveries that had profound influence on our understanding of differences in individual drug response. These events overlap with discoveries of the laws of heredity by Gregor Mendel (1822-1884) and their rediscovery around the turn of the century. A detailed historical perspective on the emergence of pharmacogenetics and a review of early experiments with several different classes of enzymes can be found in (1). Beginning in the 1950s, a combination of new technologies and more detailed knowledge of genetics allowed for better study of individual genetic differences in drug metabolism. For the first time, individual variations in response to suxamethonium, primaquine and isoniazid were investigated from the genetic standpoint. More recently, advances in laboratory automation, robotics, molecular biology, computer science, bioinformatics, statistical genetics and other related fields have allowed production of genetic data on much larger scales. Additionally, with the Human Genome Project and its privately funded competitors planning to completely sequence the human genome within this year, experiments that were unapproachable a few years ago are quickly turning into reality. These recent advances have opened the doors to developing new therapeutic and diagnostic products from genomics in timelines that are compatible with expectations from drug discovery and development.

Aims and Scope of Pharmacogenetics - The primary aim of pharmacogenetics is to discover genetic differences that affect drug action, including drug safety and efficacy. Most drugs prescribed today have an inherent problem that limits optimal efficacy: the inability to account for genetic differences in drug response among individuals. There are many examples of adverse drug reactions in the medical literature. A meta-analysis of prospective studies reported an overall incidence for adverse reactions of 6.7% in hospitalized patients (2). Their conclusion was that the incidence of serious and fatal adverse reactions in U.S. hospitals was found to be extremely high. It would not be a speculation to suggest that many people are prescribed more than one antibiotic before one works to improve their condition. It then follows that a thorough study of genetic differences in response to a particular drug can eventually help to increase a drug's efficacy. This, in turn, can lower the incidence of both non-responders and adverse reactions. Furthermore, stratifying patients according to genetic differences that predict safety or efficacy can enhance statistical power of a clinical trial and may expedite the drug approval process, resulting in substantial savings. These are some of the promises of pharmacogenetics. As we will discuss later in this chapter, pharmacogenetics approaches can be utilized in different phases of the drug development process.

Application to Drug Discovery and Clinical Therapeutics - During the drug discovery process, systematic identification of variants in the drug targets can be used to avoid new chemical entities (NCEs) that are affected by genetic variation before they enter into clinical development. Preclinical pharmacokinetic and pharmacodynamic studies will benefit when genetic effects on drug metabolism can be taken into consideration. Incorporation of genotype information in clinical trials can, in many cases, help to reduce the size of the trials (3). By reducing the number of responders necessary to achieve statistical significance, the cost and time required to achieve approval can be reduced. In addition, certain NCEs that do not exhibit statistical significance in unselected populations may show significant responses in selected populations. Given that historically 80% of compounds fail in clinical trials and the pharmaceutical industry spends $500-700 million for each new drug approval, even a small percentage of failed compounds rescued by stratification using pharmacogenetics can have a major impact in the drug industry. After drug approval, commercial molecular diagnostic tests can be devised so that medication can be prescribed more precisely to improve efficacy and minimize adverse reactions (4). By collecting relevant genetic data on participants in large-scale clinical trials, one may also be in the position to address fundamental questions about the nature and pathogenesis of the disease for which the compound was designed (5). Better understanding of the disease process will help develop future drugs for the disease. Expression profiling comparison between individuals with and without drug treatment can shed light on the drug action mechanism and aid improved the drug design process.

There are important differences between genetic effects on diseases and on drug action (6). The genetic effects on common diseases such as diabetes, and cardiovascular, inflammatory and central nervous system diseases are complex. Very often, different polymorphisms in many genes are involved and there are interactions between environmental factors and genotypes. Individual mutations that cause disease susceptibility are either low in frequency or have modest effect. As a result, the predictive or therapeutic value of any single gene, or any specific mutation in that gene, is generally limited. In contrast, the heritability of variations in drug effects can be extremely high, and there are numerous precedents for significant single gene effects on the action of common drugs. Some twin studies, used for estimating genetic contribution to a trait, indicate that genetic factors can account for much of the observed variability in important parameters of drug action (7). Thus, genetic studies of drug responses may have better potential to produce significant clinical and commercial rather results than genetic studies of diseases have.

The success of pharmacogenetics hinges upon our knowledge of genes that are related to drug responses, detection of their genetic variances (polymorphisms), and the effects of their genetic polymorphisms on safety and efficacy of different drugs. It also hinges upon the enabling technologies that allow fast and cost effective identification of an individual's genotypes of the medically related genes. The following sections will describe the recent research in two main categories of drug response related genes. The first category includes drug targets and disease susceptibility genes that are mainly related to efficacy. The second category includes genes for phase I and II drug-metabolizing enzymes (DME) that are responsible for the metabolism and disposition of drugs. DME genes are related not only to toxicity response but to efficacy as well. In many cases, DME genes are associated with disease susceptibility to several common diseases. The subsequent sections will describe different genetic and statistical approaches for studying association between genetic polymorphisms in medical-related genes and drug responses. The fast advent of genotyping, sequencing and gene expression technologies using gene chips, micro arrays and mass spectrometry is summarized. Conclusions are drawn as to how pharmacogenetics will impact future drug discovery and development and ultimately improving on health care and medical practices.

<u>Drug Targets and Disease Susceptibility Genes</u> - Many of the genes encoding drug targets have genetic polymorphisms that will alter their sensitivity to specific drugs. Recently, it has been shown that patients with mutant genotypes in the promoter region of 5-lipoxygenase (ALOX5) have a diminished clinical response to ABT-761, an asthmatic drug that is a potent and selective inhibitor of ALOX5 (8). It has been shown that patients with homozygous variants of a functional polymorphism within the promoter region of the serotonin transporter (5-HTT) showed poorer efficacy response to the antidepressant drug fluvoxamine that targets 5-HTT (9). Another study showed that substitution of phenyalanine by cysteine in position 124 of human serotonin 5-HT1B (h5-HT1B) receptor significantly affects its pharmacological properties (10). Carriers of the variant may exhibit differences in response to drugs acting on the h5-HT1B receptor or may develop side effects to such agents. It has been shown that many antipsychotic drugs have different binding affinities and potencies for the different human D2 dopamine receptor missense variants (11). Comparative pharmacological and functional analyses of variants 2 and 10 of the human dopamine D4 receptor show that there are no major discrepancies in pharmacological or functional profile between the two receptors. Moreover, there are small increases in functional potency and affinity for dopamine and quinirole at the D4.10 receptor variant compared with the D4.2 receptor variant (12). Genetic polymorphisms in angiotensin converting enzyme (ACE) have also affected the drug responses of enalapril, lisinopril, and captopril which all target ACE (13,14).

Polymorphisms in disease susceptibility genes can also have great impact on drug efficacy. The E4 allele of the ApoE gene is associated with sporadic and late-onset familial Alzheimer's disease (AD). Residual brain choline acetyltransferase activity decreases with increasing number of ApoE4 alleles. Clinical trials of Tacrine, a cholinomimetic drug, showed that > 80% of ApoE4 negative AD patients exhibited marked improvement to the drug treatment while 60% of ApoE4 carriers deteriorated (15). However, ApoE4 has the opposite effect on another drug for Alzheimers disease, S12024, with effect on vasopressinergic neutrotransmission (16). A common variant in cholesteryl ester transfer protein (CEPT) is correlated with higher CETP levels and lower levels of plasma high-density lipoproteien. Treatment with pravastatin in that group of patients slowed the progression of coronary atherosclerosis (17).

<u>Phase I and II Drug Metabolizing Enzymes</u> - Most drug-metabolizing enzymes exhibit clinically relevant genetic polymorphisms (18). Phase I enzymes catalyze the oxidation reaction of drug molecules while phase II enzymes catalyze conjugation with endogenous substituents. Major categories of phase I drug- metabolizing enzymes include different isoforms of cytochrome P450 (CYP), alcohol dehydrogenase (ADH), aldehyde dehydrogenase (ALDH), and dihydropyrimidine dehydrogenase (DPD, NADPH): quinone oxidoreductase (NQO1). Major categories of phase II drug metabolizing enzymes are uridine 5'-triphosphate glucuronosyltransferases (UGTs), sulfotransferases (STs), thiopurine methyltransferase (TPMT), N-acetyltransferases (NAT), glutathione S-transferases (GST), and catechol O-methyltransferase (COMT) (19).

In the cytochrome P450 families of genes, CYP2D6, CYP2C9, CYP3A4/5/6, CYP2C19 are the ones that contribute most to the oxidation of drugs (19). CYP3A4 is the human enzyme known to be involved in the metabolism of the largest number of medications. However, it is not as important in pharmacogenetic studies as CYP2D6, (the enzyme involved in the second largest number of medications), primarily because CYP2D6 is more polymorphic and its polymorphisms affect the therapeutic management of up to 17% of individuals in some ethnic groups. Another isoenzyme CYP2D6 is responsible for the metabolism of more than 25 commonly prescribed medications, including beta blockers, antidepressants, antipsychotics, codeine, and debrisoquin. CYP2C9 is responsible for metabolizing Tolbutamide, warfarin, phenytoin, and nonsteroidal anti-inflammatories. Pharmaceutical substrates of CYP2C19 include amitriptyline, certain barbiturates, chlorproguanil, diazepam, proguanil, and propranolol (20).

A common polymorphism in the promoter region of CYP3A4 has recently been described but no completely inactivating mutations have been found (21,22). For enzymes that apparently do not have critical endogenous substrates (for example, CYP2C19, CYP2D6, and TPMT), the genetic molecular mechanisms of inactivation include splice site mutations resulting in exon skipping (for example, CYP2C19), microsatellite nucleotide repeats (for example, CYP2D6), point mutations resulting in early stop codons (for example, CYP2D6), and amino acid substitutions that alter protein stability or catalytic activity (for example, TPMT, NAT2, CYP2D6, CYP2C19, and CYP2C9) (19). Individuals with genotypes with variant alleles are poorer metabolizers of the drug substrates and the effects can be profound toxicity for medications involved. A more complete description of the medications metabolized by the drug-metabolizing enzymes, especially CYP2D6, their genetic variants, and the altered drug responses can be found in several recent articles (1,20,23).

Recent research on the effects of genetic polymorphisms in CYP2D6 include their involvement in the cholesterol-lowering effect of simvastatin, the O-demethylation of the antidementia drug galanthamine, N-oxidation of procainamide in man, association with prostate cancer, rifampicin treatment on propafenone disposition in extensive and poor metabolizers of CYP2D6, and influence on the disposition and cardiovascular toxicity of the antidepressant agent venlafaxine in humans (24-29). New genetic polymorhphisms have also been found (30) and so has their association with poor metabolism of debrisoquine. Another study shows that 10% of North Spanish subjects carry duplicated CYP2D6 genes associated with ultrarapid metabolism of debrisoquine (31). Other recent studies on the P450 genes include the effect of genetic polymorphisms in CYP2C9 on sulphamethoxazole N-hydroxylation, and the pharmacokinetics of chlorpheniramine, phenytoin, glipizide and nifedipine in an individual homozygous for the CYP2C9*3 allele (32,33).

It has been shown that genetic polymorphisms in phase II DME thiopurine methyltransferase (TPMT) cause TPMT deficiency which will result in adverse drug reaction to the antileukemic agents mercaptopurine (MP) and thioguanine (TG) to treat cancer patients (34). Commercial genetic diagnostic tests are being developed to prospectively diagnose TPMT deficiency to minimize hematopoietic toxicity and maximize efficacy of anticancer therapy. A new diagnostic method for identifying TPMT genotypes in individuals has been described and tested in a North Portuguese population (35). Research has also been done on the mechanism for TPMT deficiency associated with the different variant alleles TPMT*2, TPMT*3A, TPMT*3B or TPMT*3C (36). Genotype-phenotype relationship studies have been done on a polymorphism in NAD(P)H: quinone oxidoreductase 1 (37). It has also been shown that N-acetyltransferase 1 and 2 (NAT1 and NAT2) are associated with susceptibility to sporadic Alzheimer's disease (38), urothelial transitional cell carcinoma, and oral/pharyngeal and laryngeal cancers (39).

Pharmacogenetic polymorphisms differ in frequency among ethnic and racial groups. A recent study shows that UDP-glucuronosyl transferases (UGTs) has great variability in allele UGT1A1 frequencies in 658 individuals from a worldwide sample of 15 aboriginal and two admixed human populations. The most common allele can vary in frequency from 33% to 91% (40). A comparative evolutionary pharmacogenetic study of CYP2D6 in Ngawbe and Embera Amerindians of Panama and Colombia shows that human CYP2D6 evolution was preferentially affected by random genetic drift and not by adaptive or purifying selection (41). There are ethnic differences in TPMT between Caucasian and Kenyan individuals (42). Xie et al. have provided a systematic overview of the population distribution of the CYP2C19 poor metabolizers phenotype and CYP2C19 alleles and genotypes in healthy Caucasians living in different geographical areas, and found a similar polymorphic pattern in the structure and expression of the CYP2C19 gene in worldwide Caucasian populations (43). The population prevalence of CYP2C19 poor metabolizers in Caucasians of European descent range from 0.9% to 7.7%. Another study shows that the majority of Pacific Islanders metabolize a wide variety of clinically important drugs to a significantly lower degree than the average European (44). The frequency of functionally important beta-2 adrenoceptor polymorphisms varies markedly among African-American, Caucasian and Chinese individuals (45).

Different Approaches of Pharmacogenetic Analysis - Human genetic studies can be divided into two main categories: family based and population based. Family based studies usually entail collecting either extended or nuclear families with the disease, traits of interest or drug responses. There are different kinds of family-based studies. One can ask the question: how much the variances in traits can be attributed to genetics? This can be addressed by a heritability study. One can do segregation analysis, which addresses the mode of inheritance of the traits. One can genotype families to determine the genetic characteristics of the individuals at different genetic markers; linkage analyses can then be done to determine whether disease status or drug responses and the alleles in genetic markers transmit together from generation to generation or how much allele sharing takes place between affected relatives as compared with allele sharing by chance. One can also do family-based association studies to determine whether a genetic marker allele is associated with the trait of interest (46). Population based studies, on the other hand, involve unrelated individuals with the disease (cases) and those without the disease but matched with the cases in other aspects (controls). The goals of population studies usually involve association of the disease status or drug response with genetic marker polymorphisms.

In pharmacogenetic studies, study populations usually come from clinical trials that enroll mostly unrelated individuals. Even when the experiments are specifically designed for genetic studies, it is difficult to collect families with patients taking the same drug. Thus most pharmacogenetic studies involve population based association studies.

Two different approaches can be used for association studies between drug responses and genetic polymorphisms. The approach used most commonly at the moment is called the "candidate gene" approach in which known polymorphisms in known genes are tested for association with the efficacy or toxic responses of the new drugs. Those genes can be the genes for drug metabolizing enzymes, the drug target or the disease susceptibility genes discussed in detail in the last section. Another approach is a more systematic approach in which genetic markers are typed at regular intervals throughout the genome to test for association, or linkage disequilibrium, in more precise human genetics terminology (47-49). That approach does not assume any a priori knowledge about the genes involved and can have greater chance of finding novel genes associated with the drug action. However, the disadvantage is that it involves very large-scale genotyping and can be very costly or cost prohibitive when using the currently available technologies.

While association studies can be based on genotypes of individual markers, a more powerful approach is based on haplotypes, which are sets of alleles received by an individual from one parent at many markers on the same chromosome (50,51). One challenge of doing haplotype based genetic analysis on population pharmacogenetics is the difficulty of statistically inferring haplotypes from individual marker data without the aid of family genetic information. Haplotypes can also be obtained experimentally by sequencing or allele-specific PCR (51). However, those experimental haplotyping methods are still under development for large-scale application. The advent of the modern molecular genetic technologies that make pharmacogenetic studies feasible for large-scale clinical trials will be discussed in the next section.

Enabling Technologies for Pharmacogenetics – The Human Genome Project and its commercial competitors are due to provide a rough draft of the entire human genome this year and completely finish it by the year 2003. Those efforts will result in identification of tens of thousands of genes, many of which are possibly relevant to various disease processes and drug responses. In addition, their biochemical and physiological actions and expression patterns, as well as the reagents and technologies necessary for assaying individuals with respect to these genes, will be known from functional genomic studies. In 1999, The Wellcome Trust Foundation and several big pharmaceutical companies founded a SNP Consortium. Their goals are to detect 300,000 single nucleotide polymorhphisms (SNPs) and map 170,000 of them throughout the entire human genome. These polymorphisms, especially those in the regulatory and coding regions of the genes, are the causes of genetic variation in human traits, susceptibility of diseases and different responses to drugs. The SNP Consortium project is due to finish in 2 years.

These projects will not be possible without the fast advance of sequencing, genotyping and mutation detection technologies. The fast and cost effective genotyping of SNPs is a very important goal of the genetic community. A variety of available technologies have the potential to transfer to high-throughput genotyping laboratories (52). These include Taqman, Oligonucleotide-ligation assays, dye-labeled oligonucleotide ligation, minisequencing, microarray technology, scorpion assay, and Invader chemistry (53-62). Bioinformatic tools to store the enormous amount of biological and genetic data, to analyze the sequences for genomic structures of novel genes, to find similarities between sequences, and to annotate the genes with the most essential information are essential for pharmacogenetic research as well. An overview of the variety of databases, programs and expert systems for human geneticists can be found in (63).

The wealth of information on human genes will enable a more informed choice of candidate genes and their SNPs in pharmacogenetic studies. As mentioned earlier, allele frequencies of SNPs vary greatly from population to population. It is important that mutation detection is carried out on the specific population under study and allele frequencies specific to that population are used in the analysis. With improvement in both speed and cost of genotyping and sequencing, the whole genome approach of identifying drug response related genes is a possibility. The major issue is the statistical power of the clinical samples to detect genetic association of drug responses at significance levels corrected for multiple testing (3,64). Advances in methodologies and software tools in statistical genetics to enhance both the power and the robustness of statistical analyses will therefore play a major role in the success of pharmacogenetic studies.

Discussion and Conclusions - Genetics and genomics are fast becoming integral parts of the drug discovery and development processes. In the discovery process, identification of disease susceptibility genes can deepen our understanding of the etiology and pathology of diseases and the pathways involved in the disease processes. That understanding can in turn provide useful targets. Knowledge of genetic polymorphisms in potential drug targets can also provide useful information for drug developers to choose the one with the least variation across individuals to start high-throughput screening for lead compounds. Genetic information applied to pharmacokinetic and pharmacodynamic studies in preclinical stages can help to understand toxicity and drug action pathways earlier on. While studying the polymorphisms of genes encoding a target may prove useful, gene expression profiling will prove to be very useful too. For example, the gene expression pattern from the liver of an animal dosed with a drug can give an indication as to whether and which gene pathways related to toxicity have been turned on. Further discussion of expression analysis by gene chips, arrays and other related technologies can be found in (65-67).

In clinical development, stratifying patients according to their genotypes can help to reduce the sample size required to achieve statistical significance in demonstrating efficacy. By identifying genes and genetic markers that are associated with efficacy and adverse effect response, diagnostic tests can be devised to tailor medicine and treatment to an individual's genetic makeup and therefore help to improve the potency and safety of the drugs. Personalized medicine can and will become a reality as a result of pharmacogenetic studies.

Reference

1. W.W. Weber in "Pharmacologenetics", Oxford Press, New York, N.Y., 1997, p 125.
2. J. Lazarou, B.H. Pomeranz and P.N. Corey, JAMA, 279; 1200 (1998).
3. L.R. Cardon, R.M. Idury, T.J.R. Harris, J.S. Witte and R.C. Elston, Pharmacogenetics, in press (2000).
4. F.S. Collins, N. Engl. J. Med., 341, 28 (1999).
5. N.J. Schork and A.B. Weder, Alzheimer Disease and Associated Disorders, 10, 22 (1996).
6. D. Housman and F.D. Ledley, Nature Biotechnology, 16, 492 (1998).
7. P. Propping, Physiol. Biochem. Pharmacol., 83, 123 (1978).
8. J.M. Drazen, C.N. Yandava, L. Dube, N. Szczerback, R. Hippensteel, A. Pillari, E. Israel, N. Schork, E.S. Silverman, D.A. Katz, J. Drajesk, Nature Genetics, 22, 168 (1999).
9. E. Smeraldi, R. Zanardi, F. Benedetti, D. Di Bella, J. Perez and M. Catalan, Molecular Psychiatry, 3, 508 (1998).
10. M. Bruss, H. Bonisch, M. Buhlen, M.M. Nothen, P. Propping and M. Gothert, Pharmacogenetics, 9, 95 (1999).
11. A. Cravchik, D.R. Sibley and P.V. Gejman, Pharmacogenetics, 9, 17 (1999).
12. V. Jovanovic, H. Guan and H. Van Tol, Pharmacogenetics, 9, 561 (1999).
13. L. O'Toole, M. Steward, P. Padfield, K. Channer, J. Cardiovasc. Pharmacol., 32, 998 (1998).
14. G.G.V. Essen et al., Lancet, 347, 94 (1996).
15. J. Poirier, M. Delisle, R. Quirion, I. Aubert, M. Farlow, D. Lahiri, S. Hui, P. Bertrand, J. Nalbantoglu, B.M. Gilix, and S. Gauthier, Neurobiology, 92, 12260 (1995).
16. F. Richard, N. Helbecque, E. Neuman, D. Guez, R. Levy, P. Amouyel, Lancet, 349, 539 (1997).
17. J.A. Kuivenhoven, J.W. Jukema, A.H. Zwinderman, P. Kniff, R. McPherson, A.V.G. Bruschke, K.I. Lie, J.J.P. Kastelein, New Eng. J. Med., 338, 86 (1998).
18. D.W. Nebert, AM. J. Hum. Genet., 60, 265 (1997).
19. W.E. Evans and M.V. Relling, Science, 286, 487 (1999).
20. N.W. Linder, R.A. Prough and R. Valdes, Clin. Chem., 43, 254 (1997).
21. T.R. Rebbeck, J.M. Jaffe, A.H. Walker, A.J. Wein, S.B. Malkowicz, J. Natl. Cancer Inst., 90, 1225 (1998).
22. C.A. Felix, A.H. Walker, T.M. Williams, N.J. Winick, N.K. Cheung, B.D. Lovett, P.C. Nowell, I.A. Blair, T.R. Rebbeck, Proc. Natl. Acad. Sci. U.S.A., 95, 13176 (1998).
23. C. Sachse, J. Brockmoller, S. Bauer and I. Roots, AM. J. Hum. Genet., 60, 284 (1997).
24. C. Nordin, M. Dahl, M. Eriksoon, S. Sjoberg, Lancet, 350, 29 (1997).
25. B. Bachus, U. Bickel, T. Thomsen, I. Roots and H. Kewitz, Pharmacogenetics, 9, 661 (1999).

26. E. Lessard, B.A. Hamelin, L. Labbe, G. O'hara, P.M. Belanger and J. Turgeon, Pharmacogenetics, 9, 683 (1999).

27. M. Wadelius, J.L. Autrup, M.J. Stubbins, S. Andersson, J. Johansson, C. Wedelius, C.R. Wolf, H. Autrup, and A. Rane, Pharmacogenetics, 9, 333 (1999).

28. K. Dilger, B. Greiner, M.F. Fromm, U. Hofmann, H.K. Kroemer and M. Eichelbaum, Pharmacogenetics, 9, 551 (1999).

29. E. Lessard, M. Yessine, B.A. Hamelin, G. O'Hara, J. Leblanc and J. Turgeon, Pharmacogenetics, 9, 435 (1999).

30. M. Chida, T. Yokoi, N. Nemoto, M. Inaba, M. Kinoshita and T. Kamataki, Pharmacogenetics, 9, 287 (1999).

31. M.L. Bernal, B. Sinues, I. Johansson, R.A. Mclellan, A. Wennerholm, M. Dahl, M. Ingelman-Sundberg and L. Bertilsson, Pharmacogenetics, 9, 657 (1999).

32. H.J. Gill, J.F. Tjia, N.R. Kitteringham, M. Pirmohamed, D.J. Back and B.K. Park, Pharmacogenetics, 9, 43 (1999).

33. R.S. Kidd, A.B. Straughn, M.C. Meyer, J. Blaisdell, J.A. Goldstein, and J.T. Dalton, Pharmacogenetics, 9, 71 (1999).

34. E.Y. Krynetski and W.E. Evans, AM. J. Hum. Genet., 63, 11 (1998).

35. S. Alves, M.J. Prata, F. Ferriera and A. Amorim, Pharmacogenetics, 9, 257 (1999).

36. H. Tai, M.Y. Fessing, E.J. Bonten, Y. Yanishevsky, A. D'Azzo, E.Y. Krynetski and W.E. Evans, Pharmacogenetics, 9, 641 (1999).

37. D. Siegel, S.M. McGuinness, S.L. Winski and D. Ross, Pharmacogenetics, 9, 113 (1999).

38. L. Rocha, C. Garcia, A. De Mendonca, J. Pedro Fil, D.T. Bishop and M.C. Lechner, Pharmacogenetics, 9, 9 (1999).

39. N. Jourenkova-Mioronova, H. Wikman, C. Bouchardy, K. Mitrunen, P. Dayer, S. Benamou and A. Hirnoven, Pharmacogenetics, 9, 533 (1999).

40. D. Hall, G. Ybazeta, G. Destro-Bisol, M.L. Petzl-Erler, and A. Di Rienzo, Pharmacogenetics, 9, 591 (1999).

41. M. Eichelbaum, L.F. Jorge, E. Griese, T. Inaba and T.D. Arias, Pharmacogenetics, 9, 217 (1999).

42. H.L. Mcleod, S.C. Pritchard, J. Githang'a, A. Indalo, M. Arneyaw, R.H. Powrie, L. Booth and E.S.R. Collie Duguid, Pharmacogenetics, 9, 773 (1999).

43. H. Xie, C.M. Stein, R.B. Kim, G.R. Wilkinson, D.A. Flockhart and A.J.J. Wood, Pharmacogenetics, 9, 539 (1999).

44. A. Kaneko, J. Kojilum, J. Yaviong, N. Takahashi, T. Ishikazi, L. Bertilsson, T. Kobayakawa and A. Bjorkman, Pharmacogenetics, 9, 581 (1999).

45. H. Xie, C.M. Stein, R.B. Kim, Z. Xiao, N. he, J. Zhou, J.V. Gainer, N.J. Brown, J.L. Haines and A. Wood, Pharmacogenetics, 9, 511 (1999).

46. R. Elston, Genet. Epidemiol., 15; 565 (1998).

47. L. Kruglyak, Nature Genetics, 22, 139 (1999).

48. R. Risch and K. Merikangas, Science, 273, 1516 (1996).

49. N. Camp, Am. J. Hum. Genet., 61, 1424 (1997).

50. R.S. Ajioka, L.B. Jorde, J.R. Gruen, P. Yu, D. Dimitrova, J. Barrow, E. Radisky, C.Q. Edwards, L.M. Griffen and J.P. Kushner, Am. J. Hum. Genet. 60, 1439 (1997).

51. A.G. Clark, K.M. Weiss, D.A. Nickerson, S.L. Tatlor, A. Buchanan, J. Stengard, V. Salomaa, E. Vartianinen, M. Perola, E. Boerwinkle, and C.F. Sing, Am. J. Hum. Genet., 63, 595 (1998).

52. U. Landegren, M. Nilsson, and P.Y. Kwok, Genome Res., 8, 769 (1998).

53. K.J. Livak, J. Marmaro, and J.A. Todd, Nat. Genet., 9, 341 (1995).

54. V.O. Tobe, S.L. Taylor, and D.A. Nickerson, Nucleic Acids Res., 24, 3728 (1996).

55. X. Chen, K.J. Livak and P.Y. Kwok, Genome Res., 8, 549 (1998).

56. X. Chen and P.Y. Kwok, Nucleic Acids Res., 25, 347 (1997).

57. T. Pastinen, A. Kurg, A. Metspalu, L. Peltonen and A.C. Syvanen, Genome Res., 7, 606 (1997).

58. P. Ross, L. Hall, I. Smirnov and L. Haff, Nat. Biotechnology, 16, 1347 (1998).

59. D. Whitcombe, J. Theaker, S.P. Guy, T. Brown and S. Little, Nat. Biotechnology, 17, 804 (1999).

60. V. Lyamichev, A.L. mast, J.G. Hall, J.R. Prudent, N.W. Kaiser, T. Takoya, R.W. Kwiatkowski, T.J. Sander, M. de Arruda, D.A. Arco, Nat. Biotechnology, 17, 292 (1999).

61. D. Ryan, B. Nuccie and D. Arvan, Mol. Diagnostics, 4, 135 (1999).

62. C.A. Mein, B.J. Barratt, M.G. Dunn, T. Siegmund, A.N. Smith, L. Esposito, S. Nutland, H.E. Stevens, A.J. Wilson, M.S. Philips N. Jarvis, .S Law, M. de Arruda and J. Todd, Genome Res., 10, 330 (2000).

63. C. Fischer, S. Schweigert, C. Spreckelsen, F. Voget, Hum. Genet. 97, 129 (1996).

64. R.C. Elston, R.M. Idury, L.R. Cardon and J.B. Lichter, Statistics in Medicine, <u>18</u>, 741 (1999).
65. T.L. Farea and P.O. Brown, Curr. Opion. Genet. Dev., <u>9</u>, 715 (1999).
66. D.J. Duggan, M. Bittner, et al., Nat. Genet., <u>21</u> (1 Suppl), 10-4 (1999).
67. E.S. Lander, Nat. Genet., <u>21</u> (1 Suppl), 3-4 (1999).

Chapter 24. Oligomerisation of G Protein-Coupled Receptors

Graeme Milligan and *Stephen Rees

University of Glasgow, Scotland,
Glaxo Wellcome Research and Development, Stevenage, UK

Introduction- Protein dimerisation is a widely employed biological regulatory mechanism, functioning in systems as wide ranging as transcriptional control and bacterial virulence (1). It is well appreciated to be a key step in the functional activation of many, if not all, single transmembrane element growth factor receptors which possess intrinsic tyrosine kinase activity. Furthermore, many of the immunoglobulin receptors of the immune system require functional interactions between multiple polypeptides to produce the receptor complexes which provide high affinity ligand recognition and functional response (2-3). In retrospect it may thus seem surprising that until relatively recently G-protein coupled receptor (GPCR) homo and hetero-dimerisation was not more clearly recognised as a likely mechanism which may contribute to their function.

GPCRs are likely to represent the single largest gene family in the human genome. Cellular responses to a vast range of stimuli are mediated through the activation of this class of receptor. Ligands that are capable of activating GPCRs include peptide and small molecule hormones, neurotransmitters, odourants, divalent cations and light. Due to the characterisation and understanding of the pharmacology of GPCR ligands, and the fact that GPCRs are cell surface proteins, GPCRs have proven to be the predominant class of targets for clinically prescribed drugs (4). With the huge increase in the number of cloned GPCR cDNA's it seems reasonable to conclude that GPCRs will continue to represent a major class of targets for drug discovery.

GPCRs are single polypeptides characterised by the possession of seven transmembrane spanning elements with extracellular N-terminal and intracellular C-terminal domains. Ligand binding to the receptor results in a conformational change in receptor structure to facilitate receptor interaction with a member of the family of heterotrimeric G proteins. G-protein activation by the receptor results in the activation of one or more of a plethora of intracellular signal transduction cascades. This in turn leads to a change in the activity of ion channels and enzymes to cause an alteration in the rate of production of intracellular second messengers such as cAMP and calcium (5-8). Until recently, it was believed that all of the structural and functional requirements of a GPCR were encoded within the single polypeptide chain which was presented as a single entity at the cell surface. However, recent studies have revealed that many GPCRs are capable of both homo- and heterodimerisation, and that dimer formation can be critical to GPCR pharmacology. Such studies have demonstrated that dimerisation may be required to facilitate agonist binding and that receptor monomerisation contributes to the desensitisation process to cause an attentaution of receptor signalling. These studies have identified specific examples in which chaperonin-type proteins are required for the appropriate folding and delivery of the GPCR to the plasma membrane. The RAMP proteins (receptor-activity modulating protein) are a family of single transmembrane domain proteins that facilitate the transport of some GPCRs to the cell surface. A single GPCR polypeptide has been demonstrated to exhibit distinct pharmacologies according to the identity of the RAMP protein involved in trafficking the receptor to the cell surface (9). A further example of chaperonin-like activity includes an example in which the transport of a single GPCR to the cell surface is enabled by hetero-dimerisation with a second, distinct, GPCR gene product. The co-immunoprecipitation of differentially epitope-tagged forms of the same (homo-dimerisation) and closely related (hetero-dimerisation) GPCRs has also indicated the possibility of differential pharmacology resulting when two GPCRs with similar ligands are co-expressed in a single cell (10).

Oligomerisation of Class I GPCRs- A range of studies, using approaches as diverse as classical ligand binding studies and target size radiation inactivation had provided data, of varying degrees of quality, that supported the presentation of oliogomeric arrays of GPCRs in membranes and intact cells. However, it was only the results of relatively recent experiments, exploring the folding of different GPCR domains, which provided clear evidence for the capacity for oligomerisation of GPCRs (11). The capacity to recover ligand binding and function following the co-expression in mammalian cells of two domains of a single GPCR, separated within the third intracellular loop of the receptor, and thus comprising transmembrane regions I-V and VI-VII, indicated that these two regions may fold and interact as two independent elements. Such studies allowed the direct analysis of the possibility of GPCR oligomerisation. Two key experimental approaches were used. In the first, chimeric GPCRs were constructed consisting of transmembrane regions I-V of the α_{2C}-adrenoceptor and transmembrane domains VI-VII of the M_3 muscarinic acetylcholine receptor and vice versa. Surprisingly coexpression of the chimeric GPCRs allowed the generation of functional receptors capable of responding to both the α_{2C}-adrenoceptor agonist and the muscarinic receptor agonist carbachol although neither chimera functioned when expressed alone (12). In the second, co-expression of two different non-functional mutants of the angiotensin II receptor resulted in the production of high affinity binding sites for this peptide ligand although neither mutant could bind ligand effectively when expressed individually (13). Such studies were best rationalised as reflecting the capacity for intermolecular interactions between the chimeric proteins to form intact GPCRs capable of responding to their natural ligand. Subsequent studies have examined the role of elements of the third intracellular loop in these interactions (14). Such studies have shown that equivalent results can be obtained with a variety of GPCRs including the prototypic family member, rhodopsin (15) and the V_2 vasopressin receptor (16-17), many distinct mutations of which produce the phenotype of diabetes insipidus. Recent evidence has built on such work to indicate that closely related GPCRs may produce hetero-dimers with pharmacological properties distinct from either parental GPCR. For example, by taking advantage of markedly different binding affinities of a chimeric M_2/M_3 muscarinic receptor and the wild type M_3 muscarinic receptor for the antagonist tripitramine (18-19) strong evidence of the capacity to form hetero-oligomeric M_2/M_3 receptors in cell membranes has been produced (20). Similar inferences have resulted from the generation of a distinct pharmacology following co-expression in mammalian cells of the κ and δ opioid receptors than observed when either of these GPCRs are expressed in isolation (21).

Although such examples provide excellent pharmacological evidence that hetero-oligomerisation can be induced in transfected cell systems, a major challenge will now be to attempt to reconcile such results with the pharmacology of native cells and tissues. Indeed the pharmacology recorded following co-expression the κ and δ opioid receptors did not appear to mimic any of the well appreciated physiologically relevant opioid receptor pharmacologies which have not been reconciled with the isolated expression of known cloned opioid receptor subtypes (21). Clearly, the potential to form hetero-oligomers requires the co-expression of related GPCRs in individual cells and the extent of such interactions will presumably depend on the levels of expression of each individual GPCR, and possible compartmentalisation of GPCR expression within that cell. Thus, pharmacological profiles in individual cells or tissues may show very subtle variation according to the type, and level of expression, of receptors in that cell or tissue. This in turn may have implications on the design of selective therapeutics. It is possible that in vivo pharmacology may alter in different cell types according to the range of receptors expressed in that cell, and this may alter in a disease state.

Evidence for GPCR oliogomerisation- Many of the key recent advances in demonstrating oligomerisation of class I GPCRs have derived from the expression of epitope-tagged forms of the receptors which allow their subsequent immunodetection. Following expression of either c-myc or HA-tagged forms of the β_2-adrenoceptor in insect Sf9 cells, detergent solubilisation of membranes and SDS-PAGE immunoblotting with appropriate antisera confirmed both the expression of the receptor as a polypeptide of some 40kDa and the presence of immunoreactivity consistent with a dimer of the GPCR (22-23). Addition of a cell permeable cross-linker prior to membrane preparation increased the proportion of the GPCR migrating as the dimer. As a peptide corresponding to a sequence of transmembrane region VI of the β_2-adrenoceptor, but not an equivalent peptide from the D_2 dopamine receptor, both limited dimerisation and inhibited agonist-mediated stimulation of adenylyl cyclase activity it appears that dimerisation/ologimerisation might either be required for, or enhance, function of this GPCR. Most importantly, however, co-expression of the c-myc and HA-tagged forms of the β_2-adrenoceptor allowed direct detection of interactions between these polypeptides by immunoprecipitation with one antibody followed by immunoblotting these fractions with the alternate antibody (10, 22). Such studies are widely applied in many areas to provide evidence of high affinity protein-protein interactions.

Similar studies have demonstrated the capacity of a range of GPCRs, including the V_2 vasopressin receptor, the δ and κ opioid receptors, the H_2 histamine receptor, the D_3 dopamine receptor, the M_3 muscarinic acetylcholine receptor, and the chemokine CCR2b and CXCR4 receptors to dimerise (22-28). Furthermore, in many of these studies immunoreactivity which might correspond to GPCR trimers, tetramers and other high order complexes have also been observed. Encouragingly, the capacity to observe oligomeric GPCR complexes has not been restricted only to cell transfection studies. For example, potential oligomers of the D_3 dopamine receptor have been immunoprecipitated from brain and of the CCR2 receptor from the human monocytic cell line Mono Mac 1 (25, 27). Moreover, dimers of the M_3 muscarinic receptor have also been immunodetected in rat brain membranes (26).

The same approaches have also been used to demonstrate hetero-oligomerisation between co-expressed κ and δ opioid receptors (21). It is clearly important in such studies that there is a lack of hetero-oligomerisation between certain GPCR pairs to provide appropriate negative controls. However, perhaps surprisingly given the sequence similarity between the μ and κ opioid receptors, a lack of capacity of co-expressed μ and κ opioid receptors to form hetero-oligomers has been reported (21). It should be noted however, that the M_3 muscarinic receptor does not appear to form hetero-oligomers with either the M_1 or M_2 muscarinic receptor subtypes in such co-immunoprecipitation studies (27). As these results are distinctly different from those obtained using M_2/M_3 muscarinic receptor chimeras it is clear that the subject of class I GPCR hetero-oligomerisation is in its infancy and that a great deal of work will be required to elucidate clear physiological relevance and thus its importance for therapeutic ligand design (20).

Ligand regulation of GPCR oligomerisation- If oligomerisation is a key feature in the function of class I GPCRs then it might be anticipated that receptor ligands would regulate the presence and extent of such oligomers. In a number of examples, but not all, this does appear to be the case.   The proportion of detected dimeric β_2-adrenoceptor following immunoprecipitation from Sf9 cell membranes is increased by addition of the β-adrenoceptor agonist isoprenaline (22). Furthermore, as the β_2-adrenoceptor possesses a degree of agonist-independent constitutive activity the β-blocker timolol, which acted as an effective inverse agonist for this GPCR in the Sf9 system, decreased the proportion of the receptor which migrated as the dimer (29, 22). Such results are entirely consistent with the concept that oligomerisation of this GPCR is required for function as are data which indicate that antibodies directed against extracellular loop elements of this receptor can have agonist function (30). However, although ligand enhanced dimerisation has also been observed for both the CCR2 and CXCR4 receptors, the cholinergic agonist carbachol has been reported to be without effect on the monomer/dimer proportions of the M_3 muscarinic receptor when expressed in COS-7 cells and the κ opioid receptor appears to exist predominantly, if not exclusively, as a dimer whether cells have been pretreated with agonist or not (27, 28, 26, 21). Perhaps most interestingly however, efficacious agonists, whether opioid peptides or alkaloids have been reported to diminish the proportion of δ opioid receptor dimers whereas both the partial agonist morphine and the classical opioid antagonist naloxone were without effect (23).

Although such results suggest that agonist regulation of class I GPCR oligomerisation and arrays may vary between receptors this is not inherently an intellectually attractive scenario. However, variation in regulation of receptor oligomerisation with agonist efficacy might provide novel insight into the mechanistic basis of this conundrum. It is also noteworthy that a photoaffinity variant of the dopaminergic antagonist spiperone has been reported to bind only to D_2 dopamine receptor monomer whereas a photoaffinity variant of nemonapride labels both monomer and dimer (31). This has been suggested to account for the greater B_{max} of [^{3}H]nemonapride compared to [^{3}H]spiperone in membranes of Sf9 cells expressing the human D_2 dopamine receptor and suggests that certain ligands may be able to discriminate between different arrays of the same GPCR (31).

Although beyond the scope of this article it is interesting to note that "dimer" or bivalent forms of a number of GPCR agonist ligands frequently have both higher affinity and higher efficacy than the equivalent monomer ligand (32). It is too early, however, to establish if such observations are directly relevant to GPCR oligomerisation.

Mechanisms of GPCR oligomerisation- Peptides derived from the transmembrane VI and VII regions of the D_2 dopamine receptor and from transmembrane region VI of the β$_2$-adrenoceptor have been shown to limit dimer formation (22,31). These effects are specific as such peptides are unable to prevent dimerisation of more distantly related receptors. Furthermore, at least in the case of the β$_2$-adrenoceptor the transmembrane VI peptide is also able to inhibit receptor (but not forskolin or direct G protein) stimulation of adenylyl cyclase. Such observations are consistent with dimer formation being dependent upon appropriate interfaces forming between transmembrane elements. However, this does not appear to the the only potential mechanism. As the majority of type I GPCRs have a highly conserved pair of cysteine residues on extracellular loops II and III it is noteworthy that mutagenesis of these residues in the M_3 muscarinic acetylcholine receptor prevents dimer formation and that treatment of rat brain membranes with dithiothreitol also prevents the detection of dimers. Furthermore, as truncated forms of certain receptors such as the V2 vasopressin receptor can act as "dominant negatives' for cell surface expression and function of the full length receptor and certain mutated, non-functional, receptors can be "rescued" by co-expression of receptor fragments there is clear evidence for receptor-receptor interactions prior to membrane insertion (33, 17). There is growing evidence from other GPCR systems for roles of both other GPCR and unrelated gene products providing a chaperonin role in appropriate membrane delivery of GPCRs and this will be discussed below.

<u>Oligomerisation of Class II GPCRs</u> - As yet there no evidence has emerged for the homo or heterodimerisation of Class II GPCRs. However one of the most surprising observations within the field of molecular pharmacology in recent years has been the identification of a family of three single transmembrane domain proteins, designated RAMP 1-3 (Receptor Activity Modulatory Protein). These proteins oligomerise with members of the calcitonin family of GPCRs to facilitate transport to the cell surface and to cause a modulation of the pharmacology of the coexpressed GPCR (34). The calcitonin family of peptides comprises five members, calcitonin, CGRP1, CGRP2, amylin and adrenomedullin. The cloning of specific receptor cDNA's for these peptides has proven to be very difficult. The calcitonin receptor was first cloned in 1991. This was followed by the cloning of a second receptor, the Calcitonin Receptor Like Receptor (CRLR), with 55% homology to the calcitonin receptor. However attempts to identify any of the calcitonin family of peptides, including CGRP, as candidate ligands for the CRLR proved unsuccessful. A further attempt to functionally clone the receptor for CGRP (Calcitonin Gene Related Peptide) resulted in the identification and characterisation of the RAMP proteins (35).

The SK-N-MC neuroblastoma cell line expresses a well characterised CGRP receptor. Hence in 1998, McLatchie et al., set out to clone the CGRP receptor from this cell line by expression cloning in Xenopus oocytes (35). In these studies the authors were looking for an enhancement of the endogenous oocyte CGRP response by cDNAs prepared from SK-N-MC cells. In this experiment a 148 amino acid single transmembrane domain protein was identified which was able to enhance the oocyte response to CGRP. This protein was designated RAMP1 (Receptor Activity Modulating Protein) (35). When coexpressed either in Xenopus oocytes or HEK 293 cells together with the CRLR, RAMP1 was found to functionally reconstitute the CGRP receptor, however expression of either protein alone did not permit CGRP reponses (35). Database searches revealed the existance of a family of three RAMP proteins (RAMP1-3) with approximately 30% identity. In recombinant cell lines expression of RAMP1 with the CRLR leads to the expression of a CGRP receptor, however coexpression of the CRLR with either RAMP2 or RAMP3 leads to the reconstitution of two pharmacologically distinct adrenomedullin receptors (35, 36). This was the first indication that coexpression of a GPCR with different accessory proteins could lead to the generation of receptors for different ligands.

RAMP proteins were also found to be capable of modulating the pharmacology of the calcitonin receptor to generate a receptor for amylin (37, 38). The calcitonin receptor exists as multiple splice varient. Cotransfection of either RAMP1 or RAMP3 with a splice varient of the human calcitonin receptor lacking a 16 amino acid domain within the first intracellular loop of the receptor, generates a receptor capable of amylin binding. This effect was not seen following cotransfection with RAMP2 (38). Binding studies revealed that the amylin receptor formed following cotransfection of RAMP1 with the calcitonin receptor is prahmacologically distinct from the receptor generated following cotransfection with RAMP3 (38). The receptor generated following cotransfection with RAMP1 has a rank order of ligand binding of salmon calcitonin>rat amylin=human CGRP>human calcitonin. In contrast the receptor generated following cotransfection with RAMP3 has a markedly reduced affinity for human CGRP (38).

How do RAMP proteins function? In the absence of RAMP1 the CRLR is expressed as an immature glycoprotein on internal membranes of transfected cells (35). Following coexpression of RAMP1, a mature glycoprotein is formed and the receptor is trafficked to the cell surface where RAMP1/CRLR complexes can be isolated by immunoprecipitation. In the presence of RAMP2 a core glycosylated protein is formed which is transported to the cell surface where it is capable of binding adrenomedullin (35). Whether the change in glycosylation state is responsible for the observed changes in pharmacology has not been proven, similarly it is not known if the RAMP protein contributes to the ligand binding site. It may be that the RAMP effect on GPCR pharmacology is simply due to a conformational alteration in the structure of the GPCR following the interaction with the RAMP protein. The precise mechanism of action of RAMP proteins, together with the effect of RAMP proteins on the function of other GPCRs, is the subject of ongoing investigation. However it is clear that RAMP proteins are able to modulate the pharmacology of a coexpressed GPCR to generate receptor complexes with distinct pharmacologies. As many cells express multiple RAMP proteins this may permit the rapid switching of responses within a particular cell type by simply modulating the function or level of expression of these proteins.

<u>Oligomerisation of Class III GPCRs</u> - Increasing evidence exists for the homodimerisation of a number of Class III GPCRs. Class III GPCRs include the family of metabotropoic glutamate receptors (mGluRs), the calcium sensing receptor and the GABAB receptor. These receptors share a common structural feature in that they possess an 800 amino acid extracellular domain which contains the ligand binding site. Receptor dimerisation may be fundamental to the function of this class of GPCR. There is evidence for homodimerisation of mGluR1a, mGluR2, mGluR3, mGluR4 and mGluR5 and for the calcium sensing receptor (39, 40). Homodimerisation within these receptors is mediated by an intermolecular disulphide bridge within the C-terminal region of the extracellular domain. There is some evidence emerging that monomerisation of these receptors results in a loss of responsiveness to agonist.

However, in recent times it has been a detailed analysis of the expression and function of the GABAB receptor which has perhaps provided the greatest insights into GPCR dimerisation (41). The cloning of the first GABAB receptor was reported in 1997 (42). The receptor was cDNA was identified by expression cloning using the high affinity antagonist ligand [^{125}I]CGP64213 and was designated GABA$_B$R1a. This was accompanied by the identification of an N-terminal splice variant which was designated GABA$_B$R1b. While the cloned GABAB receptor exhibited many of the expected propreties of the endogenous GABAB receptor in terms of tissue distribution and antagonist ligand binding, when expressed in recombinant systems it bound agonist ligands poorly and did not couple to the expected signal transduction cascades (42).

Immunocytochemistry, glycosylation studies and Fluoresceence Activated cell Sorting (FACS) studies on transfected mammalian cells demonstrated that the cloned receptor was expressed as an immature glycoprotein and expression was restricted to intracellular membranes (43). These observations led a number of groups to attempt to clone a "factor" that would permit transport of the transfected receptor to the cell surface. In 1998, no less than six groups reported the identification of a second receptor, the GABA$_B$R2 receptor, which when coexpresed with the GABA$_B$R1receptor in mammalian cells or Xenopus oocytes, allowed the reconstitution of a pharmacologically authentic GABAB receptor (43-48). Like GABA$_B$R1 the GABA$_B$R2 receptor is a family III GPCR sharing 35% sequence homology with GABA$_B$R1. Efforts to identify a ligand for the GABA$_B$R2 receptor had previously met with failure. These reports provided several lines of evidence to indicate that the functional form of the GABAB receptor consists of a heterodimer of a GABA$_B$R1 splice varient and GABA$_B$R2. First, immunocytochemistry studoes indicated that the expression of the two receptors is colocalised at the cell surface in transfected cells, and tissue distribution studies in the rat brain demonstrated that both receptors are expressed in the same neurones (44, 48). Second, when coexpressed in mammalian cells with the GABA$_B$R2 receptor the GABA$_B$R1b is expressed as a mature glycoprotein and transported to the cell surface as determined by FACS (43). This is accompanied by an increase in agonist potency in radioligand binding studies and the receptor now becomes able to regulate signal transduction pathways including the stimulation of [^{35}S]GTPγS binding and the inhibition of forskolin stimulated cAMP (43). Third, when a human brain cDNA library was screened using the yeast two hybrid system with the C-terminal domain of the GABA$_B$R1 receptor as bait the GABA$_B$R2 receptor was identified as the major interacting protein (43, 45). The interaction between the two receptors has been localised to an interaction between so-called coiled-coiled domains found within the intracellular C-terminal tails of both receptors. Colied-coiled domains are present in many proteins and have been well characterised as mediators of protein-protein interaction. Interestingly, similar studies have failed to reveal any evidence for the formation of homodimers between either GABA$_B$R1 or GABA$_B$R2. Fourth, co-immunoprecipitation experiments in transfected cells and brain tissue have shown that either receptor can be co-immunoprecipitated with the other, thus providing the direct physical evidence for an interaction between the two receptors (44, 46, 48).

These studies demonstrated that the functional GABAB receptor consists of a heterodimer between one of a number of splice variants of the GABA$_B$R1 receptor and the GABA$_B$R2 receptor. The search for further partners of either the GABA$_B$R1 or the GABA$_B$R2 receptor, partly to account for pharmacologically defined GABAB sub-types is ongoing. Within the heterodimer the GABA binding site is localised to the GABA$_B$R1 receptor. Whether the GABA$_B$R2 receptor is a functional GPCR in it's own right remains unclear. While this receptor is capable of reaching the cell surface in the absence of coexpressed GABA$_B$R1 there are no reports of the identification of a ligand specific for this receptor. However there is some evidence from recombinant systems that GABA may be able to regulate the GABA$_B$R2 receptor alone. For example GABA$_B$R2 will activate potasium channels when expressed in HEK 293/T cells (44, 45). However it fails to activate [^{35}S]GTPγS binding in this cell line (43, 46). The functional consequence of GABAB receptor hetero-dimerisation remain unclear. Unlike the other examples of GPCR dimerisation there is as yet no evidence of ligand regulation of GABAB receptor hetero-dimers.

What is the purpose of GPCR oligomerisation? - The role of GPCR oligomerisation is the subject of much investigation and is likely to vary amongst different GPCR classes. However it is already clear that GPCR dimerisation may be required to permit agonist binding and monomerisation may play a role in receptor desensitisation. The ratio of receptor monomer to dimer displayed within a cell may permit a rapid change in the sensitivity of the cell to external stimuli without the requirement for *de novo* protein synthesis or modification. It is also clear that receptor oligomerisation can permit the generation of receptor complexes with distinct pharmacologies. The consequence of this observation is that different cells may be able to very sensitively modulate their response to ligand, at different times, according to the ratio of receptors, or indeed trafficking factors such as the RAMP proteins, expressed within that cell. This observation may account for the existance of pharmacologically defined receptor sub-types that have not been reconstituted in recombinant cell lines following the cloning of receptor cDNA's. These observations may also have important consequences for drug discovery. Many drug discovery programmes rely upon the recombinant expression of GPCRs of interest, and the use of cell lines expressing such receptors, for the identification of molecules capable of modulating receptor activity. If, in the target tissue, the receptor exists as a homo-or hetero-dimer with a subtly different pharmacology then compounds identified in the laboratory may not have the desired profile *in vivo*. The existence of GPCR homo-and heterodoimers has been one of the biggest surprises within the field of molecular pharmacology. Understanding the biological relevance of the existence of such dimers will be on of the major challenges of the coming years.

References

1. H.J.Snijder, I. Ubarretxena-Belandia, M .Blaauw, K.H. Kalk, H.M. Verheij, M.R.Egmond, N. Dekker, B.W. Dijkstra, Nature 401, 717 (1999).
2. C. Chothia and E.Y. Jones, Annu. Rev. Biochem., 66, 823 (1997).
3. S.C. Garman, J-P. Kinet and T.S. Jardetzky, Annu. Rev. Immunol., 17, 973 (1999).
4. S. Wilson, D.J. Bergsma, J.K. Chambers, A.I. Muir, K.G. Fantom, C. Ellis, P.R. Murdock, N.C. Herrity, J.M. Stadel, Br. J. Pharmacol., 125, 1387 (1998).
5. J. Wess, Pharmacol. Ther., 80, 231 (1998).
6. J. Bockaert and J.P. Pin, EMBO. J., 18, 1723 (1999).
7. H. LeVine, 3rd Mol. Neurobiol., 19, 111 (1999).
8. T. Schoneberg, G. Schultz, T. Gudermann, Mol. Cell. Endocrinol. 151, 181 (1999).
9. H. Mohler and J-M. Fritschy, Trends Pharmacol. Sci. 20, 87(1999).
10. T.E. Hebert and M. Bouvier, Biochem. Cell. Biol., 76, 1 (1998).
11. N.P. Martin, L. Leavitt, C.M. Sommers and M.E. Dumont, Biochemistry, 38, 682 (1999).
12. R. Maggio, Z. Vogel and J. Wess, Proc. Natl. Acad. Sci. U S A., 90, 3103 (1993).
13. C. Monnot, C. Bihoreau, C. Conchon, K.M. Kurnow, P. Corvol and E. Clauser, J. Biol. Chem. 271, 1507 (1996).
14. R. Maggio, P. Barbier, F. Fornai, and G.U. Corsini, J. Biol. Chem. 271, 31055 (1996).
15. K.D. Ridge, S.S.J. Lee, and L.L. Yao, Proc. Natl. Acad. Sci. USA., 92, 3204 (1995).
16. T. Schoneberg, J. Liu, and J. Wess, J. Biol. Chem., Jul 28, 270, 18000 (1995).
17. T. Schoneberg, J. Yun, D. Wenkert, and J. Wess, EMBO. J. 15, 1283 (1996).
18. R. Maggio, P. Barbier, M.L. Bolognesi, A. Minarini, D. Tedeschi, and C. Melchiorre, Eur. J. Pharmacol., 268, 459 (1994).
19. P. Barbier, A. Colelli, M.L. Bolognesi, A. Minarini, V. Tumiatti, G.U. Corsini, C. Melchiorre, R. Maggio, Eur. J. Pharmacol., 355, 267 (1998).
20. R. Maggio, P. Barbier, A. Colelli, F. Salvadori, G. Demontis, and G.U. Corsini, J. Pharmacol. Exp. Ther., 291, 251 (1999).
21. B.A.Jordan, and L.A. Devi, Nature, 399, 697 (1999).
22. T.E. Hebert, S. Moffett, J-P. Morello, T.P. Loisel, D.G. Bichet, C. Barret, and M. Bouvier, J. Biol. Chem., 271, 16384(1996).
23. S. Cjevic, and L.A. Devi, J. Biol. Chem., 272, 26959 (1997).
24. Y. Fukushima, T. Asano, T. Saitoh, M. Anai, M. Funaki, T. Ogihara, H. Katagiri, N. Matsuhashi, Y. Yazaki, and K. Sugano, FEBS. Lett., 409, 283 (1997).
25. E.A. Nimchinsky, P.R. Hof, W.G.M Janssen, J.H. Morrison, C. Schmauss, J. Biol. Chem., 272, 29229 (1997).
26. F.-Y. Zeng and J. Wess, J. Biol. Chem., 274, 19487 (1999).
27. J.M. Rodriguez-Frade, A.J. Vila-Coro, A.M. de Ana, J.P. Albar, A.C. Martinez, and M. Mellado, Proc. Natl. Acad. Sci. U S A., 96, 3628 (1999).
28. A.J. Vila-Coro, J.M. Rodriguez-Frade, A. Martin De Ana, M.C. Moreno-Ortiz, A.C. Martinez, and M. Mellado, FASEB. J., 13,1699 (1999).
29. R. Leurs, M.J. Smit, A.E. Alewijnse and H. Timmerman, Trends Biochem. Sci., 23, 418 (1998).
30. D. Lebesgue, G. Wallukat, A. Mijares, C. Granier, J. Argibay, and J. Hoebeke, Eur. J. Pharmacol., 348, 123 (1998).
31. G.Y.K. Ng, B.F. O'Dowd, S.P. Lee, H.T. Chung, M.R. Brann, P. Seeman and S.R. George, Biochem. Biophys. Res. Comm., 227, 200 (1996).
32. P.J. Pauwels, D.S. Dupuis, M. Perez, and S. Halazy, Naunyn Schmiedebergs Arch. Pharmacol., 358, 404 (1998).
33. X. Zhu and J. Wess, Biochemistry 37, 15773 (1998).
34. S.M. Foord and Marshall, Trends Pharmacol. Sci., 20, 184 (1999).
35. L.M. McLatchie, N. J. Fraser, M.J. Main, A. Wise, J. Brown, N. Thompson, R. Solari, M.G. Lee, and S.M. Foord, Nature 393, 333 (1998).
36. N. Buhlmann, K. Leuthauser, R. Muff, J.A. Fischer, and W. Born, Endocrinology, 140, 2883 (1999).
37. R. Muff, N. Buhlmann, J.A. Fischer, and W. Born, Endocrinology, 140, 2924 (1999).
38. G. Christopoulos, K.J. Perry, M. Morfis, N. Tilakaratne, Y. Gao, N.J. Fraser, M. Main, S.M. Foord, and P.M. Sexton, Mol. Pharmacol., 56, 235 (1999).
39. C. Romano, W. Yang, and K. O'Malley, J. Biol. Chem.,271, 28612 (1996).
40. M. Bai, S. Trevedi and E.M. Brown, J. Biol. Chem., 273, 23605 (1998).
41. F. H. Marshall, K.A. Jones, K. Kaupmann and B. Bettler, Trends Pharmacol. Sci. 20, 396 (1999).
42. K. Kaupmann, K. Huggel, J. Heid, P.J. Flor, S. Bischoff, S.J. Mickel, G. McMaster, C. Angst, H. Bittiger, W. Forestl and B. Bettler, Nature, 386, 239 (1997).

43. J.H. Whrite, A. Wise, M.J. Main, A. Green, J. Fraser, G.H. Disney, A.A. Barnes, P. Emson, S.M. Foord, and F.H. Marshall, Nature, 396, 679 (1998).

44. K. Kaupmann, B. Malitscheck, V. Schuler, J. Heil, W. Froestl, P. Beck, J. Mosbacher, S. Bischoff, A. Kulik, R. Shigemoto, A. Karschin, and B. Bettlet. Nature 396, 683 (1998).

45. R. Kuner, G. Kohr, S. Grunewald, G. Eisenhardt, A. Bach, and H.C. Kornau. Science 283, 74 (1999).

46. G.Y. Ng, J. Clark, N. Coulombe, N. Ethier, T.E. Hebert, R. Sullivan, S. Kargman, A. Chateauneuf, N. Tsukamoto, T. McDonald, P. Whiting, E. Mezey, M.P. Johnson, Q. Liu, L.F. Jr. Kolakowski, J.F. Evans,T.I. Bonner, and G.P. O'Neill. J. Biol. Chem. 274, 7607 (1999).

47. S.C. Martin, S.J. Russek, and D.H. Farb. Mol. Cell. Neurosci., 13, 180 (1999).

48. K.A. Jones, B. Borowsky, J.A. Tamm, D.A. Craig, M.M. Durkin, M. Dai, W.-J. Yao, M. Johnson, C. Gunwaldsen, L.-Y. Huang, C. Tang, Q. Shen, J.A. Salon, K. Morse, T. Laz, K.E. Smith, D. Hagarathnam, S.A. Noble, T.A. Branchek, and C. Gerald. Nature 396, 674 (1998).

49. N.I. Tarasova, W.G. Rice, C.J. Michejda, Inhibition of G-protein-coupled receptor function by disruption of transmembrane domain interactions, J. Biol. Chem. 274, 34911 (1999).

50. Y. Djellas, K. Antonakis, G.C. Le Breton, A molecular mechanism for signaling between seven-transmembrane receptors: evidence for a redistribution of G proteins, Proc. Natl. Acad. Sci. U S A , 95, 10944 (1998).

Chapter 25. Immunomodulatory phosphorylcholine-containing proteins secreted by filarial nematodes

William Harnett
Department of Immunology, University of Strathclyde
Glasgow G4 0NR, United Kingdom

Introduction - Parasitic worms represent an enormous medical problem for humans. Current estimates as to the number of infections world-wide are greater than the global population and one species alone, *Ascaris lumbricoides*, infects more than a quarter of all humans. A characteristic of infection with these worms is its enduring nature. Infections with parasitic worms are in fact commonly life-long and indeed individual worms may survive for several decades. Such longevity is remarkable given that the worms contain a repertoire of highly immunogenic molecules and dictates that they must have evolved extremely effective mechanisms for evading host immunity. One wide-spread mechanism by which they do this involves actively modulating the host immune system such that it is rendered ineffective. It is generally accepted that an elucidation of the molecular mechanisms and associated targets underlying this "immunomodulation", will provide opportunities to develop novel strategies to eliminate these organisms. However, what is also now being realised, is that such information could additionally provide clues/insights into novel targets of imunomodulation for therapeutic purposes *per se*.

The filariids are arthropod-transmitted nematode worms which parasitise a wide range of vertebrates including man (1). As indicated above, infection in humans is long- term and the net effect of this is that hundreds of millions of people are currently infected with these parasites (2). The longevity of these worms is readily explained by the immunomodulation theory as "defects" in immune responsiveness have been revealed in infected individuals (3,4). Some discrepancy exists as to the exact nature of these defects but generally, they incorporate impairment of lymphocyte proliferation and bias in production of both cytokines, e.g., reduced production of interferon gamma (IFN-γ); increased synthesis of IL-10 (an anti-inflammatory cytokine) and immunoglobulin subclasses - greatly elevated IgG4 (an antibody of little value in eliminating pathogens due to an inability to activate complement or bind with high affinity to immune cells such as monocytes); decreases in other IgG subclasses. Overall therefore, the picture is of an immune response demonstrating a somewhat suppressed, anti-inflammatory phenotype which has recently been classified as "TH-2" ("TH" is derived from a category of T-lymphocyte referred to as "helper"). It has been speculated that such a phenotype is conducive to parasite survival.

What then is responsible for the induction and maintenance of this phenotype? Although current chemotherapy for treatment of filarial nematodes is inadequate with respect to certain species, drug treatment where successful, has often been shown to reverse the observed defects in immune responsiveness (5). This suggests that the worms may release immunomodulatory products and indeed several such molecules have been described (6,7). All filarial nematodes examined to date have been found to actively secrete proteins which contain a phosphorylcholine (PC) moiety (reviewed in 8). These molecules can be found in the bloodstream of individuals infected with filarial nematodes and were potentially of interest as there is some evidence in the literature to PC possessing immunomodulatory properties such as interference with antibody responses and driving immune responses in a TH-2 direction (9, 10, 11, 12). Arguably therefore, the PC moiety of filarial nematode molecules could be a significant contributor to the induction and maintenance of the immunological phenotype characteristic of infection. A number of recent observations strongly suggest that this is in fact the case.

IMMUNOMODULATORY PROPERTIES OF PC ON FILARIAL NEMATODE MOLECULES

Effects of PC on B- and T-lymphocyte proliferation - The first indicator that PC present on filarial molecules could interact with the immune system of the parasitised host was provided by Lal and colleagues (13). These workers showed that a soluble antigen extract prepared from adult *Brugia malayi* could suppress the normal proliferative response of human peripheral blood T lymphocytes to the mitogen, phytohemagglutinin (PHA) during co-culture and when they purified the PC-containing molecules from the extract, they obtained the same result. Moreover, replacing parasite molecules with PC chemically conjugated to bovine seum albumin (BSA) also resulted in suppression of proliferation. Subsequently Harnett and Harnett were able to show that PC on filarial molecules could also affect B cells (14). Specifically, they demonstrated that co-culture with ES-62, a PC-containing glycoprotein of the rodent filarial nematode *Acanthocheilonema viteae* inhibited the polyclonal activation of mouse small resting splenic B lymphocytes induced via the antigen receptor by anti-immunoglobulin (Ig) antibodies (15). Again it was observed that this result could be mimicked by employing PC-BSA and in addition, PC-chloride. Furthermore, splenic B cells from mice exposed to weekly injections of PC-BSA were found to be less able to proliferate in response to antigen-receptor ligation than those exposed to BSA alone (16). Attempts to undertake similar experiments with T lymphocytes were hampered by the fact that stimulation via the TCR complex in the absence of costimulatory signals does not provide a sufficient signal to induce DNA synthesis. Thus, ES-62 was tested for its ability to prevent PHA-induced proliferation of human T-lymphocytes during co-culture. In this assay however, the parasite molecule was found to have no inhibitory effect (17). The same result was obtained when an alternative T-lymphocyte mitogen, Concanavalin A (Con-A), was employed. These results were unexpected given the data produced by Lal and colleagues but may perhaps in some way, reflect the fact that purified T lymphocytes derived from tonsil tissue rather than peripheral blood mononuclear cells, were used in the ES-62 study (13). ES-62 *was* however found to inhibit proliferation of the spontaneously proliferating human leukaemic T cell line Jurkat, which was induced using antibodies against the TCR complex, in particular CD3 (17). Although not formally investigated, as will be discussed later, this is almost certainly due to PC.

PC inhibits lymphocyte proliferation by modulating associated effects on signal transduction pathways - The mechanism(s) by which PC inhibits lymphocyte proliferation was first investigated in B cells, co-cultured with anti-Ig antibodies and either, ES-62, PC-BSA or PC-chloride (14). It had been noted that unlike proliferation induced by anti-Ig antibodies, activation of B cells by lipopolysaccharide (LPS) was not influenced by the presence of PC, suggesting that it may target some aspect of signalling via the antigen receptor. All three PC-containing molecules were subsequently found to fail to inhibit surface Ig-mediated generation of the second messenger, inositol trisphosphate (IP3). Generation of IP3 is an early event during activation and results from the sIg-coupled phospholipase C-mediated, hydrolysis of phosphatidylinositol 4,5-bisphosphate. It was noted however that exposure to the PC-containing molecules was found to lead to a reduction in protein kinase C (PKC), an important downstream regulatory enzyme, with respect to both levels of protein and enzyme activity. Activation of PKC generally requires the translocation of the enzyme from the cytosol to a membrane and indeed mitogenic stimulation has been associated with nuclear translocation of PKC (18). Interestingly, pre-exposure of murine small resting splenic B cells to ES-62 has been shown to reduce the duration of nuclear translocation for certain isoforms (α,λ) of PKC in response to anti-Ig (19).

More recent analysis of the effects of PC on murine B cells has focussed on its ability to modulate the activation status of key sIg-coupled proliferative pathways such as the phosphoinositide-3-kinase (PI-3-K) and Ras/mitogen-activating protein (MAP) kinase signalling cascades. Ligation of sIg results in tyrosine phosphorylation and activation of the *src*-family protein tyrosine kinases (PTKs), Blk, Fyn, Lck and Lyn and this in turn leads to association of the ligated antigen receptors with these two cascades (20). Pre-exposure of B cells to ES-62 was found to reduce the PI-3-K, Ras and MAP kinase activity normally associated with stimulation of sIg (21). Moreover, PC-chloride was found to have similar effects to ES-62. Furthermore, pre-exposure to ES-62 was found to have essentially the same effects on these signal transduction cassettes in Jurkat T cells cultured with anti-CD3 antibodies (17). Thus, there was disruption of TCR coupling to PKC, PI-3-K and Ras/MAP kinase. Also, as observed in murine B cells, antigen receptor-mediated generation of IP3 was not affected. Again, all of these effects could be mimicked by PC-BSA or PC-chloride (16).

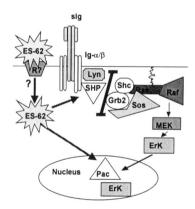

Figure Legend
ES-62 uncouples the BCR from the RasErk MAPkinase cascade.
Following ligation of the BCR the PTK, Lyn, tyrosine phosphorylates the ITAMS on the accessory transducing molecules Ig-α and Ig-β. The Ras adaptor protein, Shc binds to the phosphorylated ITAMS and in turn is phosphorylated leading to the recruitment of the Grb2/Sos complexes required for activation of Ras. Active Ras initiates the Erk MAPkinase cascade by binding and activating Raf leading to stimulation of MEK and consequent activation and nuclear translocation of Erk. ES-62/PC, binds to an unknown receptor ® and either by subversion of immune receptor signalling and/or internalisation, appears to target two major negative regulatory sites in the control of BCR-coupling to the Ras MAPkinase cascade. Firstly, ES-62 induces the activation of SHP tyrosine phosphatase to prevent initiation of BCR signalling by maintaining the ITAMS in a resting, dephosphorylated

state and hence prevents recruitment of the ShcGrb2Sos complexes required to activate Ras. Secondly, ES-62 recruits the nuclear MAPkinase dual phosphatase, Pac to terminate any ongoing Erk signals. This dual-pronged mechanism results in a rapid and profound desensitisation of BCR-coupling to the Ras ErkMAPkinase cascade.

Interestingly, although ES-62 profoundly suppresses sIg-stimulated tyrosine phosphorylation, examination of its effect on the sIg-associated PTKs in B cells, suggested that they were unlikely to be major targets in desensitization of sIg coupling to the PI-3-K and Ras/MAP kinase cascades: any effects on activity noted, tended to be minor, transient and not always moving in a negative direction (21). Desensitisation appears in fact to occur for a different reason, the targetting of two major regulatory sites in activation by ES-62 (22) (Figure): firstly, pre-exposure to the parasite protein induces recruitment of SHP-1 tyrosine phosphatase - this prevents recruitment of the Ras/MAP kinase cascade by dephosphorylating critical tyrosine residues on ITAMs (immunoreceptor tyrosine-based activation motifs) on signalling molecules associated with surface Ig (Igα, Igβ) which are necessary for early steps in activation; and secondly, the MAP kinase phosphatase Pac-1 is activated to terminate any residual MAP kinase signalling. PC has been shown to be responsible for at least the former of these effects (22): whether it is also responsible for the latter has yet to be investigated.

PC *per se* activates certain lymphocyte transduction pathways - Whilst investigating the effect of ES-62 on PKC levels/activity in B cells subjected to ligation of sIg (14), it was also observed that ES-62 alone could cause a reduction in the levels/activity found in resting B cells. Again, this was mirrored by PC-chloride. This led to the idea that the PC moiety of ES-62 might interact with certain lymphocyte signal transduction pathways and hence this was subsequently investigated. ES-62, at concentrations found to inhibit lymphocyte signal transduction due to ligation of sIg, was found with respect to murine splenic B cells to induce tyrosine phosphorylation and activation of the PTKs Lyn (535% of basal level tyrosine phosphorylation); Syk (289% basal level) and Blk (172% basal level) and activate the Erk2 isoform of MAP kinase (256% of basal level tyrosine phosphorylation) (21). These effects were again all mimicked by PC-chloride. ES-62 was also found to modulate the expression of a number of PKC isoforms in B cells (19) - for example, whereas α (31% of basal levels), β (37%), λ (41%), ζ (43%) were downregulated expression of γ (492%) and ε (185%) was upregulated. Although it was not investigated as to whether these effects on the specific isoforms was due to PC, the results relating to total PKC levels referred to earlier (14), suggest that this is highly likely. With respect to T-lymphocytes ES-62 and PC-chloride were also both found to activate certain PTKs (eg., lck, ZAP 70) and MAP kinase in Jurkat cells (16, 17).

Mechanism of action of PC? - How then is PC able to exert these effects on lymphocyte signal transduction pathways? Examination of the literature reveals clear evidence from a number of studies for PC playing a role in cellular proliferation. For example, many human tumours have elevated levels of PC and it is known that Ras-transformed cell lines produce increased levels of PC which are necessary for proliferation (23,24). Furthermore, it has recently been found that PC can exert mitogenic and co-mitogenic effects on fibroblast cell lines *in vitro* (25) and this is associated with activation of MAP kinase (26). However, evidence has been produced indicating that these latter effects can occur not only intracellularly but also following extracellular interaction with PC (25). With respect to ES-62, we cannot at this stage state whether the PC moiety acts at the cell surface or following internalisation. The ability of PC to activate certain protein tyrosine kinases which are associated with receptors found at the plasma membrane may be an argument in favour of the former but it is also worth considering that it may be possible for ES-

62 to become readily internalised via receptors for platelet activating factor (PAF) (27). Certainly this receptor is utilised by strains of *Streptococcus pneumoniae* which express PC on their surface to enter endothelial cells and it is known that PAF receptors are found on B cells (28, 29). Whatever the site of action of ES-62, the activation of MAPKinase which it induces is of interest in that it appears to take place in the absence of Ras activation (21). Recent studies indicate that ES-62/PC actually blocks Ras activation by preventing a critical phosphotyrosine-SH2 domain-dependent association between an adaptor protein Shc (this molecule becomes phosphorylated by the PTK, syk, following binding to tyrosine phosphorylated Igα/β), and the Ras guanine nucleotide exchange factor, Sos in complex with another adaptor, Grb2 (22). The alternative mechanism by which MAPkinase is activated by ES-62/PC *per se* remains to be established but interestingly, activation fails to result in lymphocyte proliferation (14,21).

PC influences antibody and cytokine profiles - Following subcutaneous injection, the murine antibody response to ES-62 is almost entirely of the IgG1 subclass (a "TH-2 antibody") (16) whereas the response to ES-62 synthesised to lack PC also contains a significant IgG2a component (a "TH-1 antibody") (30). PC thus appears to block the production of the TH-1 antibody. A role for the anti-inflammatory cytokine IL-10 in this effect is shown by the ability of IL-10 knockout mice to make an IgG2a response to native ES- 62 (30) and indeed ES-62 induces IL-10 production in naive spleen cells (16). Similarly, the PC component of filarial nematode extracts (*B. malayi*)has been shown to induce IL-10 production in BALB/c peritoneal exudate CD5$^+$ B-1 cells (31). It is possible to relate these data to how secreted PC-containing molecules might contribute to the patterns of immune responsiveness observed in humans infected with filarial nematodes. PC-containing secreted glycoproteins (PC-ES) are readily detectable in the bloodstream of people harbouring *Wuchereria bancrofti,* particularly if they are microfilaraemic (larval forms can be readily detected in their bloodstreams and often in large numbers) (9). The latter individuals are also notable in that their peripheral blood mononuclear cells have been shown to be more able to secrete IL-10, either spontaneously or in response to parasite antigens (32). This raises the possibility as to whether the PC moiety of the parasite molecules could be a contributor to the IL-10 production being observed. Elevated spontaneous and parasite antigen-induced IL-10 production has also been noted individuals infected with *Onchocerca volvulus*. Elevated spontaneous levels of IL-10, by downregulating TH-1 responses, could contribute to the TH-2 phenotype generally observed in the specific immune response of individuals harbouring filarial nematodes.

ATTACHMENT OF PC TO FILARIAL MOLECULES AS A TARGET FOR CHEMOTHERAPY

Targeting PC with respect to drug development - The immunomodulatory properties of PC appear to be important in allowing survival of filarial nematodes and for this reason the synthesis of PC-containing molecules might represent an important new target for chemotherapy. Certainly, new drugs are urgently required for treatment of filarial nematodes. With this in mind, it was decided to investigate the mode of attachment of PC to filarial nematode proteins.

PC is attached to filarial nematode proteins via N-type glycans - Early studies on the nature of the PC-containing proteins of filarial nematodes indicated that PC was present on glycosylated molecules. For example, biosynthetic labelling of filarial PC-ES products with the amino-sugar [^{3}H]-glucosamine was observed in *A. viteae*, the bovine parasite *Onchocerca gibsoni* and the feline parasite *Brugia pahangi* (15, 33, 34). Exposure of [^{3}H]-glucosamine-labelled ES-62 to N-glycosidase F, an enzyme which cleaves N-type glycans from proteins, was found to result in a total loss of radioactivity from the molecule (35). Substituting [^{3}H]-glucosamine with [^{3}H]-choline as radiolabel also resulted in a complete loss of radioactivity from ES-62 after exposure to this sugar-cleaving enzyme. These experiments provided evidence that PC is not only present on carbohydrate-containing molecules of filarial nematodes but is actually attached to the protein backbone via an N-type glycan. This was further supported by the observation that culture of *A. viteae* with tunicamycin, an antibiotic which inhibits N-linked glycosylation, resulted in production of ES-62 lacking PC (36, 37). Both of these findings, obtained using ES-62, were subsequently confirmed with PC-ES of *B. pahangi* (34,38).

The structure of N-glycans to which PC is attached is conserved amongst filarial species - By employing fast atom bombardment mass spectroscopy, it was found that the N-glycan to which PC was attached to ES-62, consisted of a trimannosyl core, with and without core fucosylation, carrying between one and four additional N-acetylglucosamine residues (39). These glycans were found to contain one PC residue although there was also some evidence suggesting that attachment of two was possible. It is most likely that the sugar to which PC is attached is one of the additional N-acetylglucosamine residues. More recently, it has been found that these particular N-glycan structures appear to be conserved amongst filarial nematodes as we have found them in both *O. gibsoni* and *O. volvulus* (40). This represents an important finding for two reasons: (i) *O. volvulus* is the filarial neatode for which new drugs are most urgently sought; (ii) any drug developed could conceivably be utilised against *all* species of filarial nematode.

Can inhibitors of PC attachment be developed for chemotherapeutic use? - The absence of PC-N-glycans from humans makes their biosynthesis an attractive option for chemotherapy. We have in fact shown that attachment of PC can be blocked by culturing filarial nematodes (*A. viteae; B. pahangi*) with 1-deoxymannojirimycin (dMM) (41, 38). dMM is a mannose analogue which inhibits mannosidase I, an enzyme involved in N-type oligosaccharide processing, in the cis Golgi (36). Although it is unlikely that this reagent could be used *in vivo* for fear of the effects on glycosylation on human cells, it raises the prospect that blocking PC addition to filarial proteins is a feasible option. Also, we have found that PC attachment can be blocked by culturing worms in the presence of hemicholinium-3 (HC-3) (42). HC-3 is a choline kinase inhibitor which would be expected to block synthesis of phosphorylcholine (43). Again, toxicity associated with this molecule would prevent its use *in vivo*, but more recently a new generation of choline kinase inhibitors has been generated, which because they are much more potent than HC-3, allows them to be utilised at much lower concentrations (44). These reagents are currently being evaluated as anti-tumour reagents in mice (many tumours have constitutively elevated levels of choline kinase) and hence it is possible that they could ultimately be consided for use in preventing PC attachment to filarial nematode glycoproteins (45, 23).

A final point here which could be of immense significance is that PC is found attached to carbohydrate in a range of other important infectious agents (e.g., *S. pneumoniae, Heamophilius influenzae*, pathogenic fungae, non-filarial nematodes) (reviewed in 11). Although attachment does not always involve interaction with an N-type glycan, it is possible that any drugs developed for treatment of filarial nematodes could have much more widespread application. At the very least, information gained in developing anti-filarial nematode drugs could be utilised in relation to targeting these other agents.

Therapeutic use of PC as an immunomodulator? - There is some evidence that concurrent infection with filarial nematodes can diminish the immune response to unrelated antigens by a mechanism likely to involve IL-10 (46). As mentioned earlier, this may relate to the secretion of PC-containing molecules and hence this raises the possibility of rather than targetting PC as in filarial nematode infection, of actually using it for therapeutic purposes. For example, PC "mimetics" could perhaps be developed for use in the treatment of inflammation-associated "TH-1 diseases" such as rheumatoid arthritis. Additionally, they could be considered as immunomodulators to drive the TH-2 phenotype in the treatment of diseases resulting from TH-1 derived pathology such as endotoxin-shock. Whether or not this turns out to be the case the unravelling at the molecular level of the effects which PC has on lymphocytes will at least suggest new targets for therapeutic intervention. There may be scope for example, in considering phosphatases such as SHP-1 and Pac-1 as routes to manipulating lymphocyte function in autoimmune diseases.

References

1.   Tropical Disease Research, 1995-96, Thirteenth Programme Report UNDP/World Bank/WHO Special Programme for Research & Training in Tropical Diseases, WHO, Geneva, 1997, p.86.
2.   P. Vanamail, K.D. Ramaiah, S.P. Pani, P.K. Das, B.T. Grenfell and D.A.P. Bundy, Trans. R. Soc. Trop. Med. Hyg. 90, 119 (1996).
3.   J.W. Kazura, T.B. Nutman, and B. Greene in "Immunology and Molecular Biology of Parasitic Infections", K.S. Warren, Ed., Blackwell Scientific Publications, Oxford, 1993, p. 473.
4.   E.A. Ottesen. J. Infect. Dis. 171, 659 (1995).
5.   R.M. Maizels, E. Sartono, A. Kurniawan, F. Partono, M.E. Selkirk and M. Yazdanbakhsh. Parasitol. Today 11, 50 (1995).
6.   A.A. Wadee, A.C. Vickery, and W.F. Piessens. Acta Trop. 44, 343 (1987).
7.   S. Hartmann, B. Kyewski, B. Sonnenburg and R. Lucius. Eur. J. Immunol. 27, 2253 (1997).
8.   W. Harnett, and R.M.E. Parkhouse in "Perspectives in Nematode Phsiology and Biochemistry," M.L. Sood and J. Kapur, Eds., Narendra Publishing House, Delhi, 1995, p. 207.
9.   W. Harnett, J.E. Bradley, and T. Garate. Parasitology 117, S59 (1998).
10.   G.F. Mitchell and H.M. Lewers. Int. Arch.Allergy Appl. Immunol. 52, 235 (1976).
11.   W. Harnett and M.M. Harnett. Immunol. Today 20, 125, (1999).
12.   G. Bordmann, W. Rudin and N. Favre. Immunology 94, 35 (1998).
13.   R.B. Lal, V. Kumaraswami, C. Steel and T.B. Nutman. Am. J. Trop. Med. Hyg. 42, 56 (1990).
14.   W. Harnett and M.M. Harnett. J. Immunol. 151, 4829 (1993).
15.   W. Harnett, M. Grainger, A. Kapli, M.J. Worms and R.M.E. Parkhouse. Parasitology 99, 229 (1989).
16.   W. Harnett, M.R. Deehan, K.M. Houston and M.M. Harnett. Parasit. Immunol. 21, 601 (1999).
17.   M.M. Harnett, M.R. Deehan, D.S. Williams and W. Harnett. Parasit. Immunol. 20, 551 (1998).
18.   H. Hug, and T.F. Sarre. Biochem. J. 291, 329 (1993).
19.   M.R. Deehan, M.M. Harnett and W. Harnett. J. Immunol. 159, 6105 (1997).
20.   J.C. Cambier, C. Pleiman and M.R. Clark. Annu. Rev. Immunol. 12, 457 (1994).
21.   M.R. Deehan, M.J. Frame, R.M.E. Parkhouse, S.D. Seatter, S.D. Reid, M.M. Harnett and W. Harnett. J. Immunol.160, 2692 (1998).
22.   W. Harnett and M.M. Harnett. Med. Asp. Immunol. in press.
23.   P.F.Daly, R.C. Lyon, P.J. Faustino and J.S. Cohen. J. Biol. Chem. 262, 14875 (1987).
24.   S. Ratnam and C. Kent. Arch. Biochem. Biophys. 323, 313 (1995).
25.   K.S. Crilly, M. Tomono and Z. Kiss. Arch. Biochem. Biophys. 352, 137 (1998).

26.   B. Jimenez, L. del Peso, S. Montaner, P.Esteve and J.P. Lacal. J. Cell. Biochem. 57, 141 (1995).

27.   C.M. Pleiman, D. D'Ambrosia and J.C. Cambier. Immunol. Today 15, 393 (1994).

28.   D.R. Cundell, N.P. Gerard, C Gerard, I. Idanpaan-Heikkila and E.I. Tuomanen. Nature 377, 435 (1995).

29.   B. Mazer, J. Domenico, H. Sawmi and E.W. Gelfand. J. Immunol. 146, 1914 (1991).

30.   K.M. Houston, E. wilson, L. Eyers, F. Brombacher, M.M. Harnett, J. Alexander and W. Harnett, Submitted for Publication.

31.   V. Palanivel, C. Posey, A.M. Horauf, W. Solbach, W.F. Piessens and D.A. Harn.   Exp. Parasitol. 84, 168 (1996).

32.   S. Mahanty and T.B. Nutman. Parasit. Immunol. 17, 385 (1995).

33.   W. Harnett, M. Patterson, D.B. Copeman and R.M.E. Parkhouse. Int. J. Parasitol. 24, 543 (1994).

34.   Z.M. Nor, K.M. Houston, E. Devaney and W. Harnett. Parasitology 114, 257 (1997).

35.   W. Harnett, K.M. Houston, R. Amess and M.J. Worms. Exp. Parasitol. 77, 498 (1993).

36.   A.D. Elbien. Ann. Rev. Biochem. 56, 497 (1987).

37.   K.M. Houston and W. Harnett. J. Parasitol. 82, 320 (1996).

38.   Z.M. Nor, E. Devaney and W. Harnett. Parasitol. Res. 83, 813 (1997).

39.   S.M. Haslem, K-H, Khoo, K.M. Houston, W. Harnett, H.R. Morris and A. Dell. Mol. Biochem. Parasitol. 85, 53 (1997).

40.   S.M. Haslam, K.M. Houston, W. Harnett, A.J Reason, H.R. Morris and A. Dell. J. Biol. Chem. 274, 20953 (1999).

41.   K.M. Houston, W.C. Cushley and W. Harnett. J. Biol. Chem. 272, 1527 (1997).

42.   K.M. Houston and W. Harnett. Parasitology 118, 311 (1999).

43.   M. Hamza, J. Lloveras, G. Ribbea, G. Soula L. Douste-Blazy. Biochem. Pharm. 32, 1893 (1983).

44.   R. Hernandez-Alcoceba, L. Saninger,J. Campos, M. Carmen Nunez, F. Khaless, M. Angel Gallo, A. Espinosa, A. and J.C. Lacal. Oncogene 15, 2889 (1997).

45.   R. Hernandez-Alcoceba, F. Fernandez, and J.C. Lacal. Cancer Res. 59,3112 (1999).

46.   P.J. Cooper, I Espinel, W. Paredes, R. H. Guderian, and T. B. Nutman. J. Infect. Dis. 178, 1133 (1998).

SECTION IV. TOPICS IN DRUG DESIGN AND DISCOVERY

Editor: George L. Trainor, DuPont Merck Pharmaceutical Company, Wilmington, Delaware

Chapter 26. Privileged Structures – An Update

Arthur A. Patchett and Ravi P. Nargund,
Merck Research Laboratories, Rahway, NJ 07065-0900

Introduction -The term "privileged structure" was introduced by B. Evans et al. (1) in describing their development of benzodiazepine-based CCK-A antagonists from the natural product lead asperlicin. In their definition a privileged structure such as a benzodiazepine "is a single molecular framework able to provide ligands for diverse receptors..." Benzodiazepines are found in several types of CNS agents and are in ligands of both ion channel and G-protein coupled receptors (GPCRs). In the latter category Evans et al. cited the analgesic tifluadom which has both analgesic activity and affinity for the CCK-A receptor. These authors pointed out that core structures from which multiple activities can be derived is a broadly recognizable phenomenon and cited a review by Ariens (2) in which a "hydrophobic double-ring system" was exemplified in a number of biogenic amine antagonists. These ring systems, many of which were 1,1-diphenylmethane variants, were considered not to bind to the same receptor sites..."to which the highly polar agonists bind. They must bind to accessory binding sites of a predominantly hydrophobic nature..." (2). Ariens (3) further elaborated on these concepts and Andrews and Lloyd (4) described a number of common topological arrangements in biogenic amine antagonists. In summarizing their successful use of benzodiazepines, Evans and colleagues concluded that "judicious modification of such structures could be a viable alternative in the search for new receptor agonists and antagonists" (1).

Privileged Structure Variations - This review will focus primarily on privileged structure based ligands of G-protein coupled receptors (GPCRs). Even with this limitation, no comprehensive reviews of GPCR privileged structures are available. Sources are quite fragmented. The papers of Ariens illustrate 1,1-diphenylmethyl units including tricyclic aromatics as in chloropromazine, phenylbenzylmethyl groups, saturated and heterocyclic variants of them and several 4-arypiperidines (2,3). A benzodiazepine and a biphenyl unit additionally appear in a review by LaBella (5). Sixteen templates useful in designing peptidomimetics were

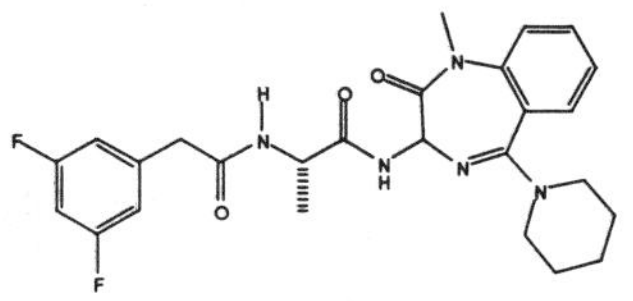

1 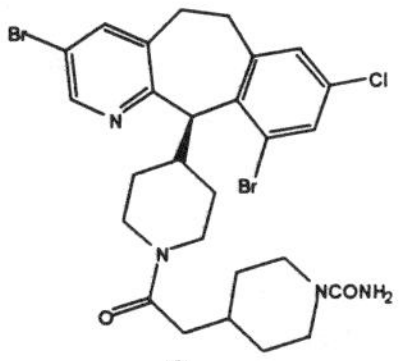   **2** 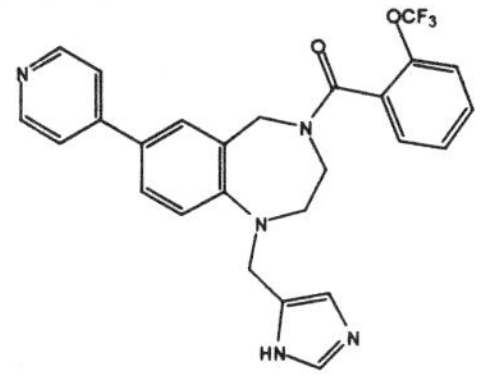   **3**

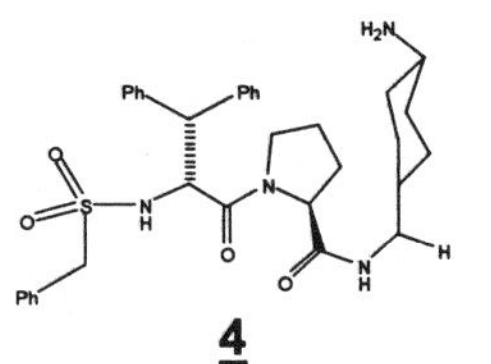

4 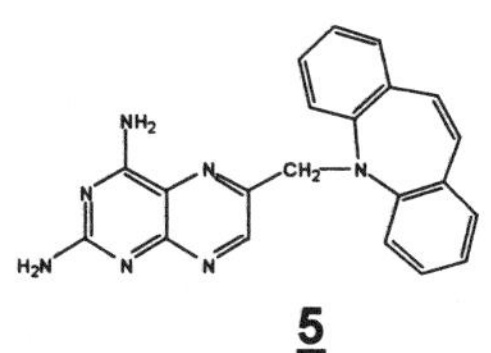   **5** 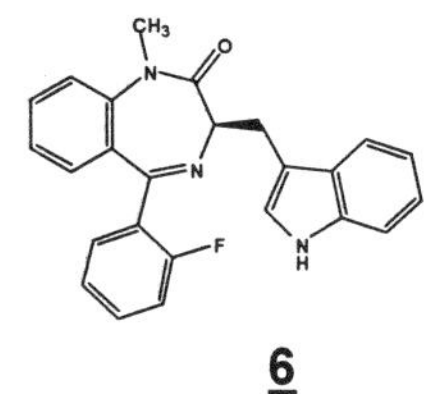   **6**

listed by Wiley and Rich in 1991 (6). These included variations of the types mentioned above with the addition of 4-aryl-1,4-dihydropyridines and certain indoles, benzimidazoles and quinazolinones. Wiley and Rich emphasized that their listed templates are characterized by a resistance to hydrophobic collapse in water. Thus, their potentially extensive hydrophobic binding energies are available to the receptor and not significantly lost in intramolecular stacking interactions.

Many of the privileged structures in GPCR ligands are also found in enzyme inhibitors. Recent examples include benzodiazepine-containing K-secretase inhibitors such as 1 and farnesyl transferase inhibitors including the tricycle-based SCH-66336 (2) and the tetrahydrobenzodiazepine 3 (7-9). A 1,1-diphenyl unit was incorporated in the potent thrombin inhibitor 4 (10) and a tricycle was used in the dihydrofolate reductase inhibitor 5 (11).

Other enzyme inhibitors containing GPCR privileged structures are listed in the peptidomimetic design review of Ripka and Rich (12). Recent examples of benzodiazepines in ion channel ligands include the delayed rectifier K+ current modulator 6 (13).

Although this review deals primarily with the use of privileged structures in the design of GPCR agonists and antagonists, their usefulness certainly extends beyond GPCR receptors. Thus binding motifs for them are to be found in ion channels and enzymes perhaps also associated with α-helical structures. Whatever the reasons for promiscuity, the use of privileged structures brings with it a concern for the selectivity of their derivatives.

Privileged Structure Based Antagonists – The discovery of nonpeptide ligands for GPCRs dramatically increased following the announcement of the natural product CCK-A antagonist asperlicin (14) and the subsequent design of benzodiazepine based CCK-A antagonists including devazepide (7) (15). A number of excellent reviews summarize these developments (16-20).

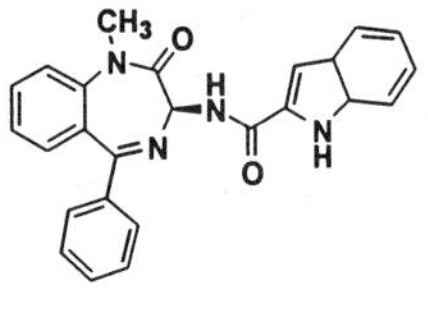

7	8	9

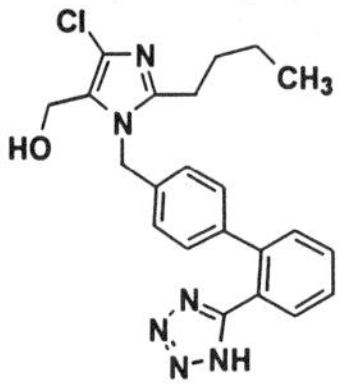

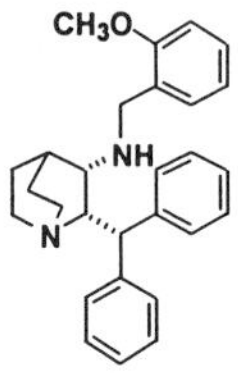

Most nonpeptide GPCR ligands are antagonists and they were derived from broad sample collection screening. In lesser number they were designed by applying pharmacophore functionality to conformationally defined scaffolds. The majority of ligands listed in the referenced reviews do not contain privileged structures as the term is used in this paper (16-20). However, a number of privileged structure derived

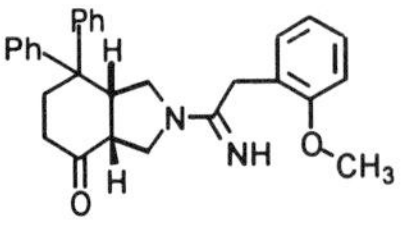

10	11	12

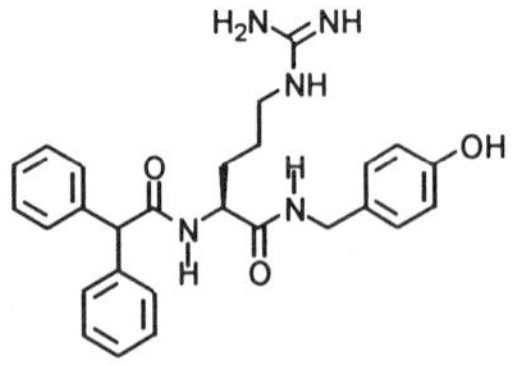

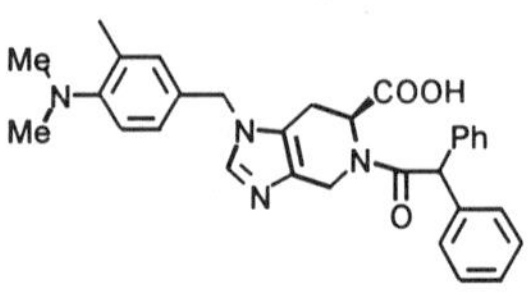

antagonists have been synthesized and some representative ones will be cited. Among them the biphenyl unit is prominently found in clinically important angiotensin II antagonists including losartan (8). 1,1-Diphenyl groups are present in the original NK-1 antagonist CP96,345 (9), in the NK-1 antagonist RP67,580 (10), in the NPY Y1 antagonist BIBP3226 (11) and in the angiotensin AT-2 antagonist PD123,319 (12) (21-24).

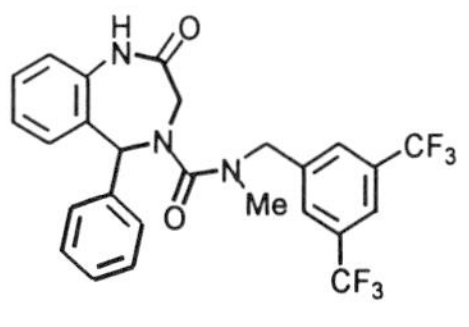

13 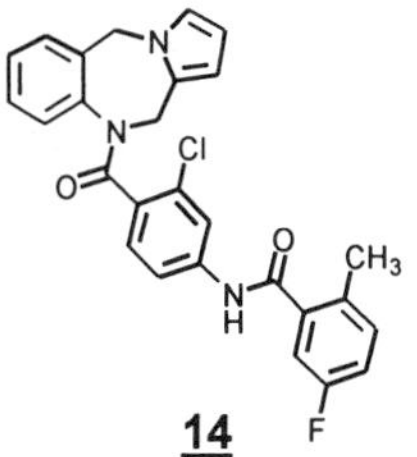   **14** 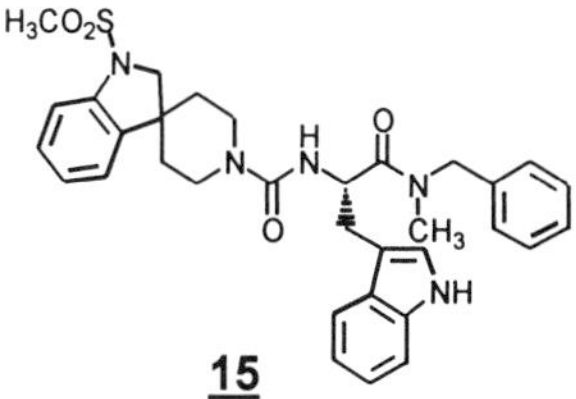   **15**

There are benzodiazepine NK-1 antagonists as in 13 and in the vasopressin antagonist 14 (25, 26). One of the larger categories of privileged structures includes 4-arylpiperidines, 4-arylpiperazines and spiro versions of the former. Examples include the neurokinin NK-2/NK-3 antagonist 15, the NK-3 antagonist 16 , the oxytocin antagonist 17, the opiate antagonist 18 and the CGRP antagonist 19 (27-31).

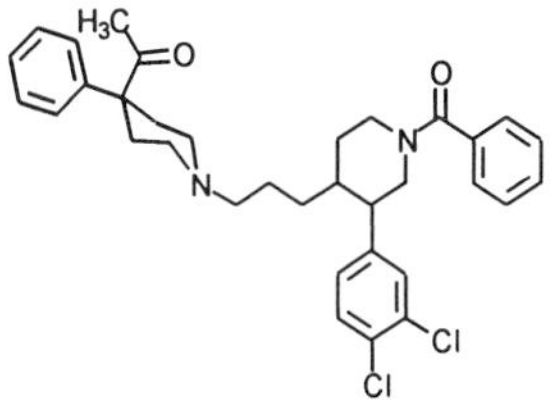

16 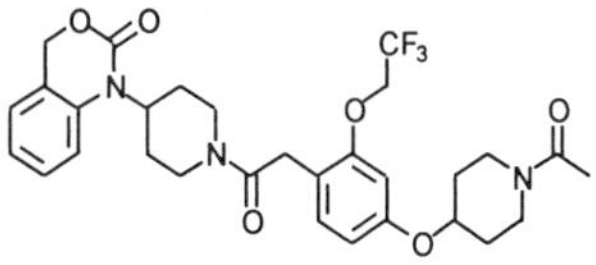   **17** 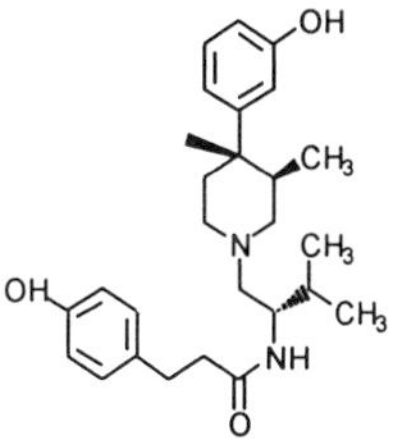   **18**

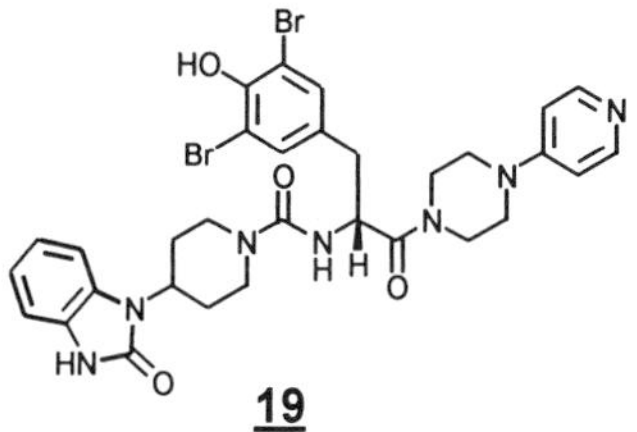

19

Peptidomimetic GPCR Agonists - Until the early 1990s morphine and related opioids were the only extensively studied peptidomimetic agonists. The benzodiazepine tifludom was a known opiate receptor agonist (32) and erythromycin a motilin agonist (33). Within the past several years, agonists for the CCK-A, bradykinin, growth hormone secretagogue and somatostatin receptors and partial agonists of the C5a and angiotensin II AT1 receptor have been identified. Broadly speaking, two strategies have emerged for the discovery of agonists: a) the screening of empirically modified ligands, for example, for the AT1, bradykinin B_2 and CCK-A receptors and b) peptidyl privileged structures (C5a, growth hormone secretagogue, somatostatin and melanocortin subtype-4 receptor).

<u>Agonists *via* Privileged Structure Modification</u> - Progress in the synthesis of agonist ligands for the opioid, angiotensin II and CCK-A receptors was reviewed by E. Sugg in 1997 (34). The addition of a single methyl group to a benzodiazepine substituent converted a CCK-A antagonist (<u>20</u>) to an agonist (<u>21</u>) (35). Similar modification of a biphenyl substituent produced partial agonism in the AII antagonist series (compounds <u>22</u> and <u>23</u>) (36). Apparently, the appended functionalities of these privileged structures bind in receptor regions which are quite sensitive to the expression of either agonist or antagonist activity. The transition from opioid agonists to antagonists by N-alkyl modification is well-known. An additional recent example of agonists and antagonists within the same chemical series is the bradykinin B$_2$ antagonist FR165649 (<u>24</u>) and the B$_2$ agonist FR190997 (<u>25</u>) (37).

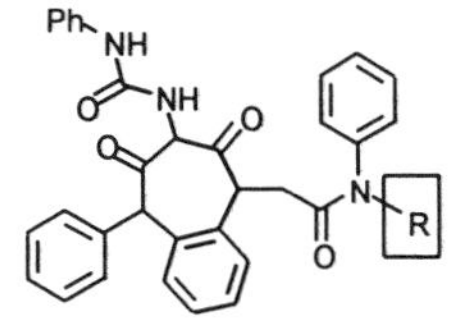

| <u>20</u> | R = CH$_3$ | (antagonist) |
| <u>21</u> | R = CH$_2$CH$_3$ | (agonist) |

| <u>22</u> | R = | (antagonist) |
| <u>23</u> | R = | (agonist) |

| <u>24</u> | R = H FR165649 | (antagonist) |
| <u>25</u> | R = | FR 196997 (agonist) |

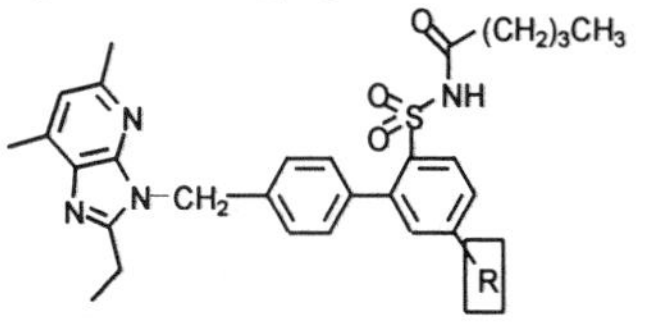

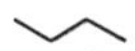

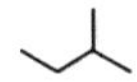

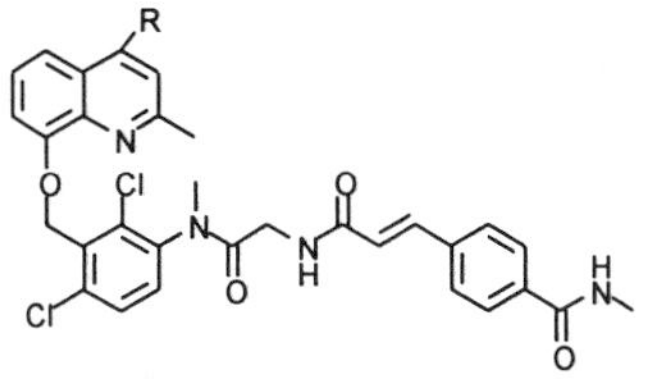

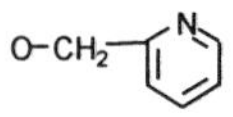

<u>Agonist Ligand Design - Peptidyl Privileged Structures</u> - In this review we are using the term "peptidyl privileged structures" to describe a design in which a privileged structure anchor is derivatized with dipeptides or capped amino acids. In synthesizing these structures it was assumed that the privileged structure bound in a hydrophobic binding site in the receptor from which essential peptide functionality could be expressed (38, 39).

<u>Examples of Peptidyl Privileged Structure Agonists</u> - Spiropiperidine growth hormone secretagogues, including orally active clinical candidate <u>5</u> (MK-0677), are the first highly active peptidyl privileged structures to have been described (38, 39). The spiroindanylpiperidine group which was used in their discovery was considered to be a privileged structure since it was present in a screening ligand

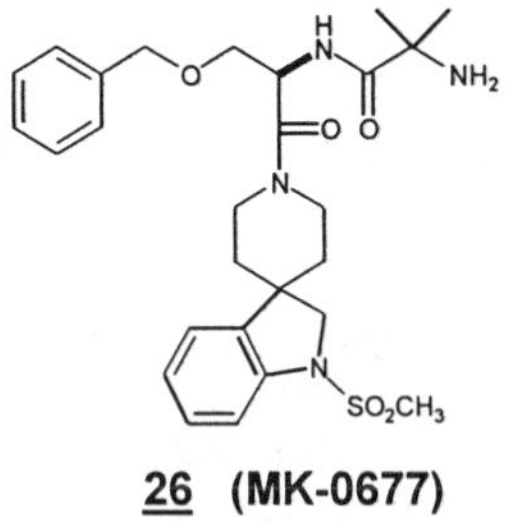

<u>26</u> (MK-0677)

that binds to the GH secretagogue and to oxytocin, vasopressin and sigma opiate receptors. It was derivatized with capped amino acids, including tryptophan since tryptophan contributes significantly to the GH-releasing activity of GHRP-6 (His-D-Trp-Ala-Trp-D-Phe-Lys-NH$_2$) and related peptides. Compound <u>5</u> is devoid of significant activities (IC$_{50s}$ > 2 µM) at a large number of G-protein-coupled receptors, ion channels and enzymes.   It was unexpected and reassuring that both high agonist potency and selectivity had been achieved by derivatizing a privileged structure with only a dipeptide (38,39).

Following the disclosure of **5**, potent orally active GH secretagogues with variations of the spiroindoline sulfonamide privileged structure have been described, including the 1,1-spiroindane 3-carboxylic acid ester <u>27</u>, the 3,3-disubstituted piperidine <u>28</u>, the fused pyrazolidinone <u>29</u> and the aminoimidazole <u>30</u> (40-43).

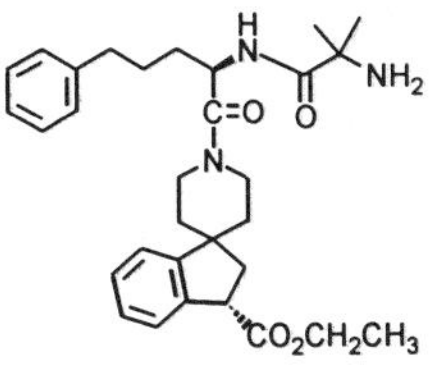

27 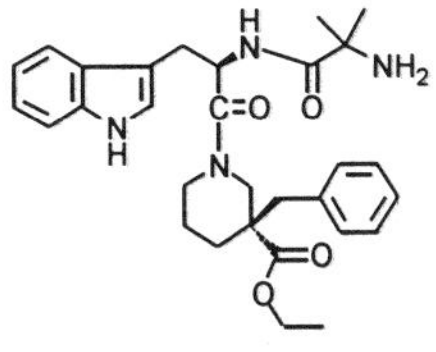   **28**

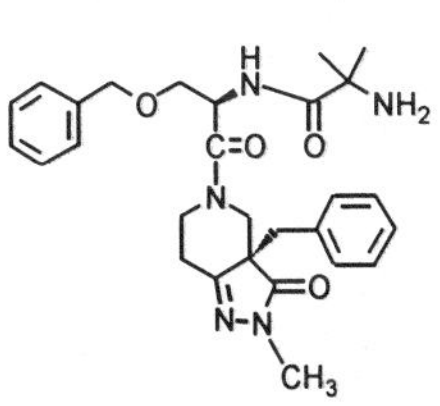

29 (CP 424,391) 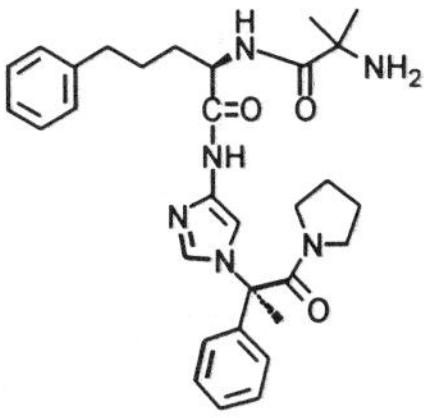   **30** (LY 444,771)

An alternate structural class of growth hormone secretagogues includes benzazepinones such as the orally active compound <u>31</u> (44).   These agonists possesses biphenyl and benzolactam part structures which are found, for example, in ligands for the angiotensin II (AT1 and AT2) and CCK-A receptors (45).   The benzolactam privileged structure is also a part structure in <u>40</u> which is a selective agonist for the somatostatin subtype-5 receptor.

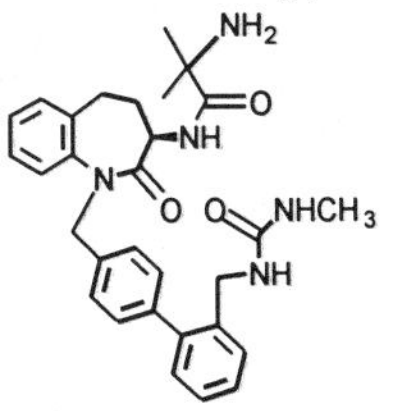

31 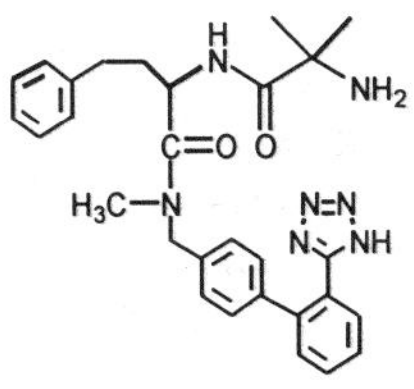   **32**

In the above compounds the benzolactam group may serve as conformationally restricted D-homophenyl alanine residue because a recent publication describes compounds such as <u>32</u> as GH secretagogues (46).

In 1996, the growth hormone secretagogue receptor la was cloned (47) and in late 1999 ghrelin, the natural ligand of this receptor, was isolated (48). It is a 28 amino acid peptide which must be octanoylated on Ser 3 to express secretagogue activity. It is remarkable that compounds such as MK-0677, which approximate tripeptides in size are able to equal ghrelin's potency in releasing growth hormone from rat pituitary cells. It will be interesting to see if a large peptide such as ghrelin activates the GH secretagogue receptor in a manner that overlaps or is independent of the MK-0677 binding site.

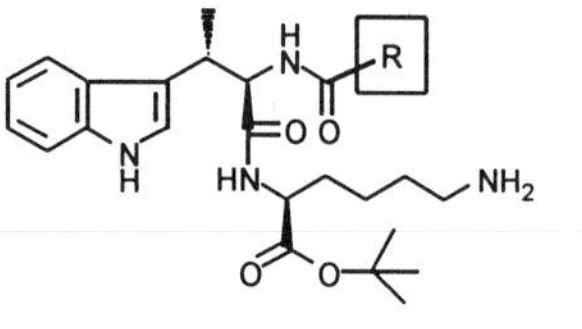

33

R =

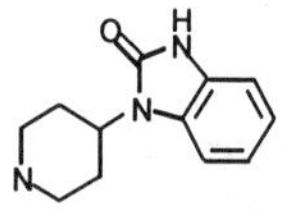

34

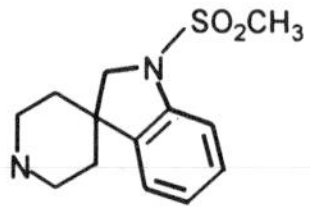

35

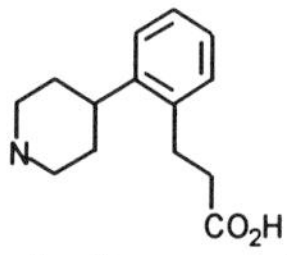

Peptidyl privileged structures <u>33</u>, <u>34</u>, and <u>35</u> are selective agonists for the human somatostatin subtype-2 receptor (sst2) (49-51). These compounds are suggested to mimic the type II' β-turn structure of Tyr-D-Trp-Lys-Thr in the somatostatin analog <u>36</u>. Compound <u>33</u> shows excellent specificity versus other G-protein coupled receptors, enzymes and ion channels (49). A recent patent application discloses 4-phenyl piperazine privileged structure compounds such as <u>37</u> as balanced sst2/sst3 agonists (52).

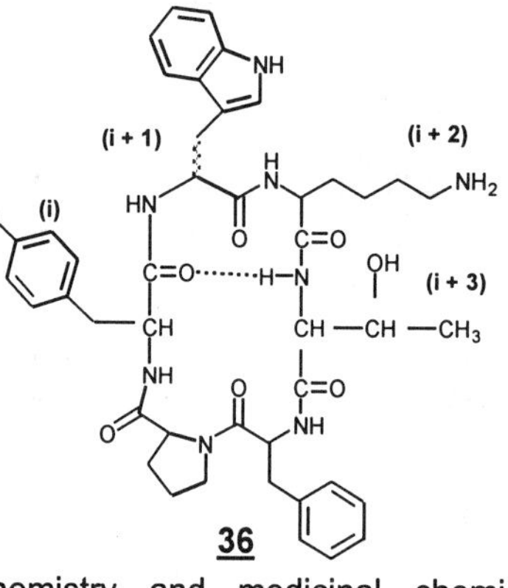

36

Combinatorial chemistry and medicinal chemistry techniques were employed for the synthesis of peptidyl privileged structures which show good selectivity towards sst1 (<u>38</u>), sst4 (<u>39</u>) and sst5 (<u>40</u>) (53).

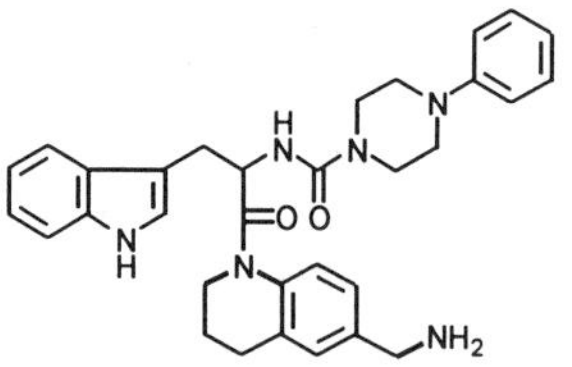

37 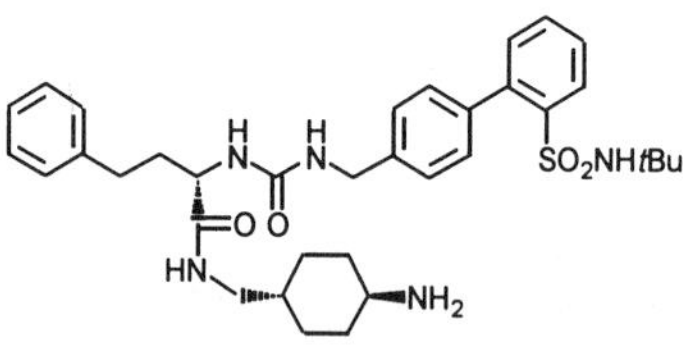   **38**

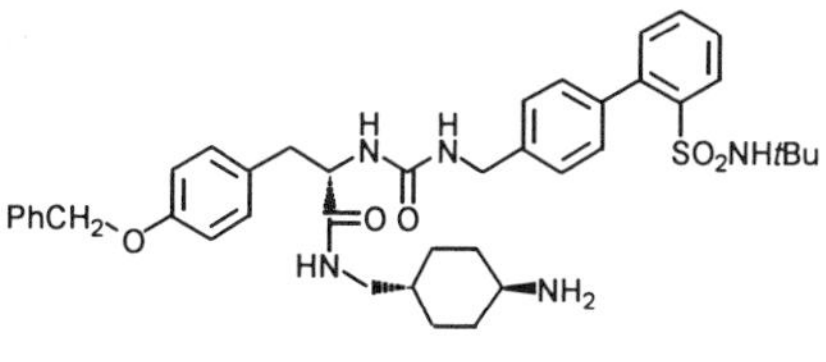

39

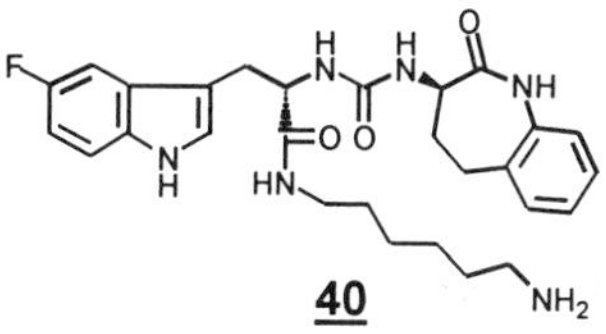

40

An early example of a modified peptidyl privileged structure partial agonist is <u>41</u> which binds to the C5a receptor with high affinity (IC_{50} = 20nM) and releases myeloperoxidase from PMNs (54). In a published patent application compounds such as <u>42</u> are claimed as agonists of the melanocortin subtype-4 receptor (55). For the vasopressin V2 receptor recent patent applications claim compounds <u>43</u> and <u>44</u> as selective agonists (56).

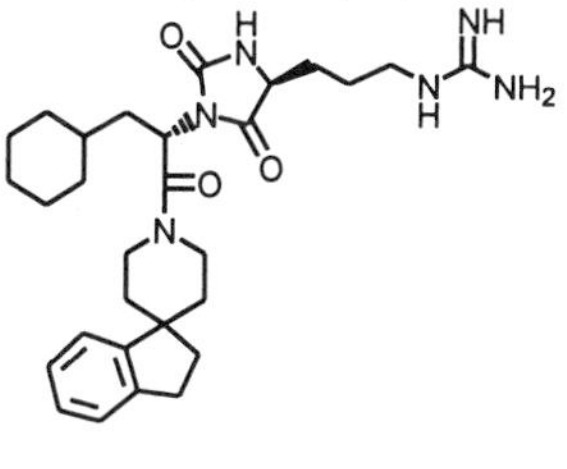

41

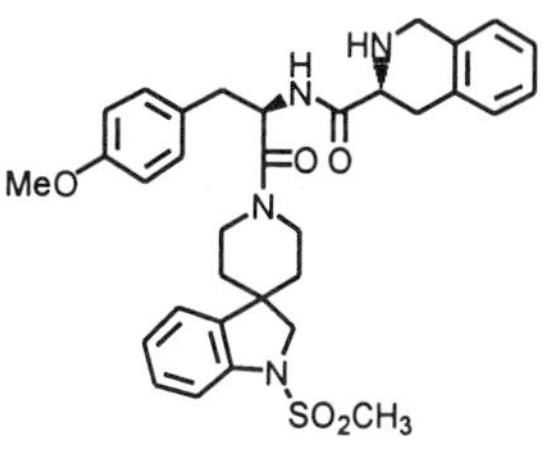

42

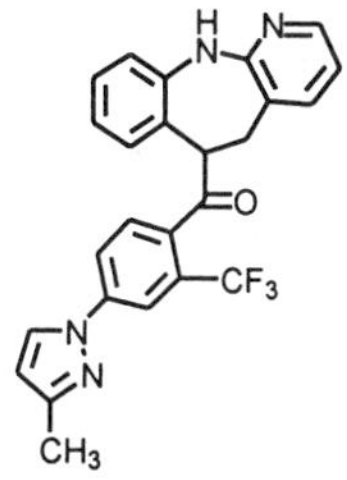

43

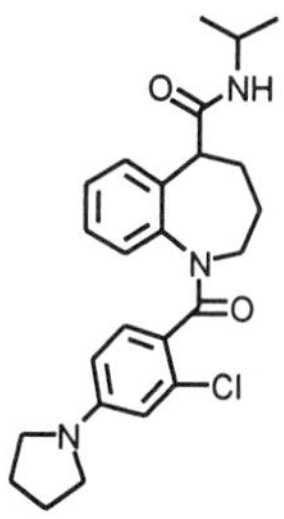

44

<u>Privileged Structure Libraries</u> – The synthesis of libraries based upon privileged structures or their incorporation in libraries as capping or modular units has been undertaken as one strategy for the creation of drug-like molecules. Early libraries of benzodiazepines and dihydropyridines were reviewed by Thomson and Ellman (57) and a comprehensive review of chemical libraries form 1992 to 1997 was made by Dolle (58) in which GPCR directed libraries based upon biphenyl, 4-phenylpiperidine and 1,1-diphenyl units are also cited. More recent biphenyl (59) and dihydropyridine (60) based libraries have been described and the somatostatin directed libraries of Berk et al. (53) made extensive use of privileged structures as modular units. Computational techniques (61) including RECAP (62) have been developed to recognize biologically active building blocks including privileged structures in databases of active molecules. Recently a 4-point pharmacophore method has been described which can be used to generate functional group diversity around privileged structures (63).

<u>Future Directions</u> – It is remarkable that potent privileged structure based agonists of GPCRs have been discovered whose molecular size is much smaller than the peptide ligands which they mimic. Determining how they bind to GPCRs will provide additional insights into the activation process. Eventually it may be possible

to select privileged structures for GPCRs with some guidance from receptor sequence and to have an improved success rate in their derivatization for either agonist or antagonist properties.

References

1. B.E. Evans, K.E. Rittle, M.G. Bock, R.M. DiPardo, R.M. Freidinger, W.L. Whitter, G.F. Lundell, D.F. Veber, P.S. Anderson, R.S.L. Chang, V.J. Loti, D.J. Cerino, T.B. Chen, P.J. Kling, K.A. Kunkel, J.P. Springer and J. Hirshfield, J. Med. Chem., 31, 2235 (1988).
2. E.J. Ariens, A.J. Beld, J.F. Rodrigues de Miranda and A.M. Simonis in The Receptors A Comprehensive Treatise, R.D. O'Brien, ed., Plenum Press, New York, 1979, pp 33.
3. E.J. Ariens, Med. Res. Rev., 7, 367 (1987).
4. P.R. Andrews and E.J. Lloyd, Med. Res. Rev., 2, 355 (1982).
5. F.S. LaBella, Biochem. Pharmacol., 42, S1 (1991).
6. R.A. Wiley and D.H. Rich, Med. Res. Rev., 13, 327 (1993).
7. J. Wu, J.S. Tung, E.D. Thorsett, M.A. Pleiss, J.S. Nissen, J. Neitz, L.H. Latimer, J. Varghese, S. Freedman, T.C. Britton, J.E. Audia, J.K. Reel, T.E. Mabry, B.A. Dressman, C.L. Cwi, J.J. Droste, S.S. Henry, S.L. McDaniel, W.L. Scott, R.D. Stucky and W.J. Porter, Patent Application WO98/28268 (1998).
8. F.G. Njoroge, A.G. Traveras, J. Kelly, S.W. Remiszewski, A.K. Mallams, R. Wolin, A. Alfonso, A.B. Cooper, D. Rane, Y.-T. Liu, J. Wong, B. Vibulbhan, P. Pinto, J. Deskus, C. Alvarez, J. Del Rosario, M. Connolly, J. Wang, J.A. Desai, R.R. Rossman, W.R. Bishop, R. Patton, L. Wang, P. Kirschmeier, M.S. Bryant, A.A. Nomeir, C.-C. Lin, M. Liu, A.T. McPhail, R.J. Doll, V. Girijavallabhan and A.K. Ganguly, J. Med. Chem., 41, 4890 (1998).
9. C.Z. Ding, R. Batorsky, R. Bhide, H.J. Chao, Y. Cho, S. Chong, J. Gullo-Brown, P. Guo, S.H. Kim, F. Lee, K. Leftheris, A. Miller, T. Mitt, M. Patel, B.A. Penhallow, C. Ricca, W.C. Rose, R. Schmidt, W.A. Slusarchyk, G. Vite, N. Yan, V. Manne and J.T. Hunt, J. Med. Chem., 42, 5241 (1999).
10. T.J. Tucker, W.C. Lumma, A.M. Mulichak, Z. Chen, A.M. Naylor-Olsen, S.D. Lewis, R. Lucas, R.M. Freidinger and L.C. Kuo, J. Med. Chem., 40, 830 (1997).
11. A. Rosowsky, V. Cody, N. Galitsky, H. Fu, A.T. Papoulis and S.F. Queener, J. Med. Chem., 42, 4853 (1999).
12. A.S. Ripka and D.H. Rich, Curr. Opin. Chem. Biol., 2, 441 (1998).
13. J.J. Salata, N.K. Jurkiewicz, J. Wang, B.E. Evans, H.T. Orme and M.C. Sanguinetti, Mol. Pharmacol., 53, 220 (1998).
14. R.S.L. Chang, V.J. Lotti, R.L. Monaghan, J. Birnbaum, E.O. Stapley, M.A. Goetz, G. Albers-Schonberg, A.A. Patchett, J.M. Liesch, O.D. Hensens and J.P. Springer, Science, 230, 177 (1985).
15. B.E. Evans, K.E. Rittle, M.G. Bock, R.M. DiPardo, W.L. Whitter, D.F. Veber, P.S. Anderson and R.M. Freidinger, Proc. Natl. Acad. Sci. USA, 83, 4918 (1986).
16. R.M. Freidinger. 1993. In Progress in Drug Research, Vol. 40, ed. E. Jucker, pp. 33. Basel: Birkhauser Verlag.
17. D.C. Rees. 1993. In Annual Reports in Medicinal Chemistry, Vol. 28, ed. J.A. Bristol, pp. 59. San Diego: Academic Press.
18. D.J. Pettibone and R.M. Freidinger, Biochem. Soc. Trans., 25, 1051 (1997).
19. C. Betancur, M. Azzi and W. Rostene, Trends Pharmacol. Sci., 18, 372 (1997).
20. R.M. Freidinger, Curr. Opin. Chem. Biol., 3, 395 (1999).
21. R.M. Snider, J.W. Constantine, J.A. Lowe III, K. P. Longo, W.S. Lebel, H.A. Woody, S.E. Drozda, M.C. Desai, F.J. Vinick, R.W. Spencer and H.-J. Hess, Science, 251, 435 (1991).
22. C. Garret, A. Carruette, V. Fardin, S. Moussaoui, J.-F. Peyronel, J.-C. Blanchard and P.M. Laduron, Proc. Natl. Acad. Sci. USA, 88, 10208 (1991).
23. K. Rudolf, W. Eberlein, W. Engel, H.A. Wieland, K.D. Willim, M. Entzeroth, W. Wienen, A.G. Beck-Sickinger and H.N. Dodds, Eur. J. Pharm., 271, R11 (1994).
24. C.J. Blankley, J.C. Hodges, S.R. Klutchko, R.J. Himmelsbach, A. Chucholowski, C.J. Connolly, S.J. Neergaard, M.S. Van Nieuwenhze, A. Sebastian, J. Quin III, A.D. Essenburg and D.M. Cohen, J. Med. Chem., 34, 3248 (1991).
25. D.-R. Armour, N.M. Aston, K.M.L. Morriss, M.S. Congreve, A.B. Hawcock, D. Marquess, J.E. Mordaunt, S.A. Richards and P. Ward, Bioorg. Med. Chem. Lett., 7, 2037 (1997).
26. J.D. Albright, M.F. Reich, E.G.D. Santos, J.P. Dusza, F.-W. Sum, A.M. Venkatesan, J. Coupet, P.S. Chan, X. Ru and H. Mazandarani, J. Med. Chem., 41, 2442 (1998).

27. H. Qi, S.K. Shah, M.A. Cascieri, S.J. Sadowski and M. MacCoss, Bioorg. Med. Chem. Lett., 8, 2259 (1998).
28. T. Harrison, M.P.G. Korsgaard, C.J. Swain, M.A. Cascieri, S. Sadowski and G.R. Seabrook, Bioorg. Med. Chem. Lett., 8, 1343 (1998).
29. P.D. Williams, M.G. Bock, B.E. Evans, R.M. Freidinger, S.N. Gallicchio, M.T. Guidotti, M.A. Jacobson, M.S. Kuo, M.R. Levy, E.V. Lis, S.R. Michelson, J.M. Pawluczyk, D.S. Perlow, D.J. Pettibone, A.G. Quigley, D.R. Reiss, C. Salvatore, K.J. Stauffer and C.J. Woyden, Bioorg. Med. Chem. Lett., 9, 1311 (1999).
30. J.B. Thomas, M.J. Fall, J.B. Cooper, R.B. Rothman, S.W. Mascarella, H. Xu, J.S. Partilla, C.M. Dersch, K.B. McCullough, B.E. Cantrell, D.M. Zimmerman and F.I. Carroll, J. Med. Chem., 41, 5188 (1998).
31. K. Rudolf, W. Eberlein, W. Engel, H. Pieper, H. Doods, G. Hallermayer and M. Entzeroth, Patent Application WO9811128A1 (1998).
32. D. Romer, H.H. Buscher, R.C. Hill, R. Maurer, T.J. Petcher, H. Zeugner, W. Benson, E. Finner, W. Milkowski, and P.W. Thies, Nature, 298, 759 (1982).
33. T.L. Peters, G. Matthijs, I. Depoortere, T. Cachet, J. Hoogmartens and G. Vantrappen, Biomed. Res., 9 (Suppl. 1), 21 (1988).
34. E. E. Sugg in Annual Reports in Medicinal Chemistry, Vol. 32, J.A. Bristol, Ed., Academic Press, San Diego, pp. 277, 1997.
35. C.J. Aquino, D.R. Armour, J.M. Berman, L.S. Birkemo, R.A.E. Carr, D.K. Croom, M. Dezube, R.W. Dougherty, Jr., G.N. Ervin, M.K. Grizzle, J.E. Head, G.C. Hirst, M.K. James, M.F. Johnson, L.J. Miller, K.L. Queen, T.J. Rimele, D.N. Smith and E.E. Sugg, J. Med. Chem., 39, 562 (1996).
36. S. Perlman, H.T. Schambye, R.A. Rivero, W.J. Greenlee, S.A. Hjorth and T.W. Schwartz, J. Biol. Chem., 270, 1493 (1995).
37. M. Asano, C. Hatori, H. Sawai, S. Johki, N. Inamura, H. Kayakiri, S. Satoh, Y. Abe, T. Inoue, Y. Sawada, T. Mizutani, T. Oku and K. Nakahara, Brit. J. Pharmacol., 124, 441 (1998).
38. A.A. Patchett, R.P. Nargund, J.R., Tata, M.-H. Chen, K.J. Barakat, D.B.R. Johnston, K. Cheng, W.W.-S. Chan, B. Butler, G. Hickey, T. Jacks, K. Schleim, S.-S. Pong, L.Y-P. Chaung, H.Y. Chen, E. Frazier, K.H. Leung, S.-H.L. Chiu and R.G. Smith, Proc. Natl. Acad. Sci. USA, 92, 7001 (1995).
39. R.P. Nargund, A.A. Patchett, M.A. Bach, M.G. Murphy and R.G. Smith, J. Med. Chem., 41, 3103 (1998).
40. J.R. Tata, Z. Lu, K. Cheng, L. Wei, W.W.-S. Chan, B. Butler, K.D. Schleim, T.M. Jacks, K. Leung, S.-H.L. Chiu, G. Hickey, R.G. Smith and A.A. Patchett, Bioorg. Med. Chem. Lett., 7, 2319 (1997).
41. L. Yang, G. Morriello, A.A. Patchett, K. Leung, T. Jacks, K. Cheng, K.D. Schleim, W. Feeney, W.W.-S. Chan, S.-H.L. Chiu and R.G. Smith, J. Med. Chem., 41, 2439 (1998).
42. P.A. Carpino, B.A. Lefker, S.M. Toler, L.C. Pan, E.R. Cook, J.N. DiBrino, W.A. Hada, J. Ithavongsay, F.M. Mangano, M.A. Mullins, D.F. Nickerson, J.A. Ragan, C.R. Rose, D.A. Tess, A.S. Wright, M.P. Zawistoski, C.M. Pirie, K. Chidsey-Frink, O.C. Ng, D.B. MacLean, J.C. Pettersen, P.A. DaSilva-Jardine and D.D. Thompson, Endocrine Soc. Mtg., New Orleans, LA, P2-188 (1999).
43. J.A. Dodge, M.D. Adrian, C.A. Alt, H.U. Bryant, M.P. Clay, D.M. Cohen, K.J. Fahey, M.L. Heiman, S.A. Jones, L.N. Jungheim, B.S. Muehl, J.J. Osborne, A.D. Palkowitz, G.A. Rhodes, R.L. Robey, K.J. Thrasher, T.A. Shepherd, L.L. Short, P.L. Surface, D.E. Seyler and T.D. Lindstrom, 218th Am. Chem. Soc. Natl. Mtg., New Orleans, LA, MEDI 135 (1999).
44. R. J. Devita, R. Bochis, A.J. Frontier, A.J. Kotliar, M.H. Fisher, W.R. Schoen, M.J. Wyvratt, K. Cheng, W.W.-S. Chan, B. Butler, T.M. Jacks, G.J. Hickey, K.D. Schleim, K. Leung, Z. Chen, S.L. Chiu, W.P. Feeney, P.K. Cunningham and R.G. Smith, J. Med. Chem., 41, 1716 (1998).
45. W.H. Parsons, A.A. Patchett, M.K. Holloway, G.M. Smith, J.L. Davidson, V.J. Lotti and R.S.L. Chang, J. Med. Chem., 32, 1681 (1989).
46. P. Lin, J.M. Pisano, W.R. Schoen, K. Cheng, W.W.-S. Chan, B.S. Butler, R.G. Smith, M.H. Fisher and M.J. Wyvratt, Bioorg. Med Chem. Lett., 9, 3237 (1999).
47. A.D. Howard., S.D. Feighner, D.F. Cully, J.P. Arena, P.A. Liberator, C.I. Rosenblum, M. Hamelin, D.L. Hreniuk, O.C. Palyha, J. Anderson, P.S. Paress, C. Diaz, M. Chou, K.K. Liu, K.K. McKee, S.-S. Pong, L.-Y.P. Chaung, A. Elbrecht, M. Dashkevicz, R. Heavens,

M. Rigby, D.J.S. Sirlnathsinghji, D.C. Dean, D.G. Melillo, A.A. Patchett, R.P. Nargund, P.R. Griffin, J.A. DeMartino, S.K. Gupta, J.M. Schaeffer, R.G. Smith, and L.H.T. Van der Ploeg, Science, <u>273</u>, 974 (1996).

48.   M Kojima, H. Hosoda, Y. Date, M. Nakazato, H. Matsuo and K. Kangawa, Nature, <u>402</u>, 656 (1999).

49.   L. Yang, S..C. Berk, S.P. Rohrer, R.T. Mosley, L. Guo, D.J. Underwood, B.H. Arison, E.T. Birzin, E.C. Hayes, S.W. Mitra, R.M. Parmar, K. Cheng, T.-J. Wu, B.S. Butler, F. Foor, A. Pasternak, Y. Pan, M. Silva, R.M. Freidinger, R.G. Smith, K. Chapman, J.M. Schaeffer and A.A. Patchett, Proc. Natl. Acad. Sci. USA, <u>95</u>, 10836 (1998).

50.   L. Yang, Y. Pan, L. Guo, G. Morriello, A. Pasternak, S. Rohrer, J. Schaeffer and A.A. Patchett in Peptides, Chemistry and Biology: Proceedings of the 16th American Peptide Symposium, ESCOM, Leiden, The Netherlands, in press.

51.   R.P. Nargund, K. Barakat in Peptides, Chemistry and Biology: Proceedings of the 16th American Peptide Symposium, ESCOM, Leiden, The Netherlands, in press.

52.   N. Suzuki, K. Kato, S. Takekawa, J. Auchi and S. Endo, PCT Patent Publication, WO9952875-A1 (1999).

53.   S. Berk, S.P. Rohrer, S.J. Degrado, E.T. Birzin, R.T. Mosley, S.M. Hutchins, A. Pasternak, J.M. Schaeffer, D.J. Underwood and K.T. Chapman, J. Comb. Chem., <u>1</u>, 388 (1999).

54.   S. E. deLaszlo, E.E. Allen, B. Li, D. Ondeyka, R. Rivero, L. Malkowitz, C. Molineaux, S.J. Siciliano, M.S. Springer, W.J. Greenlee and N. Mantlo, Bioorg. Med. Chem. Lett., <u>7</u>, 213 (1997).

55.   R.P. Nargund, Z. Ye, B. Palucki, R. Bakshi, A.A. Patchett and L. Van der Ploeg, PCT Patent Publication, WO9964002 (1999).

56.   A.A. Failli, J.S. Shumsky and R.J. Steffan, PCT Patent Publication, WO9906403-A1 (1998).

57.   L.A. Thompson and J.A. Ellman, Chem. Rev., <u>96</u>, 555 (1996).

58.   R.E. Dolle, Mol. Diversity, <u>3</u>, 199 (1998).

59.   B.R. Neustadt, E.M. Smith, N. Lindo, T. Nechuta, A. Bronnenkant, A. Wu, L. Armstrong and C. Kumar, Bioorg. Med. Chem. Lett., <u>8</u>, 2395 (1998).

60.   M.F. Gordeev, D.V. Patel, B.P. England, S. Jonnalagadda, J.D. Combs and E.M. Gordon, Bioorg. Med. Chem., <u>6</u>, 883 (1998).

61.   R.P. Sheridan and M.D. Miller, J. Chem. Inf. Comput. Sci., <u>38</u>, 915 (1998).

62.   X.Q. Lewell, D.B. Judd, S.P. Watson and M.M. Hann, J. Med. Chem. Inf. Comput. Sci., <u>38</u>, 511 (1998).

63.   J.S. Mason, I. Morize, P.R. Menard, D.L. Cheney, C. Hulme and R.F. Labaudiniere, J. Med. Chem., <u>42</u>, 3251 (1999).

Chapter 27 . *Ex Vivo* Approaches to Predicting Oral Pharmacokinetics in Humans

Barbra H. Stewart, Yi Wang, Narayanan Surendran
Parke-Davis Pharmaceutical Research, Ann Arbor MI   48105

Introduction – The competitive pharmaceutical market of the new millennium has brought an emphasis on "process" to drug discovery. Previously, customized and mechanistic studies could be conducted more-or-less routinely on a nice-to-know basis. Optimizing biological activity was the driving force and major criterion for declaration of a lead candidate. Absorption-Distribution-Metabolism-Excretion (ADME) and biopharmaceutical considerations were secondary. Poor drug delivery characteristics often resulted in a heightened failure rate, as well as expensive and lengthy development times (1). Today, lead selection may involve the testing of hundreds of thousands of candidates and speed-to-market is major factor in drug development. The numbers of compounds being tested and the need for earlier, more informed decisions with regard to drug selection has manifested in a paradigm change in drug discovery. The paradigm change is co-optimization of biological activity and drug delivery properties from very early stages of discovery. This approach enables risk analysis at the early discovery stage with subsequent selection of higher quality lead compounds that should progress through development with fewer problems.

Testing for the characteristics that make a New Chemical Entity (NCE) orally available requires a panel of *in vitro* experiments that can be conducted in relatively high throughput. After oral administration, the serial barriers to systemic availability are stability and adequate solubility in the intestinal lumen, transport across the intestinal epithelia, and first-pass through the liver. Unfavorable physicochemical (PC) properties or biological processes can limit delivery (2). Here we will describe methods for determining PC properties, as well as intestinal permeability and metabolic clearance. The progression of models is from *in vitro* to *in vivo*, or simpler to complex. The information from these tests can be integrated to be more meaningful in drug selection or risk analysis; eg, the extent of absorption as a function of both solubility in gastrointestinal fluids and permeability across intestinal membranes.

PHYSICOCHEMICAL PROPERTIES FOR PREDICTING ORAL ABSORPTION

Calculated versus measured properties – The advantages of using calculated parameters are two-fold: 1) the potential for multifactorial input and 2) the capacity for ultra-high throughput. Using computational neural networks, one- and two-dimensional molecular descriptors and large databases (eg, Comprehensive Medicinal Chemistry (CMC)), a model was developed to distinguish "drug-like" from "nondrug-like" molecules (3). Approximately 90% of molecules in the CMC library could be ascribed to a set of 1D/2D descriptors. Based on quantitative structure-property relationships (QSPRs), a nonlinear neural network model was derived that included six descriptors to estimate human intestinal absorption (HIA) (4). Applied to a set of compounds independent of the training set, this model resulted in a 16% error and linear correlations between predicted and calculated HIA. Taken with throughput limited only by computation time, calculated approaches can be employed to design combinatorial and virtual libraries. Currently, it would be the nearly unanimous view of researchers in this field that computational models are not yet adequate to design molecules *a priori* and that experimental methods will be needed for some time to come.

Calculated properties – Publication of the so-called Rule-of-Five generated widespread industrial interest in applying calculated PC properties to drug discovery. In this paradigm, poor intestinal absorption was associated with molecules possessing any two of the following properties (5): molecular weight > 500, number of hydrogen bond donors > 5, number of hydrogen bond acceptors > 10, or calculated log P > 5. These guidelines have been useful in an approximate manner; moreover, curiosity, if not acceptance of the paradigm suggests interest in a more systematic approach in drug discovery.

More recent examination of the relationship of molecular surface properties with biological performance of molecules has been revealing. Most notably, it has been demonstrated that polar surface area (PSA) has a strong correlation with transport across biological membranes (6). PSA is defined as the area of van der Waals surface that derives from oxygen or nitrogen atoms, or hydrogen atoms attached to oxygen or nitrogen atoms; thus, it is related to the hydrogen bonding capacity of the molecule. Dynamic PSA (PSA_d) is calculated as a Boltzmann-weighted average from multiple low-energy conformers for the van der Waals or solvent-accessible molecular surface. PSA_d has shown value in predicting Caco-2 membrane permeability and oral absorption in humans (7,8,9). Interpolation of the sigmoidal curve derived from a set of twenty selected compounds suggested that when PSA_d >140 A^2, incomplete absorption (<10%) resulted; furthermore, when PSA_d < 60 A^2, absorption was extensive (>90%) (10). Compared to log D (octanol:water) and immobilized liposome chromatography for an analogous series of β-blockers, PSA_d had an improved correlation with Caco-2 cell permeability (r^2 = 0.97 vs 0.72 or 0.84, respectively, in the linear model) (11).

The computation time per molecule for PSA_d is considerable (ie, hours to days), which initially rendered this approach unsuitable for higher-throughput intentions. Recently, it was demonstrated that calculation of PSA from a single conformer gave excellent correlation with intestinal absorption (12). The reduced computation time using single conformer analysis enhances the potential of this technique for application prior to synthesis. It's been suggested that passive permeation of all biological membranes is determined by similar factors (13). It follows that if PSA is useful in modeling intestinal absorption, it would have value in predicting blood-brain barrier transport. The utility of PSA was shown to discriminate molecules with the ability to traverse the blood-brain barrier and that PSA < 60 to 70 A^2 favors brain penetration (14,15,16).

Measures of lipophilicity – The lipophilicity parameter, or log P, has most direct significance in its relationship with passive membrane transport, although mathematical correlations can be complex (17,18). Lipophilicity is frequently correlated with extent of protein binding within chemical series, and factors into prediction of pharmacokinetic parameters, such as clearance (19). Advances in lipophilicity research are generally two-fold: 1) to render the calculated version, ClogP, more predictive; and 2) enhance facility, throughput, or predictability of the experimental approach (20,21). For example, an automated potentiometric titration technique was developed to measure drug partitioning in a liposomal system (22). Substitution of the liposomal partitioning parameter for log P in the calculation of Absorption Potential (AP) (23) resulted in a sigmoidal correlation of $AP_{liposome}$ with HIA, a finding not demonstrable when octanol was the partitioning phase.

Immobilized artificial membranes (IAMs) refer to a chromatographic model of the cellular membrane lipid environment where monolayers of purified phospholipids are covalently bonded to silica particle support (24,25). IAM columns are available commercially with phosphatidylcholine packing (IAM.PC.DD, Regis

Chemical, Morton Grove IL) and provide a relatively fast, simple method.   The primary characteristic that distinguishes IAM chromatography from other lipophilicity techniques is the presence of the ordered phospholipid surface that supports partitioning behavior based on both lipophilic and electrostatic interactions (26). IAM capacity factors (k'$_{IAM}$) were evaluated favorably as a model for intestinal absorption (27).   Positive correlations of k'$_{IAM}$ with uptake into T-cell suspensions and Caco-2 cell monolayers were observed in a study of HIV protease inhibitor design (28).

Measures of solubility – As in lipophilicity research, advances have been made toward higher throughput and greater predictive value, both from calculated and experimental approaches.   The turbidometric solubility measurement has become popular for use in drug discovery because it is rapid and automated, albeit yielding approximate values relative to the equilibrium measurement (5). An amended solvation energy relationship was derived for prediction of aqueous solubility (log S$_W$) from experimental data *or* chemical structure. The derived parameter, log S$_W$, could be predicted within 0.56 log units and could be corrected for Brønsted acids and Brønsted bases using a predicted pK$_a$ value (29).

PERMEABILITY MODELS IN EARLY DISCOVERY

Although a variety of models (subcellular fractions, cell monolayer models, isolated intestinal tissue, intestinal perfusion, transfected cell lines) are available, cell monolayer models and rat intestinal perfusion techniques are the most commonly used systems in the pharmaceutical industry today(30,31).

Cell monolayers – These models consist of cells grown on permeable inserts. Transport of compound across the cell monolayer can be used to quantitate the permeability of NCEs in a rapid fashion. One of the most popular cell lines is Caco-2, derived from human colon adenocarcinoma cells. The monolayers exhibit ion conductance and possess transepithelial electrical resistance indicative of fully formed tight junctions that restrict the paracellular transport of NCEs and nutrients (32,33). While Caco-2 cells are the most commonly used cells, Madin-Darby Canine Kidney (MDCK) cells are becoming more widespread in use.   The popularity of MDCK cells is due in part to the shorter culture times needed before use in experiments (4-7 days of culture vs   21-30 days for Caco-2 cells).   Excellent correlation of permeability coefficients (P$_{app}$) between MDCK and Caco-2 cells was observed for 55 compounds with known HIA (34). The availability of 3-day systems for Caco-2 (e.g., BIOCOAT) has mitigated some of these concerns (35). Regardless of the cell line used for permeability measurements, establishing correlation between P$_{app}$ and fraction of dose absorbed (F$_{abs}$) *in vivo* validates the approach (36,37,38). Correlation plots can be used to interpolate the *in vivo* F$_{abs}$ of an NCE from P$_{app}$. Cell monolayer models grown on permeable supports can be used to examine bi-directional permeability (apical-to-basolateral and basolateral-to-apical) to determine the extent and role of efflux transporters (e.g., P-gp, MRP) in the intestinal transport of NCEs.   Caco-2 clones and MDR1-transfected cell lines are the most widely studies in literature (39,40).   Transgenic animal models (mdr1a, mdr1a/b knock-out mice) serve as an additional tool to confirm the effects of efflux-mediated drug transport on the absorption and disposition of drugs *in vivo* (41). Limitations of the cell models, such as Caco-2, arise from the fact that these cells are homogenous and lack mucus; furthermore, quantitative and qualitative differences exist in the expression of transporters and metabolic enzymes compared to the *in vivo* situation.

Cytochrome P450 (CYP) metabolizing enzymes have been enhanced in Caco-2 cells by producing specific clones of Caco-2 (TC7) (42) or by increasing enzyme expression with the addition of agents such as 1,25-dihydroxy-vitamin D3 (43) Other disadvantages of the cell monolayer models are related to potential experimental difficulties for specific NCEs (poor solubility, high non-specific adsorption) and lab-to -lab variability in P_{app}. Poor solubility of NCEs can be partly addressed by incorporating co-solvents such as ethanol, DMSO or DMA in small quantities that will enable solubilization during the experiment.   Non-specific adsorption can be addressed by numerical corrections as shown by Chan et al (44). Caco-2 cells are of colonic origin, thus inherently high transepithelial electrical resistance and low ion conductance may result in underestimation of permeabilities of compounds that traverse the paracellular pathway. Caution should be exercised when examining the permeability of NCEs that are low molecular weight (< 300) and negative Log D (distribution coefficient).

In addition to modulation of CYP450 enzyme activities, alternatives or modifications of the cell monolayer system have been reported that may make it more useful for early drug discovery support. The use of artificial lipid bilayers has been used in a high-throughput fashion (96-well plate format) with UV detection to quantitate the permeability characteristics of reference compounds (45).   The efficiency of initial screening of NCEs for permeability characteristics has been enhanced by application of the "cocktail" approach for screening mixtures from combinatorial libraries (46).   The ability to use a sensitive and specific detection technique such as the LC/MS/MS makes this a reality. This approach may give confusing results if there is saturation of transporters or metabolic enzymes in the transepithelial transport of mixtures of NCEs.   The utility of LC/MS to increase throughput has been recently confirmed by assessing the transport of a peptide combinatorial library containing 375,000 individual peptides (47). By combining the above compounds into 150 pools containing 2500 tripeptide sequences, permeable sequences were rapidly identified. In another application, the intrinsic bioactivity of NCEs was used to quantitate the permeability of a series of compounds (48). To predict dissolution-absorption relationship *a priori* in humans, a dissolution/Caco-2 system has been developed (49).   The relationship between dissolution and absorption for each formulation was evaluated and found to be predictive of clinical studies.

ESTIMATED DOSE ABSORBED (EDA)

Permeability is a *component* of intestinal absorption, with the other major component being drug solubility in the gastrointestinal tract. To combine these two factors to estimate HIA, the following equation has been proposed (50):

$$EDA \text{ (mg)} = k_a * V * C_s * t_{res,} \text{ where } k_a = 2 \ P_{eff\cdot human} / r \qquad \text{Equation 1}$$

Where k_a denotes absorption rate constant and C_s is solubility.   V (volume in the GI tract), t_{res} (residence time) and r (radius of human intestine) are held constant at 250 mL, 4 h and 2 cm, respectively. For compounds that have high aqueous solubility, calculation of EDA using the above formula can be accomplished by substituting C_s with Dose/Volume.   The relationship between EDA and F_{abs} is hyperbolic (51), similar to the relationship between permeability and F_{abs}.   Using solubility in Equation 1 gave realistic estimates of EDA for poorly to moderately soluble compounds, while Dose/Volume was appropriate for compounds with high solubility.

The concept of EDA is intuitive and serves to combine two major determinants of oral absorption i.e., intestinal permeability and solubility. EDA can provide a practical guideline in the early discovery setting.   If the dose of a development candidate is projected to be greater than the calculated EDA, it serves as a catalyst to examine: 1) the potential for solubilization in the GI tract; 2) formulation variables that could ultimately result in increased solubility or dissolution during GI transit in order to increase F_{abs}.   EDA can be used to rank order compounds and structural series. In turn, this can direct medicinal chemistry efforts toward series with the greatest potential for absorption.

HEPATIC CLEARANCE ESTIMATIONS FROM *IN VITRO* SYSTEMS

Total clearance of a drug in humans is the sum of all individual organ clearances that contribute to the overall elimination. For many marketed drugs, hepatic clearance is a major contributor to removal of a drug from the body. Almost 70% of the dose of newly approved anti-AIDS drugs indinavir, ritonavir and nelfinavir is metabolized by liver enzymes (52,53). Hepatic clearance may decrease or increase because of enzyme inhibition or induction by co-administered drugs. Auto-induction or auto-inhibition of the drug may also occur after multiple dosing (54,55,56). If the drug has a narrow therapeutic window, regulation of metabolizing enzymes may play a critical role in pharmacological effect and toxicity (57).

Estimating *in vivo* hepatic clearance -- Total clearance (CL_T) is estimated from plasma concentration data after IV dosing. In the discovery scenario, this is generally conducted with cassette- or unit-dosing to rats. To obtain hepatic clearance, one needs to collect urine over a time period (at least 5 half-lives) to determine the total amount of metabolite excreted. The *in vivo* hepatic clearance is obtained from equation 2.

$$CL_H = fm \times CL_T \qquad\qquad \text{Equation 2}$$

CL_T is calculated from plasma data and f_m is the metabolite formation fraction determined from urine collection data. If there is more than one metabolite, the sum of f_m will be needed.

Prediction of *in vivo* hepatic clearance from *in vitro* systems -- Scaling factors are needed for various species to relate *in vitro* intrinsic clearance to *in vivo* intrinsic clearance. Scaling factors for man with 75 kg body weight are 1.35×10^8 cells/g liver, 45mg hepatic microsomes/g liver and 20g liver/kg body weight (58,59).

The interrelationship between hepatic clearance, hepatic blood flow (Q_H), *in vivo* intrinsic clearance and unbound fraction of a drug in the blood (f_u) has been described in Equation 3 (60).

$$CL_H = \frac{Q_H \times f_u \times CL_{int(invivo)}}{Q_H + f_u \times CL_{int(invivo)}} \qquad\qquad \text{Equation 3}$$

Based on this equation, f_u and *in vivo* $CL_{int(invivo)}$ are both needed for the estimation of hepatic clearance. $f_{u(blood)}$ can be determined using *in vitro* and *ex vivo* methods, such as ultra-filtration, and human serum albumin chromatographic column (61,62,63,64).

Determination of *in vitro* intrinsic clearance -- Two approaches have been introduced to determine this parameter. Houston proposed the use of metabolite formation, including Michaelis - Menten parameters K_m and V_{max} of the drug in question. Metabolite information is obtained from liver microsomes or hepatocytes with factoring of nonspecific protein binding (65). Another approach proposed by Obach (66,67) is to calculate *in vitro* half-life from the elimination rate constant directly obtained from microsomal incubation. The latter approach has the advantage that only the disappearance of the parent drug in the *in vitro* system is measured.

Previously, isolated hepatocytes have been used to directly calculate the elimination rate constant (68). Theoretically, the use of drug disappearance profiles to determine *in vitro* intrinsic clearance is feasible in microsomes(69).

Correlation of *in vitro* estimations with *in vivo* performance -- The correlation of hepatic clearance predicted from *in vitro* systems with *in vivo* results has been studied by numerous investigators (70,71,72,73,74,75). Recently, Iwatsubo et al have used human and dog microsomal metabolism incubation data to successfully predict hepatic clearance of TM796 in two species (76). In another report, recombinant P450 isoenzymes (CYP3A4) and hepatic microsomes have been used together to predict oral metabolic clearance, taking into account the fraction absorbed from the gut (77,78). *In vitro* microsomal studies have provided a good prediction of the *in vivo* total body clearance for caffeine (79). Moreover, Lave et al have proposed that together with human *in vitro* data, allometric scaling can accurately predict clearance in man (80,81). In this report, 80% of the predictions were within a 2-fold factor of the actual data in man.

Metabolic stability approach and *in vitro* systems -- Use of drug disappearance to determine *in vitro* intrinsic clearance is termed the metabolic stability approach, widely accepted in drug discovery.   Isolated liver perfusion, liver slices, cryopreserved or fresh isolated hepatocytes, hepatic microsomes and microsomes with cDNA-expressed CYP can be used to measure metabolic systems (82,83,84,85,86,87,88). The main attraction of *in vitro* incubations is their simplicity, but essentially, such systems are static and contrast with the *in vivo* situation, which has greater complexity and dynamic nature. The higher-throughput cellular and subcellular methods will be discussed here.

Hepatocytes contain the full complement of Phase1 and Phase 2 metabolizing enzymes including the cytochrome P450s, flavomonooxygenases, alcohol dehydrogenase, aldehyde dehydrogenase, aldehyde oxidase, diamine oxidase, acetyltransferase, sulfotransferase, methyltransferase and UDP-glucuronosyltransferases. Hepatocytes can provide unique information not only on the direct Phase 2 reactions, but also sequential metabolism. Compared to microsomal incubation, hepatocytes provide a more realistic prediction for the *in vivo* value (89). The hepatocyte model is useful in the study of induction and inhibition (90). A disadvantage of this preparation is that CYP enzyme activity in the freshly isolated hepatocytes is poorly retained (91). Cryopreserved hepatocytes are a potential alternative: they are commercially available, are a good pooled resource and maintain CYP activity (92).

Although *hepatic microsomes* do not have most Phase 2 metabolizing enzymes, the model does have special features. Notably, it can be stored at -80°C for months, even years. Generally speaking, the human liver microsome (HLM) system is first-line in metabolic stability testing in drug discovery. If the NCE is primarily cleared by Phase 1 metabolism, prediction from microsomal incubation data will be similar to *in vivo* hepatic clearance. For instance, the *in vivo* intrinsic clearance of P450 2C9 substrates phenytoin, tolbutamide (S)-ibuprofen and diclofenac was successfully predicted from microsomal incubation data (93). Metabolic stability in the microsomal system, however, does not obviate   high clearance in more enzyme-complete systems.

Using microsomes with *expressed CYP enzymes* (Gentest) to predict *in vivo* intrinsic clearance can be problematic. The level of individual CYP has been enhanced, which is not comparable to the *in vivo* liver system. Intrinsic clearance measured with a single isozyme will not be predictive for *in vivo* intrinsic clearance if multiple enzymes mediate metabolism. Conversely, if it is known that a single isozyme is responsible for clearing the drug, CYP-enhanced microsomes can be useful.

Available evidence supports using *in vitro* human metabolic stability to predict hepatic clearance in humans. For NCEs mainly cleared by hepatic metabolism, good correlation between hepatic clearance *in vivo* and estimated hepatic clearance from *in vitro* data can be expected. More predictive computer models that include physiological factors are becoming available (94).

INTEGRATING INFORMATION FOR RISK ANALYSIS

The goal of early-ADME profiling is to answer first-cut questions: Will the NCE be absorbed? Can the NCE be formulated readily? Will metabolic liability limit systemic exposure? ADME risk analysis combines experimental and derived parameters with assigned limits. For example, in order to deliver 1 mg/kg dose and assuming moderate permeability of 10 to 20x 10^{-6} cm/s, it would be necessary to achieve a solubility concentration of at least 50 microg/mL in the GI tract (95).

Examination of the individual parameters provides insight on where and how to address problem areas, whether by synthesis or formulation. Other factors, such as potency and competitive position in therapeutic class, can then be rationally balanced against ADME risk. This provides the basis for selection of quality lead candidates for development, as well as the discovery of back-up candidates with properties enhanced over the initial lead.

References

1.   [1]R.A. Lipper, Modern Drug Discovery, 2, 55 (1999).
2.   R.E. Stratford, Jr., M.P. Clay, B.A. Heinz, M.T. Kuhfeld, S.J. Osborne, K.L. Phillips, S.A. Sweetana, M.J. Tebbe, V. Vasudevan, L.L. Zornes, T.D. Lindstrom, J. Pharm. Sci., 88, 747 (1999).
3.   Ajay, W.P. Walters, M.A. Murcko, J. Med. Chem., 41, 3314 (1998).
4.   M.D. Wessel, P.C. Jurs, J.W. Tolan, S.M. Muskal, J. Chem. Ing. Conput. Sci., 38, 726 (1998).
5.   C.A. Lipinski, F., Lombardo, B.W. Dominy, P.J. Feeney, Adv. Drug Deliv. Rev., 23, 3 (1997).
6.   H. van de Waterbeemd, G. Camenisch, G. Folkers, O.A. Raevsky, Quant. Struct.-Act. Relat., 15, 480(1996).
7.   P. Stenberg, K. Luthman, P. Artursson, Pharm. Res., 16, 205 (1999).
8.   K. Palm, K. Luthman, A-I. Ungell, G. Strandlund, P. Artursson, J. Pharm. Sci., 85, 32 (1996).
9.   L.H. Krarup, I.T. Christensen, L. Hovgaard, S. Frokjaer, Pharm. Res., 15, 972 (1998).
10.   K. Palm, P. Stenberg, K. Luthman, P. Artursson, Pharm. Res., 14, 568 (1997).
11.   K. Palm, K. Luthman, A-I. Ungell, G. Strandlund, F. Berigi, P. Lundahl, P. Artursson, J. Med. Chem., 41, 5382 (1998).
12.   D.E. Clark, J. Pharm. Sci., 88, 807 (1999).
13.   H. Lennernšs, J. Pharm. Pharmacol., 49, 627 (1997).
14.   D.E. Clark, J. Pharm. Sci., 88, 815 (1999).
15.   H. van de Waterbeemd, G. Camenisch, G. Folkers, J.R. Chretien, O.A. Raevsky, J. Drug Target., 6, 151(1998).
16.   J. Kelder, P.D.J. Grootenhuis, D.M. Bayada, L.P.C. Delbressine, J-p. Ploemen, Pharm. Res., 16, 1514 (1999).
17.   N.F.J. Ho, J.Y. Park, W. Morozowich, W.I. Higuchi in "Design of Biopharmaceutical Properties through Prodrugs and Analogs," E.B. Roche, Ed., APhA/APS, Washington, D.C., 1977, p 136.
18.   M. Yazdanian, S.L. Glynn, J.L. Wright, A. Hawi, Pharm. Res., 15, 1490 (1998).
19.   A.M. Davis, P.J.H. Webborn, D.W. Salt, Drug Metab. Disposition, 28, 103 (2000).

20.   Leo, A. J. in "Molecular Design and Modeling: Concepts and Applications," Vol.
        202A of Methods in Enzymology, Part A, Proteins, Peptides and Enzymes, J. J.
        Langone, Ed., Academic Press, New York, N.Y., 1991, p 544.
21.   H. van de Waterbeemd, R. Mannhold in "Lipophilicity in Drug Action and Toxicology",
        Vol. 4 of Methods and Principles in Medicinal Chemistry; V. Pliška, B. Testa, H. van
        de Waterbeemd, Eds., VCH Publishers, Inc., New York, N.Y., 1996, p 402.
22.   K. Balon, B.U. Riebesehl, B.W. Müller, Pharm. Res., 16, 882 (1999).
23.   J.B. Dressman, G.L. Amidon, D. Fleisher, J. Pharm. Sci., 74, 588 (1985).
24.   H.L. Liu, S.W. Ong, L. Glunz, C. Pidgeon, Anal. Chem., 67, 3550 (1995).
25.   C.Y. Yang, S.J. Cai, H. Liu, C. Pidgeon, Adv. Drug Del. Rev., 23, 229 (1997).
26.   B.H. Stewart, O.H. Chan, J. Pharm. Sci., 87, 1471 (1998).
27.   C. Pidgeon, S. Ong, H. Liu, X. Qiu, M. Pidgeon, A.H. Dantzig, J. Munroe, W.J.
        Hornback, J.S. Kasher, L. Glunz, T. Szczerba, J. Med. Chem,., 38, 590 (1995).
28.   B.H. Stewart, F.Y. Chung, B. Tait, C.J. Blankley, O.H. Chan, Pharm. Res., 15, 1401
        (1998).
29.   M.H. Abraham, J. Le, J. Pharm. Sci., 88, 868 (1999).
30.   D. Fleisher in "Peptide-based Drug Design", M.D. Taylor, G.L. Amidon, Eds., ACS
        Books, American Chemical Society, Washington D.C, 1995, p 500.
31.   G. L. Amidon, P. J. Sinko, D. Fleisher, Pharm. Res., 5, 651 (1988).
32.   M. Pinto, S. Robine-Leon, M.D. Appay, M. Kedinger, N. Triadou, E. Dussaulx, B.
        Lacroix, P. Simon-Assman, K. Haffen, J. Fogh. A. Zweibaum, Biol. Cell., 4, 323
        (1983).
33.   E. Grasset, M. Pinto, E. Dussaulx, A. Zweibaum. J. F. Desjeux, Am. J. Physiol., 247,
        C260 (1984).
34.   J. D. Irvine, L. Takahashi, K. Lockhart, J. Cheong, J. W. Tolan, H. E. Selick, J. R.
        Grove, J. Pharm. Sci., 88, 1, 28 (1999).
35.   S. Chong, S.A. Dando, R. A. Morrison, Pharm. Res., 14, 1835 (1997).
36.   P. Artursson, J. Karlsson, Biochem. Biophys. Res. Comm., 175, 880 (1991).
37.   R.A. Conradi, K. F. Wilkinson, B.D. Rush, A.R. Hilgers, M. J. Ruwart, P. S. Burton,
        Pharm. Res., 10, 1710 (1993).
38.   B. H. Stewart, O. H. Chan, R. H. Lu, E. L. Reyner, H. L. Schmid, H. W. Hamilton, B.
        A. Steinbaugh, M. D. Taylor, Pharm. Res., 12, 693 (1995).
39 .   Y. Zhang, L. Z. Benet, Pharm. Res., 15, 1520 (1998).
40.   J. Alsenz, H. Steffen, R. Alex, Pharm. Res., 15, 423 (1998).
41 .   M.F. Fromm, R.B. Kim, M. Stein, G.R. Wilkinson, D.M. Roden, Circulation, 99, 552
        (1999).
42.   S.D. Raeissi, I.J. Hidalgo, J. Segura-Aguilar, P. Artursson, Pharm. Res., 16, 625
        (1999).
43.   P. Schmiedlin-Ren, K. E. Thummel, J. M. Fisher, M. F. Paine, K. S. Lown, P. B.
        Watkins, Mol. Pharmacol. 51, 741 (1997).
44.   O. H. Chan, H. L. Schmid, B-S. Kuo, D. S. Wright, W. Howson, B.H. Stewart, J.
        Pharm. Sci., 85, 3, 253 (1996).
45.   M Kansy, F. Senner, K. Gubernator, J. Med. Chem., 41, 7, 1007 (1998).
46.   E. W. Taylor, J. A. Gibbons, R. A. Braeckman, Pharm. Res., 14, 5, 572 (1997).
47.   C.L. Stevenson, P.F. Augustijns, R.W. Hendren, Int. J. Pharm., 177, 103 (1999).
48.   W. Rubas, M. E. M. Cromwell, R. J. Mrsny, G. Ingle, K. A. Elias, Pharm. Res., 13, 23
        (1996).
49.   M. J. Ginski, J. E. Polli, Int. J. Pharm., 177, 117 (1999).
50.   K.C. Johnson, A.C. Swindell, Pharm. Res., 13, 1795 (1996).
51.   N. Surendran, O.H. Chan, M.T. Whittico, C.L. Stoner, B.H.Stewart, Pharm. Res., 16s
        (1999).
52.   M. Barry, S. Gibbons, D. Back, F. Mulcahy, Clin. Pharmacokinet., 32, 194 (1997).
53.   J.H. Lin, M. Chiba, S.K. Balani, I.W. Chen, G.Y. Kwei, K.J. Vastag, J.A. Nishime,
        Drug Metab. Dispos., 24, 1111 (1996).
54.   Hsu, G.R. Granneman, G. Witt, C. Locke, J. Denissen, A. Molla, J. Valdes, J. Smith,
        K. Erdman, N. Lyons, P. Niu, J.P. Decourt, J.B. Fourtillan, J. Girault, J.M. Leonard,
        Antimicrob. Agents Chemother., 41, 898 (1997).
55.   A.T. Bowdle, I.H. Patel, R.H. Levy, A.J. Wilensky. Clin. Pharmacol. Ther., 28, 486
        (1980).
56.   S. Rohatagi, J.S. Barrett, W. Sawyers, K. Yu, R.J. Morales, Am. J. Ther., 4, 229
        (1997).

57.   K. Venkatakrishnan, L.L. von Moltke, D.J. Greenblatt, Clin. Pharmacokinet., 38,111 (2000).
58.   R.S. Obach, Drug Metab. Dispos., 27, 1350 (1999).
59.   J.B. Houston, Biochem. Pharmacol., 47, 1469 (1994).
60.   G.R. Wilkinson, Pharmacol. Rev., 39, 1 (1987).
61.   D. Colussi, C. Parisot, F. Legay, G. Lefevre, Eur. J. Pharm. Sci., 9: 9 (1999).
62.   M.J. Garrido, C. Aguirre, I.F. Troconiz, M. Marot, M. Valle, M.K. Zamacona, R. Calvo, Int. J. Clin. Pharmacol. Ther., 38, 35 (2000).
63.   F. Beaudry, M. Coutu, N.K. Brown, Biomed. Chromatogr., 13, 401(1999).
64.   Z. Liu, F, Li, Y. Huang, Biomed. Chromatogr., 13, 262 (1999).
65.   J.B. Houston, and D.J. Carlile, Drug Metab. Rev., 29, 891 (1997).
66.   R.S. Obach, Drug Metab. Dispos., 25, 1359 (1997).
67.   R.S. Obach, J.G. Baxter, T.E. Liston, B.M. Silber, B.C. Jones, F. MacIntyre, D.J. Rance, P. Wastall, J. Pharmacol. Exp. Ther., 283, 46 (1997).
68.   E. Bodd, C.A. Drevon, N. Kveseth, H. Olsen, J. Morland, J. Pharmacol. Exp. Ther., 237, 260 (1986).
69.   J.D. Carlile, A. J. Stevens, E.I.L. Ashforth, D. Waghela, and J. B. Houston, Drug Metab. Dispos., 26, 216 (1998).
70.   D.J. Carlile, K. Zomorodi, J.B. Houston, Drug Metab. Dispos., 25, 903 (1997).
71.   T. Iwatsubo, A. Hisaka, H. Suzuki, Y. Sugiyama, J. Pharmacol. Exp. Ther., 286, 122 (1998).
72.   K. Matsui, S. Taniguchi, T. Yoshimura, Xenobiotica., 29, 1059 (1999).
73.   E. Tanaka, A. Ishikawa, T. Horie, Hum. Exp. Toxicol., 18, 12 (1999).
74.   L.E. Witherow, J.B. Houston, J. Pharmacol. Exp. Ther., 290, 58 (1999).
75.   B.A. Hoener, Biopharm. Drug Dispos., 15, 295 (1994).
76.   T. Iwatsubo, H. Suzuki, Y. Sugiyama, J. Pharmacol. Exp. Ther., 283, 462 (1997A).
77.   T. Iwatsubo, N. Hirota, T. Ooie, H. Suzuki, Y. Sugiyama, Biopharm. Drug Dispos., 17, 273 (1996).
78.   T. Iwatsubo, H. Suzuki, N. Shimada, K. Chiba, T. Ishizaki, C.E. Green, C.A. Tyson, T T. Yokoi, T. Kamataki, Y. Sugiyama, J. Pharmacol. Exp. Ther., 282, 909 (1997B).
79.   K.A. Hayes, B. Brennan, R. Chenery, J.B. Houston, Drug Metab. Dispos., 23, 349 (1995).
80.   T. Lave, S. Dupin, C. Schmitt, R.C. Chou, D. Jaeck, P. Coassolo, J. Pharm. Sci., 86, 584 (1997).
81.   T. Lave, P. Coassolo, B. Reigner, Clin. Pharmacokinet., 36, 211 (1999).
82.   J.P. Villeneuve, M. Dagenais, P.M. Huet, R. Lapointe, A. Roy, D. Marleau, Can. J. Physiol. Pharmacol., 74,1327 (1996).
83.   N. Zaman, Y.K. Tam, L.D. Jewel, R.T. Coutts, J. Parenter. Enteral Nutr., 20, 349 (1996).
84.   R. Ishida, K. Suzuki, Y. Masubuchi, S. Narimatsu, S. Fujita, T. Suzuki, Biochem. Pharmacol., 44, 2281 (1992).
85.   A.B. Renwick, H. Mistry, P.T. Barton, F. Mallet, R.J. Price, J.A. Beamand, B.G. Lake, Food Chem. Toxicol., 37, 609 (1999).
86.   A.E. Vickers, R.M. Jimenez, M.C. Spaans, V. Pflimlin, R.L. Fisher, K. Brendel, Drug Metab. Dispos., 25, 873 (1997).
87.   P.D. Worboys, A. Bradbury, J.B. Houston, Drug Metab. Dispos., 24, 676 (1996A).
88.   D.J. Carlile, N, Hakooz, J.B. Houston, Drug Metab. Dispos., 27, 526 (1999).
89.   K. Zomorodi, D.J. Carlile, J.B. Houston, Xenobiotica., 25, 907 (1995).
90.   J.A. Holme, NIPH Ann., 8, 49 (1985).
91.   L.B. Tee, T. Seddon, A.R. Boobis, D.S. Davies, Br. J. Clin. Pharmacol., 19, 279 (1985).
92.   A.P. Li, C. Lu, J.A. Brent, C. Pham, A. Fackett, C.E.Ruegg, P.M. Silber, Chem. Biol. Interact., 121, 17 (1999).
93.   D.J. Carlile, N. Hakooz, M.K. Bayliss, J.B. Houston, Br. J. Clin. Pharmacol., 47, 625 (1999B).
94.   G. Schneider, P. Coassolo, T. Lave, J. Med. Chem., 42, 5072 (1999).
95.   W. Curatolo, Pharm. Sci. Technol. Today., 9, 387 (1998).

Chapter 28. Inhibition of Cysteine Proteases

Robert W. Marquis
SmithKline Beecham Pharmaceuticals
King of Prussia, PA 19406

Introduction – Several members of the cysteine protease family of enzymes have been implicated as possible causative agents in a variety of diseases (1,2). More notable examples include cathepsin K as the principle protease responsible for excessive degradation of the bone matrix (3), cathepsins L and S for MHC-II antigen presentation (4-6), the role of caspases in programmed cell death (7, 8), rhinovirus 3C protease for viral processing (9), cruzipain (10) and falcipain (11) in parasitic infections as well as the possible role played by the gingipains in periodontal disease (12). Based on the diverse role played by this class of protease in the pathology of disease, the search for specific, potent inhibitors which may serve as potential therapeutics has been expanding (recently reviewed in 13-15). A variety of cysteine protease inhibitor templates have been discovered which may be divided broadly into three mechanistically distinct groups (16). The first are a series of active site titrants whose mechanism of inhibition is based upon the irreversible alkylation of the thiol moiety of the active site cysteine of these proteases. This inhibitor class includes peptide derived α-haloketones, α-diazoketones, epoxides, the (acyloxy)methyl ketone quiescent affinity label and α, β-unsaturated ester and vinyl sulfone Michael acceptors. The second inhibitor class relies on the formation of a covalent, yet reversible, transition state like intermediate with the active site cysteine. Here, carbonyl based peptidyl aldehydes, ketones and α-keto esters and amides serve as examples of reversible transition state inhibitors. Peptidyl aldehydes have historically served as models for the development of cysteine protease inhibitors. However, due to the reactivity and metabolic liabilities associated with aldehyde based inhibitors, recent developments in this class will not be covered in this review. The third class are a series of inhibitors that form a stable thioacyl-enzyme complex with the active site cysteine which, by nature of this intermediate, is slow to hydrolyze. Aza-substituted peptides are the principle example of these slow turnover inhibitors. This review will highlight recent salient developments in the design and application of these three groups of cysteine protease inhibitors.

REVERSIBLE INHIBITORS

1,3-Bis(acyl)diaminoketone Inhibitors - An approach to the design of inhibitors of the osteoclast specific cysteine protease cathepsin K utilizing a combination of both X-ray crystallography and molecular modeling has recently been disclosed (17). The design of this series of 1,3-bis(acyl)diaminoketone inhibitors was based upon the observation that tripeptide aldehydes leupeptin **1** and Z-leu-leu-leu-CHO **2** had bound in opposite directions within the active site of papain (18, 19). Molecular modeling experiments based on these cocrystal structures generated the C2 symmetric ketone **3** which was characterized as a reversible and competitive inhibitor of cathepsin K ($K_{i,app}$ = 22 nM). Compound **3** was also a selective inhibitor of cathepsin K versus cathepsins B, L and S as well the serine proteases trypsin and chymotrypsin. The X-ray cocrystal structure of **3** bound within cathepsin K showed inhibitor binding on both the primed and unprimed sides of the active site with the ketone carbonyl having formed a hemithioketal with cysteine. Both of the Cbz-leucine groups of **3** were replaced with several peptidomimetics. Inhibitor **4**, which incorporates the 4-phenoxyphenyl sulfonamide is a 1.8 nM inhibitor of cathepsin K. Molecular modeling and X-ray crystallographic data had shown that the 4-phenoxy group contained in **4** was required in order for the inhibitor to extend into the P_3' binding pocket where it was proposed to form a critical π–π stacking interaction with tryptophan 184 of the protein. Analog **5**, in which both Cbz-leucines of **3** have been replaced by non-amino acids is a potent inhibitor of cathepsin K with $K_{i,app}$ = 1.4 nM. The 2-(3-biphenyl)-4-methylvaleryl amide has served to mimic both the hydrophobic *iso*-butyl group of leucine as well as the phenyl group of the Cbz moiety which are described as important binding elements of the P$_2$ and P$_3$ binding pockets respectively (20).   The morpholine benzofuran derivative **6** is a potent inhibitor of both human and rat cathepsin K's with $K_{i,app}$'s = 0.082 nM and 69 nM respectively (21).   Compound **6** was shown to inhibit the degradation of type I collagen in a human osteoclast-based assay of bone resorption with an IC$_{50}$ = 41 nM as well as inhibiting bone turnover in a dose dependent manner in the thyroparathyroidectomized rat model of bone loss.

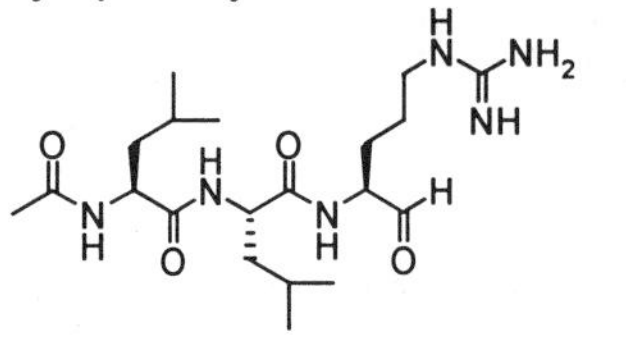

1 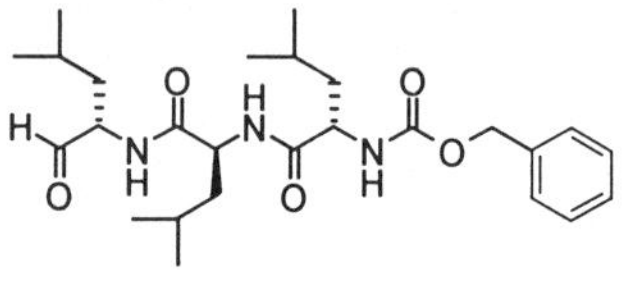   **2**

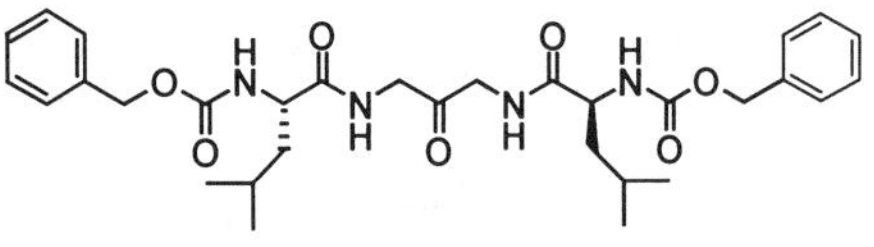

3 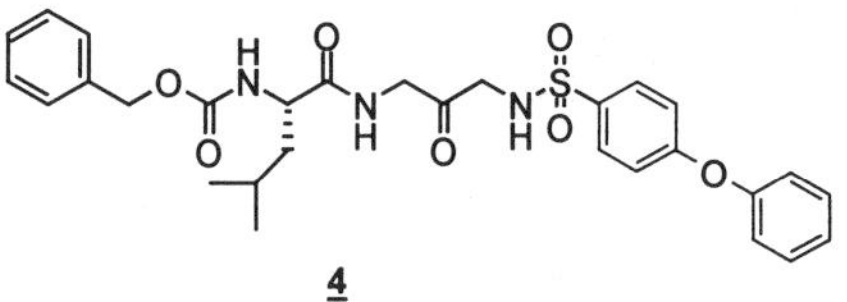   **4**

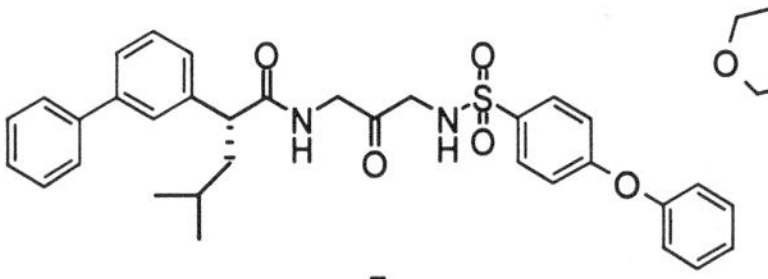

5 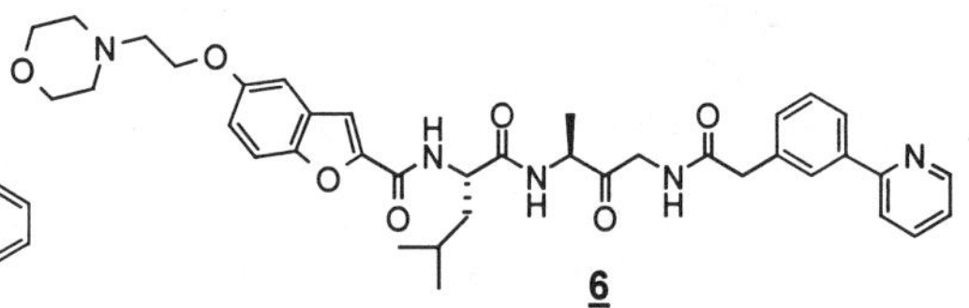   **6**

A solid phase synthesis based on the 1,3-bis(acylamino)-2-propanone template has been used to identify potent and selective inhibitors for both cathepsins K and L (22). The synthesis of this array involved the attachment of several amino acids (R^2) to a Merrifield resin bound BAL linker. The amino acid was capped with several carboxylic acids (R^1) and the protected 1,3-bis(acylamino)-2-propanone template (emboldened in structure **7**) was elaborated.   Cleavage of the inhibitors from the resin with TFA/(CH$_3$)$_2$S resulted in partial epimerization of the α-methyl group which was attributed to the increased lability of the α-chiral center upon formation of the intermediate oxonium ion during ketal hydrolysis. Analog **8** which incorporated a P$_2$ leucine capped by a benzo[b]thiophene carboxamide was a potent inhibitor of cathepsin K (K$_i$ = 1.3 nM) and was 70 fold selective over cathepsin L (K$_i$ = 90 nM). Alternatively, the P$_2$ phenylalanine derivative **9**, capped by the 6-quinoline carboxamide, shows a 5 fold selectivity for cathepsin L (K$_i$ = 36 nM) over cathepsin K (K$_i$ = 170 nM).

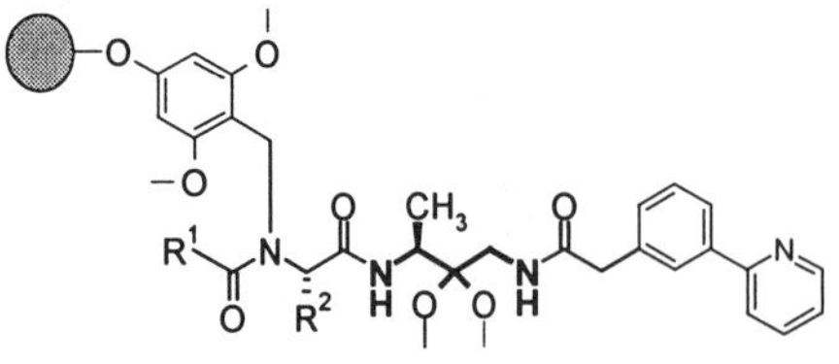

7

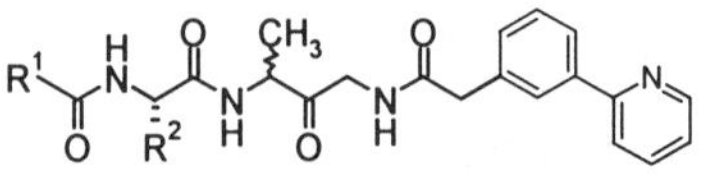

8 R$_1$ = (benzothiophene) — ; R$_2$ =CH$_2$CH(CH$_3$)$_2$

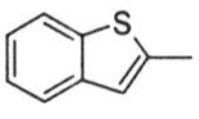

9 R$_1$ = (quinoline) — ; R$_2$ = CH$_2$Ph

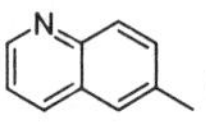

<u>Cyclic Diaminoketone Inhibitors</u> - The design of several conformationally constrained versions of **3** have been reported (23). The C2 symmetric inhibitors **10** and **11** are weak inhibitors of cathepsin K. The diastereomeric mixture of cyclopentanones **10** displayed time dependent inhibition (56 M^{-1} s^{-1}) while the individual cyclohexanone diastereomers of **11** were 16 uM and 15 uM inhibitors respectively. The diastereomeric 4-amino-pyrrolidinone and 4-amino-piperidinone inhibitors **12** and **13** were 2.3 nM and 2.6 nM inhibitors respectively. Both **12** and **13** were selective for cathepsin K versus cathepsin B ($K_{i,app}$'s = >1,000 and 440 nM) and were less selective over cathepsin L ($K_{i,app}$'s = 39 nM, and 16 nM). These ketone based analogs have been characterized as reversible and competitive inhibitors. An X-ray cocrystal structure of **12** bound in the active site of cathepsin K shows the inhibitor spans both sides of the active site with the tertiary amide nitrogen oriented on the unprimed side. The difference in potencies for **10** and **11** versus **12** and **13** was attributed to unforeseen steric interactions present in inhibitors **10** and **11** which may have hindered approach of the active site cysteine. Several peptidomimetics have been incorporated into these inhibitor templates with no significant loss in activities or selectivites relative to **12** and **13**.

10 n = 1
11 n = 2

12 n = 1
13 n = 2

α-Heteroatom Ketone Inhibitors - Based on substrate specificities (24) and the activity of the tetrapeptide aldehyde Ac-Tyr-Val-Ala-AspCHO (25) (K_i = 0.76 nM), a series of reversible alkyl and phenylalkyl inhibitors of interleukin-1β converting enzyme (ICE) have been reported (26). Ethyl ketone analog **14** was a 4 uM inhibitor while incorporation of a phenylpentyl group provided **15** which was an 18.5 nM inhibitor of ICE. The increased potency of **15** over **14** was attributed to the ability of the alkylphenyl moiety to access the hydrophobic binding pocket in the P_1' to P_2' subsites of the enzyme active site. This series of inhibitors has been extended by the incorporation of either electron withdrawing groups or heteroatoms β to the ketone carbonyl moiety (27). Aminomethylketone derivative **16** (K_i = 4.7 nM) is 25 fold more potent than the carbon analog **17** (K_i = 100 nM). Incorporation of the pyridinone based P_3-P_2 val-ala peptidomimetic has led to **18** which is a 0.37 nM inhibitor of ICE (28).

14 X = CH₂CH₃
15 X = (CH₂)₅Ph

16 X = NH
17 X = CH₂

18

A series of alkoxymethylketones have been reported as reversible inhibitors of cathepsin K (29). This class of inhibitor was designed in an effort to eliminate the metabolic liabilities associated with aldehyde based inhibitors such as **2** which had shown good *in vitro* and *in vivo* efficacy against cathepsin K (30). SAR studies in this series showed that the P_1 leucine and alanine derivatives **19** and **20** were potent inhibitors ($K_{i,app}$'s = 60, 150 nM respectively) which are selective for cathepsin K over cathepsins B and L. Inhibitor **21**, which incorporates the 4-biphenylbenzyl ether group, was a potent inhibitor with a $K_{i,app}$ = 22 nM. The increased potency of **21** was attributed to its ability to bind on both the primed and unprimed sides of the active site of the enzyme. Based on the similar potencies of the P_1 alanine and leucine derivatives **19** and **20** a series of cyclic tetrahydrofuranones and tetrahydropyranones have been designed (31, 32) Inhibitors **22** and **23** were 140 and 150 nM inhibitors of cathespin K when tested as a mixture of diastereomers. These inhibitors were characterized as reversible and competitive. Replacement of the Cbz moiety of **22** with either the benzo[b]thiophene-2-carboxamide or the quinoline-2-carboxamide provided inhibitors **24** and **25** which are 11 and 44 nM inhibitors of cathepsin K respectively.

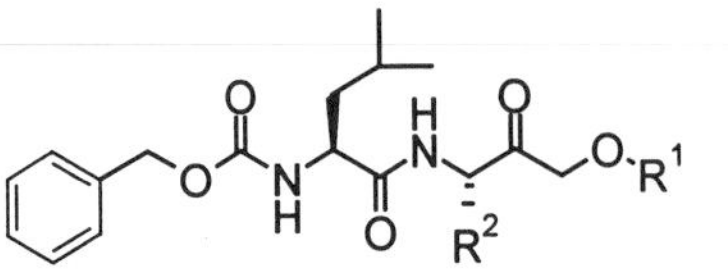

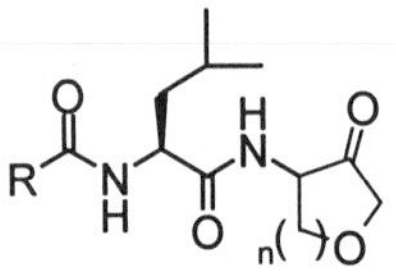

19 $R^1 = CH_3$; $R^2 = CH_2CH(CH_3)_2$
20 $R^1 = CH_3$; $R^2 = CH_3$
21 $R^1 = CH_2OPh-4-Ph$; $R_2 = CH_2CH(CH_3)_2$

22 n = 1; R = OCH_2Ph
23 n = 2; R = OCH_2Ph
24 n = 1; R = 2-benzo[b]thiophene
25 n = 1; R = 2-quinoline

A recent report has disclosed the solid phase synthesis of several ketone derived mechanism based inhibitor templates (33). This synthesis exploits the ketone carbonyl as the central point of attachment to a hydrazone based linker **26**. After appropriate functionalization of the resin bound template the inhibitor **27** may be cleaved from the resin by treatment with TFA, water, acetaldehyde and trifluoroethanol. No racemization of the chiral center α to the ketone was observed throughout these synthesis sequences. A related strategy which utilizes a semicarbazone linked aminomethyl polystyrene resin has also been reported for the synthesis of resin bound peptidyl aldehydes such as **28** (34). Here again, no racemization of the chiral center α to the aldehyde moiety was seen upon removal of the inhibitors from the resin. These methods should facilitate the synthesis and evaluation of large numbers of mechanism based inhibitors of cysteine proteases.

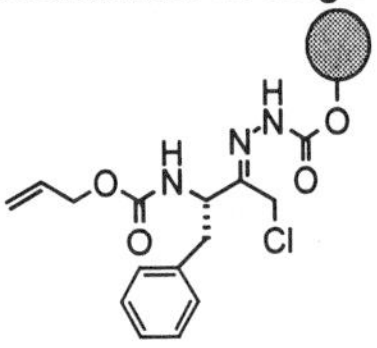

26 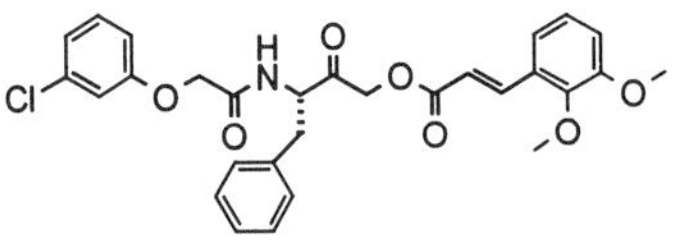   **27** 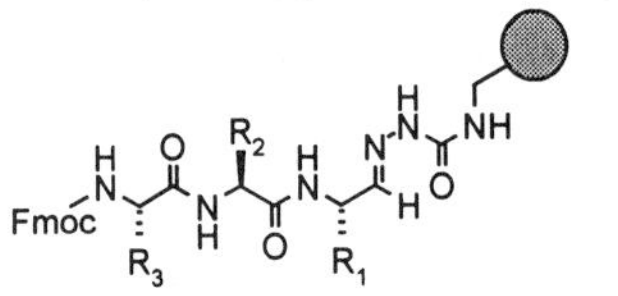   **28**

4-Heterocyclohexanones - An approach to the design of reversible inhibitors of cysteine proteases based on increased reactivity of the ketone carbonyl of the 4-heterocyclohexanone ring system (see **30-32**) has been recently reported (35). The increased susceptibility of ketones of this ring system toward reaction with nucleophiles is the result of a through space dipole-dipole repulsion between the ketone carbonyl and the heteroatom in the 4-positon of the cyclohexanone ring. Analogs **30-32** were characterized as competitive and reversible inhibitors of papain with one diastereomer significantly more active than the other. The more active diastereomer of the cyclohexanone derivative **30** was a 78 uM inhibitor of papain. Compound **30** was 20 times more potent than the acyclic derivative **29** highlighting the contribution that ring strain plays in effecting the reactivity of the carbonyl group. Incorporation of heteroatoms in the 4-position of **30** gave the tetrahydrothiopyranone **31** and 4-tetrahydropyranone **32** which were 26 uM and 11 uM inhibitors of papain respectively. The synthesis of a $^{13}C(O)$ labeled derivative of **32** provided ^{13}C NMR spectroscopic evidence for the formation of a hemithiolketal between the ketone carbonyl of **32** and the active site cysteine of papain (36). The utility of this inhibitor strategy has been extended by the synthesis of **33** which is a 6.6 mM inhibitor of the lysosomal cysteine protease cathepsin B (37). This inhibitor is reported to be capable of extending into both the S and S′ binding pockets of the active site of the enzyme.

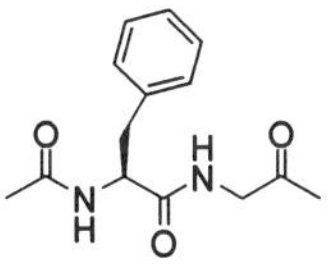

29

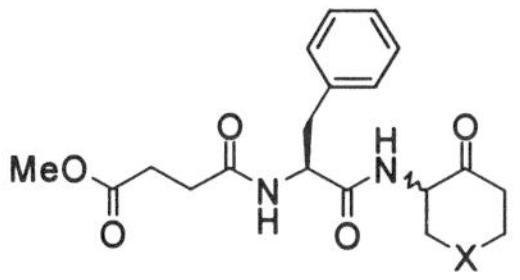

30 = CH$_2$
31 = S
32 = O

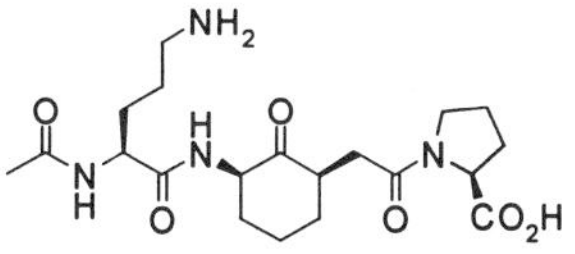

33

Aryl Ketone Based Inhibitors - A series of reversible 2-benzothiazole ketone inhibitors of human rhinovirus 3C protease have recently been disclosed (38). The 2-benzothiazole ketone **34** is a reversible inhibitor of rhinovirus 3C protease (K_i = 0.065 uM) with modest antiviral activity (EC$_{50}$ = 3.2 uM, HRV serotype-14). The related benzo[b]thiophene derivative **35** was far less potent than **34** with a K_i = 4.7 uM. This loss in activity was attributed to the inability of **35** to form a critical hydrogen bond with histidine within the active site of the protease. The 2-benzothiazole ketone electrophile of **34** was incorporated into a recently identified tripeptidyl Michael acceptor inhibitor template to provide **36** which is a 4.5 nM inhibitor of human rhinovirus C with antiviral activity versus three rhinovirus serotypes (HRV-14, EC$_{50}$ = 0.34 uM; HRV-1A, EC$_{50}$ = 0.34 uM; HRV-10, EC$_{50}$ = 0.25 uM).

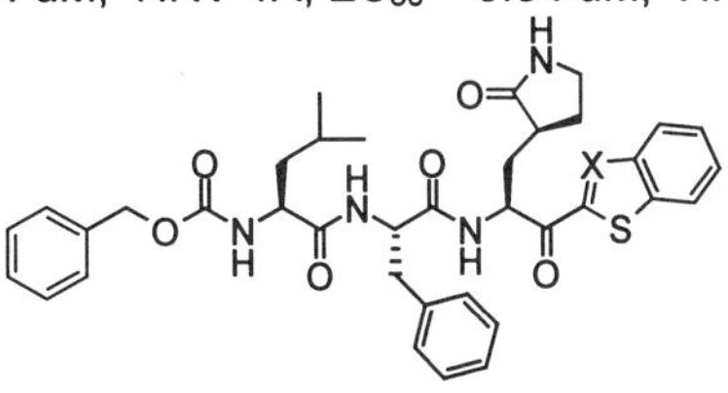

34 X = N
35 X = C

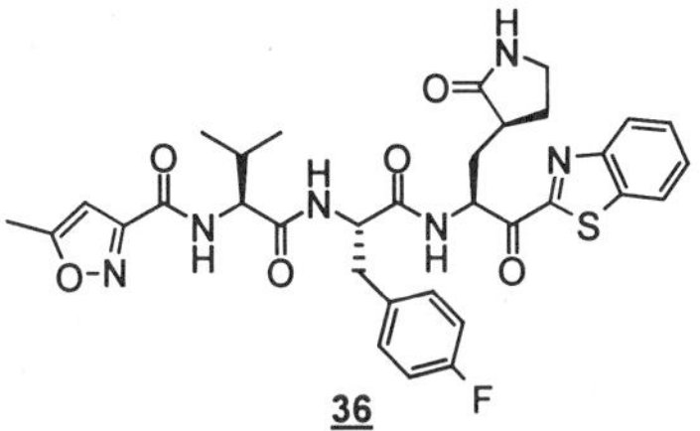

36

2, 3-Dioxindoles (Isatins) – The 2,3-dioxindole moiety has seen recent application for the inhibition of the rhinovirus 3C protease as well as the selective inhibition of caspases 3 and 7. This template was originally reported for the inhibition of the serine protease α-chymotrypsin (39). Molecular modeling and structure- based design led to the identification of 1-methylistatin-5-carboxamide **37** which is a 51 nM inhibitor of human rhinovirus 3C protease (40). Incorporation of the benzo[b]thiophene produced **38** which is a 2 nM inhibitor. The increased potency of **38** was attributed to the ability of the benzo[b]thiophene to bind more tightly in the S$_2$ pocket of the active site of the enzyme. Both **37** and **38** were selective for HRV-14 3CP versus several rhinovirus serotypes but showed little antirhinoviral activity as measured by their ability to protect against HRV infection in HI-Hela cells. The 5-nitro-1-methyl istatin **39** was identified by high throughput screening as a 500 nM inhibitor of human caspase–3 (41). Subsequent SAR focused on replacement of the 5-nitro moiety resulted in isatin sulfonamide **40** which is a selective inhibitor of caspases 3 (K_i = 15 nM) and 7 (K_i = 47 nM). The structural basis for this selectivity was shown by X-ray crystallography to involve the binding of the pyrrolidine ring of **39** in the hydrophobic S$_2$ pocket of caspase-3. Istatin **40** inhibited apoptosis in murine bone marrow neutrophils and human chondrocytes.

37   38   39   40

IRREVERSIBLE INHIBITORS

α, β-Unsaturated Esters – The α, β-unsaturated ester moiety was first reported as an irreversible inhibitor for papain in 1982 (42). Subsequent publications have highlighted utility of this approach for the irreversible inhibition of cysteine proteases (43-45). Recently, the development of a series of potent irreversible α, β-unsaturated ester inhibitors of the rhinovirus 3C protease (HRV-14 3CP) have been disclosed. Initial studies based upon the known pentapeptide substrate H$_2$N-Thr-Leu-Phe-Gln-Gly-Pro-CO$_2$H showed that the scissile amide bond between Gln and Gly could be replaced by an α, β-unsaturated ester group and that the P$_1$-P$_3$ amino acids Leu-Phe-Gln could serve as an appropriate minimal HRV-14 3CP specificity determinant. These design considerations have led to the tripeptide α, β-unsaturated ethyl ester **41** which is an irreversible inhibitor of HCV-14 3CP with a k$_{obs}$/[I] = 25,000 M^{-1}s^{-1} (46). Compound **41** inhibits several rhinovirus serotypes when tested in cell cultures with an EC$_{50}$ = 0.54 uM. A similar substrate based approach to the design of α, β-unsaturated ester inhibitors of human rhinovirus 3C protease has also been disclosed (47). The X-ray cocrystal structure of **41** bound within the active site of the protease showed that the inhibitor was attached to the protein via a 1,4-Michael addition of the active site cysteine to the β-carbon of the acrylate. Analog **43** which incorporated several preliminary modifications suggested by the HRV-14 3CP/**41** cocrystal structure was 32 times more potent than **41** and showed improved antiviral activity in cell culture versus several rhinovirus serotypes (**43**, k$_{obs}$/[I] = 800,000 M^{-1}s^{-1}; EC$_{50}$ = 0.056 uM). Structure based design has led to further refinement of the tripeptide portion of **41**. Replacement of the Leu-Phe amide nitrogen by a ketomethylene isostere provided analogs **42** and **44** with k$_{obs}$/[I] = 17,000 M^{-1}s^{-1} and 850,000 M^{-1}s^{-1} respectively. Both **42** and **44** also show improved antiviral activity in cell culture over similar amide derived analogs which was accounted for by their greater cell permeability upon removal of the hydrogen bond donor. Molecular modeling studies had also predicted that removal of an N-H hydrogen bond donor from the P$_1$ glutamine residue by the incorporation of this residue in to a conformationally constricted (S)-lactam could be accommodated by the S$_1$ binding pocket of the enzyme. The P$_1$-lactam derivative **45** (k$_{obs}$/[I] = 257,000 M^{-1} s^{-1}) is 10 times as potent as the corresponding glutamine derived analog **41**.   This modification of the glutamine amino acid has served to eliminate a hydrogen bond while maintaining the *cis*-amide bond geometry required for optimal hydrogen bonding within the S$_1$ pocket of the protein. A combination of several of the modifications discussed above has led to the identification of **46** (AG7088) which is a potent inhibitor of HRV-14-3CP with a second order rate constant for inactivation of 1,470,000 M^{-1}s^{-1} and is selective over several mammalian cysteine and serine proteases. Due to what was believed to be the hydrolysis of the ethyl ester moiety, **46** shows limited stability in rat plasma with a half life of less than 2 minutes but has a half life greater than 60 minutes in both dog and human plasma. Compound **46** has also been shown to be stable to exogenous thiol as well as chymotrypsin. Because of the combination of favorable biological and chemical properties displayed by **46**, it has recently entered clinical trials as an intranasally delivered antirhinoviral therapy for the common cold (46, 48).

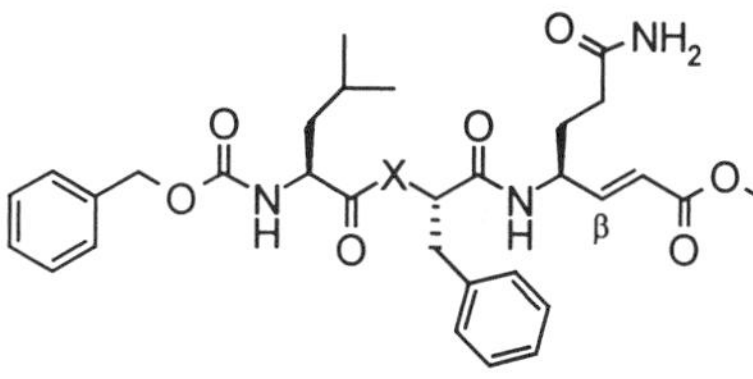

41 X = NH
42 X = CH$_2$

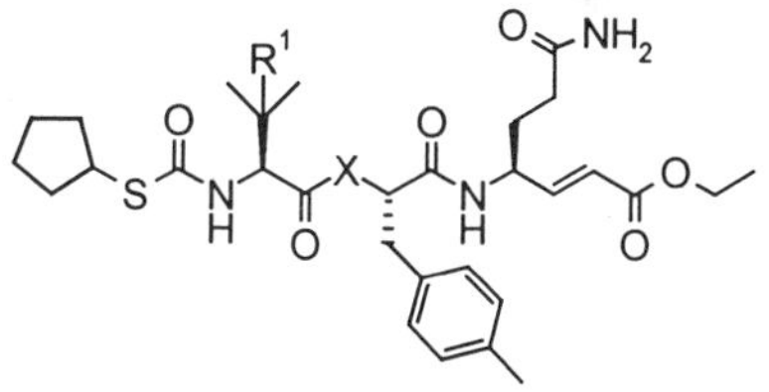

43 X = NH; R^1 = CH$_3$
44 X = CH$_2$; R^1 = H

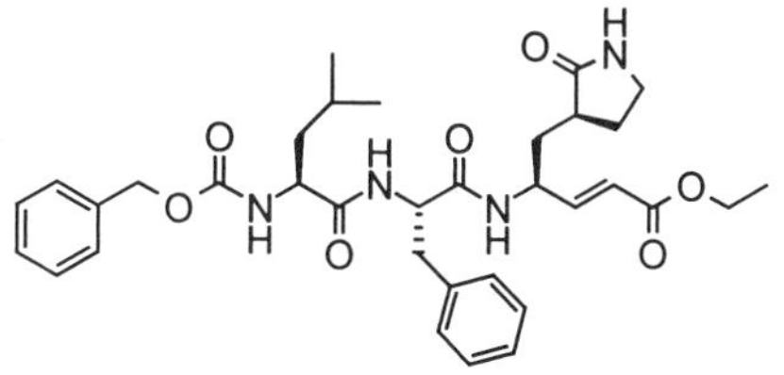

45

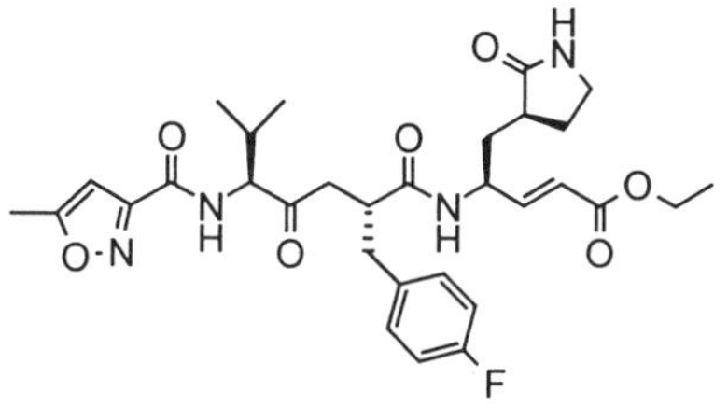

46

Peptidyl Vinyl Sulfones – An extension of the peptidyl Michael acceptor design embodied by the α, β-unsaturated ester inhibitors has been detailed (49). The design of this inhibitor template was based on both the hydrogen bonding capabilities and the polarizible nature of the vinyl sulfone moiety. This group, tethered to an appropriate peptidyl recognition sequence, has served as an effective irreversible inactivator of a variety of cysteine proteases. The prototype vinyl sulfone **47**, was reported to be a potent inactivator of cathepsin S (k_{inact}/K_i = 14,600,000 M^{-1}s^{-1}) versus cathepsins K (k_{inact}/K_i = 727,000 M^{-1}s^{-1}) and L ($k_{inact}/[I]$ = 325,000 M^{-1}s^{-1}). Variation of the P$_2$ leucine residue of **47** has

47 R = CH(CH)$_3$
48 R = Ph

permitted the comparative mapping of the S2 binding pocket of cathepsins K, L and S (50). These studies had shown that the S2 pocket of cathepsin S was the largest of the three enzymes examined, capable of accommodating several amino acid residues of varying surface area while the S2 pocket of cathepsin K demonstrated a clear preference for leucine. Sulfone **47** has recently been shown to be an effective inhibitor of cathepsin S *in vivo* (51). Upon i.p. administration of a 100 mg/kg dose of **47**, mice demonstrated altered invariant chain processing (Ii) as well as inhibition of antigen presentation. Vinyl sulfone **48**, which incorporates a P$_2$ phenylalanine, is an effective inhibitor of cruzain ($k_{inact}/[I]$ = 203,000 M^{-1}s^{-1}) and cured *T.cruzi* infected macrophages at a concentration of 20 uM (52). Evaluation of **48** in a mouse model of Chagas' disease showed that mice treated with **48** survived lethal infections with *T. cruzi* for the duration of the experiment (14-16 days after infection) while untreated mice died within 4 to 5 days. Analog **48** is also an inhibitor of falcipain (IC$_{50}$ = 0.08 uM) a cysteine protease of *Plasmodium Falciparum* which is essential for the development of malarial parasites (53). Because of the therapeutic potential of this class of inhibitor the pharmacokinetic profile of vinyl sulfone **48** has been determined. These studies show **48** to be a substrate for both CYP3A as well as p-glycoprotein (54). The sulfone **48** has low oral bioavailibility (%F = 2.9±1.4) in the male Sprague-Dawley rat at a p.o. dose of 30 mg/kg. The oral bioavailibilty of **48** was seen to increase ten fold (%F = 31.0±7.5) when co-administered with ketoconazole, a known inhibitor of p-glycoprotein and CYP3A (55).

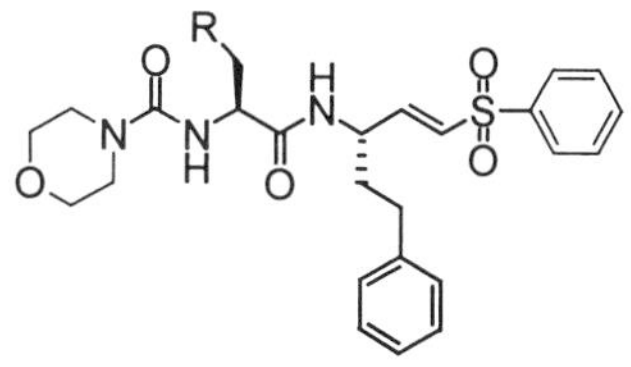

(Acyloxy)methyl Ketones – The quiescent affinity label design was first introduced in 1988 as active site, irreversible inhibitors of cathepsin B (56, 57). This template has been used for the inhibition of a variety of cysteine proteases and the reader is referred to previous reviews covering this topic (58, 59).   Briefly, inhibitors such as **49** (cathepsin B k_{inact}/K_{inact} = 1,600,000 $M^{-1}s^{-1}$) and **50** (60) (cathepsin S k_{inact}/K_{inact} = 686,000 $M^{-1}s^{-1}$) have three components which may be used in order to achieve potent and selective inhibition. First is the peptidyl recognition sequence which serves to direct the latent nucleophage to the active site of the target protease. Changes in these residues serve to enhance both the reactivity and specificity of the inhibitors. The second component is the specific and quiescent reactivity of the carbonyl moiety of these inhibitors towards the cysteine thiol group contained within the active site of the target protease. Third, is the dependence of the rate of enzyme inhibition on the pK_a of the carboxylate leaving group. Here again, the rate of protease inactivation may be increased by varying the nature of the carboxylate leaving group contained within the inhibitor. Several recent reports have expanded the utility of this inhibitor template by the incorporation of leaving groups which possess structural features required for good enzyme affinity but do not fall within the strict pK_a range initially described for potent inhibition (56, 57). Examples include the ((1-phenyl-3-(trifluoromethyl)pyrazole-5-yl)oxy)methyl ketone **51** and the α-((diphenylphosphinyl)oxy)methyl ketone **52** which inhibit ICE with a $k_{obs}/[I]$ = 280,000 and 117,000 $M^{-1}s^{-1}$ respectively. Inhibitor **53** which incorporates the P3-P2 val-ala 5-amino-pyrimidine peptidomimetic is a potent inhibitor of ICE with a second order rate constant of inactivation of 157,000 $M^{-1}s^{-1}$. This rate for inactivation compares favorably with the parent tripeptide **51**. The P3-P2 val-asp was also replaced with the pyridazinodiazepine peptidomimetic to provide **54** which is a potent inactivator of ICE with a second order rate of deactivation of 1,200,000 $M^{-1}s^{-1}$ (61). Several other leaving groups have been incorporated into this template and their mechanism of inhibition has been studied (62-64).

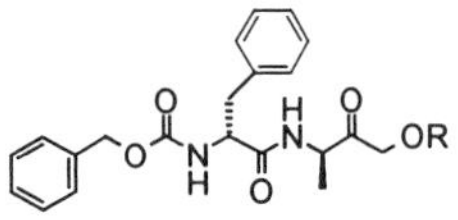

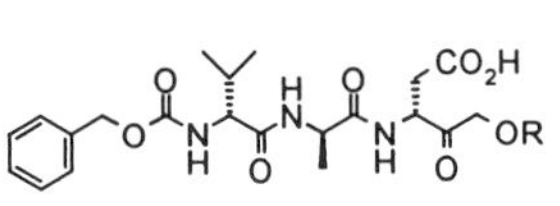

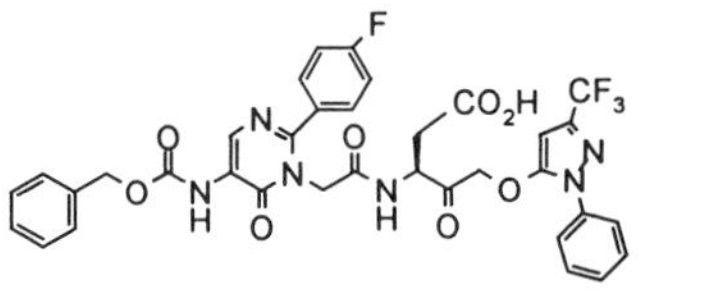

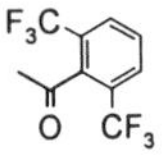

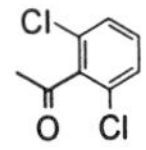

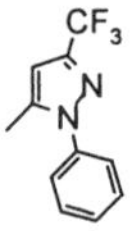

Aza-Peptides - A series of aza-peptide amides which utilize both primed and unprimed side binding elements have recently been shown to be virtually irreversible inhibitors of papain (65). These pseudo-substrates were designed so that, upon attack of the active site cysteine, they would produce a thioacyl-enzyme intermediate which would hydrolyze slowly relative to the parent peptide substrate. The incorporation of amino acid functionality which acts as a leaving group and accesses the S_1' leaving group binding pocket was seen to increase the potency of these inhibitors.

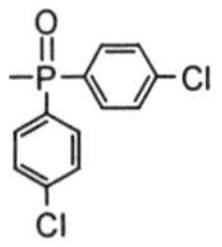

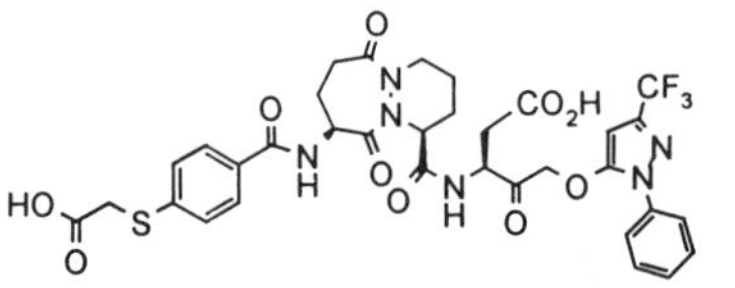

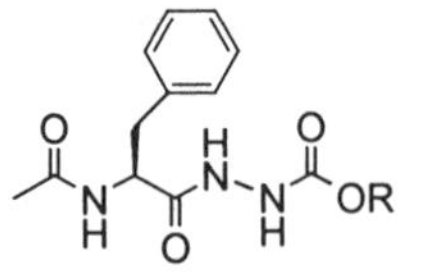

55 R = CH_3
57 R = Ph

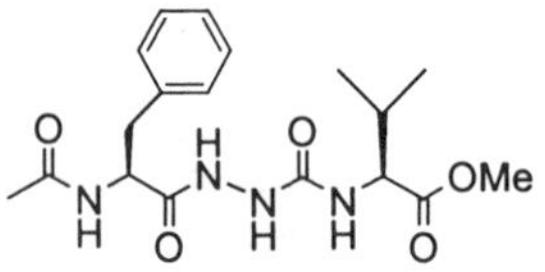

56

The rate of papain inactivation by aza-peptide **55** was reported to have a k_{on} = 13 $M^{-1}s^{-1}$ (66). Analog **56**, which incorporates a leaving group capable of interacting with the S_1' binding pocket, inactivates papain at a rate 17 times that of **55**. The thioacyl enzyme intermediate formed between papain and **56** hydrolyzed with a $t_{1/2}$ = 12 hours making these inhibitors essentially irreversible. In a similar series (67), the peptidyl carbazate ester **57** was an inactivator of papain with a second order rate constant for inactivation of 11,800 $M^{-1}s^{-1}$. The resulting carbamoylated acyl enzyme intermediate formed upon inactivation of papain by **57** was confirmed by mass spectral analysis and was shown to have a $t_{1/2}$ for hydrolysis of approximately 24 hours.

SLOW TURNOVER INHIBITORS

Peptidyl Carbazate Esters – A recent report has detailed the synthesis and evaluation of a series of peptidyl carbazate esters which are slow turnover inhibitors of human rhinovirus 3C protease (68). The carbazate ester **58** is a transient inhibitor of HRV 3CP which initially inhibits the enzyme quickly but not completely. These data suggest that the formation of the intermediate thioacyl enzyme between **58** and HRV 3CP is rapid but then hydrolyzes slowly relative to the corresponding thioacyl enzyme intermediate derived from a peptide substrate. HPLC experiments confirmed the rapid hydrolysis of inhibitor **58** in the presence of HRV 3CP.

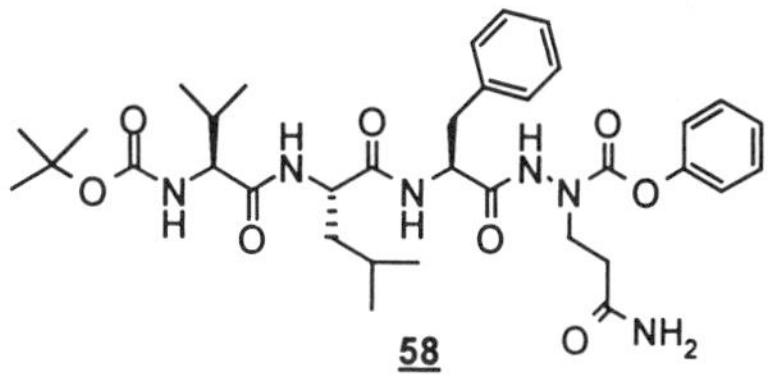

58

Acyl-Bis-Hydrazide - Nitrogen substitution of the central diaminopropanone scaffold of the reversible ketone based inhibitor **3** has led to a series of acyl-bis hydrazide inhibitors of cathepsin K (69). The parent member of this class of inhibitors **59** has been characterized as a potent time dependent inhibitor of cathepsin K with a $k_{obs}/[I] =$ 3.1 x 10^{6} $M^{-1}s^{-1}$ and is selective versus cathepsins B, L and S. The X-ray cocrystal structure of

59 R = H
60 R = $CH_2N(CH_3)_2$

61

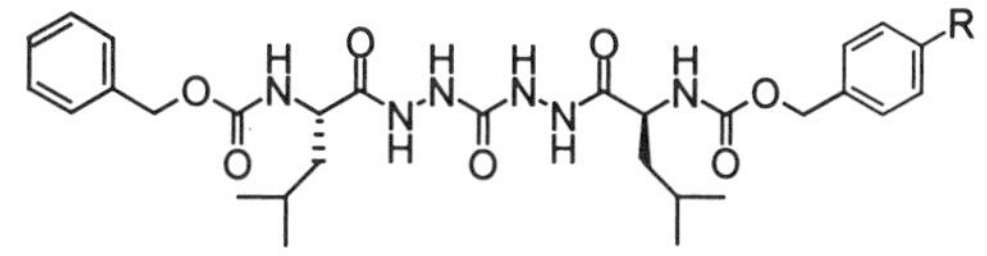

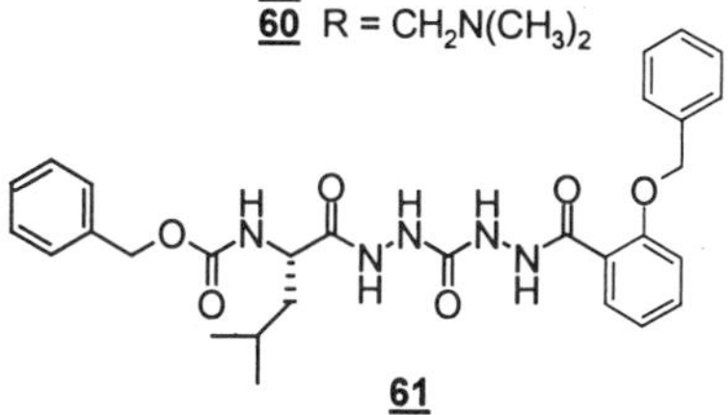

59 complexed within the active site of cathepsin K suggested that the inhibitor was attached to the enzyme via a tetrahedral thioketal complex formed between the thiol of the active site cysteine and the urea carbonyl group. The inhibitor spanned both the primed and unprimed sides of the active site, binding in a fashion analogous to diaminopropanone **3**. Studies to elucidate the mechanism of inhibition for this class of compound show **59** to have an initial, reversible association $K_{i,app}$ = 2.7 nM with a second step inactivation rate of k_{inact} = 43 s^{-1} (70). Reactivation rates are very similar after inactivation with **59** and several structurally related analogs suggestive of a common thiocarbamate intermediate. Both ^{13}C NMR and mass spectral analysis confirmed the acylation of the enzyme by the acylhydrazide Z-LeuNHNHC(O)- a substructure common within all of the inhibitors examined. Based on these studies, this class of compound has been characterized as time dependent, slow turnover inhibitors. Analog **60**, which has incorporated a water solubilizing N, N'-dimethyl moiety, is a potent and selective inhibitor of cathepsin K ($k_{obs}/[I]$ = 3.1 x 10^{6} $M^{-1}s^{-1}$). Both **59** and **60** are potent inhibitors of cathepsin K in a cell based osteoclast resorption assay with IC_{50}'s = 0.34 and 0.12 uM respectively. Analog **60** is a potent inhibitor of osteoclast mediated calcemic response to human parathyroid hormone in the thyroparathyroidectomized (TPTX) rat model of bone loss. Peptidomimetic elements have been incorporated into this inhibitor template (71). Inhibitor **61**, in which the o-benzyloxybenzoyl moiety was shown to be an effective replacement for the Z-leucine of **59**, was a potent and selective inhibitor of cathepsin K ($k_{obs}/[I]$ = 318,000 $M^{-1}s^{-1}$).

Diacyl Hydrazines – Replacement of one of the acylhydrazide moieties of **59** with a thiazole amide bond isostere provided **62** which is a potent and selective inhibitor of cathepsin K (cat K $K_{i, app}$ = 10 nM: cat B $K_{i, app}$ = 5,200 nM; cat L $K_{i, app}$ = 700 nM; cat S $K_{i, app}$ > 1,000 nM). The mode of inhibition of **62** was shown to be initially rapidly reversible with a subsequent slow turnover step[69]. Diaminopyrrolidinone **63** is a 30 nM inhibitor which has been shown to be more stable than its acyclic counterpart **62** to proteolytic processing by cathepsin K (72). The X-ray cocrystal structure of **62** within the active site of cathepsin K has led to the further incorporation of peptidomimetic elements into this inhibitor template (73). The aminothiazole **64** is a 50 nM inhibitor of cathepsin K which is 50% orally bioavailable in the male Sprague-Dawley rat (74).

          62   **63**   **64**

Conclusion – Through the application of iterative structure based design and the development of high throughput combinatorial synthesis methodologies, potent and selective inhibitors of cysteine proteases have continued to evolve. Despite this progress, pharmacokinetic and pharmacodynamic issues associated with these inhibitors have yet to be fully resolved. The further application of these methods to inhibitor design as well as a more detailed understanding of the factors which limit both inhibitor bioavailability and *in vivo* efficacy should soon produce molecules capable of fulfilling the requirements to serve as suitable candidates for clinical evaluation.

References

1.   H. A. Chapman, R. J. Riese, G-P. Shi, Ann. Rev. Physiol., 59, 63 (1997).
2.   J. H. McKerrow, M. N. G. James, Perspect. Drug Discovery Des., 6 (1996).
3.   D. S. Yamashita, R. A. Dodds, Curr. Pharm. Des., 6, 1 (2000).
4.   J. A. Villadangos, R. A. Bryant, J. Deussing, C. Driessen, A. Lennon-Dumenil, R. J. Riese, W. Roth, P. Saftig, G-P. Shi, H. A. Chapman, C. Peters, H. L. Ploegh, Immunol. Rev., 172, 109 (1999).
5.   T. Nakagawa, A. Y. Rudensky, Immunol. Rev., 172, 121 (1999).
6.   G. M. Barton, A. Y. Rudensky, Science, 283, 67 (1998).
7.   K. K. W. Wang, Curr. Opin. Drug Discovery Dev., 2, 519 (1999).
8.   D. J. Livingston, J. Cell. Biochem., 64, 19 (1997).
9.   A. K. Patick, K. E. Potts, Clin. Microbiol. Rev., 11, 614 (1998).
10. J. J. Cazzulo, V. Stoka, V. Turk, Biol. Chem. Hoppe-Seyler, 378, 1 (1997)
11. P. J. Rosenthal, K. Kim, J. H. McKerrow, J. H. Leech, J. Exp. Med., 166, 816 (1987).
12. J. Potempa, R. Pike, J. Travis, Prospect. Drug Discovery Design, 2, 445 (1995).
13. D. Leung, G. Abbenante, D. P. Fairlie, J. Med. Chem., 43, 305 (2000).
14. H-H. Otto, T. Schirmeister, Chem. Rev., 97, 133 (1997).
15. R. E. Babine, S. L. Bender, Chem. Rev., 97, 1359 (1997).
16. A. Krantz, Bioorg. Med. Chem. Lett., 2, 1327 (1992).
17. D. S. Yamashita, W. W. Smith, B. Zhao, C. A. Janson, T. A. Tomaszek, M. A. Bossard, M. J. Levy, H-J. Oh, T. J. Carr, S. T. Thompson, C. F. Ijames, S. A. Carr, M. S. McQueney, K. J. D'Alessio, B. Y. Amegadzie, C. R. Hanning, S. S. Abdel-Meguid, R. L. DesJarlais, J. G. Gleason, D. F. Veber, J. Amer. Chem. Soc., 119, 11351 (1997).
18. E. Schröder, C. Phillips, E. Garman, K. Harlos, C. Crawford, FEBS Lett. 315, 38 (1993).
19. J. M. LaLonde, B. Zhao, W. W. Smith, C. A. Janson, R. L. DesJarlais, T. A. Tomaszek, T. J. Carr, S. K. Thompson, H-J. Oh, D. S. Yamashita, D. F. Veber, S. S. Abdel-Meguid, J. Med. Chem., 41, 4567 (1998).
20. R. L. DesJarlais, D. S. Yamashita, H-J. Oh, I. N. Uzinskas, K. F. Erhard, A. C Allen, R. C. Haltiwanger, B. Zhao, W. W. Smith, S. S. Abdel-Meguid, K. D'Alessio, C. A. Janson, M. S. McQueney, T. A. Tomaszek, M. A. Levy, D. F. Veber, J. Amer. Chem. Soc., 120, 9114 (1998).

21. D. F. Veber, D. S. Yamashita, H-J. Oh, B. R. Smith, K. Salyers, M. Levy, C-P. Lee, A. Marzulli, P. L. Smith, T. Tomaszek, D. Tew, M. McQueney, G. B. Stroup, M. W. Lark, I. E. James, M. Gowen, Proceedings of the American Peptide Symposium, In Press.

22. D. S. Yamashita, X. Dong, H-J. Oh, C. S. Brook, T. A. Tomaszek, L. Szewczuk, D. G. Tew, D. F. Veber, J. Comb. Chem., 1, 207 (1999).

23. R. W. Marquis, D. S. Yamashita, Y. Ru, S. M. LoCastro, H-J. Oh, K. F. Erhard, R. L. DesJarlais, M. S. Head, W. W. Smith, B. Zhao, C. A. Janson, S. S. Abdel-Meguid, T. A. Tomaszek, M. A. Levy, D. F. Veber, J. Med. Chem., 41, 3563 (1998).

24. N. A. Thornberry, H. G. Bull, J. R. Calaycay, K. T. Chapman, A. D. Howard, M. J. Kostura, D. K. Miller, S. M. Molineaux, J. R. Weidner, J. Aunins, K. O. Elliston, J. M. Ayala, F. J. Casano, J. Chin, G. Ding, L. A. Egger, E. P. Gaffney, G. Limjuco, O. C. Palyha, S. M. Raju, A. M. Rolando, J. P. Salley, T. Yamin, T. D. Lee, J. E. Shivelly, M. MacCoss, R. A. Mumford, J. A. Schmidt, M. J. Tocci, Nature 356, 768 (1992).

25. K. T. Chapman, Bioorg. Med. Chem. Lett., 2, 613 (1992).

26. A. M. M. Majalli, K. T. Chapman, M. MacCoss, N. A. Thornberry, Bioorg. Med. Chem. Lett., 3, 2689 (1993).

27. A. M. M. Majalli, K. T. Chapman, M. MacCoss, N. A. Thornberry, E. P. Peterson, Bioorg. Med. Chem. Lett., 4, 1965 (1994).

28. G. Semple, D. M. Ashworth, A. R. Batt, A. J. Baxter, D. W. M. Benzies, L. H. Elliot, D. M. Evans, R. J. Franklin, P. Hudson, P. D. Jenkins, G. R. Pitt, D. P. Rooker, S. Yamamoto, Y. Isomura, Bioorg. Med. Chem. Lett., 8, 959 (1998).

29. R. W. Marquis, Y. Ru, D. S. Yamashita, H-J. Oh, J. Yen, S. K. Thompson, T. J. Carr, M. A. Levy, T. A. Tomaszek, C. F. Ijames, W. W. Smith, B. Zhao, C. A. Janson, S. S. Abdel-Meguid, K. J. D'Alessio, M. S. McQueney, D. F. Veber, Bioorg. Med. Chem., 7, 581 (1999).

30. B. J. Votta, M. A. Levy, A. Badger, J. Bradbeer, R. A. Dodds, I. E. James, S. Thompson, M. J. Bossard, T. Carr, J. R. Connor, T. A. Tomaszek, L. Szewczuk, F. H. Drake, D. F. Veber, M. J. Gowen, J. Bone Miner. Res., 12, 1396 (1997).

31. A. D. Gribble, A. E. Fenwick, R. W. Marquis, D. F. Veber, J. Witherington, WO 9850533 (1998).

32. R. Marquis, Y. Ru, K. F. Erhardt, A. D. Gribble, J. Witherington, A. Fenwick, B. Garnier, J. Bradbeer, M. Lark, M. Gowen, T. A. Tomaszek, D. G. Tew, B. Zhao, W. W. Smith, C. A. Janson, K. J. D'Alessio, M. S. McQueney, S. S. Abdel-Meguid, K. L. Salyers, B. R. Smith, M. A. Levy, D. F. Veber, The 218th ACS National Meeting, Division of Medicinal Chemistry, New Orleans LA, Aug. 22-26, MEDI-019, (1999).

33. A. Lee, L. Huang, J. A. Ellman, J. Amer. Chem. Soc., 121, 9907 (2000).

34. D. B. Siev, J. E. Semple, Org. Lett., 2, 19 (2000).

35. J. L. Conroy, T. C. Sanders, C. T. Seto, J. Amer. Chem. Soc., 119, 4285 (1997).

36. J. L. Conroy, C. T. Seto, J. Org. Chem., 63, 2367 (1998).

37. J. L. Conroy, P. Abato, M. Ghosh, M. I. Austermuhle, M. R. Kiefer, C. T. Seto, Tet. Lett., 39, 8253 (1998).

38. P. S. Dragovich, R. Zhou, S. E. Webber, T. J. Prins, A. K. Kwok, K. Okano, S. A. Fuhrman, L. S. Zalman, F. C. Maldonado, E. L. Brown, J. W. Meador III, A. K. Patick, C. E. Ford, M. A. Brothers, S. L. Binford, D. A. Matthews, R. A. Ferre, S. T. Worland, Bioorg. Med. Chem. Lett., 10, 45 (2000).

39. R. A. Iyer, P. E. Hanna, Bioorg. Med. Chem. Lett., 5, 89 (1995).

40. S. E. Webber, J. Tikhe, S. T. Worland, S. A. Fuhrman, T. F. Hendrickson, D. A. Matthews, R. A. Love, A. K. Patick, J. W. Meador, R. A. Ferre, E. L. Brown, D. M. DeLisle, C. E. Ford, S. L. Binford, J. Med. Chem. 39, 5072 (1996).

41. D. Lee, S. A. Long, J. L. Adams, G. Chan, K. S. Vaidya, T. A. Francis, K. Kikly, J. D. Winkler, C-M. Sung, C. Debouck, S. Richardson, M. A. Levy, W. E. DeWolf, Jr., P. M. Keller, T. Tomaszek, M. S. Head, M. D. Ryan, R. C. Haltiwanger, P-H. Liang, C. A. Janson, P. J. McDevitt, K. Johanson, N. O. Concha, W. Chan, S. S. Abdel-Meguid, A. M. Badger, M. W. Lark, D. P. Nadeau, L. J. Suva, M. Gowen, M. E. Nuttall, J. Biol. Chem. In Press.

42. A. J. Barrett, A. A. Kembhavi, M. A. Brown, H. Kirschke, C. G. Knight, M. Tamai, K. Hanada, Biochem. J., 201, 189 (1982).

43. R. P. Hanzlik, S. A. Thompson, J. Med. Chem., 27, 711 (1984).

44. S. A. Thompson, P. R. Andrews, R. P. Hanzlik, J. Med. Chem., 29, 104 (1986).

45. S. Liu, R. P. Hanzlik, J. Med. Chem., 35, 1067 (1992).

46. D. A. Matthews, P. S. Dragovich, S. E. Webber, S. A. Fuhrman, A. K. Patick, L. S. Zalman, T. F. Hendrickson, R. A. Love, T. J. Prins, J. T. Marakovits, R. Zhou, J. Tikhe, C. E. Ford, J. W. Meador, R. A. Ferre, E. L. Brown, S. L. Binford, M. A. Brothers, D. M. DeLisle, S. T. Worland, Proc. Natl. Acad. Sci., 96, 11000 (1999) and references cited therein.

47. J-s. Kong, S. Venkatraman, K. Furness, S. Nimkar, T. A. Shepherd, Q. M. Wang, J. Aubé, R. P. Hanzlik, J. Med. Chem., 41, 2579 (1998).
48. A. K. Patick, S. L. Binford, M. A. Brothers, R. L. Jackson, C. E. Ford, M. D. Diem, F. Maldonado, P.S. Dragovich, R. Zhou, T. J. Prins, S. A. Fuhrman, J. W. Meador, L. S. Zalman, D. A. Matthews, S. T. Worland Antimicrob. Agents Chemother., 43, 2444 (1999).
49. J. T. Palmer, D. Rasnick, J. L. Klaus, D. Brömme, J. Med. Chem., 38, 3193 (1995).
50. D. Brömme, J. L. Klaus, K. Okamoto, D. Rasnick, J. T. Palmer, Biochem. J., 315, 85 (1996).
51. R. J. Riese, R. N. Mitchell, J. A. Villadangos, G-P. Shi, J. T. Palmer, E. R. Karp, D. T. DeSanctis, H. L. Ploegh, H. A. Chapman, J. Clin. Invest., 101, 2351 (1998).
52. J. C. Engel, P. S. Doyle, I. Hsieh, J, H. McKerrow, J. Exp. Med., 188, 725 (1998).
53. P. J. Rosenthal, J. E. Olson, G. K. Lee, J. T. Palmer, J. L. Klaus, D. Rasnick, Antimicrob. Agents Chemother., 40, 1600 (1996).
54. Y. Zhang, X. Guo, E. T. Lin, L. Z. Benet, Drug Metab. Dispos., 26, 360 (1998).
55. Y. Zhang, Y. Hsieh, T. Izumi, E. T. Lin, L. Z. Benet, J. Pharmcol. Exp. Ther., 287, 246 (1998).
56. R. A. Smith, L. J. Copp, P. J. Coles, H. W. Pauls, V. J. Robinson, R. W. Spencer, S. B. Heard, A. Krantz, J. Amer. Chem. Soc., 110, 4429 (1988).
57. A. Krantz, L. J. Copp, P. J. Coles, R. A. Smith, S. B. Heard, Biochemistry, 30, 4678 (1991).
58. A. Krantz, Methods in Enzymology, 47, 656 (1994).
59. A. Krantz, Advances in Medicinal Chemistry, 1, 235 (1992).
60. D. Brömme, R. A. Smith, P. J. Coles, H. Kirschke, A. C. Storer, A. Krantz, Biol. Chem. Hoppe-Seyler, 375, 343 (1994).
61. R. E. Dolle, C. V. C. Prasad, C. P. Prouty, J. M. Salvino, M. M. A. Awad, S. J. Schmidt, D. Hoyer, T. M. Ross, T. L. Graybill, G. J. Speier, J. Uhl, B. E. Miller, C. T. Helaszek, M. A. Ator, J. Med. Chem., 40, 1941 (1997) and references cited therein.
62. T. L. Graybill, C. P. Prouty, G. J. Speier, D. Hoyer, R. E. Dolle, C. T. Helaszek, M. A. Ator, J. Uhl, J. Strasters, Bioorg. Med. Chem. Lett., 7, 41, 1997.
63. L. C. Dang, R. V. Talanian, D. Banach, M. C. Hackett, J. L. Gilmore, S. J. Hays, J. A. Mankovich, K. D. Brady, Biochemistry 35, 14910 (1996).
64. K. D. Brady, Biochemistry, 37, 8508 (1998).
65. R. Baggio, Y-Q. Shi, Y-Q. Wu, R. H. Abeles, Biochemistry, 35, 3351 (1996).
66. J. McGrath, R. H. Abeles, J. Med. Chem., 35, 4279 (1992).
67. R. Xing, R. P. Hanzlik, J. Med. Chem., 41, 1344 (1998).
68. S. Venkatraman, J-s. Kong, S. Nimkar, Q. M. Wang, J. Aubé, R. P. Hanzlik, Bioorg. Med. Chem. Lett., 9, 577 (1999).
69. S. K. Thompson, S. M. Halbert, M. J. Bossard, T. A. Tomaszek, M. A. Levy, B. Zhao, W. W. Smith, S. S. Abdel-Meguid, C. A. Janson, K. J. D'Alessio, M. S. McQueney, B. Y. Amegadzie, C. R. Hanning, R. L. DesJarlais, J. Briand, S. K. Sarkar, M. J. Huddleston, C. F. Ijames, S. A. Carr, K. T. Garnes, A. Shu, J. R. Heys, J. Bradbeer, D. ZembryKi, L. Lee-Rykaczewski, I. E. James, M. L. Lark, F. H. Drake, M. Gowen, J. G. Gleason, D. F. Veber, Proc. Natl. Acad. Sci., USA 94, 14249 (1997).
70. M. J. Bossard, T. A. Toamszek, M. A. Levy, C. F. Ijames, M. J. Huddleston, J. Briand, S. Thompson, S. Halbert, D. F. Veber, S. A. Carr, T. D. Meek, D. G. Tew, Biochemistry, 38, 15893 (1999).
71. S. K. Thompson, W. W. Smith, B. Zhao, S. M. Halbert, T. A. Tomaszek, D. G. Tew, M. A. Levy, C. A. Janson, K. J. D'Alessio, M. S. McQueney, J. Kurdyla, C. S. Jones, R. L. DesJarlais, S. S. Abdel-Meguid, D. F. Veber, J. Med. Chem., 41, 3923 (1998).
72. K. J. Duffy, L. H. Ridgers, Renee L. DesJarlais, T. A. Tomaszek, M. J. Bossard, S. K. Thompson, R. M. Keenan, D. F. Veber, Bioorg. Med. Chem. Lett., 9, 1907 (1999).
73. S. K. Thompson, S. M. Halbert, R. L. DesJarlais, T. A. Tomaszek, M. A. Levy, D. G. Tew, C. F. Ijames, D. F. Veber, Bioorg. Med. Chem., 7, 599 (1999).
74. S. K. Thompson, E. Michaud, S. M. Halbert, T. A. Tomaszek, D. G. Tew, B. Zhao, W. W. Smith, K. A. D'Alessio, M. S. McQueney, J. Kurdyla, C. S. Jones, S. S. Abdel-Meguid, C. F. Ijames, R. L. DesJarlais, K. L. Salyers, B. R. Smith, M. A. Levy, D. F. Veber The 217th ACS National Meeting, Division of Medicinal Chemistry, Anahiem, CA Mar. 21-25, MEDI-009, (1999).

Chapter 29. Principles for Multivalent Ligand Design

Laura L. Kiessling, Laura E. Strong, Jason E. Gestwicki
Departments of Chemistry and Biochemistry
University of Wisconsin, Madison

Multivalent interactions control a wide variety of cellular processes including cell surface recognition events (1). Examples of specific cell – cell binding events can be found in diverse processes, such as inflammation, tumor metastasis, and fertilization. An understanding of the mechanistic principles that underlie multivalent binding events would facilitate the generation of new classes of therapeutic agents and biomaterials. Synthetic multivalent ligands can be used to illuminate and exploit biological processes that benefit from multipoint contacts. This review focuses on the principles for designing synthetic multivalent ligands and the interplay between ligand structure and biological activity

Antibodies are the perhaps most widely used tools for studying multivalency in biological systems. Because of their quaternary structures, antibodies have multiple recognition sites. There are many antibody isotypes, that vary in size, shape, orientation of binding sites, and valency. Although IgG, IgD, and IgE are dimeric **1**, IgA and IgM form higher order oligomers. IgM is a decamer **3**, and IgA ranges from a tetramer **2** to an octamer. These differences in antibody size and quaternary structure directly influence their resulting biological activities (2).

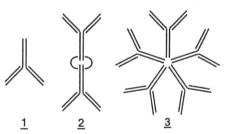

1   **2**   **3**

The use of antibodies can provide information about the involvement of multivalent interactions in a particular process; however, this information is limited. Understanding the structural requirements for multivalent ligand activity requires a wider variety of multivalent ligands whose structure, including size and valency, can be controlled and tailored. It is chemical synthesis that provides access to molecules that possess a variety of sizes, shapes, and valencies.

The flexibility offered by chemical synthesis is illustrated by design strategies that have been used to create small molecule drug candidates. Given a natural product lead, chemical synthesis can be used to install critical molecular features, thereby creating small molecule analogs of a more complex structure. The display of functional groups and other features (i.e. hydrophobicity and flexibility) can be optimized to create ligands with higher affinity or specificity for the chosen target. The principles of small molecule drug design have been reviewed extensively, including a recent discussion of peptidomimetic design (3).

Small molecule ligand design can be analyzed to emphasize common principles important for multivalent ligand design. A primary recognition epitope, which may contain important binding features identified in a natural product, or be derived from a common pharmacophore, or act as a transition state analog, serves as the lead structure (square). In addition, ancillary functional groups that exploit interactions not initially identified in the lead structure or that

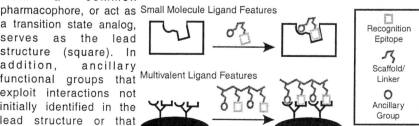

Small Molecule Ligand Features

Multivalent Ligand Features

Recognition Epitope

Scaffold/ Linker

Ancillary Group

restrict the overall ligand conformation often lead to ligands with significantly increased potencies (circle). The features of the scaffold (or the linker) that presents the groups that occupy the binding site(s) on the protein target also are important determinants of productive binding. These are basic design principles that can be used to guide the preparation of larger, multivalent ligands.

In multivalent displays, the identity of the monovalent recognition epitopes is dictated by the biological interaction to be studied. These epitopes can be derived from a natural ligand or from identification of small molecule agonists or antagonists. After selecting the recognition epitope, the choice of scaffold or backbone from which these epitopes are displayed is a critical design parameter. This choice determines many of the overall ligand features including size, three-dimensional shape, and valency. The relative sizes of the scaffolds most commonly used to study multivalent interactions can be estimated using structures derived from X-ray crystallography, molecular modeling, or electron microscopy. Models and scaled schematic depictions of scaffolds discussed in this review are depicted (4). In addition to their differences in size, scaffolds also vary in three-dimensional shape, as exemplified in the comparison of linear polymers and more spherical dendrimers. The number of attachment points available on a scaffold determines the maximum valency, or number of recognition epitopes, that can be displayed. This feature, combined with the size and shape of the scaffold, dictates the spacing of recognition epitopes. Valency, spacing, and the flexibility and hydrophilicity of the framework can have significant effects on the biological activity of the ligand. The linker, which also can influence activity, must be attached to the recognition epitope in such a way that receptor-ligand binding is not disrupted. Finally, the incorporation of ancillary groups, which contribute binding energy but are not found in the original recognition epitope, can be exploited for multivalent ligand design as with small molecule targets. The interplay of all of these ligand features can result in molecules with unique biological activities. A change in ligand structure can alter the mechanisms by which a ligand functions. For example, changing the maximum distance between terminal binding epitopes on a synthetic scaffold can influence the number of receptors that can bind simultaneously to a single ligand, thereby altering biological responses to the ligand. The mechanisms that underlie multivalent ligand activity have been discussed elsewhere (1,5).

To illustrate the influence of multivalent ligand design features, we present examples in which one or more multivalent ligand design features have been systematically varied and the biological activity of the resulting ligands explored. Two

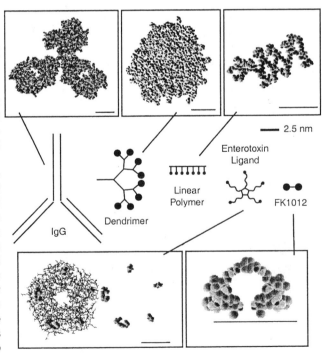

examples of defined divalent molecules will be discussed as will additional examples of higher valency ligands. Dimeric ligands are discussed separately because the strategy used to display the recognition epitopes is different. Specifically, dimeric ligands often consist of recognition elements connected by a linker, as opposed to higher valency compounds that generally display recognition epitopes from a scaffold. Descriptions of strategies for multivalent ligand synthesis have appeared elsewhere, including a discussion of bioactive polymer assembly (6).

Defined divalent molecules have been generated to explore protein dimerization and the signal transduction events that result. The chemical induced dimerization (CID) strategy developed by Schreiber, Crabtree and coworkers uses cell permeable

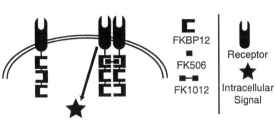

synthetic dimers of natural products with known protein receptors (7). A target receptor is fused to the natural product-binding domain, and the divalent natural product ligand is able to oligomerize the target receptor. A cell line transfected with the fusion protein can be treated with synthetic dimers and the biological effects of receptor oligomerization measured. This controlled association can mediate intracellular events, such as cytoplasmic calcium concentration, gene transcription, or kinase activity (8, 9).

The first demonstration of the CID strategy utilized FK506, a cell permeable immunosuppressant. FK506, **4**, binds to the peptidyl-prolyl cis-trans isomerase FK506-binding protein (FKBP12); this complex is, in turn, a ligand for the serine/threonine phosphatase calcineurin (10). The binding of the FK506-FKBP12 complex to calcineurin is responsible for the immunosuppressive activity of FK506. The CID strategy exploited the interaction between FK506 and FKBP12 while minimizing the interaction of the FK506-FKBP complex with calcineurin. The functional groups of FK506 that contribute to FKBP12 and calcineurin binding are known, and sites involved in calcineurin binding were chosen as points of modification for dimerization of FK506 (11). Two of the FK506 functional groups that do not appear to be required for FKBP12 binding are the hydroxyl group (*) and the allyl group (*). Linkers appended through hydroxyl group modification obstructed binding to FKBP12 (12). Modification of the alkene did not interfere with the interaction of FK506 and FKBP12; therefore, the terminal alkene was dihydroxylated, the diol intermediate was oxidatively cleaved, and the resulting aldehyde was reduced to a primary alcohol (7). The alcohol was elaborated to afford a group of FK506 dimers, referred to as FK1012s, carbamates **5-7**. Dimers also were assembled by cross metathesis of the terminal methylene group with the Grubbs catalyst (13), a strategy that produced **8-10**. All of the divalent ligands can mediate oligomerization of the FKBP12 fusion proteins. Their biological activities do not differ significantly; consequently, the specific structural features of the linker group (rigidity, hydrophobicity, and length) have little influence on ligand function. Thus, it is the point of linker attachment that is critical; the attachment preserves interactions with the target protein (FKBP) and minimizes unwanted interactions with calcineurin.

Linker elements, which are generally not relevant in the design of monovalent compounds, are often crucial to the successful generation of multivalent ligands. Although the identity of the linker in the FK506 strategy was not critical, the activity of many multivalent ligands depends on the nature of the linker. For example, both linker length and conformational flexibility play important roles in the interaction of a series of various divalent N-acetyl neuraminic acid (referred to here as sialic acid) derivatives with the influenza virus hemagglutinin (HA).

Divalent sialic acid derivatives were designed to bind to HA, a homotrimeric protein with three sialic acid binding sites (14). HA mediates the binding of virus to the cell surface sialic acids. Monovalent sialic acid derivatives are only weak inhibitors of HA-mediated cell adhesion (IC_{50} values ca. 3 mM). Divalent displays might have increased potency because they have the potential to occupy more than one binding site. Three spacers, ethylene glycol (**11**), glycine (**12**), and piperazine (**13**), were selected, and linkers of a variety of lengths were generated. The tethers were chosen because they differ in conformational flexibility. The ability of the sialic acid dimers to inhibit HA-mediated hemagglutination was measured (hemagglutination inhibition assay; HIA). None of the anomeric linkers interfered with HA binding, but alterations in linker structure had dramatic effects on biological activities. When ethylene glycol linkers are employed, the maximum increase in potency was only 20-fold greater than monovalent sialic acid, in contrast to the 100-fold increase observed for the derivative with the less flexible glycine spacer. Interestingly, series **13**, which possesses the most rigid piperazine linker, showed no enhanced potencies, even when the tether length, as estimated by molecular modeling, was similar to that of the linker of the most potent glycine dimer. Inter- and intra-receptor binding may contribute to a ligand's activity for a receptor with multiple ligand binding sites, such as HA. When the ability of the most potent glycine dimer to bind free trimeric HA in solution was measured, it was found that the dimer was no more potent than monomeric sialic acid. The potency of active **12**, therefore, is likely due to its ability to interact with two HA trimers on the viral surface.

Both the FK506 and sialic acid dimers illustrate the importance of design in the construction of divalent ligands. The FK506 examples demonstrate that the point of covalent attachment is critical for biological activity; changing the site of modification can result in dramatically different biological results. The HA example highlights the necessity to design

Schematic Ligand Structure: ●~~~~~~●

linkers based on the biological system where they will be studied or ultimately used. Changing the nature or length of the linker may allow different binding modes to be accessed, resulting in altered biological activities. The point of modification and the structure of the linking unit are the two key features involved in the synthesis of dimeric ligands, however there are other ligand features that must be considered in the design of compounds with greater valency.

　　　　Compounds with higher valencies have been generated to explore systematically the effects of various multivalent ligand parameters on biological activity. The design of these materials requires consideration of additional ligand features. The parameters outlined in the examples discussed here include the three-dimensional shape and size of the multivalent scaffold, the length and nature of the linkers that connect the recognition epitope and the incorporation of ancillary functional groups.

Size and three dimensional shape

　　　　Different synthetic approaches can be used to vary the size and shape of multivalent ligands. Even when the identity of the recognition epitope is held constant, variations in size and shape of the scaffold can result in ligands with dramatically different biological activities. A protein that has been used as a tool to study multivalent interactions is concanavalin A (Con A), a tetrameric, mannose-binding lectin. A number of multivalent ligands that display mannose have been generated in different shapes and sizes. For example, a trivalent mannose display based on a conformationally defined macrocycle, **14**, was synthesized and determined to have a 30- to 40-fold higher affinity for Con A than monomeric mannose in a surface plasmon resonance assay. Fluorescence resonance energy transfer (FRET) experiments revealed that one of the mechanisms responsible for the increase in potency is the ability of the ligand to form soluble Con A clusters (15). Dendrimers bearing mannose, **15**, also are ligands for Con A and these are more potent inhibitors of hemagglutination than monovalent mannose. Tetrameric **15** showed a 31-fold enhancement over mannose on a saccharide residue basis. The activity of the dendrimer was suggested by titration microcalorimetry and X-ray crystallographic studies to be due in part to aggregation and precipitation of Con A from solution (16). A cyclic scaffold, β-cyclodextrin, was used to generate a heptavalent display of mannose residues, **16**. Like the dendrimers, this ligand elicits efficient Con A precipitation (17). Mannose-substituted linear polymers, **17**, generated using ring-opening metathesis polymerization (ROMP) have shown increased potency in HIAs as well as in surface plasmon resonance assays. As the size of the polymer increases, the ligands exhibit

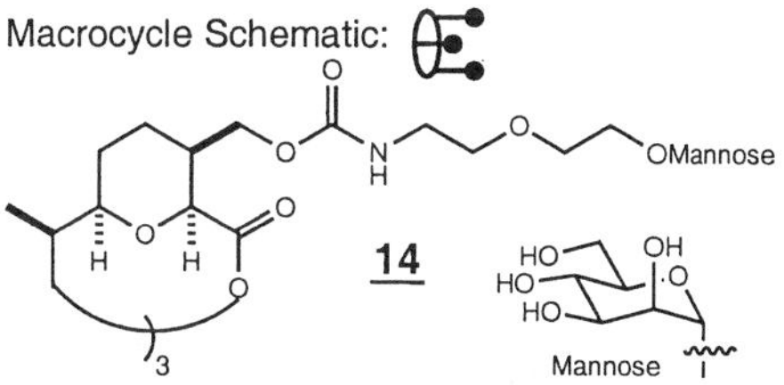

Macrocycle Schematic:

14

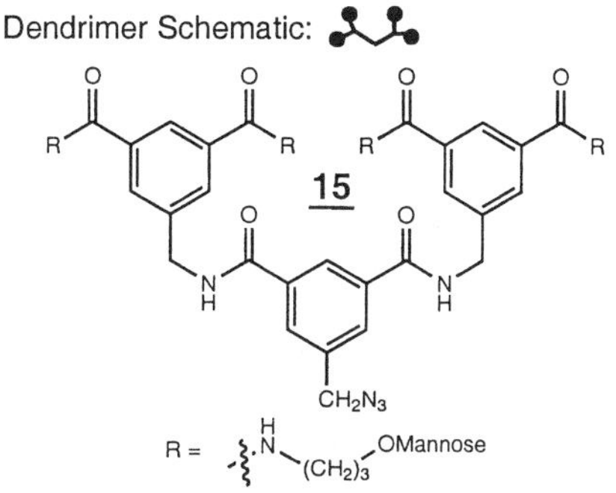

Mannose

Dendrimer Schematic:

15

CH_2N_3

$R = $ Mannose

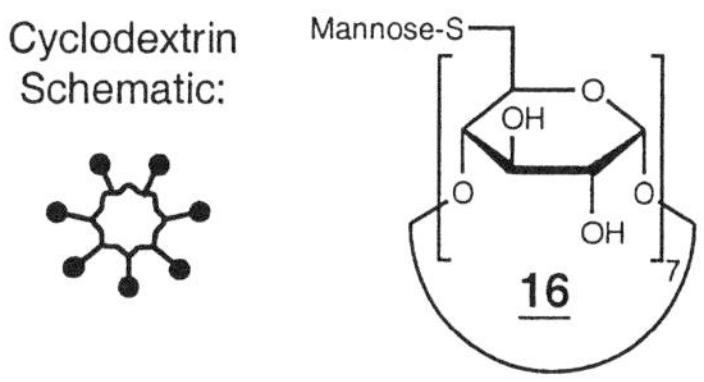

Cyclodextrin Schematic:

16

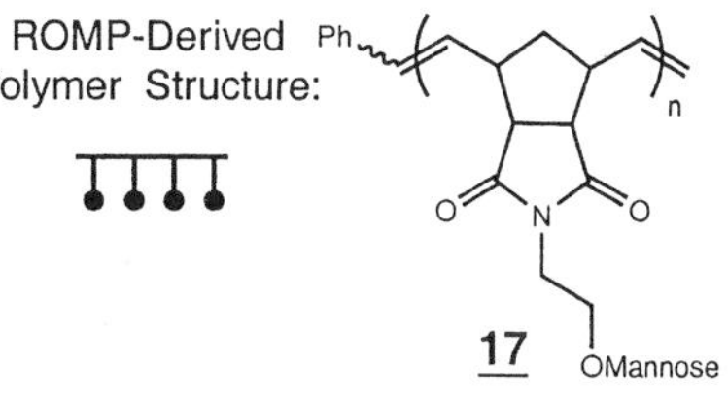

ROMP-Derived Polymer Structure:

17 OMannose

potencies significantly greater than many of the smaller templates described above (18-20). The use of different assays for each of these ligand types, however, makes direct comparisons of their activities difficult. Still, based on the degree of potency enhancement and the mechanistic information from compounds **14** and **15**, the differences in potency of these ligands can likely be attributed to changes in their mechanisms of action. The modes of binding available to these ligands are dictated by their sizes and shapes.

　　　Two commonly used multivalent scaffolds are linear polymers, such as polyacrylamide polymers, and wedge-shaped or spherical dendrimers, such as PAMAM dendrimers. Baker and coworkers have combined polymer and dendrimer chemistry to generate different ligand architectures (21). The products displayed carbohydrate recognition epitopes and were tested against four different viral strains in a hemagglutination assay. Sialic acid-substituted, linear acrylamide polymers **18** were compared to linear polymers displaying sialic acid-bearing dendrimers along the polymer backbone (dendron-displaying linear polymer, **19**). The biological activity of the linear polymer and dendron-displaying linear polymer was target dependent. Specifically, the linear polymers **18** were more effective against one viral strain (influenza A H2N2 mouse) than was the equivalent dendron-displaying polymer **19**. The activity of polymer **19**, however, was greater against another strain (X31), and both materials had essentially the same activity towards a third virus (Sendai). Two additional polymer architectures, termed comb-branched (**20**) and dendrigraft (**21**) also were examined. Comb-branched refers to a linear polymer displaying sialic acid-bearing linear polymers, and dendrigraft refers to a similar, but more highly branched, structure. Unlike the previous comparison, there were no significant differences in the hemagglutination inhibition activities of **20** and **21** for any of the four viral strains tested. Thus, for high valency ligands, backbone structure can play a role in biological activity, although the trends are difficult to predict.

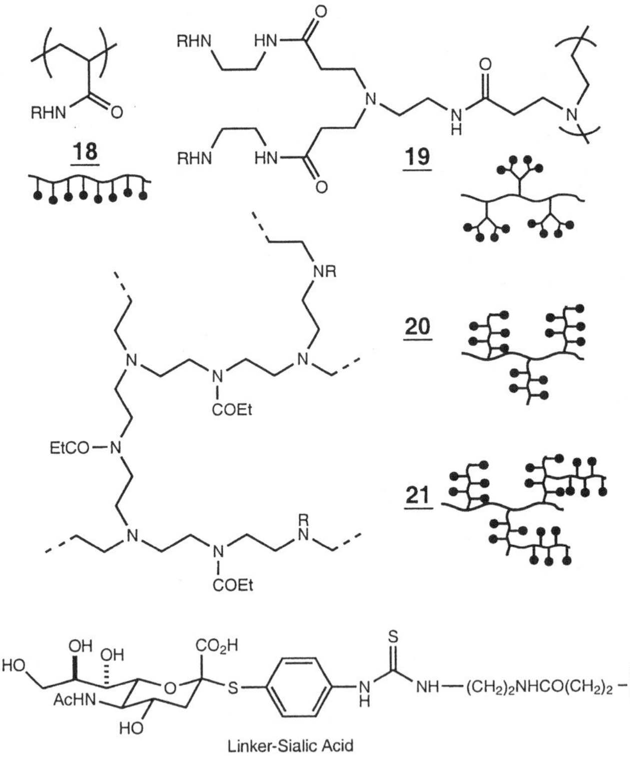

Flexibility of the Scaffold

　　　The flexibility of a scaffold also can affect the biological activity of a multivalent display, as observed by Kobayashi and coworkers (22). These researchers explored the relative potencies of two classes of linear saccharide substituted polymers, which vary in backbone flexibility. Saccharide-bearing poly(phenyl isocyanides) (PPI) **22** were generated, and these materials have a rigid, helical structure. For comparison, carbohydrate-substituted phenylacrylamide polymers (PAPs) **23** were synthesized because they are more flexible and can adopt a more extended conformation. The activities of the substituted PPI and PAP materials were assessed using two different lectins, ricin agglutinin (RCA$_{120}$) and Con A. When the activities of galactose-bearing PPI and PAP materials in an assay with RCA$_{120}$, were compared, the flexible PAP materials exhibited significantly higher potencies. Con A, which binds both glucose and mannose, was tested with glucose-bearing PAP and PPI displays.

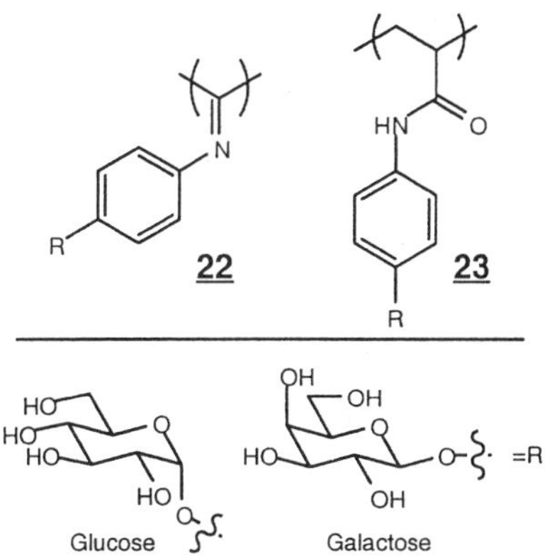

22   **23**

Glucose   Galactose

The backbone again affected the biological activity; the more flexible PAP materials had greater activity in assays with Con A. These data indicate that too rigid a scaffold can obstruct protein binding to recognition elements. It should be noted, however, that increasing backbone flexibility too much could diminish activity because of unfavorable contributions due to increased conformational entropy.

Ligand Valency

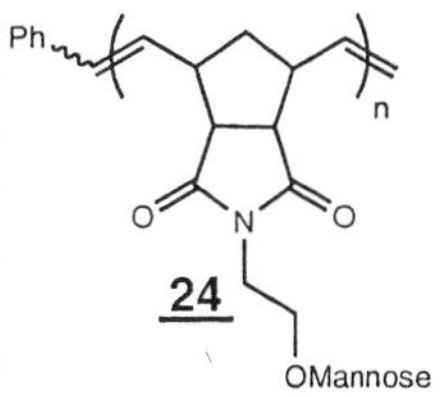

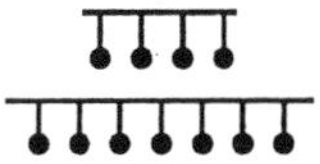

Dissecting the effects due to changes in valency from those due to changes in presentation is difficult. One investigation that addressed this issue took advantage of the features of ROMP. This synthetic approach was used to generate mannose-bearing materials **24** that varied in valency. A key feature of this strategy is that valency can be changed, while holding other ligand features, such as epitope spacing and backbone flexibility, constant. The inhibitory activities of the ligands in a Con A-mediated HIA depended on their valency (18). Higher valency ligands had increased potency. Recently, the synthesis of defined, linear displays via ROMP was

24

OMannose

generalized so that libraries of displays can be generated (20). The ability to access different mechanisms by varying valency may provide new opportunities for the development of multivalent scaffolds with a variety of biological effects.

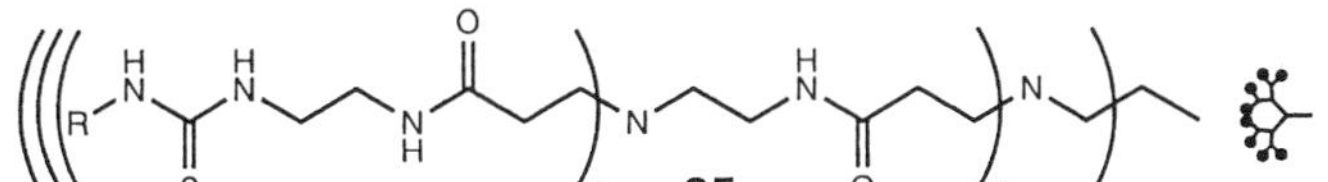

25

Several studies have explored simultaneous variations in scaffold size and the number of attachment sites for recognition epitopes (23). In one example, a series of dendrimers, with 2, 4, 8 (**25**), 16, 32, 64, and 128 lactose residues, was generated and the activities of these molecules in assays with a panel of proteins was measured. The proteins were chosen because they differ in their number of saccharide binding sites and their orientation of binding sites. The authors compared potency on a saccharide residue basis. In general, increasing the valency of the dendrimer increased protein-dendrimer binding for all four proteins. However, the magnitude of this effect is dependent on the structure of the target protein. This study demonstrates the importance of tailoring the structure of a multivalent ligand to the target protein; the design of ligands should account for the number, orientation, and spacing of protein binding sites.

Linkage to the Scaffold

Another critical design feature is the length of the linker or tether that connects the recognition element to the scaffold. This importance is highlighted in a search for multivalent inhibitors of the bacterial toxin enterotoxin (24). Enterotoxin is a pentameric protein with five saccharide-binding sites, one per monomer. Enterotoxin binds the GM1 ganglioside, which possesses a terminal galactose residue. A pentacyclen core was used to display five galactose residues, and these residues were tethered using linkers that range from 60 to 180 Å in length (26). The four ligands were tested for their ability to inhibit enterotoxin binding to its target ganglioside. The activity of each depended on the length of the linker employed. The compound with the longest linker was the most potent; it was 10^4-fold more effective than the monomeric galactose control. Dynamic light scattering experiments indicate that 1:1 protein-ligand complexes are formed, suggesting the efficacy of the ligand may be due to the chelate effect. In a related study, a decameric ligand was found to be a highly active inhibitor of Shiga toxin, which has 15 galactose binding sites (25). A structure of the complex reveals that this ligand dimerizes the toxin and each saccharide residue of the complex occupies a saccharide-binding site. Together, these studies demonstrate the importance of linker length and its impact on the biological activity and mechanism of action of the ligand. Moreover, they are outstanding examples of the importance and potential of multivalent ligand design.

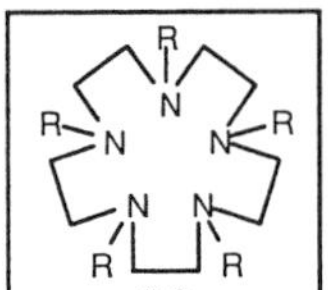

26

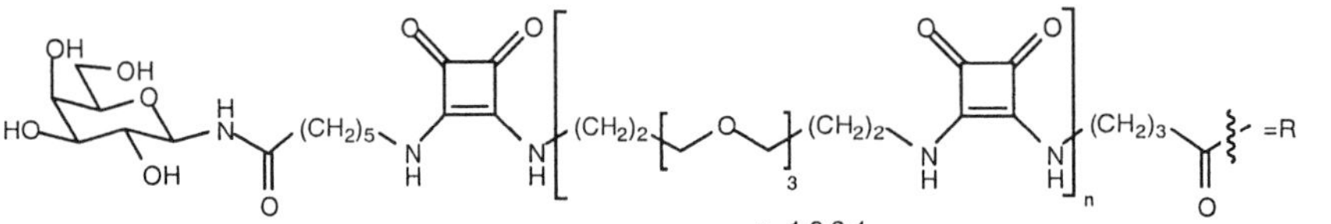

n=1,2,3,4

Schmidt and coworkers explored the importance of linker length in the context of a surface. Using sialyl Lewis x (sLe^x) glycolipid derivatives, they examined the structural requirements for selectin-mediated cell rolling (26). The selectins are a family of carbohydrate binding proteins that facilitate the rolling of leukocytes on the endothelial cell wall, the first step in the inflammatory response. SLe^x is a tetrasaccharide known to bind to the selectins, but is generally displayed on a protein backbone at the end of a complex carbohydrate chain. Determination of the optimal spacing between sLe^x and a multivalent scaffold may lead to ligands with greater biological activity. In this example, an ethylene glycol-based linker was installed between the recognition epitope, sLe^x, and the lipid necessary for incorporation on the surface, **27**. The structure of the linker was designed to mimic the "length" of a carbohydrate. The cell rolling assays suggest that there is a minimum requirement for accessibility of the recognition epitope; the incorporation of six or nine ethylene glycol units resulted in compounds with significantly higher activities than were derivatives with zero and three ethylene glycol units. These results highlight the importance of considering the physiological presentation of a ligand when designing the linker for synthetic macromolecular conjugates.

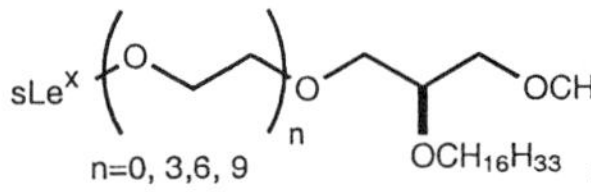

n=0, 3,6, 9   **27**

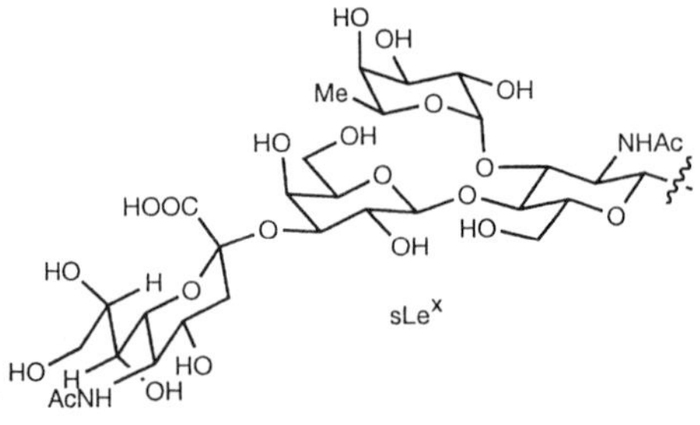

sLe^x

Incorporation of Ancillary Functional Groups

Most examples of multivalent ligand design focus on increasing the specific interactions between a recognition epitope and the target receptor. As with small molecules, however, incorporation of a variety of functional groups may contribute to biological activity. Applying this concept to multivalent displays, Whitesides and coworkers incorporated a range of functional groups into sialic acid-bearing polyacrylamide displays, **28**, and tested the ability of the resulting materials to inhibit viral agglutination (27). The inhibitory concentrations were determined based on the amount of sialic acid appended onto polymer backbone; the concentration of each ancillary functional group incorporated was kept constant. Inclusion of many of these groups resulted in materials with increased inhibitory potencies. Two types of ancillary groups were of particular interest due to their similar structures yet distinct biological activities. Incorporation of D-2-amino-2-deoxymannose resulted in a 40-fold greater activity than D-2-amino-2-deoxyglucose. In addition, the use of 1-amino-1-cyclopropanecarboxylic acid, 1-amino-1-cyclopentanecarboxylic acid, and 1-amino-1-cyclohexanecarboxylic acid, resulted in materials with increased potencies of 5- to 20-fold. These results suggest that ancillary groups can exploit unknown yet specific secondary binding interactions. This strategy could be useful in increasing the affinity of multivalent ligands for their targets.

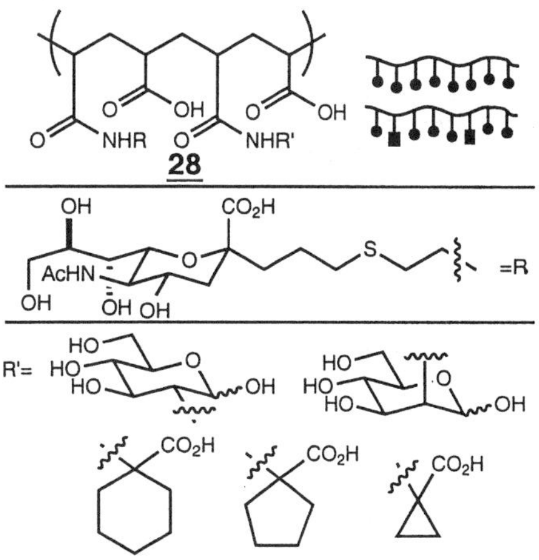

Conclusion: Design of Multivalent Scaffolds

By assessing the effects of structural variations on multivalent ligand activities, structure-activity relationships (SAR) can be established, in analogy to SAR studies with small molecules. In both cases, the final three-dimensional display of recognition epitopes, determined by the features of the scaffold and linking groups, is critical for proper recognition by the appropriate target. Ligand activity can be modulated by incorporation of functional groups that may participate in specific or non-specific interactions. There are, however, critical differences that distinguish multivalent ligand design. Because multivalent ligands can generally bind multiple binding sites, there are a wide variety of mechanisms multivalent ligands can exploit. In addition, the size of ligands needed to bind discrete, spatially separated targets often is significantly greater than that needed for ligands that bind a single site. The examples described here emphasize the importance of controlling multivalent ligand structure with synthesis. We have focused on design features, because we anticipate that an understanding of the issues associated with these unconventional ligands will facilitate the development of new general approaches for their synthesis Advances in chemical synthesis undoubtedly will continue to facilitate systematic studies of ligand parameters. Such ligands could be used to elucidate and manipulate a diversity of cellular processes.

References

1. M. Mammen, S.-K. Choi, G. M. Whitesides, Angew. Chem. Int. Ed. Engl., 37, 2755 (1998).
2. D. M. Crothers, H. Metzger, Immunochem., 9, 341 (1972).
3. A. S. Ripka, D. H. Rich, Current Opinion in Chemical Biology, 2, 441 (1998).

4. IgG (PDB accession number 1IGT) and enterotoxin (PDB accession number 1EEF) are structures generated from X-ray crystallographic studies. Images were generated in RasMol 2.7.1. See www.rcsb.org/pdb. The FK1012 structure is based on PDB accession number 1A7X. The dendrimer is an eighth generation PAMAM dendrimer modeled by the Caltech Molecular and Process Simulation Center. The linear polymer is a mannose-substituted 10mer generated by ROMP and modeled by C.W. Cairo (UW-Madison). The enterotoxin protein is shown on the left and was removed from the right hand image to clarify the location of the saccharides. The scale bars indicate 2.5 nm.

5. L. L. Kiessling, N. L. Pohl, Chem. Biol., 3, 71 (1996).

6. L. L. Kiessling, L. E. Strong, Top. Organomet. Chem., 1, 199 (1998).

7. D. M. Spencer, T. J. Wandless, S. L. Schreiber, G. R. Crabtree, Science, 262, 1019 (1993).

8. J. D. Klemm, S. L. Schreiber, G. R. Crabtree, Annu. Rev. Immunol., 16, 569 (1998).

9. D. J. Matthews, R. S. Topping, R. T. Cass, L. B. Giebel, Proc. Natl. Acad. Sci. USA, 93, 9471 (1996).

10. J. Liu, J. D. J. Farmer, W. S. Lane, J. Friedman, I. Weissman, S. L. Schreiber, Cell, 66, 807 (1991).

11. S. L. Schreiber, Science, 251, 283 (1991).

12. G. D. Van Duyne, R. F. Standaert, P. A. Karplus, S. L. Schreiber, J. Clardy, Science, 252, 839 (1991).

13. S. T. Diver, S. L. Schreiber, J. Am. Chem. Soc., 119, 5106 (1997).

14. G. D. Glick, P. L. Toogood, D. C. Wiley, J. J. Skehel, J. R. Knowles, J. Biol. Chem., 266, 23660 (1991).

15. S. D. Burke, Q. Zhao, M. C. Schuster, L. L. Kiessling, J. Am. Chem. Soc., 121, 4518 (2000).

16. S. M. Dimick, S. C. Powell, S. A. McMahon, D. N. Moothoo, J. H. Naismith, E. J. Toone, J. Am. Chem. Soc., 121, 10286 (1999).

17. J. J. Garcia-Lopez, F. Hernandez-Mateo, J. Isac-Garcia, J. M. Kim, R. Roy, F. Santoyo-Gonzalez, A. Vargas-Berenguel, J. Org. Chem., 64, 522 (1999).

18. M. Kanai, K. H. Mortell, L. L. Kiessling, J. Am. Chem. Soc., 119, 9931 (1997).

19. D. A. Mann, M. Kanai, D. J. Maly, L. L. Kiessling, J. Am. Chem. Soc., 120, 10575 (1998).

20. L. E. Strong, L. L. Kiessling, J. Am. Chem. Soc., 121, 6193 (1999).

21. J. D. Reuter, A. Myc, M. M. Hayes, Z. Gan, R. Roy, D. Qin, R. Yin, L. T. Piehler, R. Esfand, D. A. Tomalia, J. R. Baker Jr., Bioconjugate Chem., 10, 271 (1999).

22. T. Hasegawa, S. Kondoh, K. Matsuura, K. Kobayashi, Macromolecules, 32, 6595 (1999).

23. S. Andre, P. J. Cejas Ortega, M. Alamino Perez, R. Roy, H.-J. Gabius, Glycobiology, 9, 1253 (1999).

24. E. Fan, Z. Zhang, W. E. Minke, Z. Hou, C. L. M. J. Verlinde, W. G. J. Hol, J. Am. Chem. Soc., 122, 2663 (2000).

25. P. I. Kitov, J. M. Sadowska, G. Mulvey, G. D. Armstrong, H. Ling, N. S. Pannu, R. J. Read, D. R. Bundle, Nature, 403, 669 (2000).

26. C. Gege, J. Vogel, G. Bendas, U. Rothe, R. R. Schmidt, Chem.-Eur. J., 6, 111 (2000).

27. S. K. Choi, M. Mammen, G. M. Whitesides, J. Am. Chem. Soc., 119, 4103 (1997).

SECTION VII. TRENDS AND PERSPECTIVES

Editor : Annette M. Doherty
Institut de Recherche Jouveinal/Parke-Davis

Chapter 30. To Market, To Market – 1999
Bernard Gaudillière and Patrick Berna
Institut de Recherche Jouveinal/Parke-Davis
Fresnes, France

In 1999, the trend observed in the previous year was repeated : among the new active entities marketed worldwide for human therapeutic use, the new biological entities (NBEs) that are biotechnology-based drugs (mainly recombinant proteins) represent a significant proportion ; moreover, they represent innovative approaches to the treatment of several important human diseases. In this chapter, new chemical entities (NCEs) and NBEs will be treated together as new therapeutic entities.

A total of 35 new therapeutic entities were introduced into their first markets last year, making 1999 more productive than previous years 1995-1998 (1–5). Moreover, the launch of drugs considered as priority drugs, which represent significant therapeutic gains over the existing drugs for the patients and the medical community ranked particularly high.

With 15 launches of new drugs, the US maintained their top rank in terms of new introductions, followed by Europe and Japan with 8 and 7 respectively. In terms of pharmaceutical companies, no company really emerged apart from Akzo Nobel, Boehringer Ingelheim and Ligand with 2 new entities originating at each.

With 7 new launches, antiinfectives were one of the most active groups : Anti-viral agents were first represented last year with two new drugs for the treatment of HIV infections : Agenerase (Amprenavir), the first new protease inhibitor to be approved for AIDS and marketed in more than two years and then Ziagen (Abacavir), one of the most potent inhibitors of HIV-replication which as a nucleoside reverse transcriptase inhibitor is highly synergistic with protease inhibitors. First in the new class of neuraminidase inhibitors, Tamiflu (Oseltamivir) and Relenza (Zanamivir) were introduced to treat a large variety of influenza strains. Synercid, a two component semi-synthetic mixture, was the first in a new class of antibiotics with a broad spectrum of activity, in particular effective against gram-positive bacteria responsible for life-threatening infections resistant to all other antibiotics. Two new-generation quinolones with an improved spectrum of antibacterial activity, Avelox (moxifloxacin) and Tequin (gatifloxacin) were launched for the treatment of refractory respiratory tract infections and are expected to usefully complete the available arsenal against community-acquired infections.

LYMErix is a vaccine against Lyme's disease that was introduced for the first time last year for the prevention of this tick borne disease.

In the field of oncology, Panretin (alitretinoin) was the first topical therapy for the AIDS-related Kaposi's sarcoma. A new potent antiproliferative. Arglabin, was developed in several countries as an ethical pharmaceutical for cancer. Temodal (temozolomide) was the first new drug in 20 years for the treatment of certain malignant gliomas. Valstar (Valrubicin) and Sunpla (SKI-2053R) were both launched as new antineoplastic agents, against bladder and stomach carcinoma respectively, with significant advantages over existing therapies. Ontak (Denileukin diftitox), produced by genetically fusing protein from the diphtheria toxin to IL-2, was marketed for the treatment of persistent or

331

recurrent cutaneous T-cell lymphoma. Octin (OCT-43), a recombinant variant of IL-1beta, was approved for treatment of skin tumors. The first recombinant tumor necrosis factor marketed for cancer treatment, Beromun (tasonermin) was used for soft tissue sarcoma of the limbs and malignant melanoma as an adjunct to surgery.

Major high profile launches of last year were those of the « super-aspirins » Celebrex (celecoxib) and Vioxx (rofecoxib). These selective COX-2 inhibitors exhibit potent antiinflammatory activity without gastrointestinal toxicities and are already considered as breakthrough medicines for the relief of symptoms in osteoarthritis and rheumatoid arthritis.

Another success has been the introduction of several new drugs for type-2 diabetes. Actos (pioglitazone) and Avandia (rosiglitazone) have been shown to significantly improve the glycemic control in type-2 diabetes with a lower occurrence of hepatic toxicity than for Rezulin, the first drug in this class. In another mechanistic class, Starsis (nateglinide) was launched last year with a unique pharmacological profile and pharmacokinetic properties.

Avishot (naftopidil) was marketed for the treatment of dysurea associated with benign prostatic hypertrophy, demonstrating a selective urodynamic effect.

Hectorol (Doxercalciferol) is a new vitamin D2 analog for the management of secondary hyperparathyroidism in patients with end-stage renal failure. Cetrotide (cetrorelix) was the first luteinizing hormone-releasing hormone approved worldwide in the treatment of female infertility.

In the cardiovascular field, Colforsin daropate, a water-soluble activator of adenylate cyclase, was launched as the first in its class for the treatment of acute heart failure. Shinbit (nifekalant) was a novel class III anti-arrythmic agent with specific characteristics of K+ current blockade conferring some differentiation from other class III agents. Micardis (telmisartan) was the sixth antihypertensive agent representing the angiotensin II antagonist class of drugs. Integrilin (Eptifibatide) should have been included last year in this chapter because this significant new agent was actually introduced on the market in 1998 ; this short-acting antagonist of the glycoprotein IIb/IIIa complex completes the arsenal of drugs for the treatment of acute coronary syndrome.

Migsis (lomerizine) was introduced as the first dual sodium and calcium channel blocker for the treatment of migraine. Sonata (zaleplon) was the first in a new chemical class with a unique profile for the treatment of insomnia.

Finally, 12 other biological entities were launched in 1999, although they are not considered as NBEs:
- natural products such as Calfactant from Forest (Infasurf™, an extracted calf lung surfactant indicated for prevention and treatment of respiratory distress syndrome in premature infants), Dermatan sulfate from Mediolanum (Mistral™, a natural glycosaminoglycan extracted from pig and indicated for the prevention of deep vein thrombosis in bedridden patients and in patients undergoing certain types of surgery), combined hepatitis A and typhoid vaccine (Hepatyrix™, vaccine containing natural antigenic extracts), or meningococcal group C conjugate vaccine (Meningitec™, vaccine containing natural antigenic extracts) ;
- existing products launched for new indications such as hepatitis B immune globulin (Nabi-HB™, indicated for treatment following exposure to hepatitis B virus) or Infliximab (Remicade™, the anti-TNF-alpha monoclonal antibody launched for use in combination with methotrexate, for reducing the symptoms of rheumatoid arthritis);

launched for use in combination with methotrexate, for reducing the symptoms of rheumatoid arthritis);

- new formulation or new production mode of already existing products such as Moroctocog alfa (ReFacto™, a second-generation B-domain deleted recombinant factor VIII albumin-free formulation for the treatment of hemophilia A), somatropin (Nutropin depot™, first long-acting dosage form of recombinant human growth hormone), recombinant Glucagon (for the treatment of severe hypoglycemia in patients with diabetes), Insulin aspart (NovoRapid™, new rapid-acting insulin analog for the treatment of diabetes mellitus), recombinant human insulin (Insuman™, insulin for the treatment of type-1 and type-2 diabetes), or Pamiteplase (Solinase™, a second generation tissue plasminogen activator as thrombolytic agent in acute myocardial infarction).

Abacavir sulfate (Antiviral) (6-12)

Country of Origin : UK
Originator : Glaxo Wellcome
First Introduction : US
Introduced by : Glaxo Wellcome
Trade Name : Ziagen
CAS Registry No : 188062-50-2
Molecular Weight : 335.34

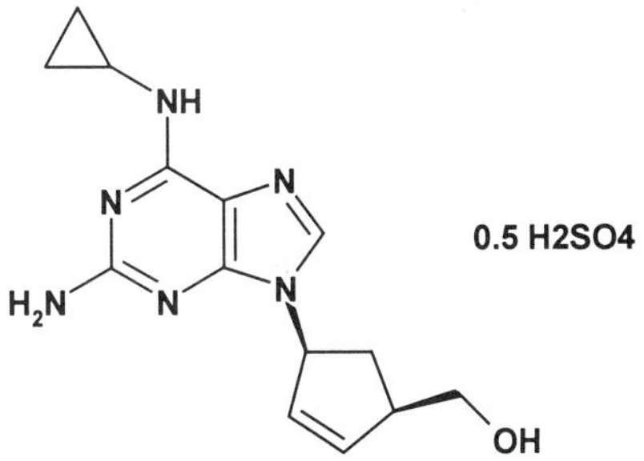

0.5 H2SO4

Abacavir sulfate was first launched as Ziagen in the US for the treatment of human immunodeficiency virus (HIV) infection, in combination with other antiretroviral drugs. Abacavir is a carbocyclic nucleoside reverse transcriptase inhibitor (nRTI) ; it is one of the most potent anti-HIV agents to date. The compound can be prepared by an enantioselective synthesis involving palladium-catalyzed coupling of a chloropurine with a carbocyclic allylic diacetate. In vitro, Abacavir is a potent and selective inhibitor of HIV-1 and HIV-2 replication. Resistance to Abacavir develops more slowly than for other anti-HIV agents. Abacavir is highly synergistic with protease inhibitors such as Amprenavir. In clinical trials for HIV infections in adults, it produced durable suppression in viral load. Combinations with different protease inhibitors such as Nelfinavir, Saquinavir or Indinavir markedly reduced plasma viral load to undetectable levels for at least 48 weeks, and significantly raised CD4+ cell counts in adults with HIV infection, especially nRTI-naive patients. Abacavir has a good oral availability and its penetration into CSF is much more significant than for other anti-HIV drugs. The two major metabolites identified in humans were the 5'-carboxylate and the 5'-glucuronide, mainly excreted via the renal route.

Alitretinoin (Anticancer) (13-19)

Country of Origin : US
Originator : Ligand
First Introduction : US
Introduced by : Ligand

Trade Name : Panretin
CAS Registry No : 5300-03-8
Molecular Weight : 300.44

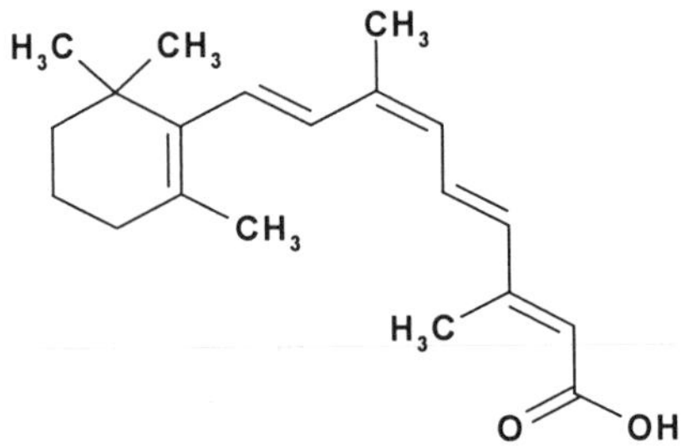

Alitretinoin was introduced in the US for the topical treatment of cutaneous lesions in patients with Kaposi's sarcoma (KS), the most frequent malignancy observed in AIDS patients. It is a derivative of 9-cis-retinoic acid (9-cis-RA) identified as an endogenous hormone in mammalian tissues. Alitretinoin binds to all isoforms of the intracellular retinoid X (RXR) and retinoid A (RAR) receptors thus inducing cell differentiation, increasing cell apoptosis and inhibiting cellular proliferation in experimental models of human cancer. In an international phase III trial in patients with KS, 42% of 82 patients treated with 0.1% gel formulation experienced complete or partial responses cf 7% for controls. Alitretinoin is the first topical therapy for KS and is expected to be a new option to its traditional management.

Amprenavir (Antiviral)   (20-25)

Country of Origin : US
Originator : Vertex Pharm
First Introduction : US
Introduced by : Glaxo Wellcome
Trade Name : Agenerase
CAS Registry No : 161814-49-9
Molecular Weight : 505.64

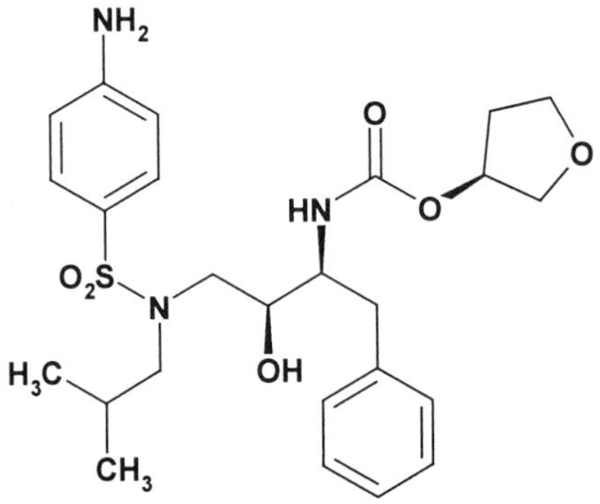

Amprenavir was launched as Agenerase in the US for the treatment of AIDS patients in combination with approved agent antiretroviral nucleoside analogs. It is the fifth non-peptidic inhibitor of HIV-1 protease to be marketed in this indication after the last approved Neflinavir. Amprenavir, designed via a structure-based process, is the smallest molecule in the « navir » class and exhibits a reduced peptidic character. An improved process for preparation comprising four steps from a (1S, 2R)-2-hydroxy-3-aminopropylcarbamate has been developed. Amprenavir is a potent inhibitor of HIV-1 aspartyl protease (Ki = 0.6nM), an enzyme required by the virus to cleave pro-form polyproteins to structural proteins during the last stage in the replication process. The compound displays good oral bioavailability in humans and penetrates the CNS, which is an important advantage in long-term treatment. Its plasma half-life is approximately 10h. Treatment with Amprenavir in combination with nucleoside analog reverse transcriptase inhibitors considerably decreases viral load and restores CD4+ T-cell counts in patients with HIV infection.

Arglabin (Anticancer)   (26-31)

Country of Origin : Kazakstan
Originator : NuOncology Labs
First Introduction : Russia
Introduced by : NuOncology Labs
Trade Name :   ?
CAS Registry No : 84692-91-1
Molecular Weight : 246.31

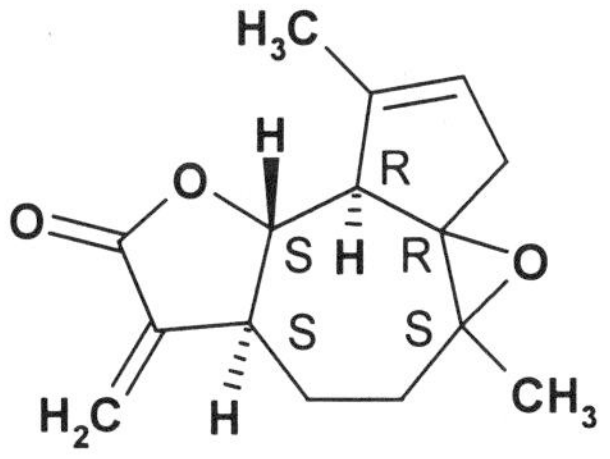

Arglabin is a new antineoplastic agent launched in Russia for the treatment of a variety of cancers. It is a sesquiterpene gamma-lactone extracted from Artemisia glabella, a plant from Kazakstan ; it was first developed in Kazakstan followed by US and other countries as an ethical pharmaceutical. Arglabin can be obtained by resin extraction from *A.glabella*, purification by chromatography and finally recrystallization. Arglabin is a potent and selective inhibitor of farnesyl transferase, an enzyme critical to the function of the Ras oncogene in cancer cell reproduction. It showed a strong Inhibitory effect against tumor cells proliferation in in vitro models of transformed mouse hepatocytes and splenocytes. This cytotoxic effect against tumor cells was 50-100 times higher than that on intact cells. In many patients with various cancers, Arglabin was shown to stop or inhibit the tumour development, alone or in combination with other anticancer agents, and was reportedly well tolerated with few side-effects. Promising response rates were also reported with Arglabin after treating other « difficult-to-treat cancers » such as breast, ovarian, lung or colon cancers.

Celecoxib (Antiarthritic)   (32-38)

Country of Origin : US
Originator : Searle
First Introduction : US
Introduced by : Searle / Pfizer
Trade Name : Celebrex
CAS Registry No : 169590-42-5
Molecular Weight : 381.37

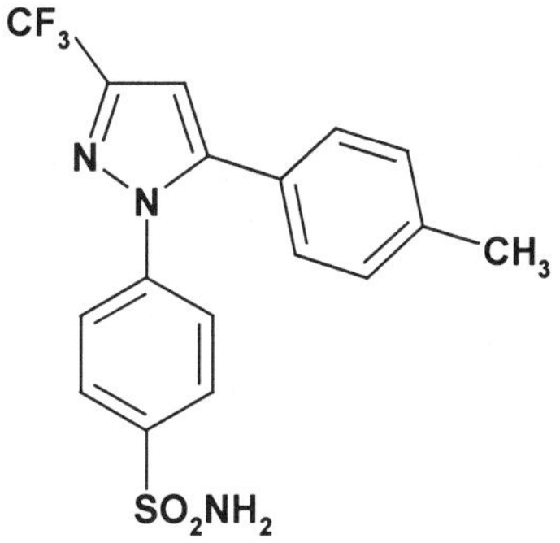

Celecoxib is a nonsteroidal antiinflammatory drug (NSAID) first launched as Celebrex in the US for the treatment of symptoms in patients with rheumatoid arthritis (RA) and osteoarthritis (OA). Celecoxib belongs to a new class of 1,5-diarylpyrazoles and can be synthesized by heat-promoted heterocyclization of a trifluoro-1,3-dione with appropriate arylhydrazine. Celecoxib is a highly selective inhibitor of COX-2, the inducible form of cyclooxygenase expressed during inflammatory processes ; it does not block the constitutive form COX-1, thus suppressing the gastric and intestinal toxicity of most non-selective NSAIDs. The potency ratio COX1/COX2 on purified human enzymes was about 400. In several in vivo models of acute and chronic inflammation, Celecoxib demonstrated potent antiinflammatory activity without affecting gastric or urinary prostaglandin PGE2. In several clinical studies performed with patients suffering from osteoarthritis or rheumatoid arthritis, Celecoxib was shown to be well tolerated and to relieve pain and inflammation more efficiently compared with other standard NSAIDs ; the gastrointestinal safety profile was significantly better than that of other NSAIDs. Interestingly, Celecoxib was approved before the end of last year for another indication in patients with familial adenomatous polyposis (FAP). A six-month clinical trial demonstrated a 28% reduction in the number of colorectal polyps with Celecoxib, compared to a 5% reduction with placebo.

Cetrorelix   (39-46)
(Female Infertility)

Country of Origin :
Germany
Originator : Asta
Medica
First Introduction :
Germany
Introduced by :
Asta Medica
Trade Name :
Cetrotide
CAS Registry No :
120287-85-6 ;
130289-71-3
Molecular Weight :
1431.07

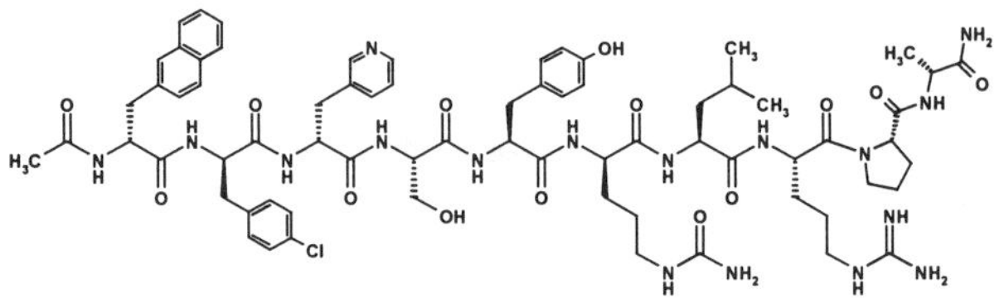

Cetrorelix was launched last year in Germany for the treatment of female infertility. It is a decapeptidic analog of luteinizing hormone-releasing hormone (LH-RH) bearing structural modifications in the crucial positions 1, 2, 3, 6 and 10 : [Ac-D-Nal1, D-4-Cl Phe-2, D-Pal3, D-Cit6, D-Ala10]-GnRH. Cetrorelix is an extremely potent and long acting gonadotrophin releasing hormone (GnRH) antagonist and thus blocks gonadotrophins and sex steroid secretion immediately after administration. Moreover, it has a low histamine-releasing potency. Cetrorelix will be the first LH-RH antagonist approved worldwide. In several Phase III clinical trials, female patients receiving Cetrorelix had a successful controlled ovulation thus avoiding a premature LH-surge. Cetrorelix could be a first-choice in-vitro fertilization (IVF) treatment without the complications of current controlled ovarian hyperstimulation protocols. Cetrorelix is currently under clinical investigation for the treatment of diverse sex hormone dependent disorders such as benign prostate hypertrophy, breast, ovarian or prostate cancers and diverse gynaecological disorders.

CHF-1301 (Antiparkinsonian)   (47-49)

Country of Origin : Italy
Originator : Chiesi Farmaceutici SpA
First Introduction : Italy
Introduced by : Chiesi
Trade Name : Levomet
CAS Registry No : 7101-51-1
Molecular Weight : 211.22

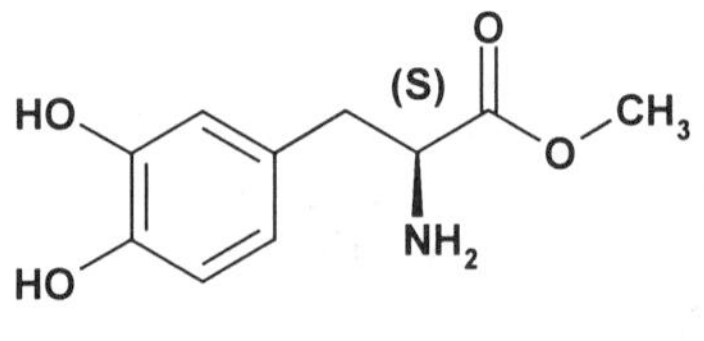

CHF-1301 was launched as an injection in Italy for the treatment of patients with Parkinson's disease, particularly those with complications in their clinical responses. CHF-1301 is the methyl ester of L-Dopa (levodopa), the immediate precursor of dopamine. Thus, this prodrug is more soluble in water at physiological pH, better absorbed and slowly transformed into levodopa in vivo ; it consequently improves the duration of action and prevents the rapid fluctuations in response or involuntary movements (dyskinesia) observed during chronic treatment with levodopa. In patients with a delayed effect of levodopa (on-phenomenon) or a severe resistance to L-Dopa (off-phenomenon), CHF-1301 significantly improves the onset and the stability of the response.

Colestimide (Hypolipidaemic)   (50-56)

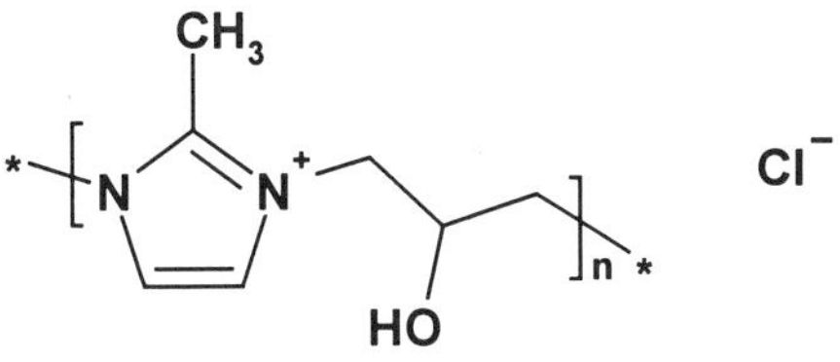

Country of Origin : Japan
Originator : Mitsubishi
First Introduction : Japan
Introduced by : Tokyo Tanabe, Yamanouchi
Trade Name : Cholebine
CAS Registry No : 95522-45-5

Colestimide was launched in Japan for the treatment of hyperlipidemia and hypercholesterolemia. It is a newly developed methyl imidazole-epichlorhydrin copolymer ; this anion exchange resin works as a powerful bile acid sequestrant, thus increasing the rate of conversion of cholesterol to bile acids in the liver which may lead to enhancement of the rate of LDL removal. In animal models and humans, Colestimide proved to be more potent than cholestyramine and seems to be better tolerated. It reduced plasma levels of total and LDL cholesterol and did not alter HDL and triglyceride levels. Colestimide can be used in monotherapy or co-administered with HMG-CoA reductase inhibitors.

Colforsin Daropate Hydrochloride   (57-61)
(Cardiotonic)

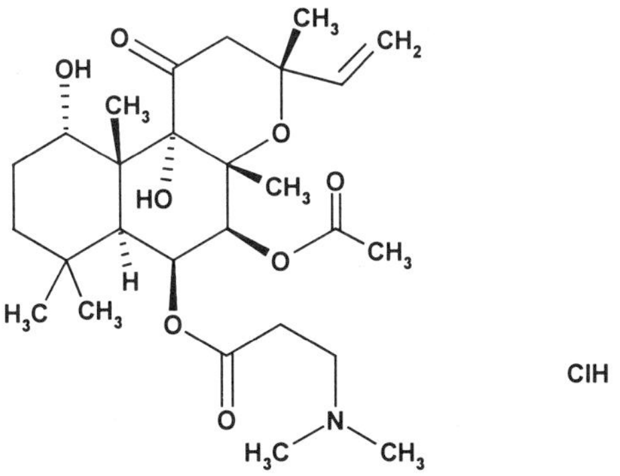

Country of Origin : Japan
Originator : Nippon Kayaku
First Introduction : Japan
Introduced by : Nippon Kayaku
Trade Name : Adele ; Adehl
CAS Registry No: 138605-00-2
Molecular Weight : 546.11

Colforsin daropate was launched in Japan as a treatment for acute heart failure. It is a water-soluble and orally active 3-dimethylaminopropionyl (daropate) derivative of forskolin, a labdane-type diterpene isolated from the root of *Coleus forskohlii*. It is obtained from forskolin in 7 steps based upon a series of protections and deprotections of the different OH groups and a key 7,8-transacylation. Colforsin daropate is the first in its class of adenylate cyclase activators to be marketed in the treatment of heart failure. As forskolin, it directly stimulates adenylate cyclase thus increasing intracellular concentration of cAMP and inhibiting calcium mobilization. As a direct consequence, it induces a significant vasodilatation, so provides a moderate positive inotropic and chronotropic effect. In contrast with well-established generators of cAMP synthesis (dobutamine), Colforsin daropate maintains its action on cardiac muscle in the severe stage even when the beta-adrenoceptor is down-regulated in myocardium. In several clinical trials in patients with severe congestive heart failure (CHF), Colforsin daproate improved hemodynamic symptoms of heart failure patients without serious adverse effects.

Dalfopristin/Quinupristin (Antibiotic)   (62-68)

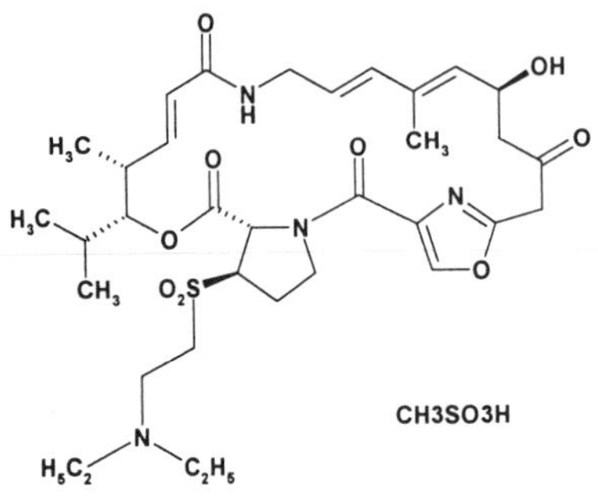

CH3SO3H

dalfopristin (methanesulfonate)

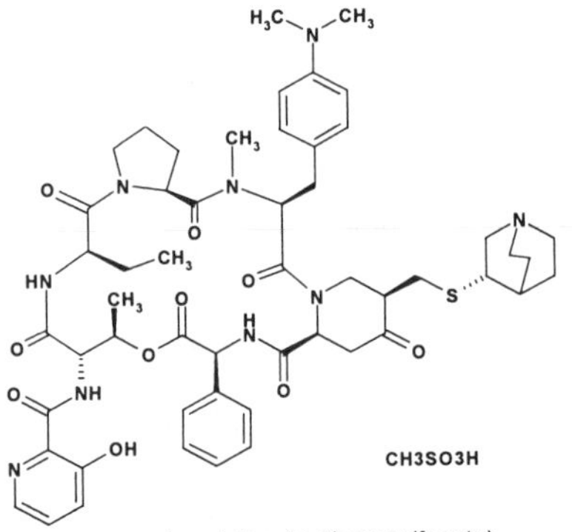

CH3SO3H

quinupristine (methanesulfonate)

Country of Origin : France
Originator : Rhone-Poulenc-Rorer
First Introduction : UK
Introduced by : Aventis

Trade Name : Synercid
CAS Registry No : 126602-89-9
MW Dalfopristin (CH3SO3H) : 786.97
MW Quinupristin (CH3SO3H) : 1118.35

Dalfopristin and Quinupristin are two well-defined semi-synthetic antibacterials belonging to the class of streptogramins. Together, they form Synercid, an injectable formulation in a ratio 70 : 30 that was launched last year first in the UK followed by the US. The two distinct antibiotic agents, pristinamycins, are bacteriostatic in their own right but, when they are in association, act synergistically at the ribosomal level to inhibit protein synthesis. Synercid is the first antibiotic in its class to reach the market. It appears to be a sound treatment alternative for patients with severe or life-threatening infections, such as the most problematic forms of gram-positive nosocomial sepsis including vancomycin-resistant Enterococcus faecium (VREF) or methicillin-resistant Staphylococcus aureus (MRSA).

Denileukin diftitox (Anticancer)   (69-71)

Country of Origin : US
Originator : Ligand
First Introduction : US
Introduced by : Ligand
Trade Name : Ontak
CAS Registry No : 173146-27-5

Class : Recombinant protein
Type : receptor targeted fusion toxin
Molecular Weight : 58 kDa
Expression system : E. Coli
Manufacturer : Seragen

Denileukin diftitox was first launched as Ontak in the US for the treatment of patients with persistent or recurrent cutaneous T-cell lymphoma (CTCL) whose malignant cells express the CD25 component of the IL-2 receptor. Denileukin diftitox is a diphtheria toxin fragment A - fragment B genetically fused to a human interleukin-2 (IL-2) fragment. In vivo the IL-2 fragment seeks out activated T-cells through their IL-2 receptor. After binding, the fusion protein enters the cell by endocytosis and the toxin kills the cell by inhibiting protein synthesis. Affinity, potency and half-life of this product were greatly improved by deletion of 97 amino acids on the C-terminus of the diphtheria toxin. It is, to date, one of the most extensively evaluated fusion proteins that combine a toxin with a targeting molecule. The fusion protein is expressed in *Escherichia coli* strain and purified by a series of diafiltration steps and finally by reverse phase chromatography. Antibodies against Ontak were present in 92% of patients after two courses but there was no relationship between the presence of antibodies and the likelihood of responding to the drug. In a randomized, double blind, phase III trial in 71 patients with recurrent CTCL, injection for 5 days every three weeks over 8 courses produced >50% reduction in tumor burden lasting up to 4 months in 30% of patients. Other potential indications for Denileukin diftitox are for treatment of psoriasis, HIV infection, atopic dermatitis, alopecia aerata, multiple sclerosis and inflammatory bowel disease.

Doxercalciferol   (72-74)
(Vitamin D prohormone)

Country of Origin : US
Originator : Bone Care Int.
First Introduction : US
Introduced by : Bone Care Int.
Trade Name : Hectorol
CAS Registry No : 54573-75-0
Molecular Weight : 428.66

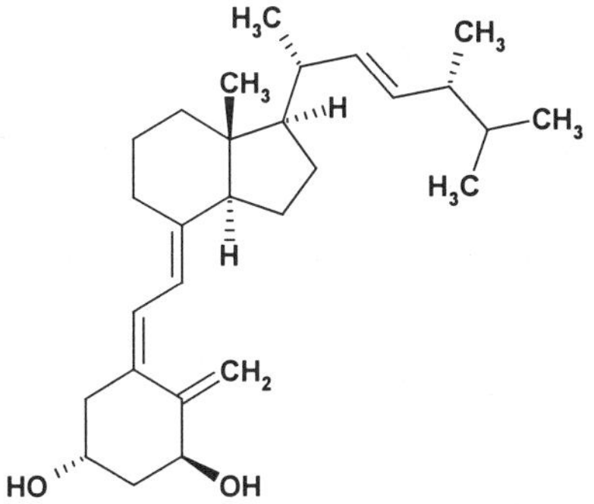

Doxercalciferol is a synthetic vitamin D2 analog introduced in the US as Hectorol for the treatment of secondary hyperparathyroidism (SHPT) in patients with end-stage renal failure. It is actually 1-alpha-hydroxyvitamin D2 (one-alpha-D2) and can be synthesized by direct hydroxylation of vitamin D2 tosylate or in 5 steps from ergosterol. Doxercalciferol itself is inactive until metabolized by the liver into active 1,24-dihydroxyvitamin D2, showing a pharmacokinetic profile similar to calcitriol. During placebo-controlled clinical trials, patients with moderate to severe SHPT were treated with 1-alpha-D2 (10 μg doses) for 16 to 24 weeks ; in 9 patients out of 10, blood levels of parathyroid hormone were considerably reduced, so maintaining a good control of the parathyroid hyperplasia. Moreover, only a moderate tendency to raise calcium and phosphate was observed. Doxercalciferol is orally-active and less toxic than other vitamin D analogs like alfacalcidol ; it shows differentially controlled effects and has an improved safety profile compared to calcitriol.

Eptifibatide (Antithrombotic)   (75-82)

Country of Origin : US
Originator : Cor Therapeutics
First Introduction : US
Introduced by : Cor
Ther./Schering-Plough
Trade Name : Integrilin
CAS Registry No : 188627-80-7
Molecular Weight : 831.98

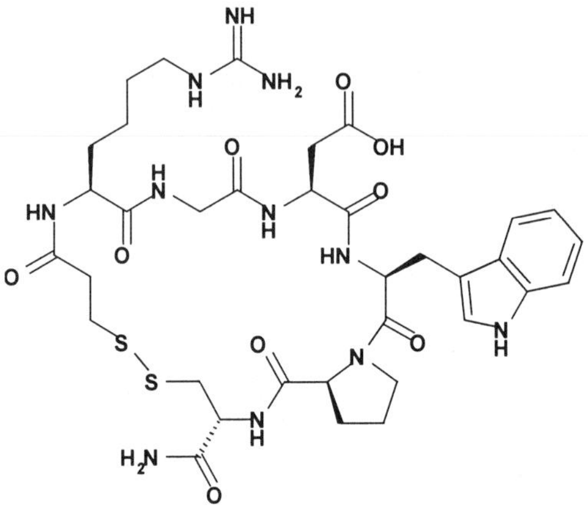

Eptifibatide was first launched in the US (1998) as Integrilin for the iv treatment of acute coronary syndrome, in particular for patients at risk for abrupt vessel closure during or after coronary angioplasty. It is a synthetic cyclic heptapeptide based on rattlesnake venom disintegrin peptides and containing the RGD sequence (arginyl-glycyl-aspartyl) involved in the adhesive function of platelets. It can be prepared by classical solid-phase peptide synthesis or by fragment synthesis in solution. Eptifibatide is a reversible antagonist of the glycoprotein IIb/IIIa complex, a specific platelet adhesion receptor that plays a central role in the cascade of thrombus formation by allowing mediators such as fibrinogen or von Willbrand factor to cross-link adjacent platelets and to give rise to aggregation. In diverse animal experimental models of arterial thrombosis, treatment with Eptifibatide resulted in an enhanced lysis of occlusive thrombus and a restoration of arterial blood flow. In clinical trials involving patients with acute coronary syndromes, Eptifibatide demonstrated a significant decrease in the incidence of death or nonfatal myocardial infarction at 30 days. In other trials in patients undergoing percutaneous coronary intervention (PCI), it showed a positive trend. Eptifibatide has the advantage of being short acting, its antiplatelet effect being rapidly reversible.

Gatifloxacin (Antibiotic)   (83-88)

Country of Origin : Japan
Originator : Kyori
First Introduction : US
Introduced by : Bristol-
Myers Squibb
Trade Name : Tequin
CAS Registry No : 160738-
57-8
Molecular Weight : 375.40

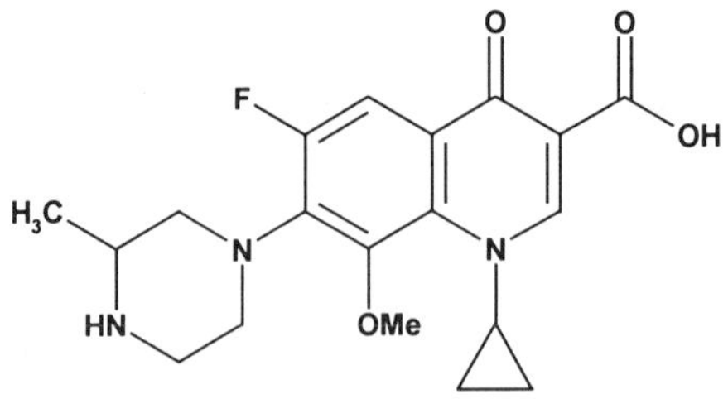

Gatifloxacin is a novel orally-active fluoroquinolone antibiotic with a potent and broad spectrum of activity against gram-positive (*S.pneumonia*, *S.aureus*) and gram-negative bacteria. It shows good activity against strains resistant to established antituberculosis agents. Gatifloxacin was marketed for the treatment of respiratory tract and urinary infections, particularly the treatment of certain community-acquired infections (such as bronchitis, pneumonia and common sexually-transmitted diseases). Comparative studies with ciprofloxacin, ofloxacin and sparfloxacin against fluoroquinolone-resistant organisms indicated a better or equivalent effectiveness with less phototoxic adverse effects.

Kinetin (Dermatologic)   (89-91)

Country of Origin : UK
Originator : Senetek
First Introduction : US
Introduced by : Osmotics Corp.
Trade Name : Kinerase
CAS Registry No: 525-79-1
Molecular Weight : 215.22

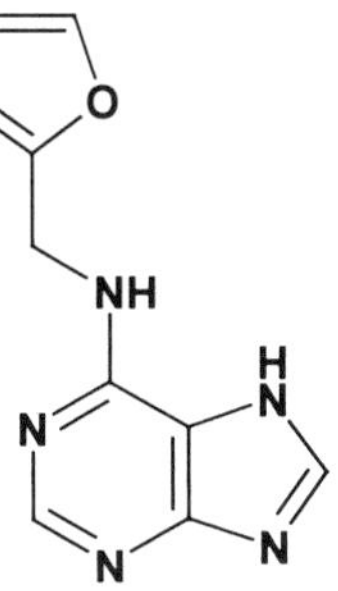

Kinetin (or Vivakin) was introduced as Kinerase in the US as a new ingredient for the treatment of age related photodamage of skin. This 6-furfurylaminopurine is a synthetic cytokinin, a family of plant growth factors, and was shown to be a highly potent growth factor. In vitro, it was able to delay or prevent the onset of age-related changes in skin cells without affecting cellular lifespan. In a double-blind clinical trial, Kinetin (0.005%) partially reversed the clinical signs of photodamaged skin and demonstrated a good safety profile.
It could have potential in psoriasis as well as in other proliferative skin disorders.

Levalbuterol Hydrochloride   (92-98)
(Antiasthmatic)

Country of Origin : US
Originator : Sepracor
First Introduction : US
Introduced by : Sepracor
Trade Name : Xopenex
CAS Registry No : 34391-04-3
Molecular Weight : 239.32

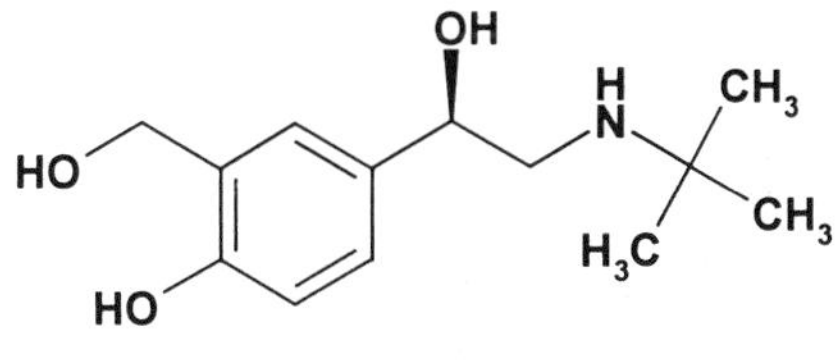

Levalbuterol was launched last year in the US for the treatment or prevention of bronchospasm in patients with reversible obstructive airway disease. It is the single R-enantiomer version of racemic albuterol (salbutamol) marketed for more than 30 years as a mainstay in the treatment of asthma. The R-isomer can be obtained with an excellent optical purity by enantiomeric purification based on the separation of diastereomeric tartrates. This isomer retains solely the desired bronchodilating effect of the racemic mixture due to a potent agonistic effect on ß2-adrenoceptors, with a lower incidence of ß-mediated side effects such as pulse rate increase, tremor and decrease in blood glucose and potassium levels. A pivotal clinical trial with two doses of levalbuterol and racemic albuterol, given by nebulization, demonstrated a greater improvement in lung function for the pure enantiomer levalbuterol.

Lomerizine Hydrochloride   (99-102)
(Antimigraine)

Country of Origin : Netherlands
Originator : Akzo Nobel
First Introduction : Japan
Introduced by : Kanebo
Trade Name : Teranas ; Migsis
CAS Registry No : 101477-54-7
Molecular Weight : 541.47

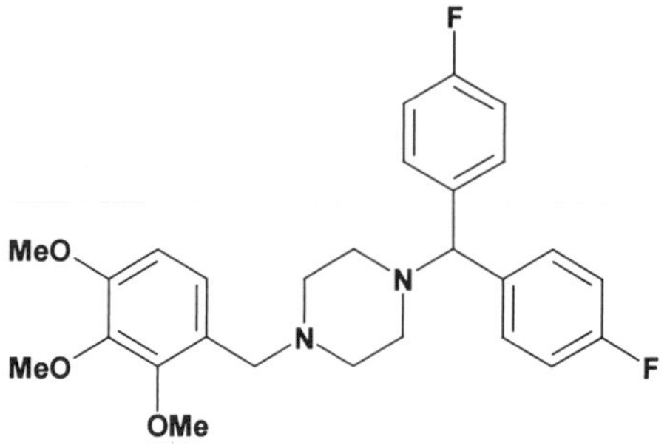

2 HCl

Lomerizine hydrochloride was introduced as Teranas and Migsis in Japan for the treatment of migraine. It is the first in its class of dual sodium and calcium channel blockers to be marketed for this indication. It can be synthetically obtained by reductive amination of 2,3,4-trimethoxybenzaldehyde with the appropiate benzhydryl piperazine. In dogs, Lomerizine exerts a potent, selective and long-lasting vasodilation of cerebral arteries related to a combination of mechanisms, especially a functional block of L-type voltage-sensitive calcium channels (L-VSCCs). Lomerizine increases cerebral blood flow compared to peripheral blood flow with only weak effects on systemic arterial blood pressure. Other mechanisms involved could be blockade of other VSCC and sodium channels, 5-HT2 and alpha-1 receptors. As a reducing agent of cortical spreading depression and neurogenic inflammation, Lomerizine was shown to be useful in migraine. In an open clinical study, it demonstrated efficacy in the treatment of cluster headache. Moreover, it may have utility in other neurological diseases such as cerebrovascular ischemia or cerebral infarction.

Lyme disease vaccine (Vaccine)   (103-105)

Country of Origin : US
Originator : Yale university
First Introduction : US
Introduced by : SmithKline Beecham
Trade Name : LYMErix
CAS Registry No : 147519-65-1

Class : Recombinant protein
Type : Antibody inducing vaccine
Molecular Weight : 30 kDa
Expression system : E. Coli
Manufacturer : SmithKline Beecham
Biologicals

LYMErix was introduced in the US as a vaccine against Lyme disease. Lyme disease (LD) is a tick borne disease caused by infection with the spirochete, *Borrelia burgdorferi,* which manifests itself with rashes, arthritis, cardiac and neurological disorders. This first vaccine against LD was critical as the spirochete has an incomplete response to antibiotic treatment. The vaccine is based on a lipidated outer-surface protein (L-OspA) of the spirochete, which is prepared by expressing the protein with a post-translational addition at the N-terminal of the lipid moiety in *E. coli* strain. After injection of the vaccines, patients develop antibodies to OspA which either inhibit the growth of the spirochete or are bactericidal, facilitate phagocytosis and block the attachment of the organism to host cells. Antibodies produced against this vaccine are cross-reactive for European and North American strains but not for Japanese strains. A phase III trial involving 11,000 people in the US showed an overall vaccine efficacy after three injections of 76% against definite Lyme disease and 100% against asymptomatic infection. LYMErix was generally well tolerated.

Mivotilate (Hepatoprotectant)   (106-109)

Country of Origin : South Korea
Originator : Yuhan Corp
First Introduction : South Korea
Introduced by : Yuhan Corp
CAS Registry No : 130112-42-4
Molecular Weight : 330.45

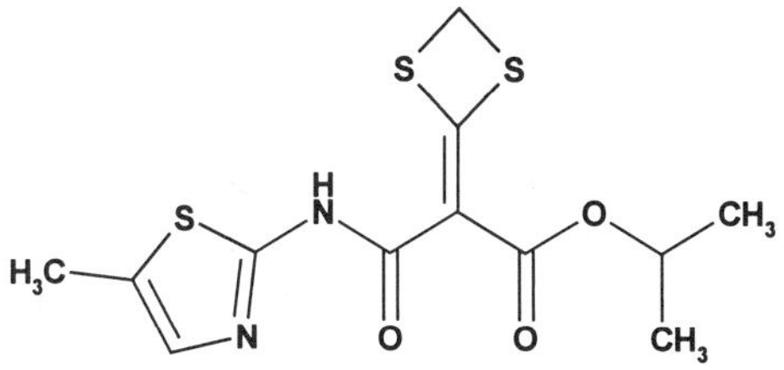

Mivotilate is an orally-active hepatoprotective agent introduced last year for the treatment of liver cirrhosis and hepatitis-B infection. It is a dithietanylidene malonate analogue of malotilate and is without the haemotological adverse effects associated with the latter. Mivotilate can be obtained by partial saponification of the corresponding diisopropylmalonate, conversion of the acid function into mixed anhydride then to the amide with an aminothiazole. It displays potent hepatoprotective activity in rat and mice liver injury models by a mechanism involving the inactivation of Kupffer cells. Mivotilate was shown to exert multiple effects on the hepatic cytochrome P450 system, particularly to inhibit CYP2E1 expression and to up-regulate the CYP1A1 expression. The low oral bioavailability of Mivotilate in rats (F < 0.22%) could be primarilly attributed to poor absorption and considerable hepatic and gastrointestinal first-pass effects. Mivotilate prevents mutagenesis caused by agents such as benzo[a]pyrene and reduces skin tumors induced by these agents.

Moxifloxacin Hydrochloride   (110-115)
(antibiotic)

Country of Origin : Germany
Originator : Bayer
First Introduction : Germany
Introduced by : Bayer
Trade Name : Avelox
CAS Registry No : 186826-86-8
Molecular Weight : 437.94

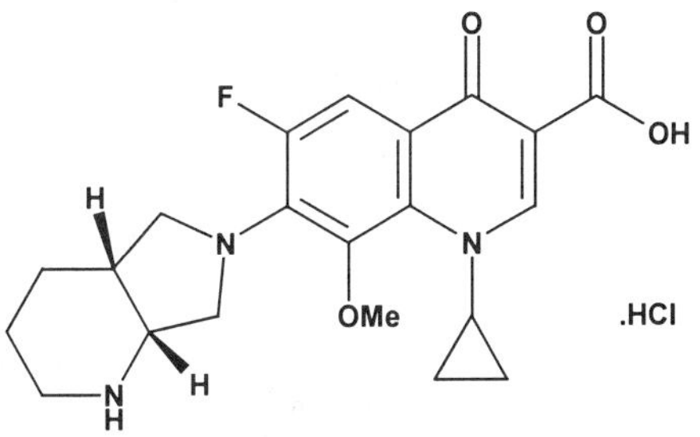

Moxifloxacin hydrochloride is one of the two new fluoro-quinolonecarboxylic acid antibiotics introduced in 1999 for the treatment of respiratory tract infections such as community-acquired pneumonia, acute exacerbations of bronchitis or acute sinusitis. Moxifloxacin can be synthesized through a 12 step sequence from the classical 4-quinolone-3-carboxylic acid template. Some advantages of Moxifloxacin over Ciprofloxacin, another of Bayer's launched quinolones, have been shown, i.e. an enhanced activity against Gram-positive bacteria (*Streptococcus pneumoniae, Clostridium pneumoniae*), a favorable pharmacokinetic profile (good tissue penetration and plasma concentrations above MICs) and a lack of phototoxicity (UVA irradiation).

Naftopidil (Dysuria) (116-120)

Country of Origin : Germany
Originator : Boehringer Mannheim
First Introduction : Japan
Introduced by : Kanebo ; Asahi
Trade Name : Avishot ; Flivas
CAS Registry No : 57149-07-2
Molecular Weight : 392.50

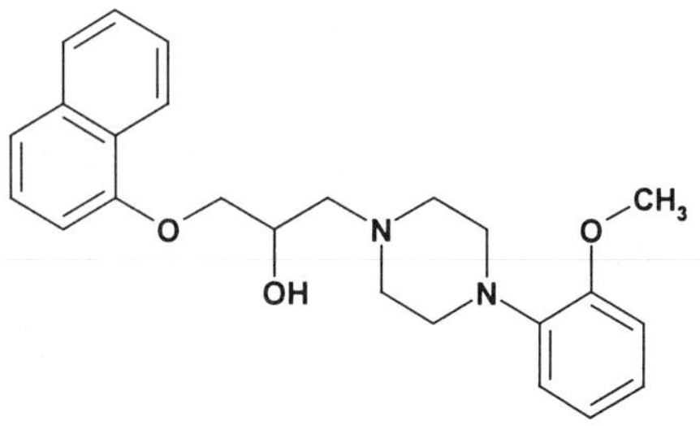

Naftopidil was launched in Japan for the treatment of dysuria associated with benign prostatic hypertrophy (BPH). It can be prepared by a two step route starting with α-naphthol. Naftopidil is a potent postsynaptic-selective alpha-1-antagonist with a slightly higher affinity for the human prostatic than for the aortic alpha-adrenoceptor. It also shows a 5-HT1A agonistic effect, as well as a weak calcium antagonistic activity, but no alpha-2 or beta-adrenoreceptor affinity. In experiments with rats or rabbits, Naftopidil was shown to be more potent and selective for the urodynamic effect than the hypotensive effect. Aromatic or aliphatic hydroxylation are the major routes of metabolism, producing metabolites with a profile similar to the parent compound.

Nateglinide (antidiabetic) (121-124)

Country of Origin : Japan
Originator : Ajinomoto
First Introduction : Japan
Introduced by : Ajinomoto/
Yamanouchi/Roussel
Trade Name : Fastic ; Starsis
CAS Registry No : 105816-04-4
Molecular Weight : 317.43

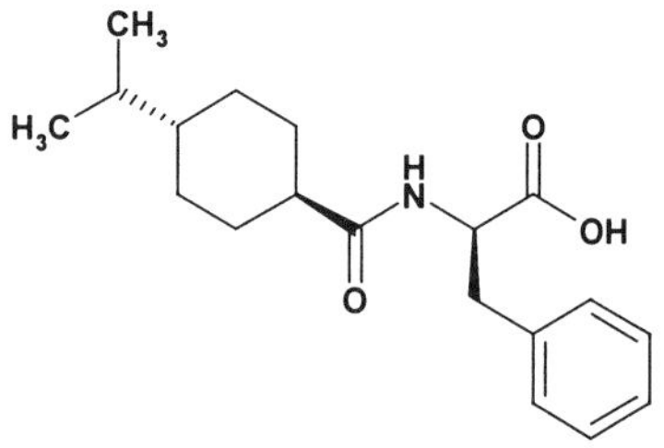

Nateglinide is a N-acylated D-phenylalanine marketed last year in Japan as novel orally active insulinotropic agent for the treatment of type-2 diabetes mellitus. It belongs to the class of nonsulfonylureas and shows some structural similarity to repaglinide, the only other representative in this family. In single pancreatic beta-cells isolated from rats, Nateglinide was found to specifically block the ATP-sensitive K+ channel resulting in an increase in intracellular calcium concentration. This primary action would underlie the mechanism by which Nateglinide markedly stimulates or potentiates, depending on glucose concentrations, insulin secretion from pancreatic beta-cells. Clinical studies demonstrated a good safety profile with a low potential for hypoglycemia. The pharmacokinetic profile was consistent with the changes of the blood glucose and plasma insulin level. Interestingly, Nateglinide exerts a rapid onset and short duration of action due to a rapid absorption and clearance. Unlike other similar agents, Nateglinide suppresses postprandial glucose elevations.

Nifekalant Hydrochloride   (125-129)
(antiarrhythmic)

Country of Origin : Japan
Originator : Mitsui
First Introduction : Japan
Introduced by : Mitsui
Trade Name : Shinbit
CAS Registry No : 130656-51-8
Molecular Weight : 405.46

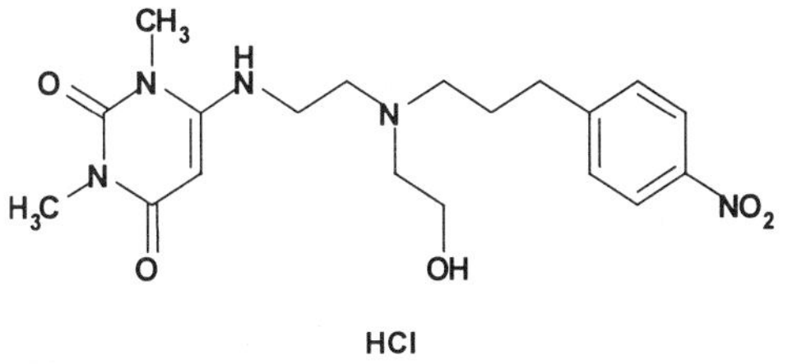

HCl

Nifekalant is a novel and pure class III anti-arrhythmic agent introduced in Japan as Shinbit injection for the treatment of serious ventricular arrhythmias. It can be synthesized in 3 steps with a final assembly of two ethanolamine blocks. Nifekalant is a non-selective blocker of myocardial repolarizing potassium currents and is completely devoid of any ß-adrenergic blocking effect. In hearts from different species, Nifekalant significantly prolonged action potential duration, atrial and ventricular refractory periods in in situ and in isolation studies. In intact animals, hemodynamic parameters were not affected by doses up to 1mg/kg. The specific characteristics of its voltage- and time-dependent K+ current blockade might differentiate Nifekalant from other class III agents currently in use.

OCT-43 (Anticancer)   (130-132)

Country of Origin : Japan
Originator : Otsuka
First Introduction : Japan
Introduced by : Otsuka
Trade Name : Octin
CAS Registry No : 159074-77-8

Class : Recombinant protein
Type : Immunostimulant
Molecular Weight : 17 kDa
Expression system : **?**
Manufacturer : Otsuka

Oct-43 is a recombinant variant of Interleukin 1 beta (IL-1beta) launched in Japan for the treatment of mycosis fungoides. In this variant cysteine has been substituted for serine at position 71. Recent literature on this macromolecule is scarce. Apart from the antitumor effect in malignant skin tumors, Oct-43 was also found to be clinically useful in the treatment of aplastic anemia and myelodysplastic syndrome.

Oseltamivir Phosphate (Antiviral)   (133-139)

Country of Origin : US
Originator : Gilead
First Introduction : Switzerland
Introduced by : Roche
Trade Name : Tamiflu
CAS Registry No : 204255-11-8
Molecular Weight : 410.40

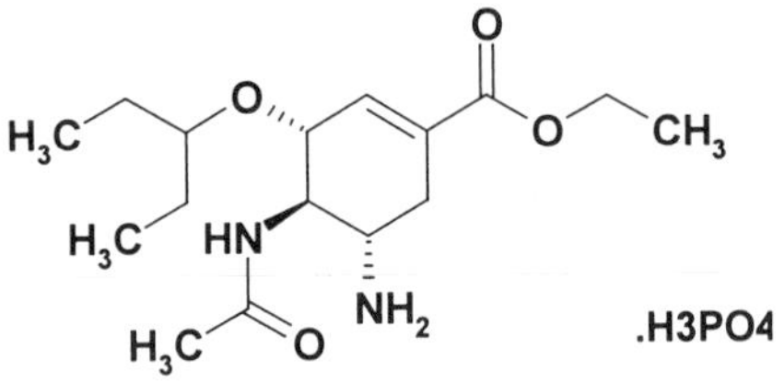

.H3PO4

Oseltamivir was launched in the US and Switzerland for the treatment of influenza infections by all common strain viruses. It can be obtained by at least 2 different ways including a novel 12-step synthesis from (-)-quinic acid. Oseltamivir is the ethyl ester prodrug of GS-4071, the corresponding acid, which is one of the most potent inhibitors of both influenza A and B virus neuraminidase (sialidase) isoenzymes ; these glycoproteins are expressed on the virion surface and are essential for virus replication for both A and B strains. Oseltamivir emerged as one of the first two neuraminidase inhibitors to reach the market last year. GS-4071 demonstrated a low (< 5%) oral bioavailability in animals due to a poor absorption from the gastrointestinal barrier ; by incorporating a more lipophilic ester group, the oral bioavailabilty can reach 30 to 100% in mice, rats and dogs. Following oral administration of Oseltamivir in rats, a similar concentration of GS-4071 was found in the bronchoalveolar lining fluid and the plasma which indicated a good penetration of the active compound into the lower respiratory tract. In mice, chickens and ferrets, orally administered Oseltamivir was found to have significant inhibitory effects on A and B influenza infections in protecting against a lethal challenge of virus and lessening virus titer in the lungs or nasal washings. In several clinical trials with patients receiving oral capsules daily, Oseltamivir was shown to be effective in reducing significantly the duration and severity of the clinical symptoms, including fever, cough and general malaise, in both early treatment and prevention.

Pioglitazone Hydrochloride   (140-145)
(Antidiabetic)

Country of Origin : Japan
Originator : Takeda
First Introduction : US
Introduced by : Eli Lilly
Trade Name : Actos
CAS Registry No : 112529-15-4
Molecular Weight : 401.97

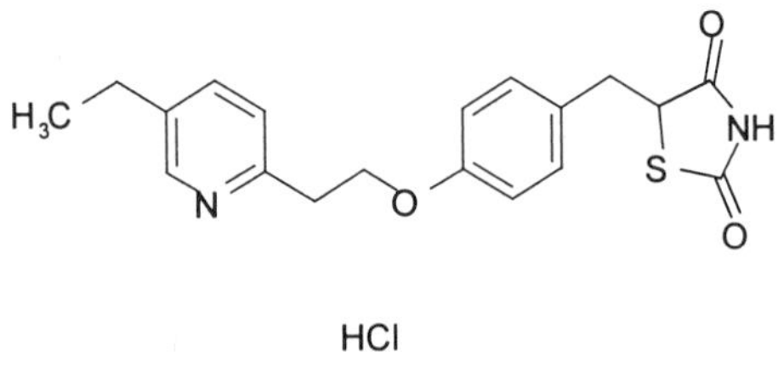

HCl

Pioglitazone is a new orally active thiazolidinedione (TZD) launched last year in the US for the treatment of non-insulin dependent diabetes mellitus (NIDDM). It can be synthesized in 4 steps, the last one transforming an alpha-bromoester into thiazolidine with thiourea. As with other representatives in this class, it potently activates the nuclear receptor peroxisome proliferator-activated receptor gamma which is believed to be involved in the regulation of insulin resistance and adipogenesis. In several obese and obese diabetic animal models, treatment with Pioglitazone resulted in reductions in plasma glucose and serum lipids. In clinical studies, Pioglitazone at a once daily oral dose of 15-

45 mg, as monotherapy or in combination with non-TZDs or insulin, was shown to significantly improve glycemic control in type-2 diabetes and demonstrated a beneficial effect on insulin resistance and other clinically relevant parameters as plasma levels of triglycerides or HDL-cholesterol. Pioglitazone is reported to be safe and well tolerated and is said to have a lower occurrence of hepatic toxicity as well as a low probability for drug interaction.

Rapacuronium Bromide   (146-150)
(Muscle Relaxant)

Country of Origin : Netherlands
Originator : Akzo Nobel
First Introduction : US
Introduced by : Akzo Nobel
Trade Name : Raplon
CAS Registry No :156137-99-4
Molecular Weight : 677.81

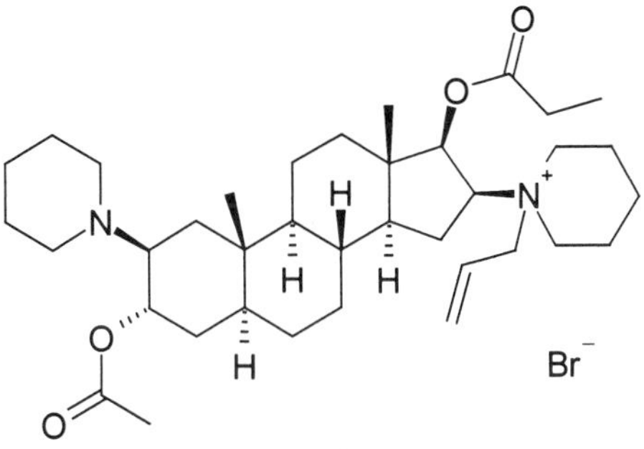

Rapacuronium bromide was introduced in the US as a parenterally administered adjunct to general anesthesia during surgical procedures. It is a new steroidal derivative belonging to the same class as vecuronium bromide. This 2,16-bispiperidinylandrostane derivative can be synthesized in 3 steps from the 3-hydroxy-androstan-17-one analog and finally quaternized with allyl bromide. Rapacuronium bromide is a nondepolarizing neuromuscular blocking drug with a rapid onset and a short duration of action. With a relatively low potency (ED$_{90}$ = 1.15 mg/kg), Rapacuronium is expected to be substituted for the polarizing succinylcholine with very short duration of action.

Rofecoxib (Antiarthritic)   (151-156)

Country of Origin : US
Originator : Merck
First Introduction : Mexico
Introduced by : Merck
Trade Name : Vioxx
CAS Registry No : 162011-90-7
Molecular Weight : 314.36

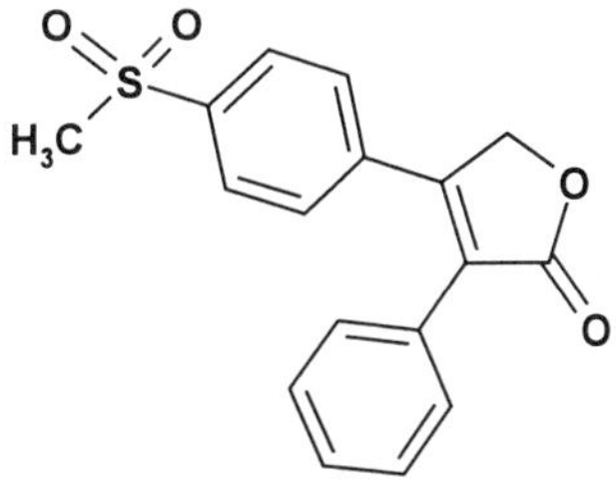

Rofecoxib is a non-steroidal anti-inflammatory drug (NSAID) launched in Mexico, its first market, for the management of acute pain and the treatment of osteoarthritis (OA) and primary dysmenorrhea. Rofecoxib can be obtained by several different ways ; one example is by arylation of a 4-bromofuranone with a phenylboronic acid under Suzuki conditions. Rofecoxib is a highly selective inhibitor of COX-2, the inducible isoform of cyclooxygenase and therefore exhibits a potent antiinflammatory activity without concomitant gastric or renal toxicities linked to the non-specific COX-1/2 inhibitors. In several clinical studies in patients with knee or hip osteoarthritis, Rofecoxib was evaluated at 12.5-50 mg doses once daily : it demonstrated efficacy for all primary and secondary endpoints at doses considerably weaker than those for classical non-specific NSAIDs, with good tolerance and less adverse effects. Selective COX-2 inhibitors potentially have a large spectrum of activity including new indications such as Alzheimer's disease, colorectal cancer, irritable bowel disease or urinary incontinence.

Rosiglitazone Maleate   (157-162)
(Antidiabetic)

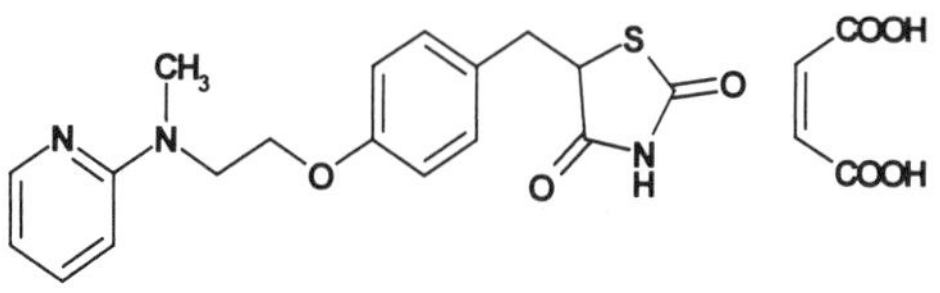

Country of Origin : US
Originator : SmithKline Beecham
First Introduction : US ; Mexico
Introduced by : SmithKline Beecham
Trade Name : Avandia
CAS Registry No : 155141-29-0
Molecular Weight : 473.50

Rosiglitazone maleate, belongs to a novel class of thiazolidine diones launched for the treatment of non-insulin-dependent diabetes mellitus (NIDDM), a disease characterized by a pancreatic ß-cell defect and insulin resistance in the liver and peripheral tissues. The racemic base can be obtained by Knövenagel condensation between 2, 4-thiazolidinedione and the corresponding 4-substituted benzaldehyde (itself prepared in 2 steps from 2-chloropyridine), followed by reduction of the benzylidene. Rosiglitazone was shown to be a potent agonist of peroxisome proliferator activated receptor-gamma (PPAR-gamma), a nuclear receptor involved in the differentiation of adipose tissue, without activating liver PPAR-alpha receptors. This activation of PPAR-gamma could mediate the down-regulation of leptin gene expression. In animal models, Rosiglitazone has been shown to normalize glucose metabolism and reduce the exogenous dose of insulin needed to achieve glycemic control. In patients with Type II diabetes, daily doses (4 or 8 mg) of Rosiglitazone significantly improved blood sugar control without affecting cardiac structure or function.

SKI-2053R (Anticancer)   (163-166)

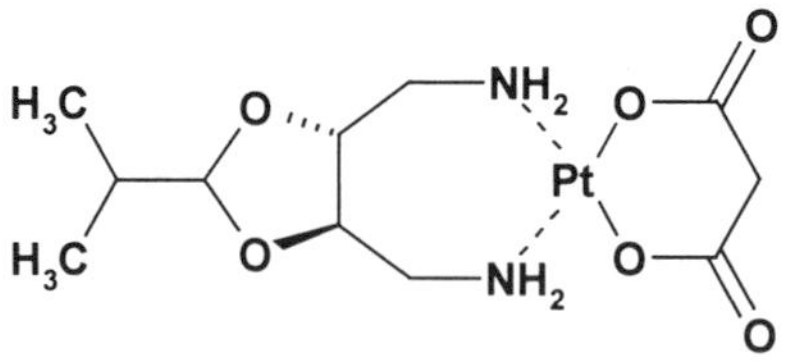

Country of Origin : Korea
Originator : SK Pharma
First Introduction : Korea
Introduced by : SK Pharma
Trade Name : Sunpla
CAS Registry No : 146665-77-2
Molecular Weight : 471.38

SKI-2053R was launched in Korea for the treatment of patients with advanced gastric adenocarcinoma. It is a third generation platinum complex alkylating agent with less nephrotoxicity than cisplatin. In in vitro experiments, SKI-2053R was highly active against various cell lines including cisplatin-resistant tumor cell lines. In mice implanted with L1210 cells, SKI-2053R showed activity comparable or superior to cisplatin or carboplatin. In mice implanted s.c. with human tumor xenograft KATO III (stomach adenocarcinoma), SKI-2053R achieved growth-inhibition rates comparable with those of cisplatin. General pharmacological studies in animals did not display any severe adverse effects on the different systems at antitumour doses. Pharmacokinetic studies after a single iv administration in rats demonstrated that the compound was extensively and rapidly distributed into all tissues except the CNS. In patients with unresectable or metastatic gastric adenocarcinoma, SKI-2053R was active without significant hematological toxicity. Phase III trials with SKI-2053R are currently ongoing for the treatment of various other cancers including lung, head and neck.

Tasonermin (Anticancer)   (167-168)

Country of Origin : US   Class : Recombinant protein
Originator : Genentech   Type : Tumor necrosis factor
First Introduction : Germany   Molecular Weight :   17 kDa
Introduced by : Boehringer Ingelheim   Expression system : E. Coli
Trade Name : Beromun   Manufacturer : Boehringer Ingelheim
CAS Registry No : 94948-59-1

　　　Tasonermin was launched last year in Germany for the treatment of soft tissue sarcoma of the limbs. Recent literature is scarce for this compound. It is the first tumor necrosis factor (TNF) launched for cancer treatment to date and works through destruction of tumor blood vessels. This recombinant protein is expressed in *E. coli* and purified. Tasonermin is too toxic for systemic use and consequently is being used in isolated limb perfusion (ILP) for sarcoma and malignant melanoma. The product has shown notable efficacy when used in combination with melphalan to prevent amputation. In four European clinical trials involving 260 patients destined for amputation or mutilating surgery, Beromun enabled durable limb salvage in 80% of the patients. Due to the complexity of ILP, the application of tasonermin is restricted to clinical centers with sufficient experience and facilities including radionucleide-based continuous monitoring of leakage from the isolated circuit during ILP and postoperative intensive care surveillance.

Telmisartan (Antihypertensive)   (169-175)

Country of Origin : Germany
Originator : Boehringer Ingelheim
First Introduction :US
Introduced by : Boehringer Ingelheim
Trade Name : Micardis
CAS Registry No : 144701-48-4
Molecular Weight : 514.63

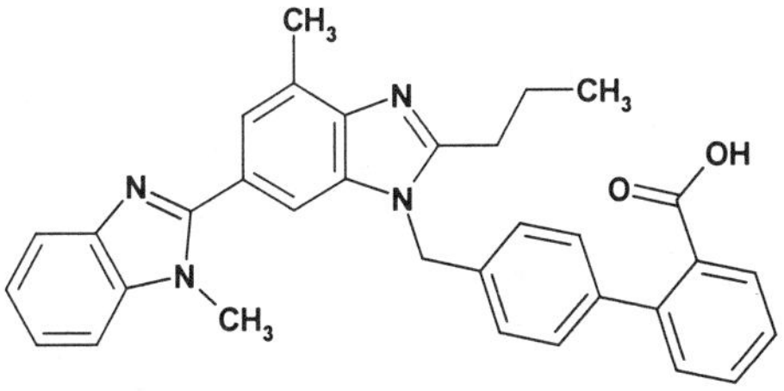

　　　Telmisartan was launched in the US for the treatment of hypertension. It can be prepared in eight steps starting with methyl 4-amino-3-methyl benzoate ; the first and second cyclization into a benzimidazole ring occur at steps 4 and 6 respectively. Telmisartan blocks the action of angiotensin II (Ang II), the primary effector molecule of the renin-angiotensin-aldosterone system (RAAS). It is the sixth of this class of « sartans » to be marketed after the lead compound Losartan. Its long lasting effect (24h half-life) could be the main difference with other angiotensin II antagonists. Unlike several other agents in this category, its activity does not depend upon transformation into an active metabolite, the 1-O-acylglucuronide being the principal metabolite found in humans. Telmisartan is a potent competitive antagonist of AT1 receptors that mediate most of the important effects of angiotensin II while lacking affinity for the AT2 subtypes or other receptors involved in cardiovascular regulation. In several clinical studies, Telmisartan, at a once daily dosage, produced effective and sustained blood-pressure lowering effects with a low incidence of side effects (particularly treatment-related cough associated with ACE inhibitors in elderly patients).

Temozolomide (Anticancer)   (176-180)

Country of Origin : UK
Originator : CRC Technology
First Introduction : UK
Introduced by : Schering-Plough
Trade Name : Temodal
CAS Registry No : 85622-93-1
Molecular Weight : 194.15

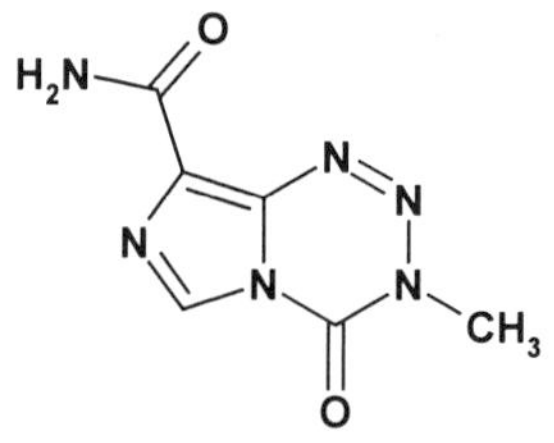

Temozolomide was launched for the first time in the UK for the treatment of patients with glioblastoma multiforme showing recurrence or progression after standard therapy. It can be considered as a cyclic variant of highly reactive triazenes producing a cascade of ionic or radical antitumoral species. Temozolamide is a 3-methyl analog of mitozolomide and was shown to be converted to cytotoxic triazene MCTIC. It can be prepared in 2 steps from 5-aminoimidazole-4-carboxamide by diazotization then cyclization with methylisocyanate. Mechanistically, the depletion of O6-alkylguanine-DNA alkyltransferase (OGAT) in cells or tumors was shown to be correlated with the cytotoxicity of temozolomide which is a potent inhibitor of this enzyme involved in DNA repair activity.
Further approval applications for the treatment of other malignant gliomas such as relapsed anaplastic astrocytoma, advanced metastatic melanoma were submitted.

Valrubicin (Anticancer)   (181-183)

Country of Origin : US
Originator : Anthra Pharm.
First Introduction : US
Introduced by : Medeva
Trade Name : Valstar
CAS Registry No : 56124-62-0
Molecular Weight : 723.66

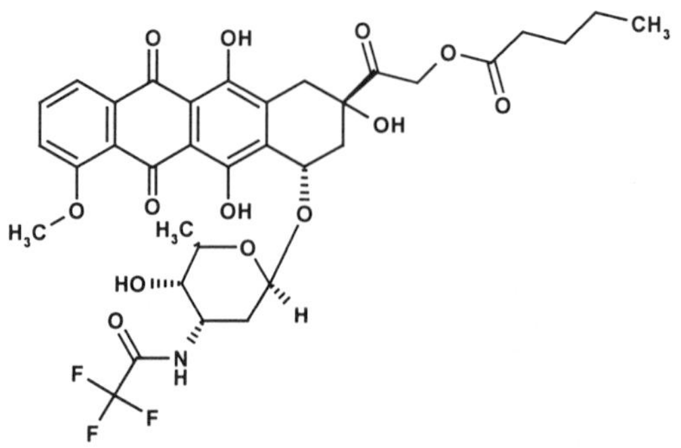

Valrubicin was launched as a new chemotherapeutic agent for the treatment of bladder cancer, particularly in patients with BCG-refractory carcinoma in situ (CIS) of the bladder for whom immediate cystectomy is unacceptable. It belongs to the class of anthracyclines, the widest used in human cancers, and is a N-trifluoroacetyl 14-valerate derivative of doxorubicin.

Valrubicin can be obtained in 3 steps from daunomycin by N-trifluoroacetylation of the sugar moiety then iodination of the 2-acetyl group and introduction of a valerate residue. A proposed mechanism involved in the cytotoxicity of Valrubicin coud be the blockade of SV40 large T antigen helicase ; this cellular enzyme is involved in the formation of a ternary complex with DNA to maintain the topographic structure of DNA during transcription. In patients with CIS of bladder refractory to front line and second line therapies, intravesical instillation of Valrubicin resulted in a complete response with a significant rate and allowed a delay in cystectomy. Systemic absorption was minimal and accordingly produced a lower incidence of cardiotoxicity compared with doxorubicin.

Zaleplon (Hypnotic)   (184-187)

Country of Origin : US
Originator : American Home Products
First Introduction : Sweden, Denmark
Introduced by : American Home Products
Trade Name : Sonata
CAS Registry No : 15319-34-5
Molecular Weight : 332.32

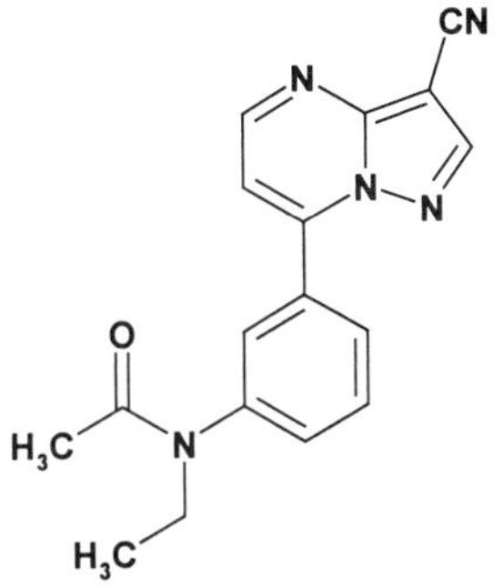

Zaleplon was introduced in Sweden and Denmark as a new treatment for insomnia, particularly in patients who have difficulty in falling asleep. Zaleplon is a non-benzodiazepine compound and is the first in a new generation belonging to the pyrazolopyrimidine class, showing therefore fewer benzodiazepine-like side effects. It can be synthesized in 3 steps from the corresponding acetophenone, the key step being the cyclization of the appropriate enaminone with 3-aminopyrazole-4-carbonitrile. Biochemically, Zaleplon is a full agonist at the benzodiazepine $\omega 1$ site of the gaba-A receptor complex, but its behavioural profile remains distinct from both benzodiazepine (e.g. Lorazepam) or non-benzodiazepine (e.g. Zopiclone or Zolpidem) sedative-hypnotic drugs. Clinical pharmacokinetic analysis showed rapid absorption and elimination. In man, the main metabolic route was oxidative giving the major metabolites 5-oxo Zaleplon and its N-desethyl analog. Both were shown to have no effect at central benzodiazepine receptors and to be rapidly excreted as glucuronides. In patients with chronic insomnia, Zaleplon at 5 and 10 mg/kg significantly reduced sleep latency and improved the quality of sleep compared with placebo without altering the normal sleep architecture. Given its short half-life, the next-day residual effects such as hangover are minimized. It may have some advantages over benzodiazepines regarding unwanted amnesic effects and psychomotor impairment. There was no evidence for the occurrence of rebound insomnia at 10 mg/kg.

Zanamivir (Antiviral)   (188-194)

Country of Origin : Australia
Originator : Biota Scientific
Management
First Introduction : Australia
Introduced by : Glaxo Wellcome
Trade Name : Relenza
CAS Registry No : 139110-80- 8
Molecular Weight : 332.32

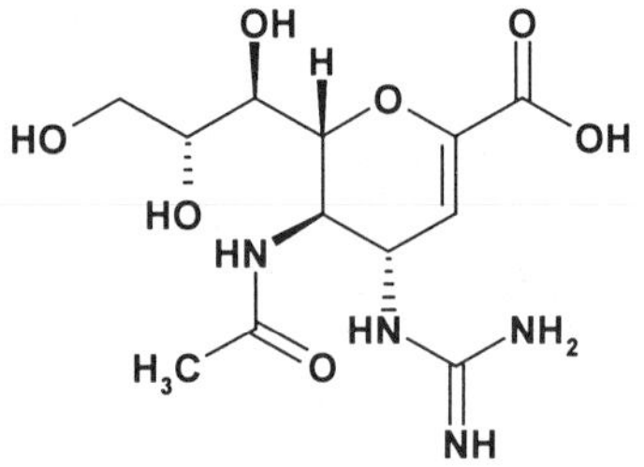

Zanamivir was launched as Relenza in Australia for treatment of human influenza A and B virus infections. Zanamivir (4-guanidino-Neu5Ac2en) can be obtained by several similar ways, for instance in seven step synthesis starting from N-acetyl-D-neuraminic acid. Mechanistically, Zanamivir is a potent and specific inhibitor of neuraminidase (or sialidase), a key viral surface glycohydrolase essential for viral replication and disease progression by catalyzing the cleavage of terminal sialic acid residues from the glycoprotein. The in vitro activity of Zanamivir against a wide variety of influenza A and B strains was demonstrated in different model systems ; its activity against clinically relevant isolates of influenza virus, with IC_{50} values ranging from 0.005 to 15 µM was superior to those of amantadine and rimantadine.

Acknowledgments : the authors sincerly thank Claude Barrere for providing library searches and literature references.

References

1.   The selection of new therapeutic entities launched for the first time in 1999 is based on information extracted from the following selected sources : (a) A.I.Graul, Drug News Perspect. 13, 37 (2000). (b) Scrip Magazine, 63, January 2000. (c) Pharmaprojects. (d) Drug Topics, W. M. Davis, I. W. Waters, M. C. Vinson, "New Drug Approvals of 1999", Part 1, 82, Feb. 2000. (e) ib. Part 2, 103, March 2000. (f) Pharmaceutical Research and Manufacturers of America, « New Drug Approvals In 1999 », January 2000.

2.   B. Gaudillière, Ann. Rep. Med. Chem., 34, 317 (1999).

3.   P. Galatsis, Ann. Rep. Med. Chem., 33, 327 (1998).

4.   P. Galatsis, Ann. Rep. Med. Chem., 32, 305 (1997).

5.   X.-M. Cheng, , Ann. Rep. Med. Chem., 31, 337 (1996).

6.   A. Graul, P. A. Leeson, J. Castañer, Drugs Future, 23(11), 1155 (1998) ; 24(11), 1250 (1999).

7.   R. H. Foster and D. Faulds, Drugs, 55(5), 729 (1998).

8.   S. M. Roberts, Idrugs, 1(8), 896 (1998).

9.   B. M. Barry, F. Mulcahy, C. Merry, S. Gibbons, D. Back, Clin. Pharmacokinetics, 36(4), 289 (1999).

10.   S. Staszewski, Int. J. Clin. Pract., Suppl. 103, 35, (1999).

11.   J.-P. Vartanian, Curr. Opin. Anti-Infect. Invest. Drugs, 1(2), 227 (1999).

12.   J. G. Moyle, Curr. Opin. Infect. Dis., 13(1), 19 (2000).

13.   B. F. Tate, A. A. Levin, J. F. Grippo, Trends Endocrinol. Metab., 5(5), 189 (1994).

14.   M. M. Gottardis, K. Malewicz, W. W. Lamph, Proc. Am . Assoc. Cancer Res., 36, 514 (1995).

15.   J. L. Napoli, Clin. Immunol. Immunopathol., 80(3, Pt. 2), S52-S62 (1996).

16. M. M. Gottardis, W. W. Lamph, D. R. Shalinsky, A. Wellstein, R. A. Heyman, Breast Cancer Res. Treat. $\underline{38}$(1), 85 (1996).

17. C. P. F. Redfern, Adv. Organ. Biol., $\underline{3}$, 35 (1997).

18. Z. Szondy, U. Reichert, L. Fesus, Cell Death Differ., $\underline{5}$(1), 4 (1998).

19. X.-C. Xu, R. Lotan, Handb. Exp. Pharmacol., $\underline{139}$, 323 (1999).

20. G. R. Painter, S. Ching, D. Reynolds, M. Clair, B. Sadler et al., Drugs Future, $\underline{21}$(4), 347 (1996).

21. J. C. Adkins, D. Faulds, Drugs, $\underline{55}$(6), 837 (1998).

22. J. Gatell, J. Clin. Pract., Suppl., $\underline{103}$, 42 (1999).

23. A. Billich, Idrugs, $\underline{2}$(5), 466 (1999).

24. P. Reddy, J. Ross, Formulary, $\underline{34}$(7), 567 (1999).

25. B. Conway, S. D. Shafran, Expert Opin. Invest. Drugs, $\underline{9}$(2), 371 (2000).

26. C. Bottex-Gauthier, D. Vidal, F. Picot, P. Potier, F. Menichini, G. Appendino, Biotechnol. Ther., $\underline{4}$(1-2), 77 (1993).

27. S. M. Adenekov, B. B. Rakhimova, K. A. Dzhazin, V. B. Alikov, Z. K. Shaushekov, Z. R. Toregozhina, K. M. Turdybekov, Fitoterapia, $\underline{66}$(2), 142 (1995).

28. T. E. Shaikenov, Zdravookhr. Kaz. (1), 52 (1997).

29. T.E. Shaikenov, S.M. Adekenov, K. D. Rakhimov, E. M. Neldibaiev, K. Z. Musulmanbekov, Dokl. Minist. Nauki-Akad. Nauk. Resp. Kaz., (3), 69 (1997).

30. T.E. Shaikenov, S.M. Adekenov, S. Basset, M. Trivedi, L. Wolfinbarger, Dokl. Minist. Nauki-Akad. Nauk. Resp. Kaz., (5), 64 (1998).

31. A. S. Fazylova, K. I. Itzhanova, A. T. Kulyyasov, K. M. Turdybekov, S. M. Adekenov, Chem. Nat. Compd. $\underline{35}$(3), 305 (1999).

32. A. Graul, A. M. Martel, J. Castañer, Drugs Future, $\underline{22}$(7), 711 (1997) ; $\underline{23}$(7), 782 (1998) ; $\underline{24}$(7), 793 (1999).

33. E. G. Boyce, G. A. Breen, Formulary, $\underline{34}$(5), 405 (1999).

34. J. Wallace, B. Chin., Curr. Opin. Anti-Inflammatory Immunomodulatory Invest. Drugs, $\underline{1}$(2), 100 (1999).

35. H. B. Fung, H. L. Kirschenbaum, Clin. Ther., $\underline{21}$(7), 1131 (1999).

36. M. M. Goldenberg, Clin. Ther., $\underline{21}$(9), 1497 (1999).

37. R. Infante, R. G. Lahita, Geriatrics, $\underline{55}$(3), 30 (2000).

38. A. A. Schuna, C. Megeff, Am. J. Health-Syst. Pharm., $\underline{57}$(3), 225 (2000).

39. T. Reissmann, J. Engel, B. Kutscher, M. Bernd, P. Hilgard, M. Peukert, I. Szelenyi, S. Reichert, D. Gonzales-Barcena, E. Nieschiag, A.M. Comaru-Schally, A.V. Schally, Drugs Future, $\underline{19}$(3), 228 (1994).

40. T. Reissmann, R. Felderbaum, K. Diedrich, J. Engel, A. M. Comaru-Schally, A. V. Schally, Human Reprod. $\underline{10}$(8), 1974 (1995).

41. A. V. Schally, A. M. Comaru-Schally, Adv. Drug Delivery Rev., $\underline{28}$(1), 157 (1997).

42. R. Felderbaum, K. Dietrich, Human Reprod. 14 Suppl., 1207 (1999).

43. U. Brunner, B. M. Gensthaler, Pharm. Ztg, $\underline{144}$(22), 1782 (1999).

44. A. V. Schally, Gynecol. Endocrin., $\underline{13}$(6), 401 (1999).

45. A. V. Schally, Peptides, $\underline{20}$(10), 1247 (1999).

46. G. Emons, K.-D. Schultz, Recent Res. Cancer Res., $\underline{153}$, 83 (2000).

47. M. Brunner-Guenat, P.-A. Carrupt, G. Lisa, B. Testa, S. Rose, K. Thomas, P. Jenner, P. Ventura, J. Pharm. Pharmacol., $\underline{47}$, 861 (1995).

48. F. Stocchi, L. Barbato, L. Bramante, G. Nordera, L. Vacca, S. Ruggieri, J. Neurol., $\underline{243}$(5), 377 (1996).

49. M. De Ceballos, Curr. Opin. Cent. Peripher. Nerv. Syst. Invest. Drugs, $\underline{1}$(5), 649 (1999).

50. G. M. Benson, D. M. Hickey, Drugs Future, $\underline{18}$(1), 15 (1993).

51. Y. Homma et al., Nutr. Metab. Cardiovasc. Dis., $\underline{6}$(4), 211 (1996).

52. C. M. Hayward, M. J. Bamberger, Ann. Rep. Med. Chem., $\underline{32}$, 101 (1997).

53. W. H. Mandeville, D. I . Goldberg , Curr. Pharm. Design, $\underline{3}$(1), 15 (1997).

54. Y. Homma, T. Kobayashi, H. Yamaguchi, H. Ozawa, H. Sakane, H. Nakamura, Atherosclerosis, $\underline{129}$(2), d241 (1997).

55. K. Pettersson, Curr. Opin. Cardiovasc., Pulm. Renal Invest. Drugs, $\underline{1}$(2), 296 (1999).

56. W. H. Mandeville, C. Arbeeny, Idrugs, $\underline{2}$(3), 237 (1999).

57. J. Castañer, R. M. Castañer, J. Prous, Drugs Future, $\underline{18}$(2), 134 (1993).

58. Y. Toya, C. Schwencke, Y. Ishikawa, J. Mol. Cell. Cardiol., $\underline{30}$(1), 97 (1998).

59. R. Miyake, H. Yoshida, K. Tanonaka, Y. Miyamoto, H. Hayashi, H. Kajirawa, S. Takeo, Can J . Physiol. Pharmacol., $\underline{77}$(4), 225 (1999).

60. S. Momomura, Nippon Byoin Yakuzaishikai Zasshi, $\underline{35}$(7/8), 985 (1999).

61. M. Hosono, Nippon Yakurigaku Zasshi, $\underline{114}$(2), 83 (1999).

62. C. Chant, M. J. Ribak, Ann. Pharmacother., $\underline{29}$(10), 1022 (1995).

63. M. W. Griswold, B. M. Lomaestro, L. Briceland, Am. J. Health-Syst. Pharm., $\underline{53}$(17), 2045 (1996).

64. D. H. Bouanchaud, J. Antimicrob. Chemother., 39 (Suppl.A), 15 (1997).
65. D. H. Bouanchaud, Infect. Dis. Ther., 21, 51 (1997).
66. C. Carbon, Clin. Infect. Dis., 27(1), 28 (1998).
67. M. J. Woods, J. Chemother., 11(6), 446 (1999).
68. H. M. Lamb, D. P. Figgitt, D. Faulds, Drugs, 58(6), 1061 (1999).
69. J. Murphy, J. Vanderspek, Semin. Cancer Biol., 6(5), 259 (1995).
70. J. Nichols, F. Foss, T. Kuzel, C. Lemaistre, L. Platanias, M. Ratain, A. Rook, M. Saleh, G. Schwartz, Eur. J. Cancer, 33(1), S64 (1997)
71. M. Potter, Curr. Opin. Oncol. , Endocr. Metab. Invest. Drugs, 1(3), 291 (1999).
72. A. U. Tan, B. S. Levine, R. B. Mazess, D. M. Kyllo, C. W. Bishop, J. C. Knutson, K. S. Kleinman, J. W. Coburn, Kidney Int., 51(1), 317 (1997).
73. E. Slatopolsky, A. Brown, A. Dusso, Kidney Int., Suppl., 73, S14-9 (1999).
74. J. Cunningham, Kidney Int., Suppl., 73, S59-S64 (1999).
75. D. R. Phillips, R. M. Scarborough, Am. J. Cardiol., 80(4A), 11B-20B (1997).
76. R. M. Scarborough, Drugs Future, 23(6), 585 (1998).
77. A. Dunn, M. S. Chow, Formulary, 33(7), 632, 635, 639, 647 (1998).
78. E. J. Topol, T. V. Byzova, E. F. Plow, Lancet, 353(9148), 227 (1999).
79. K. L. Goa, S. Noble, Drugs, 57(3), 439 (1999).
80. J. C. O'Shea, J. E. Tcheng, Expert Opin. Invest. Drugs, 8(11), 1893 (1999).
81. R. M. Scarborough, Am. Heart J., 138(6, Pt.1), 1093 (1999).
82. J. E. Tcheng, Am. Heart J., 139(2, Pt.2), S38-S45 (2000).
83. J. Castañer, R. M. Castañer, J. Prous, Drugs Future, 18(3), 203 (1993).
84. A. Dalhoff, Exp. Opin. Invest. Drugs, 8(2), 123 (1999).
85. U. Holzgrabe, P. Heisig, Pharm. Unserer Zeit, 28(1), 30 (1999).
86. J. M. Blondeau, J. Microb. Chemother., 43(Suppl.B), 1 (1999).
87. R. Wise, D. Honeybourne, Eur. Respir. J., 14(1), 221 (1999).
88. C. M. Perry, J. A. Balfour, H. M. Lamb, Drugs, 58(4), 683 (1999).
89. S. I. Rattan, B. F. Clark, Biochem. Biophys. Res. Comm., 201, 665 (1994).
90. M. West, Arch. Dermatol., 130, 87 (1994).
91. J. Barciszewski, S. Rattan, G. Siboska, B. Clark, Plant. Sci., 148(1), 37 (1999).
92. J. R. McCullough, D. A. Handley, T. P. Jerussi, T. E. Rollins, P. Koch, S. DeGraw, M. Munsey, D. Kellerman, P. D. Rubin, Respir. Drug Delivery VI, Int. Symp. 6th , 113-118, Publ. Interpharm Press, Inc., Buffalo Grove (1998).
93. D. A. Handley, J. Morley, L. Vaickus, Expert Opin. Invest. Drugs, 7(12), 2027 (1998).
94. D. Ormrod, C. M. Spencer, BioDrugs, 11(6), 431 (1999).
95. C. Page, J. Morley, J. Allergy Clin. Immunol., 104(2, Pt.2), S31-S41 (1999).
96. J. Costello, J. Allergy Clin. Immunol., 104(2, Pt.2), S61-S68 (1999).
97. D. Handley, J. Allergy Clin. Immunol., 104(2, Pt.2), S69-S76 (1999).
98. H. S. Nelson, J. Allergy Clin. Immunol., 104(2, Pt.2), S77-S84 (1999).
99. M. N. Serradell, J. Castañer, R. M. Castañer, Drugs Future, 13(4), 312 (1988).
100. H. Hara, T. Morita, T. Sukamoto, F. M. Cutrer, CNS Drug Rev.,1(2), 204 (1995).
101. H. Hara, M. Shimazawa, M. Hashimoto, T. Sukamoto, Nippon Yakurigaku Zasshi. Folia Pharmacol. Japonica, 112 Suppl., 1138P-142P (1998).
102. D. Dooley, Curr. Opin. Cent. Peripher. Nerv. Syst. Invest. Drugs, 1(1), 116 (1999).
103. T. Haupl, S. Land, P. Netusil, N. Biller, C. Capiau, P. Desmons, P. Hauser, G. Burmester, Fems Immunol. Med. Microbiol., 19(1), 15 (1997).
104. M. Hayney, M. Grunske, L. Boch, C. Da Camada, M. Perreault, Ann. Pharmacother., 33(6), 723 (1999).
105. W. Thanassi, R. Schoen, Ann. Intern. Med., 132(8), 661 (2000).
106. B. Y. Lee, Hwahak Sekye, 35(8), 39 (1995).
107. Y.-J. Surth, M. Shlyankevitch, J. W. Lee, J.-K. Yoo, Mutat. Res., 367(4), 219 (1996).
108. W. H. Yoon, J. H. Park, W. I. Lee, J. W. Lee, C.-K. Shim, M. G. Lee, Drug Stab., 1(2), 106 (1996).
109. S. G. Kim, Y.-J. Surth, Y. Sohn, J.-K. Yoo, J. W. Lee, A. Liem, J. A. Miller, Carcinogenesis, 19(4), 687 (1999).
110. J. Castañer, R. M. Castañer, J. Prous, L. Martin, Drugs Future, 24(2), 193 (1999).
111. J. M. Blondeau, D. Felmingham, Clin. Drug Invest.), 18(1), 57 (1999).
112. R. Wise, Clin. Drug Invest., 17(5), 365 (1999).
113. A. P. Alasdair, Expert Opin. Invest. Drugs, 8(2), 181 (1999).
114. J. A. Balfour, H. M. Lamb, Drugs, 59(1), 115 (2000).
115. C. H. Nightingale, Pharmacotherapy, 20(3), 245 (2000).
116. M. N. Serradell, J. Castañer, R. M. Castañer, Drugs Future, 12(1), 31 (1987).
117. H. M. Himmel, Cardiovasc. Drug Rev., 12(1), 32 (1994).
118. S. Yamada, C. Tanaka, R. Kimura, K. Kawabe, Life Sci., 54(24), 1845 (1994).

119. N. Pescalli, V. Sala, G. Sponer, R. Ceserani, Pharmacol. Res., 34(3-4), 121 (1996).
120. R. J. Thurlow, Emerging Drugs, 3, 225 (1998).
121. J. Prous, A. Graul, J. Castañer, Drugs Future, 18(6), 503 (1993).
122. C. J. Tack, P. Smits, Neth. J. Med., 55(5), 209 (1999).
123. M. Kikuchi, Jap. J. Clin. Med., 57(3), 702 (1999).
124. Y. Iwamoto, Saishin Igaku, 55(3), 331 (2000).
125. T. Katakami, T. Yokoyama, M. Michihiko, H. Mori, N. Kawauchi, T. Nobori, K. Sannohe, T. Kaiho, J. Kamiya, J. Med. Chem., 35(18), 3325 (1992).
126. J. Castañer, R. M. Castañer, J. Prous, Drugs Future, 18(3), 226 (1993).
127. M. Ishii, J. Kamiya, K. Hashimoto, Drug Dev Res., 35(2), 61 (1995).
128. H. Nakaya, H. Uemura, Cardiovasc. Drug. Rev., 16(2), 133 (1998).
129. A. Zaza, Curr. Opin. Cardiovasc., Pulm. Renal Invest. Drugs, 2(1), 86 (2000).
130. A. Ozaki, Y. Toba, M. Umezato, S. Fujita, E. Hosoki, N. Nakagiri, K. Ikezono, M. Ishikawa, T. Maehara, S. Shintani, Jpn Pharmacol. Ther., 22(6), 199(1994).
131. K. Kitamura, F. Takaku, K. Takakura, S. Kitamura, Y. Aso, M. Mizuno, S. Ikeda, T. Taguchi, Biotherapy, 9/11, 1381 (1995).
132. M. Elkordy, M. Crump, J. Vredenburgh, W. Petros, A. Hussein, P. Rubin, M. Ross, C. Gilbert, C. Modlin, B. Meisenberg, D. Coniglio, J. Rabinowitz, M. Laughlin, J. Kurtzberg, W. Peters, Bone Marrow Transplant., 19(4), 315 (1997).
133. A. Billich, Idrugs, 1(1), 122 (1998).
134. C. U. Kim, Med. Chem. Res., 8(7/8), 392 (1998).
135. C. U. Kim, X. Chen, D. B. Mendel, Antiviral Chem. Chemother., 10(4), 141 (1999).
136. A. Bardsley-Elliot, S. Noble, Drugs, 58(5), 851 (1999).
137. A. Graul, P. A. Leeson, J. Castañer, Drugs Future, 24(11), 1189 (1999).
138. M. D. Khare, M. Sharland, Expert Opin. Pharmacother., 1(3), 367 (2000).
139. L. V. Gubareva, L. Kaiser, F. G. Hayden, Lancet, 355(9206), 827 (2000).
140. J. Castañer, R. M. Castañer, Drugs Future, 15(11), 1080 (1990).
141. S. L. Grossman, J. Lessem, Expert Opin. Invest. Drugs, 6(8), 1025 (1997).
142. A. J. Evans, A. J. Krentz, Drugs R&D, 2(2), 75 (1999).
143. M. Paul, Arzneim.-Forsch., 49(10), 835 (1999).
144. R. J. Jha, Clin. Exp. Hypertension, 21(1-2), 157 (1999).
145. G.P.Samraj, L. Kuritzky, D. M. Quillen, Hospital Practice(Office Edition), 35(1), 123 (2000).
146. J. Prous, A. Graul, J. Castañer, Drugs Future, 19(10), 916 (1994).
147. J. M. Wierda, Int. Congr. Ser., 1164, 43 (1998).
148. S. V. Onrust, R. H. Foster, Drugs, 58(5), 887 (1999).
149. G. E. Larijani, A. Zafeiridis, M. E. Goldberg, Pharmacotherapy, 19(10), 1118 (1999).
150. A. N. Plowman, Idrugs, 2(7), 711 (1999).
151. L. A. Sorbera, P. A. Leeson, J. Castañer, Drugs Future, 23(12), 1287 (1998) ; 24(12), 1374 (1999).
152. F. Kamali, Curr. Opin. Anti-Inflammatory Immunomodulatory Invest. Drugs, 1(2), 111 (1999).
153. F. Kamali, Curr. Opin. Centr. Peripher. Nerv. Syst. Invest. Drugs, 1(1), 126 (1999).
154. B. Kaplan-Machlis, B. S. Klostermeyer, Ann. Pharmacother., 33(9), 979 1999).
155. L. J. Scott, H. M. Harriet, Drugs, 58(3), 499 (1999).
156. H. B. Fung, H. L. Kirschenbaum, Clin. Ther., 21(7), 1131 (1999).
157. L. A. Sorbera, X. Rabasseda, J. Castañer, Drugs Future, 23(9), 977 (1998) ; 24(9), 1038 (1999).
158. P. V. Amato, D. Domenichini, Formulary, 34(10), 825 (1999).
159. D. A. Greene, Expert Opin. Invest. Drugs, 8(10), 1709 (1999).
160. J. A. Balfour, G. L. Plosker, Drugs, 57(6), 921 (1999).
161. R. Jones, Curr. Opin. Oncol., Endocr. Metab. Invest. Drugs, 1(1), 65 (1999).
162. P. Tontonov, L. Nagy, Curr. Opin. Lipidol., 10(6), 485 (1999).
163. K. H. Kim, Drugs Future, 20(11), 1128 (1995).
164. D.-K. Kim, J. Gam, Y.-B. Cho, H. T. Kim, K. H. Kim,Biorg. Med. Chem. Lett., 6(6), 647 (1996).
165. N. K. Kim, S.-A. Im, D.-W. Kim, M. H. Lee, C. W. Jung, E. K. Cho, J. T. Lee, J. S. Ahn, D. S. Heo, Y.-J. Bang, Cancer, 86(7), 1109 (1999).
166. K. S. Kang, K. H. Lee, J. S. Lee, J. H. Lee, W. K. Kim, S. H. Kim, W. D. Kim, D. S. Kim, J. H. Kim, B. S. Kim, Y. B. Cho, D. K. Kim, K. H. Kim, Am. J. Clin. Oncol., 22(5), 495 (1999).
167. K. Anderson, Y. Lie, M. Low, E. Fennie, Antiviral Res., 21(4), 343 (1993).
168. G. Parenteau, G. Doherty, G. Peplinski, K. Tsung, J. Norton, Ann. Surg., 221(5), 572 (1995).

169. U. J. Ries, G. Mihm, B. Narr, K. M. Hasselbach, H. Wittenben, M. Entzeroth, J. C. A. Van Meel, W. Wienen, N. H. Hauel, J. Med. Chem. 36(25), 4040 (1993).
170. M. Merlos, A. Casas, J. Castañer, Drugs Future, 22(10), 1112 (1997).
171. K. J. McClellan, A. Markham, Drugs, 56(6), 1039 (1998).
172. P. A. Meredith, Am. J. Cardiol., 84(2A), 7K-12K (1999).
173. H. Siragy, Am. J. Cardiol., 84(10A), 3S-8S (1999).
174. P. B. Timmermans, Hypertens. Res., 22(2), 147 (1999).
175. I. Gavras, H. Gavras, Cardiovasc. Rev. Rep., 21(2), 76 (2000).
176. K. M. Hvizdos, K. L. Goa, CNS Drugs, 12(3), 237 (1999).
177. J. Prous, A. Graul, J. Castañer, Drugs Future, 19(8), 746 (1994).
178. E. S. Newlands, M. F. Stevens, S. R. Wedge, R . T. Wheelhouse, C. Brock, Cancer Treat. Rev., 23(1), 35 (1997).
179. N. G. Avgeroupolos, T. Batchelor, Oncologist, 4(3), 209 (1999).
180. U. Brunner, B. M. Gensthaler, Pharm. Ztg, 144(13), 1040 (1999).
181. S. M. Albonico, Drugs Future, 5(4), 171 (1980).
182. N. R. Bachur, L. Lur, P. M. Sun, C. M. Trubey, E. Elliott, M. J. Egorin, L. Malkas, R. Hickey, Biochem. Pharmacol., 55(7), 1025 (1998).
183. S. V. Onrust, H. M. Lamb, Drugs Aging, 15(1), 69 (1999).
184. N. Mealy, J. Castañer, Drugs Future, 21(1), 37 (1996).
185. B. Beer, D. E. Clody, R. Mangano, M. Levner, P. Mayer, J. E. Barrett, CNS Drug Rev., 3(3), 207 (1997).
186. J. Wagner, M. L. Wagner, W. A. Hening, Ann. Pharmacother., 32(6), 680 (1998).
187. M. Hurst, S. Noble, CNS Drugs, 11(5), 387 (1999).
188. R. Fromtling, J. Castañer, Drugs Future, 21(4), 375 (1996).
189. M. J. Milton, M. von Itzstein, Prog. Med. Chem., 36, 1 (1999).
190. J. N. Varghese, Drug Dev. Res., 46(3/4), 176 (1999).
191. B. Freund, S. Gravenstein, M. Elliott, I. Miller, Drug Saf., 21(4), 267 (1999).
192. C. Dunn, K. L. Goa, Drugs, 58(4), 761 (1999).
193. C. A. Silagy, K. Campion, Ann. Med., 31(5), 313 (1999).
194. J. S. Oxford, Idrugs, 3(4), 447 (2000).

Chapter 31. Genetically Modified Crops as a Source for Pharmaceuticals

Véronique Gruber and Manfred Theisen
MERISTEM THERAPEUTICS
8, rue des Frères Lumière, 63100 Clermont-Ferrand, France

Introduction - The use of plants as bioreactors for recombinant protein production was made possible through the development of genetic engineering technologies and plant transformation procedures that allow introduction of foreign genes which are stably expressed and transmissible through generations into plant cells. This concept is based on the capability to generate transformed cells which are able to regenerate into plants. Adapted techniques for DNA delivery, selection of transgenic tissues and regeneration of transgenic plants have been developed for different plant species. DNA is delivered either via Agrobacterium-mediated gene transfer (1), plant viral vectors (2), or direct gene transfer through chemical (3, 4), electrical (5) or physical methods such as microprojectile bombardment (6, 7). Transgenic plants can be considered as a safe production system with respect to lack of human and animal pathogens. They are a eukaryotic expression system that allows the production of complex mammalian proteins in an active form. Production scale-up is easy and costs are often low. An unlimited supply of plant material is possible by using different geographic production sites or adapted storage conditions for the biomass. Depending on the species, only particular plant tissues are industrially exploitable such as tobacco leaves, potato tubers, rape and maize seeds, tomato fruits. From a practical point of view, storage of maize seeds is easier than storage of fresh leaves from tobacco plants. Nevertheless, advantages of different plant species depend on their own features and properties, the stability of the recombinant product, the administrative route of the drug and the process for recombinant protein recovery.

Figure 1 represents a general scheme for the production of recombinant protein drugs in plants. An efficient plant production system requires an expression vector recognisable by the plant host cell. The choice of regulatory elements controlling the expression of the recombinant protein is an important factor for product yields. One of the major elements is the promoter which can be constitutive, as for instance the 35S promoter from cauliflower mosaic virus (8-10), tissue-specific (leaf, seed, tuber, root) or inducible. The promoter is involved in control of the expression level with respect to the harvesting stage, which is an important parameter for the stability and quality of the product.

Improved expression levels can also be obtained through plant species adapted codon usage which will avoid mis-splicing and polyadenylation, and promote efficient recognition by the translational machinery of the plant. Furthermore, protein stability and biological activity are dependent on post-translational processing, including for instance proteolytic cleavage, protein folding with the assistance of molecular chaperones, oligomerisation, disulfide bond formation and glycosylation. Thus, the targeting of the recombinant protein to a subcellular compartment remains an important criteria. The choice of a cell compartment is dependant on the signal sequences. For example, the fusion of a signal peptide to a mature protein sequence will direct the protein to the extracellular compartment, whereas the presence of an additional propeptide signal will target the protein to the vacuole. The choice of the cell compartment will also determine the stability of the recombinant protein, extraction and purification conditions and the type of glycans found on a recombinant glycoprotein (11, 12). Differences in glycosylation between plants and mammals do not necessarily have a negative impact on a protein function. It depends on the nature of the recombinant protein and on the administrative route chosen for the drug. Functionality is a crucial factor for the exploitation of recombinant proteins.

Transformed plants are selected via the presence of a selectable marker gene and screened for expression of the target protein using adapted biochemical procedures. Plants expressing the recombinant protein at high expression levels are multiplied in the greenhouse to allow laboratory-scale purification and characterisation of the recombinant protein. Production of large quantities of plant biomass is assured in field trials. The plant biomass is then handled for scale-up extraction and purification procedures on the semi-industrial pilot scale. The purified protein is biochemicaly characterised. Toxicology, pharmacology as well as pre-clinical and clinical trials will decide if the protein will be a new recombinant drug.

Figure 1. Production of commercially valuable plant-derived recombinant protein drugs

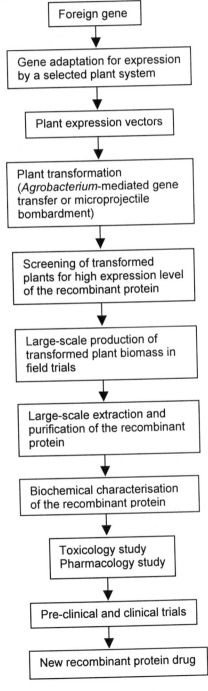

The potential of transgenic plants to function as efficient bioreactors for production of recombinant proteins has generated enthusiasm for this alternative production system. Today, several human proteins are produced in transgenic crops. In this chapter, we will focus on a review of pharmaceuticals produced in transgenic plants and discuss the impact of the plant system on the development of new pharmaceuticals.

PRODUCTION IN TOBACCO-BASED EXPRESSION SYSTEM

Today, most plant-derived recombinant pharmaceuticals are produced in tobacco. This model species was the first one to be stably transformed and regenerated (13, 14). Transformation to seed harvesting takes only six months. Tobacco is not only an excellent leaf biomass producer, (30 tons of leaf fresh weight per hectare) but also a prolific seed producer (seeds from 4 to 5 plants are sufficient for a one hectare culture). Nevertheless, large-scale production of recombinant proteins from tobacco has several considerations including storage of the leaves, elimination of nicotine and interference of alkaloids with the purification procedure.

Many mammalian proteins for different disease targets have been successfully produced at the laboratory-scale level in tobacco, such as blood proteins, hormones, growth factors, enzymes, milk protein, structural protein, antigens, vaccines and antibodies (see Table 1 for some examples).

Table 1. Some examples of pharmaceutical proteins produced in the tobacco plant system

Protein	Potential use	Expression level[a]	Glycosylation	Ref.
Human albumin	Shock treatment, burns, co-adjuvant	0.02 %	no	15
Human haemoglobin	Blood substitute	0.05 %	no	16
Protein C	Anticoagulant	0.002 %	no	17
γ-interferon	Phagocyte activator	1 %	yes	18
Cytokine CM-CSF	Leukopoiesis in bone marrow transplants	NR	yes	19
Epidermal growth factor	Mitogen	0.001 %	yes	20
Human erythropoietin	Mitogen, blood cells	0.0026 %	yes	21

NP1 defensin	Antibiotic	NR	no	18
α-galactosidase	Fabry disease	12.1 mg/kg tissue	yes	18
Glucocerebrosidase	Gaucher disease	1 - 10 %	yes	17
Glutamic acid decarboxylase	Diabetes	0.4 %	no	22
Human lactoferrin	Multifunction iron binding protein	0.3 %	yes	23
Human collagen	Multiple uses	100 mg/kg tissue	yes	24
Hepatitis B (surface antigen)	Vaccine	0.01 %	yes	25
Enterotoxin B (E. coli)	Vaccine (Traveller's diarrhea)	0.001 %	no	26
Cholera toxin	Vaccine	NR	no	27
Malaria antigen	Vaccine	NR	no	28
IgG	Antibodies	1.3 %	yes	29
Secretory antibodies (IgG-A hybrid)	Antibodies for mucosal immuno-therapy	500 µg/g fresh weight of leaves	yes	30
Fab	Antibodies against human creatine kinase	0.044 %	no	31
single chain Fv (scFv)	Antibodies against human creatine kinase	0.01 %	no	32
Monoclonal antibodies	Antibodies	variable	yes	33 – 35

^{a} : % of extracted proteins
NR : not reported

Table 1 clearly demonstrates that glycoproteins and complex proteins such as human haemoglobin, protein C, human collagen and full-length antibodies, can be produced in tobacco even if complete structural conformity to the native protein is not obtained. Moreover, 35S or enhanced 35S promoters from cauliflower mosaic virus have been used in most of the examples noted in Table 1. It is known that the 35S promoter is a strong constitutive promoter in tobacco. Nevertheless, rather low expression levels were obtained for the majority of recombinant proteins (see Table 1) and improvements are required. The level of expression is a very crucial point for further development of recombinant proteins at an industrial level. The impact of this parameter is nevertheless influenced by the type of recombinant protein produced and the actual amount of protein needed to satisfy market demands.

Plants offer an opportunity for large-scale production of antibodies. In topical immunotherapy, mucosal passive immunisation through secretory antibodies could be considered as an effective prophylactic measure for some bacterial, viral or fungal diseases. Several authors have shown that topically applied antibodies can prevent colonisation by pathogenic bacteria in a highly specific manner (36-38). As an example, monoclonal antibodies against the cell surface adhesin from *Streptococcus mutans* have been applied directly to human teeth. A long-term protection against *Streptococcus mutans* colonisation has been obtained (38). Moreover, limited data indicate that *in vivo* half-life is similar to that of a monoclonal antibody produced in mammalian cells (39).

An other interesting application of antibodies is the production of novel polypeptides such as antibodies fused to enzymes, biological response modifiers or toxins which has been also widely exploited (40-44).

OTHER RECOMBINANT PROTEIN PRODUCTION SYSTEMS

The concept of seeds used as storage containers for recombinant proteins has been widely tested. This concept has been based on the fact that seed production retains the potential to capture the full value of a plant system such as flexibility in scale-up, culturing, long term storage of seeds, shipping of large quantities of seeds and processing of a dried product. Nevertheless, a defined time of seed development has to be determined to ensure high expression level and stability of recombinant proteins.

It is worth noting that an extraction and purification procedure based on the use of oleosins, which are proteins associated with oil bodies in all oil seeds, has been developed (45, 46). These oleosins accumulate in oil seeds at quite high levels and act as carriers for recombinant proteins. For example, this procedure was applied to the production of the anticoagulant hirudin. The hirudin gene was fused to the *Arabidopsis* oleosin gene and introduced into rape. Between both genes, a recognition site for the protease Factor Xa was engineered, enabling cleavage and recovery of the recombinant hirudin from the fusion protein in the course of rape seed purification (47).

The concept of using seeds, tubers or fruit as a medium with which to deliver oral vaccines has been widely tested. For some vaccine antigens, transgenic plants may provide an ideal expression system. Plant material can be fed directly without purification to subjects as their oral dose of recombinant vaccine. In 1995, Haq *et al.* have published the first proof of concept for edible vaccines by demonstrating that mice fed doses of transgenic tubers expressing the B subunit of enterotoxin produced both serum and mucosal antibodies against the enterotoxin B protein (26). Thus, antibody delivery through foodstuff has been successful in the oral cavity and it has been demonstrated that recombinant antigens can be delivered to the gastrointestinal tract *via* raw plant material to induce an immune response.

Studies with vaccine epitopes delivered on the surface of plant viruses have also shown exciting potential. Dalsgaard *et al.* have demonstrated a functional and potent vaccine produced in cowpea mosaic virus (48).

In conclusion, the concept of vaccines has been essentially tested through animal trials. The balance between an effective dose range of vaccine and a level of antigens remains to be elucidated and will probably be dependent on the antigens.

DEVELOPMENT STATUS FOR PHARMACEUTICALS PRODUCED IN PLANTS

To date there are no recombinant pharmaceutical products obtained from transgenic plants currently on the market.

Table 2. Development status for pharmaceuticals produced in plants

Protein	Development status	Ref.
Enterotoxin B	Phase I trials	49
Chimeric mouse-human therapeutic antibodies	Pre-clinical trials	50, 51
Antibodies against HSV-2	Pharmacology study Pre-clinical trials	52
sIgA against *Streptococcus mutans*	Clinical trials	52, 53

Previous human trials with the B subunit of cholera toxin used up to 1 mg per oral dose. To use the same amount of plant produced enterotoxin B, useful in Traveller's diarrhea, volunteers ingested 50-100 g raw tuber per dose in phase I trials started in October 1997 at the University of Baltimore (see Table 2). Thus, Tacket *et al.* have demonstrated that recombinant antigens delivered orally in raw potatoes without adjuvant elicit immune responses in human subjects (49).

So far, human trials have focused essentially on antibodies. A few recombinant antibodies have been evaluated in pre-clinical or clinical trials (see Table 2). Production of chimeric mouse-human therapeutic antibodies in plants for pre-clinical trials has been described (50, 51). As reported by Fischer *et al.*, transgenic soybeans that expressed antibodies against herpes simplex virus 2 (HSV-2) have been produced (52). These antibodies have been shown to be efficient in preventing vaginal HSV-2 transmission in mice. EPIcyte began pre-clinical trials in 1999 with monoclonal antibodies produced in corn against herpes simplex virus 1 and 2 as well as with contraceptive antibodies. The first clinical trial with plant-based immunotherapy was reported by Planet Biotechnology (52). The novel drug CaroRx^{TM} is based on sIgA antibodies that have been produced in transgenic tobacco plants and is designed to eliminate *Streptococcus mutans* contributing to dental carries (35). Planet Biotechnology has demonstrated that CaroRx ^{TM} can effectively eliminate the bacteria and clinical trials are underway (53).

The impact of recombinant antibodies on human health will be important and quantities of protein required are significant (54). As an example, for cancer therapy, each patient will require 10-200 mg of recombinant anti-tumour antibody, which could create a demand for up to 130 kg per year in the USA alone. Thus, large amounts of monoclonal antibodies are required for *in vivo* passive immunisation in order to administer sufficient amounts of antibodies at the site of disease, to overcome the rapid rate of clearance from the body and because repeated treatments are usually required. Transgenic plants appear as one of the most likely systems to meet this demand.

Conclusions - The successful exploitation of the plant production system is dependent on the recombinant target protein, its biochemistry and the capability of the plant cell to produce the target molecule in an active form.

This will largely influence the post-translational modifications such as glycosylation of the target protein and thus determine its pharmacological activity and applicability in human therapy.

The creation of modified plants that allow the specific design of humanised glycan structures is one of the key aspects of research in the field of molecular pharming. Several recombinant proteins from plants are today in large-scale production for clinical trials and the first products will be on the market with the next couple of years.

The steps that will largely determine the future of this production technology will be the development of feasible large-scale industrialisation processes, the proof of efficiency and safety of plant derived proteins in medical trials, the acceptance of plant-derived medical proteins by the regulatory authorities and finally the acceptance of these recombinant proteins by the medical community and the patients. Plant-based medication has accompanied man since its origins. We are now able to use modern biotechnology to specifically produce targeted medical products in plants.

References

1. M.W. Bevan, Nucl.Acids Res., 22, 8711 (1984).
2. B. Gronenborn in "Plant DNA infectious elements", T. Hohn and J. Schell, Eds., Spinger-Verlag, Vienna, 1987, p. 1.
3. J. Draper, M.R. Davey, J.P. Freeman, E.C. Cocking, and B.J. Cox, Plant Cell Physiol., 23, 451 (1982).
4. F.A. Krens, L. Molendijk, G.J. Wullems, and R.A. Schilperoort, Nature, 296, 72 (1982).
5. H. Jones, M.J. Tempelaar, and M.G.K. Jones, Oxf. Surv. Plant Mol. Cell Biol., 4, 347 (1987).
6. J.C. Sanford, T.M. Klein, E.D. Wolf, and N. Allen, J.Part.Sci.Tech., 5, 27 (1987).
7. J.C. Sanford, Trends Biotechnol., 6, 299 (1988).
8. M.H. Harpster, J.A. Townsend, J.D.G. Jones, J. Bedbrook, and P. Dunsmuir, Mol.Gen.Genet., 212, 182 (1988).
9. R. Kay, A. Chan, M. Daly, and J. McPherson, Science, 236, 1299 (1987).
10. A. Franck, H. Guilley, G. Jonard, K. Richards, and L. Hirth, Cell, 21, 285 (1980).
11. L. Faye, K.D. Johnson, A. Sturm, and M.J. Chrispeels, Physiol.Plant, 75, 309 (1989).
12. P. Lerouge and L. Faye, Plant Physiol.Biochem., 34, 263 (1996).
13. R.B. Horsch, R.T. Fraley, S.G. Rogers, P.R. Sanders, A. Loyd, and N. Hoffmann, Science, 223, 496 (1984).
14. R.B. Horsch, J.E. Fry, N.L. Hoffmann, D. Eichholtz, S.G. Rogers, and R.T. Fraley, Science, 227, 1229 (1985).
15. P.C. Sijmons, B.M.M. Dekker, B. Schrammeijer, T.C. Verwoerd, P.J.M. van den Elzen, and A. Hoekma, Biotechnology, 8, 217 (1990).
16. W. Dieryck, J. Pagnier, C. Poyart, M.C. Marden, V. Gruber, P. Bournat, S. Baudino, and B. Merot, Nature, 385, 29 (1997).
17. C.L. Cramer, D.L. Weissenborn, K.K. Oishi, E.A. Grabau, S. Bennett, E. Ponce, G.A. Grabowski, and D.N. Radin in "Engineering plants for commercial products and applications", G.B. Collins and R.J. Shepherd, Eds., New York Academy of Sciences, New York, 1996, p. 62.
18. L.K. Grill in "IBCs 3rd Annual International Symposium on Producing the Next Generation of Therapeutics: Exploiting Transgenic Technologies", West Palm Beach, 1997, 5-6 Feb.
19. P.R. Ganz, A.K. Dudani, E.S. Tackaberry, R. Sardana, C. Sauder, X. Cheng, and I. Altosaar in "Transgenic plants: a production system for industrial and pharmaceutical proteins", M.R.L. Owen and J. Pen, Eds., Wiley, Chichester, 1996, p. 281.
20. K. Higo, Y. Saito, and H. Higo, Biosci Biotechnol.Biochem., 57, 1477 (1993).
21. S. Matsumoto, K. Ikura, M. Ueda, and R. Sasaki, Plant Mol.Biol., 27, 1163 (1995).
22. S. Ma, D. Zhao, A. Yin, R. Mukherjee, B. Singh, H. Qin, C.R. Stiller, and A.M. Jevnikar, Nat.Med., 3, 793 (1997).

23. V. Salmon, D. Legrand, M.C. Slomianny, I. El Yazidi, G. Spik, V. Gruber, P. Bournat, B. Olagnier, D. Mison, M. Theisen, and B. Merot, Protein Express.Purif., 13, 127 (1998).
24. F. Ruggiero, J-Y. Exposito, P. Bournat, V. Gruber, S. Perret, J. Comte, B. Olagnier, R. Garrone, and M. Theisen, FEBS Let., 469, 132 (2000).
25. H.S. Mason, D.M. Lam, and C.J. Arntzen, Proc.Natl.Acad.Sci.USA, 89, 11745 (1992).
26. T.A. Haq, H.S. Mason, J.D. Clements, and C.J. Arntzen, Science, 268, 714 (1995).
27. M.B. Hein, T.C. Yeo, F. Wang, and A. Sturtevant in "Engineering plants for commercial products and applications", G.B. Collins and R.J. Shepherd, Eds., New York Academy of Sciences, New York, 1996, p. 50.
28. T.H. Turpen, S.J. Reinl, Y. Charoenvit, S.L. Hoffman, V. Fallarme, and L.K. Grill, Bio/Technol., 13, 53 (1995).
29. A.C. Hiatt, R. Cafferkey, and K. Bowdish, Nature, 342, 76 (1989).
30. J.K.C. Ma, A. Hiatt, M. Hein, N.D. Vine, F. Wang, P. Stabila, C. Van Dolleweerd, K. Mostov, and T. Lehner, Science, 268, 716 (1995).
31. M. De Neve, M. De Loose, A. Jacobs, H. Van Houdt, B. Kaluza, U. Weidle, M. Van Montagu, and A. Depicker, Transgenic Res., 2, 227 (1993).
32. A.M. Bruyns, G. De Jaeger, M. De Neve, C. De Wilde, M. Van Montagu, and A. Depicker, FEBS Letters, 386, 5 (1996).
33. K. Düring, S. Hippe, F. Kreuzaler, and J. Schell, Plant Mol.Biol., 15, 281 (1990).
34. J.K.C. Ma, T. Lehner, P. Stabila, C.I. Fux, and A. Hiatt, Eur.J.Immunol., 24, 131 (1994).
35. J.K. Ma, B.Y. Hikmat, K. Wycoff, N.D. Vine, D. Chargelegue, L. Yu, M.B. Hein, and T. Lehner, Nat.Med., 4, 601 (1998).
36. T. Lehner, J. Caldwell, and R. Smith, Infect.Immun., 50, 796 (1985).
37. D. Bessen and V.A. Fischetti, A.J.Exp.Med., 167, 1945 (1988).
38. J. K-C. Ma, M. Hunjan, R. Smith, C. Kelly, and T. Lehner, Infect.Immun., 58, 3407 (1990).
39. L. Zeitlin, S.S. Olmsted, T.R. Moench, M.S. Co, B.J. Martinell, V.M. Paradkar, D.R. Russell, C. Queen, R.A. Cone, and K.J. Whaley, Nat.Biotechnol., 16, 1361 (1998).
40. P.D. Senter, M.G. Sautmer, G.J. Schreiber, D.L. Hirschberg, J.P. Brown, I. Hellström, and K.E. Hellström, Proc.Natl.Acad.Sci.USA, 85, 4842 (1988).
41. M.A. Bookman, Semin.Oncol., 25, 381 (1998).
42. B. Gerstmayer, M. Hoffmann, U. Altenschmidt, and W. Wels, Cancer Immunol.Immunother., 45, 156 (1997).
43. S.U. Shin, A. Wright, and S.L. Morrison, Int.Rev.Immunol., 10, 177 (1993).
44. U. Brinkmann and I. Pastan, Biochem.Biophys.Acta, 1198, 27 (1994).
45. A.H.C. Huang, Ann. Rev. Plant Physiol.Plant Mol.Biol., 43, 177 (1992).
46. G.J.H. Van Rooijen and M.M. Moloney, Bio/Technol., 13, 72 (1995).
47. D.L. Parmenter, J.G. Boothe, and M.M. Moloney in "Transgenic plants: a production system for industrial and pharmaceutical proteins", M.R.L. Owen and J. Pen, Eds., Wiley, Chichester, 1996, p. 261.
48. K. Dalsgaard, A. Uttenthal, T.D. Jones, F. Xu, A. Merryweather, W.D.O. Hamilton, J.P.M. Langeveld, R.S. Boshuizen, S. Kamstrup, G.P. Lomonossoff, C. Porta, C. Vela, J.I. Casal, R.H. Meloen, and P.B. Rodgers, Nat.Biotechnol., 15, 248 (1997).
49. C.O. Tacket, H.S. Mason, G. Losonsk, J.D. Clements, M.M. Levine, and C.J. Arntzen, Nat.Med., 4, 607 (1998).
50. L. Zeitlin, S.S. Olmsted, T.R. Moench, M.S. Co, B.J. Martinell, V.M. Paradkar, D.R. Russell, C. Queen, R.A. Cone, and K.J. Whaley, Nat.Biotechnol., 16, 1361 (1998).
51. C. Vaquero-Martin, M. Sack, J. Chandler, J. Drossard, F. Schuster, M. Monecke, S. Schillberg, and R. Fisher, Proc.Natl.Acad.Sci.USA, 96, 11128 (1999).
52. R. Fischer, Y-C. Liao, K. Hoffmann, S. Schillberg, and N. Emans, Biol.Chem., 380, 825 (1999).
53. J.W. Larrick, L. Yu, J. Chen, S. Jaiswal, and K. Wycoff, Res.Immunol., 149, 603 (1998).
54. R. Fischer, J. Drossard, U. Commandeur, S. Schillberg, and N. Emans, Biotechnol.Appl.Biochem., 30, 101 (1999).

CUMULATIVE NCE INTRODUCTION INDEX, 1983–1999

GENERIC NAME	INDICATION	YEAR INTRO.	ARMC VOL., PAGE
abacavir sulfate	antiviral	1999	35, 333
acarbose	antidiabetic	1990	26, 297
aceclofenac	antiinflammatory	1992	28, 325
acetohydroxamic acid	hypoammonuric	1983	19, 313
acetorphan	antidiarrheal	1993	29, 332
acipimox	hypolipidemic	1985	21, 323
acitretin	antipsoriatic	1989	25, 309
acrivastine	antihistamine	1988	24, 295
actarit	antirheumatic	1994	30, 296
adamantanium bromide	antiseptic	1984	20, 315
adrafinil	psychostimulant	1986	22, 315
AF-2259	antiinflammatory	1987	23, 325
afloqualone	muscle relaxant	1983	19, 313
alacepril	antihypertensive	1988	24, 296
alclometasone dipropionate	topical antiinflammatory	1985	21, 323
alendronate sodium	osteoporosis	1993	29, 332
alfentanil HCl	analgesic	1983	19, 314
alfuzosin HCl	antihypertensive	1988	24, 296
alglucerase	enzyme	1991	27, 321
alitretinoin	anticancer	1999	35, 333
alminoprofen	analgesic	1983	19, 314
alpha-1 antitrypsin	protease inhibitor	1988	24, 297
alpidem	anxiolytic	1991	27, 322
alpiropride	antimigraine	1988	24, 296
alteplase	thrombolytic	1987	23, 326
amfenac sodium	antiinflammatory	1986	22, 315
amifostine	cytoprotective	1995	31, 338
aminoprofen	topical antiinflammatory	1990	26, 298
amisulpride	antipsychotic	1986	22, 316
amlexanox	antiasthmatic	1987	23, 327
amlodipine besylate	antihypertensive	1990	26, 298
amorolfine HCl	topical antifungal	1991	27, 322
amosulalol	antihypertensive	1988	24, 297
ampiroxicam	antiinflammatory	1994	30, 296
amprenavir	antiviral	1999	35, 334
amrinone	cardiotonic	1983	19, 314
amsacrine	antineoplastic	1987	23, 327
amtolmetin guacil	antiinflammatory	1993	29, 332
anagrelide HCl	hematological	1997	33, 328
anastrozole	antineoplastic	1995	31, 338
angiotensin II	anticancer adjuvant	1994	30, 296
aniracetam	cognition enhancer	1993	29, 333
APD	calcium regulator	1987	23, 326
apraclonidine HCl	antiglaucoma	1988	24, 297
APSAC	thrombolytic	1987	23, 326
aranidipine	antihypertensive	1996	32, 306
arbekacin	antibiotic	1990	26, 298
arglabin	anticancer	1999	35, 335

GENERIC NAME	INDICATION	YEAR INTRO.	ARMC VOL., PAGE
argatroban	antithromobotic	1990	26, 299
arotinolol HCl	antihypertensive	1986	22, 316
artemisinin	antimalarial	1987	23, 327
aspoxicillin	antibiotic	1987	23, 328
astemizole	antihistamine	1983	19, 314
astromycin sulfate	antibiotic	1985	21, 324
atorvastatin calcium	dyslipidemia	1997	33, 328
atovaquone	antiparasitic	1992	28, 326
auranofin	chrysotherapeutic	1983	19, 314
azelaic acid	antiacne	1989	25, 310
azelastine HCl	antihistamine	1986	22, 316
azithromycin	antibiotic	1988	24, 298
azosemide	diuretic	1986	22, 316
aztreonam	antibiotic	1984	20, 315
balsalazide disodium	ulcerative colitis	1997	33, 329
bambuterol	bronchodilator	1990	26, 299
barnidipine HCl	antihypertensive	1992	28, 326
beclobrate	hypolipidemic	1986	22, 317
befunolol HCl	antiglaucoma	1983	19, 315
benazepril HCl	antihypertensive	1990	26, 299
benexate HCl	antiulcer	1987	23, 328
benidipine HCl	antihypertensive	1991	27, 322
beraprost sodium	platelet aggreg. inhibitor	1992	28, 326
betamethasone butyrate propionate	topical antiinflammatory	1994	30, 297
betaxolol HCl	antihypertensive	1983	19, 315
bevantolol HCl	antihypertensive	1987	23, 328
bicalutamide	antineoplastic	1995	31, 338
bifemelane HCl	nootropic	1987	23, 329
binfonazole	hypnotic	1983	19, 315
binifibrate	hypolipidemic	1986	22, 317
bisantrene HCl	antineoplastic	1990	26, 300
bisoprolol fumarate	antihypertensive	1986	22, 317
bopindolol	antihypertensive	1985	21, 324
brimonidine	antiglaucoma	1996	32, 306
brinzolamide	antiglaucoma	1998	34, 318
brodimoprin	antibiotic	1993	29, 333
bromfenac sodium	NSAID	1997	33, 329
brotizolam	hypnotic	1983	19, 315
brovincamine fumarate	cerebral vasodilator	1986	22, 317
bucillamine	immunomodulator	1987	23, 329
bucladesine sodium	cardiostimulant	1984	20, 316
budipine	antiParkinsonian	1997	33, 330
budralazine	antihypertensive	1983	19, 315
bunazosin HCl	antihypertensive	1985	21, 324
bupropion HCl	antidepressant	1989	25, 310
buserelin acetate	hormone	1984	20, 316
buspirone HCl	anxiolytic	1985	21, 324

GENERIC NAME	INDICATION	YEAR INTRO.	ARMC VOL., PAGE
cetrorelix	female infertility	1999	35, 336
chenodiol	anticholelithogenic	1983	19, 317
CHF-1301	antiparkinsonian	1999	35, 336
choline alfoscerate	nootropic	1990	26, 300
cibenzoline	antiarrhythmic	1985	21, 325
cicletanine	antihypertensive	1988	24, 299
cidofovir	antiviral	1996	32, 306
cilazapril	antihypertensive	1990	26, 301
cilostazol	antithrombotic	1988	24, 299
cimetropium bromide	antispasmodic	1985	21, 326
cinildipine	antihypertensive	1995	31, 339
cinitapride	gastroprokinetic	1990	26, 301
cinolazepam	hypnotic	1993	29, 334
ciprofibrate	hypolipidemic	1985	21, 326
ciprofloxacin	antibacterial	1986	22, 318
cisapride	gastroprokinetic	1988	24, 299
cisatracurium besilate	muscle relaxant	1995	31, 340
citalopram	antidepressant	1989	25, 311
cladribine	antineoplastic	1993	29, 335
clarithromycin	antibiotic	1990	26, 302
clobenoside	vasoprotective	1988	24, 300
cloconazole HCl	topical antifungal	1986	22, 318
clodronate disodium	calcium regulator	1986	22, 319
clopidogrel hydrogensulfate	antithrombotic	1998	34, 320
cloricromen	antithrombotic	1991	27, 325
clospipramine HCl	neuroleptic	1991	27, 325
colestimide	hypolipidaemic	1999	35, 337
colforsin daropate HCl	cardiotonic	1999	35, 337
cyclosporine	immunosuppressant	1983	19, 317
cytarabine ocfosfate	antineoplastic	1993	29, 335
dalfopristin	antibiotic	1999	35, 338
dapiprazole HCl	antiglaucoma	1987	23, 332
defeiprone	iron chelator	1995	31, 340
defibrotide	antithrombotic	1986	22, 319
deflazacort	antiinflammatory	1986	22, 319
delapril	antihypertensive	1989	25, 311
delavirdine mesylate	antiviral	1997	33, 331
denileukin diftitox	anticancer	1999	35, 338
denopamine	cardiostimulant	1988	24, 300
deprodone propionate	topical antiinflammatory	1992	28, 329
desflurane	anesthetic	1992	28, 329
dexfenfluramine	antiobesity	1997	33, 332
dexibuprofen	antiinflammatory	1994	30, 298
dexrazoxane	cardioprotective	1992	28, 330
dezocine	analgesic	1991	27, 326
diacerein	antirheumatic	1985	21, 326
didanosine	antiviral	1991	27, 326
dilevalol	antihypertensive	1989	25, 311

GENERIC NAME	INDICATION	YEAR INTRO.	ARMC VOL., PAGE
dirithromycin	antibiotic	1993	29, 336
disodium pamidronate	calcium regulator	1989	25, 312
divistyramine	hypocholesterolemic	1984	20, 317
docarpamine	cardiostimulant	1994	30, 298
docetaxel	antineoplastic	1995	31, 341
dolasetron mesylate	antiemetic	1998	34, 321
donepezil HCl	anti-Alzheimer	1997	33, 332
dopexamine	cardiostimulant	1989	25, 312
dornase alfa	cystic fibrosis	1994	30, 298
dorzolamide HCL	antiglaucoma	1995	31, 341
doxacurium chloride	muscle relaxant	1991	27, 326
doxazosin mesylate	antihypertensive	1988	24, 300
doxefazepam	hypnotic	1985	21, 326
doxercalciferol	vitamin D prohormone	1999	35, 339
doxifluridine	antineoplastic	1987	23, 332
doxofylline	bronchodilator	1985	21, 327
dronabinol	antinauseant	1986	22, 319
droxicam	antiinflammatory	1990	26, 302
droxidopa	antiparkinsonian	1989	25, 312
duteplase	anticougulant	1995	31, 342
ebastine	antihistamine	1990	26 302
ebrotidine	antiulcer	1997	33, 333
ecabet sodium	antiulcerative	1993	29, 336
efavirenz	antiviral	1998	34, 321
efonidipine	antihypertensive	1994	30, 299
emedastine difumarate	antiallergic/antiasthmatic	1993	29, 336
emorfazone	analgesic	1984	20, 317
enalapril maleate	antihypertensive	1984	20, 317
enalaprilat	antihypertensive	1987	23, 332
encainide HCl	antiarrhythmic	1987	23, 333
enocitabine	antineoplastic	1983	19, 318
enoxacin	antibacterial	1986	22, 320
enoxaparin	antithrombotic	1987	23, 333
enoximone	cardiostimulant	1988	24, 301
enprostil	antiulcer	1985	21, 327
entacapone	antiparkinsonian	1998	34, 322
epalrestat	antidiabetic	1992	28, 330
eperisone HCl	muscle relaxant	1983	19, 318
epidermal growth factor	wound healing agent	1987	23, 333
epinastine	antiallergic	1994	30, 299
epirubicin HCl	antineoplastic	1984	20, 318
epoprostenol sodium	platelet aggreg. inhib.	1983	19, 318
eprosartan	antihypertensive	1997	33, 333
eptazocine HBr	analgesic	1987	23, 334
eptilfibatide	antithrombotic	1999	35, 340
erdosteine	expectorant	1995	31, 342
erythromycin acistrate	antibiotic	1988	24, 301
erythropoietin	hematopoetic	1988	24, 301

GENERIC NAME	INDICATION	YEAR INTRO.	ARMC VOL., PAGE
esmolol HCl	antiarrhythmic	1987	23, 334
ethyl icosapentate	antithrombotic	1990	26, 303
etizolam	anxiolytic	1984	20, 318
etodolac	antiinflammatory	1985	21, 327
exifone	nootropic	1988	24, 302
factor VIIa	haemophilia	1996	32, 307
factor VIII	hemostatic	1992	28, 330
fadrozole HCl	antineoplastic	1995	31, 342
famciclovir	antiviral	1994	30, 300
famotidine	antiulcer	1985	21, 327
fasudil HCl	neuroprotective	1995	31, 343
felbamate	antiepileptic	1993	29, 337
felbinac	topical antiinflammatory	1986	22, 320
felodipine	antihypertensive	1988	24, 302
fenbuprol	choleretic	1983	19, 318
fenoldopam mesylate	antihypertensive	1998	34, 322
fenticonazole nitrate	antifungal	1987	23, 334
fexofenadine	antiallergic	1996	32, 307
filgrastim	immunostimulant	1991	27, 327
finasteride	5α-reductase inhibitor	1992	28, 331
fisalamine	intestinal antiinflammatory	1984	20, 318
fleroxacin	antibacterial	1992	28, 331
flomoxef sodium	antibiotic	1988	24, 302
flosequinan	cardiostimulant	1992	28, 331
fluconazole	antifungal	1988	24, 303
fludarabine phosphate	antineoplastic	1991	27, 327
flumazenil	benzodiazepine antag.	1987	23, 335
flunoxaprofen	antiinflammatory	1987	23, 335
fluoxetine HCl	antidepressant	1986	22, 320
flupirtine maleate	analgesic	1985	21, 328
flurithromycin ethylsuccinate	antibiotic	1997	33, 333
flutamide	antineoplastic	1983	19, 318
flutazolam	anxiolytic	1984	20, 318
fluticasone propionate	antiinflammatory	1990	26, 303
flutoprazepam	anxiolytic	1986	22, 320
flutrimazole	topical antifungal	1995	31, 343
flutropium bromide	antitussive	1988	24, 303
fluvastatin	hypolipaemic	1994	30, 300
fluvoxamine maleate	antidepressant	1983	19, 319
follitropin alfa	fertility enhancer	1996	32, 307
follitropin beta	fertility enhancer	1996	32, 308
fomepizole	antidote	1998	34, 323
fomivirsen sodium	antiviral	1998	34, 323
formestane	antineoplastic	1993	29, 337
formoterol fumarate	bronchodilator	1986	22, 321
foscarnet sodium	antiviral	1989	25, 313
fosfosal	analgesic	1984	20, 319
fosphenytoin sodium	antiepileptic	1996	32, 308

GENERIC NAME	INDICATION	YEAR INTRO.	ARMC VOL., PAGE
fosinopril sodium	antihypertensive	1991	27, 328
fotemustine	antineoplastic	1989	25, 313
fropenam	antibiotic	1997	33, 334
gabapentin	antiepileptic	1993	29, 338
gallium nitrate	calcium regulator	1991	27, 328
gallopamil HCl	antianginal	1983	19, 319
ganciclovir	antiviral	1988	24, 303
gatilfloxacin	antibiotic	1999	35, 340
gemcitabine HCl	antineoplastic	1995	31, 344
gemeprost	abortifacient	1983	19, 319
gestodene	progestogen	1987	23, 335
gestrinone	antiprogestogen	1986	22, 321
glatiramer acetate	Multiple Sclerosis	1997	33, 334
glimepiride	antidiabetic	1995	31, 344
glucagon, rDNA	hypoglycemia	1993	29, 338
GMDP	immunostimulant	1996	32, 308
goserelin	hormone	1987	23, 336
granisetron HCl	antiemetic	1991	27, 329
guanadrel sulfate	antihypertensive	1983	19, 319
gusperimus	immunosuppressant	1994	30, 300
halobetasol propionate	topical antiinflammatory	1991	27, 329
halofantrine	antimalarial	1988	24, 304
halometasone	topical antiinflammatory	1983	19, 320
histrelin	precocious puberty	1993	29, 338
hydrocortisone aceponate	topical antiinflammatory	1988	24, 304
hydrocortisone butyrate	topical antiinflammatory	1983	19, 320
ibandronic acid	osteoporosis	1996	32, 309
ibopamine HCl	cardiostimulant	1984	20, 319
ibudilast	antiasthmatic	1989	25, 313
ibutilide fumarate	antiarrhythmic	1996	32, 309
idarubicin HCl	antineoplastic	1990	26, 303
idebenone	nootropic	1986	22, 321
iloprost	platelet aggreg. inhibitor	1992	28, 332
imidapril HCl	antihypertensive	1993	29, 339
imiglucerase	Gaucher's disease	1994	30, 301
imipenem/cilastatin	antibiotic	1985	21, 328
imiquimod	antiviral	1997	33, 335
incadronic acid	osteoporosis	1997	33, 335
indalpine	antidepressant	1983	19, 320
indeloxazine HCl	nootropic	1988	24, 304
indinavir sulfate	antiviral	1996	32, 310
indobufen	antithrombotic	1984	20, 319
insulin lispro	antidiabetic	1996	32, 310
interferon alfacon-1	antiviral	1997	33, 336
interferon, β-1a	multiple sclerosis	1996	32, 311
interferon, β-1b	multiple sclerosis	1993	29, 339
interferon, gamma	antiinflammatory	1989	25, 314
interferon, gamma-1α	antineoplastic	1992	28, 332

GENERIC NAME	INDICATION	YEAR INTRO.	ARMC VOL., PAGE
interferon gamma-1b	immunostimulant	1991	27, 329
interleukin-2	antineoplastic	1989	25, 314
ipriflavone	calcium regulator	1989	25, 314
irbesartan	antihypertensive	1997	33, 336
irinotecan	antineoplastic	1994	30, 301
irsogladine	antiulcer	1989	25, 315
isepamicin	antibiotic	1988	24, 305
isofezolac	antiinflammatory	1984	20, 319
isoxicam	antiinflammatory	1983	19, 320
isradipine	antihypertensive	1989	25, 315
itopride HCl	gastroprokinetic	1995	31, 344
itraconazole	antifungal	1988	24, 305
ivermectin	antiparasitic	1987	23, 336
ketanserin	antihypertensive	1985	21, 328
ketorolac tromethamine	analgesic	1990	26, 304
kinetin	skin photodamage/ dermatologic	1999	35, 341
lacidipine	antihypertensive	1991	27, 330
lamivudine	antiviral	1995	31, 345
lamotrigine	anticonvulsant	1990	26, 304
lanoconazole	antifungal	1994	30, 302
lanreotide acetate	acromegaly	1995	31, 345
lansoprazole	antiulcer	1992	28, 332
latanoprost	antiglaucoma	1996	32, 311
lefunomide	antiarthritic	1998	34, 324
lenampicillin HCl	antibiotic	1987	23, 336
lentinan	immunostimulant	1986	22, 322
lepirudin	anticoagulant	1997	33, 336
lercanidipine	antihyperintensive	1997	33, 337
letrazole	anticancer	1996	32, 311
leuprolide acetate	hormone	1984	20, 319
levacecarnine HCl	nootropic	1986	22, 322
levalbuterol HCl	antiasthmatic	1999	35, 341
levobunolol HCl	antiglaucoma	1985	21, 328
levocabastine HCl	antihistamine	1991	27, 330
levodropropizine	antitussive	1988	24, 305
levofloxacin	antibiotic	1993	29, 340
lidamidine HCl	antiperistaltic	1984	20, 320
limaprost	antithrombotic	1988	24, 306
lisinopril	antihypertensive	1987	23, 337
lobenzarit sodium	antiinflammatory	1986	22, 322
lodoxamide tromethamine	antiallergic ophthalmic	1992	28, 333
lomefloxacin	antibiotic	1989	25, 315
lomerizine HCl	antimigraine	1999	35, 342
lonidamine	antineoplastic	1987	23, 337
loprazolam mesylate	hypnotic	1983	19, 321
loprinone HCl	cardiostimulant	1996	32, 312
loracarbef	antibiotic	1992	28, 333

GENERIC NAME	INDICATION	YEAR INTRO.	ARMC VOL., PAGE
muzolimine	diuretic	1983	19, 321
mycophenolate mofetil	immunosuppressant	1995	31, 346
nabumetone	antiinflammatory	1985	21, 330
nadifloxacin	topical antibiotic	1993	29, 340
nafamostat mesylate	protease inhibitor	1986	22, 323
nafarelin acetate	hormone	1990	26, 306
naftifine HCl	antifungal	1984	20, 321
naftopidil	dysuria	1999	35, 344
nalmefene HCl	dependence treatment	1995	31, 347
naltrexone HCl	narcotic antagonist	1984	20, 322
naratriptan HCl	antimigraine	1997	33, 339
nartograstim	leukopenia	1994	30, 304
nateglinide	antidiabetic	1999	35, 344
nazasetron	antiemetic	1994	30, 305
nebivolol	antihypertensive	1997	33, 339
nedaplatin	antineoplastic	1995	31, 347
nedocromil sodium	antiallergic	1986	22, 324
nefazodone	antidepressant	1994	30, 305
neflinavir mesylate	antiviral	1997	33, 340
neltenexine	cystic fibrosis	1993	29, 341
nemonapride	neuroleptic	1991	27, 331
neticonazole HCl	topical antifungal	1993	29, 341
nevirapine	antiviral	1996	32, 313
nicorandil	coronary vasodilator	1984	20, 322
nifekalant HCl	antiarrythmic	1999	35, 344
nilutamide	antineoplastic	1987	23, 338
nilvadipine	antihypertensive	1989	25, 316
nimesulide	antiinflammatory	1985	21, 330
nimodipine	cerebral vasodilator	1985	21, 330
nipradilol	antihypertensive	1988	24, 307
nisoldipine	antihypertensive	1990	26, 306
nitrefazole	alcohol deterrent	1983	19, 322
nitrendipine	hypertensive	1985	21, 331
nizatidine	antiulcer	1987	23, 339
nizofenzone fumarate	nootropic	1988	24, 307
nomegestrol acetate	progestogen	1986	22, 324
norfloxacin	antibacterial	1983	19, 322
norgestimate	progestogen	1986	22, 324
OCT-43	anticancer	1999	35, 345
octreotide	antisecretory	1988	24, 307
ofloxacin	antibacterial	1985	21, 331
olanzapine	neuroleptic	1996	32, 313
olopatadine HCl	antiallergic	1997	33, 340
omeprazole	antiulcer	1988	24, 308
ondansetron HCl	antiemetic	1990	26, 306
orlistat	antiobesity	1998	34, 327
ornoprostil	antiulcer	1987	23, 339
osalazine sodium	intestinal antinflamm.	1986	22, 324

GENERIC NAME	INDICATION	YEAR INTRO.	ARMC VOL., PAGE
oseltamivir phosphate	antiviral	1999	35, 346
oxaliplatin	anticancer	1996	32, 313
oxaprozin	antiinflammatory	1983	19, 322
oxcarbazepine	anticonvulsant	1990	26, 307
oxiconazole nitrate	antifungal	1983	19, 322
oxiracetam	nootropic	1987	23, 339
oxitropium bromide	bronchodilator	1983	19, 323
ozagrel sodium	antithrombotic	1988	24, 308
paclitaxal	antineoplastic	1993	29, 342
parnaparin sodium	anticoagulant	1993	29, 342
panipenem/betamipron	carbapenem antibiotic	1994	30, 305
pantoprazole sodium	antiulcer	1995	30, 306
paricalcitol	vitamin D	1998	34 327
paroxetine	antidepressant	1991	27, 331
pefloxacin mesylate	antibacterial	1985	21, 331
pegademase bovine	immunostimulant	1990	26, 307
pegaspargase	antineoplastic	1994	30, 306
pemirolast potassium	antiasthmatic	1991	27, 331
penciclovir	antiviral	1996	32, 314
pentostatin	antineoplastic	1992	28, 334
pergolide mesylate	antiparkinsonian	1988	24, 308
perindopril	antihypertensive	1988	24, 309
picotamide	antithrombotic	1987	23, 340
pidotimod	immunostimulant	1993	29, 343
piketoprofen	topical antiinflammatory	1984	20, 322
pilsicainide HCl	antiarrhythmic	1991	27, 332
pimaprofen	topical antiinflammatory	1984	20, 322
pimobendan	heart failure	1994	30, 307
pinacidil	antihypertensive	1987	23, 340
pioglitazone HCL	antidiabetic	1999	35, 346
pirarubicin	antineoplastic	1988	24, 309
pirmenol	antiarrhythmic	1994	30, 307
piroxicam cinnamate	antiinflammatory	1988	24, 309
pivagabine	antidepressant	1997	33, 341
plaunotol	antiulcer	1987	23, 340
polaprezinc	antiulcer	1994	30, 307
porfimer sodium	antineoplastic adjuvant	1993	29, 343
pramipexole HCl	antiParkinsonian	1997	33, 341
pramiracetam H_2SO_4	cognition enhancer	1993	29, 343
pranlukast	antiasthmatic	1995	31, 347
pravastatin	antilipidemic	1989	25, 316
prednicarbate	topical antiinflammatory	1986	22, 325
prezatide copper acetate	vulnery	1996	32, 314
progabide	anticonvulsant	1985	21, 331
promegestrone	progestogen	1983	19, 323
propacetamol HCl	analgesic	1986	22, 325
propagermanium	antiviral	1994	30, 308
propentofylline propionate	cerebral vasodilator	1988	24, 310

GENERIC NAME	INDICATION	YEAR INTRO.	ARMC VOL., PAGE
propiverine HCl	urologic	1992	28, 335
propofol	anesthetic	1986	22, 325
pumactant	lung surfactant	1994	30, 308
quazepam	hypnotic	1985	21, 332
quetiapine fumarate	neuroleptic	1997	33, 341
quinagolide	hyperprolactinemia	1994	30, 309
quinapril	antihypertensive	1989	25, 317
quinfamide	amebicide	1984	20, 322
quinupristin	antibiotic	1999	35, 338
rabeprazole sodium	gastric antisecretory	1998	34, 328
raltitrexed	anticancer	1996	32, 315
raloxifene HCl	osteoporosis	1998	34, 328
ramipril	antihypertensive	1989	25, 317
ramosetron	antiemetic	1996	32, 315
ranimustine	antineoplastic	1987	23, 341
ranitidine bismuth citrate	antiulcer	1995	31, 348
rapacuronium bromide	muscle relaxant	1999	35, 347
rebamipide	antiulcer	1990	26, 308
reboxetine	antidepressant	1997	33, 342
remifentanil HCl	analgesic	1996	32, 316
remoxipride HCl	antipsychotic	1990	26, 308
repaglinide	antidiabetic	1998	34, 329
repirinast	antiallergic	1987	23, 341
reteplase	fibrinolytic	1996	32, 316
reviparin sodium	anticoagulant	1993	29, 344
rifabutin	antibacterial	1992	28, 335
rifapentine	antibacterial	1988	24, 310
rifaximin	antibiotic	1985	21, 332
rifaximin	antibiotic	1987	23, 341
rilmazafone	hypnotic	1989	25, 317
rilmenidine	antihypertensive	1988	24, 310
riluzole	neuroprotective	1996	32, 316
rimantadine HCl	antiviral	1987	23, 342
rimexolone	antiinflammatory	1995	31, 348
risedronate sodium	osteoporosis	1998	34, 330
risperidone	neuroleptic	1993	29, 344
ritonavir	antiviral	1996	32, 317
rivastigmin	anti-Alzheimer	1997	33, 342
rizatriptan benzoate	antimigraine	1998	34, 330
rocuronium bromide	neuromuscular blocker	1994	30, 309
rofecoxib	antiarthritic	1999	35, 347
rokitamycin	antibiotic	1986	22, 325
romurtide	immunostimulant	1991	27, 332
ronafibrate	hypolipidemic	1986	22, 326
ropinirole HCl	antiParkinsonian	1996	32, 317
ropivacaine	anesthetic	1996	32, 318
rosaprostol	antiulcer	1985	21, 332
rosiglitazone maleate	antidiabetic	1999	35, 348

GENERIC NAME	INDICATION	YEAR INTRO.	ARMC VOL., PAGE
roxatidine acetate HCl	antiulcer	1986	22, 326
roxithromycin	antiulcer	1987	23, 342
rufloxacin HCl	antibacterial	1992	28, 335
RV-11	antibiotic	1989	25, 318
salmeterol hydroxynaphthoate	bronchodilator	1990	26, 308
sapropterin HCl	hyperphenylalaninemia	1992	28, 336
saquinavir mesvlate	antiviral	1995	31, 349
sargramostim	immunostimulant	1991	27, 332
sarpogrelate HCl	platelet antiaggregant	1993	29, 344
schizophyllan	immunostimulant	1985	22, 326
seratrodast	antiasthmatic	1995	31, 349
sertaconazole nitrate	topical antifungal	1992	28, 336
sertindole	neuroleptic	1996	32, 318
setastine HCl	antihistamine	1987	23, 342
setiptiline	antidepressant	1989	25, 318
setraline HCl	antidepressant	1990	26, 309
sevoflurane	anesthetic	1990	26, 309
sibutramine	antiobesity	1998	34, 331
sildenafil citrate	male sexual dysfunction	1998	34, 331
simvastatin	hypocholesterolemic	1988	24, 311
SKI-2053R	anticancer	1999	35, 348
sobuzoxane	antineoplastic	1994	30, 310
sodium cellulose PO4	hypocalciuric	1983	19, 323
sofalcone	antiulcer	1984	20, 323
somatomedin-1	growth hormone insensitivity	1994	30, 310
somatotropin	growth hormone	1994	30, 310
somatropin	hormone	1987	23, 343
sorivudine	antiviral	1993	29, 345
sparfloxacin	antibiotic	1993	29, 345
spirapril HCl	antihypertensive	1995	31, 349
spizofurone	antiulcer	1987	23, 343
stavudine	antiviral	1994	30, 311
succimer	chelator	1991	27, 333
sufentanil	analgesic	1983	19, 323
sulbactam sodium	β-lactamase inhibitor	1986	22, 326
sulconizole nitrate	topical antifungal	1985	21, 332
sultamycillin tosylate	antibiotic	1987	23, 343
sumatriptan succinate	antimigraine	1991	27, 333
suplatast tosilate	antiallergic	1995	31, 350
suprofen	analgesic	1983	19, 324
surfactant TA	respiratory surfactant	1987	23, 344
tacalcitol	topical antipsoriatic	1993	29, 346
tacrine HCl	Alzheimer's disease	1993	29, 346
tacrolimus	immunosuppressant	1993	29, 347
talipexole	antiParkinsonian	1996	32, 318
tamsulosin HCl	antiprostatic hypertrophy	1993	29, 347

GENERIC NAME	INDICATION	YEAR INTRO.	ARMC VOL., PAGE
tandospirone	anxiolytic	1996	32, 319
tasonermin	anticancer	1999	35, 349
tazobactam sodium	β-lactamase inhibitor	1992	28, 336
tazanolast	antiallergic	1990	26, 309
tazarotene	antipsoriasis	1997	33, 343
teicoplanin	antibacterial	1988	24, 311
telmesteine	mucolytic	1992	28, 337
telmisartan	antihypertensive	1999	35, 349
temafloxacin HCl	antibacterial	1991	27, 334
temocapril	antihypertensive	1994	30, 311
temocillin disodium	antibiotic	1984	20, 323
temozolomide	anticancer	1999	35, 349
tenoxicam	antiinflammatory	1987	23, 344
teprenone	antiulcer	1984	20, 323
terazosin HCl	antihypertensive	1984	20, 323
terbinafine HCl	antifungal	1991	27, 334
terconazole	antifungal	1983	19, 324
tertatolol HCl	antihypertensive	1987	23, 344
thymopentin	immunomodulator	1985	21, 333
tiagabine	antiepileptic	1996	32, 319
tiamenidine HCl	antihypertensive	1988	24, 311
tianeptine sodium	antidepressant	1983	19, 324
tibolone	anabolic	1988	24, 312
tilisolol HCl	antihypertensive	1992	28, 337
tiludronate disodium	Paget's disease	1995	31, 350
timiperone	neuroleptic	1984	20, 323
tinazoline	nasal decongestant	1988	24, 312
tioconazole	antifungal	1983	19, 324
tiopronin	urolithiasis	1989	25, 318
tiquizium bromide	antispasmodic	1984	20, 324
tiracizine HCl	antiarrhythmic	1990	26, 310
tirilazad mesylate	subarachnoid hemorrhage	1995	31, 351
tirofiban HCl	antithrombotic	1998	34, 332
tiropramide HCl	antispasmodic	1983	19, 324
tizanidine	muscle relaxant	1984	20, 324
tolcapone	antiParkinsonian	1997	33, 343
toloxatone	antidepressant	1984	20, 324
tolrestat	antidiabetic	1989	25, 319
topiramate	antiepileptic	1995	31, 351
topotecan HCl	anticancer	1996	32, 320
torasemide	diuretic	1993	29, 348
toremifene	antineoplastic	1989	25, 319
tosufloxacin tosylate	antibacterial	1990	26, 310
trandolapril	antihypertensive	1993	29, 348
tretinoin tocoferil	antiulcer	1993	29, 348
trientine HCl	chelator	1986	22, 327
trimazosin HCl	antihypertensive	1985	21, 333

GENERIC NAME	INDICATION	YEAR INTRO.	ARMC VOL., PAGE
trimetrexate glucuronate	*Pneumocystis carinii* pneumonia	1994	30, 312
troglitazone	antidiabetic	1997	33, 344
tropisetron	antiemetic	1992	28, 337
trovafloxacin mesylate	antibiotic	1998	34, 332
troxipide	antiulcer	1986	22, 327
ubenimex	immunostimulant	1987	23, 345
unoprostone isopropyl ester	antiglaucoma	1994	30, 312
valaciclovir HCl	antiviral	1995	31, 352
valrubicin	anticancer	1999	35, 350
valsartan	antihypertensive	1996	32, 320
venlafaxine	antidepressant	1994	30, 312
vesnarinone	cardiostimulant	1990	26, 310
vigabatrin	anticonvulsant	1989	25, 319
vinorelbine	antineoplastic	1989	25, 320
voglibose	antidiabetic	1994	30, 313
xamoterol fumarate	cardiotonic	1988	24, 312
zafirlukast	antiasthma	1996	32, 321
zalcitabine	antiviral	1992	28, 338
zaleplon	hypnotic	1999	35, 351
zaltoprofen	antiinflammatory	1993	29, 349
zanamivir	antiviral	1999	35, 352
zileuton	antiasthma	1997	33, 344
zidovudine	antiviral	1987	23, 345
zinostatin stimalamer	antineoplastic	1994	30, 313
zolpidem hemitartrate	hypnotic	1988	24, 313
zomitriptan	antimigraine	1997	33, 345
zonisamide	anticonvulsant	1989	25, 320
zopiclone	hypnotic	1986	22, 327
zuclopenthixol acetate	antipsychotic	1987	23, 345

CUMULATIVE NCE INTRODUCTION INDEX, 1983–1999, (BY INDICATION)

GENERIC NAME	INDICATION	YEAR INTRO.	ARMC VOL., PAGE	
gemeprost	ABORTIFACIENT	1983	19,	319
mifepristone		1988	24,	306
lanreotide acetate	ACROMEGALY	1995	31,	345
nitrefazole	ALCOHOL DETERRENT	1983	19,	322
tacrine HCl	ALZHEIMER'S DISEASE	1993	29,	346
quinfamide	AMEBICIDE	1984	20,	322
tibolone	ANABOLIC	1988	24,	312
mepixanox	ANALEPTIC	1984	20,	320
alfentanil HCl	ANALGESIC	1983	19,	314
aminoprofen		1983	19,	314
dezocine		1991	27,	326
emorfazone		1984	20,	317
eptazocine HBr		1987	23,	334
flupirtine maleate		1985	21,	328
fosfosal		1984	20,	319
ketorolac tromethamine		1990	26,	304
meptazinol HCl		1983	19,	321
mofezolac		1994	30,	304
propacetamol HCl		1986	22,	325
remifentanil HCl		1996	32,	316
sufentanil		1983	19,	323
suprofen		1983	19,	324
desflurane	ANESTHETIC	1992	28,	329
propofol		1986	22,	325
ropivacaine		1996	32,	318
sevoflurane		1990	26,	309
azelaic acid	ANTIACNE	1989	25,	310
emedastine difumarate	ANTIALLERGIC	1993	29,	336
epinastine		1994	30,	299
fexofenadine		1996	32,	307
nedocromil sodium		1986	22,	324
olopatadine hydrochloride		1997	33,	340
repirinast		1987	23,	341
suplatast tosilate		1995	31,	350
tazanolast		1990	26,	309
lodoxamide tromethamine	ANTIALLERGIC OPHTHALMIC	1992	28,	333
loteprednol etabonate		1998	34,	324

GENERIC NAME	INDICATION	YEAR INTRO.	ARMC VOL., PAGE	
donepezil hydrochloride	ANTI-ALZHEIMERS	1997	33,	332
rivastigmin		1997	33,	342
gallopamil HCl	ANTIANGINAL	1983	19,	319
cibenzoline	ANTIARRHYTHMIC	1985	21,	325
encainide HCl		1987	23,	333
esmolol HCl		1987	23,	334
ibutilide fumarate		1996	32,	309
moricizine hydrochloride		1990	26,	305
nifekalant HCl		1999	35,	344
pilsicainide hydrochloride		1991	27,	332
pirmenol		1994	30,	307
tiracizine hydrochloride		1990	26,	310
celecoxib	ANTIARTHRITIC	1999	35,	335
meloxicam		1996	32,	312
leflunomide		1998	34,	324
rofecoxib		1999	35,	347
amlexanox	ANTIASTHMATIC	1987	23,	327
emedastine difumarate		1993	29,	336
ibudilast		1989	25,	313
levalbuterol HCl		1999	35,	341
montelukast sodium		1998	34,	326
pemirolast potassium		1991	27,	331
seratrodast		1995	31,	349
zafirlukast		1996	32,	321
zileuton		1997	33,	344
ciprofloxacin	ANTIBACTERIAL	1986	22,	318
enoxacin		1986	22,	320
fleroxacin		1992	28,	331
norfloxacin		1983	19,	322
ofloxacin		1985	21,	331
pefloxacin mesylate		1985	21,	331
pranlukast		1995	31,	347
rifabutin		1992	28,	335
rifapentine		1988	24,	310
rufloxacin hydrochloride		1992	28,	335
teicoplanin		1988	24,	311
temafloxacin hydrochloride		1991	27,	334
tosufloxacin tosylate		1990	26,	310
arbekacin	ANTIBIOTIC	1990	26,	298
aspoxicillin		1987	23,	328
astromycin sulfate		1985	21,	324
azithromycin		1988	24,	298

GENERIC NAME	INDICATION	YEAR INTRO.	ARMC VOL., PAGE	
RV-11		1989	25,	318
sparfloxacin		1993	29,	345
sultamycillin tosylate		1987	23,	343
temocillin disodium		1984	20,	323
trovafloxacin mesylate		1998	34,	332
meropenem	ANTIBIOTIC,	1994	30,	303
panipenem/betamipron	CARBAPENEM	1994	30,	305
mupirocin	ANTIBIOTIC, TOPICAL	1985	21,	330
nadifloxacin		1993	29,	340
alitretinoin	ANTICANCER	1999	35,	333
arglabin		1999	35,	335
denileukin diftitox		1999	35,	338
letrazole		1996	32,	311
OCT-43		1999	35,	345
oxaliplatin		1996	32,	313
raltitrexed		1996	32,	315
SKI-2053R		1999	35,	348
tasonermin		1999	35,	349
temozolomide		1999	35,	350
topotecan HCl		1996	32,	320
valrubicin		1999	35,	350
angiotensin II	ANTICANCER ADJUVANT	1994	30,	296
chenodiol	ANTICHOLELITHOGENIC	1983	19,	317
duteplase	ANTICOAGULANT	1995	31,	342
lepirudin		1997	33,	336
parnaparin sodium		1993	29,	342
reviparin sodium		1993	29,	344
lamotrigine	ANTICONVULSANT	1990	26,	304
oxcarbazepine		1990	26,	307
progabide		1985	21,	331
vigabatrin		1989	25,	319
zonisamide		1989	25,	320
bupropion HCl	ANTIDEPRESSANT	1989	25,	310
citalopram		1989	25,	311
fluoxetine HCl		1986	22,	320
fluvoxamine maleate		1983	19,	319
indalpine		1983	19,	320
medifoxamine fumarate		1986	22,	323
metapramine		1984	20,	320

GENERIC NAME	INDICATION	YEAR INTRO.	ARMC VOL., PAGE	
milnacipran		1997	33,	338
mirtazapine		1994	30,	303
moclobemide		1990	26,	305
nefazodone		1994	30,	305
paroxetine		1991	27,	331
pivagabine		1997	33,	341
reboxetine		1997	33,	342
setiptiline		1989	25,	318
sertraline hydrochloride		1990	26,	309
tianeptine sodium		1983	19,	324
toloxatone		1984	20,	324
venlafaxine		1994	30,	312
acarbose	ANTIDIABETIC	1990	26,	297
epalrestat		1992	28,	330
glimepiride		1995	31,	344
insulin lispro		1996	32,	310
miglitol		1998	34,	325
nateglinide		1999	35,	344
pioglitazone HCl		1999	35,	346
repaglinide		1998	34,	329
rosiglitazone maleate		1999	35,	347
tolrestat		1989	25,	319
troglitazone		1997	33,	344
voglibose		1994	30,	313
acetorphan	ANTIDIARRHEAL	1993	29,	332
fomepizole	ANTIDOTE	1998	34,	323
dolasetron mesylate	ANTIEMETIC	1998	34,	321
granisetron hydrochloride		1991	27,	329
ondansetron hydrochloride		1990	26,	306
nazasetron		1994	30,	305
ramosetron		1996	32,	315
tropisetron		1992	28,	337
felbamate	ANTIEPILEPTIC	1993	29,	337
fosphenytoin sodium		1996	32,	308
gabapentin		1993	29,	338
tiagabine		1996	32,	320
topiramate		1995	31,	351
centchroman	ANTIESTROGEN	1991	27,	324
fenticonazole nitrate	ANTIFUNGAL	1987	23,	334
fluconazole		1988	24,	303
itraconazole		1988	24,	305

GENERIC NAME	INDICATION	YEAR INTRO.	ARMC VOL., PAGE	
lanoconazole		1994	30,	302
naftifine HCl		1984	20,	321
oxiconazole nitrate		1983	19,	322
terbinafine hydrochloride		1991	27,	334
terconazole		1983	19,	324
tioconazole		1983	19,	324
amorolfine hydrochloride	ANTIFUNGAL, TOPICAL	1991	27,	322
butenafine hydrochloride		1992	28,	327
butoconazole		1986	22,	318
cloconazole HCl		1986	22,	318
flutrimazole		1995	31,	343
neticonazole HCl		1993	29,	341
sertaconazole nitrate		1992	28,	336
sulconizole nitrate		1985	21,	332
apraclonidine HCl	ANTIGLAUCOMA	1988	24,	297
befunolol HCl		1983	19,	315
brimonidine		1996	32,	306
brinzolamide		1998	34,	318
dapiprazole HCl		1987	23,	332
dorzolamide HCl		1995	31,	341
latanoprost		1996	32,	311
levobunolol HCl		1985	21,	328
unoprostone isopropyl ester		1994	30,	312
acrivastine	ANTIHISTAMINE	1988	24,	295
astemizole		1983	19,	314
azelastine HCl		1986	22,	316
ebastine		1990	26,	302
cetirizine HCl		1987	23,	331
levocabastine hydrochloride		1991	27,	330
loratadine		1988	24,	306
mizolastine		1998	34,	325
setastine HCl		1987	23,	342
alacepril	ANTIHYPERTENSIVE	1988	24,	296
alfuzosin HCl		1988	24,	296
amlodipine besylate		1990	26,	298
amosulalol		1988	24,	297
aranidipine		1996	32,	306
arotinolol HCl		1986	22,	316
barnidipine hydrochloride		1992	28,	326
benazepril hydrochloride		1990	26,	299
benidipine hydrochloride		1991	27,	322
betaxolol HCl		1983	19,	315
bevantolol HCl		1987	23,	328
bisoprolol fumarate		1986	22,	317

GENERIC NAME	INDICATION	YEAR INTRO.	ARMC VOL., PAGE	
bopindolol		1985	21,	324
budralazine		1983	19,	315
bunazosin HCl		1985	21,	324
candesartan cilexetil		1997	33,	330
carvedilol		1991	27,	323
celiprolol HCl		1983	19,	317
cicletanine		1988	24,	299
cilazapril		1990	26,	301
cinildipine		1995	31,	339
delapril		1989	25,	311
dilevalol		1989	25,	311
doxazosin mesylate		1988	24,	300
efonidipine		1994	30,	299
enalapril maleate		1984	20,	317
enalaprilat		1987	23,	332
eprosartan		1997	33,	333
felodipine		1988	24,	302
fenoldopam mesylate		1998	34,	322
fosinopril sodium		1991	27,	328
guanadrel sulfate		1983	19,	319
imidapril HCl		1993	29,	339
irbesartan		1997	33,	336
isradipine		1989	25,	315
ketanserin		1985	21,	328
lacidipine		1991	27,	330
lercanidipine		1997	33,	337
lisinopril		1987	23,	337
losartan		1994	30,	302
manidipine hydrochloride		1990	26,	304
mebefradil hydrochloride		1997	33,	338
moexipril HCl		1995	31,	346
moxonidine		1991	27,	330
nebivolol		1997	33,	339
nilvadipine		1989	25,	316
nipradilol		1988	24,	307
nisoldipine		1990	26,	306
perindopril		1988	24,	309
pinacidil		1987	23,	340
quinapril		1989	25,	317
ramipril		1989	25,	317
rilmenidine		1988	24,	310
spirapril HCl		1995	31,	349
telmisartan		1999	35,	349
temocapril		1994	30,	311
terazosin HCl		1984	20,	323
tertatolol HCl		1987	23,	344
tiamenidine HCl		1988	24,	311
tilisolol hydrochloride		1992	28,	337

GENERIC NAME	INDICATION	YEAR INTRO.	ARMC VOL., PAGE	
trandolapril		1993	29,	348
trimazosin HCl		1985	21,	333
valsartan		1996	32,	320
aceclofenac	ANTIINFLAMMATORY	1992	28,	325
AF-2259		1987	23,	325
amfenac sodium		1986	22,	315
ampiroxicam		1994	30,	296
amtolmetin guacil		1993	29,	332
butibufen		1992	28,	327
deflazacort		1986	22,	319
dexibuprofen		1994	30,	298
droxicam		1990	26,	302
etodolac		1985	21,	327
flunoxaprofen		1987	23,	335
fluticasone propionate		1990	26,	303
interferon, gamma		1989	25,	314
isofezolac		1984	20,	319
isoxicam		1983	19,	320
lobenzarit sodium		1986	22,	322
loxoprofen sodium		1986	22,	322
nabumetone		1985	21,	330
nimesulide		1985	21,	330
oxaprozin		1983	19,	322
piroxicam cinnamate		1988	24,	309
rimexolone		1995	31,	348
tenoxicam		1987	23,	344
zaltoprofen		1993	29,	349
fisalamine	ANTIINFLAMMATORY,	1984	20,	318
osalazine sodium	INTESTINAL	1986	22,	324
alclometasone dipropionate	ANTIINFLAMMATORY,	1985	21,	323
aminoprofen	TOPICAL	1990	26,	298
betamethasone butyrate propionate		1994	30,	297
butyl flufenamate		1983	19,	316
deprodone propionate		1992	28,	329
felbinac		1986	22,	320
halobetasol propionate		1991	27,	329
halometasone		1983	19,	320
hydrocortisone aceponate		1988	24,	304
hydrocortisone butyrate propionate		1983	19,	320
mometasone furoate		1987	23,	338
piketoprofen		1984	20,	322
pimaprofen		1984	20,	322
prednicarbate		1986	22,	325

GENERIC NAME	INDICATION	YEAR INTRO.	ARMC VOL., PAGE	
pravastatin	ANTILIPIDEMIC	1989	25,	316
artemisinin	ANTIMALARIAL	1987	23,	327
halofantrine		1988	24,	304
mefloquine HCl		1985	21,	329
alpiropride	ANTIMIGRAINE	1988	24,	296
lomerizine HCl		1999	35,	342
naratriptan hydrochloride		1997	33,	339
rizatriptan benzoate		1998	34,	330
sumatriptan succinate		1991	27,	333
zomitriptan		1997	33,	345
dronabinol	ANTINAUSEANT	1986	22,	319
amsacrine	ANTINEOPLASTIC	1987	23,	327
anastrozole		1995	31,	338
bicalutamide		1995	31,	338
bisantrene hydrochloride		1990	26,	300
camostat mesylate		1985	21,	325
capecitabine		1998	34,	319
cladribine		1993	29,	335
cytarabine ocfosfate		1993	29,	335
docetaxel		1995	31,	341
doxifluridine		1987	23,	332
enocitabine		1983	19,	318
epirubicin HCl		1984	20,	318
fadrozole HCl		1995	31,	342
fludarabine phosphate		1991	27,	327
flutamide		1983	19,	318
formestane		1993	29,	337
fotemustine		1989	25,	313
gemcitabine HCl		1995	31,	344
idarubicin hydrochloride		1990	26,	303
interferon gamma-1α		1992	28,	332
interleukin-2		1989	25,	314
irinotecan		1994	30,	301
lonidamine		1987	23,	337
mitoxantrone HCl		1984	20,	321
nedaplatin		1995	31,	347
nilutamide		1987	23,	338
paclitaxal		1993	29,	342
pegaspargase		1994	30,	306
pentostatin		1992	28,	334
pirarubicin		1988	24,	309
ranimustine		1987	23,	341
sobuzoxane		1994	30,	310

GENERIC NAME	INDICATION	YEAR INTRO.	ARMC VOL., PAGE	
toremifene		1989	25,	319
vinorelbine		1989	25,	320
zinostatin stimalamer		1994	30,	313
porfimer sodium	ANTINEOPLASTIC ADJUVANT	1993	29,	343
masoprocol	ANTINEOPLASTIC,	1992	28,	333
miltefosine	TOPICAL	1993	29,	340
dexfenfluramine	ANTIOBESITY	1997	33,	332
orlistat		1998	34,	327
sibutramine		1998	34,	331
atovaquone	ANTIPARASITIC	1992	28,	326
ivermectin		1987	23,	336
budipine	ANTIPARKINSONIAN	1997	33,	330
CHF-1301		1999	35,	336
droxidopa		1989	25,	312
entacapone		1998	34,	322
pergolide mesylate		1988	24,	308
pramipexole hydrochloride		1997	33,	341
ropinirole HCl		1996	32,	317
talipexole		1996	32,	318
tolcapone		1997	33,	343
lidamidine HCl	ANTIPERISTALTIC	1984	20,	320
gestrinone	ANTIPROGESTOGEN	1986	22,	321
cabergoline	ANTIPROLACTIN	1993	29,	334
tamsulosin HCl	ANTIPROSTATIC HYPERTROPHY	1993	29,	347
acitretin	ANTIPSORIATIC	1989	25,	309
calcipotriol		1991	27,	323
tazarotene		1997	33,	343
tacalcitol	ANTIPSORIATIC, TOPICAL	1993	29,	346
amisulpride	ANTIPSYCHOTIC	1986	22,	316
remoxipride hydrochloride		1990	26,	308
zuclopenthixol acetate		1987	23,	345
actarit	ANTIRHEUMATIC	1994	30,	296
diacerein		1985	21,	326

GENERIC NAME	INDICATION	YEAR INTRO.	ARMC VOL., PAGE	
octreotide	ANTISECRETORY	1988	24,	307
adamantanium bromide	ANTISEPTIC	1984	20,	315
cimetropium bromide	ANTISPASMODIC	1985	21,	326
tiquizium bromide		1984	20,	324
tiropramide HCl		1983	19,	324
argatroban	ANTITHROMBOTIC	1990	26,	299
defibrotide		1986	22,	319
cilostazol		1988	24,	299
clopidogrel hydrogensulfate		1998	34,	320
cloricromen		1991	27,	325
enoxaparin		1987	23,	333
eptifibatide		1999	35,	340
ethyl icosapentate		1990	26,	303
ozagrel sodium		1988	24,	308
indobufen		1984	20,	319
picotamide		1987	23,	340
limaprost		1988	24,	306
tirofiban hydrochloride		1998	34,	332
flutropium bromide	ANTITUSSIVE	1988	24,	303
levodropropizine		1988	24,	305
benexate HCl	ANTIULCER	1987	23,	328
ebrotidine		1997	33,	333
ecabet sodium		1993	29,	336
enprostil		1985	21,	327
famotidine		1985	21,	327
irsogladine		1989	25,	315
lansoprazole		1992	28,	332
misoprostol		1985	21,	329
nizatidine		1987	23,	339
omeprazole		1988	24,	308
ornoprostil		1987	23,	339
pantoprazole sodium		1994	30,	306
plaunotol		1987	23,	340
polaprezinc		1994	30,	307
ranitidine bismuth citrate		1995	31,	348
rebamipide		1990	26,	308
rosaprostol		1985	21,	332
roxatidine acetate HCl		1986	22,	326
roxithromycin		1987	23,	342
sofalcone		1984	20,	323
spizofurone		1987	23,	343
teprenone		1984	20,	323

GENERIC NAME	INDICATION	YEAR INTRO.	ARMC VOL., PAGE	
tretinoin tocoferil		1993	29,	348
troxipide		1986	22,	327
abacavir sulfate	ANTIVIRAL	1999	35,	333
amprenavir		1999	35,	334
cidofovir		1996	32,	306
delavirdine mesylate		1997	33,	331
didanosine		1991	27,	326
efavirenz		1998	34,	321
famciclovir		1994	30,	300
fomivirsen sodium		1998	34,	323
foscarnet sodium		1989	25,	313
ganciclovir		1988	24,	303
imiquimod		1997	33,	335
indinavir sulfate		1996	32,	310
interferon alfacon-1		1997	33,	336
lamivudine		1995	31,	345
neflinavir mesylate		1997	33,	340
nevirapine		1996	32,	313
oseltamivir phosphate		1999	35,	346
penciclovir		1996	32,	314
propagermanium		1994	30,	308
rimantadine HCl		1987	23,	342
ritonavir		1996	32,	317
saquinavir mesylate		1995	31,	349
sorivudine		1993	29,	345
stavudine		1994	30,	311
valaciclovir HCl		1995	31,	352
zalcitabine		1992	28,	338
zanamivir		1999	35,	352
zidovudine		1987	23,	345
alpidem	ANXIOLYTIC	1991	27,	322
buspirone HCl		1985	21,	324
etizolam		1984	20,	318
flutazolam		1984	20,	318
flutoprazepam		1986	22,	320
metaclazepam		1987	23,	338
mexazolam		1984	20,	321
tandospirone		1996	32,	319
flumazenil	BENZODIAZEPINE ANTAG.	1987	23,	335
bambuterol	BRONCHODILATOR	1990	26,	299
doxofylline		1985	21,	327
formoterol fumarate		1986	22,	321
mabuterol HCl		1986	22,	323

GENERIC NAME	INDICATION	YEAR INTRO.	ARMC VOL., PAGE	
oxitropium bromide		1983	19,	323
salmeterol hydroxynaphthoate		1990	26,	308
APD	CALCIUM REGULATOR	1987	23,	326
clodronate disodium		1986	22,	319
disodium pamidronate		1989	25,	312
gallium nitrate		1991	27,	328
ipriflavone		1989	25,	314
dexrazoxane	CARDIOPROTECTIVE	1992	28,	330
bucladesine sodium	CARDIOSTIMULANT	1984	20,	316
denopamine		1988	24,	300
docarpamine		1994	30,	298
dopexamine		1989	25,	312
enoximone		1988	24,	301
flosequinan		1992	28,	331
ibopamine HCl		1984	20,	319
loprinone hydrochloride		1996	32,	312
milrinone		1989	25,	316
vesnarinone		1990	26,	310
amrinone	CARDIOTONIC	1983	19,	314
colforsin daropate HCL		1999	35,	337
xamoterol fumarate		1988	24,	312
cefozopran HCL	CEPHALOSPORIN, INJECTABLE	1995	31,	339
cefditoren pivoxil	CEPHALOSPORIN, ORAL	1994	30,	297
brovincamine fumarate	CEREBRAL VASODILATOR	1986	22,	317
nimodipine		1985	21,	330
propentofylline		1988	24,	310
succimer	CHELATOR	1991	27,	333
trientine HCl		1986	22,	327
fenbuprol	CHOLERETIC	1983	19,	318
auranofin	CHRYSOTHERAPEUTIC	1983	19,	314
aniracetam	COGNITION ENHANCER	1993	29,	333
pramiracetam H_2SO_4		1993	29,	343
carperitide	CONGESTIVE HEART FAILURE	1995	31,	339

GENERIC NAME	INDICATION	YEAR INTRO.	ARMC VOL., PAGE	
nicorandil	CORONARY VASODILATOR	1984	20,	322
dornase alfa	CYSTIC FIBROSIS	1994	30,	298
neltenexine		1993	29,	341
amifostine	CYTOPROTECTIVE	1995	31,	338
nalmefene HCL	DEPENDENCE TREATMENT	1995	31,	347
azosemide	DIURETIC	1986	22,	316
muzolimine		1983	19,	321
torasemide		1993	29,	348
atorvastatin calcium	DYSLIPIDEMIA	1997	33,	328
cerivastatin		1997	33,	331
naftopidil	DYSURIA	1999	35,	343
alglucerase	ENZYME	1991	27,	321
erdosteine	EXPECTORANT	1995	31,	342
cetrorelix	FEMALE INFERTILITY	1999	35,	336
follitropin alfa	FERTILITY ENHANCER	1996	32,	307
follitropin beta		1996	32,	308
reteplase	FIBRINOLYTIC	1996	32,	316
rabeprazole sodium	GASTRIC ANTISECRETORY	1998	34,	328
cinitapride	GASTROPROKINETIC	1990	26,	301
cisapride		1988	24,	299
itopride HCL		1995	31,	344
mosapride citrate		1998	34,	326
imiglucerase	GAUCHER'S DISEASE	1994	30,	301
somatotropin	GROWTH HORMONE	1994	30,	310
somatomedin-1	GROWTH HORMONE INSENSITIVITY	1994	30,	310
factor VIIa	HAEMOPHILIA	1996	32,	307
pimobendan	HEART FAILURE	1994	30,	307

GENERIC NAME	INDICATION	YEAR INTRO.	ARMC VOL., PAGE	
anagrelide hydrochloride	HEMATOLOGIC	1997	33,	328
erythropoietin	HEMATOPOETIC	1988	24,	301
factor VIII	HEMOSTATIC	1992	28,	330
malotilate	HEPATROPROTECTIVE	1985	21,	329
mivotilate		1999	35,	343
buserelin acetate	HORMONE	1984	20,	316
goserelin		1987	23,	336
leuprolide acetate		1984	20,	319
nafarelin acetate		1990	26,	306
somatropin		1987	23,	343
sapropterin hydrochloride	HYPERPHENYLALANINEMIA	1992	28,	336
quinagolide	HYPERPROLACTINEMIA	1994	30,	309
cadralazine	HYPERTENSIVE	1988	24,	298
nitrendipine		1985	21,	331
binfonazole	HYPNOTIC	1983	19,	315
brotizolam		1983	19,	315
butoctamide		1984	20,	316
cinolazepam		1993	29,	334
doxefazepam		1985	21,	326
loprazolam mesylate		1983	19,	321
quazepam		1985	21,	332
rilmazafone		1989	25,	317
zaleplon		1999	35,	351
zolpidem hemitartrate		1988	24,	313
zopiclone		1986	22,	327
acetohydroxamic acid	HYPOAMMONURIC	1983	19,	313
sodium cellulose PO4	HYPOCALCIURIC	1983	19,	323
divistyramine	HYPOCHOLESTEROLEMIC	1984	20,	317
lovastatin		1987	23,	337
melinamide		1984	20,	320
simvastatin		1988	24,	311
glucagon, rDNA	HYPOGLYCEMIA	1993	29,	338
acipimox	HYPOLIPIDEMIC	1985	21,	323
beclobrate		1986	22,	317
binifibrate		1986	22,	317

GENERIC NAME	INDICATION	YEAR INTRO.	ARMC VOL., PAGE	
ciprofibrate		1985	21,	326
colestimide		1999	35,	337
fluvastatin		1994	30,	300
meglutol		1983	19,	321
ronafibrate		1986	22,	326
modafinil	IDIOPATHIC HYPERSOMNIA	1994	30,	303
bucillamine	IMMUNOMODULATOR	1987	23,	329
centoxin		1991	27,	325
thymopentin		1985	21,	333
filgrastim	IMMUNOSTIMULANT	1991	27,	327
GMDP		1996	32,	308
interferon gamma-1b		1991	27,	329
lentinan		1986	22,	322
pegademase bovine		1990	26,	307
pidotimod		1993	29,	343
romurtide		1991	27,	332
sargramostim		1991	27,	332
schizophyllan		1985	22,	326
ubenimex		1987	23,	345
cyclosporine	IMMUNOSUPPRESSANT	1983	19,	317
gusperimus		1994	30,	300
mizoribine		1984	20,	321
muromonab-CD3		1986	22,	323
mycophenolate mofetil		1995	31,	346
tacrolimus		1993	29,	347
defeiprone	IRON CHELATOR	1995	31,	340
sulbactam sodium	β-LACTAMASE INHIBITOR	1986	22,	326
tazobactam sodium		1992	28,	336
nartograstim	LEUKOPENIA	1994	30,	304
pumactant	LUNG SURFACTANT	1994	30,	308
sildenafil citrate	MALE SEXUAL DYSFUNCTION	1998	34,	331
telmesteine	MUCOLYTIC	1992	28,	337
interferon ß-1a	MULTIPLE SCLEROSIS	1996	32,	311
interferon ß-1b		1993	29,	339
glatiramer acetate		1997	33,	334

GENERIC NAME	INDICATION	YEAR INTRO.	ARMC VOL., PAGE	
afloqualone	MUSCLE RELAXANT	1983	19,	313
cisatracurium besilate		1995	31,	340
doxacurium chloride		1991	27,	326
eperisone HCl		1983	19,	318
mivacurium chloride		1992	28,	334
rapacuronium bromide		1999	35,	347
tizanidine		1984	20,	324
naltrexone HCl	NARCOTIC ANTAGONIST	1984	20,	322
tinazoline	NASAL DECONGESTANT	1988	24,	312
clospipramine hydrochloride	NEUROLEPTIC	1991	27,	325
nemonapride		1991	27,	331
olanzapine		1996	32,	313
quetiapine fumarate		1997	33,	341
risperidone		1993	29,	344
sertindole		1996	32,	318
timiperone		1984	20,	323
rocuronium bromide	NEUROMUSCULAR BLOCKER	1994	30,	309
fasudil HCL	NEUROPROTECTIVE	1995	31,	343
riluzole		1996	32,	317
bifemelane HCl	NOOTROPIC	1987	23,	329
choline alfoscerate		1990	26,	300
exifone		1988	24,	302
idebenone		1986	22,	321
indeloxazine HCl		1988	24,	304
levacecarnine HCl		1986	22,	322
nizofenzone fumarate		1988	24,	307
oxiracetam		1987	23,	339
bromfenac sodium	NSAID	1997	33,	329
lornoxicam		1997	33,	337
alendronate sodium	OSTEOPOROSIS	1993	29,	332
ibandronic acid		1996	32,	309
incadronic acid		1997	33,	335
raloxifene hydrochloride		1998	34,	328
risedronate sodium		1998	34,	330
tiludronate disodium	PAGET'S DISEASE	1995	31,	350

GENERIC NAME	INDICATION	YEAR INTRO.	ARMC VOL., PAGE	
beraprost sodium	PLATELET AGGREG.	1992	28,	326
epoprostenol sodium	INHIBITOR	1983	19,	318
iloprost		1992	28,	332
sarpogrelate HCl	PLATELET ANTIAGGREGANT	1993	29,	344
trimetrexate glucuronate	*PNEUMOCYSTIS CARINII* PNEUMONIA	1994	30,	312
histrelin	PRECOCIOUS PUBERTY	1993	29,	338
gestodene	PROGESTOGEN	1987	23,	335
nomegestrol acetate		1986	22,	324
norgestimate		1986	22,	324
promegestrone		1983	19,	323
alpha-1 antitrypsin	PROTEASE INHIBITOR	1988	24,	297
nafamostat mesylate		1986	22,	323
adrafinil	PSYCHOSTIMULANT	1986	22,	315
finasteride	5α-REDUCTASE INHIBITOR	1992	28,	331
surfactant TA	RESPIRATORY SURFACTANT	1987	23,	344
kinetin	SKIN PHOTODAMAGE/ DERMATOLOGIC	1999	35,	341
tirilazad mesylate	SUBARACHNOID HEMORRHAGE	1995	31,	351
APSAC	THROMBOLYTIC	1987	23,	326
alteplase		1987	23,	326
balsalazide disodium	ULCERATIVE COLITIS	1997	33,	329
tiopronin	UROLITHIASIS	1989	25,	318
propiverine hydrochloride	UROLOGIC	1992	28,	335
Lyme disease	VACINE	1999	35,	342
clobenoside	VASOPROTECTIVE	1988	24,	300
paricalcitol	VITAMIN D	1998	34,	327

GENERIC NAME	INDICATION	YEAR INTRO.	ARMC VOL., PAGE	
doxercalciferol	VITAMIN D PROHORMONE	1999	35,	339
prezatide copper acetate	VULNERARY	1996	32,	314
cadexomer iodine	WOUND HEALING AGENT	1983	19,	316
epidermal growth factor		1987	23,	333

American Chemical Society

Division of Medicinal Chemistry

Dear Colleague:

As a scientist involved in drug research and development, you are well aware of the need for viewing your specific area of interest from an interdisciplinary perspective. To assist you, we cordially invite you to join the Division of Medicinal Chemistry of the American Chemical Society (ACS).

Each year the Division publishes a review of the previous year's literature in a volume entitled "**Annual Reports in Medicinal Chemistry**" in which all important aspects of the drug discovery process are covered. The chapters are compiled by chemists, pharmacologists, biochemists, microbiologists and others from industrial and academic settings. This annual volume is an excellent source of information for structures of novel medicinal agents, biological activity, and current concepts. **Annual Reports** is furnished to division members at no extra cost (publisher's price ca. $65).

Other membership benefits include divisional abstracts of all ACS meetings, mixers at national meetings, a semi annual divisional newsletter and biannual medicinal chemistry symposia. In addition a directory of the division, with addresses and phone numbers of members is published periodically. For younger scientists, travel grants to attend national ACS meetings and research fellowships are awarded annually.

Dues are outlined below for regular and student ACS members, as well as foreign and affiliate (non ACS) members. We look forward to your participation in a large and stimulating scientific discipline.

APPLICATION FOR NEW MEMBERSHIP
DIVISION OF MEDICINAL CHEMISTRY
AMERICAN CHEMICAL SOCIETY

Mail to: David Rotella (phone 609-818-5398, fax 609-818-3450)
            Bristol-Myers Squibb Pharmaceutical Research Institute
            P. O. Box 5400
            Princeton, NJ 08543-5400

[check or money order, drawn on a US bank, payable to the Division of Medicinal Chemistry]

_____ ACS member, US, Canada, Mexico ($19)   Charge my: [ ] Visa/Mastercard [ ] Amex
_____ ACS member, foreign ($20)   card # _____
_____ Non-ACS member, US, Canada, Mexico ($20)   Exp. date _____
_____ Non-ACS member, foreign ($23)   name on card _____
_____ Student ($10)   Signature _____

Name (Dr., Mr., Ms.) _____

Affiliation _____

Address   Street_____
[use same
mailing address   City _____
as national ACS
membership]   State/Country _____ Zip/Postal Code _____

Signature _____ Phone/Fax _____
[membership applications received after October 1 will be applied to membership for the following calendar year]